WOOD-PLASTIC COMPOSITES

The Wiley Bicentennial–Knowledge for Generations

Each generation has its unique needs and aspirations. When Charles Wiley first opened his small printing shop in lower Manhattan in 1807, it was a generation of boundless potential searching for an identity. And we were there, helping to define a new American literary tradition. Over half a century later, in the midst of the Second Industrial Revolution, it was a generation focused on building the future. Once again, we were there, supplying the critical scientific, technical, and engineering knowledge that helped frame the world. Throughout the 20th Century, and into the new millennium, nations began to reach out beyond their own borders and a new international community was born. Wiley was there, expanding its operations around the world to enable a global exchange of ideas, opinions, and know-how.

For 200 years, Wiley has been an integral part of each generation's journey, enabling the flow of information and understanding necessary to meet their needs and fulfill their aspirations. Today, bold new technologies are changing the way we live and learn. Wiley will be there, providing you the must-have knowledge you need to imagine new worlds, new possibilities, and new opportunities.

Generations come and go, but you can always count on Wiley to provide you the knowledge you need, when and where you need it!

William J. Pesce
President and Chief Executive Officer

Peter Booth Wiley
Chairman of the Board

WOOD-PLASTIC COMPOSITES

ANATOLE A. KLYOSOV

WILEY-INTERSCIENCE
A JOHN WILEY & SONS, INC., PUBLICATION

Published by John Wiley & Sons, Inc., Hoboken, New Jersey
Published simultaneously in Canada

Wiley Bicentennial Logo: Richard J. Pacifico

Library of Congress Cataloging-in-Publication Data:

Klysov, A. A. (Anatolii Alekseevich)
Wood-plastic composites / Anatole A. Klyosov.
p. cm.
Includes index.
ISBN 978-0-470-14891-4 (cloth)
1. Plastic-impregnated wood. 2. Engineered wood. 3. Strength of materials. I. Title.
TA418.9.C6K5838 2007
620.1'2–dc22 2007001698
Printed in the United States of America

10 9 8 7 6 5 4 3 2 1

The book is targeted for multidisciplinary scientists and engineers dealing with wood-plastic composites, material science, cellulose, polymers, minerals, strain and stress, flammability, microbiology, rheology, plastic technology, as well as graduate-level students in these disciplines.

CONTENTS

PREFACE

This book is by no means a comprehensive review of wood-plastic composites. Such an imaginary gigantic volume, or, rather, a set of volumes would be extremely boring, overloaded with little and unnecessary details, and would largely duplicate a great many books and papers on plastics, particularly in descriptions of compounders, extruders, downstream equipment, and other machinery.

My initial goal was to present a series of assays in the field of wood-plastic composites, which were supposed to bridge a gap between the laboratory-based research and testing and the real world, real decks, and real railing systems made of wood-plastic composite (WPC) materials. I was fortunate to spend a number of—I hope—productive years being a head of Research and Development division and Vice President, R&D of a company which first had a name Thermo Fibergen, then Kadant Composites, and then LDI Composites, manufacturing GeoDeck WPC decking and railing systems. We have gone both through the highs of our professional achievements, such as when our product was awarded the "Best Buy" status by a major U.S. magazine, and through the lows, when several years ago we voluntarily recalled many trackloads of suspect composite boards from distributors of our product. We have a thing or two to say about wood-plastic composites, about a road to success and a road to failure.

A manufacturing company like ours generates a wealth of knowledge, particularly when works along with many other experts in the field. We have all four great components in our work in order to absorb, digest, and generate knowledge: we research and develop WPC composites in the lab, we manufacture them, sell them, and install them in the real world. And, as a reward, we have a feedback from the field, on how our product performs. If an academic researcher, reading it, is jealous,

I can understand this. I myself was an academic researcher most of my life, including many good years at Harvard University.

So, having an academic experience in the fields of chemistry and biochemistry, and then having acquired an experience in engineering, strain and stress, material science, and, first and foremost, in keeping focus on priorities of the market, hence, on priorities of manufacturing, troubleshooting and problemsolving, and translation of these priorities back into research and development, I decided to share these experiences with a wider audience. A wider, compared with the audience I rather frequently meet at professional meetings, conferences, symposia in the area.

This book is focused on "substance," that is, on wood-plastic composites and their properties, their behavior, rather than on means to manufacture them. I have soberly decided not to describe machinery, "hardware" in WPC manufacturing, because there are countless volumes in which the machinery is described in detail. Generally, as I see it, there is no principal difference between that machinery in the plastic industry and that of wood-plastic composites. Who thinks otherwise, seven feet under the keel and a favorable wind, and welcome to write a textbook or a monograph on the subject.

The choice of chapters for the book was very simple—all the seventeen topics were those which were of a great interest to colleagues of mine and myself when we were working on wood-plastic composites. All these topics seriously determine aesthetics, properties, performance, durability of wood-plastic composite products, and/or processabilty of the material.

By the time the book was submitted for publication, there was not a single volume on the market which would have the subject covered. The material is either scattered in multiple proceedings of WPC and related conferences and symposia, or published as separate papers in professional and semiprofessional editions. Hence, this might be the first try to collect the topics under a single cover, and I am fully responsible for it.

As a reader would notice, many features of wood-plastic composites are illustrated with GeoDeck decking and railing products. This is not a sales pitch but a reality because, as one knows, manufacturing companies very seldom publish data on their products, particularly in comparison with the competition. I have collected as many data published on commercial WPC as I could, including data provided by the manufacturer in their commercial literature and on their websites, and they are described in this book. We generated other data in our laboratory, using commercially available wood-plastic composite products. In many cases, when publication of data could have hurt the image of the manufacturer, I did not indicate the brand name and hid it under a number. I testify that the purpose of the book is not to show which material and product is better (there is no a universally better WPC product compared to competition), but to show a range of properties and explain why such a range exists.

The book would not appear without close cooperation with my co-workers who are too many to name them all. I would like to specially thank here Alan James, Dr. Tatyana Samoylova, David Leeman, Dr. Yiannis Monovoukas, Jonathan Painter, Steve Anderson, Mikiko Kubala, Matt Beachler, Brian Betz, Brent Gwatney, Lanny

Jass, John Long, Burl Boone, Tim Lusk, all from LDI composites. In the process of writing the book I have discussed the material with a good number of experts in the wood-plastic composite area; many of them have read separate chapters and were making suggestions, corrections, and sharing with me their data for inclusion into the book. Their contributions are referenced in the respective chapters. I truly appreciate inputs of the following individuals to this book: Velichko Hristov (McMaster University), Yash P. Khanna (Imerys), Rick Mann (KibbeChem), Thomas Kelley (Dover Chemical Corporation), Zhenghong Tao (University of Massachusetts, Lowell), David Dean (DuPont Packaging & Industrial Polymers), William Sigworth (Chemtura), Jonas Burke (Ferro), Shawn Mealey (Dow Corning).

Last, but not least, I must thank my wife Gail, who had to tolerate my spending so many days, evenings and weekends, not to forget whole vacation time spans spent between as close to Boston as in Berkshires and as far away as in Tahiti and Bora-Bora, cloistered with computers, books, and papers in order to prepare this book.

ANATOLE A. KLYOSOV
December 20, 2006

1

FOREWORD-OVERVIEW: WOOD–PLASTIC COMPOSITES

Let us take a look at a generic wood–plastic composite (WPC) deck, preferably of a premium quality. What should be done in order to avoid the deck owner complaints and, god forbid, a lawsuit? Which properties of the deck should we consider, in order to extend its lifetime as much as possible, preferably longer than that of a common pressure-treated lumber deck? In other words, what is required to make a material that is both durable enough to meet the warranty guidelines and at the same time cost-efficitive to be competitive in the marketplace? What can happen to the WPC deck in use, and how to prevent it? Which properties of the composite material should we aim at, what should we study in that regard, what should we test and how, what should we optimize in order to make a premium product, or, at least—for a less ambitious manufacturer—to pass the building code?

These are the questions considered in this book.

Decking is defined as a platform either attached to a building, or unattached, as in the case of boardwalks, walkways, piers, docks, and marinas. The decking market includes deck boards, railing systems (consisting of a top rail, balusters, bottom rail, and posts), and accessories, such as stairs and built-in benches. According to Principia Partners, U.S. demand for decking (wood and WPC) in 2005 reached $5.1 billion, or approximately 4.0 billion board feet, and projected to grow to $5.5 billion and 4.2 billion lineal feet in 2006 [1].

Wood-Plastic Composites, by Anatole A. Klyosov

Now, let us consider a WPC deck. In a simple case, it is assembled with boards, made of a composite material. The boards can be solid or hollow, or of an "opened," engineered design (see Figs. 1.1–1.25), which can be extruded (in a common case) or compression molded. Typically, but not always, WPC boards have a width of 5½ in. (139.7 mm), height (thickness) of 1¼, 1.00, 15/16, or 13/16 in. (31.75–20.64 mm), and—for standard boards—12, 16, or 20 ft. in length. The board's surface can be smooth (unbrushed), brushed, embossed, or having an "exotic" pattern, such as streaks, simulated wood texture, among others.

Percentage of Decking Demand By Material (Sources: [1, 2, 5])

		Share of (%)		
Year	Market ($ billion)	Wood	Neat plastic	WPCs
1992	2.3	97	1	2
2002	3.4	91	2	7
2005	5.1	77	4	19
2006	5.5	73	5	22
2011 (forecast)	6.5	66	4	30

Note: According to The Freedonia Group, composite and plastic lumber decking demand in 1999, 2004, and 2009 (forecast) was/will be (in million) $317, $662, and $1,370, respectively. According to Principia Partners, only composite decking was sold in North America in 2004 for $670 million.

The boards can be made of plastic of any kind. However, the majority (if practically not all) of WPC boards, manufactured and sold today, are based on polyethylene (PE), polypropylene (PP), or polyvinyl chloride (PVC), see

Figure 1.1 Fading of a composite deck. The "Welcome" mat is just removed.

Figure 1.2 Pressure-treated lumber (as a reference deck board).

Figure 1.3 Trex.

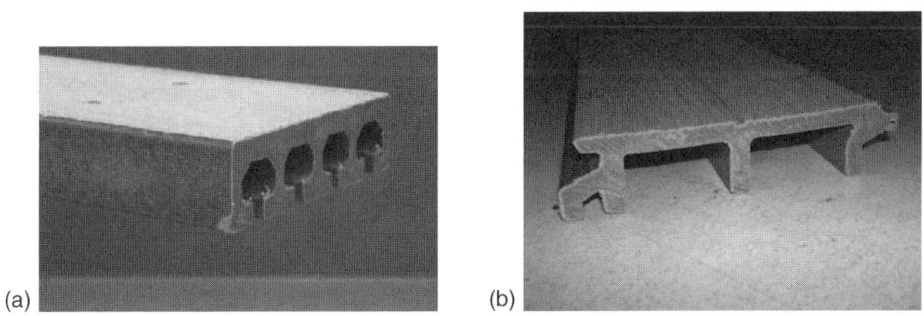

Figure 1.4 TimberTech.

Figure 1.5 Fiberon.

Figure 1.6 WeatherBest^{EHP} (hollow).

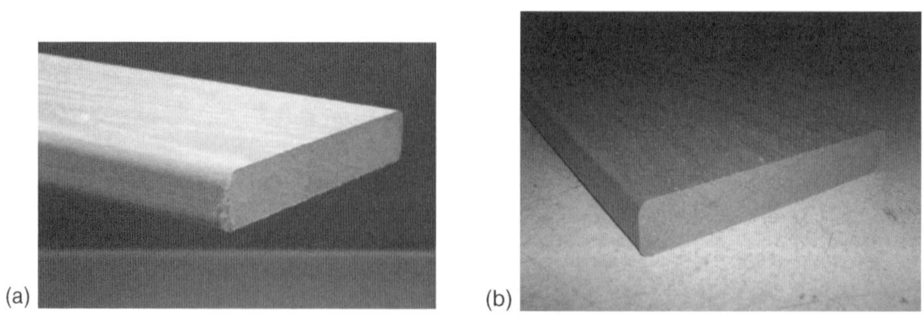

Figure 1.7 WeatherBest^{SP} (solid).

Figure 1.8 ChoiceDek.

Figure 1.9 Nexwood. See color insert.

Figure 1.10 Rhino Deck. See color insert.

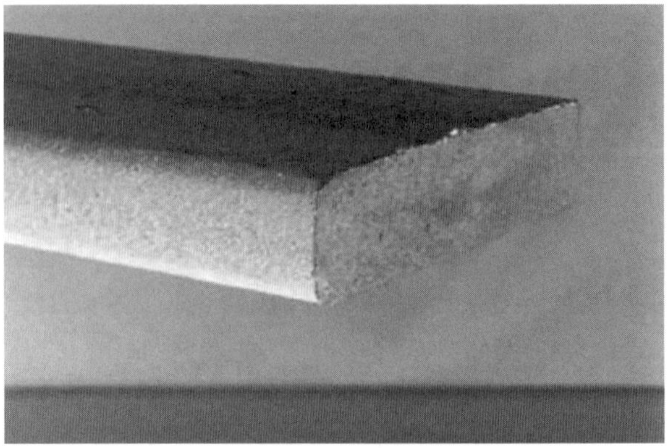

Figure 1.11 SmartDeck. See color insert.

Figure 1.12 Geodeck, Tongue, and Groove.

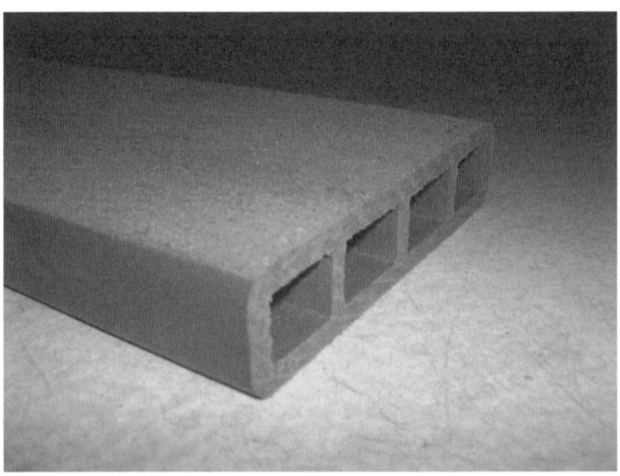

Figure 1.13 Geodeck, Traditional.

Figure 1.14 Geodeck, Heavy Duty (Commercial).

(a) (b)

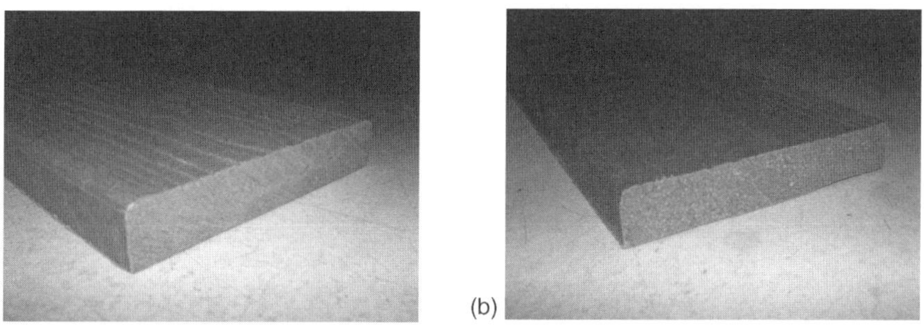

Figure 1.15 Evergrain/Epoch.

Figure 1.16 Ultradeck. See color insert.

Figure 1.17 Boardwalk.

Figure 1.18 CorrectDeck.

Figure 1.19 USPL (Carefree).

Figure 1.20 Millenium.

Figure 1.21 Xtendex.

Figure 1.22 Life Long. See color insert.

Figure 1.23 Cross Timbers

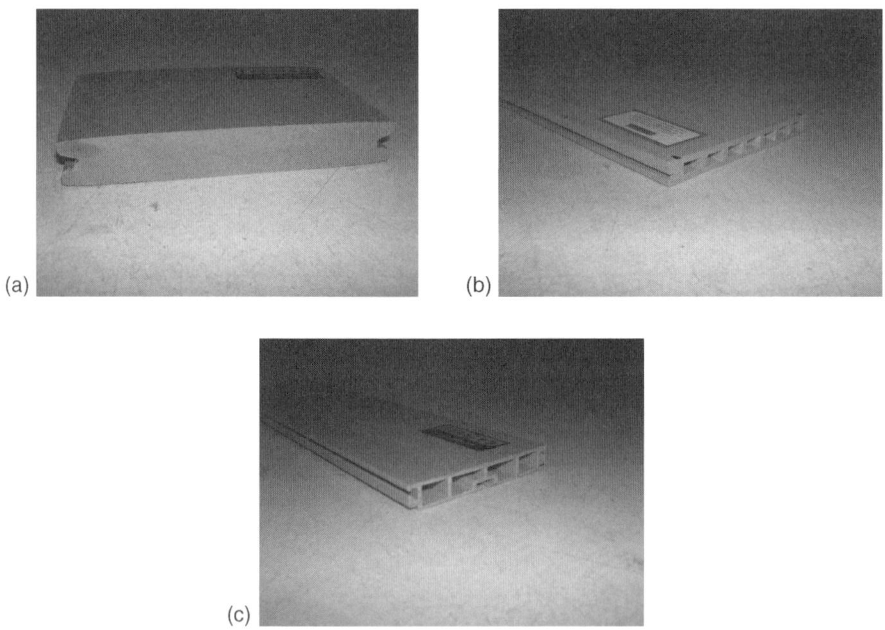

Figure 1.24 Fasalex (Austria)

Table 1.1. The reason is simple: as WPC boards are competing in the market with common lumber, their price should be in the same ballpark. In practical terms, their cost should be no more than 2–3 times higher than that of wooden boards, and that increase should be justified by, say, aesthetics (good looks and

(a)  (b) 

Figure 1.25 An experimental composite board.

the absence of knots, splinters, warping, and checking), acceptable mechanical properties, good durability, low maintenance, lack of microbial degradation, resistance to termites, and possibly even fire resistance. Many customers would pay a premium price to have such a material on their decks. So far, only three plastics named above (PE, PP, and PVC) can fit into the respective pricing category, at the same time having properties necessary for the WPC material to pass the building code.

Types of wood used for decking include pressure-treated lumber, redwood, cedar, and other imported wood. Pressure-treated lumber encompasses for about 78% of wood demand of decking in 2006. WPC building materials consist of a blend of cellulosic fibers and industrial grade polymers, such as polyethylene, polypropylene, and polyvinyl chloride. "Cellulosic fiber" or "wood" in this context is (ligno)cellulosic fiber, such as wood flour, rice hulls, and so on, typically in the form of milled wood products or particles of waste lumber, bleached cellulose fiber or natural fiber of different grades and origins. WPC materials are made by mixing (compounding) plastic and (ligno)cellulose fiber with additives (lubricants, coupling agents, pigments, antioxidants, UV stabilizers, antimicrobial agents, etc.), and manufacturing, using a high volume process such as extrusion or compression or injection molding.

WPC: PRICING RESTRICTIONS

In WPC decking and railing, plastic is filled with natural fiber, such as wood flour, rice hulls and by-product residues from the papermaking industry. Again, there are countless types of natural fiber, obtainable from countless plant sources, however, either a scale is not there, or an availability is restricted, and/or price is too high. Rice hulls cost is about 3¢/lb, wood flour about 3–5¢/lb, bleached fiber by-product (as a blend with minerals) of paper mills between 3 and 9 ¢/lb

TABLE 1.1 WPC decking and railing systems and manufacturing companies, as described in manufacturer's ICC-ES reports

No.	Deck	Manufacturer	ICC-ES report number and date	Cellulosics	Solid or hollow
		Polyethylene-based products			
1	ChoiceDek, Dreamworks, LifeCycle, MoistureShield, A.E.R.T.	Advanced Environmental Recycling Technologies; Weyerhaeuser	NER-596 2/1/2006	Not reported (52% Wood fibers)	Solid
2	Epoch/Evergrain	Epoch Composite Products	ESR-1625 6/1/2005 NER-630 (Legacy)4/1/2006	50% Wood fibers	Solid
3	EverGreen	Integrated Composite Technologies	N/A (referenced by Principia Partners, 2006)	Wood flour	Solid and Hollow
4	EverX, Latitudes, Veranda	Universal Forest Products Ventures II	ESR-1573 6/1/2005	50% Wood flour	Solid
5	Fiberon, Perfection, Veranda	Fiber Composites; LMC	22–41 (Legacy) 10/1/2004	50% Wood fibers	Solid
6	GeoDeck	LDI Composites	ESR-1369 6/1/2005	Rice hulls, Biodac	Hollow
7	Lakeshore	Bluelinx	ESR-1573 6/1/2005	50% Wood flour	Solid
8	Life Long	Brite Manufacturing	ESR-1279 6/1/2005	50% Wood flour	Hollow
9	Monarch	Green Tree Composites	ESR-1084 2/1/2005	55% Wood flour	Solid
10	Nexwood	Nexwood Industries	BOCA 99-8.1 (January, 2000) not current	60% Rice hulls	Hollow
11	Oasis	Alcoa Home Exteriors	ESR-1425 6/1/2005	55% Wood flour	Solid
12	Premier	Composatron Manufacturing	ESR-1481 6/1/2005	50% Wood flour	Solid
13	Rhino Deck	Master Mark Plastic Products	ESR-1461 6/1/2005	50% wood fibers	Solid
14	SmartDeck	SmartDeck Systems	N/A Not current		
15	Tendura	HB&G Building Products; Tendura Industries	N/A (referenced by Principia Partners, 2006)	60% Wood flour	Solid
16	TimberTech	TimberTech	ESR-1400 6/1/2005	50% Wood flour	Solid and open profiles

		Polyethylene-based products			
17	Trex	Trex Company	ESR-1190 6/1/2005	50% Wood flour	Solid
18	UltraDeck	Midwest Manufacturing Extrusion	ESR-1674 11/1/2004	60% Wood flour	Hollow and Solid
19	XTENDEX, E-Deck	Carney Timber	NER-695 (Legacy) 11/1/2004	60% Rice hulls	Hollow
20	WeatherBest, LP Composite, Veranda	Louisiana Pacific	ESR-1088 6/1/2005	60% Wood flour	Hollow and solid
		Polypropylene-based products			
21	CorrectDeck	Correct Building Products	ESR-1341 6/1/2005	60% Wood fibers	Solid
22	Cross Timbers	Elk Composite Building Products	ESR-1590 6/1/2005	Wood flour	Solid
23	K-Decking	Kenaf Industries	N/A (referenced by Principia Partners, 2006)		
		PVC-based products			
24	Boardwalk	CertainTeed	NER-576 3/1/2004	35–45% Hardwood fiber	Solid
25	Millenium	Millenium Decking	ESR-1603 6/1/2006	Wood fiber	Hollow
26	Procell	Procell	ESR-1667 11/1/2006	Flax fiber	Solid

(processed). There is no other natural fiber known to be used in common WPC deck boards on commercial scale, except—on a limited scale—flax, obtained with a good discount.

In order to increase stiffness of composite boards, some WPC manufacturers add mineral fillers to a composition. Why and how minerals increase stiffness, we will consider later. At this point, we would only notice that there are only a few industrially available minerals, which are not prohibitively expensive, for the same reason we have mentioned earlier. Only a few manufacturers of WPC materials use such fillers as calcium carbonate (6–10 ¢/lb), talc (15–20 ¢/lb), and, again, mineral-filled by-products of paper mills (3–9 ¢/lb, processed). In general, all ingredients, combined, should keep material cost of WPC at about 30–40 ¢/lb, or \$0.60–1.10 per lineal foot of a deck board because the weight of WPC deck boards is commonly in a range of 1.8–2.8 lb/ft. With manufacturing costs often to be approximately equal to material costs, WPC boards cost about 60–80 c/lb, or \$1.20–\$2.20 per lineal foot, and in fact are sold for about \$2.20–2.80 per lineal foot. As one can see, there is not much room in WPC costs for expensive plastics, fillers, and additives. Compared to retail prices of common 2 × 6 pressure-treated lumber boards, which are around \$0.90–1.20 per lineal foot, one can see a challenge that WPC boards face in the market.

Consumer Reports magazine (July 2004, p.22) has quoted the following price figures for 100 ft^2 of a deck (just boards; the cost of railings, stairs, supporting structures, and labor is not included):

- Wood, \$190–450. It corresponds to \$0.95–2.25 per lineal foot (2″ × 6″ board).
- WPC, \$300–720. It corresponds to \$1.50–3.60 per lineal foot.
- Plastic or vinyl, \$400–1,000. It corresponds to \$2.00–5.00 per lineal foot.

Note: As a concrete example, a local lumberyard (Newton, MA) sells in October 2006 2 × 6 pressure-treated boards for \$14.24/16 ft. (\$0.89/ft.) and Trex deck boards for \$41.60/16 ft. (\$2.60/ft.). Almost a 3-fold price difference.

Now, regardless of the shape of the WPC board, the materials it consist of, or the aesthetics of the board, the deck board should pass the building code requirements, which are "bind" with respect to materials used. That is, the very same code is for both common wood boards or rails and WPC boards or rails. The principal Acceptance Criteria (AC) for decking and railing systems, the ICC-ES (International Code Council Evaluation Service) AC 174 "Acceptance Criteria for Deck Board Span Ratings and Guardrail Systems (Guards and Handrails)," effective since July 1, 2006, does not differentiate between different kinds of materials decking and railing systems are made from. Wood, steel, concrete, or WPCs, the final product should pass the same building code. This is quite a challenging task for WPC materials.

What are the criteria? Let us come back to our generic WPC deck and boards the deck assembled with.

WPC: BRANDS AND MANUFACTURERS

WPC decking and railing systems brands and manufacturers are given in Table 1.1.

Consumer Reports magazine (July 2004, p.24) has described its investigation of 19 deck materials, three of them were wood (pressure treated, natural cedar, and tropical hardwood), four plastic lumber (Eon, CertainTeed Ever New Vinyl, Brock, and Carefree), one aluminum (LockDry), and 11 WPC boards (Veranda, ChoiceDek, Evergrain, WeatherBest, Trex, Boardwalk, GeoDeck, Timbertech, ChoiceDek, CorrectDeck, and Monarch). Five of them were granted the Best Buy status (board cost per 100 ft^2 is indicated, by the magazine data):

- Eon (plastic lumber), $440
- Veranda, $320
- ChoiceDek, $300
- WeatherBest, $440
- GeoDeck, $430

The magazine has also noted that the following two deck board brands provide with the greatest range in styles:

- Evergrain (four colors and three sizes), $460
- Trex (five colors and sizes), $330.

The rest of the deck board materials were (in the order of residual rating)

- Certainteed Ever New Vinyl (plastic lumber), $1,000
- Brock (plastic lumber), $700
- Boardwalk, $400
- LockDry (aluminum), $700
- TimberTech, $500
- Carefree (plastic lumber), $420
- CorrectDeck, $720.
- Monarch, $590
- ACQ pressure treated lumber, $190
- Cedar natural wood, $320
- Ipe tropical hardwood, $340.

FLEXURAL STRENGTH

The most obvious requirement is that the deck should not collapse under a certain reasonable weight (load). What is a reasonable weight though? The code specifies it as service load and employs a "fail" term rather than "collapse." The ICC requirement

A typical hot tub for, say, five persons has dimensions of 93″ × 78″, hence, it occupies an area of 50.375 ft^2. Total weight of the hot tub consists of its own weight (1100 lb), water (350 gal, or 2,900 lb), and people (5 × 200 lb = 1000 lb), for a rather heavy scenario, total 5,000 lb. Therefore, the hot tub produces a uniformly distributed load of about 100 lb/ft^2.

is a uniformly distributed load of 100 lb/ft.^{2} of a deck. This roughly correlates to a load that a common hot tub, filled with water and having five adult occupants in it, would uniformly distribute on its support.

However, the ICC code also requires a 2.5× safety factor, on top of the 100 lb/ft.^{2} requirement, that is, a deck should hold a live uniform load of 250 lb/sq.ft. Pretty stringent, isn't it?

What about WPC deck boards? Chapter 7, in this book covers this issue in detail. As a brief example, let us consider two WPC boards—Trex and GeoDeck. Trex has reported that flexural strength of their boards (solid boards of 5.5″ width and 1.25″ thickness) is 1423 psi. It means that a Trex board placed on two joists at 16″ span would have a break load derived from the formula

$$S = PLh/8I$$

where
S = flexural strength (1423 psi in this case)
P = break load, or a center point break load (lb)
L = span (16″ in this case)
h = board height/thickness (1.25″ in this case)
I = moment of inertia, equal to $bh^3/12$ in this case of a solid board, with b = board width, 5.5″. For a standard Trex board, the moment of inertia is equal to 0.895 in^4.

From the above equation, a break load (an ultimate load) for a standard Trex board equals to 509 lb. This would translate to an ultimate uniformly distributed load of 1667 lb/ft^2.

The latter value was calculated using a standard formula for an ultimate uniformly distributed load:

$$W = 16 \times 144 \times S \times I/bhL^2$$

where
W = uniformly distributed load
b = board width

and other factors are defined above.

As one can see, a Trex deck is able to hold 1667 lb/sq.ft. which is more than six times higher than the ICC required load including the necessary safety factor.

Similar calculations for GeoDeck deck show that at flexural strength of the board (hollow boards of 5.5″ width and 1.25″ thickness and moment of inertia of 0.733 in.^{4}) of 2782 psi, a break load at 16″ span (center point load) would be 816 lb. This would translate to an ultimate uniformly distributed load of 2670 lb/ft^2, which is more than 10 times higher than the ICC required load including the necessary safety factor.

These examples illustrate that the flexural strength of composite deck boards is quite satisfactory. It is several times higher than the respective building code requirements. Indeed, out of hundreds of thousands of composite decks installed in the United States, none is known to be collapsed during service from the beginning of composite boards appearance.

How strong a WPC deck board can be? We know that wood is very strong, at least for the same purposes WPCs are intended. As it is shown in Chapter 7, flexural strength of wood can reach 20,000 psi. In WPC, wood fiber is blended with a much weaker polymer matrix, which for high-density polyethylene has flexural strength of about 1400 psi (Chapter 2). In a very simplified case, when, say, 50% plastic–50% wood fiber is *ideally* blended into the WPC, and wood fiber is oriented along the flow, that is, longitudinally, the flexural strength would be equal to a symmetrical superposition of the flexural strength of the matrix and the fiber, which is about 10,700 psi.

In reality, flexural strength of wood–HDPE composites is of 1500–4400 psi for commercial deck boards, up to 5000 psi for laboratory WPC, obtained at carefully controlled conditions, and up to 9000 psi, obtained in laboratory conditions and in the presence of coupling agents. At the lower end of this range are Trex boards, which, according to the manufacturer's data, have a flexural strength of 1423 psi. According to the author's data, Trex boards have a flex strength of 1900–2200 psi (Table 7.13). At the highest end of this range are wood flour (pine, 61–63%) filled HDPE composites, obtained in finely optimized and carefully controlled conditions, using best available lubricants and having flexural strength of 4670 ± 90 psi (without coupling agents) and 9100 ± 150 psi (in the presence of 3% Polybond 3029) (Jonas Burke, Ferro Corporation, private communication). Hence, in the last case flexural strength of the WPC reaches 85% of the theoretical maximum of 10,700 psi.

FLEXURAL MODULUS AND DEFLECTION

If flexural strength is directly related to a break load of a board (in this context) placed on supports, flexural modulus is directly related to a deflection of a board, placed on supports, under a certain load. Unlike the flexural strength of composite boards, typically significantly exceeding building code requirements at commonly accepted spans (such as 16 in. on center), flexural modulus of plastic-based composite boards often puts certain restrictions for their installation.

There are two main situations concerning deflection of boards that may not pass the building code requirements: deck boards at a certain span (distance between neighboring joists) and stair tread at a certain span. Let us consider these situations using the same examples: Trex composite deck boards and GeoDeck composite deck boards. These examples would illustrate general shortcomings of plastic-based composite deck boards in terms of their flexibility and deflection.

Deck Boards

The building code requires that the maximum load at certain deflection of the test span shall be recorded (ASTM D 7042, Section 5). A common load requirement for measuring deflection of deck boards is uniformly distributed live load of 100 lb/ft^2.

If one is to choose common requirements for flooring, a deflection shall not exceed 1/360 of the span (The BOCA® National Building Code/1999, Section 1604.5.4).

Deflection under uniformly distributed load is determined by the following formula:

$$D = 5WbL^4/384 \times 144 \times EI$$

where

W = uniformly distributed load (100 lb/sq.ft in this case)

D = deflection, in inches, at the load W (should not exceed 16″/360 = 0.0444″ in this case)

b = board width (5.5 in. in this case)

L = support span, in inches

E = flexural modulus, psi

I = moment of inertia, $in.^4$

For Trex boards (E = 175,000 psi as reported by Trex, with $I = 0.895\ in.^4$), deflection under uniformly distributed load of 100 $lb/in.^2$ at 16-in. span would be 0.021 in., which is within the building code requirements. However, at support span of 24 in., deflection at the same conditions would be 0.105 in. that significantly exceeds the allowable limitation (24″/360 = 0.0667 in.).

For GeoDeck boards (E = 374,000 psi, $I = 0.733\ in.^4$), deflection under uniformly distributed load of 100 $lb/in.^2$ at 16-in. span would be 0.012 in., which is within the building code requirements. Furthermore, at support span of 24 in., deflection at the same conditions would be 0.060 in., which is also within the allowable limitation (0.0667 in.).

As a result, for Trex boards (5/4 × 6) maximum decking span at 16 in. at 100 lb/ft^2 is allowed (ICC-ES Report ESR-1190), and for Geodeck (5/4 × 6) at 24 in. is allowed (ICC-ES Report ESR-1369). Only two WPC deck boards, GeoDeck and TimberTech (ICC-ES Report ESR-1400) are allowed to employ 24-in. span on decks; three more WPC commercial deck boards are allowed to have 19–20-in. span; 14 WPC commercial deck boards on ICC-ES record have 16-in. allowable span; and one WPC board has allowed only 12-in. span on decks.

These records show that flexural modulus of commercial WPC deck boards (and the respective span on decks) certainly has room for improvements. This in turn will improve quality of WPC boards and save money and material on deck joists. This conclusion is supported by consideration of support spans for stair treads (see below).

Stair Treads

The building code requires that the maximum deflection of deck boards used as stair treads under concentrated load of 300 lb placed at midspan shall be 1/8 in. (3.2 mm) or 1/180th of the span (AC 174, Section 4.1.1; 2000 International Building Code, Section 1607.1). For 16-in. span, the allowed deflection is either 0.125 in. or 16″/180 = 0.089 in.

At a span of 16 in. on center, deflection of stair tread under 300 lb of load will be approximately defined by the following equation:

$$D = PL^3/48 \times EI$$

where
D = deflection, inches
P = 300 lb, center point load
L = span, 16 in.
E = flexural modulus
I = moment of inertia.

For Trex solid board (see above) deflection at a span of 16 in. would be equal to 0.163″. It is too much for both criteria, which is 0.125″ and 0.089″ allowable deflection. The span would not pass.

Even for a span of 12 in., with the allowed deflection of 12″/180 = 0.067″, the deflection for this solid board under concentrated load of 300 lb would be 0.069″, which is slightly higher than the allowed one (L/180). Indeed, in ICC-ES Report ESR-1190 maximum stair tread span for Trex boards is listed as 10.5 in.

For hollow GeoDeck, a calculated deflection at a span of 16 in. would be equal to 0.093 in., which is slightly higher than 16″/180 = 0.089″, but within the allowed 1/8 (0.125) in. Direct experiments with GeoDeck boards as stair treads showed that the 16″/180 deflection was reached at an average 301 lb, which is satisfactory compared with the designated 300 lb.

Overall, for 12 WPC deck board brands for which allowable stair tread span is on ICC-ES record (published in the respective ICC-ES reports), only two (CorrectDeck and GeoDeck) have allowable span of 16 in., six have allowable span of 12 in., and four have allowable span of 10.5″, 9″, or even 8 in.

This again shows that stiffness of commercial WPC deck boards (and the respective span on decks and stair treads) certainly can and should be improved. This in turn will improve the quality of WPC boards and bring them closer in this regard to stiffness of real wood. This is one of the most challenging tasks for WPC materials.

In a similar manner, as it was discussed in the preceding section, we can ask—how stiff a WPC deck board can possibly be, if not filled with mineral fillers? We know that wood is very stiff, at least in applications WPCs are intended for. As it is shown in Chapter 7, flexural modulus of wood is about 1,500,000 psi. Polymers are much more flexible, and flexural modulus for HDPE is at best at 150,000 psi (Chapter 2). Again, in a very simplified case, for 50% HDPE – 50% wood fiber composites, in which both principal ingredients are ideally mixed and wood fiber is oriented along the flow, that is, longitudinally, the flexural modulus would be equal to a symmetrical superposition of the flexular moduli of the matrix and the fiber, which is about 825,000 psi.

In reality, for industrial WPCs, exemplified again with Trex, it is 175,000 psi, which is about five times less. For best laboratory WPCs, flexural modulus is close to 700,000 psi in the absence of coupling agents (696,000 ± 30,000 and 717,000 ± 33,000 psi for a wood flour-filled HDPE in the presence of two different lubricants) and slightly higher in the presence of coupling agents (727,000 ± 25,000 and 773,000 ± 13,000 psi, respectively; Jonas Burke, Ferro Corporation, personal communication). Hence, in the last case flexural modulus of the WPC reaches 88–94% of the alleged theoretical maximum of 825,000 psi. It fits rather well with a similar 85% figure for flexural strength of experimentally available WPC with respect to the alleged theoretical maximum (see the preceding section).

Composite decking has become a large established category of building products, particularly for decking and railing systems and related outdoor structures, such as deck stairs. Composite building products present significant advantages over traditional materials, and, as a result of it, composites are one of the most rapidly growing segments of the building products industry. The WPC deck boards segment in particular is estimated in 2005 at $766 million, and in 2006 (forecast) at $929 million. In addition, composite railing systems contributed $190 million in 2005 and $271 million in 2006 (forecast). It makes a total of $956 million in 2005 and $1.2 billion in 2006 (forecast) [1].

THERMAL EXPANSION–CONTRACTION

This is a rather unpleasant phenomenon of decks made of WPC boards, hence, a very important area for R&D of WPC. Almost exclusively (except specially engineered and aerospace-designed materials), all solid materials expand almost linearly (in every direction) with increasing temperature and contract with decreasing temperature. It is this degree of expansion–contraction that can make the phenomenon an unpleasant one, and at the same time challenging for designers with plastic and composite decking. Would consumer like it if the ends of deck boards would quite visibly stick out of the deck frame for extra several inches on a hot day and completely disappear under the deck frame on a chilly night? Well, may be not that much as several inches, but··· how much?

The coefficient of linear expansion–contraction (CTE, for coefficient of thermal expansion) is a measure of the "how much." In fact, the coefficient numerically describes a fraction of the board length that would be added to (expansion) or subtracted from (contraction) per 1°C temperature. If, for example, a 20-ft WPC board is elongated by 1/2 in. when the board surface temperature increased from 70 to 130°F, the coefficient of linear expansion is 0.5″/240″/60 deg $= 3.47 \times 10^{-5}$ 1/deg. This, by the way, is in the neighborhood of a very typical value for expansion–contraction of WPC boards.

Wait a minute, one would say—is not the 130°F a little bit too high a temperature, even on a hottest day?

No, it is not too high for some situations. Commonly, on a summer afternoon a deck surface temperature is higher than the air temperature. To be more specific, it is 40° higher in the North and 50° higher in the South. Hence, if the air temperature increases from 70° in the morning to 90° in the afternoon, by about 2 P.M. a deck surface temperature will be about 130° (North) and 140° (South).

For neat plastics, the CTE is about twice as much compared with WPC boards, which are about 50% filled with nonplastic materials, which is wood fiber and sometimes minerals. As the coefficients of expansion–contraction of both wood fiber and minerals are about ten times lower than those for WPC materials, hence, the reduction in the coefficient's value for filled WPC. In reality, the picture is somewhat more complicated because it is the expansion–contraction of wood *along the grain* that is 10 times lower compared to common WPC. Expansion–contraction of wood across the grain is close to that of WPC. That is, an orientation of wood fiber in a WPC material can increase or decrease the coefficient of expansion–contraction.

The longer the fiber (the higher is the fiber aspect ratio) and the more it is oriented longitudinally, along the deck board, the lower is the coefficient of thermal expansion–contraction. Overall, for different commercial WPC deck boards the coefficient is in the range of 2×10^{-5} to 5×10^{-5} 1/°F. In other words, some commercial WPC boards can expand–contract by 250% higher than others. These "overexpanded" decks are very noticeable and sometimes cause complaints from the deck owners.

> The WPC decking products have realized an annual growth sales at an explosive rate of over 30% per year over the last 10 years [3]. WPCs were projected to grow at a rate of 23% in 2006 in terms of volume, reaching 608 million lineal feet and approximately $1.2 billion in market, value [1]. By 2011, the market for composite decking is expected to surpass $2 billion, or a third of the overall decking market according to estimates from The Freedonia Group [2]. Among the advantages being recognized by consumers for the last decade are lower (compared to wood) maintenance requirements, including no need for staining, sealing, and painting, higher resistance to termites and wood-destroying microbes, the absence of knots and splinters, and environmentally friendly characteristics compared to preservative-treated lumber.

There are two principal ways to decrease a magnitude of expansion–contraction of WPC deck boards. First, to change the formulation of WPC material (less plastic, different fillers, higher fiber aspect ratio) and/or the extrusion regime (the faster the extrusion speed, the more longitudinal is the orientation of the fiber). Second, to better restrain boards on the deck, employing more powerful nails or screws. The moving forces of expanding–contracting boards can be neutralized and blocked by powerful fasteners. Certainly, in these cases the stress has to go somewhere, and it can be expected that at some point the restraint would manifest itself into torsion damage in the joist substructure underneath, or into damage of the boards themselves. However, for these WPC boards (exemplified by GeoDeck) that were observed to be resrained enough on a real deck and not to thermally expand–contract, such damages have not been noticed. However, when fasteners (screws) were being removed, these boards were producing sounds like a guitar string. Hence, they indeed "held" a good deal of stress.

Overall, values of expansion–contraction of WPC boards are largely unpredictable and represent highly empirical values. To make composite deck boards with truly minimized coefficients of thermal expansion–contraction is a very challenging task, not resolved as yet in the industry.

SHRINKAGE

Unlike linear thermal expansion–contraction, which is a completely reversible phenomenon, shrinkage of WPC boards is a one-way, irreversible, though limited process. If contraction of deck boards on a chilly night or during winter seasons opens a gap (sometimes 1/8 to 1/4 of an inch on long decks), the gap is typically closed on a warm day or during summer seasons. However, when boards shrink, the gap is never closed back (Figs. 1.26 through 1.29).

Shrinkage happens when a plastic-based board, extruded and pulled from the die, cool too fast. Too fast means that stretched long polymer molecules, coming from the die, do not get enough time to settle, to come back to their thermodynamically favorable coiled form. They are "trapped" in the board solidified matrix in an unsettled, stretched shape.

To be exact, these "distorted in space" polymeric molecules continue to get rearranged into their energetically minimized shape, but at ambient temperature rates of this rearrangement are too slow, about 100 million times slower than those at hot melt temperature. If it would take 5 s for a polymer molecule to coil from its stretched shape at hot melt temperature, at ambient temperature it would take about 16 years.

However, on a deck on a hot summer day, it might take only a few weeks. In the North, it might take a year or two (see the insert).

How were those temperature-dependent figures for deck shrinkage obtained?

Let's take hot melt temperature (HDPE-based WPC) as 300°F (about 150°C). The temperature coefficient for polymer molecules conformational rearrangements, which is a change in speed of the process by each 10°C, approximately equals to 4. This value for so-called cooperative processes is significantly higher than common temperature coefficients, typically between 2 and 3.

In this case, a temperature drop from 300°F (about 150°C) to ambient 70°F (about 20°C) would result in 4^{13} slower rate of the polymer molecules rearrangements, which is approximately 10^8, or 100,000,000 times.

Five hundred million seconds approximately equal to 139,000 h, which is 5800 days, or 16 years.

Increase of temperature from 70°F (about 20°C) to 140°F (60°C) on a deck would accelerate the rearrangement of polymer molecules in $4^4 = 256$ times, which is from 16 years to 23 days of hot temperature on the deck.

At 200°F (about 90°C), deck boards would further accelerate the rearrangement of polymer chains in $4^3 = 64$ times faster than that at 60°C (see above), which is in about 9 h. This is a common annealing time period for WPC deck boards.

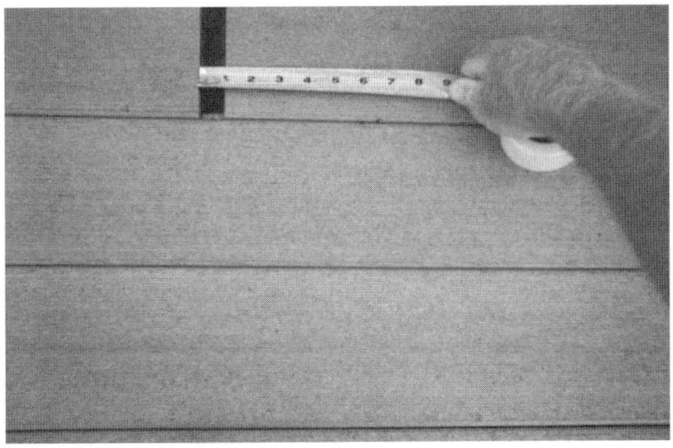

Figure 1.26 An 1-in. gap due to shrinkage of composite deck boards on a deck.

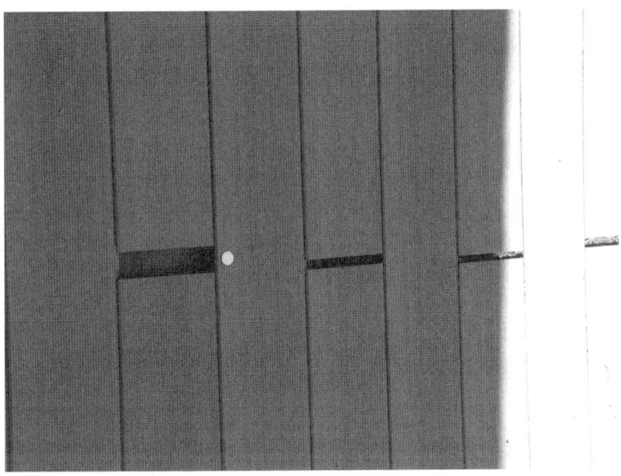

Figure 1.27 Shrinkage of WPC board.

Figure 1.28 Shrinkage of WPC boards.

Figure 1.29 Shrinkage of WPC boards.

In order to eliminate shrinkage, WPC boards are treated by annealing them in a chamber between 180 and 200°F for about a day (see color insert).

It should be noted here that shrinkage is observed, and a respective annealing is required, as a rule, only for profile (hollow) WPC boards. Solid boards, because of their mass, are cooled much slower than that of hollow boards, hence cooling time for solid boards is typically long enough to have stretched polymer molecules to settle in their coiled form. Therefore, shrinkage often is an issue only for hollow WPC boards.

The amount of postmanufacturing shrinkage has several variables and depends on the WPC formulation (especially the percentage of plastic in formulation), the extrusion speed, the cooling regime, the density of the resulting board, and the downstream pooling (and the rate of pool). While still hot, the rate of shrinkage is rapid, so the faster the cooling rate, the higher the postmanufacturing shrinkage. In its worst case, postmanufacturing (in-service) shrinkage reaches 0.3–0.5% of the board length, which is 3/4″–1¼″ for a 20-ft long board. For shorter boards, shrinkage is proportionally smaller.

SLIP RESISTANCE

Slippage on a deck is a very serious matter. A broken limb can financially devastate a good company, particularly if it is not an isolated case.

Generally, WPC deck boards are more slippery than the wood boards. It is easy to verify using a simple experimental setup. Take a 4-ft. conditioned (not wet) board, fix it at a certain angle, place onto the board a leather-sole shoe with a chunk of a heavy metal in it (to increase the weight of the shoe for its stability on the board), and slowly (or step-wise) incline the board until the shoe starts to slide down. With a wood board (such as pressure-treated lumber), it will happen at an angle of about

29° (at the ratio of an opposite side of the triangle to the adjacent side, which is at the tangent ratio of about 0.55). With WPC boards, the same shoe will start sliding down at an angle of about 16–26° for different WPC materials with brushed or unbrushed board surface (the tangent ratio between 0.28 and 0.48). These tangent ratio values in a simplified case are called the coefficient of friction of the board

An Angle of the Triangle (Deg) Versus the Tangent Ratio Table (the Coefficient of Friction)

16	0.29
17	0.31
18	0.33
19	0.34
20	0.36
21	0.38
22	0.40
23	0.42
24	0.45
25	0.47
26	0.49
27	0.51
28	0.53
29	0.55
30	0.58
31	0.60
32	0.62
33	0.65
34	0.67
35	0.70
36	0.73
37	0.75
38	0.78

The coefficients of friction should be determined in more controlled conditions and using professional equipment, as it will be shown below in this book in Chapter 11, but for illustrative purposes the experiment described above would be good enough. It will show that WPC boards are commonly more slippery than wood boards, that some WPC boards are more slippery than others, and that wet boards, both wood and WPC, are less slippery than dry boards. The last statement sounds counterintuitive, however, thanks to the capillary effect of wood and WPC materials, it is, as a rule, true. For the shoe to slip down, wet wood and WPC boards should be inclined up to 34–36° (the coefficient of friction of 0.67–0.73).

At more controlled laboratory conditions, the coefficient of friction for dry wood boards is about 0.70–0.90 and that for dry WPC materials is typically between 0.40 and 0.65.

There is a common perception, but not supported by building code documents (or supported by some outdated documents) that the coefficient of friction for any materials made for walking surfaces should be not less than 0.50, in order to be safe. Not all WPC deckboards would satisfy this (unofficial) criterion.

In order to minimize slippage, some WPC manufacturers texture the surface of their material (typically brushing or deep embossing). It is known that some types of plastic, for example, low-density polyethylene, are noticeably less slippery (have higher coefficient of friction) than other plastics (for example, high-density polyethylene). However, making of WPC boards with predetermined and controlled traction properties is generally not among WPC manufacturers concerns as yet.

WATER ABSORPTION, SWELL, AND BUCKLING

WPC materials will absorb variable amounts of moisture, some more, some less. Why so, it will be discussed later. When immersed into water, they absorb moisture typically between 0.7 and 3% by weight after 24 h of the immersion. This can be compared to water absorption by wood, such as pressure-treated lumber, which absorbs about 24% water by weight after 24 h of immersion. When immersed into water for much longer time, commercial WPC materials absorb up to 20–30% of water, wood more than 100% by weight.

Water absorption by WPC materials may lead to a number of unpleasant events. One is board distortions, swelling, and buckling (Figs. 1.30 and 1.31). Another is mold propagation. Also, saturation of WPC boards with water sometimes decreases flexural modulus of the boards, hence, results in a higher deflection under load. Besides, water absorption leads to a faster board deterioration, oxidation (water is a catalyst of plastic oxidation), and other negative consequences.

WPC materials absorb water due to their porosity. The base plastic material of WPC, such as neat HDPE, practically does not absorb water. However, after being filled with

Figure 1.30 WPC deckboards distortions, swelling, and buckling.

Figure 1.31 A typical appearance of a composite deck in case of buckling or thermal expansion of WPC boards.

cellulose fiber, minerals, and pigment additives (which often contain free metals, serving as effective catalysts of plastic oxidation), and during processing at high temperatures, plastic undergoes rather noticeable degradation, depolymerization, which leads to VOC (volatile organic compounds) formation. Along with it, moisture in cellulose fiber is converted to steam at hot melt temperatures and also adds to microbubbling in the hot melt. Steam and VOC make the material foamed, with noncontrolled porosity. This noticeably decreases the density of the final WPC product. For example, Trex's specific gravity (density) theoretically should be 1.10 g/cm^3, in reality it is reportedly 0.91–0.95 g/cm^3 (Trex data). Even the very fact that the range of density is listed indicated that this parameter is poorly controllable. These densities indicate that porosity of Trex material is between 16 and 21%. When the material is immersed, water fills this void volume.

How the theoretical specific gravity (density) of WPC can be calculated?

Trex example (50% HDPE, 50% wood flour).

100 g of the composite material contain 50 g of HDPE (d = 0.96 g/cm^3) and 50 g of wood flour (d = 1.30 g/cm^3). Each of these components takes the following volume:

HDPE 50 g/0.96 g/cm^3 = 52.083 cm^3,

Wood flour 50 g/1.30 g/cm^3 = 38.462 cm^3.

Therefore, total volume of the 100 g of the composite will be 90.545 cm^3. Hence, specific density of the composite is 100 g/90.545 cm^3 = 1.104 g/cm^3.

Water absorption accelerates mold growth because water is a necessary component for microbial life. Typically, materials that have moisture content of 19% or lower do not support the growth of mold. This amount of moisture can be retained in the very thin upper layer of WPC profiles in humid, moist areas, with inadequate deck ventilation, for an indefinitely long time. In cases where installation instructions are violated and deck boards are installed too close to the ground, or they can be installed high enough, but the deck is "boxed" and completely isolated underneath, creating a perfect "greenhouse," which is moist and wet—in these cases moisture content in WPC deckboards can exceed 20–25% and stay long enough at that level. These are very favorable conditions for mold growth and possibly creating the respective health issues.

Most composite decking manufacturers utilize high-density polyethylene, (HDPE), polypropylene, or polyvinylchloride (PVC) as polymer matrix, and wood flour or rice hulls as the principal filler for their products. Some manufacturers also add mineral fillers, such as talc. These and other changes in compositions make composite materials to vary in their appearance, shape, strength, deflection, moisture absorption, fade resistance, microbial resistance, slip resistance, flammability, and other properties, which will be discussed later in this book.

That is why installation instructions for many composite decks prescribe a deck to be installed at least 12 in., and preferably 24 in. from the grade or rooftop, or provide a wider space between boards (such as 3/16″ or even 1/4″). Some installation instructions say that failure to adhere to proper ventilation may void the warranty.

When WPC boards absorb water, they swell. When the boards are in close contact with each other, a very high pressure can develop in the area of contact, reaching several thousand pounds (Fig. 1.32). This may lead to boards buckling.

Composite decking materials are manufactured under heat and pressure to encase the cellulosic fiber in the plastic, resulting in a product with high resistance to weathering, moisture, insect infestation, and decay. In reality, the encasement is never complete, which leaves room for water absorption, thermal- and UV-induced oxidation with oxygen from the air, and microbial contamination.

Typically, for WPC boards to be buckled they should be contacted with water for a long time, days and weeks. However, the lower the board density and higher the swell, the more likely boards would buckle after their shorter exposure to water. Buckling typically results from an improper installation of a composite deck—causing a prolonged contact with water (from outside or from inside of deckboards, such as for hollow boards), lack of proper gapping, and so on.

In order to minimize water absorption by WPC boards, they should have as high a density that their formulation allows. To achieve this goal, a proper amount of antioxidants should be introduced to the formulation.

Antioxidants slow down the plastic degradation under high temperature, attrition, and so on, hence, minimize the VOC formation and the respective decrease of density. Moisture in the ingredients also leads to a decrease of the final density of the material, hence, cellulose fiber should be dried, if necessary. Last, but not the least, vented extruders remove VOCs and steam from hot melt and greatly increase density of the final product.

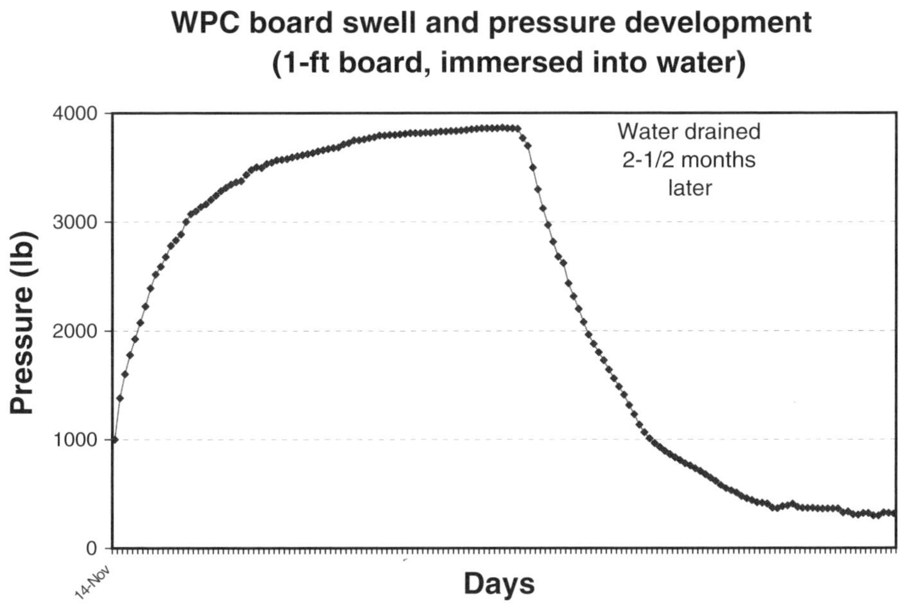

Figure 1.32 Pressure development by a composite deck board (13.5-in. long and 5.5-in. wide) immersed into water, for 2-1/2 month. The initial pressure, caused by holding clamps, was 1000 lb. After 80 days, water was drained and the board was getting dry with the resulting pressure release for the following 2-1/2 months .

MICROBIAL DEGRADATION

On May 28, 2004, the Superior Court of New Jersey certified a nationwide class action in a case originally filed in 2000 against Trex Company, Inc. and ExxonMobil Corp. The case alleged that the Trex product was defective. A press release by the Law Offices of Marc B. Kramer, P.C., which announced the class action on June 2, 2004, said: "In addition, although the Company claims that the product does not need sealants, after the product exhibits mold, the Company allegedly recommends that consumers apply sealants."

While we are not going to discuss here merits of the class action and history of the case, we just point out at the words "after the product exhibit mold···" apparently recognized by the manufacturer.

Mold of the product was one of the main reasons of the class action against Trex. This is how serious it can be. Deck Forum (http://www.mrdeck.com/deck_forum_trex.html) contains many complaints about mold on Trex decks, such as

> ··· had a black mold problem that was coming up from the inside ··· Even after stripping and acid wash, this material would still have the black mold come back up in a few months (10/02).

> ··· It was molding from the inside out (11/03).

··· terrible mold and stains (02/04).

··· The deck is absolutely covered with dark spots which do not come off with cleaning (03/04).

··· I built a Trex deck in 2003 that gets full sun all day long. This past summer it developed mold spots all over it (02/05).

The issue of where mold is coming from and in which form is covered in detail in Chapter 13 in this book. In this introduction overview, we just mention that appearance of mold on some WPC decks and stairs (Fig. 1.33) can be made more likely by certain types of a WPC formulation (and less likely by other compositions of WPC), by improper deck installation, and by climatic conditions.

WPC formulations that invite mold are those with a relatively high porosity (typically made using moist wood fiber) and, hence, having lower density that it might be in the final product. Particularly, it happens if the WPC profile is extruded in the absence or with not enough amounts of antioxidants. Typically, these WPC materials absorb more water than other WPC products in the market. Formulations that make mold on the deck is less likely, contain not only antioxidants but also minerals, which create a natural barrier for microbial degradation of WPC materials. Obviously, biocides and other antimicrobial agents in the formulation help to prevent or slow down mold on decks.

Figure 1.33 Mold stains on a WPC.

Figure 1.34 Black mold on a WPC material.

As an example, a WPC post sleeve is shown in Figure 1.34. It was made with no added antioxidant, unlike regular post sleeves. As a result, it absorbed water in the amount of 3% per bulk material (after 24 h under water) compared with a regular value of 1%, and after some outdoor exposure developed black mold.

Improper installation of WPC decks is associated typically with lack of ventilation at the bottom of the deck and/or deck level too close to the ground, particularly when ground is wet. Water in these decks is retained for a long time and that in turn creates more favorable conditions for mold to grow. Naturally, in wet areas rain water absorbed by decks dries out much more slower than in dry areas, which may lead to mold on decks (see, for example, Fig. 1.35).

Figure 1.36 illustrates a curious case, when WPC rails, quite resistant to mold, were wrapped into corrugated carboard for shipping. Being stored in a wet and warm place, the cardboard was heavily infested and almost completely destroyed by mold. Otherwise intacted composite handrails were soiled with metabolic products of the mold.

Unfortunately, antimicrobial components (often as much as $15–50/lb) are often too expensive to be affordable by WPC deck manufacturers. Biocides for plastics

Figure 1.35 Mold on a composite deck board.

are commonly designed aiming at a quite a different, higher, price structure, which takes place in, say, small plastic-made biomedical devices. If there is a cost, for example, \$10 for a two ounce device and \$50/lb for a biocide, the latter cost is still affordable at 0.1% biocide load. This would increase cost of the device by 0.6¢, or by 0.06% of the total cost of the product. However, 0.1% of the \$50/lb biocide in WPC deck boards that cost otherwise \$0.30/lb would increase the cost of boards by 5¢, that is, by 17% of the cost.

Realistically, at 0.2% of an effective antimicrobial agent in a WPC formulation and the allowable price of the formulation to be increased by 1¢ (cost of materials), cost of the biocide should not exceed \$5/lb.

Figure 1.36 Black-colored metabolic products of *Gonatobotryum* and *Epicoccum* on the surface of composite handrails.

TERMITE RESISTANCE

In the list of homeowners' problems, termites rate ranks very high. According to the Boston Globe, which in turn refers to Bay Colony Home Inspections, between 20 and 25% of the homes sold in most areas of New England have termites or have had them in the past. Toward the South of the United States problem is higher. And, of course, termites not just live around the house. In many cases, termites eat as much as 80% and more of all the structural components of a house, including its deck, if it has one. According to Home Inspection data, in 70% of the above cases the termites have been treated and returned.

There are several main types of termites. Some of them require elevated moisture content, such as dampwood termites (Fig. 1.37). Some live deep inside wood, such as drywood termites (Fig. 1.38). Some live in colonies in the ground and build tunnels, using wood as their food (Fig. 1.39). These pictures were provided with permission by Specialty Termite, Inc. (Pleasanton, CA).

WPC materials are commonly very resistant to termites. Despite that wood fibers are not completely—as a rule—encapsulated into the plastic matrix, and form a sort of continuing chains across WPC materials (unless the ratio of plastic to fiber is really high, more than 80%), termites cannot get into the plastic matrix. At best, termites can only slightly trim cellulose fiber at the WPC surface.

Figure 1.37 Dampwood termites on wood (© Specialty Termites, Inc.).

Figure 1.38 Drywood termites on a wood post base (© Specialty Termites, Inc.).

As a result, weight loss of WPC materials by termites is negligible, if anything. Let us consider GeoDeck composite board as an example. It was subjected to termites collected from a colony of subterranean termites *Reticulitermes flavipes,* according to a procedure given in ASTM D3345-74. Five of 1.00 × 1.00 × 0.25 in. blocks of Southern Yellow Pine sapwood and five blocks cut from WPC board were exposed to termites for 8 weeks. With the wood samples, weight loss due to termite action was of 9.1 ± 0.7%. With the WPC samples, two out of five samples were practically untouched (no weight loss), and an overall average weight loss was 0.2 ± 0.2%.

Figure 1.39 Subterranean termite damage in wood (© Specialty Termites, Inc.).

Here are few examples of termite resistance ratings, showed in the respective company records:

- Trex, rating 9.6
- GeoDeck—No attack, rating 10
- Nexwood—No damage, rating 10.

One can see that commercial WPC deck boards are dramatically more termite resistant than wood lumber.

FLAMMABILITY

Polyethylene and polypropylene-based WPC materials are flammable (see, for example, Fig. 1.40).

Flammability of materials is characterized by many different ways, one of them is the flame spread index (FSI). As reference values, FSI for inorganic reinforced cement board surface is arbitrarily set as 0, and for select grade oak surface as 100 under the specified conditions. FSI for ordinary wood species is typically between 100 and 200, and for some special cases it is as low as 60–70. An average FSI for about 30 different wood species is 125 ± 45.

(a) (b)

Figure 1.40 WPC deck boards before and a few minutes after ignition (by permission from the University of California Forest Products Laboratory).

For comparison, wood fiber-filled HDPE hollow boards have an FSI around 150, solid boards about 80–100, WPC hollow boards containing minerals have an FSI around 100, and PVC-based wood-filled deck boards typically have an FSI between 25 and 60.

Both ordinary wood species and most of WPC deck boards belong to Class C category of flammability in terms of flame spread. There are four basic categories, or classes, for flame spread index: Class A, with FSI between 0 and 25; Class B, with FSI between 26 and 75; Class C, with FSI between 76 and 200; and below Class C, with FSI above 200 (unclassified materials). Classes A, B, and C sometimes are called Classes I, II, and III.

Until recently, the flammability of WPC decks was not even a concern. Decks were not supposed to be inflammable. What is a point if the house would burn and the deck would stay, right? Then it was recognized that brushfires often ignite a house via the deck. Now legislatures of several states, California first, are working on a new law, according to which decks should be fireproof to some extent. This poses a new challenge for WPC decks. The new law is scheduled to be effective in the state of California starting January 2008.

Technically to make a WPC deck of a low flammability is not difficult. Principally, there are two ways to go—either to load a WPC formulation with flame retardant components or to employ PVC (or other low-flammable plastics) as a base plastic for WPC. Chapter 14 considers these issues.

As always, optimization is a name of the game. PVC is not considered as an environmentally friendly material. When ignited, the resin releases hydrogen chloride (HCl), a toxic and volatile strong acid. If not stabilized properly, PVC can release HCl under direct sunlight, at high temperature of a hot deck surface on a sunny summer day. Some flame retardants, particularly polybrominated diphenyl esters, are also far from being benign. Mineral flame retardants, such as aluminum trihydrate and magnesium hydroxide, are required at a high loading level (up to 40–50% w/w) to be effective.

Considering that plastic often takes 40–50% w/w of flame retardants, there is no room for wood filler in WPC, which will not be WPC anymore but rather a mineral-filled plastic. At any rate, replacement of wood fiber of 3–5 ¢/lb with mineral flame retardant of 20–30 ¢/lb would significantly increase the cost of the resulting material. All these questions pose a great challenge to WPC manufacturers aiming at fireproof composite deck boards.

OXIDATION AND CRUMBLING

One of the most unpleasant, damaging, and unexpected features of some WPC materials happened to be their elevated vulnerability to oxidation, leading to board crumbling (Figs. 1.41–1.44). In the progress of crumbling, the WPC board shows tiny and then developing cracks, its surface becomes dustier and softer, until one can easily scratch it, leaving deep tracks. Eventually, the board can collapse under its own weight.

Figure 1.41 An intermediate step of crumbling of Cedar tongue-and-groove composite deck board (Arizona).

There are number of factors leading to accelerated WPC oxidation, and lack of antioxidants (in the initial, incoming plastics), and/or insufficient amounts added to the formulation is the most important of them. Adding antioxidants aims both at preserving the plastic during the processing at high temperatures and protecting the WPC profile during service on a deck under the damaging effects of sunlight, air oxygen, water, pollutants, and other elements.

Figure 1.42 An advanced step of crumbling of Mahogany tongue-and-groove composite deck board (Arizona).

Figure 1.43 An advanced step of crumbling of Mahogany tongue-and-groove composite deck board (Arizona).

Briefly, antioxidants quench free radicals that are formed in the process of plastic degradation by oxygen and initiated by temperature and UV light, and assisted by moisture, stress, presence of metals, and other catalysts of plastic oxidation. If not intercepted by antioxidants, the polymeric plastic is degraded (depolymerized) so much that it loses its integrity and ceases to be a plastic anymore. It is converted to a loose powderous material, mainly a filler.

Figure 1.44 Catastrophic failure of Mahogany tongue-and-groove composite deck board due to an oxidative degradation (Arizona).

Other factors that accelerate WPC oxidation, hence, decrease durability and shorten a deck's lifetime, are decreased density (specific gravity) of the board compared with the maximum density for the same board, presence of metals (in pigments, lubricants, and other additives), moisture content, and unsettled stress in boards. Decreased density is the result of an increased porosity of the boards, due to moisture presence in the initial ingredients of the WPC (of wood fibers first of all) and plastic degradation during the processing (due to overheating, excessive shear and/or lack of antioxidants). An excessive porosity allows oxygen to permeate into the WPC material "from inside," significantly increasing the accessible surface area, along with the rate of oxidation. Metals, particularly free metals, often are efficient catalysts of plastic oxidation. Moisture is also an effective catalysts of plastic oxidation.

Until recently, the effects of these factors and their quantitative manifestation were practically unknown and not even recognized, neither in the WPC industry nor in academic research in the area. That is why the acute deterioration and crumbling of some WPC boards turned out to be quite unexpected and puzzling, and resulted in some cases in an avalanche of warranty claims. These cases will be considered in detail in chapter 15. It turned out that the progressive deterioration and crumbling have resulted from WPC oxidation, and as soon as it was recognized, measures were taken. The OIT (oxidative induction time) parameter was introduced into characterization of WPC products and evaluation of their lifetime in the real world, on real decks.

Essentially, the OIT value quantitatively describes a lifetime of a composite (or actually any organic-based) material during its accelerated oxidation in pure oxygen at an elevated temperature, such as 190°C. For example, for unstabilized (without added antioxidants) WPC materials the OIT can be as low as 0.3–0.5 min. The lifetime of such WPC boards in the South (Arizona, Texas, Florida) can be as low as only several months. The "lifetime" in this context is a time period by the end of which the consumer can see there is something wrong with the deck and calls for help. In real terms, the deck owner contacts with the manufacturer and files a warranty claim.

For partially stabilized WPC materials, the OIT can be between 1 and 10 min (Fig. 1.45). A number of commercial WPC deck boards being sold in the market falls into this range. Depending on the deck profile (solid or hollow) and the board density, and, of course, on location/geography/climatic conditions, the lifetime of the deck can vary, but there is a risk that these boards would not live long enough to see the end of their warranty time period, particularly in the South.

For well-stabilized WPC board, the OIT can be in the range of stabilized plastics (15–100+ min).

Figure 1.45 shows the OIT values for commercial wood–plastic decking boards. Twelve of them have OIT lower than 10 min. This is a troubling observation because these boards can be time bombs for the manufacturers. Boards with OIT above 15–20 min are not going to suffer from their deterioration and crumbling due to oxidation, at least caused by hot summer season temperature and UV light. Certainly, these boards can be damaged by other mechanisms (water, mold, bacteria, or algae),

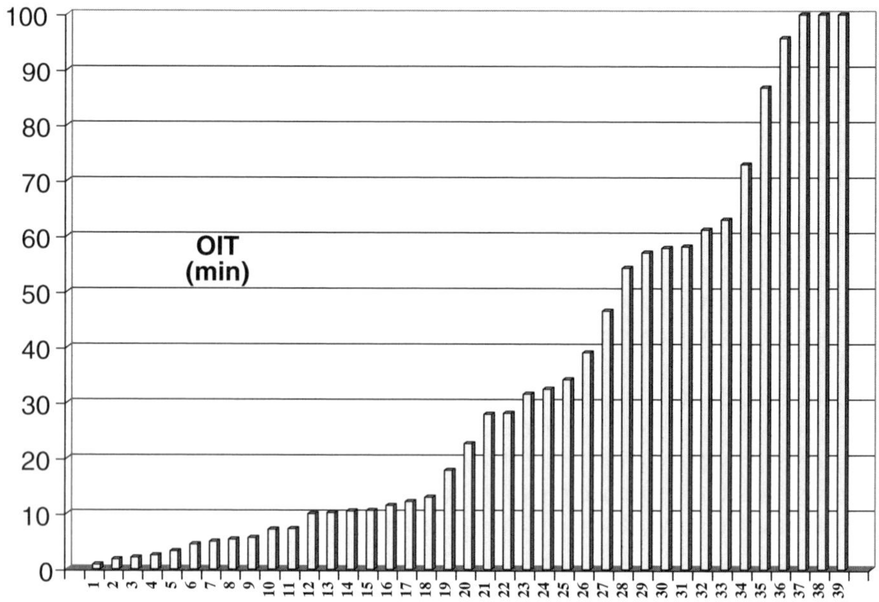

Figure 1.45 The oxidative induction time (OIT) values for commercially available WPC deck boards. The manufacturers and board names are numbered in the order of increasing of the OIT values.

or can be broken by force, or burned by fire (which is a very rapid oxidation), but there are means to minimize each of these effects as well. All these aspects are also covered in this book.

PHOTOOXIDATION AND FADING

Fading is generally an accepted feature of WPCs, probably because people get used to it with common wooden decks; hence, this phenomenon has a kind of grandfather status. However, some composite materials fade less than others, and some much more (Fig. 1.46).

Clearly, customers generally prefer to have their deck not fading at all. However, they are either not informed on the prospective fading or do not know that some WPC practically do not fade, or accept the fading as it is. When the sun irradiation on their deck is uniform throughout the day, it does not create a problem. However, in many cases just after several months the difference in color of their deck is too noticeable (see Fig. 1.1).

Figure 1.46 shows a difference in fading (in terms of lightness) of 32 commercially available WPC deck boards after 1000 h of the accelerated weathering. A difference between ΔL (on the Hunter Lab color scale) is between 0.4 and 35 L units. In a simplified manner, one unit is the first shade difference that the naked eye can

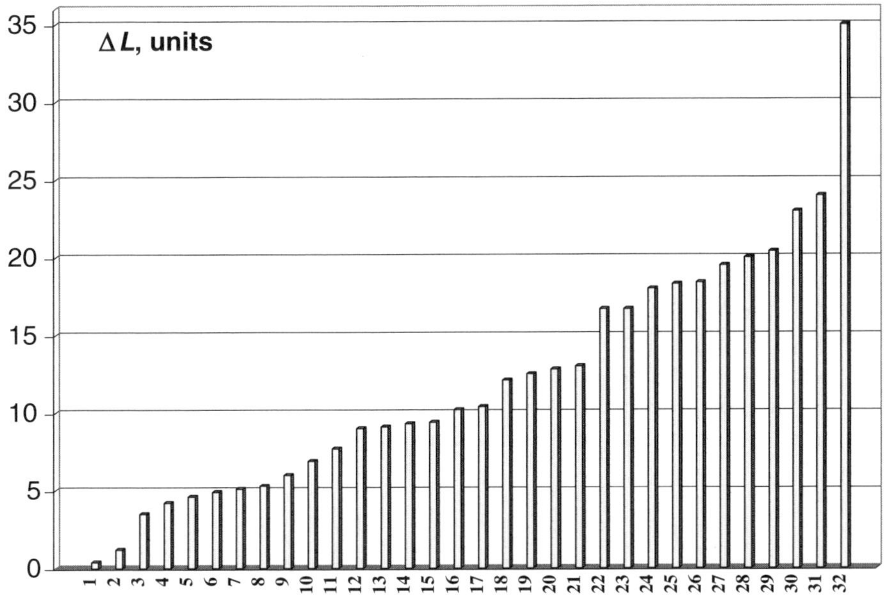

Figure 1.46 Fading of commercial WPC boards in terms of lightness (L in the Hunter Lab color scale) shift after 1000 h of accelerated weathering (vertical axis) in Q Sun-3000 weathering chamber at 0.35 W/m^2 at 340 nm, 102:18 min (UV light: UV light + water spray) cycle, 63°C black panel temperature. ΔL values vary from 0.4 to 35 units/1000 h, that is in almost 100 times, for the boards available on the market. Since some boards can be outdated compared to current manufacturing, they are not named but numbered.

normally detect. That is, a difference in lightness by 0.4 units one cannot detect, while 35 units of fading from, say, the initial $L = 53$ to $L = 88$ results in the final lightness of almost a white sheet of paper.

It is rather difficult, if possible at all, to quantitatively translate the fading in the weathering box to the real world. However, some very approximate comparisons can be made. Depending on the material color, one day in the weathering box under "standard" conditions (340 nm, 0.35 W/m^2, 102:18 cycle, 63°C black panel temperature) often corresponds to 9 ± 4 days of natural weathering in the U.S. Midwest and New England. This figure is often called the "acceleration factor." In Arizona and Florida, the acceleration factor is about 40% of the above, that is, around 3.5.

Hence, a detectable level of lightness change ($\Delta L = 1$) for the WPC board with the lowest degree of fading in Figure 1.46 (0.4/1000 h) will be reached in Midwest and New England after about two and a half years, and a level of $\Delta L = 5$ (a surely noticeable change of lightness) will be reached after at least 15 years, taking into account some slowing down of the fading process with time. On the contrary, a detectable level of lightness ($\Delta L = 1$) for the WPC board with the highest degree of fading in Figure 1.46 (35/1000 h) will be reached in these geographical areas after about 10 days, and a level of $\Delta L = 5$ will be reached after about 2 months. This is confirmed on direct observations.

Indeed, if one is to place some commercial composite deck boards outside under direct sunlight, their fading would be noticeable after only a couple of weeks. This can be observed for boards in Figure 1.46 with $\Delta L = 20/1000$ h. Approximate calculations show that the "theoretical" figure, based on the acceleration factor of 9 ± 4 (see above), would be equal to 19 ± 8 days, which is close to the observed time period.

Fading of composite materials depend on many factors, some of them are related to the WPC composition (wood fiber content, type of cellulosic fiber, amount of UV stabilizers and antioxidants and amount and type of colorants) and some to the outdoor conditions (covered or open deck and amount of moisture on the deck and other climatic conditions). It does not appear that processing of WPC and the profile manufacturing noticeably affect the material fading.

WOOD–PLASTIC COMPOSITES—PRODUCTS, TRENDS, MARKET SIZE AND DYNAMICS, AND UNSOLVED (OR PARTIALLY SOLVED) PROBLEMS

WPC Products

At present, a lion's share of WPCs goes for decking and railing systems (deck boards, stairs, posts and post sleeves, handrails and bottom rails, post caps, balusters, and other small accessories), and similar structures attached to the exterior of dwellings, as well as boardwalks. A relatively small amount of commercially produced WPC goes for siding, fencing, pallets, roofing tiles, and window frame lineals. Other products, such as pilings, railroad ties, marinas, window blinds, and sound barriers, are rather experimental, not commercial as yet, or sold at a very small fraction compared to principal WPC products. Automotive products (interior panels, trunk liners, spare tire covers, package trays, etc.) form a separate category of composite products, often use long cellulose fiber, and fall into a quite different price category. They will not be considered here.

The Public View: Perception

The public view of WPCs is hard to evaluate more or less objectively. Many have never heard of WPCs. Many prefer "real wood," and they are hard to blame. Wood is an excellent material, far exceeding WPC in many properties, first of all in strength and stiffness, in slip resistance, and in many types of wood—in fire resistance (except PVC-based wood composites). Common wood, however, is an inferior material compared to WPC with respect to water absorption, microbial degradation, and durability. There are exceptional types of wood that satisfy a taste of a sophisticated customer, but these types are too expensive for a general market.

Overall, great many customers have gladly accepted the appearance of WPC in the market, though many are doubtful and many have rejected them outright. Nevertheless, WPC-based building products are capturing the market with pretty high speed for the last 10 years.

What is so attractive in WPC products? This question is fair only with respect to WPC decks because other products have not attracted enough attention in the market. Well, WPC decks, as a rule, look pretty good. I have one. Anyone can walk on them barefoot without any precautions to get a splinter in the foot. There are no splinters whatsoever. Then, WPC decks indeed require minimum maintenance. And maintenance with wood decking means—first of all—regular staining and painting. WPC decks do not require them because they are colored for life, when contain colorants, and do not fade. Though, very few WPC deck boards do not fade. In the South, decks often require treatment with antimicrobial and antitermite chemicals. WPC boards do not require it, as they are much more resistant to biological degradation.

It should be noticed here that bioresistance of WPC deck boards is diminished with the increase of wood fiber content (above 40%) and increased with mineral content (silica, calcium carbonate, talc, etc.).

WPC decks require, though, normal washing, cleaning, and other care, as conventional wood decks do. It is obvious that a barbeque on a deck would unavoidably lead to grease and fat stains; when potato salad is dropped upside down on a deck (and there is no other way for potato salad to drop, as everyone can testify), made of either wood or WPCs, leaves stains, which are not easy, though possible, to remove. In fact, it is much easier to remove grease from WPC deck than from a wood deck.

Overall, a WPC deck is much more durable than a wooden deck and requires much less work in a long run. This is certainly attractive for some people. However, it requires a steep payment upfront. This is repulsive for other people. Both features of WPC affect the public acceptance, and both are considered as a practicality issue.

A key issue in public perception regarding a new product in building industry is an appreciation of the product by both builders and homeowners. It seems that WPCs hit the right spot. Deck installers commonly like WPC deckboards as safe to work with due to lack of splinters, easiness to cut, saw, nail, and screw (except polypropylene-based WPC deck boards that are too tough, but, however, this problem is generally solved with development of special fastening systems). These properties of WPC deckboards result in an ultimate goal of any installer for hire to be accomplished: a good speed of deck installing, hence, a faster and a better pay.

Another important factor in the success in the the builder's market is the market accessibility for the product. Technically it means a speedy way from the plant's warehouse to a lumberyard, to distributors, dealers, suppliers, retailers, and to the end user. This is called "strong channel position to access the market" and "distribution channels." This leads to a competitive advantage of some manufacturers compared to others.

WPC Market Size and Dynamics

The U.S. market for two major WPC products, that is, decking and railing components, amounted $1.3 billion in 2006 (projection), approximately 22% of total decking and railing (wood, plastic lumber, vinyl, WPC). Fifteen years back, in the beginning of 1990s, total decking and railing dollar market was about two times smaller, and a share of wooden decks was as much as 97% (compared with the present 73%), while a share of WPC decks was 2% (compared to the present 22%). The rest was and is the only plastic-made decks and rails (1% in 1992, and 5% in 2006).

Figures on composite decking and railing systems, particularly forecasts, vary a great deal among analysts. For example, the Freedonia Group has forecasted in 2002 that WPC decking and railing sales will be of $680 million in 2006 [6]. According to Principia Partners, this volume was significantly exceeded as soon as in 2004 ($820 million), further increased in 2005 ($956 million), and was projected at $1,195 million for 2006 (Table 1.2), which is almost 80% higher than the Freedonia Group projection. Clearly, all these figures, particularly when they show the precision of up to 0.1% (as shown above), have a rather limited value and depend on many

TABLE 1.2 Total (all materials) and WPC decking and railing market size in North America [1–5]

	2004	2005	2006 (projection)
Product	Dollar value (million)		
WPC decking	670	766	929
WPC railing	150	190	271
WPC decking and railing	820	956	1,200
Total decking (all material)	2,570	2,960	3,170
Total railing (all materials)	1,860	2,150	2,280
Total decking and railing (all materials)	4,430	5,110	5,450
	Lineal feet (million)		
WPC decking	450	479	590
WPC railing	12	14	18
WPC decking and railing	462	493	608
Total decking (all materials)	3,650	3,760	3,990
Total railing (all materials)	220	230	240
Total decking and railing (all materials)	3,870	3,990	4,230

different and variable factors. What is undisputable is that WPC building materials, first of all WPC decking and railing, are steadily displacing conventional wooden decking and railing from the market. From 2001 to 2011, WPC decking expenditures are forecasted to grow at a compound annual growth rate of 22%. Similarly, a share of WPC decking in the market is forecasted to grow from 7% in 2002 to 14% in 2007 and to more than 30% in 2011.

The composite decking segment has realized compound annual growth rates (CAGR) in excess of 20% over the last 10 years. It was predicted that a CAGR will further grow for composite decking at 26% from 2002 to 2011. According to another set of numbers, wood–plastic decking represented approximately 7% of the overall decking market in 2001 and is expected to represent almost 14% in 2007. Among other factors, the wood-decking segment stands to be significantly affected by the withdrawal of chromated copper arsenate (CCA) preservaties for residential pressure-treated wood market, as it is described in Chapter 13. CCA is no longer used for consumer application effective from December 31, 2003.

The above growth figures reflect both "physical" (lineal) growth and cost of materials and labor. For example, "physical" growth of decking and railing, in lineal feet, from 2004 to 2005 was about 4% (total, for all materials), whereas dollar growth was twice as much. Overall, "physical" amount of total decking and railing (all materials) in 2004–2005 is related to their dollar amount of $1.14 and $1.28 per lineal foot, respectively.

According to Principia Partners, an industrial consulting firm, specializing in building products (among other materials and manufactured goods), WPC decking and railing reached $956 million in market value and 493 million lineal feet in 2005, and $1.2 billion and 608 million lineal feet in 2006 (projected).

One can see that the "physical" amount of WPC decking and railing is related to their dollar amount of $1.77 in 2004 and $1.94 in 2005, and $1.97 in 2006 (projected) per lineal foot, which is 52–55% higher than that of total decking and railing (all materials). Compared to 4 and 8% in lineal and dollar growth for total decking and railing, respectively (all materials), from 2004 to 2005, growth of WPC decking and railing represented 6 and 17%, respectively. In 2006, WPC decking and railing are expected to grow by 24 and 26%, respectively. Of the total of $956 million for WPC decking and railing systems in 2005, boards were worth $766 million and railing systems $190 million (Principia Partners).

The very recent data by Principia Partners show that annual growth in WPC decking in 2005 and 2006 was 14% and 21%, respectively, and that in railing systems was 27% and 43%, respectively.

Regarding a combined North American and European WPC market, in 2002 it was of 685,000 metric tons, that is, 1.51 billion lb [7].

Competition on the WPC Market

Principal players on WPC market are described in Table 1.1, and their financial standing is outlined very briefly in Table 1.3.

TABLE 1.3 The competitive landscape ([8, 9] with additions)

Company (years)	Revenues ($ million)	Comments
Trex		Public company; market leader; supplier to Home Depot
2003	210	
2004	254	
2005	294	
TimberTech		Private company; early entrant; only 30% of revenues is by WPC
2003	65	
2004	83	
2005	112	
Fiber Composites		Private company; supplier to Home Depot
2003	45	
2004	72	
2005	90	
A.E.R.T. (Weyerhaeuser)		Public company; sells its ChoiceDek through Lowe's stores nationwide
2003	44	
2004	49	
2005	70	
Louisiana Pacific Specialty Products		Public company; only 5% of revenues is by WPC; 2005 sales data are taken from (Natural & Wood Fiber Composites, v. 5, No. 3, 2006, p. 2)
2003	41	
2004	67.1	
2005	70.5	
Nexwood Industries		Went out of business in 2005
2003	35	
2004	25	
2005	5	
Epoch Composite Products		Private company; only 10% of revenues is by WPC; makes compression-molded Evergrain composite decking
2003	30	
2004	55	
2005	82	
Mikron Industries		Private company; window lineal producer; only 15% of revenues is by WPC
2003	25	
2004	(N/A)	
2005	(N/A)	
Certainteed		Public company; less than 1% of revenues is by WPC
2003	20	
2004	15	
2005	12	
Kadant composites		Subsidiary of a public company; 112% sales growth between 2002 and 2003, then sales leveled off because of shrinkage and crumbling problems. In 2003 and 2004, both
2003	17	
2004	17	
2005	19	

TABLE 1.3 (*Continued*)

Company (years)	Revenues ($ million)	Comments
		problems were solved at the plant. In October of 2005 was acquired by LDI, formed LDI Composites, a private company
Correct Building Products		Private company; polypropylene-based WPC decking
2003	15	
2004	16	
2005	20	
Elk composite Building products		Polypropylene-based WPC decking
2004	15	
2005	27	
UFP		Supplier of its brand Veranda composite decking to Home Depot
2004	33	
2005	42	
Green Tree Composites		A private company
2004	15	
2005	18	
Master Mark Plastics		A private company
2004	15	
2005	18	
Brite Manufacturing		A Canadian company
2004	12	
2005	12	
Composatron		A private company. Produces more WPC railing products compared to WPC decking
2004	11	
2005	18	
Procell		A private company. A relatively new WPC entrant with PVC-based flax-filled composite decking
2004	7	
2005	14	
Alcoa Home Exteriors		Sold to Ply Gem Industries at the end of 2006
2004	3	
2005	9	
Integrated Composite Technologies		A private company
2004	5	
2005	10	

Unsolved (or Only Partially Solved) R & D Problems

A recent meeting of a few dozen of manufacturers of WPCs and their R & D representatives had—as a central event—a brainstorming session. That session had as a principal goal to identify the most "burning" issues in WPCs to be solved in years to come.

The indentified issues are as following, in no particular order. I am reproducing the list below, first, to show the "burning" issues as they are identified by WPC manufactures; second, to indicate that most of them, if not all, are discussed in the following chapters in this book; and the third, in order to illustrate how many issues are considered to be important and not solved for the WPCs:

- Fundamental research on accelerated weathering
- Effects of wood extractives on the look and properties of WPCs
- Effects of recycled resins on properties of WPCs, and quantitative characterization of recycled resins compared to virgin ones
- Plastics for structural (engineering, load-bearing) WPC materials
- Long-term creep issues in WPC decking
- Stain resistance of WPC products
- Fade resistance of WPC products
- How to make WPC products superior to wood
- How to reduce density of WPC products in controlled way, without the presence of moisture in raw materials
- Polymer alloys to improve properties of WPCs
- Fire resistance of WPCs
- Surface biocides as an economical way to increase microbial resistance
- Consistency in mechanical properties of commercial WPCs
- Simple ways to measure rheology of WPC hot melts to characterize and predict performance of WPC products
- Simple ways to measure durability of WPC; clear criteria of durability
- New low-density fillers for WPC materials
- Modeling of material properties of WPC products
- Improved ways of fiber dispersion in plastic matrix
- Decrease thermal expansion–contraction of WPC products
- Assessment of UV stabilizers in WPC products
- Effective flame retardants for WPC products
- Development of WPC products for ground contact applications
- Antioxidants and UV stabilizers for WPC roofing shingles, tiles, and slates
- Plastic and cellulose fiber degradation during extrusion: qualitative evaluation and countermeasures
- Fasteners for WPC deck boards: short- and long-term issues

- Abrasion resistance of WPC deck boards
- Slip resistance of WPC deck boards: science and practical measures to increase it
- Recycled nylon for WPC products

The above list shows that WPCs face a long way to go in order to realize their potential, have their properties improved, and replace wood decking, railing, and roofing materials providing their benefits both from structural and aesthetic, and environmental point of view.

REFERENCES

1. *Composite Decking, 2004 vs. 2005*, Principia Partners, Jersey City, NJ, 2006.
2. *Composite& Plastic Lumber*, The Freedonia Group, Cleveland, OH, 2006.
3. *Composite Decking & Railing*, Principia Partners, Jersey City, NJ, 2004.
4. J. Morton. Wood-plastic composites: taking it to the next level. In: *WPC Conference 2004 "Realizing the Full Potential"*, Principia Partners, Baltimore, MD, October 11–12, 2004.
5. *Focus on Decking*, The Freedonia Group, Cleveland, OH, 2003.
6. *Composite and Plastic Lumber*, The Freedonia Group, Cleveland, OH, 2002, p. 68.
7. W. Sigworth, L. Walp, R. Bacaloglu, and P. Kleinlauth. The role of additives in formulating WPC. In: *The Global Outlook for Natural Fiber & Wood Composites 2003*, New Orleans, LA, December 3–5, 2003. Principia Partners, *7th International Conference of Wood-Fibre Plastic Composites*, Madison, WI, May 19–20, 2003.
8. N. Beare, J. Courtney, and S. Ratledge. New ways of assessing capital and building the value of your WPC business. *WPC Conference 2004 "Realizing the Full Potential"*, Baltimore, MD, October 11–12, 2004.
9. *Natural & Wood Fiber Composites*, Vol. 4, No. 10, Principia Partners, Cleveland, OH, 2005, p. 3.

2

COMPOSITION OF WOOD–PLASTIC COMPOSITE DECK BOARDS: THERMOPLASTICS

INTRODUCTION

Wood–plastic composite (WPC) deck boards are extruded or molded products of a specified shape and, by definition, represent plastic filled with cellulose fiber and other ingredients. In this context "wood" is a proxy for fibrous materials of plant origin. These materials will be considered in the next chapter. This chapter deals with a plastic component of composite deck boards, with an understanding that the plastic is thermoplastic.

Only those thermoplastics are applicable in wood–plastic composites which can be processed at temperatures below 400°F (about 200°C). These limitations are due to limited thermal stability of wood. This temperature limitation, hence, rather narrow selection of plastics, is not absolute, as (a) delignified cellulosics can be used in WPC, and lignin is the most temperature-sensitive fraction of woody materials; (b) cellulose fibers can be partly thermoinsulated by blending them with minerals, such as in Biodac® (see the next chapter); and (c) residence time for cellulosic materials in compounders and extruders can be significantly reduced by increasing processing speed and by other means to shorten contact time of cellulosics with hot melt. Therefore, other, high–processing temperature plastics could be used to make WPC with superior properties. On that reason we briefly consider here Nylon 6 (melting temperature 216°C, or 421°F) [Nylon 6/6, 255–265°C] and ABS as examples of a rather nonconventional (regarding WPC) plastic.

Wood-Plastic Composites, by Anatole A. Klyosov

Manufacturers of wood–plastic composites in North America have used about 600 million pounds of thermoplastics in 2005, of which polyethylene accounts for about 90% of the volume and polypropylene and PVC for the remaining 10%. Of these, reclaimed resin is about 35–40% of the total resin demand, virgin plastics is 60–65% [1].

Table 1.1 lists 26 commercially available brands of wood–plastic composite deck boards. Twenty of them are made of filled polyethylene, three of filled polypropylene, and three of filled PVC. In fact, only one brand of each of the two latter categories has established itself in the market (CorrectDeck and Boardwalk). Of ten the most sellable brands of composite deck boards (Trex, TimberTech, Fiberon, ChoiceDeck, WeatherBest, Evergrain, Monarch, GeoDeck, EverX/Latitudes, RhinoDeck) all ten are polyethylene-based products.

Why so?

This is a subject of this chapter, among some other things related to thermoplastics.

This chapter is not a comprehensive description of thermoplastics because those readings are abundant in the literature. This chapter aims at a brief description of plastic properties directly related to behavior of wood–plastic composites and at a comparison of these properties for thermoplastics currently used in making WPC. Besides, this comparison is extended to Nylon and ABS as prospective—apparently—plastics for future brands of WPC possessing superior properties.

POLYETHYLENE

Polyethylene (PE) is the largest volume plastic produced in the world. It has a relatively low-melting temperature (typically between 106 and 130°C, depending on density/branching of PE) and can be produced in a very wide range of viscosity of its melts. The melts mix well with fillers, and the low melting point allows the use of cellulose fiber as a filler without much risk of significant thermal degradation.

Polyethylene is a semicrystalline polymer. It means that at ambient temperatures the polymer consists of two rather distinct fractions, or phases—crystalline and amorphous. The amorphous part of polyethylene, which is a sort of rubbery at ambient temperatures, becomes a glass-like at a certain transition temperature, the so-called glass transition point. For polyethylene the glass transition point varies from very low to low (from –130 to −20°C), thus making the plastic ductile at common temperatures. The lower glass transition point (γ-transition) is always present in the range of −130 to −100°C, the higher one (β–transition, at −20°C) is manifested not in all PE materials. To complicate the picture even more, we can notice that there is one more transition in polyethylenes, called α-transition, commonly found between 10 and 70°C, and it is associated with crystallinity of PE. For WPC the last two transitions (α- and β-) are of little importance.

Polyethylene is rather soft, making PE-based composite deck boards easier to nail, screw, cut, and saw. Polyethylene, as well as polypropylene, shows near-zero moisture absorption (typically below 0.02% after 24 h underwater immersion) and very high resistance to chemicals, including strong acids, such as sulfuric, hydrochloric, and nitric. Only fuming acids can produce some staining of polyethylene, but those acids are not common in households.

Polyethylene shows a relatively high resistance to oxidation compared to other polyolefins (polypropylene among them, see below), hence, requires lesser amount of antioxidants for processing and for the following service outdoors.

On the contrary, polyethylene is rather flexible and not very strong. Its high flexibility forces PE-based deck boards to be installed at a maximum clear span of 16″–24″ on center (o.c.), with installations typically at 16″ o.c., and also as low as 12″ o.c. Its high flexibility does not allow it to be used for composite handrails without special reinforcements, such as metal inserts. Its relatively low strength is generally acceptable for composite deck boards with a standard code requirement to withstand a uniformly distributed load of 100 lb/sq.ft. (with the additional safety factor of 2.5), but it is generally not enough to make safe railing system posts and handrails without reinforcements.

Compared to wood, polyethylene shows a higher coefficient of thermal expansion–contraction. A 20-ft. long unrestrained HDPE-based composite deck board expands or contacts (lengthwise) by 3/8″ to one inch in the temperature range between 50 and 130°F, depending on the amount and type of fillers. Fasteners on a real deck often restrict that movement, though it is still typically larger than that of wood.

Polyethylene is manufactured in various polymeric forms, differing by their molecular weight and "linearity," or presence of irregularities, or branches, unsaturations, and so on. This in turn determines the density, or specific gravity of the polymer, which is used as the principal classification feature of polyethylenes. The main forms of polyethylenes are as follows:

- High-density PE (HDPE)
- High-molecular weight HDPE (HMW-HDPE)
- Ultra high-molecular-weight HDPE (UHMW-HDPE)
- Low-density PE (LDPE)
- Linear low-density PE (LLDPE)
- Very low-density PE (VLDPE).

Naturally, there are all possible combinations of intermediate forms. Weight average molecular weight of UHMW-HDPE is typically higher than three million, whereas HMW-HDPE is in the range of 200,000–500,000. A pattern of molecular weight of polyethylenes (and of most polymers) is rather complicated, as a result of their various molecular weight distributions. In a real estate market, for example, it does not have much sense to talk about "average" cost of houses, if there are, say, ten \$100,000 houses and one \$10,000,000 house. Formally speaking, the average price of these houses is 1.1 million dollars each. It surely does not give much information to a potential buyer of any one of the houses.

Such an approach to polymers gives a "weight-average" molecular weight and does not provide much of information either, particularly when the polymer has a wide distribution of molecular weights of its individual chains. "Number-average" molecular weight is a median, which is equal to the molecular weight at which half of polymeric chains are smaller in size (molecular weight), and half is larger than the "number-average" figure. It shows which is the molecular weight of majority of polymers in the polymeric mix. For the above real estate example, the number-average is $100,000.

Finally, "viscosity-average" molecular weight is that of the highest molecular weight fraction. For example, if one studies the polymeric mix using viscometry, he would not "see" small molecules, as they do not add to viscosity of the system compared with long polymers. He would "see" practically only long molecules. For the above real estate example, the "viscosity-average" is $10,000,000 house.

Hence, for a better and more adequate picture, polymers are characterized by all three "average" figures—weight-average, number-average, and viscosity-average. Some particular examples will be given below in this book.

Table 2.1 shows the density and melt flow index values for low-, medium-, and high-density polyethylenes. By definition, these polymers are characterized by a certain range of density. Each of them can have practically any melt flow index, hence, MFI is independent of the respective density category. Table 2.1 shows only the most applicable range of MFI, though it varies between 0.01 and 100 and wider.

Many users do not consider medium density PE as a separate category and divide polyethylenes into low- and high- density materials by the benchmark density of 0.94 g/cm^3. Generally, none of these subdivisions of PE were scientifically defined, though they are largely (but not specifically) determined by methods of their synthesis. The classification is commonly accepted just for convenience.

Generally, the density of a 100% amorphous PE sample is considered to be 0.85 g/cm^3, whereas that of a 100% crystalline PE sample is 1.0 g/cm^3. The degree of crystallinity in HDPE is typically in the range of 60–80%, and 40–50% for LDPE. 50% crystallinity (in LDPE) corresponds to about two branches per hundred carbon atoms in the chain, and 60–90% crystallinity (in HDPE) corresponds to about 0.5 to practically zero branches per hundred carbon atoms. However, in polyethylenes usually the density rather than the crystallinity is referred to because both are connected by a certain linear relationship [4], and density can be faster and more precisely determined experimentally.

Polyethylene resins, employed in wood–plastic composite materials, typically belong to HDPE, and only in one case (Trex deck boards) it was LDPE (or LDPE/HDPE), originated from grocery sacks and stretch film, used as reclaimed plastic source.

TABLE 2.1 Polyethylene, molded or extruded [2, 3]

Polyethylene	Density (g/cm^3)	Melt flow index (g/10 min)
High-density	0.941–0.965	0.2–30
Medium-density	0.926–0.940	1–20
Low-density	0.915–0.925	0.3–26
Linear low-density	0.915–0.925	0.1–100
Very low-density	0.870–0.914	0.02–10

There are many grades of polyethylene available (see Chapter 7 for flexural modulus values for LDPE, LLDPE, and HDPE, supplied by Chevron Phillips Chemical Company). Flexural modulus values vary from 30,000 to 50,000 psi for LDPE, equal to 60,000 psi for one particular LLDPE material (in fact, they can vary from about 40,000 psi to 130,000 psi), and from 125,000 to 240,000 psi for HDPE. Hence, HDPE is much stiffer compared with LDPE and LLDPE.

Similar in kind, comparisons of flex strength are difficult because most of PE samples do not break when bent at conditions of ASTM D 790. Some data indicate that HDPE shows flex strength around 1400 psi. Filling polyethylene with wood fiber, rice hulls, and other plant fiber material increases flexural strength of resulting composites—to about 1600–2200 psi (for Trex composites) and higher, for HDPE-based composite materials up to about 3000 psi. Further increase of flex strength of wood–HDPE composites, to about 3800 psi and higher, is typically achieved by using coupling and cross-linking agents (see Chapter 5).

Low-Density Polyethylene (LDPE)

As the number of branch points in PE chains increases, PE density decreases. The amount of unsaturations in PE also increases with the decrease in density. For example, an average number of methyl branches per 1000 carbons in LDPE increased from 21 at 0.922 g/cm^3 to 43 at 0.916 g/cm^3. An average number of vinylidene unsaturations/branches per 1000-Da (molecular weight) segments of LDPE increased from 16 at 0.922 g/cm^3 to 35 at 0.912 g/cm^3 [5]. Hence, sensitivity of PE to oxidation increases as its density decreases. LDPE is more vulnerable to oxidation compared to HDPE. On the contrary, LDPE is oxidized rather uniformly compared to HDPE, in which amorphous areas are oxidized faster than crystalline ones.

LDPE typically has long side-chain branching off the main molecular chain and therefore is a more amorphous polymer. As a result, it shows lower shrinkage compared to a more crystalline HDPE, in which many of the polymer molecules are packed closely together.

Linear low-density PE (LLDPE) has a density similar to LDPE, but the linearity of HDPE. Branches of LLDPE are comparatively short. None of WPC manufacturers has reported that they make composite deck boards based on LLDPE, as well as on VLDPE (very low-density PE) and ULDPE (ultra low-density PE). The two latter polyethylenes have extremely high flexibility, which would make them inappropriate for composite deck boards or railing systems.

LDPE can be easily scratched by a thumbnail, HDPE can be scratched with difficulty, and polypropylene can hardly be scratched at all. This is also related to the respective WPC. Apparently, this is why unbrushed polypropylene-based WPC feels more like a plastic. This property is directly related to hardness, which is the resistance of a material to deformation, indentation, scratching, and to abrasion resistance. The latter is lower for LDPE compared to that of HDPE (10–15 mg/100 cycles and 2–5 mg/100 cycles, respectively) [6].

Maximum operating temperature of LDPE is considered to be 71°C (160°F), compared to 82°C (180°F) for HDPE.

Medium-Density Polyethylene (MDPE)

MDPE is typically a mixture of LDPE and HDPE and will not be considered here as a separate type of PE. None of WPC manufacturers have reported that they use "medium-density polyethylene" to make their products. Trex, apparently, can get into this category, but they report that they use "polyethylene" for the manufacturing of WPC (ICC-ES Report ESR-1190, June 1, 2005).

High-Density Polyethylene (HDPE)

Due to its higher crystallinity compared to LDPE, HDPE is stronger and stiffer than LDPE, but is more prone to warpage. It shows a higher shrinkage, due to formation of crystalline, packed areas upon transition from melted state to solid one. Its tensile strength is two to three times that of LDPE and it has a reasonably good compressive strength (4600 psi), whereas LDPE typically does not break on compression. Both of them generally have a very good impact resistance, though some HDPE-made materials with a high degree of crystallinity could be rather brittle.

Some polyethylenes are supplied as so-called bimodal grades: a mix of high- and low-molecular weight components in similar concentrations. They often show quite different processing characteristics compared to "normal," monomodal (unmodified) polyethylenes. Bimodal grades PE often have higher die swell.

Ultra high molecular weight PE (UHMWPE) is a linear homopolymer, structurally similar to HDPE, but having average molecular weight of 10–100 times higher than that of standard grades of HDPE. None of WPC manufacturers have reported that they use "ultra high molecular weight" PE.

As crystallinity of HDPE is generally higher than that of LDPE, the following HDPE properties, depending on crystallinity, are higher than that of LDPE: strength, modulus, density, shrinkage, creep resistance, wear resistance, and hardness. On the contrary, lower crystallinity often offers better processability and impact resistance. Degree of crystallinity and a character of crystal areas often depend on how rapidly the profile is cooled; this in turn can effect postmanufactured shrinkage and brittleness of the product, as well as stresses and—as a result of it—a sensitivity of the product to oxidation and to a thermal expansion–contraction.

Permeability by gases is significantly lower for HDPE compared to that for LDPE. As the permeability (P) for gases is determined as $P = D \times S$, where D = diffusion constant and S = solubility coefficient, the units of permeability are expressed in the amount of gases (cm^3) through a film of a unit thickness (mm) per unit area (cm^2) per time (s) per pressure difference (cm Hg), that is (cm$^3 \cdot$ mm)/(cm$^2 \cdot$ s $\cdot$ cm Hg). In these units, the permeability of HDPE and LDPE (at 25°C) for nitrogen is 1.4 and 9.7, for helium 11 and 49, for oxygen 4 and 29, for CO_2 3.6 and 126, respectively. At 30°C these values are as follows: for nitrogen 2.7 and 19, for oxygen 11 and 55, for CO_2 35 and 352 [7]. Obviously, a higher rate of penetration for oxygen would lead to a faster oxidative degradation of LDPE compared to HDPE.

Linear PE (more crystalline) is more stable to oxidation than branched (more amorphous) PE. When the polymer is melted (140°C, or 284°F), oxidation rates are similar for both crystalline and amorphous PE. The amorphous regions are oxidized faster compared to crystalline regions of polyethylene because of two reasons: higher reactivity of PE at branched points and a higher oxygen diffusion into amorphous domains compared with that of crystalline ones. However, the oxidation of amorphous areas (such as at outdoor exposure) in partially crystalline PE leads to a noticeable increase of its brittleness due to an increased fraction of the crystalline regions.

Generally, the increase of density and the respective transition from LDPE to HDPE leads to increase in abrasion resistance, chemical resistance, hardness, strength, stiffness, decrease in gas and water permeability, thermal expansion, and impact strength.

In terms of creep—the gradual deformation of a sample under prolonged loading—LDPE is inferior compared with HDPE. In creep, once a sample has yielded, the deformation is practically irreversible. Creep is less pronounced in samples with a higher crystallinity, hence, it is less pronounced in HDPE than in LDPE. Generally, the higher the density of PE, the lower the creep. Overall, polyethylenes are poor plastics compared to polypropylene and PVC, and even more so compared to engineering plastics, such as Nylon. Creep of polyethylenes is a major obstacle to make them structural plastics, and, therefore, to make WPC based on PE structural composite materials.

Oriented plastic samples are less prone to creep compared to isotropic samples. Apparently, on the same reason extruded WPC composites, such as deck boards, which show clear signs of anisotropicity, might produce less creep compared to the base plastic, even considering a contribution of the fillers.

An important parameter for deck boards is coefficient of friction at their surface. This subject is described in detail in Chapter 11, "Slip resistance and coefficient of friction of composite deck boards." To a certain extent friction determines safety of the deck and a liability of the manufacturer. A slippery deck could literally bankrupt the manufacturer. Regarding friction, polyethylenes vary in a wide range, by about 500%, with the lowest friction of HDPE of the highest density (coefficient of friction of 0.1), and the highest friction of LDPE of the lowest density (coefficient of friction of 0.5) [8]. These are values for PE sliding against itself. Values of the coefficient of friction of WPC against a leather sole are given in Chapter 11.

POLYPROPYLENE

The share of polypropylene-based WPC decking in the market is relatively small. There are only two recognized brands—CorrectDeck and Cross Timbers—that totally account for about 10% sales of WPC materials. For example, reported sales of PE-based Trex composite deck boards in 2005 were $294 million out of about $956 million total for WPC, whereas sales of PP-based Cross Timbers (Elk Composite Building Products) in 2005 were $27 million.

In a number of properties polypropylene is superior compared to polyethylene. It is lighter, stronger, stiffer; it shows improved creep resistance, less wear, and less slippery. However, it is more brittle than polyethylene, particularly at low temperatures, and so stiff that it is difficult to fasten using nails or screws. That is why PP-based WPC are installed using special fastening systems, recommended by the manufacturers. For a comparison, using common nails and screws makes polyethylene-based WPC boards so easy to install. Also, PP-based boards are much harder to cut and saw at a job site, compared to PE-based boards.

Unlike polyethylene, which exists—in an ideal case—in a form of a monotonous (— CH_2 — CH_2 — CH_2 —) chain, polypropylene has a "microbranched" chemical structure

$$-\underset{\displaystyle CH_3}{\underset{|}{CH}}-CH_2-\underset{\displaystyle CH_3}{\underset{|}{CH}}-CH_2-\underset{\displaystyle CH_3}{\underset{|}{CH}}-CH_2-\underset{\displaystyle CH_3}{\underset{|}{CH}}-CH_2-\underset{\displaystyle CH_3}{\underset{|}{CH}}-CH_2-$$

This structure largely determines chemical and other properties of polypropylene, for example, faster oxidation compared to polyethylene.

Description of polypropylene in the literature is often accompanied by terms "atactic," "isotactic," or "syndiotactic." The origin of the terms is as follows. Side methyl groups in polypropylene chains can be all on the same side of plane (as it is shown above in a very simplified manner, as in reality carbon atoms in the chain are in *zig-zag* stereoconfiguration), on alternate sides, or in a random arrangement with respect to plane of carbon atom chain. These forms of PP are called isotactic (fiber-forming), syndiotactic, and atactic, respectively. All these forms are related to homopolymers of polypropylene. PP used for common applications, including WPC, is atactic and will be referred to here as polypropylene.

Polypropylenes are subdivided to homopolymers and copolymers. Homopolymers are more crystalline, have a rather well-defined melting temperature at 161–165°C (322–329°F), softens at about 155°C (311°F), and have a rather narrow molecular weight distribution. Copolymers typically contain some amount of ethylene comonomer and in turn are subdivided to random and block copolymers. Their melting points are in the range of 140–155°C (284–311°F). Polypropylenes have a specific gravity (density) of 0.90–0.91 g/cm^3, which is approximately equal to that of very low-density polyethylene and lower than that of majority of polyethylenes, particularly HDPE (0.941–0.965 g/cm^3). PP homopolymers are stiffer than copolymers, with their flexural modulus of 165,000–290,000 psi and 130,000–175,000 psi, respectively. PP homopolymers, in turn, are generally stiffer compared to HDPE, which has flex modulus in the range of 125,000–240,000 psi.

For polypropylene homopolymer, the glass transition temperature varies from −20 to −18°C (−4 to 0°F). However, in a number of cases it was recorded at −4°C (25°F), or between −1 and 5°C, that is, between 30 and 41°F [9]. What it all means in practical terms is that between −4 and 40°F polypropylene becomes brittle.

Flexural strength for polypropylene (6000–7000 psi) is much higher than that for polyethylene (around 1400 psi, if can be measured because of its high flexibility). Compressive strength for PP is also higher than that for HDPE, and for some specific examples they equal to 6720 and 4570 psi, respectively.

As it was mentioned above, polypropylenes are more prone to oxidation, hence, requiring significantly higher amounts of antioxidants and UV stabilizers compared to PE. It was shown that oxygen intake is much faster in polypropylene compared to that in PE [10]. The primary reason is in the "microbranched" chemical structure of PP (see above), containing tertiary hydrogens that makes formation of hydroperoxides in PP much easier compared to that in polyethylenes. Overall, the mechanisms of oxidation (both photo- and thermooxidation) in PP and PE are quite different. For example, the termination reaction rates for oxidation in PE are 100–1000 times faster compared to PP [11].

Typical polypropylenes used for extrusion of WPC have melt flow index (MFI) of 2–5 g/10 min. However, standard MFI for PP cannot be directly compared to standard MFI for PE, as these are measured at different temperatures (190 and 230°C, respectively), however under the same constant load (commonly 2.16 kg).

Polypropylene, as well as polyethylene, shows negligible water absorption (less than 0.01% after 24 h underwater submersion; some data show 0.008% for PP, and 0.03% for HDPE).

Thermal expansion–contraction of plastics will be considered in detail in Chapter 10, "Temperature-driven expansion–contraction of wood–plastic composites. Linear coefficient of thermal expansion–contraction." Here it can be briefly mentioned that this property is about the same with HDPE, polypropylene, PVC, ABS, and Nylons 6 and 6/6, and the respective coefficients of thermal expansion are all overlapping in the range of $2\text{–}7 \times 10^{-5}$ 1/°F ($4\text{–}13 \times 10^{-5}$ 1/°C). Only with LDPE the coefficient is noticeably higher and equal to $6\text{–}12 \times 10^{-5}$ 1/°F (10–22 1/°C) [12].

According to some other data the coefficient of expansion–contraction for HDPE is $8\text{–}11 \times 10^{-5}$ 1/°F, for LDPE 13.1×10^{-5} 1/°F, and for homopolymer PP $3.8-5.8 \times 10^{-5}$ 1/°F. That is, PP-based profiles can thermally move more compared with PE-based ones.

POLYVINYL CHLORIDE

In the WPC decking market, the total share of PVC-based materials is even smaller than those based on PP. There are only three commercially available brands—Boardwalk, Millenium, and Procell, the last two are only entering the market. According to a Saint Gobain's data, CertainTeed reported sales of their WPC boards Boardwalk for 1,361,000 euros in Q1 2005, which approximately corresponds (if sales are at the same level through the year) to less than $8 million in 2005. According to Principia Partners, sales of Boardwalk boards in 2005 reached $12 million. For a comparison, as it was indicated above, sales of PE-based Trex composite deck boards in 2005 was $294 million out of about $956 million total for WPC, and sales of PP-based Cross Timbers in 2005 was $27 million.

PVC is a thermoplastic polymer of a chemical structure

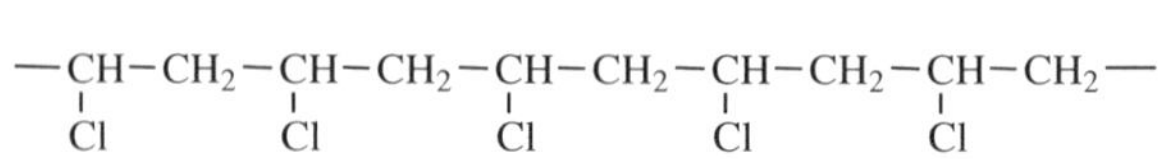

As with polypropylene (see above), PVC can be syndiotactic (chlorine atoms in the polymer chains on the same side of plane, as it is shown above in a very simplified manner because in reality chlorine atoms in the chain are in *zig-zag* stereoconfiguration), isotactic (on an alternate sides), and atactic (in a random arrangement with respect to plane of carbon atom chain). PVC is mostly atactic, but some portions of the polymer can be syndiotactic, making PVC crystalline at some regions (often as low as 5% crystallinity). Average molecular weight of PVC is often in the range of 100,000–200,000, with its number-average molecular weight of 45,000–64,000.

There are two principal forms of PVC, rigid and plasticized, or flexible. Rigid, unmodified PVC is stronger and stiffer than polyethylene and polypropylene. Credit cards are made of rigid PVC. Flexible, modified, plasticized PVC has a rubbery behavior more suited for tubing and shower curtains and is not employed in composite deck boards manufacturing. In order to reduce the cost of PVC-based WPC and to avoid difficulties with installation (sawing, nailing, and screwing) of tough PVC-based composite deck boards, most of PVC-based WPC are currently foamed.

PVC is the heaviest material compared to polyethylenes and polypropylenes. Specific gravity (density) for PVC is 1.32–1.44 g/cm^3, compared to HDPE (0.94–0.96 g/cm^3) and polypropylene (0.90–0.91 g/cm^3).

Compared to polyethylene and polypropylene, PVC has some inherent disadvantages, among them low thermal stability and high brittleness (for example, compared to HDPE). The high brittleness of PVC at ambient temperatures is caused by a relatively high glass transition temperature of PVC, which is typically in the range of 70–90°C (approx. 160–190°F). For crystalline regions of polymers at temperatures between the glass transition temperature and melting temperature, chain mobility is constrained by the crystalline regions of the polymer. The crystalline structures are not fully mobile until the temperature exceeds the crystalline melting point. Practically, below the glass transition temperature the polymer is more brittle.

One of principal beneficial properties of PVC is that it is inherently flame resistant. PVC contains about 57% chlorine. Flame spread index for a PVC-based WPC (Boardwalk) is 25, whereas HDPE-based WPC is in the range of 50–170, depending mainly on a shape of the board (solid or hollow) and content of mineral fillers. However, the self-ignition point for PVC-based composites is slightly lower than that for HDPE-based WPC (e.g., 345°C for Boardwalk, 395°C for Trex, and 436°C for EverX). Flash ignition points are about the same for all three said materials: 361°C for Boardwalk, 370°C for Trex, and 355°C for EverX (see Chapter 14, "Flammability and fire rating of wood–plastic composites," Table 14.3).

When it burns, PVC releases toxic hydrogen chloride, HCl, hence, it is commonly considered as the most environmentally damaging among thermoplastics. Even at temperatures above 70°C (158°F), PVC, if not stabilized, can start to degrade and release HCl. As this temperature can be reached in the south (the average July high temperature in Phoenix, AZ is 109°F, so the composite deck surface reaches a temperature of 150–160°C), the use of PVC-based composite boards in the south might be of a certain concern.

That is why PVC-based building materials, including WPC, are often considered as "not environmentally preferable" materials. Environmentally oriented studies place in that category such PVC-based products (containing wood or neat PVC) as Boardwalk (CertainTeed), Millenium (Wood Composite Technologies), AmeriDeck (American Composite Building Products), Procell (Procell Decking Systems), Country Estate (Nebraska Plastics), Deck Lok (Royal Crown), Deck/Dock (Westech Fencing), Dream Deck (Thermal Industries), EverNew, Bufftech (CertainTeed), Forever-Wood (Forever Wood), Oasis PVC Deck (Alcoa Home Exteriors), Sheerline (L.B. Plastics), Synboard (Synboard America), VEKAdeck (VEKA), and vinyl decking (Poly Vinyl Creations). However, as Principia has noticed [13], there is no discernible recognition among distributors or end users of any pushback against PVC. As Principia has estimated, PVC-based WPC make 4% of decking (that has not changed much between 2002 and 2005) and 22% of railing products [13].

There is another aspect of the thermally induced toxic and corrosive fumes of HCl relating to the processing PVC-based materials. PVC decomposes at 148°C (298°F). This causes corrosion in processing equipment and requires corrosion-resistant metals and coatings. Besides, PVC has a very narrow processing window and undergoes fast degradation at even a slight overheating. This thermal degradation results in a color change and malodor in the extrudate. The result of this color shift and thermal instability along with environmental concerns is that PVC recycling is often unrealistic.

Photodegradation of PVC outdoors takes place, naturally, in the upper layer of the product, with a thickness of the degraded layer in the range of 0.2–0.3 mm [14].

Flexural strength for PVC (6000 to 10,000–16,000 psi) is higher than that for polypropylene (6000–7000 psi) and HDPE (which is too flexible to break, or sometimes breaks around 1400 psi). It is interesting, though, that a PVC-based solid composite board "Boardwalk" (CertainTeed) has flex strength of only about 2700 psi—apparently, because of foaming of the material. For a comparison, commercial wood–PP composites have flex strength in the range of 5300–6200 psi. There might be some inherent problem in making Boardwalk, particularly in the light of a recent announcement on a voluntary recall of Boardwalk HFS™ Planks (on July 27, 2005) due to some incidents of planks breaking on decks [15]. However, the relatively low-flex strength of Boardwalk can be explained by just its foaming.

Flexural modulus for PVC (350,000–600,000 psi) is also higher compared to polypropylene (165,000–250,000 psi) and HDPE (125,000–240,000 psi).

Compressive strength for PVC (close to 11,000 psi) is higher compared to that of polypropylene (6700 psi) and HDPE (4600 psi).

PVC exhibits very low water absorption—about 0.1% after underwater submersion for 24 h. This places it close to polypropylene and polyethylene, which both show negligible water absorption (less than 0.01% after 24 h underwater submersion, see above).

In terms of thermal expansion–contraction, PVC is practically overlapping with other thermoplastics (HDPE, PP, ABS, Nylons) in the range of $2–7 \times 10^{-5}$ 1/° F ($4–13 \times 10^{-5}$ 1/°C). This will be discussed in more detail in Chapter 10,

"Temperature-driven expansion–contraction of wood–plastic composites. Linear coefficient of thermal expansion–contraction."

ACRYLONITRILE–BUTADIENE–STYRENE COPOLYMER (ABS)

This is a group of tough, rigid thermoplastics, in which all three monomers can vary in a fractional amount to tailor properties of the final copolymer. The acrylonitrile component contributes strength, heat resistance, and chemical resistance. The butadiene component contributes impact resistance, toughness, and flexibility. The styrene component contributes rigidity and processability. Disadvantages of ABS include poor durability/weatherability, low fire resistance, high density/weight, and a relatively high cost compared to polyolefines.

An attractive property of ABS as a plastic for WPC is its relatively low melt point of 100–110°C (212–230°F). This keeps wood fiber from burning during processing. However, a recommended processing temperature for neat ABS is typically in the range of 177–260°C (351–500°F), which makes WPC manufacturing rather difficult.

There is apparently only one commercial ABS-based WPC product, containing two-thirds ABS and one-third wood flour. The product is a railing system. Compared to HDPE, the ABS material behaved quite favorably with respect to temperature. Flex strength for ABS-based bottom rails at 70°F and 130°F were 5200 ± 200 psi and 4400 ± 200 psi, respectively. That is, "high temperature loss" was 800 ± 400 psi (15% only). For the balusters the figures were 6000 ± 100 psi and 4800 ± 80 psi, respectively, and "high temperature loss" was 1200 psi (20% only). For HDPE-based WPC this "high temperature loss" often reaches 50% (see below).

Flex modulus values for ABS-based bottom rails were 560,000 ± 40,000 psi and 500,000 ± 70,000 psi at 70 and 130°F, respectively, hence, the "high temperature loss" was 11%, and for balusters the respective figures were 460,000 ± 20,000 and 390,000 ± 30,000 psi, hence, the "high temperature loss" was 15%. The highest figure, 20%, was taken for the temperature adjustment, unlike 50% for HDPE-based railing system (see Chapter 7). This allowed to reduce the code requirement: (a) for in-fill from 187.5 lb for HDPE-based material to 150 lb for ABS-based one, (b) for uniform load on the top rail from 1125 lb for HDPE-based rail to 900 lb for ABS-based rail, and (c) for concentrated load test (at top of the rail) from 750 to 600 lb. As a result, the ABS-based railing system has met all the code requirements, unlike the corresponding HDPE-based railing system.

The above data also show how superior ABS-based WPC is when compared to HDPE-based WPC in terms of flexural strength and flexural modulus. Under the same conditions (ambient temperature), flex strength for HDPE-based WPC was 2200 psi, whereas that for ABS-based WPC it was 5200–6000 psi. Flexural modulus was 360,330 ± 33,600 psi (for HDPE-based railing) and 560,000 ± 40,000 psi (for ABS-based railing).

Overall, flex strength for neat ABS is 4300–6400 psi compared to about 1400 psi for HDPE, 6000–7000 psi for polypropylene, and 6000 to 10,000–16,000 psi for PVC. Flex modulus is 130,000–420,000 psi for ABS, 125,000–240,000 for HDPE, 165,000–250,000 psi for polypropylene, and 350,000–600,000 psi for PVC.

Tensile strength for ABS is around 6800 psi compared to 3500–5300 psi for polypropylene, 3000–5000 psi for PVC, and 10,150–10,875 for Nylon 6 and Nylon 66.

Compressive strength for ABS is 6750 psi compared to 4600 psi for HDPE, 6700 psi for polypropylene, and about 11,000 psi for PVC.

ABS shows higher water absorption (0.3% after 24 h underwater, 0.7% at saturation) compared to HDPE, PP, and PVC (less than 0.01, 0.01, and 0.1%, respectively). However, an ABS-based WPC, containing 50% w/w of 80-mesh maple wood flour, showed a very reasonable water absorption performance, namely 5.5% after 10 days, 8% after 20 days, and 9.5% after a month [16]. These values are similar with those for many HDPE-based WPC products.

In terms of thermal expansion–contraction, ABS is practically overlapping with other thermoplastics (HDPE, PP, PVC, Nylons) in the range of $2\text{–}7 \times 10^{-5}$ 1/° F ($4\text{–}13 \times 10^{-5}$ 1/°C). This will be discussed in more detail in Chapter 10.

An increased interest in manufacturing WPC deck boards from ABS might be observed soon as a result of a project aiming by using recycled ABS from discarded computers [17].

It was mentioned above that disadvantages of ABS include poor durability/weatherability, low fire resistance, high density/weight, and a higher material cost. Besides, it was observed that processing of ABS-based WPC is a rather difficult task, primarily due to high viscosity of hot melt and, hence, very high torque (it is very easy to get torque values near 80% of maximum torque); die pressure fluctuations (die surging); severe melt fracture, particularly at a high wood fiber content, such as 50% w/w; difficulties with pelletizing extruded strands (extrusion instability) [16].

NYLON 6 AND OTHER POLYAMIDES

Currently there are no commercial wood–nylon deck boards or other wood-filled composite materials. Melting temperatures of Nylon 6, 216–223°C (421–433°F) and Nylon 6/6, 255–265°C (491–509°F) are seemingly too high to avoid burning wood. It is generally considered that only those thermoplastics are applicable in WPCs that can be processed at temperatures below 200°C (about 400°F). However, using delignified cellulosics, either as it is or blended with thermo insulating minerals, and/or decreasing a residence time for cellulosic materials in compounders and extruders by increasing processing speed and by other means to shorten contact time of cellulosics with hot melt, can make cellulose–Nylon composites possible. These composites could possess superior properties, such as high strength and high modulus, as Nylons are structural plastics.

Nylon is a family of polyamide polymers characterized by the presence of amide group–CONH. Nylon 66 comprises approximately 75% of the U.S. consumption of polyamide polymers, with Nylon 6 comprising most of the remainder. Except the difference in melting point (see above), their properties are almost identical. Density (specific gravity) of both the Nylons is 1.13–1.14 g/cm^3. Other types of Nylons are 4, 9, 11, and 12. As Nylon 6 has a lower melting point compared to Nylon 66, it is more suitable for making—potentially—WPC materials.

Each nylon has a very sharp melting point. The rather wide range of melting points (see above) reflects different polyamide materials. Nylons also have low melt viscosity, convenient for injection molding, but difficult to handle in extrusion. That is why in extrusion processing a wide molecular weight distribution of nylon is often preferred, along with using a screw with a short compression zone and an ability to reduce temperature at the exit to increase melt viscosity. Nylon is often processed at 232–271°C (450–520°F).

Nylon is a crystalline polymer with high strength, modulus, and impact resistance. The general chemical structure of Nylon is as follows:

$$-CH_2-\underset{\underset{O}{\|}}{C}-NH-CH_2-CH_2-CH_2-CH_2-CH_2-\underset{\underset{O}{\|}}{C}-NH-CH_2-CH_2-$$

The type of nylon, such as Nylon 6, is identified by a number of carbon atoms in the repeating block. The repeating block in Nylon 6 is

$$-NH-CH_2-CH_2-CH_2-CH_2-CH_2-\underset{\underset{O}{\|}}{C}-$$

This is defined by the number of carbon atoms in the monomers used in the polymerization. When two different monomers are used in the preparation of the polymer, the nylon is identified using two numbers, such as Nylon 6/6 or Nylon 66.

Nylon properties are greatly affected by the amount of crystallinity, which in turn is controlled, to a great extent, by the cooling of the polymer. A slowly cooled nylon will have greater crystallinity, in the range of 50–60%. A rapidly cooled nylon will have crystallinity as low as 10%. In addition, slowly cooled nylon contains larger crystals than when rapidly cooled.

Due to the presence of the polar groups (CO-NH), nylon attracts water. Hence, water absorption by nylon is relatively high, and higher than by any other thermoplastics, considered in this book. Compared to water absorption (24 h) of HDPE and PP (less than 0.01%), PVC (0.1%) and ABS (0.3%), Nylon 6 absorbs 1.2% water in 24 h (underwater), and 9% water when reached saturation. The lower the degree of crystallinity of nylon, the higher the water absorption. Due to this hygroscopicity, nylons must be dried before melt processing.

Because of its hygroscopicity, the moisture content of nylon affects the glass transition temperature of the polymer. Dried nylon has the glass transition temperature near 50°C (122°F), though other data place it between 45 and 57°C (113 and 135°F), whereas wet nylon can have it from 20°C (68°F) to 0°C (32°F). The last figure apparently refers to water freezing temperature in wet nylon. At any rate, dry nylon is more brittle at ambient temperature compared to that of wet nylon.

Flexural strength of Nylon 6 is from 9700 to 14,000–16,500 psi, that is higher than for all other thermoplastics considered in this book, except PVC, which generally has a similar flex strength (6000 to 10,000–16,000 psi). For a comparison, flex strength of HDPE is around 1400 psi (when can be measured), PP 6000–7000 psi, ABS 4300–6400 psi, sometimes to 12,000 psi [18].

Flexural modulus of Nylon 6 is 100,000–464,000 psi, which is in the same range with that of PVC and ABS, and generally higher than that of HDPE and polypropylene.

Compressive strength of Nylon 6 is 12,000 psi, higher than that of PVC (10,830 psi), ABS (6750 psi), polypropylene (6720 psi), and HDPE (4570 psi).

Coefficient of friction of Nylon 6 against itself (0.36) is generally between that of HDPE (0.10–0.23) and those of LDPE and PVC (0.50) and polypropylene (0.67) [19].

Nylon is commonly considered as an engineering thermoplastic. In terms of wear, it is one of the most superior among other thermoplastics. The most wear-sensitive among common thermoplastics is polyethylene, following by polypropylene and PVC [20].

Nylon 6 has a high flame-resistance and is a self-extinguishing material. Generally, nylon is not considered as a weather-stable material.

As examples of commercial product line, Ultramid® polyamides, supplied by BASF, can be considered. They are molding compounds based on Nylon 66 (Ultramid A) and Nylon 6 (Ultramid B), and also on their copolymer (Ultramid C). There are many brands available—more than 10 variants of Ultramid B of "general purpose," plus flame retarded PA6 (Ultramid® 8232 G HS FR BK-102), impact modified PA6 (16 versions), mineral reinforced PA6 (30- and 40%-mineral reinforced, three versions), glass reinforced (26 versions, from 14- to 63%-glass fiber-filled material), and so on. There are eight variants of Ultramid A available for "general purpose," besides those of flame retarded, impact modified, and reinforced brands. These materials are employed in almost all fields of engineering and for many special applications, due to their high strength, high rigidity, and thermal stability. Many of them afford good impact resistance even at low temperatures.

Two of the most common concerns associated with nylon polymers are brittleness at low temperatures and moisture absorption. For example, a nylon material dried to a moisture content of 0.20% may flow 30–40% faster than the same material dried 10 times better, to 0.02% [21]. In some cases it could be taken as a sign of the polymer degradation.

However, it is moisture absorption that gives nylon its toughness. Upon initial injection molding manufactured nylon profiles are typically very hard and brittle. After they left to cure, they absorb moisture from the air and become very impact resistant. The curing (or "softening") period for nylon profiles takes 2–3 weeks. Clearly, at outdoors conditions it usually takes much less time.

CONCLUSION

Table 2.2 compares some important properties of thermoplastics used in making WPC materials on a commercial scale, or, as in case of Nylon 6, can be used in a near future.

TABLE 2.2 Some important properties of thermoplastics [12, 21–23]

Material	Specific gravity (density) (g/cm^3)	Flexural properties		Compressive strength (psi)	Water absorption after 24 h (%)	Coefficient of thermal expansion–contraction, $\times 10^{-5}$ (1/°F)	Weather stability
		Strength, psi	Modulus, psi				
LDPE	0.910–0.925	No break	30,000–50,000	No break	< 0.01	6–12 (13.1)	Medium
HDPE	0.941–0.965	~1400 or no break	125,000–240,000	4570	< 0.01	3–6 (8–11)	Medium
PP (homopolymer)	0.90–0.91	6000–7000	165,000–290,000	6720	0.008	3.8–5.8	Medium w/UV stabilizers
PVC (rigid)	1.32–1.44	6000 to 10,000–16,000	350,000–600,000	10,830	0.1	2.8–3.3	Good for light colors
ABS	1.01–1.08	4300–6400	130,000–420,000	6750	0.3	5.3	Poor
Nylon 6	1.12–1.14	9700 to 14,000–16,500	100,000–464,000	12,000	1.2	4.8	Poor

Despite unfavorable mechanical properties of polyethylene compared to polypropylene, PVC, ABS, and Nylon, HDPE is the most popular plastic in WPCs. Polypropylene is too tough and makes difficult to use nails and screws as fasteners at a deck installation. Polypropylene-based composites require special fastening systems. PVC is typically considered as not environmentally friendly. Other polymers are not weather stable, brittle, or expensive.

Globally, manufacturing and sales of HDPE, LLDPE, polypropylene, and some engineering resins, such as nylon, are likely to grow above average growth rates of plastics (5% or better) annually from 2007 to 2009, whereas PVC and LDPE will likely show below-average growth (4% or less annually) [24]. It is expected that PVC will be gradually replaced by polyolefins and other resins, spurred by environmental concern. This concern might not be always justified scientifically, but public perception is a powerful thing, which should not be underestimated.

Obviously, the resin price is a very serious factor in choosing plastics for WPC. However, in a long run a situation with prices is largely unpredictable. As of November 27, 2006, for example, resin-pricing chart looked as follows (Table 2.3)

TABLE 2.3 ***Plastic News*** **resin pricing chart, November 27, 2006. Prices are given for annual volumes greater than 20 million pounds, in U.S. cents per pound for prime resin or as indicated, unfilled, natural color, FOB supplier**

Resin/grade	Price for prime resin	Lowest price in the plastic grades	Price for recycled plastics, pellets
HDPE, HMW, extrusion	83–85 37–39[a]	69–70	55–58 31–36[a]
LDPE, fractional melt, extrusion	76–78	76–78	51–55
Polypropylene, profiles, extrusion	81–84 35–38[a]	77–79	40–44 20–25[a]
ABS, general purpose, extrusion	92–94	80–81	50–60 (mixed colors)
PVC, general purpose, homopolymer	63–67 27–29[a]	63–65	28–34 13–21[a]
Engineering plastics, annual volume greater than 1 million pounds			
ABS, extra high impact	106–116	106–116	—
PVC/ABS (alloy)	119–124	119–124	—
Nylon 6	165–175	165–175	—
Nylon 66	172–180	172–180	—
Nylon/ABS (alloy)	164–174	164–174	—

[a]Plastic News, May 7, 1998 (as a reference).

ADDENDUM: ASTM TESTS COVERING DEFINITIONS OF TECHNICAL TERMS AND THEIR CONTRACTIONS USED IN PLASTIC INDUSTRY AND SPECIFICATIONS OF PLASTICS

The following ASTM standards explain terminology used in this chapter.

ASTM D 883 "Standard Terminology Relating to Plastics"

Block copolymer: an essentially linear copolymer in which there are repeated sequences of polymeric segments of different chemical structure.

Branched polyethylene plastics: those containing significant amount of both short-chain and long-chain branching and having densities in the 0.910–0.940 g/cm^3 range.

Crosslinking: the formation of a three-dimensional polymer by means of interchain reactions resulting in changes in physical properties.

Ethylene plastics: plastics based on polymers of ethylene or copolymers of ethylene with other monomers, the ethylene being in the greatest amount by mass.

Glass transition: the reversible change in an amorphous polymer or in amorphous regions of a partially crystalline polymer from (or to) a viscous or rubbery condition to (or from) a hard and relatively brittle one.

Glass transition temperature (T_g): the approximate midpoint of the temperature range over which the glass transition takes place.

High-density polyethylene plastics (HDPE): those linear polyethylene plastics, having a standard density of 0.941 g/cm^3 or greater.

Homopolymer: a polymer resulting from polymerization involving a single monomer.

Isotactic: pertaining to a type of polymeric molecular structure containing a sequence of regularly spaced asymmetric atoms arranged in like configuration in a polymer chain.

Linear low-density polyethylene plastics (LLDPE): those linear polyethylene plastics, having a standard density of 0.919–0.925 g/cm^3.

Linear medium-density polyethylene plastics (LMDPE): those linear polyethylene plastics, having a standard density of 0.926–0.940 g/cm^3.

Linear polyethylene plastics: those containing insignificant amounts of long-chain branching but which may contain significant amounts, by design, of short-chain branching.

Low-density polyethylene plastics (LDPE): those branched polyethylene plastics, having a standard density of 0.910–0.925 g/cm^3.

Medium-density polyethylene plastics (MDPE): those branched polyethylene plastics, having a standard density of 0.926–0.940 g/cm^3.

Melt temperature: the temperature of the molten plastic.

Nylon plastics: plastics based on resins composed principally of a long-chain synthetic polymeric amide that has recurring amide groups as an integral part of the main polymer chain.

Polyethylene: a polymer prepared by the polymerization of ethylene as the sole monomer.

Polypropylene: a polymer prepared by the polymerization of propylene as the sole monomer.

Poly(vinyl chloride): a polymer prepared by the polymerization of vinyl chloride as the sole monomer (vinyl chloride content in monomer not less than 99%;).

Propylene plastics: plastics based on polymers of propylene or copolymers of propylene with other monomers, the propylene being in the greatest amount by mass.

Rigid plastic: for purpose of general classification, a plastic that has a modulus of elasticity, either in flexure or in tension, greater than 700 MPa (100,000 psi) at 23°C and 50%; relative humidity when tested in accordance with Test Methods D 747, D 790, D 638, or D 882.

Note: ASTM methods D 790 is described in Chapter 7, D 638 in Chapter 8.

Thermoplastic: a plastic that repeatedly can be softened by heating and hardened by cooling through a temperature range characteristic of the plastic, and that in the softened state can be shaped by flow into articles by molding or extrusion.

ASTM D 1600 "Standard Terminology for Abbreviated Terms Relating to Plastics"

Acrylonitrile–butadiene–styrene plastics	ABS
High-density polyethylene plastics	HDPE
Linear low-density polyethylene plastics	LLDPE
Linear medium-density polyethylene plastics	LLDPE
Low-density polyethylene plastics	LDPE
Medium-density polyethylene plastics	MDPE
Nylon (see also polyamide)	PA
Poly(vinyl chloride)	PVC
Polyamide (Nylon)	PA
Polyamide 6	PA6
Polyamide 66	PA66
Polyethylene	PE
Polypropylene	PP
Ultra-high molecular weight polyethylene	UHMWPE

ASTM D 1784 "Standard Specifications for Rigid Poly(Vinyl Chloride) PVC Compounds and Chlorinated Poly(Vinyl Chloride) (CPVC) Compounds"

Of properties, related to the subject of this book, the following specifications can be quoted with respect to PVC homopolymer (all—the minimum property values):

- Impact resistance, ASTM D 256, <34.7 J/m of notch, or <0.65 ft. lb/in. of notch
- Tensile strength, ASTM D 638, <5000 psi
- Modulus of elasticity in tension, ASTM D 638, <280,000 psi
- Flammability, ASTM D 635, <25 mm; average time of burning, <10 s.

ASTM D 1972 "Standard Practice for Generic Marking of Plastic Products"

The following are examples of plastic products marking:

- For example, for products made from acrylonitrile–butadiene–styrene polymer:

>ABS<

- For a polypropylene containing 30 mass percentage of mineral powder:

>PP–MD30<

Note: abbreviations for mineral powder (MD), flame retardants (FR), glass fiber (GF). Presence of a flame retardant is indicated by "FR." The "FR" should be followed by the code number to identify the flame retardant.

- For a polyamide 66 containing a mixture of 15 mass percentage of mineral powder (MD) and 25 mass percentage of glass fiber (GF):

>PA66–(GF25 + MD15)<

or:

>PA66–(GF + MD)40<

ASTM D 4066 "Standard Classification System for Nylon Injection and Extrusion Materials (PA)"

General-purpose Nylon plastics (minimum requirements)

Nylon 6

- Density 1.12–1.15 g/cm3
- Tensile strength 10,150–10,875 psi
- Flexural modulus 319,000–348,000 psi
- Impact resistance 3–4 kJ/m2.

Nylon 66

- Density 1.13–1.15 g/cm3
- Tensile strength 10,150 psi

- Flexural modulus 333,500 psi
- Impact resistance 3.3 kJ/m^2.

ASTM D 4101 "Standard Specification for Polypropylene Injection and Extrusion Materials"

General-purpose polypropylene homopolymer (minimum requirements)

- Density 0.910 g/cm^3
- Tensile strength 3480–3988 psi
- Flexural modulus 116,000–152,250 psi
- Izod impact resistance 12–32 J/m.

ASTM D 4216 "Standard Specification for Rigid Poly(Vinyl Chloride) (PVC) and Related PVC and Chlorinated Poly(Vinyl Chloride) (CPVC) Building Products Compounds"

Of properties, related to the subject of this book, the following specifications can be quoted with respect to rigid PVC (all—the minimum property values):

- Impact resistance, ASTM D 256, >34.7 J/m of notch, or >0.65 ft. lb/in. of notch
- Impact resistance, drop dart. ASTM D 4226, >4450 J/m, or >1.0 ft. lb/in.
- Tensile strength, ASTM D 638, >5000 psi
- Modulus of elasticity in tension, ASTM D 638, >290,000 psi
- Coefficient of linear expansion, $<4 \times 10^{-5}$ 1/C, or $<2.2 \times 10^{-5}$ 1/F.

ASTM D 4396 "Standard Specification for Rigid Poly(Vinyl Chloride) (PVC) and Chlorinated Poly(Vinyl Chloride) (CPVC) Compounds for Plastic Pipe and Fittings Used in Nonpressure Applications"

Of properties, related to the subject of this book, the following specifications can be quoted with respect to rigid PVC (all—the minimum property values):

- Impact resistance, ASTM D 256, 40.0 J/m of notch, or 0.65 ft. lb/in. of notch
- Tensile strength, ASTM D 638, 3000 psi
- Modulus of elasticity in tension, ASTM D 638, 280,000 psi.

ASTM D 4673 "Standard Classification System for Acrylonitrile–Butadiene-Styrene (ABS) Plastics and Alloys Molding and Extrusion Materials"

Of properties, related to the subject of this book, the following specifications can be quoted with respect to ABS materials suitable for extrusion and injection molding (all—the minimum property requirements).

ABS—medium impact

- Impact resistance, ASTM D 256, 80–150 J/m (Izod), 3–8 kJ/m^2 (Charpy)
- Tensile strength, ASTM D 638, 5075–7250 psi
- Flexural modulus of elasticity, ASTM D 790, 290,000–377,000 psi.

ABS—high impact

- Impact resistance, ASTM D 256, 200–440 J/m (Izod), 13–34 kJ/m^2 (Charpy)
- Tensile strength, ASTM D 638, 3625–5075 psi
- Flexural modulus of elasticity, ASTM D 790, 261,000–348,000 psi.

ASTM D 4976 "Standard Specification for Polyethylene Plastics Molding and Extrusion Materials"

All PE plastics–fractional melt (melt index 0.4–1.0 g/10 min at 190°C/2.16 kg). The figures below are minimum requirements.

Branched LDPE, density 0.910–0.925 g/cm^3

- Tensile strength at yield, ASTM D 638, 1378 psi
- Elongation at break, ASTM D 638, 300%
- Secant modulus at 2% strain, 18,125 psi.

Branched MDPE, density >0.925–0.940 g/cm^3

- Tensile strength at yield, ASTM D 638, 1595 psi
- Elongation at break, ASTM D 638, 200%
- Secant modulus at 2% strain, 36,250 psi.

Linear LDPE, density 0.910–0.925 g/cm^3

- Tensile strength at yield, ASTM D 638, 1450 psi
- Elongation at break, ASTM D 638, 400%
- Secant modulus at 2% strain, 50,750 psi.

Linear MDPE, density >0.925–0.940 g/cm^3

- Tensile strength at yield, ASTM D 638, 2175 psi
- Elongation at break, ASTM D 638, 200%
- Secant modulus at 2% strain, 87,000 psi.

Linear HDPE, density >0.940–0.960 g/cm^3

- Tensile strength at yield, ASTM D 638, 2755 psi
- Elongation at break, ASTM D 638, 400%
- Secant modulus at 2% strain, 72,500 psi.

Linear HDPE, density .0.960 g/cm^3

- Tensile strength at yield, ASTM D 638, 4060 psi
- Elongation at break, ASTM D 638, 300%
- Secant modulus at 2% strain, 130,500 psi.

ASTM D 5203 "Standard Specification for Polyethylene Plastics Molding and Extrusion Materials from Recycled Postconsumer (HDPE) Sources"

Fractional melt indexes below are measured as g/10 min at 190°C/2.16 kg. The figures below (Table 2.4) are minimum specification requirements.

ASTM D 6263 "Standard Specification for Extruded Rods and Bars Made from Rigid Poly(Vinyl Chloride) (PVC) and Chlorinated Poly(Vinyl Chloride) (CPVC)"

Of properties, related to the subject of this book, the following specifications can be quoted with respect to PVC materials suitable for extrusion (all—the minimum property requirements).

- Impact resistance, ASTM D 256, 35J/m of notch (Izod), or <0.65 ft. lb/in. of notch
- Tensile strength, ASTM D 638, 7000 psi
- Tensile modulus, ASTM D 638, 400,000 psi
- Flexural modulus of elasticity, ASTM D 790, 375,000 psi.

TABLE 2.4 Specification values

Source of HDPE	Melt index (g/10 min)	Density (g/cm^3)	Tensile strength (psi)	Secant modulus (psi)
Blow molded or thermoformed containers for chemicals, food, personal care	0.2–0.6	<0.959	2030	90,000
Containers from milk, juice, water	0.4–1.0	>0.958	2900	97,000
Spunbonded materials	0.4–0.9	>0.955	2900	97,000
Injection molded articles from food and beverage base cups	>20	0.956–0.962	2500	90,000
Injection molded articles from housewares	4–20	0.956–0.962	2500	80,000

ASTM D 6779 "Standard Classification System for Polyamide Molding and Extrusion Materials (PA)"

General-purpose Nylon plastics (minimum requirements)

Nylon 6

- Density 1.12–1.15 g/cm^3
- Tensile strength 10,150–10,875 psi
- Tensile modulus 319,000–348,000 psi
- Charpy impact resistance 3–4 kJ/m^2.

Nylon 66

- Density 1.13–1.15 g/cm^3
- Tensile strength 10,150 psi
- Tensile modulus 333,500 psi
- Charpy impact resistance 3.3 kJ/m^2.

REFERENCES

1. *Natural & Wood Fiber Composites*, Vol. IV, No. 10, Principia Partners, Cleveland, OH, 2005, p. 1.
2. J. F. Shakelford and W. Alexander. *Materials Science and Engineering Handbook*, 3rd edition, CRC Press, Boca Raton, 2001, p. 98.
3. C. Vasile and M. Pascu. *Practical Guide to Polyethylene*, Rapra Technology Ltd., UK, 2005, p. 19.
4. C. Vasile and M. Pascu. *Practical Guide to Polyethylene*, Rapra Technology Ltd., UK, 2005, p. 38.
5. G. Wypych (Ed.), *Handbook of Material Weathering*, 3rd edition, ChemTec Publishing, Toronto, 2003, p. 361.
6. C. Vasile and M. Pascu. *Practical Guide to Polyethylene*, Rapra Technology Ltd., UK, 2005, p. 67.
7. A.J. Peacock (Ed.), *Handbook of Polyethylene: Structures, Properties, and Applications*, Marcel Dekker, New York, 2000, p. 194.
8. A.J. Peacock (Ed.), *Handbook of Polyethylene: Structures, Properties, and Applications*, Marcel Dekker, New York, 2000, p. 203.
9. V. Thirtha, R. Lehman, and T. Nosker. Glass transition phenomena in melt-processed polystyrene/polypropylene blends. In: *Polymer Engineering and Science*, Vol. 45, No. 9, Thompson Gale, Farmington Hills, MI, 2005, pp. 1187–1193.
10. G. Wypych (Ed.), *Handbook of Material Weathering*, 3rd edition, ChemTec Publishing, Toronto, 2003, p. 382.
11. G. Wypych (Ed.), *Handbook of Material Weathering*, 3rd edition, ChemTec Publishing, Toronto, 2003, p. 383.

12. A.J. Peacock (Ed.), *Handbook of Polyethylene: Structures, Properties, and Applications*, Marcel Dekker, New York, 2000, p. 181.
13. *Natural & Fiber Composites*, Vol. IV, No. 9, Principia Partners, Cleveland, OH, 2005.
14. G. Wypych (Ed.), *Handbook of Material Weathering*, 3rd edition, ChemTec Publishing, Toronto, 2003, p. 425.
15. *Natural & Fiber Composites*, Vol. IV, No. 9, Principia Partners, Cleveland, OH, 2005, p. 4.
16. S.-K. Yeh, S. Agarval, and R.K. Gupta. Process development for ABS-based wood-plastic composites. In: *Progress in Wood-Fibre Plastic Composites 2006 International Conference*, Toronto, Canada, May 1–2, 2006.
17. *Plastics News*, A Crain Publication, Crain Communications Inc., Chicago, IL, 2006, p. 7.
18. A.J. Peacock (Ed.), *Handbook of Polyethylene: Structures, Properties, and Applications*, Marcel Dekker, New York, 2000, pp. 681, 683, 685.
19. A.J. Peacock (Ed.), *Handbook of Polyethylene: Structures, Properties, and Applications*, Marcel Dekker, New York, 2000, p. 205.
20. A.J. Peacock (Ed.), *Handbook of Polyethylene: Structures, Properties, and Applications*, Marcel Dekker, New York, 2000, p. 200.
21. M. Sepe. The material analyst, Part 72: When investigations change direction. *Injection Molding Magazine*, Canon Communications, Los Angeles, CA, April 2006 (on-line edition).
22. J.F. Shakelford and W. Alexander (Eds.). *Material Science and Engineering Handbook*, 3rd edition, CRC Press, Boca Raton, 2001, pp. 93, 96, 681, 683, 685, 799–801, 1619.
23. A.J. Peacock. *Handbook of Polyethylene: Structures, Properties, and Applications*. Marcel Dekker, New York, 2000, Chapter 5, pp. 135, 138.
24. J. Fordyce. Additives market on the mend. In: *Plastics Engineering*, Vol. 62, No. 11, Society of Plastic Engineers, Brookfield, CT, 2006, pp. 10–12.

3

COMPOSITION OF WOOD–PLASTIC COMPOSITES: CELLULOSE AND LIGNOCELLULOSE FILLERS

INTRODUCTION

The term "wood–plastic composites" (WPCs) refers to wood as a proxy for fibrous materials of plant origin. It can be wood flour or sawdust, or agricultural plant residues, typically cut, milled, or ground, or other types of natural fiber, such as hemp, jute, and kenaf, commonly as a by-product of the respective industrial process.

Delignified cellulose is used as a commercial WPC filler in only one form, namely as a constituent of Biodac®, in GeoDeck deck boards. It is delignified by virtue of chemical treatment of wood pulp for papermaking applications. All other cellulose-based fillers for WPC, such as wood flour, sawdust, and rice hulls, are natural materials, containing cellulose, hemicellulose, and lignin.

This composition has practically unavoidable impact on processing and properties of the resulting composite product. Briefly, cellulose provides positive effect on mechanical and other properties of the composite material (such as decreased coefficient of thermal expansion–contraction, etc.); lignin generally makes the product weaker, easily burns in the course of processing and releases CO_2 and other gaseous products, making the product density lower, and greatly accelerates fading of the WPC after outdoor exposure. Wood extractives (terpenes, pinenes, tannins, carbonyl compounds, etc.) produce volatile organic compounds (VOC), also making the product density lower. Hemicellulose easily decomposes at plastic melt temperatures, particularly at sharp changes of pressure, and forms acetic acid,

Wood-Plastic Composites, by Anatole A. Klyosov

TABLE 3.1 WPC deck boards, manufacturing companies, and wood and natural fiber as a filler as described in manufacturer's ICC-ES reports

No.	Deck	Manufacturer	Natural fiber	ICC-ES report number and date
Polyethylene-based products				
1	ChoiceDek, Dreamworks, LifeCycle, MoistureShield, A.E.R.T.	Advanced Environmental Recycling Technologies; Weyerhaeuser	Not reported(50% wood fiber)	NER-596 2/1/2006
2	Epoch/Evergrain	Epoch Composite Products	50% wood fibers	ESR-1625 (6/1/2005) NER-630 (Legacy) 4/1/2006
3	EverGreen	Integrated Composite Technologies	Wood flour	N/A (referenced by Principia Partners, 2006)
4	EverX, Latitudes, Veranda	Universal Forest Products Ventures II	50% wood flour	ESR-1573 6/1/2005
5	Fiberon, Perfection, Veranda	Fiber Composites; LMC	Wood fibers	22-41 (Legacy) 10/1/2004
6	GeoDeck	LDI Composites	Rice hulls, Biodac	ESR-1369 6/1/2006
7	Lakeshore	Bluelinx	50% wood flour	ESR-1573 6/1/2005
8	Life Long	Brite Manufacturing	50% wood fiber	ESR-1279 6/1/2005
9	Monarch	Green Tree Composites	55% wood flour	ESR-1084 2/1/2005
10	Nexwood	Nexwood Industries	60% rice hulls	BOCA 99-8.1 (January 2000) Not current
11	Oasis	Alcoa Home Exteriors	55% wood floor	ESR-1425 6/1/2005
12	Premier	Composatron Manufacturing	50% wood flour	ESR-1481 6/1/2005
13	Rhino Deck	Master Mark Plastic Products	50% wood fibers	ESR-1461 6/1/2005
14	SmartDeck	SmartDeck Systems	—	N/A Not current
15	Tendura	HB&G Building Products; Tendura Industries	60% wood flour	N/A (referenced by Principia Partners, 2006)
16	Timberlast (railing system)	Kroy Building Products	50% wood flour	ESR-1422 6/1/2005
17	TimberTech	TimberTech	Wood flour	ESR-1400 6/1/2005
18	Trex	Trex Company	50% wood fiber	ESR-1190 6/1/2005

TABLE 3.1 ***(Continued)***

No.	Deck	Manufacturer	Natural fiber	ICC-ES report number and date
19	UltraDeck	Midwest Manufacturing Extrusion	60% wood flour	ESR-1674 11/1/2004
20	XTENDEX, E-Deck	Carney Timber	60% rice hulls	NER-695 (Legacy) 11/1/2004
21	WeatherBest, LP Composite, Veranda	Louisiana Pacific	60% wood flour	ESR-1088 6/1/2005
Polypropylene-based products				
22	CorrectDeck	Correct Building Products	60% wood fibers	ESR-1341 6/1/2005
23	Cross Timbers	Elk Composite Building Products	60% wood fibers	ESR-1590 6/1/2005
24	K-Decking	Kenaf Industries	—	N/A (referenced by Principia Partners, 2006)
PVC-based products				
25	Boardwalk	CertainTeed	35–45% hardwood fiber	NER-576 3/ 1/2004
26	Millenium	Millenium Decking	Wood flour	ESR-1603 6/1/2006
27	Procell	Procell	Flax fiber	ESR-1667 11/1/2006

thereby causing significant corrosion of the equipment. This effect is particularly expressed at some moisture content in the lignocellulosic material and is called "steam explosion" of lignocellulosics.

Table 3.1 lists WPC deck boards (and one railing system), commercially available and registered by the ICC in 2006 or before (with two exceptions). Most of the materials contain wood flour (or "wood fiber"), typically in 50–60% amount by weight. Only four products contain rice hulls as a cellulosic filler. One product—GeoDeck—contains delignified cellulose fiber as a part of its principal filler Biodac® (which in turn contains about 50% delignified cellulose and 50% minerals).

Pure cellulose fiber has a relatively low density, such as 1.0–1.1 g/cm^3. However, being penetrated with lignin and hemicellulosics, its density, such as in wood flour or rice hulls, reaches 1.3–1.5 g/cm^3. When ashed at 525°C (977° F), wood flour leaves 0.13–0.40% of ash and rice hulls—as much as 18.8% of ash. This is because rice extracts much of silica from soil with water through its capillary system. When stored, particularly in summer time, the moisture content of wood flour and rice hulls can reach 10%. Multiple measurements of rice hulls moisture content at the Kadant Composites plant in Green Bay, WI gave 9.3 ± 0.6% of moisture. In a direct contact with water, moisture content in cellulose can easily exceed its own weight.

Pore sizes in cellulose fiber are about 10 nm (0.01 μm). Only the smallest HDPE chains, with molecular weight smaller than 10,000 Da, can enter pores of cellulose. Cellulose fiber diameters are typically between 3 and 5 μm.

Properties of cellulose fiber will be considered in more detail in the next section and at the description of specific plant materials.

A BRIEF HISTORY OF CELLULOSE FILLERS IN WPC IN U.S. PATENTS

Materials now known as WPCs first appeared as thermosetting molding compounds in the 1960s. Other thermosetting wood composites had appeared much earlier (see below). This section provides a brief look at the development of WPC for the last 40 years or so, not pretending to quote all relevant patents and publications. I would restrict the consideration only by U.S. (and occasionally other) patents, citing the most significant (to my personal viewpoint) developments.

Five periods in the development of WPC can be tentatively identified:

1. Development of thermosetting composites. These types of materials are mentioned here only because they are still called by some "wood composites." In fact, they are much older than the 1960s and have entered the industrial applications after World War II. The earliest of them, Bakelite was invented in the beginning of 1900s and used wood flour in 1920s. Thermoset wood composites created a base for the appearance of WPCs.
2. Transition of adhesives in thermoset "wood composites" to thermoplastic polymeric materials and realization that a key point would be a uniform mixing of cellulose fiber with the plastic.
3. Improving of mechanical properties of WPC. Of course, other properties of WPC were recognized as important ones, but mechanical properties always were considered as key properties for building, construction materials, such as deck boards.
4. Improving the compatibility of the filler with the polymer matrix, by providing an interaction between the filler and the polymer, using coupling agents.
5. Further improving the properties of WPC by using plastics other than polyethylene, for example, polypropylene, polyvinyl chloride (PVC), and acrylonitrile-butadiene-styrene (ABS).

The newest history of WPCs in the current understanding of this term has begun in the early 1990s, when advanced environmental recycling technologies (AERT) and a division of Mobil Chemical Company that later was named Trex began to make solid WPC deck boards (and other WPC items) consisting of 45–55% wood fiber in polyethylene. At about the same time, Andersen Corporation began to make PVC-based wood composite products such as door profiles and then window frames. Since the mid of 1990s, other WPC manufacturers have begun to expand the market.

These WPC milestones are marked with U.S. patents granted to AERT in 1992 (Nos. 5,082,605; 5,088,910; 5,096,046) [1–3], Andersen Corporation in 1996 (Nos. 5,486,553; 5,539,027; 5,585,155) [4–6], Strandex Corporation in 1996 (No. 5,516,472) [7], Trex in 1998 (Nos. 5,746,958 and 5,851,469) [8, 9], Crane in 1998 (No. 5,827,462) [10] and 1999 (No. 5,866,264) [11], Nexwood in 1999 (No. 5,863,480) [12], CertainTeed in 2003 (No. 6,590,014) [13], and Kadant Composites in 2004 (No. 6,758,996) [14]. More patents are listed below in this chapter.

Currently, long-fiber cellulosic fiber attracts more and more attention as a measure to further improve WPC. Then, nanoparticles are claimed to be a promising additive to address some shortcomings in WPC, such as their relativelty high flammability (except those based on PVC). Foamed WPC attract attention of researchers and developers in the course of the recent several years. Biodegradable plastics, such as polylactic acid, are considered by many as a promising direction in WPC, mainly for interior applications. The last four directions—long fiber, nanoparticles, foamed WPC, and biodegradable WPC—cannot be considered as yet as a mainstream in WPC development, though they should be mentioned in this context.

Patents obtained by leading manufacturing companies are marked with the names of those companies.

Beginning of WPC: Thermosetting Materials

The first WPC—in the 1960s—were thermosetting molding compounds containing cellulose fiber as filler.

For example, U.S. Pat. No. 3,367,917 [15] describes a thermosetting melamine–formaldehyde–benzoguanamine resinous molding composition containing a fibrous filler, such as α-cellulose pulp, in an amount between 25 and 42% by weight.

U.S. Pat. Nos. 3,407,154 [16] and 3,407,155 [17] describe thermosetting urea–formaldehyde and aminoplast resinous molding composition comprising fusible reactive urea–formaldehyde and aminotriazine–formaldehyde resin, respectively, and purified α-cellulose fibers (14–25% by weight) as a filler.

The next innovations in this area are described in U.S. Pat. Nos. 4,282,119 [18] and 4,362,827 [19]. In accordance with these patents, particleboards were produced employing a binding agent that was a combination of a polyisocyanate with an aminoplast resin containing 0.25–0.65 mole of formaldehyde per mole equivalent of amino groups.

U.S. Pat. No. 3,546,158 [20] describes a flooring composition involving a terpolymer, a nonfibrous filler (calcium carbonate, silica, clay, kaolin, and carbon black) and a fibrous filler, such as wood flour, cellulose fibers, and asbestos. Compositions may include 25–53% by weight of nonfibrous filler and 17–40% by weight of fibrous filler, with the fillers accounting for 50–80% of the total composite weight.

U.S. Pat. No. 3,720,641 [21] describes a process of blending an aromatic polyamide molding resin with reinforcing fillers, such as glass fibers, asbestos, cellulose fibers, and cotton fabric paper, with the fillers ranging from 2 to 70% by weight based on the total molding composition.

U.S. Pat. No. 5,288,775 [22] describes moldable thermoset acrylic polymer composites containing 3–15% of cellulose fibers, fillers, and water; the composite is a hard, high molecular crosslinking type that requires a chemical hardener.

U.S. Pat. No. 5,767,177 [23] describes a thermosetting composition comprising 33–43% of a thermosetting polyester resin, 5–15% of cellulose fiber of wood or cotton origin, 15–21% of mineral fillers (calcium carbonate or hydrated alumina), and 12.5–22.5% of reinforcement fibers such as glass, carbon, or Kevlar.

U.S. Pat. No. 5,767,178 [24] describes a thermoset (or a cold-set) composition of a phenol–aldehyde, a urea–aldehyde, or a polyurethane, mixed with cellulose fibers and latex, containing magnesium oxychloride or magnesium oxysulfate to improve fire resistance of the resulting composite.

A thermosetting composition, comprising both PVCs and polyester resins along with sawdust and mineral fillers, is also described in U.S. Pat. No. 5,783,125 [25] (by Crane Plastics Co.).

U.S. Pat. No. 6,291,558 [26] discloses a composite material prepared by spraying a urea–formaldehyde resin mixed with a thermosetting lignosulfonate-based resin material to wood particles, to assist in the adhesion of the wood particles.

U.S. Pat. No. 6,702,969 [27] discloses a composite material prepared by mixing a thermoset resin, such as phenolic resin, urea resin, melamine resin, epoxy resin, urethane resin, and mixture thereof, to bind wood pieces and a filler, such as natural and synthetic graphites, metal, carbon, and other similar compounds and their mixtures.

Cellulose as a Reinforcing Ingredient in Thermoplastic Compositions

Earlier attempts—in the 1970s—to make cellulose-filled thermoplastic compositions had identified a serious obstacle. It became recognized that fillers, particularly cellulose fibers, do not disperse easily throughout the plastic formulations during compounding and molding. Accordingly, the finished products typically do not exhibit the desirable physical characteristics ordinarily associated with fiber-reinforced plastic composites. This problem has been dealt with in a number of patents.

U.S. Pat. No. 3,856,724 [28] describes a composite based on polypropylene or low-density polyethylene (LDPE) (density 0.92) and 5–45%, preferably 20%, by weight of α-cellulose (100-mesh flock) along with some additives.

U.S. Pat. No. 3,875,088 [29] describes a composite material comprising 50–75% of a thermoplastic resin binder (ABS or rubber-modified polystyrene) and 20–40% of wood flour (40 mesh and 100 mesh), with the ratio of plastic to wood flour being between 1.5 and 3.0.

U.S. Pat. No. 3,878,143 [30] describes a composite material comprising 63% by weight of PVC or polystyrene or ABS, and 30% of wood flour along with some minor additives.

U.S. Pat. No. 3,943,079 [31] describes a composite material comprising thermoplastic PVC polymer and cellulose fiber as major components, the cellulose fiber being wood pulp or cotton linters in amounts ranging from 16 to 30% by weight of the total.

U.S. Pat. No. 4,165,302 [32] describes filled thermoplastic resin compositions comprising LDPE, polypropylene and other resins (in amounts ranging from 95 to 50% by weight), organic fillers (such as wood flour), and inorganic fillers (such as fly ash or calcium carbonate). This patent is concerned primarily with increasing the melt flow index of filled thermoplastic resin compositions rather than their mechanical properties.

U.S. Pat. No. 4,250,064 [33] describes usage, along with low-density organic fibers (such as polyester fiber or cellulosic fiber), of a combination of coarse and fine inorganic filler such as calcium carbonate (20–50% by weight), which makes the organic filler more easily and uniformly dispersed in a plastic matrix (preferably chlorinated polyethylene or a vinyl chloride/vinyl acetate copolymer), avoiding visible clumps of fiber.

U.S. Pat. No. 4,339,363 [34] describes a thermoplastic resin composition involving crushed wastepaper (40–60% by weight), polyethylene, polypropylene or other thermoplastic resin and their combinations, and optionally an inorganic filler, such as calcium carbonate, talc, barium sulfate, or the like (8–12% by weight). This patent indicates that such compositions provide higher heat resistance, flame retardancy, and mechanical strength compared with those made of synthetic resins, pure or blended with woodmeal or having incorporated inorganic filler.

U.S. Pat. No. 4,343,727 [35] describes a thermoplastic compounded injection molding material comprising PVC and cellulosic fiber (Solca-Floc, 5–20% by weight) along with epoxidized soybean oil, hydrocarbon extenders, and stearic acid as a lubricant.

Another method of improving the dispersibility of cellulose fibers in a thermoplastic matrix is described in U.S. Pat. No. 4,414,267 [36], according to which cellulose fibers (hardwood kraft, from 1 to 40% by weight of the final composite) are pretreated by slurrying them in water, contacting them with an aqueous suspension of a mixture of a vinyl chloride polymer and a plasticizer, and drying the thus-treated fibers.

Another approach to improving cellulose filler dispersibility is described in U.S. Pat. No. 4,559,376 [37], according to which cellulose or lignocellulose material is subjected to a hydrolytic pretreatment using diluted hydrochloric or sulfuric acid. Essentially, the treatment converts cellulose or lignocellulose material to a fine powder of microcrystalline cellulose, which exhibits better dispersibility in a polymer matrix such as high-density polyethylene (HDPE).

U.S. Pat. No. 5,474,722 [38] describes a composite material 20–80% of which is a cellulosic material (ground wood, sawdust, wood flour, rice hulls, etc.) and polyethylene.

U.S. Pat. Nos. 5,746,958 [8] and 5,851,469 [9] (both—by Trex Company) disclose production of a wood–thermoplastic composite material comprising a wood component (40–65% by weight, and preferably 52%) and polyethylene (60–35%), preferably HDPE and preferably 48% by weight.

U.S. Pat. No. 6,610,232 [39] describes a process for preparing a moldable thermoplastic composite composition comprising 20–50% by weight of a reinforcing lignocellulose fiber filler having a length of at least about 15 mm and a diameter of less than 0.5 mm, and including a significant percentage by weight of "hair-like" fibers.

U.S. Pat. No. 6,743,507 [40] discloses cellulose-fiber-reinforced composites comprising a matrix polymers such as polyethylene, polypropylene, copolymers, terpolymers and mixture thereof in an amount ranging from about 25 to 99% by weight, and cellulose pulp having an α-cellulose purity of greater than about 80, 90, or 98% by weight.

U.S. Pat. No. 6,758,996 [14] (by Kadant Composites, now LDI Composites) describes cellulose-reinforced thermoplastic composite in which delignified cellulose is agglomerated along with minerals, calcium carbonate and kaolin clay, in wet papermaking sludge, into lightweight dried small porous granules. These granules easily and uniformly mix with molten plastic, and due to hydrophobicity of the granules (their oil absorption capacity is 150% by weight) and their porosity, they form a good bond with the polymer matrix.

U.S. Pat. No. 6,833,399 [41] discloses flax bast fibers and flax shives as reinforcing agent for a thermoplastic resin in a composite comprising 15–70% by weight of flax portion and the thermoplastic resin such as polyethylene, polypropylene, PVC, styrene, and other polymers.

U.S. Pat. No. 6,929,841 [42] discloses a plastic-based composite product comprising a thermoplastic polymer such as polypropylene and polyethylene, and two types of wood or natural fiber particles, that is, small particles of a length of 0.2–2 mm, and large particles being larger than the small particles and having a length of 2–6 mm. The wood- or nonwood fiber particles are present in the plastic mass in amounts of 50–70% by weight. Fibers of natural cellulose are selected from flax, jute, hemp, sisal, coconut, and other nonwood materials. Nonwood fibers can include glass fibers with aspect ratio of 300–400 and in amounts of 3–25%.

U.S. Pat. No. 6,939,496 [43] discloses a system and method of extruding plastic–cellulose fiber composite material, comprising 60–95% by weight of a thermoplastic polymer such as polypropylene, polyethylene, and PVC, 20–30% of cellulose fiber obtained from such materials as wood sawdust (60 mesh particle size), hemp, flax, straw and wheat, and a binding agent mixed with the filler.

Improving Mechanical and Other Properties of WPC

In the 1980s, more attention has been increasingly paid to improving the physical properties, such as mechanical strength, stiffness, resistance to thermal deformation, and so forth, of the fiber–plastic composite products. It was recognized that moisture in cellulosic fiber leads to steam formation in the compounder and the extruder, increases porosity of the final product and decreases its density, accelerates the oxidation of the hot melt during processing, and makes the final product weaker and less durable.

U.S. Pat. Nos. 4,687,793 [44] and 4,783,493 [45] describe elimination of moisture from cellulosic fiber (wood flour, rice hulls, waste paper, pulp, etc.) before blending them with a thermoplastic polymer (polypropylene, polyethylene, ABS, PVC, etc.), by heat-treating (80–150°C, that is, 176–300° F) the cellulosic material with glyoxal (0.5–12% by weight of the untreated cellulose). The patentees indicate that this treatment produces a cellulosic filler largely free of moisture, results in molded products free of the cavities or blisters typically caused by moisture-generated steam at molding temperatures of 160°C (320° F.) and higher, and improves the physical properties of molded products (containing 10–65% of this cellulose filler by weight).

U.S. Pat. No. 4,833,011 [46] describes an approach to reducing the water content of cellulose fibers (bleached chemical or thermomechanical pulp) prior to mixing them with the thermoplastic resin (particularly an unsaturated polyester resin, such as an epoxy resin, with a crosslinking accelerator). The disclosed drying method involves thermal treatment (heating the fibers) or the chemical treatment (treating the fibers with alcohols, such as ethanol, polyvinyl alcohol, butanol, or heptanol). The mix is heated and pressed in a mold, causing its crosslinking and curing.

U.S. Pat. No. 4,717,743 [47] discloses polyolefin-based resin compositions comprising newspaper sheets cut to small pieces (3–6 mm in size, 20–60% by weight of total composite, with 20–40% by weight given in the examples) and thermoplastic polymer such as low-density or HDPE or polypropylene.

U.S. Pat. No. 4,737,532 [48] describes a thermoplastic resin composition comprising PVC, an ABS resin, and a high amount of wood flour (up to 60% by weight, as given in the examples), pressed at 200°C (392° F) to obtain composite sheets.

U.S. Pat. No. 6,368,529 [49] discloses a lignocellulose–thermoplastic resin composition comprising calcium borate to make the material resistant to attack by wood-destroying fungi and insects. The preferred calcium borates are calcium polytriborates and calcium hexaborates, in amounts from about 0.1 to 4% by weight. The lignocellulosic material is selected from the group consisting of wood, flax, hemp, jute, bagasse, and straw.

U.S. Pat No. 6,527,532 [50] (by Trex Company) discloses a method of making a wood–thermoplastic composite material, according to which the extruded profile is quenched in a nonoxidizing environment, such as water.

U.S. Pat No. 6,578,368 [51] discloses a method for cooling extruded and molded cellulosic–polymer composites, among other plastics and plastic compositions; the said method uses cryogenic fluid with a temperature below about −250° F, such as liquid oxygen, liquid nitrogen, liquid neon, liquid hydrogen, and liquid helium.

U.S. Pat No. 6,590,014 [13] (by Certainteed Corporation) discloses a thermoplastic or thermosetting composite, based on PVC, polyethylene, polypropylene, nylon, polyesters, epoxies, and so on, filled with wood flour, particularly with oak flour, containing tannins, and including chemical compounds capable to react with tannins and thereby preventing water staining of the composite profile. According to the patent, the chemical compounds are those having at least two functional groups, such as succinic anhydride, succinic acid, maleic acid, maleic anhydride, and so on.

U.S. Pat No. 6,627,676 [52] discloses an antimicrobial biocidic fiber-plastic composite comprising a cellulose fiber material, an industrial waste plastic, and biocides such as sodium hypochlorite, vitamin E, and citric acid. Before adding biocides, lignin is removed from the fiber material through a delignification process such as the one utilizing steam explosion, and after adding the biocides, lignin is added back to the mixture to allow the cellulose fibers to bind with the plastic material.

U.S. Pat No. 6,827,995 [53] discloses a single coextrusion process of making a WPC as a hollow profile comprising a weatherable outer layer made of a first polymeric material, a core layer made of a thermoplastic polymeric foamed composition including a wood component, and an inner layer made of a third thermoplastic material. The first and second polymeric materials are PVC or acrylonitrile–styrene–acrylic polymer or a combination thereof. The third polymeric material is PVC.

U.S. Pat. No. 6,844,040 [54] discloses a coextrusion process of making reinforced WPC structural members comprising hollow profiles formed from a thermoplastic–cellulose fiber composite material and reinforcing sections bonded to the hollow profile.

U.S. Pat. No. 6,844,049 [55] discloses PVC–wood composite profiles having a natural wood grain finish by applying multiple paint transfers onto the composite.

U.S. Pat. Nos. 6,881,367 [56] and 6,890,637 [57] (both by Elk Composite Building Products) disclose a process of forming of a wood–thermoplastic composite material, comprising 25% by weight of 20–50 mesh wood fiber, 40% of plastic such as polyethylene or polypropylene, and 35% of chopped fiberglass of about 0.25–1 in. in length.

U.S. Pat. No. 6,903,149 [58] discloses a plastic–wood composite comprising 100 parts by weight of a thermoplastic resin such as PVC, polyolefin, polystyrene, a polyester resin, and a polyamide resin, 5–400 parts by weight of wood flour, and 0.05–20 parts by weight of a dibasic polyol ester having at least one hydroxyl group esterified with fatty acid.

U.S. Pat. No. 6,958,185 [59] (by Crane Plastics Co., TimberTech) describes a wood–plastic composition comprised of a layer of a first composition that is secured to a layer of a second composition, with the resulting multilayer component having an improved combination of appearance, strength, durability, weight, weatherability, and other characteristics.

Improving the Compatibility of the Filler with the Polymer Matrix: Coupling Agents

Attention has also been paid to improving the compatibility of the filler with the polymer matrix, by providing an interaction between the filler and the polymer. "Compatibilizers" can be chemically attached either to cellulose fibers, or to a polymer, or can form a covalent chemical bond between the two, or can help to form a sort of uniform "alloy" between two or several polymeric components ordinarily not very compatible with each other. Or compatibilizers can improve adhesion between fiber and plastic by other means. Generally, compatibilizers markedly improve physical properties of the polymeric composite, its weatherability, and overall performance.

U.S. Pat. No. 4,376,144 [60] discloses a thermoplastic composite in which cellulose fiber (hardwood pulp, from 1 to 40% by weight) is attached to the PVC matrix with an isocyanate bonding agent (a cyclic trimer of toluene diisocyanate).

A similar approach that uses alkyl isocyanates to bind cellulose fibers (from 1 to 50% by weight of the composite) to thermoplastic polymers such as polyethylene is described in U.S. Pat. No. 4,791,020 [61]. In both cases the patentees suggest that cellulose fibers treated in such a way have improved dispersability into the polymer and improved adhesion thereto. As the last patent indicates, the precise mechanism by which the bonding occurs is not known, but the active isocyanate moieties in the bonding agent apparently react with the hydroxyl groups on the cellulose fibers, forming a chemical bonding with the latter.

Another compatibilizer, a silylating agent, is described in U.S. Pat. Nos. 4,820,749 [62] and 5,008,310 [63], along with a series of isocyanate bonding agents, such as polymethylene polyphenylisocyanate, 1,6-hexamethylene di-isocyanate, and others. These patents disclose a reinforced composite material comprising a thermoplastic or thermosetting polymer, such as low-density polyethylene, polypropylene or polystyrene, ground wood pulp (10–40% by weight), inorganic filler, such as clay, calcium carbonate, asbestos, and glass fibers (10–30% with respect to wood pulp), maleic anhydride (0–5%), and γ-aminopropyltriethoxysilane or similar silylating agents (0.1–8% by weight) or isocyanates (0.1–20% by weight). The patentees indicate that the silane or isocyanate grafting and bonding process results in better flowability of molten polyethylene and better physical properties of the finished composite product.

U.S. Pat. No. 5,120,776 [64] also discloses using of maleic anhydride (0–4%) to improve the bonding and dispersibility of an extruded mixture of 60–90% of HDPE and 40–10% of wood flour (or chemithermomechanical pulp).

U.S. Pat. No. 4,851,458 [65] describes a composite material comprising PVC and cellulose filler (cellulose fiber or wood meal, preferably from 3 to 20% by weight of total) along with an "adhesion promoter" of undisclosed chemical structure or composition; the adhesion promoter is applied (coated or sprayed) to the fibers prior to incorporation thereof into the polymer. The patentees claim that the disclosed composite exhibits less shrinkage at elevated temperatures compared with PVC itself or filled with calcium carbonate; in particular, in a test at 110°C for 1 h, the patentees indicate that PVC exhibits shrinkage of 2–4% of the original length, whereas the disclosed PVC–cellulose composite shows shrinkage below 1% at the same conditions.

U.S. Pat. No. 5,153,241 [66] discloses composites made of thermoplastic polymers such as low-density polyethylene, polypropylene or polystyrene (90–60%) and wood pulp or sawdust (10–40% by weight) grafted with a titanium coupling agent (isopropyltri[n-ethylaminoethylamino] titanate) in acetone, along with some inorganic fillers, such as calcium carbonate and Portland cement.

U.S. Pat. No. 5,516,472 [7] (Strandex Corporation) discloses a composite having approximately 26% HDPE and 65% wood flour, extruded in the presence of zinc stearate (2%) as a lubricant along with phenolic resin and polyurethane as minor additives and crosslinking agents (4–1.3%, respectively).

U.S. Pat. No. 5,952,105 [67] describes a thermoplastic composition comprising sheared poly-coated paper (50–70% by weight) and polyethylene (30–50%). An example provided in the patent describes making an 80-g batch of a compression-molded composite comprising HDPE (39%), a poly-coated paper (scrap milk cartons, 59%), and a coupling agent (Polybond 3009, 2%).

U.S. Pat. No. 5,973,035 [68] by the same authors describes a similar thermoplastic composition comprising sheared paper (50–70% by weight) and polyethylene (30–50%). An example provided in this patent describes production of an 80-g batch of a compression-molded composite comprising HDPE (39%), sheared scrap newspapers or magazines (59%), and a coupling agent (Fusabond 100D, 2%).

U.S. Pat. No. 6,207,729 [69] describes a similar thermoplastic composition comprising shredded and sheared cellulosic materials (33–59% by weight) such as old newspapers, magazines, kenaf, kraftboard, and so on, HDPE (33–50%), calcium carbonate (11–17%), and a coupling agent (Fusabond 100D, 2%).

U.S. Pat. No. 5,981,631 [70] describes achieving compatibility of the ingredients in a thermoplastic composition by extruding a polymer (polyethylene, polypropylene, or PVC), wood fibers, and a coupling agent containing fatty acids and rosin acids, both having at least 16 carbon atoms.

U.S. Pat. No. 6,939,903 [71] discloses a process for preparing a WPC material comprising treatment of natural fiber such as wood flour, wood fiber, hemp, flax, and kenaf, with a reactive organosilane, mixing the treated fiber at its weight amount of 20–85% with HDPE, adding a functionalized polyolefin coupling agent such as maleic anhydride modified polypropylene or polyethylene (molecular weight between 20,000–300,000, and a ratio of maleic anhydride to mole of polymer is 0.6–310 for the both polymers).

U.S. Pat. No. 6,942,829 [72] discloses a method of forming a polymer–wood composite material comprising 20–80% by weight of cellulose filler such as hardwood fiber, softwood fiber, hemp, jute, rice hulls, and wheat straw, 20–80% of a thermoplastic polymer such as polypropylene, polyethylene, polyamides, polyesters, and other polymers, 0.1–10% of a blend of a nonionic compatibilizer and a lubricant.

U.S. Pat. No. 6,983,571 [73] discloses a composite roofing panel comprising polyethylene, polypropylene, or a combination thereof in an amount of 20–40% by weight, and natural plant fiber in an amount of 40–75%, selected from the group consisting of wood flour, sugarcane bagasse, hemp, coconut coir, jute, kenaf, sisal, flax, coir pith, rice hulls, and cotton.

U.S. Pat. No. 7,041,716 [74] describes a thermoplastic composite comprising a polyolefin such as polyethylene or polypropylene in an amount of 20–90% by weight, a cellulose filler such as wood flour in an amount of 30–80% by weight, a basic reactive filler such as glass fibers in an amount of 5–25% by weight, and a carboxylic acid anhydride graft polyolefin such as maleic anhydride grafted polypropylene in an amount of 1–5%.

Plastics Beyond HDPE in Wood–Plastic Composite Materials

Attention in 1990s was paid to alternative thermoplastic materials (e.g., other than polyethylene).

U.S. Pat. Nos. 5,082,605, 5,088,910, and 5,096,046 (all three—by Advanced Environmental Recycling Technologies, Inc.) disclose a composite made of 40–60% of plastic (LDPE, or a combination of 60% LDPE and 40% HDPE, or having 10–15% of polypropylene of the total amount of plastic) and about 60–40% of wood fiber.

U.S. Pat. No. 5,288,772 [75] discloses a cellulose-fiber-reinforced thermoplastic composition for compression molding, where thermoplastic material is polypropylene, or a mixture of polypropylene, polystyrene, and polyethylene (40–90% of plastic by weight), and the cellulosic material (10–60% by weight) was milled scrap newspaper with an initial moisture content of at least 30% by weight. The patentees suggest that lignin present in the cellulosic scrap provides a coherent mass of thermoplastic and cellulosic material.

Canadian Patent Nos. 2,100,320, 2,178,036, 2,242,326 [76–78], and U.S. Pat. Nos. 5,486,553, 5,539,027, 5,585,155, 5,827,607, 5,948,524, 6,015,611 [4–6, 79–81], and 6,015,612 [82] (all seven—by Andersen Corporation) describe a composite material comprising a PVC polymer (50–70%) and wood fiber (sawdust, 30–50%).

Canadian Patent No. 2,250,161 [83] and U.S. Pat. No. 5,738,935 [84] disclose a thermoplastic composite comprising PVC (45–90%), cellulose fiber (10–55%), a porosity aid (0.01–5%), and an interfacial agent (0.01–2%).

U.S. Pat. No. 5,480,602 [85] discloses a composite comprising polypropylene, polyethylene, or their combination along with lignocellulosic particles (50–70% by weight) and a polyurethane coupling agent (15–3% by weight of the mixture).

U.S. Pat. No. 5,539,027 [5] (by Andersen Corporation) discloses a composite structural member comprising PVC and wood fiber (30–50% by weight) and having the coefficient of thermal expansion of $1.5–2.5 \times 10^{-5}$ 1/F.

U.S. Pat. Nos. 5,635,125 [86] and 5,992,116 [87] describe a molded composite of wood sawdust particles (34–44%) and ground-up recycled PVC particles (55–65%).

U.S. Pat. Nos. 5,827,462 [10] and 5,866,264 [11] (both—Crane Plastics Co., TimberTech) describe extruded thermoplastic composites comprising 20–40% HDPE or PVC and 50–70% of wood flour.

U.S. Pat. No. 5,863,480 [12] (by SRP Industries Ltd., Nexwood) discloses a thermoplastic composite of polyethylene, polypropylene, vinyls or other extrudable plastics, cellulosic fiber such as sawdust, wood flour, ground rice hulls, and so on, fillers, and lubricants. The patentees describe the extrusion occurring through a die at a temperature below the melting point of the polymer, so that the deformation of the polymer takes place in the solid phase, making the product biaxially oriented.

Canadian Patent No. 2,278,688 [88] discloses a thermoplastic composite material 50–60% of which is polyethylene or polypropylene, 10–30% of which is wood powder, and 10–35% of which is a silicate (mica).

U.S. Pat. Nos. 6,011,091, 6,248,813 [89, 90] (both—Crane Plastics Co., TimberTech) disclose a cellulosic composite comprising 50% to about 75% by weight of a cellulosic material, 25–50% by weight of PVC, and 0–4% by weight of a thermosetting material.

U.S. Pat. No. 6,617,376 [91] (Crane Plastics Co., TimberTech) discloses a cellulosic composite comprising about 20% to about 55% by weight of a cellulosic material, 45–80% by weight of chlorinated polyethylene, and melt-processing rubbers.

U.S. Pat. Nos. 6,117,924 [92] and 6,984,676 [93] (both by Crane Plastics Co., TimberTech) describe an extrusion process of making a cellulose–plastic composite material; the process includes a significantly higher compression ratio through which the extrusion product must pass. Plastics include polyethylene, polypropylene, PVC, LDPE, and other commercial polymers.

U.S. Pat. No. 6,171,688 [94] describes a composite material comprising wood flour having a particle size between 50 and 400 μm, dried natural fibers (e.g., kenaf fiber) of width of 40–60 μm and length of 0.6–2.5 cm (aspect ratio of 15–60), hollow microspheres of a copolymer of vinylidene chloride and acrylonitrile having a diameter between about 15 and 50 μm, hollow glass microspheres having a diameter between about 50 and 200 μm, and a polyester polymer as cured polymer matrix. According to the patentees, the polymer microspheres reduce voids and prevent cracking of the glass microspheres upon compression molding.

U.S. Pat. Nos. 6,265,037, 6,680,090, and 6,682,789 [95–97] (all three by Andersen Corporation) disclose a composite structural member comprising 25–40% by weight of polypropylene, 50–75% by weight of wood fiber of an aspect ratio of 1:2–1:5, and 0.1–5% by weight of maleic anhydride grafted polypropylene.

U.S. Pat. No. 6,270,883 [98] describes reinforced composites containing less than 50% by weight of cellulose pulp fibers dispersed in a thermoplastic matrix comprising nylons, such as Nylon 6, Nylon 12, Nylon 66, or mixtures thereof, and the cellulose pulp comprising fibers with a lignin content less than 2% by weight.

U.S. Pat. No. 6,274,248 [99] discloses an impact-resistant composite material comprising an olefin series plastic (70–82% by weight), wood cellulose filler (10–43% by weight), and mica (10–50% by weight) with the aspect ratio of 60–70.

U.S. Pat. No. 6,284,098 [100] discloses a composite material comprising a thermoplastic polymer (50–80% by weight) and long lignocellulose fibers as a reinforcing filler (20–50% by weight), of at least about 20% of which are fibers with length greater than 15 mm.

U.S. Pat. No. 6,337,138 [101] (by Crane Plastics Company, TimberTech) discloses a cellulosic, inorganic-filled plastic composite, comprising 25–40% of polyethylene, 30–70% of cellulosic material, such as wood fiber, seed husks, rice hulls, newspaper, kenaf, coconut shells, and 1–20% (by weight) of talc.

U.S. Pat. Nos. 6,344,504 [102] and 6,498,205 [103] (both by Crane Plastics Company, TimberTech) disclose a thermoplastic-based composite, comprising at least 16% by weight of a powdered thermoplastic material, at least 50% by weight of cellulosic material, and a lubricant, a phenolic resin, and a compound containing one or more isocyanate groups, each of at least 1% by weight. A powdered thermoplastic material comprises polyethylene, LDPE, polypropylene, ethyl–vinyl acetate (EVA), and polyethylene copolymers.

U.S. Pat. No. 6,511,757 [104] (by Crane Plastics Company, TimberTech) discloses a thermoplastic-based composite, obtained by compression molding, and comprising about 20–40% by weight of a thermoplastic, such as polyethylene or PVC, about 50–70% by weight of a cellulosic material such as wood flour, and between 0 and 30% by weight of additives, such as crosslinking agents, lubricants, accelerators, inhibitors, enhancers, compatibilizers, and blowing agents.

U.S. Pat. Nos. 6,586,503 [105] and 6,737,006 [106] (both by Correct Building Products) disclose a polypropylene-based composite, comprising polypropylene, preferably reactor flake polypropylene, in amounts of 20–80% by weight, and cellulose fibrous material, such as sawdust, newspaper, alfalfa, straw, cotton, rice hulls, kenaf, and other cellulosics, in amounts of 20–80% by weight.

U.S. Pat, Nos. 6,122,877 [107] and 6,682,814 [108] (both by Andersen Corporation) disclose a cellulosic fiber–polymeric composite comprising 45–70% of thermoplastic polymers such as PVC, polyethylene and its copolymers, polystyrene, polyacrylate, polyester and their mixtures, and 30–65% of wood fiber, such as sawdust.

U.S. Pat. No. 6,780,359 [109] (by Crane Plastics Company, TimberTech) discloses a cellulosic fiber–polymeric composite comprising mixing a cellulosic material, such as wood fiber, with a plastic material, such as HDPE, LDPE, PVC, chlorinated PVC, polypropylene, EVA, ABS, and polystyrene, to form a cellulosic reinforced plastic composite.

U.S. Pat. No. 6,971,211 [110] (by Crane Plastics Company, TimberTech) discloses a cellulosic fiber–polymeric composite comprising a cellulosic filler in an amount of 30–70% by weight, a plastic material, such as PVC and polypropylene in an amount of 30–70% by weight. The PVC material may include 1–10 parts of a stabilizer, 2–12 parts of a lubricant, and 0.5–8 parts of a process aid per 100 parts of the PVC resin. The polypropylene material includes 10–20 parts of a lubricant per 100 parts of the polypropylene resin.

U.S. Pat. No. 7,022,751 [111] describes a fiber-reinforced composite plastic material comprising thermoplastic polymers such as HDPE, LDPE, polypropylene, PVC, and polystyrene; a high melting point waste polymer fiber material such as polyethylene terephthalate and nylon, an inorganic filler, such as glass and other material, and an organic filler such as wood or particles of a thermoset plastic, such as rubber and polyurethane foam.

Cellulose–Polyolefin Composite Pellets

Another innovation in the 1990s involved compounding thermoplastic polymers and cellulosic materials along with other ingredients into a feedstock in the form of durable, easy to transport, and durable pellets.

U.S. Pat. No. 5,938,994 [112] describes a WPC material produced as feedstock pellets in a twin screw extruder, and comprising wood flour (about 20–80%) and polyethylene (80–20%) as a preferred plastic.

U.S. Pat. Nos. 5,948,524, 6,015,612 [80, 82] and 6,210,792 [113] (all—by Andersen Corporation) disclose a wood–plastic composition manufactured as feedstock pellets comprising PVC and wood fiber.

U.S. Pat. No. 6,255,368 [114] describes plastic cellulosic composite pellets comprising 20–60% by weight of polyethylene, polypropylene or polystyrene, 40–80% of cellulosic fiber (jute, kenaf, sisal, bamboo, rice hulls, corn husks, wood fiber, and wood flour) with an aspect ratio of between 2 and 20 and a trace of mineral coating (talc) dispersed on the surface of the pellet.

U.S. Pat. No. 6,632,863 [115] (by Crane Plastics Company, Timber-Tech) discloses a wood–plastic composition manufactured as feedstock pellets comprising 55–90% cellulosic material such as wood flour and wood fiber, 10–40% of polyolefin such as HDPE, LDPE, and polypropylene, and 0–35% total of additive(s), such as lubricants and inorganic fillers, such as talc and mica.

U.S. Pat. Nos. 6,756,114 [116] and 7,052,640 [117] disclose a moldable pellet comprising a thermoplastic polymer(s) and a synthetic cellulosic fiber such as yarn or tow form such as Rayon or Lyocell in an amount of 2–80% by weight.

Foamed Wood–Plastic Composite Materials

A new and a challenging direction in making composite materials more lightweight and economical became visible by the middle of the 1990s. Foaming was, of course, well known in neat plastic materials, particularly having a high melt strength. A high content of cellulose fiber in plastics, particularly in polyethylene, makes the composite material much less suitable for foaming and significantly decreases the composite's flexural strength and flexural modulus. Therefore, foaming of WPC often requires reinforcing, coupling agents, or employing a strong plastic, such as PVC.

U.S. Pat. No. 6,153,293 [118] discloses a thermoplastic composite comprising wood fiber (40–60% by weight) and polyethylene (60–40%) along with a powdered endothermic foaming agent such as bicarbonate of soda. The patentees suggest that the foaming agent causes a greater degree of expansion of the composite, conferring a lower specific gravity of between 0.8 and 1.2 g/cm^3 with no significant loss of overall strength (flexural strength of 1676 psi).

U.S. Pat. No. 6,342,172 [119] (by Andersen Corporation) discloses a method of making a foamed wood–plastic composition comprising PVC, wood fibers, and a foaming agent, which is a nitrogen generating blowing agent or a carbon dioxide generating blowing agent or mixtures thereof.

U.S. Pat. No. 6,344,268 [120] (by CertainTeed Corporation) discloses a foamed fiber composite comprising 35–73% by weight of PVC, 27–65% wood fiber, and an acrylic modifier to improve the melt strength. The composite comprised 5–40% porosity and had specific gravity of 0.5–1.07 g/cm^3. The composite was made using a blowing agent in amounts of 1–1.5% by weight of PVC.

U.S. Pat. No. 6,590,004 [121] (by Crane Plastics Company, TimberTech) discloses a foamed fiber composite comprising 40–80% by weight of PVC or polypropylene, 20–60% by weight of cellulosic fiber, and additives such as stabilizer(s) in an amount of 1–10% per the polymer resin, lubricant(s) in an amount of 1–12% per the polymer resin, process aid(s) in an amount of 0.25–5% per the polymer resin.

U.S. Pat. No. 6,784,216 [122] (by Crane Plastics Company, TimberTech) discloses a foamed fiber-ABS-based composite comprising about 100 parts of ABS, 10–80 parts of cellulosic material, 2–10 parts of foam modifier, such as an acrylic foam modifier and styrene–acrylonitrile polymer, 0.5–2 parts of blowing agent such as hydrocerol, and 1–4 parts of a lubricant.

U.S. Pat. No. 6,863,972 [123] (by Crane Plastics Company, TimberTech) discloses a component for building or construction purposes comprising a composite PVC or polypropylene–cellulosic layer bonded to foamed PVC layer, with the composite layer comprising 20–70% by weight of PVC or polypropylene and 20–70% by weight of cellulosic filler, and the foamed layer comprising PVC or polypropylene (100 parts), stabilizer (1.5–7 parts), lubricant (3–10 parts), process aid (6–12 parts), and blowing agent (0.3–1 part). Bonding can be chemical, adhesive, or mechanical (fasteners) by nature.

U.S. Pat. No. 6,936,200 [124] discloses a process for producing foamed plastic-wood fiber composite, in which a blowing agent is introduced during a mixing step, and the blowing agent is a chemical or a physical blowing agent, the physical blowing agent being such as CO_2, N_2, He, Ar, air, or their mixture.

Biodegradable Wood–Plastic Composites

Ultimately, all WPCs are degradable and biodegradable in the grand scheme of time. This is particularly related to WPCs, having low amounts of antioxidants. "Biodegradable" WPCs are composites specially designed to have a short lifetime outdoors, especially when disposed of after the end of their intended service. Often, biodegradable WPCs are made for indoor applications, such as flooring.

A biodegradable thermoplastic composition comprising polyvinyl alcohols, polyurethanes or polyacrylates, cellulose fibers, and chitosan is disclosed in U.S. Pat. No. 5,306,550 [125].

Biodegradable thermoplastic composites, comprising aliphatic polyester urethanes or polyester amides and wood flour as a reinforcing material, are described in U.S. Pat. No. 5,827,905 [126].

U.S. Pat. No. 6,274,652 [127] discloses a biodegradable composite material comprising bacterial cellulose in a powdery state and a polymeric material such as polyhydroxybutyrate, polyhydroxyvalerate, polycaprolactone, polybutylenesuccinate, polyethylenesuccinate, polylactic acid, polyvinylalcohol, cellulose acetate, starch, and other biodegradable polymers.

U.S. Pat. No. 6,479,164 [128] discloses an extrudable or moldable biodegradable composite material comprising cellulosic fiber such as wood, wood chips, or cotton, and the starch-based biodegradable binder matrix.

It should be noted that although the developments include various thermoplastic composites based on waste materials, few of these materials are readily available on a widespread and cost-effective basis. Moreover, their proportion in the overall composite mixture tends to be somewhat low, because high proportions of nonplastic components can compromise mechanical properties. As a result, the environmental benefits offered by these compositions are limited.

Compairing the above claims with data of Table 3.1, one can see that despite versatile research into cellulose fillers, not many of them end up in commercial WPC materials, such as decking boards. There are still two principal cellulosic fillers in commercial deck boards—wood flour and rice hulls. Papermaking sludge, exemplified with its granulated form (Biodac®), is used in just one brand of deck boards, GeoDeck. Long natural fiber, such as kenaf, is still too expensive to be used in composite deck boards. Unless the long fiber is a nonutilized by-product with a cost well below 10 ¢/lb, it is hardly applicable in WPC. Typically, the long fiber costs in the range between 25 and 45 ¢/lb, when the whole composite materials typically cost around 25 ¢/lb. Hence, in order for long fiber to find way into WPC deck boards, its cost should be reduced at least twice, and better 10 times.

GENERAL PROPERTIES OF LIGNOCELLULOSIC FIBER AS FILLERS

Chemical Composition

The three principal components of plant materials are cellulose, lignin, and hemicellulosics. Cellulose and hemicellulosics are polysaccharides. Cellulose is a highly regular structure, crystalline polymer, made up of thousands of glucose residues, covalently bound "head-to-tail" (Fig. 3.1).

Hemicellulosics form much shorter branched chains consisting of five- and six-carbon ring sugars. These chains play a role of amorphous soft fillers, wrapping cellulose regions. "Hemicellulosics" is a collective term for a great many structures of heteropolysaccharides of plant origin, forming plant cell walls along with cellulose.

Lignin is a phenol propane-based amorphous solidified resin, filling the spaces between the polysaccharide fibers. Lignin is not just a "concrete" but also a highly engineered chemical structure (Fig. 3.2).

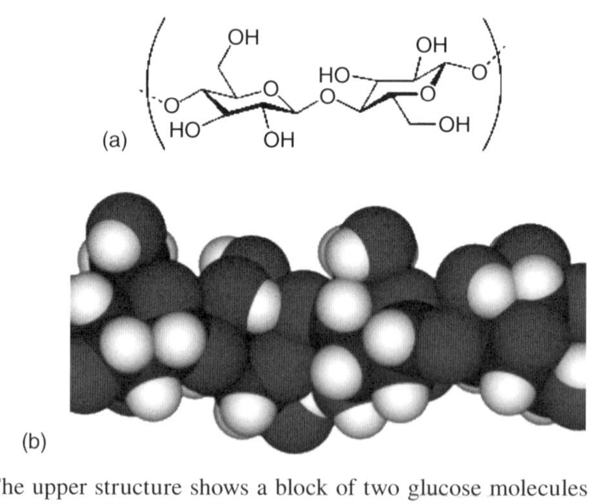

Figure 3.1 The upper structure shows a block of two glucose molecules in cellulose that is repeated thousands of times to make a more or less complete cellulose chain. Bundles of those chains form a cellulose fiber. The lower structure is a three-dimensional model of the same block of two glucose molecules.

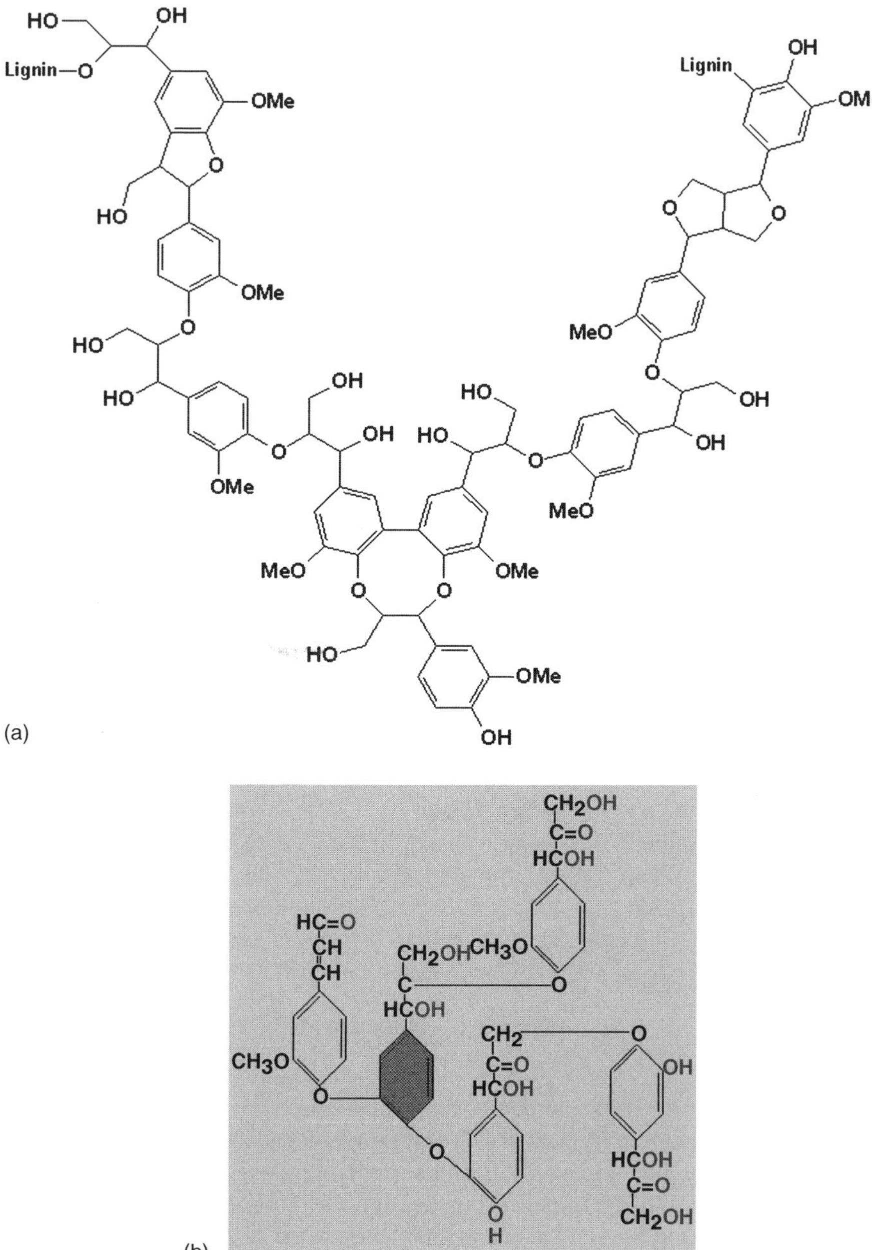

Figure 3.2 Several possible presentations (out of an almost infinite number of them) of structures of lignin.

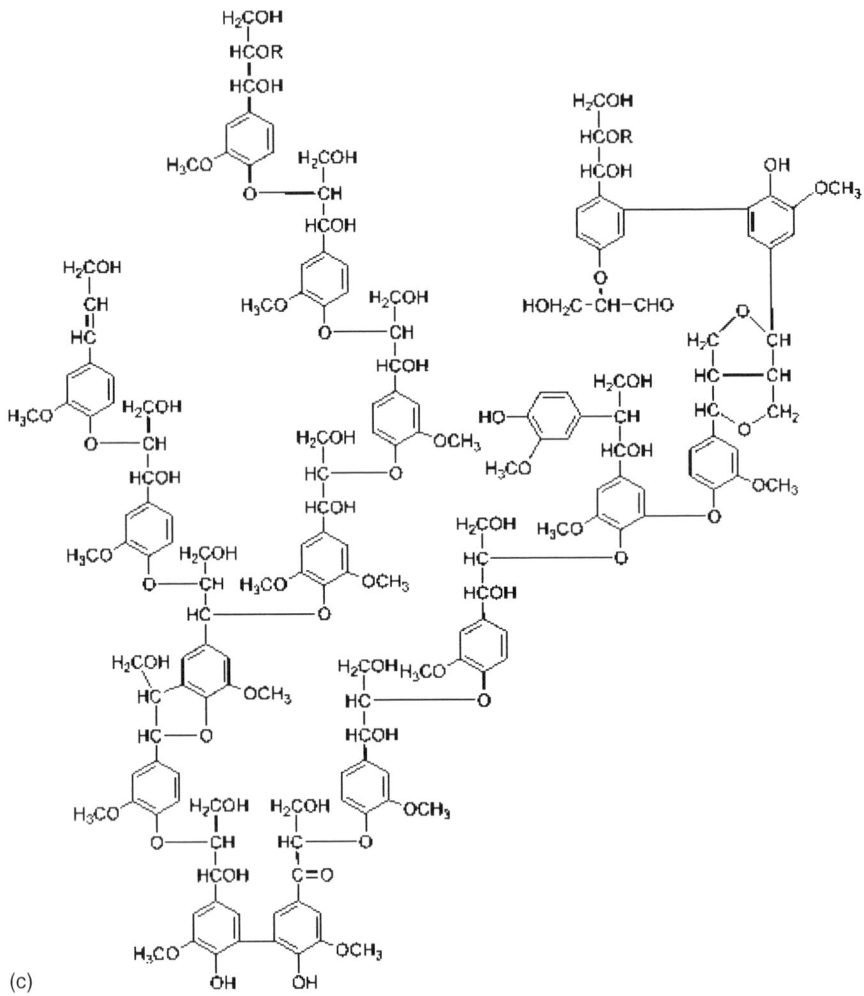

(c)

Figure 3.2 (*Continued*)

Besides these three principal components, plant materials contain nonstructural components—extractives (typically 4–10%), inorganic ash (between 0.2 and 0.5% in wood and up to 19% in rice hulls and some other plant materials), and water. Examples of extractives are terpenes, pinenes, tannins, carbonyl compounds, and so on. They contribute to wood odor, can diffuse to the wood surface during drying and can affect adhesion, and some of them possess antimicrobial properties.

The term "lignocellulose," often used for description of plant materials, does not properly describe their main components. Besides cellulose and lignin, the third main component, as indicated above, is hemicellulose, or, more correctly, hemicellulosics. The plural form here reflects a set of various heteropolymeric saccharides, the so-called matrix plant polysaccharides. These hemicellulosics mainly form amorphous three-dimensional structures, surrounding cellulose fibers. Hemicellulosics include xylans, arabinoxylans, glucuronoxylans, glucomannans, galactomannans, xyloglucans, and so on.

Lignocellulose was apparently singled out as an individual term because it represents a combination of structural components that strengthens plant cells in woody materials.

Lignin is a highly polymeric material, crosslinked, highly aromatic structure. Lignin is considered to be largely responsible for strength and durability of wood. In fact, trees stand upright because lignin supports their integrity. Besides, lignin slows down greatly accessibility of wood to cellulolytic microbes. Lignin can be defined as the residue left after the plant material is hydrolyzed with strong acids. Lignin can be considered as one huge polymeric molecule penetrating wood matrix. When one cuts wood, one cuts the single lignin molecule in parts. The same is related to hemicellulose and—with some reservations—to cellulose that has a regular structure, unlike lignin and hemicellulose.

Generally, plant materials, including wood, contain a third to half of cellulose, a third or less of lignin, and a third or less of hemicellulosics by weight. In wood, for example, there is 40–45% of cellulose and 25–35% of hemicellulosics (Table 3.2). On some data, aspen contains 18.2% of lignin, maple 22.5%, and spruce 27.6% of lignin, though such precise figures are doubtful; besides, they are changed depending on age of the wood. Lignin is largely removed from wood during chemical pulping operations; hence, papermaking sludge typically has only a little of lignin, often between 1 and 5%.

Typically, wood begins to undergo some dehydration at temperatures below or at the polyethylene melting point of 110–130°C (230–266°F). This is accompanied by the decrease of degree of polymerization of all three main components of plant materials, which is accelerated in the presence of moisture. Lignin begins to thermally degrade at about 150°C (300°F), and hemicellulosics begin to degrade at about 160°C (320° F). All these processes release volatiles, which in turn increase porosity and reduce density of the resulting composite material, unless vented extruders are used.

Detrimental Effects of Lignin

Lignin is a photosensitive material, and its degradation under UV light is believed to be largely responsible for fading of wood and WPC materials. Under UV light lignin is transformed from a brownish material to a gray one.

TABLE 3.2 A typical chemical composition of hardwoods, softwoods, and rice hulls

Component	Hardwoods (%)	Softwoods (%)	Rice hulls (%)
Cellulose	44 ± 3	42 ± 2	38 ± 8
Hemicellulose	32 ± 5	26 ± 3	25 ± 3
Lignin	18 ± 4	29 ± 4	14 ± 2
Minerals (silica)	0.2–0.8	0.2–0.8	19

Lignin generally makes the product weaker, because it easily burns in the course of processing and releases CO_2, making the product density lower, and greatly accelerates fading of the WPC after outdoor exposure.

Detrimental Effects of Hemicellulosics: Steam Explosion

As lignin, hemicellulosics easily burn in the course of processing, though they release mainly VOC, leading to a lower density of the WPC product. However, it is not the only problem associated with hemicellulosics under high pressure and temperature. During the processing of hot melt in the extruder under high pressure, moisture in wood particles does not boil even at high temperatures. For example, at 1000 psi in the extruder, water boils only at 546° F (286°C). At 2000 psi, water boils at 636° F (336°C). In other words, at normal extrusion temperatures at high pressure, water is liquid and does not boil. However, when the material passes zones where pressure sharply drops, water suddenly and violently boils, producing steam.

This effect of instantaneous decompression and sharp boiling across the whole wet wood particle causes the so-called steam explosion. The result of it is that many lignin–hemicellulose bonds are severed; hemicellulosics fragments dissolve in water trapped in the matrix and instantly decompose, with its many acetylated residues forming acetic acid. At high temperature in the extruder, even small amounts of acetic acid are highly reactive, leading to progressive corrosion of the equipment (see, for example Figs. 3.3 and 3.4).

Figure 3.3 View of the screw in the metering section of a 24:1 nonvented extruder. The hot melt flows around the downstream side of the mixing pins, where a sharp drop of pressure takes place. The area between the mixing pins and the flight is corroded as a result of steam explosion and acetic acid formation from hemicellulosics of the material.

Figure 3.4 View of the blister of the screw of a 30:1 vented extruder. The area of a large pressure drop is shown, between the blister and the flight. The screw is designed to melt and pressurize the material, then drop the pressure to below atmospheric pressure where the gases are evacuated. The pressure drop occurs on the downstream side of the blister part of the screw, and the chrome finish on the screw fails due to corrosion in this area. This area is corroded as a result of steam explosion and acetic acid formation from hemicellulosics of the material.

Aspect Ratio

Aspect ratio in this context is a ratio of the fiber length to fiber thickness. For wood flour this ratio is often about 3:1–4:1 (Table 3.3). Generally, long fibers, oriented along the flow, render a composite material with improved mechanical properties, compared to short-fiber-filled composite material. In other words, a higher aspect ratio leads to better flexural properties. As a result, in WPCs the cellulose fiber is the main load-bearing component, and the more the fibers are oriented along the flow, the higher are the flexural properties of the material.

TABLE 3.3 Aspect ratios of wood fiber [129]

	Aspect ratio	
Particle size range	Hardwoods	Softwoods
20 mesh (850 μm = 0.85 mm)	4.6	3.5
40 mesh (425 μm = 0.425 mm)	4.4	3.4
60 mesh (250 μm = 0.25 mm)	4.4	4.2
80 mesh (180 μm = 0.18 mm)	4.2	4.5

The figures for aspect ratio are rounded up in a more realistic manner than in the original publication [129] (in which aspect ratios were given with a precision to a second decimal).

More specifically, some WPCs were made using wood flour derived from pine, with an average aspect ratio of 4.0; from salt cedar, with an average aspect ratio of 3.2; and from juniper, with an average aspect ratio of 4.4 [130]. Wood fiber typically has higher aspect ratio, such as 10:1–25:1. For example, many wood fibers have a length of 3 mm and width of 0.2 mm (aspect ratio of 15) or a length of 10 mm and width of 0.4 mm (aspect ratio of 25).

Commercial wood fiber from aspen, birch, maple, and spruce typically has a length of 0.4–3.5 mm, that is, 400–3500 μm (spruce is the longest one), and width 50–27 μm. Average aspect ratio values for fiber from these species are 35 (maple), 60 (aspen), 100 (birch), and 130 (spruce).

Extrusion processes result in the decrease of aspect ratio of cellulose fiber, that is, in their shortening at the same fiber thickness. For example, the processing of bleached sulfite cellulose fiber along with polypropylene in a corotating twin screw extruder at 100 and 300 rpm resulted in the shifting of fiber length to shorter length, so that a fraction of the smallest particles (around 50 μm in length) increased from 3 to 5%. The 100-μm fraction, that is, the most abundant fraction, contributing about 12% of all particles after 100-rpm extrusion, reached 16% of all particles after 300-rpm extrusion. After a second extrusion at 300 rpm, this 100-μm fraction has increased from 16 to 26% of the total [131].

Density (Specific Gravity)

Specific gravity of lignified cellulose fiber (wood flour, saw dust, and rice hulls) in WPCs typically is between 1.3 and 1.5 g/cm^3. Rice hulls lignocellulosic matrix in WPC materials is of about the same density as that of wood flour, and they are equal to about 1.35 g/cm^3 and 1.30–1.35 g/cm^3, respectively. However, because rice hulls contain 19% of silicates (specific gravity approximately 2.8 g/cm^3), the resulting specific gravity of rice hulls as a filler is around 1.50 g/cm^3.

Long fiber cellulosics, such as flax, hemp, jute, ramie, coir, sisal, and cotton, all have density in WPC materials also in the 1.3–1.5 g/cm^3 range.

Density (specific gravity) of WPCs does not depend on the particle size of the wood flour. It appears that during compounding and extrusion, cellulose fiber is compressed to the maximum density of 1.3–1.5 g/cm^3.

A simple formula for the calculation of specific gravity of a composite material is as follows. If we take 100 g of a composite material, containing, say, 50% w/w of HDPE (d = 0.96 g/cm^3), 30% of wood flour (d = 1.30 g/cm^3), and 20% of talc (d = 2.8 g/cm^3), each of these components take the following volume:

$$\text{HDPE } 50\text{ g}/0.96\text{ g/cm}^3 = 52.083\text{ cm}^3,$$
$$\text{Wood flour } 30\text{ g}/1.30\text{ g/cm}^3 = 23.077\text{ cm}^3,$$
$$\text{Talc } 20\text{ g}/2.8\text{ g/cm}^3 = 7.143\text{ cm}^3.$$

Therefore, total volume of the 100 g of the composite will be 82.303 cm^3. Hence, density of the composite is 100 g/82.303 cm^3 = 1.215 g/cm^3.

For 85% wood-flour-filled polypropylene calculations of density (specific gravity) would be as follows:

$$\text{PP } 15 \text{ g}/0.91 \text{ g/cm}^3 = 16.484 \text{ cm}^3,$$
$$\text{Wood flour } 85 \text{ g}/1.30 \text{ g/cm}^3 = 65.385 \text{ cm}^3,$$

Total volume of the composite will be 81.869 cm^3. Hence, density of the composite is 100 g/81.869 cm^3 = 1.22 g/cm^3.

If density of wood flour is 1.35 g/cm^3, the above two figures become equal to 1.23 g/cm^3 (the HDPE-based composite) and 1.26 g/cm^3 (PP-based composite), respectively. A published figure for a density of 85%-wood-flour-filled polypropylene is 1.26 g/cm^3 [132], that is, equal to the last figure.

Particle Size

There are two different systems for the evaluation of wood size and natural fiber particles and fibers—either as effective "particle size," often measured as mesh size, or as fiber length, particularly in case of particles/fibers with a high aspect ratio. Typically, the particle size of fillers (including cellulose) for WPC is chosen by mesh screening to be in the range between 40 and 80 mesh (approximately 0.35 and 0.18 mm). Particles larger than 40 mesh are considered as "oversize" and smaller than 80 mesh are considered as "fines." Fines in wood flour can be between 10 and 100 μm in size. Length of cellulose fibers, used as fillers, often is in the range of 20–300 μm (0.02–0.3 mm). More data on cellulose fiber length are given in section "Aspect ratio".

Particle size analysis of commercial wood flour grades has shown that size varies quite significantly. For example, four samples of wood flour gave the following particle size ranges (in microns) [133]:

- ~5–100
- ~5–150
- ~5–250
- ~5–430

There was a broad particle size distribution within each wood flour grade.

More recent data from American Wood Fibers company (Columbia, MD) show that in three different wood flour grades—"Standard grades" (4010 and 4020), "X-mesh" (4017 and 4037), and "M-series" (40M3 and 40M7)—more than 80% of all particles belong to a range of 40–80, 40–120, and 20–100 mesh, respectively [129].

Particle Shape

Cellulose particles used in WPC are typically fibers, with aspect ratio between 3 and 4, or longer, particularly in case of long cellulose fiber. Hardwoods wood flour particles are commonly cylindrical, whereas softwoods wood flour particles are commonly fragmented.

Particle Size Distribution

For pine, salt cedar, and juniper, with aspect ratio 4.0, 3.2, and 4.4, respectively, image analysis using more than 500 particles per species showed the following particle size (in terms of area, in square microns) distribution [130]:

Small (around 0.1 × 0.5 mm):	
0.05 mm^2	15–23%
Medium (around 0.2 × 0.8 mm):	
0.10 mm^2	26–32%
0.15 mm^2	16–18%
0.20 mm^2	10–13%
Total medium	52–63%
Large (around 0.3 × 1.2 mm):	
0.25 mm^2	6–8%
0.30 mm^2	2–3%
0.35 mm^2	2–3%
0.40 mm^2	2%
0.45 mm^2	2%
0.50 mm^2	1%
Total large	15–19%

Particle Surface Area

Specific surface area of cellulose fiber is not large, compared to that of fine mineral fillers. Although the latter often have it in dozens of meter square per grams, cellulose fiber has it around 1 m^2/g. For example, hardwood has particle surface area of 1.01 m^2/g, softwood—1.34 m^2/g. Flax fiber has surface area between 0.31 and 0.88 m^2/g, depending on a mode of its treatment [134]. Furthermore, specific surface area principally depends on which tool was used for its measurements—a gas, water, mercury, or large organic molecules. The larger the molecule used for the measurement, the lower the specific surface area. For example, specific surface area of cellulose fiber accessible for water in wet state is 100–200 m^2.

Moisture Content, the Ability to Absorb Water

Neat plastics (except nylons), which are used in composite materials, practically do not absorb water (hundredth or thousandth percent by weight). Incorporation of cellulose fiber into plastic significantly increases water absorption (0.5–2% for 24 h underwater), which is still much lower compared to wood itself (typically more than 20% w/w for 24 h underwater).

Cellulose and lignocellulose fiber absorbs water in amount up to 200% and higher. Ground rice hulls particles absorb 220% of water by weight.

Moisture content in cellulose fiber typically reaches 2–12%, depending on origin of the fiber and air humidity at the storage. Moisture content of ground rice hulls supplied for WPC manufacturing varies between 8.5 and 9.5% throughout the year.

Ramie and cotton fibers reportedly can reach their moisture absorption capacity in ambient conditions up to 17 and 25%, respectively; however, those conditions are often not clearly defined.

The Ability of Filler to Absorb Oil

Cellulose fiber is a unique material in that it can absorb equally well both water and oil. "Equally" does not mean quantitatively equal, it means that cellulose fiber absorbs both liquids well. For instance, ground rice hulls absorbs 240% of motor oil by weight. As it was indicated above, it absorbs 200% water by weight. Biodac® absorbs 120% of water and 150% of motor oil, both by weight.

Flammability

Cellulose fiber is flammable. Flammability of various types of wood is described in Chapter 14.

Effect on Mechanical Properties of the Composite Material

Cellulose fiber is a good reinforcing filler. In fact, this is one of the two major factors of the very existence of WPC materials: (a) to make the composite material less expensive and (b) to obtain material with overall better properties compared to neat plastic, on the one hand, and wood, on the other. For example, tensile modulus of a particular sample of neat polypropylene was 203,000 psi, whereas for the same polypropylene filled with 40% of jute it was 1,030,000 psi. For a comparison, for the same polypropylene filled with 40% glass fiber it was 1,100,000 psi. Tensile modulus for natural fiber itself is in the range of 3,800,000–17,400,000 psi [135]. Table 3.4 shows data in more detail.

Effect of a different filler, wood flour, on mechanical properties of polypropylene is shown in Table 3.5.

Comparison of Tables 3.4 and 3.5 shows that hemp as a filler results in a much higher flexural strength increase compared to wood flour (57% compared to 12–15%); however, increase of flexural modulus is about the same with the two fillers (2.2 times with 33% of hemp, compared to 1.6 and 2.5 times with 20 and 40% of wood flour, respectively).

Another set of data, obtained in a wider range of wood flour content as polypropylene filler (Table 3.6), shows that flexural strength of the composite reached maximum at 40% filler, and then decreases. Apparently, there is not enough plastic to provide good adhesion for all filler particles. Flexural modulus, however, is increased

TABLE 3.4 Mechanical properties of filled polypropylene composites [135]

Filler	Flexural Strength (psi)	Flexural Modulus (psi)	Tensile Strength (psi)	Tensile Modulus (psi)	Impact resistance, Izod, ft-lb/in. (J/m)
No (neat polypropylene)	4,160	200,000	4,000	203,000	0.30 (16)
33% hemp	6,525	436,000	5,800	609,000	0.83 (45)
33% hemp woody core	10,100	465,000	8,300	626,000	0.78 (42)
30% glass fiber	16,700	589,999	7,700	709,000	1.2 (65)

TABLE 3.5 Mechanical properties of polypropylene composites, filled with wood flour [136, 137]

Filler	Flexural Strength (psi)	Flexural Modulus (psi)	Tensile Strength (psi)	Tensile Modulus (psi)	Izod impact energy, notched (J/m)
No (neat polypropylene)	5,550	172,600	4,130	222,000	20.9
20% wood flour	6,235	267,000	3,680	323,000	19.5
40% wood flour	6,410	439,000	3,680	561,000	22.2
20% hardwood fiber	6,220	273,000	3,800	329,000	20.8
40% hardwood fiber	6,950	470,000	4,100	609,000	23.2

monotonously with increase of the filler content. Tensile modulus also increases, whereas tensile strength decreases.

Reinforcing effect of cellulose fiber on Nylon is shown in Table 3.7

Often (but not always), the higher the aspect ratio of wood fiber, the higher the flexural strength and flexural modulus of filled WPC (Table 3.8).

TABLE 3.6 Mechanical properties of polypropylene composites, filled with 40-mesh wood flour (data by M. Berger and N. Stark, 1997, cited in [138])

Wood flour in the composite, (%)	Flexural Strength (psi)	Flexural Modulus (psi)	Tensile Strength (psi)	Tensile Modulus (psi)
No (neat polypropylene)	5,075	145,000	4,060	189,000
20	6,090	283,000	3,640	290,000
30	6,235	377,000	3,625	464,000
40	6,380	464,000	3,700	551,000
50	6,100	544,000	3,335	609,000
60	5,510	595,000	2,900	667,000

TABLE 3.7 Mechanical properties of filled Nylon 6 composites [135]

	Flexural		Tensile		Impact resistance,
Filler	Strength (psi)	Modulus (psi)	Strength (psi)	Modulus (psi)	Izod, ft-lb/in. (J/m)
No (neat Nylon 6)	9,700	406,000	9,600	480,000	0.92 (50)
33% cellulose fiber	17,100	930,000	12,800	970,000	0.44 (24)

TABLE 3.8 Mechanical properties of 60%-wood-fiber-filled polypropylene composites [139]

	Flexural		Tensile		Impact resistance,
Wood fiber, aspect ratio (length, mm)	Strength (psi)	Modulus (psi)	Strength (psi)	Modulus (psi)	Charpy (kJ/m^2)
No (neat PP)	—	1000	21	1150	55
3 (5)	32	3200	36	3800	4
1 (0.4)	54	4600	27	3700	6
15 (3)	71	5200	40	5000	10
25 (10)	78	5000	26	4200	10

It is often observed that the higher the fiber content, the higher the flexural strength and modulus of the WPC. For example, increase of wood fiber content from 20 (w/w) to 40, 50, and 60% (w/w) in polypropylene led to a systematic increase in flex strength and modulus, resulting in their overall increase by more than 200% [139].

Effect on Fading and Durability of Plastics and Composites

The higher the amount of the cellulose fiber in the composite, the higher the fading, at all other conditions being equal (see, for example, Table 3.9).

TABLE 3.9 Effect of wood flour content on fading of wood-flour (40-mesh pine)-filled HDPE

Wood flour content, % (w/w)	Shift in ΔL (%)
10	2
20	16
30	33
40	38
50	51
60	102

Weathering was conducted under a 2-h cycle (108 min UV + 12 min UV/water spray) [140].

TABLE 3.10 Effect of rice hulls on the oxidative induction time (the OIT) of an HDPE, processed on the Brabender laboratory extruder

Amount of rice hulls, % (w/w)	OIT (min)
0 (neat HDPE)	4.4
Traces	7.4
10	15
30	22
60	37

The OIT was measured at 190°C. The OIT values show the resistance of the material to oxidation.

However, both rice hulls and wood flour increase the stability of composite materials to oxidative degradation (Tables 3.10 and 3.11).

One can see that an increasing amount of cellulose filler on the one hand increases fading (Table 3.9), on the other hand increases stability of the material to oxidation (Tables 3.10 and 3.11). Obviously, an increase in the overall stability of a material can be accompanied by an increase of its fading.

Tables 3.10 and 3.11 also show that wood flour provides better antioxidative properties compared to rice hulls.

Effect on Hot Melt Viscosity

Cellulose fiber, obviously, increases viscosity of plastic hot melt. Furthermore, fiber from different species affects viscosity differently, even at the same particle size. For example, pine, juniper, and salt cedar (particle size 40–60 mesh), each at 50% amount w/w in HDPE, resulted in melt flow index (MFI) of 0.2, 0.6, and 1.6, respectively. In other words, pine flour increased viscosity the most. Viscosity continued to increase with smaller wood flour particles (Table 3.12).

TABLE 3.11 Effect of rice hulls and wood flour on the OIT of HDPE, filled with Biodac of two different sizes (20–50 mesh and 100 mesh, Biodac 1 and Biodac 2, respectively)

Compositions (principal ingredients)	OIT (min)
Neat HDPE (reference)	4.4
HDPE + Biodac 1 + rice hulls	10.6
HDPE + Biodac 1 + wood flour	20.6
HDPE + Biodac 2 + rice hulls	6.6
HDPE + Biodac 2 + wood flour	12.2

The composites were processed on the Brabender laboratory extruder. The OIT was measured at 190°C. The OIT values show the resistance of the material to oxidation: The higher the OIT, the higher the resistance. The amount of Biodac was 28% in all cases. The amount of rice hulls and sawdust was 32% in all the cases. The amount of HDPE was 40% in all the cases.

TABLE 3.12 MFI of HDPE filled with 50% (w/w) of various wood flour types [130]

	Melt flow index for particle size, mesh			
Wood	−40/+60	−60/+80	−80/+100	−100
Pine	0.2	0.3	—	—
Juniper	0.6	0.4	0.4	0.2
Salt cedar	1.6	1.4	1.3	0.9

Table 3.12 shows that the melt flow index of HDPE in the presence of 50% wood flour decreases rather significantly, from 1.6 g/10 min for salt cedar to 0.2 g/10 min in the presence of pine. Furthermore, if we consider the initial MFI value for the neat plastic, a drop in MFI is quite significant. For example, a polypropylene with MFI 23 g/10 min (MFI was 29 g/10 min after processing in the extruder at 100 or 300 rpm) was loaded with 20% bleached sulfite cellulose fibers, and after extrusion at 100 rpm MFI dropped to 2 g/10 min. With 30% loading with the fiber, MFI further decreased to 0.5 g/10 min [131].

Effect on Mold Shrinkage

There is a belief that increased particle size provides lower mold shrinkage because of a better flow (lower viscosity). However, in reality a connection between the two is more complex. It was shown that when average wood flour particle size increased from about 50–60 μm to about 200 μm, mold shrinkage decreased from 0.58 ± 0.02% to 0.47 ± 0.02%, and with further increase to 500 μm, mold shrinkage either stayed the same or even slightly increased, to about 0.49 ± 0.02% [133].

WOOD FIBER

Wood Flour

Wood flour is finely divided ground wood having a flour-like appearance. Wood flour that is typically used in WPC has mesh size of about 40, that is, about 400 μm. However, wood flour grades are classified in particular size ranges of 50–150 μm, 100–200 μm, 200–450 μm, and 250–700 μm. Increase of particle size of wood flour typically provides better flow of molten composite, lower mold shrinkage, and higher flexural modulus.

Bulk density of wood flour is typically around 0.1–0.3 g/cm^3. Aspect ratio of wood flour (length to thickness of fibers) is typically between 3:1 and 5:1. Specific gravity (density) of wood is about 1.3–1.4 g/cm^3. When ashed (commonly at temperatures between 525 and 575°C), wood often leaves 0.25–0.50% of mineral residues. For example, ashing of Trex boards, containing reportedly 50% of wood flour, produced 0.55 ± 0.02% of ash.

Wood often contains tannins, and higher levels of tannins are commonly associated with oak. Tannins are water-soluble phenolic and polyphenolic compounds, which can form dark color complexes with iron salts (ferric salts). That is why some WPC deck boards, containing wood fiber, particularly oak fiber, can form dark stains around nails. Sometimes those stains are referred to as "tannin stains." These stains can be removed using oxalic acid-based (or other dicarbonyl acid-based) cleaning agents.

Wood flour and sawdust do not melt but rather decompose above 190°C. This is, again, due to a more susceptible to temperature lignin and hemicellulosics, and because pure cellulose decomposes above 240°C.

Sawdust

Sawdust is essentially the same thing as wood flour except it is not ground but formed as a by-product of wood sawing. Sawdust typically consists of particles of 30–600 μm and larger in size. Sometimes sawdust is classified as particles of 20 μm–5 mm (5000 μm). Cost of wood flour and sawdust is typically of 6–8 ¢/lb, but it can vary depending on local conditions.

Rice Hulls

Rice hulls are by-products of rice mills, where lighter rice hulls are separated from more heavy husked rice through aspiration. In some rice mills, hulls are ground and sold according to their sizes.

Because rice hulls make about 20% of rough harvested weight of rice, rice mills across the United States, from California to Florida, with Arkansas, Missouri, Louisiana, and Texas in between, produce large amounts of rice hulls, namely more than 2.5 billion pounds (1.25 million metric tons) a year. One of them is Riceland Foods, or Rice Hull Specialty Products (Stuttgart, AR), the world's largest miller of rice and a major supplier of milled rice hulls. Rice hulls are often sold for about $40/ton (2 cents a pound), though cost and availability generally vary from producer to producer.

As it was mentioned above, chemically rice hulls are close to wood and contain the same principal components, that is, cellulose (28–48%), lignin (12–16%), and hemicellulosics (23–28%), plus about 19% of silica (18.8 ± 0.8% from analysis of many batches of rice hulls). There is only 0.14 ± 0.02% of calcium in rice hulls and 0.74 ± 0.09% of calcium in the silica in rice hulls, that is, less than one hundredth of the silica.

However, lignin in rice hulls significantly differs from lignin in wood, and this fact along with a much higher content of minerals makes rice hulls much more resistant to microbial degradation compared to wood flour, and more resistant to moisture penetration. Rice hulls even in ground form rarely absorb more than 10% of moisture. At the same time rice hulls are much more abrasive, due to its high silica content.

Of the three main types of milled rice hulls supply by a particular supplier, "bulky," "medium" and "fine," names related to average particle size, all three had the same moisture content, measured in a February (in fact, measurements in summertime gave similar results), 9.3 ± 0.6%. Bulk density figures were 6.3, 14.2, and 18.6 lb/cft, respectively, for these three types of milled rice hulls. Particle sizes were as follows:

TABLE 3.13 Flexural strength and modulus of composite 2 × 6 deck boards, filled with rice hulls (57% by weight) and those filled with rice hulls (32% w/w) and Biodac (29% w/w)

Principal filler	Flexural strength (psi)	Flexural modulus (psi)
Rice hulls	1,910 ± 50	297,000 ± 25,000
Rice hulls and Biodac	2,320 ± 70	374,000 ± 39,000

Third-point load, 20-in. span.

TABLE 3.14 Flexural strength and modulus of composite 2 × 6 deck boards

Principal filler	Flexural strength (psi)	Flexural modulus (psi)
Rice hulls	2,120 ± 160	276,000 ± 7,000
Rice hulls and Biodac	2,460 ± 50	371,000 ± 43,000

Composition is given in Table 3.13. Three-point load, 20-in. span.

- for “bulky,” 64% of particles (by weight) were larger than 14 mesh, and 6.4% were at or smaller than 30 mesh;
- for “medium” (20–80 mesh size), the largest fraction, 48%, was retained between 30 and 20 mesh size, and 7.0% were at or smaller than 50 mesh;
- for “fine” (40–80 mesh size), the largest fraction, 42%, was retained between 40 and 30 mesh size, 46% were at or smaller that 50 mesh size, and 6.5% were smaller than 60 mesh. There were no particles retained on 20 mesh size screen.

Blending of HDPE with rice hulls certainly improves the material strength and stiffness, however, less compared with Biodac®, blend of cellulose fiber with minerals (Tables 3.13 through 3.16). These tables show flexural properties of commercial

TABLE 3.15 Flexural strength and modulus of composite 2 × 6 tongue and groove deck boards

Principal filler	Flexural strength (psi)	Flexural modulus (psi)
Rice hulls	1,670 ± 140	266,000 ± 36,000
Rice hulls and Biodac	2,030 ± 80	389,000 ± 65,000

Composition is given in Table 3.13. Third-point load, 20-in. span.

TABLE 3.16 Flexural strength and modulus of composite 2 × 6 tongue and groove deck boards

Principal filler	Flexural strength (psi)	Flexural modulus (psi)
Rice hulls	1900 ± 220	226,000 ± 25,000
Rice hulls and Biodac	2190 ± 90	373,000 ± 45,000

Composition is given in Table 3.12. Three-point load, 20-in. span.

and experimental boards measured in two different modes of load application, that is, center-point load (three-point load) and third-point load.

One can see that a transition from rice hulls filled boards (about 11% of a natural mineral filler presented in rice hulls) to Biodac®/rice hulls filled boards (about 21% of combined mineral fillers) leads to 19 ± 4% increase in flexural strength and 43 ± 17% increase in flexural modulus on average. This increase may be attributed not only to increase of minerals but also to morphology of Biodac® porous granules and different in kind cellulose fibers in Biodac® (delignified and differently packed into the filler).

VOC from Rice Hulls When heated, rice hulls release gaseous products. This is characteristic of any plant materials, including even delignified cellulose, at least as a constituent of Biodac®. In fact, the amount of gaseous products released by rice hulls and Biodac® under heating is about the same (by weight). Data are given in Tables 3.17, 3.18, 3.21, and 3.22.

In the experiment, 41.106 g of rice hulls having a conventional amount of moisture (typically 7–10% by weight) were heated in an oven, starting at 80°C (176°F) and consequently increasing temperature up to 193°C (380°F), for a total time period of 514 h. The rice hulls sample lost 7.9% of moisture (between 80 and 116°C, that is 176 and 240°F), and between 127 and 193°C (260 and 380°F) it lost 32%

TABLE 3.17 Dynamics of weight loss of ground rice hulls (20–80 mesh size) in the course of heating in an oven

Temperature, °C (° F)	Heating time (h)	Weight loss, mg (%)	Weight loss rate (mg/h)	VOC, mg (% of dry weight)
Ambient	0	0	0	0
Loss of moisture				
80° (176)	16	2,880 (7.01)	180	—
85° (185)	96	3,029 (7.37)	1.6	—
93° (200)	23	3,059 (7.44)	1.3	—
105° (221)	26	3,161 (7.69)	3.9	[102]
Same	50	3,241 (7.88)	1.6	[182]
116° (240)	46	3,253 (7.91)	0.26	[194]
Loss of VOC				
127° (260)	36	3,412 (8.30)	4.4	159 (0.42%)
Same	13	3,430 (8.34)	1.4	177 (0.47%)
138° (280)	7	3,495 (8.50)	9.3	242 (0.64%)
Same	24	3,642 (8.86)	6.1	389 (1.0%)
149° (300)	12	3,848 (9.36)	17	595 (1.6%)
Same	60	4,692 (11.4)	14	1,439 (3.8%)
160° (320)	48	6,556 (16.0)	39	3,303 (8.7%)
171° (340)	24	8,557 (20.8)	83	5,304 (14%)
182° (360)	24	12,775 (31.1)	176	9,522 (25%)
193° (380)	9	15,274 (37.2)	278	12,021 (32%)

Initial weight of the sample 41.106 g.

TABLE 3.18 Dynamics of weight loss of ground rice hulls (20–80 mesh size) in the course of heating in an oven at 170°C (338 ° F)

Heating time (h:min)	VOC, mg (% of dry weight)	Weight loss rate (mg/h)
0:20	46 (0.16%)	153
1:10	140 (0.48%)	113
1:50	170 (0.59%)	45
2:20	191 (0.66%)	42
2:55	214 (0.66%)	40
4:10	262 (0.90%)	38
5:10	300 (1.03%)	38
6:10	336 (1.17%)	36
8:20	400 (1.38%)	35

Initial weight of the sample was 29 g.

of volatiles. This amount is close to a total amount of lignin and hemicellulosics in rice hulls; however, the gasification process (Table 3.17) was continuing at high speed when the test was terminated. After it rice hulls became much darker than the initial sample.

If we take an average molecular weight of VOC as about 200 Da (main VOC from Biodac® were identified as naphthalates, see below) and that 100 kg of rice hulls release 32 kg of VOC at the conditions of the above experiment, this amount of VOC corresponds to 160 moles. If 1 mole of VOC occupies the volume of 22.4 L (at standard conditions), 100 kg of dry rice hulls produce 3584 L of VOC. At 50% (w/w) loading of rice hulls into HDPE, 200 kg of the composite material may be produced, with a specific gravity of 1.14 g/cm^3 and total volume of 175 L. That is, VOC in this case would take a volume 20-fold larger than the volume of the composite material. This would be a highly porous (density about 0.06 g/cm^3) material, if it can be made (as a continuous material) at all.

In reality an amount of VOC produced due to decomposition of rice hulls (or wood flour for that matter) in the extruder hardly exceeds 0.25–0.50%. However, even in this case porosity of the composite material, and decrease of its density (and, hence, flexural strength and modulus) can be significant. In a modified above example, 200 kg of the composite material, containing 50% (w/w) rice hulls and having a total volume of 175 L, would accumulate 500–1000 g of VOC, that is, 2.5–5.0 moles of VOC, which would occupy 56–112 L of volume. That is, density of the composite material would drop by 30–60% and instead of the theoretical 1.14 g/cm^3 would become 0.41–0.77 g/cm^3. This would be a great "foamed composite," except its pores would be open, irregular, and the material would be very weak and flexible.

The following experiment shows the release of VOC from dried rice hulls at 170°C (338°F) during 8 h of exposure in the oven (Table 3.18).

One can see that after only 20 min at 338°F, rice hulls released 0.16% of VOC. After 1 h this amount was close to 0.5%, and after 8 h 20 min the amount of VOC released was 1.38%. Our estimates show that during extrusion, VOC from rice hulls (or plant materials in general) can be about 0.25–0.5%, taking into account a high pressure effects, higher local temperatures, and effects of moisture that can serve as a catalyst of decomposition of lignocellulosics and hemicellulosics.

LONG NATURAL FIBER

There are many kinds of natural fibers, such as bast fibers (flax, hemp, jute, kenaf, ramie, nettle, and mesta), leaf fibers (sisal, henequen, pineapple, abaca, oil palm, and screw pine), seed fibers (cotton), fruit fibers (coconut husk, or coir), and stalk fibers (straw of various kinds). They are not used for commercial WPC, primarily on economical reasons (except maybe Procell, see Table 1.1). Most of these fibers have found applications in established industries, such as textile industry (cotton, flax, jute, ramie, hemp, and sisal) and paper industry (straw).

Long natural fibers are considered as excellent reinforcing fillers for WPCs. As an example, comparison of Tables 3.4 and 3.5 shows that hemp-filled plastic results in a higher flexural strength and tensile strength and modulus compared with those for wood flour-filled plastic. As drawbacks of long fiber as a filler, high moisture uptake and swelling are often considered. Also, it is not easy to uniformly mix long cellulose fiber with molten plastic. Besides, long cellulose fiber of a good quality costs much more than rice hulls and wood flour; hence, offstream long cellulose fiber should be identified, which would cost not 25–40 ¢/lb but 10 times less. Due to high cost, long cellulose fiber finds an application not in composite building materials but in automotive applications.

A widely cited example of a long cellulose fiber is hemp. Hemp has specific gravity of 1.5 g/cm^3, cellulose content about 80%, and lignin content only 4%. Average fiber diameters for long cellulose fibers are typically from 20 (hemp) to 28 mc (coir).

Examples of length of long natural cellulose fiber are shown in Table 3.19. Cross-sectional area for long cellulose fiber, listed in Table 3.19, vary within error margin (standard deviation) for all species, so that the average cross-sectional area for all of them can be expressed as 273 ± 109 mc^2, or, more realistically, 300 ± 100 mc^2. From those, hemp is the strongest one (in terms of tensile strength) and the stiffest one (in terms of tensile modulus) (Table 3.20).

TABLE 3.19 Long cellulose fiber length [141]

Fiber	Range in fiber length (mm)	Mean fiber length (mm)
Hemp	3–47	22 ± 11
Linseed	2–28	7 ± 5
Milkweed	2–11	7 ± 2
Nettle	2–16	6 ± 3
Flax	2–15	5 ± 3

TABLE 3.20 Mechanical (tensile) properties of long cellulose fiber (modified from [141])

Fiber	Tensile strength (psi)	Tensile modulus (psi)
Hemp	157,000 ± 46,000	1,300,000 ± 400,000
Nettle	133,000 ± 109,000	1,400,000 ± 1,100,000
Milkweed	106,000 ± 31,000	1,400,000 ± 300,000
Flax	94,000 ± 41,000	2,800,000 ± 1,700,000
Linseed	54,000 ± 35,000	1,500,000 ± 500,000

PAPERMAKING SLUDGE

Pulp and paper sludge (a by-product of primary pulping operations, recycle streams, or waste paper pulping and the like) represents an environmental and disposal problem for manufacturers of pulp and paper. Generally, pulp and paper sludge is unsuitable for paper making, although it generally contains the same components—cellulose, lignin, hemicellulose, calcium carbonate, clay, and other inorganic components—as those present in the paper pulp itself. The main reason why papermaking sludge is unsuitable to bring it back for paper making is that it contains predominantly short cellulose fiber, which decreases the quality of paper.

Paper sludge has traditionally been disposed of by landfilling, composting, incorporation into cement, and incineration. The latter option, in turn, creates another problem, namely, disposal of the resulting ash, which often makes up to 50% (and sometimes as much as 80% or higher) of the volume of the sludge itself.

The principal components of ash are calcium carbonate—in the form of precipitated calcium carbonate (PCC) or ground calcium carbonate (GCC)—that typically constitutes 20% and up to 75% of dry sludge content, and clay. These two minerals are typically loaded into paper as a coating and filler to improve the mechanical characteristics as well as the appearance of paper. The resulting papermaking sludge, particularly mixed office paper sludge, consists primarily of two major components, that is, fiber and minerals finely mixed with each other.

A typical recycling mill processes 600 tons of wastepaper per day, yielding 450 tons of pulp and producing 150 tons of papermaking sludge. The 228 mills currently under operation in North America produce 9 million tons of pulp residue, approximately 5 million tons of which is cellulose. The 154 European pulp and paper mills produce about 8 million tons of pulp residue, approximately 4 million tons of which is cellulose. The conversion of such waste material into value-added products has, therefore, long been desired. One such product was developed by Kadant Grantek (Green Bay, WI), which manufactures controlled size dust-free granules, made of pulp and paper sludge, under the brand name Biodac® (see below).

The granules are a tight blend of organic and inorganic materials, that is, cellulose fiber and minerals, and possess a developed porous structure. This granulated papermaking sludge is described, for example, in U.S. Pat. No. 5,730,371 [142]. The granules have a controlled size and possess a developed porous structure; they are composed of organic and inorganic materials, that is, cellulose fiber and minerals. It is also found that granulated pulp and paper sludge, compared with loose cellulose fiber, improves the properties of fiber–plastic composites. It has been found that granulation of pulp and paper sludge, approximately half of which is typically cellulose fiber, reduces fiber-to-fiber interaction prior to incorporating it into the matrix and improves the mechanical properties of the composite.

BIODAC®

Biodac® is an agglomerated granulated papermaking sludge (see Chapter 4). Chemical composition of Biodac® slightly varies, following the composition of the papermaking sludge it is prepared from. It typically contains 50% of largely delignified cellulose (amount of lignin between 1 and 4%) and 50% of minerals, about half-and-half of calcium carbonate and kaolin clay. In one particular analysis Biodac® contained 46.3% of cellulose, 23.5 ± 0.3% of calcium carbonate, and 30.2% of kaolin clay and other admixtures. It contained 9.4 ± 0.1% of calcium, from which the amount of $CaCO_3$ was calculated.

Biodac® is supplied by Kadant Grantek (Green Bay, WI) for about $180/ton (9 cents a pound). Its major application is as an agricultural carrier for herbicides and pesticides, animal bedding, cat litter, oil absorbents, filler for GeoDeck composite deck boards, and components of GeoDeck composite railing systems.

VOC from Biodac®

When heated, Biodac®, as well as rice hulls (see above, Tables 3.17 and 3.18), releases gaseous products. It is interesting that despite the principal differences in their chemical composition, the amounts of VOC produced by both the materials under heating are very close to each other. Data are given in Tables 3.21 and 3.22.

In the experiment, 93.290 g of Biodac® having a conventional amount of moisture (typically 2–3% by weight) was heated in an oven, starting at 80°C (176°F) and consequently increasing temperature up to 193°C (380°F), for a total time period of 514 h. The Biodac® sample lost 2.3–2.5% of moisture (between 80 and 116°C, that is, 176 and 240°F), and between 127 and 193°C (260 and 380°F) it lost 16.7% of volatiles. Biodac® practically does not contain lignin and hemicellulosics, because they were chemically removed during pulping, and VOC were released by cellulose fiber.

It is not clear whether it was a final phase of a loss of moisture (making a total moisture content in Biodac® of 2.58%) or beginning of decomposition of Biodac® at 105°C with the releasing of VOC (225 mg, or 0.25% of dry weight of Biodac®). As one can see, it is not very important compared with the total VOC amount released at higher temperatures.

Total weight loss after 9 days of elevated temperature was 17.666 g, or 18.9% of the initial weight of Biodac®. About 2.5–2.7% of this weight was moisture, and

TABLE 3.21 Dynamics of weight loss of Biodac® (20–50 mesh size) in the course of heating in an oven

Temperature, °C (°F)	Heating time (h)	Weight loss, mg (%)	Weight loss rate (mg/h)	VOC, mg (% of dry weight)
Ambient	0	0	0	0
Loss of moisture				
80° (176)	16	1,996 (2.14)	125	—
85° (185)	96	2,154 (2.31)	1.7	—
93° (200)	23	2,178 (2.33)	1.0	—
105° (221)	26	2,291 (2.46)	4.3	[113]
Same	50	2,403 (2.58)	2.2	[112]
116° (240)	46	2,549 (2.73)	3.2	[146]
Loss of VOC				
127° (260)	36	2,813 (3.02)	7.3	264 (0.29%)
Same	13	2,960 (3.17)	11	411 (0.45%)
138° (280)	7	3,098 (3.32)	20	549 (0.61%)
Same	24	3,308 (3.55)	8.8	759 (0.84%)
149° (300)	12	3,733 (4.12)	35	1,184 (1.3%)
Same	60	5,180 (5.55)	24	2,631 (2.9%)
160° (320)	48	7,614 (8.16)	51	5,065 (5.6%)
171° (340)	24	10,004 (10.7)	100	7,455 (8.2%)
182° (360)	24	15,564 (16.7)	232	13,015 (14.3%)
193° (380)	9	17,666 (18.9)	234	15,117 (16.7%)

Initial weight of the sample was 93.290 g.

16.7% of the dry weight were VOC. With rice hulls, 32% of the dry weight loss were VOC (see above). The difference is not surprising, because in Biodac® half of its weight is taken by mineral fillers (calcium carbonate and kaolin), which do not decompose at those heating conditions ($CaCO_3$ starts to loose weight because of CO_2 release at about 825°C, or 1,517°F). This indicates that both cellulose in

TABLE 3.22 Dynamics of weight loss of dried Biodac (20–50 mesh size) in the course of heating in an oven at 170°C (338°F)

Heating time (h:min)	VOC, mg (% of dry weight)	Weight loss rate (mg/h)
0:15	49 (0.16%)	196
0:45	108 (0.36%)	118
1:25	163 (0.54%)	83
2:00	207 (0.69%)	75
3:15	287 (0.96%)	64
4:15	337 (0.90%)	50
5:15	377 (1.12%)	40
6:15	413 (1.38%)	36

Initial weight of the sample was 30 g.

Biodac®, on the one hand, and cellulose, lignin, and hemicellulosics in rice hulls, on the other hand, produce similar amounts of VOC (about one third of the whole plant material) and with about the same rate, when heated between 105 and 190°C (220 and 380°F).

What are VOC from Biodac®? The following list shows the results of some volatile and semivolatile organic compounds quantitative identification and chemical analysis (only those compounds that are formed in amounts higher than 100 ppb are shown, provided that they are above their limit of detection and/or limit of quantification):

- bis(2-Ethylhexyl)phthalate 19 ppm
- Butylbenzylphthalate 5.3 ppm
- Di-n-butylphthalate 3.3 ppm
- Diethylphthalate 1.6 ppm
- Acetone 1.3 ppm

These data can be fully assigned to the papermaking sludge, from which Biodac® was made.

The following experiment shows the release of VOC from dried Biodac® at 170°C (338°F) during 8 h of exposure in the oven (Table 3.22).

One can see that after only 15 min at 338°F, Biodac® released 0.16% of VOC. After 1 h this amount is close to 0.5%, and after 6 h 15 min the amount of VOC released was 1.38%. Our estimates show that during extrusion, VOC from Biodac®, as well as from rice hulls (see above) can be about 0.25–0.5%.

Rice Hulls and Biodac® as Antioxidants in WPC

Ground rice hulls, being added to HDPE, remarkably increases oxidative induction time (OIT) of the plastic (Table 3.23). Biodac® also increases the OIT of

TABLE 3.23 Effect of rice hulls and Biodac® on the OIT (at 190°C) of an HDPE, processed on the Brabender laboratory extruder

Filler of HDPE	OIT (min)
Neat HDPE	4.4 ± 0.5
Rice hulls (traces)	7.4 ± 0.3
Rice hulls (10%)	15 ± 1
Rice hulls (30%)	22 ± 2
Rice hulls (60%)	37 ± 4
Biodac (traces)	5.4 ± 0.4
Biodac (10%)	7.5 ± 0.3
Biodac (30%)	7.1 ± 0.4
Biodac (60%)	6.3 ± 0.3

HDPE but not to that extent. Apparently, the shearing effect of Biodac® in the extruder and the respective decrease of the OIT compensates the positive effect of the filler.

REFERENCES

1. U.S. Pat. No. 5,082,605 (January 21, 1992). J.G. Brooks, B.D. Goforth, and C.L. Goforth. Method for making composite material.
2. U.S. Pat. No. 5,088,910 (February 18, 1992). B.D. Goforth, C.L. Goforth, and J.G. Brooks. System for making synthetic wood products from recycled materials.
3. U.S. Pat. No. 5,096,046 (March 17, 1992). B.D. Goforth, C.L. Goforth, and J.G. Brooks. System for making synthetic wood products from recycled materials.
4. U.S. Pat. No. 5,486,553 (January 23, 1996). M.J. Deaner, G. Puppin, and K.E. Heikkila. Advanced polymer/wood composite structural member.
5. U.S. Pat No. 5,539,027 (July 23, 1996). M.J. Deaner, G. Puppin, and K.E. Heikkila. Advanced polymer/wood composite structural member.
6. U.S. Pat. No. 5,585,155 (December 17, 1996). K.E. Heikkila and A.L. Garofalo. Fiber reinforced thermoplastic structural member.
7. U.S. Pat. No. 5,516,472 (May 14, 1996). T.C. Laver. Extruded synthetic wood composition and method for making same.
8. U.S. Pat. No. 5,746,958 (May 5, 1998). K.-A. Gustafsson, J.J. Muller, and R.A. Wittenberg. Method of producing a wood-thermoplastic composite material.
9. U.S. Pat. No. 5,851,469 (December 22, 1998). J.J. Muller and R.A. Wittenberg. Process for making a wood-thermoplastic composite.
10. U.S. Pat. No. 5,827,462 (October 27, 1998). J.R. Brandt, W.G. Taylor, and J.M. Miller. Balanced cooling of extruded synthetic wood material.
11. U.S. Pat. No. 5,866,264 (February 2, 1999). B.E. Zehner and J.R. Brandt. Renewable surface for extruded synthetic wood material.
12. U.S. Pat. No. 5,863,480 (January 26, 1999). D. Suwanda. Process for making a filler reinforced thermoplastic composites having biaxially oriented components.
13. U.S. Pat. No. 6,590,014 (July 8, 2003). P. Ruede. Non-staining polymer composite product.
14. U.S. Pat. No. 6,758,996 (July 6, 2004). Y.A. Monovoukas, S.J. Anderson, D. Leeman, A.A. Klyosov, and G.P. Philippidis. Cellulose-reinforced thermoplastic composite and methods of making same.
15. U.S. Pat. No. 3,367,917 (February 6, 1968). N.A. Granito. Melamine–formaldehyde–benzoguanamine resin and process for preparing the same.
16. U.S. Pat. No. 3,407,154 (October 22, 1968). G.S. Casebolt and D.E. Cordier. Acid amide modified urea formaldehyde alpha-cellulose filled molding compound.
17. U.S. Pat. No. 3,407,155 (October 22, 1968). G.S. Casebolt. Aminotriazine decorative, dimensionally stable molding compound containing a polymerized aromatic monomer and hydroxyl-containing vinyl monomers copolymer.
18. U.S. Pat. No. 4,282,119 (August 4, 1981). A. Tinkelenberg, H.W.L.M. Vaessen, K.W. Suen, and A.J. Van Doorn. Manufacture of chipboard having high strength and reduced

formaldehyde emission, using a minor amount of protein in combination with low formaldehyde:urea resins.
19. U.S. Pat. No. 4,362,827 (December 7, 1982). A. Tinkelenberg, H.W.L.M. Vaessen, and K.W. Suen. Manufacture of chipboard.
20. U.S. Pat. No. 3,546,158 (December 8, 1970). W.C. Champion. Flooring composition.
21. U.S. Pat. No. 3,720,641 (March 13, 1973). F.M. Berardinelli. Process for producing an improved wholly automatic polyamide molding resin.
22. U.S. Pat. No. 5,288,775 (February 22, 1994). R.K. Bischoff and E.A. Witzleben. Moldable building material composites.
23. U.S. Pat. No. 5,767,177 (June 16, 1998). P. Omente and L. Landru. Injection mouldable thermosetting composition especially for motor vehicle bodies, method of production and method or recycling.
24. U.S. Pat. No. 5,767,178 (June 16, 1998). W.R. Kolker and L. Bach. Water and fire resistant materials and methods for making the same.
25. U.S. Pat. No. 5,783,125 (July 21, 1998). A.L. Bastone, L.E. Lloyd, C.D. Newby, P.R. Pelfrey, T.E. Philips, B.E. Zehner, and J.R. Brandt. Reinforced extrusion products and method of making same.
26. U.S. Pat. No. 6,291,558 (September 18, 2001). M. Raskin, L.O. Ioffe, A.Z. Pukis, and M.H. Wolf. Composition board binding material.
27. U.S. Pat. No. 6,702,969 (March 9, 2004). L.M. Matuana and J.A. King. Method of making wood-based composite board.
28. U.S. Pat. No. 3,856,724 (December 24, 1974). J.J. O'Connor and J.T. Nolan. Reinforced thermoplastic compositions.
29. U.S. Pat. No. 3,875,088 (April 1, 1975). I.J. Arons, R. Eller, and R.E. Merrill. Pencil sheath compositions.
30. U.S. Pat. No. 3,878,143 (April 15, 1975). H. Baumann, A. Kriisa, and H.E. Grahn. Method of preventing corrosion in connection with extrusion of mixtures containing polyvinyl chloride and wood flour or similar cellulosic material, and analogous mixtures containing polystyrene or acrylonitrile-butadiene-styrene resin, respectively.
31. U.S. Pat. No. 3,943,079 (March 9, 1976). P. Hamed. Discontinuous cellulose fiber treated with plastic polymer and lubricant.
32. U.S. Pat. No. 4,165,302 (August 21, 1979). C.F. Armenti and J.V. De Juneas. Filled resin compositions containing atactic polypropylene.
33. U.S. Pat. No. 4,250,064 (February 10, 1981). H. Chandler. Plastic formulation reinforced with organic fibers.
34. U.S. Pat. No. 4,339,363 (July 13, 1982). Y. Nakagima. Composite material compositions using wastepaper and method of producing same.
35. U.S. Pat. No. 4,343,727 (August 10, 1982). A. Cohen. Polyvinyl chloride shoe sole molding composition containing cellulosic fibrous flock.
36. U.S. Pat. No. 4,414,267 (November 8, 1983). A.Y. Coran and L.A. Lloyd. Method for treating discontinuous cellulose fibers characterized by specific polymer to plasticizer and polymer-plasticizer to fiber ratios, fibers thus treated and composites made from the treated fibers.
37. U.S. Pat. No. 4,559,376 (December 17, 1985). J. Kubat and T.C.F. Klason. Method of producing plastic composites filled with cellulose or lignocellulosic materials.

38. U.S. Pat. No. 5,474,722 (December 12, 1995). R.T. Eoodhams. Oriented thermoplastic and particulate matter composite material.
39. U.S. Pat. No. 6,610,232 (August 26, 2003). W.W. Jacobsen. Lignocellulose fiber filler for thermoplastic composite compositions.
40. U.S. Pat. No. 6,743,507 (June 1, 2004). F. Barlow, Y. Khanna, D.B. Pietsch, and J. Underwood. Cellulose fiber reinforced composites having reduced discoloration and improved dispersion and associated methods of manufacture.
41. U.S. Pat. No. 6,833,399 (December 21, 2004). M. Khavkine and B. Isman. Flowable flax bast fiber and flax shive blend useful as reinforcing agent.
42. U.S. Pat. No. 6,929,841 (August 16, 2005). D. Van Dijk and F.B.A. De Vries. Plastic-based composite product and method and apparatus for manufacturing same.
43. U.S. Pat. No. 6,939,496 (September 6, 2005). F.W. Maine and W.R. Newson. Method and apparatus for forming composite material and composite material therefrom.
44. U.S. Pat. No. 4,687,793 (August 18, 1987). T. Motegi, K. Aoki, and K. Kimura. Thermoplastic resins containing gloxal heat treated cellulosic fillers.
45. U.S. Pat. No. 4,783,493 (November 8, 1988). T. Motegi, K. Aoki, and K. Kimura. Thermoplastic resins with cellulosic filler.
46. U.S. Pat. No. 4,833,011 (May 23, 1989). K. Horimoto. Synthetic pulp and absorbent comprising the same.
47. U.S. Pat. No. 4,717,743 (January 5, 1988). H. Wakabayashi, F. Kato, T. Matsubara, and Y. Ishikawa. Polyolefin-base resin composition.
48. U.S. Pat. No. 4,737,532 (April 12, 1988). T. Fujita and F. Nagoshi. Thermoplastic resin composition containing wood flour.
49. U.S. Pat. No. 6,368,529 (April 9, 2002). J.D. Lloyd, M.J. Manning, and F.M. Ascherl. Lignocellulose composite.
50. U.S. Pat. No. 6,527,532 (March 4, 2003). J.J. Muller, R.A. Wittenberg, and B.S. Yarmoska. Apparatus for making a wood-plastic profile.
51. U.S. Pat. No. 6,578,368 (June 17, 2003). J.R. Brandt, M.F. Kollar, and H.L. Hutchison. Cryogenic cooling of extruded and compression molded materials.
52. U.S. Pat. No. 6,627,676 (September 30, 2003). R. George, M.A. Saide, and J.G. Saide. Antimicrobial biocidic fiber-plastic composite and method of making same.
53. U.S. Pat. No. 6,827,995 (December 7, 2004). R.E. Hughes, K. Fidati, J. Schiff, J. Cuttle, and G. Vuduris. Composites useful as fence and decking components and methods for producing same.
54. U.S. Pat. No. 6,844,040 (January 18, 2005). A.A. Pabedinskas and W.K.H. Gregori. Reinforced composite structural members.
55. U.S. Pat. No. 6,844,049 (January 18, 2005). H. Amin-Javaheri. Polyvinyl chloride/wood composite having a natural wood grain finish and a method for creating the finish.
56. U.S. Pat. No. 6,881,367 (April 19, 2005). C. Baker. Composite materials, articles of manufacture produced therefrom, and methods for their manufacture.
57. U.S. Pat. No. 6,890,637 (May 10, 2005). C. Baker. Composite materials, articles of manufacture produced therefrom, and methods for their manufacture.
58. U.S. Pat. No. 6,903,149 (June 7, 2005). N. Yasuda and H. Nemoto. Composition for wood-polymer composite and wood-polymer composite made from the composition.

59. U.S. Pat. No. 6,958,185 (October 25, 2005). B.E. Zehner. Multilayer synthetic wood component.
60. U.S. Pat. No. 4,376,144 (March 8, 1983). L.A. Goettler. Treated fibers and bonded composites of cellulose fibers in vinyl chloride polymer characterized by an isocyanate bonding agent.
61. U.S. Pat. No. 4,791,020 (December 13, 1988). B.V. Kokta. Bonded composites of cellulose fibers polyethylene.
62. U.S. Pat. No. 4,820,749 (April 11, 1989). A.D. Beshay. Reinforced polymer composites with wood fibers grafted with silanes.
63. U.S. Pat. No. 5,008,310 (April 16, 1991). A.D. Beshay. Polymer composites based cellulose-V.
64. U.S. Pat. No. 5,120,776 (June 9, 1992). G. Raj and B.V. Kokta. Process for chemical treatment of discontinuous cellulosic fibers and composites of polyethylene and treated fibers.
65. U.S. Pat. No. 4,851,458 (July 25, 1989). S. Hopperdietzel. Use of cellulose fibers for structurally modifying polyvinyl chloride articles.
66. U.S. Pat. No. 5,153,241 (October 6, 1992). A.D. Beshay. Polymer composites based cellulose-VI.
67. U.S. Pat. No. 5,952,105 (September 14, 1999). M. Medoff and A. Lagace. Poly-coated paper composites.
68. U.S. Pat. No. 5,973,035 (October 26, 1999). M. Medoff and A. Lagace. Cellulosic fiber composites.
69. U.S. Pat. No. 6,207,729 (March 27, 2001). M. Medoff and A. Lagace. Texturized cellulosic and lignocellulosic materials and compositions and composites made therefrom.
70. U.S. Pat. No. 5,981,631 (November 9, 1999). C.P. Ronden and J.C. Morin. Process for the production of composites of co-mingled thermoset resin bonded wood waste blended with thermoplastic polymers.
71. U.S. Pat. No. 6,939,903 (September 6, 2005). W.D. Sigworth, R.A. Cheney, L.M. Panzer, and H.L. Mei. Natural fiber-filled polyolefin composites.
72. U.S. Pat. No. 6,942,829 (September 13, 2005). G.W. Drabeck, J. Bravo, M. DiPierro, A.C. Andrews, J.M. McKinney, B. Hollo, and D. Chundury. Polymer-wood composites and additive system therefor.
73. U.S. Pat. No. 6,983,571 (January 10, 2006). C.C. Felton. Composite roofing panel.
74. U.S. Pat. No. 7,041,716 (May 9, 2006). M.-T. Ton-That, F. Perrin-Sarazin, and J. Denault. Cellulose filled thermoplastic composites.
75. U.S. Pat. No. 5,288,772 (February 22, 1994). D.N.-S. Hon. Pre-treated cellulosic materials for producing molded composite articles therefrom and process.
76. Canadian Pat. No. 2,100,320 (March 1994). M.J. Deaner, G. Puppin, and K.E. Heikkila. Advanced polymer wood composite.
77. Canadian Pat. No. 2,178,036 (December 1996). K.V. Seethamraju, N.J. Beaverson, K.E. Heikkila, and M.J. Deaner. Advanced compatible polymer wood fiber composite.
78. Canadian Pat. No. 2,242,326 (July 1997). M.J. Deaner and K.V. Seethamraju. Advanced engineering resin and wood fiber composite.
79. U.S. Pat. No. 5,827,607 (October 27, 1998). M.J. Deaner, G. Puppin, and K.E. Heikkila. Advanced polymer wood composite.

80. U.S. Pat. No. 5,948,524 (September 7, 1999). K.V. Seethamraju and M.J. Deaner. Advanced engineering resin and wood fiber composite.
81. U.S. Pat. No. 6,015,611 (January 18, 2000). M.J. Deaner, J. Puppin, and K.E. Heikkila. Advanced polymer wood composite.
82. U.S. Pat. No. 6,015,612 (January 18, 2000). M.J. Deaner, J. Puppin, and K.E. Heikkila. Polymer wood composite.
83. Canadian Pat. No. 2,250,161 (August 28, 1997). O. Grill and D.L. Turk. Interfacial blending agent for natural fiber composites.
84. U.S. Pat. No. 5,738,935 (April 14, 1998). D.L.Turk and O. Grill. Process to make a composite of controllable porosity.
85. U.S. Pat. No. 5,480,602 (January 2, 1996). L. Nagaich. Extruded particle board.
86. U.S. Pat. No. 5,635,125 (June 3, 1997). T.J. Ternes, K.K. Kimberly, and B.L. West. Method for forming simulated shake shingles.
87. U.S. Pat. No. 5,992,116 (November 30, 1999). T.J. Ternes, K.K. Ternes, and B.L. West. Simulated shake shingle.
88. Canadian Pat. No. 2,278,688 (June 17, 1999). Y. Kasahara and F. Goto. Thermoplastic composite composition reinforced with mica and wooden fiber filler.
89. U.S. Pat. No. 6,011,091 (January 4, 2000). B.E. Zehner. Vinyl based cellulose reinforced composite.
90. U.S. Pat. No. 6,248,813 (June 19, 2001). B.E. Zehner. Vinyl based cellulose reinforced composite.
91. U.S. Pat. No. 6,617,376 (September 9, 2003). A.F. Korney. Flexible wood composition.
92. U.S. Pat. No. 6,117,924 (September 12, 2000). J.R. Brandt. Extrusion of synthetic wood material.
93. U.S. Pat. No. 6,984,676 (January 10, 2006). J.R. Brandt. Extrusion of synthetic wood material.
94. U.S. Pat. No. 6,171,688 (January 9, 2001). J. Zheng, R.J. Morgan, and R. Jurek. Material and method for the preparation thereof.
95. U.S. Pat. No. 6,265,037 (July 24, 2001). S. Godavarti, R.K. Williams, and M.J. Deaner. Polyolefin wood fiber composite.
96. U.S. Pat. No. 6,680,090 (January 20, 2004). S. Godavarti, R.K. Williams, and M.J. Deaner. Polyolefin wood fiber composite.
97. U.S. Pat. No. 6,682,789 (January 27, 2004). S. Godavarti, R.K. Williams, and M.J. Deaner. Polyolefin wood fiber composite.
98. U.S. Pat. No. 6,270,883 (August 7, 2001). K.D. Sears. R.E. Jacobson, D.F. Caulfield, and J. Underwood. Composites containing cellulosic pulp fibers and methods of making and using the same.
99. U.S. Pat. No. 6,274,248 (August 14, 2001). F. Goto and Y. Kasahara. Thermoplastic composite composition reinforced with mica and wooden fiber filler.
100. U.S. Pat. No. 6,284,098 (September 4, 2001). W.W. Jacobsen. Lignocellulose fiber filler for thermoplastic composite compositions.
101. U.S. Pat. No. 6,337,138 (January 8, 2002). B.E. Zehner and B.K. Buhrts. Cellulosic, inorganic-filled plastic composite.
102. U.S. Pat. No. 6,344,504 (February 5, 2002). B.E. Zehner and T.C. Laver. An extruded wood simulative material is described which includes a high degree of cellulosic material content and begins with powdered thermoplastic material.

103. U.S. Pat. No. 6,498,205 (December 24, 2002). B.E. Zehner. Extrusion of synthetic wood material using thermoplastic material in powder form.
104. U.S. Pat. No. 6,511,757 (January 28, 2003). J.R. Brandt and B.E. Zehner. Compression molding of synthetic wood material.
105. U.S. Pat. No. 6,586,503 (July 1, 2003). M. Grohman. Composite products comprising cellulosic materials and synthetic resins and methods of making the same.
106. U.S. Pat. No. 6,737,006 (May 18, 2004). M. Grohman. Composite products comprising cellulosic materials and synthetic resins and methods of making the same.
107. U.S. Pat. No. 6,122,877 (September 26, 2000). G.L. Hendrickson, K.E. Heikkila, T.P. Murphy, and M.N. Goeser. Fiber-polymeric composite siding unit and method of manufacture.
108. U.S. Pat. No. 6,682,814 (January 27, 2004). G.L. Hendrickson, K.E. Heikkila, T.P. Murphy, and M.N. Goeser. Fiber-polymer composite siding unit and method of manufacture.
109. U.S. Pat. No. 6,780,359 (August 24, 2004). B.E. Zehner and J.R. Brandt. Synthetic wood composite material and method for molding.
110. U.S. Pat. No. 6,971,211 (December 6, 2005). B.E. Zehner. Cellulosic/polymer composite material.
111. U.S. Pat. No. 7,022,751 (April 4, 2006). C. Zhang and B. Casciato. Composite plastic materials produced from waste materials and method of producing same.
112. U.S. Pat. No. 5,938,994 (August 17, 1999). B.W. English and K.P. Gohr. Method for manufacturing of plastic wood-fiber pellets.
113. U.S. Pat. No. 6,210,792 (April 3, 2001). K.V. Seethamraju, N.J. Beaverson, K.E. Heikkila, and M.J. Deaner. Advanced compatible polymer wood fiber composite.
114. U.S. Pat. No. 6,255,368 (July 3, 2001). B.W. English and K.P. Gohr. Plastic wood-fiber pellets.
115. U.S. Pat. No. 6,632,863 (October 14, 2003). H.L. Hutchison and J.R. Brandt. Cellulose/polyolefin composite pellet.
116. U.S. Pat. No. 6,756,114 (June 29, 2004). F.R. Cichocki, J.L. Thomason, and T.L. Cheney. Moldable pellet based on the combination of synthetic cellulose fibers and thermoplastic polymers.
117. U.S. Pat. No. 7,052,640 (May 30, 2006). F.R. Cichocki, J.L. Thomason, and T.L. Cheney. Moldable pellet based on the combination of synthetic cellulose fibers and thermoplastic polymers.
118. U.S. Pat. No. 6,153,293 (November 28, 2000). M.E. Dahl, R.G. Rottinghaus, and A.H. Stephens. Extruded wood polymer composite and method of manufacture.
119. U.S. Pat. No. 6,342,172 (January 29, 2002). M.D. Finley. Method of forming a foamed thermoplastic polymer and wood fiber profile and member.
120. U.S. Pat. No. 6,344,268 (February 5, 2002). D.J. Stucky and R. Elinski. Foamed polymer-fiber composite.
121. U.S. Pat. No. 6,590,004 (July 8, 2003). B.E. Zehner. Foam composite wood replacement material.
122. U.S. Pat. No. 6,784,216 (August 31, 2004). B.E. Zehner and S.R. Ross. ABS foam and method of making same.
123. U.S. Pat. No. 6,863,972 (March 8, 2005). C.C. Burger, J.R. Brandt, D.W. Smith, J.P. Frechette, and K.T. Burt. Synthetic wood component having a foamed polymer backing.

124. U.S. Pat. No. 6,936,200 (August 30, 2005). C.B. Park, G.M. Rizvi, and H. Zhang. Plastic wood fiber foam structure and method of producing same.

125. U.S. Pat. No. 5,306,550 (April 26, 1994). M. Nishiyama, J. Hosokawa, K. Yoshihara, T. Kubo, K. Kanaoka, K. Kondo, S. Maruyama, K. Tateishi, and A. Ueda. Biodegradable composition and shaped article obtained therefrom.

126. U.S. Pat. No. 5,827,905 (October 27, 1998). E. Grigat, H.P. Muller, W. Schulz-Schlitte, A.E. Sayed, and R. Timmermann. Biodegradable plastics filled with reinforcing materials.

127. U.S. Pat No. 6,274,652 (August 14, 2001). M. Uryu and K. Tokura. Composite polymer materials and process for producing the same.

128. U.S. Pat No. 6,479,164 (November 12, 2002). J. Lorcks, W. Pommeranz, K. Klenke, H. Schmidt, and J. Heuer. Method for preparing composite materials from renewable raw materials.

129. S. Faehner. The economic impact of wood flour consistency in WPC applications. In: *The Global Outlook for Natural Fiber & Wood Composites 2005*, Lake Buena Vista, FL, November 14–16, 2005.

130. C. Clemons. Use of invasive wood species in wood plastic composites: compositional differences and their effects on processing and performance. In: *Progress in Wood and BioFibre Plastic Composites 2006*, Toronto, Canada, May 2006.

131. M. Le Baillif and K. Oksman. The influence of the extrusion process on bleached pulp fiber and its composites. In: *Progress in Wood and BioFibre Plastic Composites 2006*, Toronto, Canada, May 2006.

132. M.A. Svoboda. Property profiles and structure-property relationships of polypropylene-wood composites with high wood content. In: *Wood-Plastic Composites, A Sustainable Future. The International Conference*, Vienna, Austria, May 14–16, 2002.

133. N.M. Stark and M.J. Berger. Effect of particle size on properties of wood-flour reinforced polypropylene composites. In: *Fourth International Conference on Woodfiber-Plastic Composites*, Madison, WI, May 12–14, 1997, pp. 134–143.

134. A. Bismark, S. Mishra, and T. Lampke. Plant fibers as reinforcement for green composites. In: A.K. Mahanty, M. Misra, and L.T. Drzal (Eds.), *Natural Fibers, Biopolymers, and Biocomposites*, Taylor & Francis, CRC, Boca Raton, 2005, p. 65.

135. K. Williams. Automotive industry uses of natural fiber reinforced composites. In: *The Global Outlook for Natural Fiber & Wood Composites 2003*, Intertech, New Orleans, LA, December 2003.

136. L. Magdzinski. Quality-certified specialty cellulose and wood fibers for composites from market pulp producers. In: *Progress in Wood and BioFibre Plastic Composites 2006*, Toronto, Canada, May 2006.

137. M.M. Stark and R.E. Roland. Effects of wood fiber characteristics on mechanical properties of wood/polypropylene composites. *Wood J Fiber Sci.*, 2003, **35**(2), 167–174.

138. M.A. Svoboda and R.W. Lang. Property profiles and structure-property relationships of polypropylene-wood-composites with high wood content. In: *Wood-Plastic Composites, A Sustainable Future. The International Conference*, Vienna, Austria, May 14–16, 2002.

139. K. Gehrmann, M. Busch, S. Meinicke, R. Starke, and A. Krombholz. The influence of fibre geometry of different natural fibres and different PP grades on key properties of composites and extruded hollow profiles. First experiences with the industrial profile

extrusion line. In: *Fifth Global Wood and Natural Fibre Composites Symposium*, Kassel, Germany, April 27–28, 2004.

140. N.M. Stark. The effect of weathering variables on the lightness of HDPE/WF composites. In: *Eighth International Conference on Woodfiber Composites*, Madison, WI, May 23–25, 2005.

141. S. Snell, J. Hague, and L. Groom. Characterizing agrofibers for use in composite materials. In: *The Fourth International Conference on Woodfiber-Plastic Composites*, Madison, WI, May 1997, pp. 5–11.

142. U.S. Pat. No. 5,730,371 (March 24, 1998). P.A. Dangieux, S.J. Anderson, and R. L. Yoder. Delumper.

4

COMPOSITION OF WOOD–PLASTIC COMPOSITES: MINERAL FILLERS

INTRODUCTION

Minerals, such as calcium carbonate, talc, silica, are quite common fillers in plastic industry. They, often at about 6–15 cent/lb, replace a much more expensive plastic, increase stiffness of the filled product, and render the plastic more flame resistant. The world filler market for plastics is dominated by carbon black and calcium carbonate. Of about 15 billion pounds of filler in America and Europe, about half the filler volume goes into elastomers, a third into thermoplastics, and the reminder into thermosets. About 15% of all manufactured plastics contain fillers.

However, only few wood–plastic composites (WPCs) use mineral fillers along with cellulose fiber. Table 4.1 lists these composites. Some of them contain minerals only because they employ rice hulls as a principal filler, as rice hulls typically contain $19 \pm 1\%$ of silicates. Hence, 50% of rice hulls (by weight) in WPC results in about 9.5% of minerals. In comparison, wood flour typically contains about 0.25% of minerals by weight, and 50% of wood filler would result in only 0.125% (w/w) of minerals in WPC.

This chapter is not a comprehensive description of fillers in plastics and of the respective properties of the resulting materials. Books are written on this subject (see, for example, Ref. [1]). This chapter aims at a brief description of minerals which are either used as fillers in WPC or can be used readily, taking into account their cost and value-added properties. Besides, this chapter gives some examples of properties of WPCs after mineral fillers were added to the composition.

Wood-Plastic Composites, by Anatole A. Klyosov

TABLE 4.1 WPC deck boards, containing minerals in amounts higher than their natural content in wood flour or sawdust, and manufacturing companies, as described in manufacturer's ICC-ES reports

No.	Deck	Manufacturer	Mineral content and density (specific gravity)	ICC-ES report number and date
		Polyethylene-based products		
1	GeoDeck	LDI Composites	20% ($CaCO_3$ and aluminosilicates) d=1.24 g/cm^3	ESR-1369 6/1/2006
2	EverGreen	Integrated Composite Technologies	>15% talc	N/A
3	Nexwood	Nexwood Industries	12% (aluminosilicates) d=1.17 g/cm^3	BOCA 99-8.1 (January, 2000) Not current
4	TimberTech	TimberTech	16% (talc) d=1.22 g/cm^3	ESR-1400 6/1/2005
5	UltraDeck	Midwest Manufacturing Extrusion	4% (talc) d=1.22 g/cm^3	ESR-1674 11/1/2004
6	WeatherBest, LP Composite, Veranda	Louisiana Pacific	7% (talc) d=1.20 g/cm^3	ESR-1088 6/1/2005

Besides cost, the following properties of minerals are typically considered (or should be considered) for using them as fillers in composite materials (in no particular order of the properties):

- Chemical composition
- Aspect ratio
- Density (specific gravity)
- Particle size
- Particle shape
- Particle size distribution
- Particle surface area
- Moisture content, the ability to absorb water
- The ability to absorb oil
- Flame retardant properties
- Effect on mechanical properties of the composite material
- Effect on hot melt viscosity
- Effect on hot melt shrinkage

- Thermal properties
- Color, optical properties
- Effect on fading and durability of plastics and composites
- Health and safety.

Some of these properties are considered in detail in the following sections of this chapter.

Before that, let us give a few preliminary general descriptions, which will be detailed below with specific examples of mineral (and mixed) fillers.

GENERAL PROPERTIES OF MINERAL FILLERS

Chemical Composition

Fillers can be inorganic, organic, or mixed ones, such as in the case of Biodac®, as described above. Biodac® is a granular mix of cellulose fiber, calcium carbonate, and kaolin (clay). Typical inorganic fillers can be simple salts, such as calcium carbonate ($CaCO_3$) or wollastonite ($CaSiO_3$), with a well-defined chemical structure; complex inorganic materials, such as talc [hydrated magnesium silicate, $Mg_3Si_4O_{10}(OH)_2$] or kaolin (hydrated aluminum silicate, $Al_2O_3{\cdot}2SiO_2{\cdot}2H_2O$), or can be compounds with an undefined or variable composition, such as mica, clay, and fly ash. The latter can be considered as aluminum silicate with inclusions of other elements.

Aspect Ratio

It is the length of a particle divided by its diameter. For spherical or cubical particles, the aspect ratio equals to one. For calcium carbonate particles, the aspect ratio is typically 1–3. For talc, the aspect ratio is typically in the range of 5–20. For milled glass fiber, it is between 3 and 25. For mica, it is 10–70. For wollastonite, it is between 4 and 70. For chopped glass fiber, it is between 250 and 800. For natural fibers, such as cellulose, the aspect ratio can be from 20–80 to thousands. Low aspect ratio is below 10. However, the above figures are given for fillers not processed in a compounder and/or an extruder. Upon processing, aspect ratio can decrease from dozens and hundreds to as low as 3 – 10.

Density (Specific Gravity)

Though specific gravity of mineral fillers can vary in a wide range, those of fillers used (or likely to be used) in WPCs are all high, around 2.1–2.2 (fly ash) and 2.6–3.0 g/cm^3 (calcium carbonate, talc, kaolin, mica, clay). Biodac®, a granular blend of calcium carbonate with kaolin and cellulose fiber, has specific gravity of 1.58 g/cm^3.

TABLE 4.2 Effect of specific gravity of fillers on density of the filled plastic. Cellulose fibers (wood flour, rice hulls) typically have specific gravity of 1.3 g/cm³; calcium carbonate and talc typically have specific density of 2.8 g/cm³.

Plastic (density)	Filler	Content of the filler, (% w/w)	Specific gravity of the filled plastic, calculated (g/cm3)
HDPE (0.96 g/cm³)	Cellulose fibers	20	1.01
		40	1.07
	$CaCO_3$ or talc	20	1.105
		40	1.30
Polypropylene (0.90 g/cm³)	Cellulose fibers	20	0.96[a]
		40	1.03[a]
	$CaCO_3$ or talc	20	1.04
		40	1.24[a]

[a] The respective experimental data for the filled polypropylene are as follows: with 20% cellulose fibers, 0.98 – 1.00 g/cm³; with 40% cellulose fibers, 1.08 – 1.10 g/cm³; with 40% calcium carbonate or talc, 1.23 – 1.24 g/cm³ [2].

Table 4.2 shows how minerals affect the density of the filled plastics compared with wood fiber.

One can see that the presence of 20–40% of mineral fillers significantly increases density of filled HDPE and polypropylene compared with the plastics filled with cellulose fiber.

Note: These calculations can be done as shown with the following example. For HDPE, filled with 20% of calcium carbonate, 100 g of the filled plastic contains 20 g of $CaCO_3$ and 80 g of the plastic. The respective volume fractions are equal to 20 g/2.8 g/cm³ = 7.1429 cm³ for $CaCO_3$ and 80 g/0.96 g/cm³ = 83.3333 cm³ for HDPE. Total volume of the filled plastic is 7.1429 cm³ + 83.3333 cm³ = 90.4762 cm³. As the weight of this sample is 100 g, specific gravity of the filled plastic is 100 g/90.4762 cm³ = 1.105 g/cm³.

Note: *How* not *to calculate specific gravity of a composite material.* A common mistake is to mix volume and weight fractions in calculations. For example, in the above case of HDPE filled with 20% of calcium carbonate, a wrong calculation of the resulting specific gravity would be 0.2×2.8 g/cm³ + 0.8×0.96 g/cm³ = 1.328 g/cm³. The right answer, as we know, is 1.105 g/cm³ (see above). The mistake was to take volume fractions 0.2 and 0.8 as weight fractions of the resulting composite.

Particle Size

For the purpose of this discussion, fillers can be divided into coarse particles (above 0.1–0.3 mm, 20–150 mesh), large particle size (around 0.1 mm, or 100 μm, 150–200 mesh), medium particle size (around 10 μm, 250 mesh), small particle size (around 1 μm), fine particle size (around 0.1 μm), and nanoparticles (exfoliated—thickness

1 nm, or 0.001 μm, and length 200 nm, or 0.2 μm; intercalated—thickness 30 nm, length 200 nm). Nanoparticles are not considered as fillers, but rather as additives. Examples of the above particle sizes are provided with Biodac® (coarse particles), ground calcium carbonate (large particle size), clay (medium particle size), precipitated $CaCO_3$ (small particle size), some special kinds of silica (fine particle size), exfoliated layered organoclays particles. Cost of these fillers increases very significantly from coarse and large to small and fine particles, and especially for nanoparticles. Hence, only coarse and large particles of fillers can result in cost savings when replacing plastics, unless fillers provide the composite material with really beneficial properties justifying the incremental cost.

Mesh conversions to microns (0.001 mm) and mils (0.001 of in.)

Mesh	Microns	Mils
4	4700	185
6	3330	131
8	2360	92.9
10	1680	66.1
12	1410	55.5
14	1190	46.9
16	1000	39.4
20	841	33.1
24	707	27.8
28	595	23.4
32	500	19.7
35	420	16.5
42	354	13.9
48	297	11.7
60	250	9.8
65	210	8.3
80	177	7.0
100	149	5.9
115	125	4.9
150	105	4.1
170	88	3.5
200	74	2.9
250	63	2.5
270	53	2.1
325	44	1.7
400	37	1.5

Particle Shape

This property is related in part to aspect ratio of particles, but not completely. At the same aspect ratio, equal to 1.0, particles can be spherical or cubic, and spherical particles (such as carbon black, titanium dioxide, zinc oxide) improve flowability and decrease viscosity of molten plastics and provide a uniform distribution of stress in the solidified profile, whereas cubic particles (calcium hydroxide) give good reinforcement to the profile. Flakes (kaolin, mica, talc) facilitate orientation of plastics. Elongated particles such as wollastonite, glass fiber and cellulose fiber, wood flour (fiber) reduce shrinkage and thermal expansion–contraction, and particularly reinforce the solid material.

Particle Size Distribution

Particles can be monodisperse, or possess a certain distribution of sizes—wide, narrow, bimodal, and so on. Distribution can be irregular, typically a mix of particles of different sizes. This property of a particle mix largely depends on milling technology and classification (screening) of particles. A wide distribution or a bimodal distribution of particles of a mineral filler can be beneficial because it can provide a better packing density of particles in the matrix. Particle size distribution can affect viscosity of the hot melt.

Particle Surface Area

If is directly connected with "topography" of the surface and porosity of the filler. It is measured in square meters per gram of the filler and can vary from fractions of m^2/g to hundreds of m^2/g. For example, specific surface area for wollastonite varies between 0.4 and 5 m^2/g, silica 0.8 and 3.5 m^2/g, cellulose fibers around 1 m^2/g, talc 2.6 and 35 m^2/g, calcium carbonate 5 and 24 m^2/g, kaolin 8 and 65 m^2/g, clay 18 and 30 m^2/g, titanium dioxide 7 and 162 m^2/g, precipitated silica 12 and 800 m^2/g ([1], p. 253). Particle specific surface area depends very much on the procedure that is employed to measure the area. The smaller the molecule utilized for the measurements, the larger the specific surface area obtained per gram of the material. However, when mixed with molten plastic, small molecular size pores of a mineral filler are irrelevant. Large open pores, on the contrary, can provide not only adhesion area for the molten plastic, but also an additional physical interaction area between the filler and the plastic after its solidification.

Moisture Content: The Ability to Absorb Water

These two properties go hand in hand and are connected in a way with "hygroscopicity" of the filler. However, moisture content normally reflects the weight (percent) of water per unit weight of the filler under given circumstances (for instance, after or in the process of drying), whereas the ability to absorb water often means the maximum achievable moisture content or the moisture content after an apparent equilibrium at ambient conditions is attained. Moisture content of bulk rice hulls in summer months can be around 9.5% (w/w). Moisture content of dried rice hulls can be 0.2–0.5%. High moisture content in the filler leads to steam formation in the course of compounding and extrusion, which may result in a high porosity (and low density) of the final extruded profile. This in turn decreases its strength and stiffness, and increases rate of oxidation during service, hence, decreases durability.

Low moisture content of fillers are typically observed in calcium carbonate and wollastonite (0.01–0.5%), talc and aluminum trihydrate, mica (0.1–0.6%). Medium moisture content can be observed in titanium hydroxide (to 1.5%), clay (to 3%), kaolin (1–2%), and Biodac® (2–3%). High moisture content is often seen in cellulose fiber (5–10%), wood flour (to 12%), and fly ash (to 20%). Biodac® absorbs 120% of water under direct contact with an excess of water.

The Ability to Absorb Oil

This property can be beneficial for hydrophobic plastics, such as polyolefins, as hydrophobic fillers can show a good interaction with the matrix. Also, hydrophobic fillers can very significantly influence the viscosity of the matrix, hence, its rheology and flowability. Fillers typically absorb oil in much higher quantities compared to water. Calcium carbonate absorbs 13–21% of oil, aluminum trihydrate absorbs 12–41% of oil, titanium dioxide 10–45%, wollastonite 19–47%, kaolin 27–48%, talc 22–57%, mica 65–72%, and wood flour 55–60%. Biodac® absorbs 150% of oil by weight.

Typically, if oil absorption is low, the filler does not change the hot melt viscosity much. Because of this, oil absorption test is often used to characterize the effect of fillers on rheological properties of filled plastics.

Flame Retardant Properties

"Active" flame retardants, such as aluminum trihydrate or magnesium hydroxide, cool down the burning area by releasing water at and above certain temperature (see Chapter 14). Many inert fillers, such as calcium carbonate, talc, clay, glass fiber, and so on, can slow down flame spread by just "removing fuel" for flame propagation or slow heat generation. However, they do not significantly change the ignition point. They act rather by dilution of the fuel in the solid (plastic) phase. Calcium carbonate evolves inert gases (carbon dioxide) at about 825°C, which is too high to dilute flammable gaseous phases that were ignited well below that.

Effect on Mechanical Properties of the Composite Material

Minerals generally improve both flexural strength and flexural modulus of filled plastics and WPCs (Table 4.3), however, the extent of improvement is different for flex strength and flex modulus. Effect on flex strength is often not more than 10–20%. Effect on flex modulus can reach 200–400%, and it often depends on particle size of the filler and its aspect ratio. The higher the filler amount and the aspect ratio, the larger the effect of filler on flexural modulus (though not always, particularly that related to the filler amount).

Based on effects of fillers on strength of filled plastics, fillers can be subdivided into just "fillers" and "reinforcing fillers." "Fillers," such as wood flour, calcium

TABLE 4.3 Effect of inorganic fillers and wood flour on flexural strength and flexural modulus of polypropylene (homopolymer) [2]

Filler	Percentage of the filler	Flexural strength (psi)	Flexural modulus (psi)
None	0	5902	174,000
$CaCO_3$	40	5771	261,000
Talc	40	7917	479,000
Glass fiber	40	8497	508,000
Wood (pine) flour	20	6337	247,000
	40	6482	450,000

carbonate, often leave strength almost unchanged, typically within ±10% of the unfilled plastic. With reinforcing fillers, such as wood fiber with a high aspect ratio, glass fiber, strength of the filled plastic always increases.

That is, some mineral fillers increase flexural strength of polypropylene up to 30–45%, whereas wood flour increases flex strength of the same polymer by only 7–10%. Effect of fillers on stiffness of plastics is much more pronounced, and the mineral fillers increase flexural modulus of polypropylene up to 300% and wood flour increases flex modulus of the same polymer by 150–250%.

Tensile strength of neat and filled polypropylene is about the same, or slightly decreased when the plastic is filled with wood flour (Table 4.4).

Glass fiber increases tensile strength of polypropylene up to 15%; talc shows almost no change; calcium carbonate and wood flour decrease tensile strength of the same polymer by 15–30%. Regarding tensile modulus, increase of strength was up to 3.6 times (talc, glass fiber) and by 1.6–2.6 times (wood flour, calcium carbonate).

It is hard to predict quantitatively how strength and modulus of WPC will be affected by added mineral fillers, as the nature and amount of lubricants can interfere (Table 4.5).

TABLE 4.4 Effect of inorganic fillers and wood flour on tensile strength and tensile modulus of polypropylene (homopolymer) [2]

Filler	Percentage of the filler	Tensile strength (psi)	Tensile modulus (psi)
None	0	4481	203,000
$CaCO_3$	40	3161	319,000
Talc	40	4684	725,000
Glass fiber	40	5133	740,000
Wood (pine) flour	20	3828	334,000
	40	3538	537,000

TABLE 4.5 Effect of talc on flexural strength and modulus of wood-flour-polypropylene composite material in the presence of different amounts of a lubricant (data were provided by Luzenac America)

Ratio of talc/ wood flour	Lubricant (% w/w)	Flexural strength (psi)	Flexural modulus (psi)
10/90	1	3100	595,000
	3	2390	513,000
	5	2020	431,000
50/50	1	4000	806,000
	3	3290	724,000
	5	2915	641,000

Table 4.5 shows that although flexural strength and modulus increase with the increase of the amount of talc with respect to that of wood flour, the lubricant decreases the effect.

Effect on Hot Melt Viscosity

It depends on particle size, particle shape, aspect ratio, specific gravity of the filler, and other properties of fillers. The following example illustrates this "overall" property of fillers. When polypropylene, having melt flow index of 16.5 g/10 min, was filled with some mineral and cellulosic fillers, its MFI (in g/10 min) was as follows [2]:

- 40% $CaCO_3$ 15.1;
- 40% talc 12.2;
- 40% glass fiber 9.6;
- 20% wood (pine) flour 8.6;
- 40% wood flour 1.9.

Obviously, wood flour produced far more significant effect on melt viscosity compared to the inorganic fillers.

Effect on Mold Shrinkage

It apparently depends on the amount of the fillers (hence, amount of the plastic) and ability of the fillers to interfere with crystallization of the plastic. The less the crystallites in the filled plastic, the less the shrinkage. The less the plastic in the filler composite, the less the shrinkage. At the same filler loading, fillers with nucleation effects lead to lesser mold shrinkage. For example, when polypropylene, showing mold shrinkage of 1.91%, was filled with some mineral and cellulosic fillers, its mold shrinkage was as follows [2]:

- 40% $CaCO_3$ 1.34%;
- 20% wood flour 0.94%;

- 40% talc 0.89%;
- 40% wood flour 0.50%;
- 40% glass fiber 0.41%.

As one can see, all the fillers decrease the mold shrinkage, with wood flour showing better results compared with calcium carbonate and talc, but a higher shrinkage compared to glass fiber.

Thermal Properties

Thermal expansion–contraction of inorganic fillers is much lower compared with that of plastics. Therefore, the higher the filler content, the lower the coefficient of expansion–contraction of the composite material (see Chapter 10). Many inorganic nonmetallic fillers decrease thermal conductivity of the composite material. For example, compared with thermal conductivity of aluminum (204 W/deg Km) to that of talc is of 0.02, titanium dioxide of 0.065, glass fiber of 1, and calcium carbonate of 2–3. Therefore, nonmetallic mineral fillers are rather thermal insulators than thermal conductors. This property of the fillers effects flowability of filled plastics and plastic-based composite materials in the extruder.

Color: Optical Properties

Color of fillers should be certainly taken into consideration at their high content, particularly when a light-colored profile is to be made. However, composite materials typically contain enough colorants to overcome color of fillers, except very dark ones, such as carbon black. Fillers effect opacity of the product, which is a negligible factor in colored composite materials.

Effect on Fading and Durability of Plastics and Composites

Minerals often contain admixtures (such as free metals) which catalyze thermo- and/or photooxidation of the filled plastic. This subject will be considered in more detail in Chapter 15. Here we give just two examples of fading of $CaCO_3$-filled HDPE and polypropylene, with 76 and 80% w/w of the filler, respectively. The matrix had the melt flow index equal to 1 g/10 min (HDPE) and 8 g/10 min (polypropylene). Ashing of the both filled plastics at 525°C gave ash content of 76.0 ± 0.1% (HDPE–$CaCO_3$) and 79.9 ± 0.1% (PP–$CaCO_3$). After 250 h in the weathering box (Q-SUN 3000, UV filter: daylight, UV sensor: 340, 0.35 W/m^2, black panel 63°C, ASTM G155−97, cycle 1: light 1:42, light+spray 0:18), the lightness coefficient increased from 83.7 to 84.3 (ΔL=+0.6) [HDPE–$CaCO_3$ 76%] and from 85.6 to 88.8 (ΔL=+ 3.2) [PP–$CaCO_3$ 80%]. As calcium carbonate in this experiment was of the same origin, the increased fading should be attributed to a higher sensitivity of polypropylene to thermo- and/or photooxidation in the surface layer.

Another example that now shows the effects of mineral fillers on oxidation of WPCs (in terms of the OIT, that is, the oxidative induction time) is the durability of GeoDeck experimental deck boards made with talc and mica, in addition to the conventional composition. GeoDeck board without added antioxidants had the OIT of 0.50 min. In the presence of 3 and 10% talc the OIT values were 0.51 and 0.46 min, respectively. In the presence of 12.5 and 28.5% mica, the OIT values were 0.17 and 0.15 min, respectively. It means that in the last two examples mica practically eliminated the resistance (though very low one) of the composite material to oxidation.

Health and Safety

Some fillers are hazardous materials and require special handling and processing. Below are listed some fillers that are used or can readily be used in composite materials, classified according to the principal characteristics accepted in the industry. The indexes mean: no hazard, 0; slight hazard, 1; moderate, 2; severe, 3; extreme hazard, 4. Storage area codes: general, orange; special, blue; hazardous, red.

- *Health*: fly ash and wood flour, not classified; calcium carbonate, kaolin, 0; aluminum hydroxide, clay, glass fiber, magnesium hydroxide, mica, silica, talc, wollastonite, 1.
- *Flammability*: fly ash and wood flour, not classified; all other listed above, 0.
- *Reactivity*: fly ash and wood flour, not classified; all other listed above, 0.
- *Storage color code*: wood flour, not classified; all other listed above, orange.
- *Toxicity* (mg/kg): all listed above, not classified; except aluminum hydroxide, 150.
- *Carcinogenicity*: all listed above, no (except talc—if contains asbestos).
- *Silicosis*: calcium carbonate, clay, mica, yes; all other listed above, no.
- *Time weighted averages* (TWA, an average value of exposure over the course of an 8 h work shift), in mg/m^3: talc, 2; mica, 3; fly ash, calcium carbonate, glass fibers, kaolin, silica, wood flour, 10; aluminum hydroxide, clay, magnesium hydroxide, wollastonite, not classified.

As one can see, listed fillers are generally considered as reasonably safe, except where indicated.

FILLERS

Calcium Carbonate ($CaCO_3$)

Apparently, none of WPC manufacturers adds calcium carbonate as a filler in their products. LDI Composites, which use Biodac® that consists of about 25% $CaCO_3$ and 25% of kaolin (clay), also did not use individual minerals as fillers. Nevertheless, there are many publications, mainly by suppliers of minerals and university researchers, describing benefits of calcium carbonate in WPCs.

Specific gravity (density) of calcium carbonate is typically 2.7–2.9 g/cm^3, Mohs hardness 3–4. The linear coefficient of thermal expansion of $CaCO_3$ is from 2 to 6 × 10^{-6} 1/F, which is about 10–20 times lower than that of HDPE. Hence, HDPE filled with $CaCO_3$ typically has proportionately lower value of the coefficient of expansion. Particles of calcium carbonate have irregular shapes. Particle sizes of commercial $CaCO_3$ vary for different brands from 0.2 to 30 μm. Precipitated $CaCO_3$ can have smaller particles, such as down to 0.02 μm. Oil absorption is between 13 and 21 g/100 g. Specific surface area is between 5 and 24 m^2/g ([1], p.48).

There are three principal forms of $CaCO_3$ used as fillers in plastics—milled, precipitated, and coated. More than 90% of calcium carbonate is used as milled mineral.

At about 825°C (1517°F) and higher temperatures $CaCO_3$ is decomposed, releasing CO_2 and converting to CaO. If we were to add some water to the resulting calcium oxide, temperature sharply jumps up due to the exothermic chemical reaction of the formation of calcium hydroxide. However, ashing of Biodac® at 850–950°C gives a solid residue, adding water to which does not lead to any noticeable increase of temperature, not to a single degree. This is despite of Biodac® that contains about a quarter of $CaCO_3$, and the heating (ashing) is accompanied by the release of CO_2. However, the resulting CaO chemically interacts with another mineral ingredient of Biodac®, that is, kaolin (clay), forming a complex aluminosilicate, not reactive with water.

Apparently the most remarkable property of plastics changed by filling with calcium carbonate is its flexural modulus. For example, HDPE has flexural modulus typically between 125,000 and 240,000 psi (Chapter 2). HDPE filled with 76% of $CaCO_3$ gave the flex modulus of 738,000 psi, which is 3–6 times higher. Similarly, when polypropylene (flex modulus typically between 165,000 and 250,000 psi) was filled with 80% $CaCO_3$, its flex modulus increased to 579,500 psi, that is, by 230–350%. With another HDPE and PP samples, both filled with 58% $CaCO_3$, the respective flexural modulus values were 348,500 ± 4,400 and 365,000 ± 44,000 psi. Flex strength values were 3840 ± 280 and 4430 ± 270 psi, respectively. Hollow 2 × 2 pickets made from HDPE filled with 58% calcium carbonate (moment of inertia of 0.606 in^4.) showed flex strength (with various lubricants used) close to 3000 psi and flex modulus in the range of 276,000–312,000 psi. That is, added $CaCO_3$ was a beneficial filler in terms of both flex strength and modulus.

Generally, the same trend is typically observed when plastic, filled with wood fiber, is additionally filled with calcium carbonate.

In a research conducted by Yash Khanna (Imerys, Performance Minerals North America, Roswell, GA) [private communication], it was shown that calcium carbonate (median particle size 7 μm) in a filling range of 12–32% (w/w) increased flexural modulus of HDPE-based WPC up to 64%, from 280,000 psi (WPC, no minerals) to 460,000 psi (32% $CaCO_3$), as shown in Figure 4.1.

Talc (median particle size of 5 μm) in the amount of 10–15% w/w of the composition did not show a clear pattern (it rather decreased flex modulus in this range, see Fig. 4.1); however, the author concluded that these two minerals were very similar

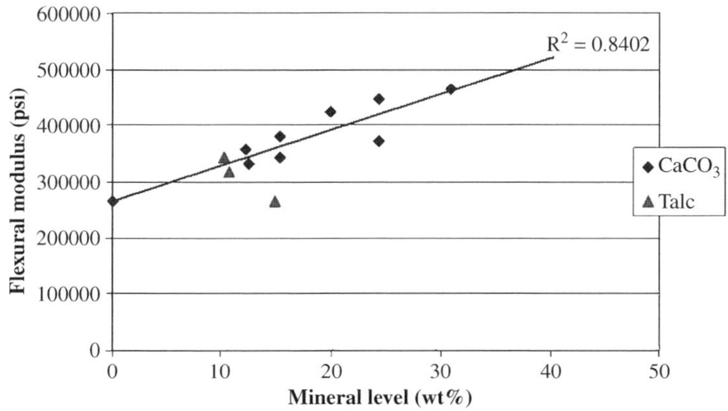

Figure 4.1 Effect of calcium carbonate on flexural modulus of HDPE-based WPC, small laboratory-scale samples (courtesy by Yash Khanna, Imerys). Calcium carbonate was of Imerys Arbocarb 7000, median particle size 7 μm. $CaCO_3$ was used in a filling range of 12–32% (w/w), replacing HDPE. MFI of the HDPE was 0.4; amount of wood fiber, 40-mesh size, was 50%. The WPC also contained 2% of Polybond 3029 and 3% lubricant TPW-113. Some data on talc are also shown for a comparison.

(see also Fig. 4.2). Figure 4.2 shows that calcium carbonate in the range of 12 – 32% reduces the impact resistance of the WPC by about 25%.

Data in Table 4.6 show that both calcium carbonate and talc increase flexural modulus by about 40–60%, decrease impact resistance and energy to break by about 30–40%, and slightly increase moisture absorption. All these effects are generally known in the composite research and development, and slightly vary quantitatively by changing the particle size of minerals, a type of plastic, a type of cellulose (wood) fiber, density (specific gravity) of the matrix, and so on.

Considering that calcium carbonate has better impact performance than many other mineral fillers, one can expect that other fillers might give more dramatic drop in impact resistance, than shown in Figure 4.2.

Effect of particle size of calcium carbonate on mechanical properties of WPCs is not pronounced. Flexural modulus for 40-mesh wood flour (27.5% w/w) -filled nonoriented polypropylene was about the same with both no added minerals, and with 200-mesh and 325-mesh calcium carbonate at 27.5% w/w loading level. Flex strength for the same systems increased from 5600–6500 to 6600–7500 psi in the presence of 200-mesh $CaCO_3$, but came back to the initial WPC (no added minerals) in the presence of 325-mesh calcium carbonate. Dolomite at 10, 20, and 30% loading either did not change flex modulus or slightly decreased it (for 16.7% wood flour content in oriented polypropylene). Flex strength was not changed in this system, being at 5800–7300 psi. As it was noticed, carbonates "maintain MOR and MOE relative to the control" [3].

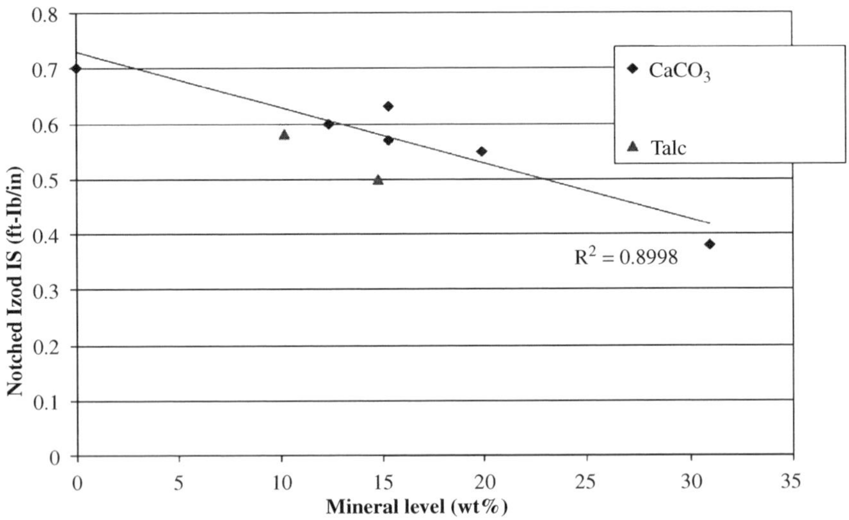

Figure 4.2 Effect of calcium carbonate (median particle size 7 μm) and talc (median particle size 5 μm) on impact strength (Notched Izod) of HDPE-based WPC, small laboratory-scale samples (courtesy by Yash Khanna). The WPC and conditions of the experiments are described in Figure 4.1 and Table 4.6.

TABLE 4.6 Effect of calcium carbonate (Imerys Arbocarb 7000, median particle size 7 μm) and talc (median particle size of 5 μm) on a WPC (MFI of the HDPE 0.4; 50% of wood fiber, 40-mesh size; 2% of Polybond 3029; 3% lubricant TPW-113; $CaCO_3$ or talc replaced HDPE). Composite boards 3.5″ × 1.5″ were made on Coperion's ZSK-58mm counter rotating twin-screw machine at a throughput of about 800 lbs/h. Flex modulus data were determined using ASTM D790, impact strength—ASTM D256, energy to break—ASTM D638. Moisture uptake was determined by immersing 33-in. long boards in a water bath. Data provided by Yash Khanna (Imerys).

Formulation	Flexural modulus(psi)	Notched Izod impact strength (ft/lbs)	Energy to break (ft/lb)	Percentage moisture pick-up in 2 weeks (another set of data)
HDPE (45%) – control	342,000 ± 22,000	0.48 ± 0.24	11.9 ± 1.5	1.4(0.6)
HDPE (35%) / $CaCO_3$ (10%)	457,000 ± 27,000	0.36 ± 0.25	9.5 ± 0.5	7.1
HDPE (30%) / $CaCO_3$ (15%)	498,000 ± 17,000	0.37 ± 0.25	9.2 ± 0.8	9.0
HDPE (25%) / $CaCO_3$ (20%)	476,000 ± 12,000	0.33 ± 0.25	8.5 ± 0.5	8.9(1.3)
HDPE (25%) / Talc (20%)	538,000 ± 7,000	0.31 ± 0.24	6.9 ± 0.5	7.8(2.0)

Talc

Talc is hydrated magnesium silicate, a nonmetallic mineral, white-colored, chemically inert. Unlike many other minerals, its particles have a distinct platy shape. It has a natural affinity to oil and, therefore, serves as a good filler for hydrophobic plastics, such as polyethylenes and polypropylene. Platy particles of talc are structurally not uniform; they have a layered composition, in which a brucite (magnesium-based, tetrahedron-cell atomic structure) sheet is sandwiched between two silica (octahedron-cell atomic structure) sheets. The "elementary" sheet is of 7Å (0.7 nm) thick.

Talc is complex magnesium silicate hydroxide, $Mg_3Si_4O_{10}(OH)_2$. It contains SiO_2 (45–65%), MgO (25–30%), CaO (0.5–13%), minor inclusions of Al_2O_3 and Fe_2O_3, and trace elements. Specific gravity (density) of talc is 2.7–2.9 g/cm^3, Mohs hardness 1–1.5, specific surface area 3–35 m^2/g, moisture content is typically low (0.1–0.6%), median particle size 1.5–15 μm (Tables 4.7 and 4.8).

In general, talc improves many properties of polyolefines when used as a filler. It increases stiffness (flexural modulus) and strength (tensile and flexural), see Tables 4.3 and 4.4, reduces mold shrinkage, water absorption, and the coefficient of thermal expansion–contraction—the latter not specifically, just by replacing the matrix. It saves cost by virtue of replacing more expensive resin. Talc is most commonly used as a filler for polypropylene. It is typically compounded with plastics at 10–40% by weight, but for some applications the loading can vary from 0.1% (for nucleation, such as in Nylons) to 70% (in reinforced elastomers). In polypropylene, at 40% loading, talc increases the softening point from about 151 to about 212°F.

Like many minerals, mined from natural sources, talc typically contains some admixtures that can cause plastic oxidation and degradation during processing and at the end-use service. Hence, when talc is added to a polymer, it requires more antioxidants (compared with neat plastic), a necessary amount of which can vary with different talcs.

TABLE 4.7 Effect of particle size and content of talc on Izod impact strength of talc-filled polypropylene [4]

Particle size (μm)	Talc content (% w/w)	Impact resistance (ft-lbs/in) (J/m)
None (neat PP)	None (neat PP)	1.02 (55)
1.7	10	0.80 (43)
3.5	10	0.70 (38)
10.5	10	0.80 (43)
1.7	20	0.52 (28)
3.5	20	0.56 (30)
10.5	20	0.48 (26)
1.7	30	0.40 (22)
3.5	30	0.40 (22)
10.5	30	0.40 (22)

TABLE 4.8 Talc products offered by Luzenac America (Centennial, CO). Source — Luzenac Technical Bulletin.

Product name	Medium particle size (μm)	Topsize (μm)
Cimpact® 699	1.5	10–15
JetFill® 700C	1.5	10–15
Cimpact® 710	1.7	10–15
Nicron® 674	1.9	10–15
Mistron® ZSC	2.0	20–25
Arctic Mist®	2.0	20–25
JetFill® 625C	2.2	20–25
Cimpact® 610	3.2	20–25
JetFill® 575C	3.4	25–30
Mistron® AB	3.5	25–30
Mistron® NT	3.5	25–30
Mistron® 400C	4.0	25–30
Silverline® 403[a]	4.8	25–30
Stellar® 510	5.0	25–30
JetFill® 350	7.5	45–55
Mistrofil® P403	7.8	45–55
Vertal® UA40	9.6	45–55
Stellar® 410	10	45–55
Stellar® 420	12	50–60
Silverline® 002[b,c]	12.5	40–50
Vertal® 97	15	65–75

[a] Oil absorption 35 g/100 g talc; surface area 12 m^2/g; specific gravity 2.8 g/cm^3; composition >97% talc, <3% dolomite. Source: Silverline® 403 Data Sheet.

[b] Used in commercial WPCs.

[c] 99.5% passing 200 mesh screen; oil absorption 29 g/100 g talc; surface area 10.5 m^2/g; specific gravity 2.8 g/cm^3; composition 97% talc, 2.5% dolomite. Source: Silverline® 002 Data Sheet.

The most remarkable effect of talc on plastics is the increase of stiffness. In general, the flexural modulus of a polyolefin increases by 150–200% with every 20% of talc added. The more platy the talc particles, the greater the effect.

Effect of talc on flexural strength and modulus of plastics is often higher compared with that of calcium carbonate. For example, polypropylene filled with 58% $CaCO_3$ had flex strength and modulus of 4430 ± 270 and 365,000 ± 44,000 psi, respectively, but the same polypropylene filled with 20% $CaCO_3$ and 30% talc had flex strength and modulus of 6125 ± 100 and 486000 ± 18000 psi, respectively.

Effect of talc on mechanical properties of polyolefins typically depends on particle size of the filler. It is generally considered that the finer the talc particles, the better its effect on impact resistance, which generally suffers with addition of minerals. However, data in Table 4.7 show that it is not always so. Although introduction of talc indeed reduced impact resistance of the filled polypropylene regardless of particle size, and the higher the talc content (from 10 to 30% w/w) the lower the impact resistance, particle size itself was not a significant factor.

On the contrary, finer talc particles increased flexural modulus of wood-flour (27.5% w/w) – filled nonoriented polypropylene (27.5% w/w) from 430,000–580,000 psi (no added talc) to 610,000–760,000 psi (10 μm talc). In the same system, but with larger particle sizes (45 and 70 μm talc), flex modulus increase was barely noticeable: 460,000–580,000 and 460,000–620,000 psi, respectively. Flex strength in the same system generally decreased with the decrease of talc particle size [5].

There is a broad selection of talcs of their particle sizes offered by suppliers. For example, Table 4.8 shows some of the talc products provided by Luzenac America (Centennial, CO).

There is a belief (apparently, justified) in the industry that talcs used as fillers in thermoplastics should be less than 10 μm in size, and when used along with wood fiber they should be about 7 – 10 μm in size. For example, with talk particles of this size (7.5 μm, Mistrofil P403, see Table 4.8), maximum flexural strength and modulus of a WPC were reached when 27% (w/w) of the talc was loaded along with 40% HDPE and 33% of wood fiber (4104 and 687,300 psi, respectively). At a slightly higher amount of talc in the same system (30% w/w), both flex strength and modulus were either the same or slightly lower. In comparison, at 6% (w/w) talc in the same system (with 54% of wood fiber), the respective figures were noticeably lower, namely 3248 and 514,750 psi [6–8].

Talc also positively improves retention of flexural properties of WPC after long underwater exposure. For example, after 210 days of underwater exposure of an HDPE-based WPC with 60% filler content (wood flour and talc combined), drop in physical strength was 26% (6% talc), 20% (20% talc), 17% (27% talc), and 16% (30% talc). The corresponding figures for drop in flexural modulus were as follows: 51% (6% talc), 45% (20% talc), 40% (27% talc), and 37% (30% talc) [6–8].

Water absorption showed a rather complex quantitative pattern for composites of HDPE, wood flour, and talc. Evidently, the higher the plastic content and the talc content, the lower the water absorption. However, in the triple system when talc also replaced plastic, and wood fiber content increased, the relationship with water absorption was not that simple, particularly when weight and volume percents of the ingredients were considered. For example, after 4000 h of water immersion, the composition of 44% HDPE, 27% wood flour, and 27% talc (the balance was a lubricant) absorbed 6% water (w/w). A slight increase of HDPE content to 47%, with a concurrent increase of wood flour to 40% and decrease of talc content to 10% gave 11% of water absorption. A sharp decrease of HDPE content to 25%, with both a concurrent decrease of wood flour (36%) and increase in talc (36%) resulted in 13% of water absorption. Finally, a composition with 28% of HDPE, 54% of wood flour, and 14% of talc absorbed 20% of water [6–8].

Besides, talc noticeably reduced warpage and creep resistance [6].

Impact resistance typically suffers progressively with an increased content of a mineral filler. For example, 40% of talc in polypropylene decreases unnotched impact

TABLE 4.9 Effect of talc (Mistrofil P403, median particle size of 7.8 μm) on unnotched Izod impact resistance of an HDPE (MFI 0.4)-based composite material containing a 40-mesh pine wood flour (up to 55% content w/w) and the talc [8]

HDPE				
Wt.%	Vol.%	Wood flour (%)	Talc (%)	Izod impact (kJ/m)^2
42	55.3	49	6	2.1
37	50.1	54	6	1.8
37	51.1	48	12	1.9
37	53.8	33	27	2.0
36	50.0	49	12	1.8
34	50.3	37	27	1.8

strength by four to five times. However, it depends on a specific system. Table 4.9 shows, for instance, that increase of talc content from 6 to 27% insignificantly reduces impact resistance of HDPE-based composite material [8].

Effect of talc on the coefficient of linear thermal expansion–contraction was not pronounced (Table 4.10) and was actually superimposed with that of wood flour. In other words, the principal effect was just a displacement of plastic with a filler, regardless whether it was wood flour or talc. It certainly makes sense because both wood and mineral fillers have their own coefficients of thermal expansion–contraction by an order of magnitude lower than that of HDPE (see Chapter 10).

Talc is the softest industrial material; hence, the wear on processing equipment is minimal, unlike that of calcium carbonate or, more than that, mica (Table 4.11).

Abrasion of talc is so low that it is compared with that of wood flour. For example, abrasion on a brass mixer blade, operating at 2000 rpm for 1.5 h, was practically undetecteable in a study in which talc (Silverline 002, see Table 4.8), wood flour, and polyethylene were employed. It was also undetectable even when the volumetric loading of talc was three times that of the wood flour but was quite noticeable with rice hulls containing silica [8].

TABLE 4.10 Effect of talc (Mistrofil P403, median particle size of 7.8 μm) on the coefficient of linear thermal expansion–contraction (CTE) of an HDPE (MFI 0.4)-based composite material containing a 40-mesh pine wood flour (up to 55% content w/w) and the talc [8]

HDPE				
Wt.%	Vol.%	Wood flour (%)	Talc (%)	CTE (1/F)
37	50.1	54	6	1.4
37	51.1	48	12	1.6
36	50.5	49	12	1.5
34	50.3	36	27	1.5
32	45.5	53	12	1.4

TABLE 4.11 Abrasion of some minerals (the test is used in the paper industry to evaluate wear) [8]

Mineral	Weight loss of bronze wire (g/m^2)
Talc	12
Air float clay	25
Calcined clay	37
Ground calcium carbonate	38
Mica	>700

Partial replacement of wood flour with talc, or the addition of talc to WPC, resulted in a decrease of the melt viscosity of WPC and increase of the volumetric output through the die. The substituting of 10% of wood flour with talc increased output by 15%, and replacing 50% of wood flour with talc increased output by 36%. It was proposed that talc minimized the fluidization of the low bulk-density feed and, thus, enhanced solid conveying [8].

Biodac® (a Blend of Cellulose and Mineral Fillers)

Biodac® is a granulated papermaking sludge, manufactured by Kadant Grantek (Green Bay, Wisconsin). Its ingredients—bleached cellulose, lignin (1–4% w/w), calcium carbonate, clay, and other organic and inorganic components, such as traces of ink—are those that are present in the paper pulp itself. They were described in Chapter 3.

The principal mineral components of Biodac® are calcium carbonate—in the form of precipitated calcium carbonate (PCC) or ground calcium carbonate (GCC)—that typically constitutes 20% and up to 75% of dry sludge content, and clay. These two minerals are typically loaded into paper as a coating and filler to improve the mechanical characteristics as well as the appearance of paper. The resulting papermaking sludge, particularly mixed office paper sludge, consists primarily of two major components, that are fiber and minerals finely mixed with each other.

Kadant Grantek manufactures controlled-size dust-free granules, made of pulp and paper sludge, under the brand name Biodac®. The granules are a tight composite of organic and inorganic materials, that is, cellulose fiber and minerals, and possess a developed porous structure. Biodac® absorbs oil or other hydrophobic fluids to a high extent (about 150% w/w).

It has been found that Biodac®, compared with loose cellulose fiber and minerals, greatly improves the properties of fiber–plastic composites. It is well known that the incorporation of cellulose fiber into polymer hot melt is difficult. Intensive prolonged mixing is ordinarily required to disperse the fiber. It is particularly difficult to obtain high-strength fiber–plastic composites because fibers typically possess a high degree of fiber–fiber interaction, tending to stick together in bundles of fibers and resisting dispersion of the individual fibers. It has been found, however, that granulation of pulp and paper sludge into Biodac®, approximately half of which is typically cellulose fiber, reduces fiber-to-fiber interaction prior to incorporating it into the matrix and improves the mechanical properties of the composite.

Compared to calcium carbonate and talc, and to other conventional fillers having specific gravities around 2.8 g/cm^3, Biodac® has a lower density, 1.58 g/cm^3. Biodac® has a spherical shape, which allows to load mixing machinery at high rates and without the risk of dusting or exposure in production area, which are usual issues for cellulose or fine mineral fillers. Biodac® does not change the size of granules during moisture absorption or drying. It has a relatively low abrasiveness in the processing equipment due to its porous structure. About 13% of Biodac® by volume is air, 56% (v/v) is a very low abrasive cellulose, and 31% (v/v) is taken by $CaCO_3$ and clay. By weight, Biodac® contains about 50% of paper cellulose fiber, approximately 25% of $CaCO_3$, and 25% kaolin clay, though the two last ingredients can vary from 25 to 34% and from 14 to 25%, respectively.

Under the microscope, Biodac® appears as small rocks attached to cellulose fibers and bundled into granules. Because of short cellulose fibers on the surface of Biodac®, the filler provides a mechanical adhesion (interlocking) with the plastic matrix without any surface treatment. The thermal expansion–contraction coefficient of Biodac® appears to be much closer to that of plastics, compared to the coefficients of common mineral fillers. Hence, the bond between Biodac® and the plastic after weathering of the composite material becomes loose at lesser extent compared to inorganic fillers.

The following properties of Biodac®--filled LDPE can be considered advantageous compared to those of $CaCO_3$- or talc-filled LDPE:

- Lower density (specific gravity) compared to both $CaCO_3$ and talc
- Higher tensile modulus compared to $CaCO_3$
- Higher Izod impact resistance compared to talc and $CaCO_3$
- Lower abrasion resistance compared to both talc and $CaCO_3$
- Lower mold shrinkage compared to both talc and $CaCO_3$

The following properties of Biodac®-filled LDPE can be considered as disadvantageous:

- Lower tensile properties compared to those of talc
- Lower flexural strength compared to talc
- Higher water absorption compared to both talc and $CaCO_3$
- More noticeable coloring of the final product compared to that with both talc and $CaCO_3$

Table 4.12 provides some illustrations of the advantageous and disadvantageous properties mentioned.

TABLE 4.12 Tensile properties of filled LDPE (Petrothene, Equistar Inc., MFI 13.5, density 0.918 g/cm³). Biodac 20/50 mesh size (Kadant Grantek Inc.), specific gravity 1.58 g/cm³, cost FOB 8.75 c/lb. Talc Mistrofit P403 (Luzenac America Inc.), specific gravity 2.8 g/cm³, median particle size 7.6 μm, cost 11 c/lb. $CaCO_3$ coarse (Marble Elite Alpha, J.M. Huber Corp.), specific gravity 2.7 g/cm³, median particle size 90 μm, cost 7 c/lb. $CaCO_3$ fine (HuberCarb G 325, J.M. Huber Corp.), specific gravity 2.7 g/cm³, median particle size 9 μm, cost 6.5 c/lb

Filler (% w/w)	Tensile modulus (psi)	Tensile strength at break (psi)	Elongation at break (mm/mm)	Yield strength (psi)	Toughness, (J/m³)
Neat LDPE, virgin (control)	16,300 ± 200	1,460 ± 10	0.90 ± 0.01	1,520 ± 10	8.4 ± 0.1
Neat LDPE, recycled	17,400 ± 200	1,520 ± 10	0.88 ± 0.01	1,560 ± 10	8.5 ± 0.2
Talc, 30%	52,200 ± 600	1,800 ± 20	0.41 ± 0.01	1,940 ± 20	4.9 ± 0.2
$CaCO_3$ coarse, 30%	30,800 ± 800	1,580 ± 100	0.37 ± 0.01	1,600 ± 10	3.6 ± 0.1
$CaCO_3$, fine, 30%	28,700 ± 600	1,670 ± 10	0.46 ± 0.01	1,730 ± 10	4.8 ± 0.2
Biodac, 18%	30,700 ± 200	1,420 ± 20	0.29 ± 0.01	1,420 ± 20	2.5 ± 0.2
Biodac, 30%	40,100 ± 1,000	1,320 ± 20	0.15 ± 0.01	1,350 ± 10	1.2 ± 0.1

One can see that Biodac® as a filler significantly increases tensile modulus of LDPE— more than $CaCO_3$ but less than talc (by the same weight amount), does not increase tensile strength at break, decreases elongation at break, decreases yield strength, and decreases toughness.

Similarly, Biodac® significantly increased flexural strength of molded samples of filled neat LDPE, but showed less strength than talc-filled LDPE or similar with that of $CaCO_3$-filled LDPE (Table 4.13).

Generally, Izod impact resistance was decreased with the addition of fillers, including Biodac®, or stayed the same (Table 4.14).

Similarly, falling weight test showed that impact resistance of the filled LDPE samples was lower compared with the neat LDPE (Table 4.15).

TABLE 4.13 Flexural strength of filled LDPE, molded samples. For materials—see Table 4.12

Filler % (w/w)	Apparent flexural strength (psi)
Neat LDPE, virgin (control)	11,800 ± 300
Neat LDPE, recycled	13,300 ± 130
Talc, 30%	29,000 ± 600
$CaCO_3$ coarse, 30%	20,700 ± 100
$CaCO_3$, fine, 30%	23,500 ± 300
Biodac, 18%	18,200 ± 200
Biodac, 30%	21,800 ± 300

TABLE 4.14 Izod impact resistance (notched test) of filled LDPE. For materials—see Table 4.12

	Near gate		Far gate	
Filler, (% w/w)	Impact Energy (kJ/m^2)	Failure type	Impact Energy (kJ/m^2)	Failure type
Neat LDPE, virgin (control)	16.7 ± 1.0	Non-break	16.3 ± 1.6	Non-break
Neat LDPE, recycled	14.6 ± 0.9	Non-break	16.1 ± 0.9	Non-break
Talc, 30%	9.3 ± 0.2	Hinge break	9.2 ± 0.2	Hinge break
$CaCO_3$ coarse, 30%	14.3 ± 0.3	Hinge break	13.0 ± 0.5	Hinge break
$CaCO_3$, fine, 30%	17.7 ± 0.3	Hinge break	17.2 ± 0.3	Hinge break
Biodac, 18%	17.6 ± 0.5	Partial break	15.5 ± 1.0	Partial break
Biodac, 30%	12.8 ± 0.4	Partial break	10.9 ± 0.4	Partial break

TABLE 4.15 Falling weight test (impact resistance) of filled LDPE, molded samples. For materials—see Table 4.12

Filler (% w/w)	Total energy (J)
Neat LDPE, recycled	12.9 ± 0.4
Talc, 30%	8.9 ± 0.1
$CaCO_3$ coarse, 30%	5.5 ± 0.2
$CaCO_3$, fine, 30%	10.1 ± 0.1
Biodac, 18%	5.0 ± 0.4
Biodac, 30%	3.8 ± 0.1

Adding fillers, such as talc and $CaCO_3$ to LDPE increases abrasion of the filled plastic. However, adding of Biodac® to LDPE increases abrasion much less (Table 4.16).

Mold shrinkage of LDPE is increased as a result of incorporation of calcium carbonate and decreased by talc. The most significant decrease is caused by filling of LDPE with Biodac® (Table 4.17).

TABLE 4.16 Abrasion resistance of filled LDPE. For materials—see Table 4.12

Filler (% w/w)	Average weight loss (%)
Neat LDPE, virgin (control)	0.32
Neat LDPE, recycled	0.32
Talc, 30%	0.36
$CaCO_3$ coarse, 30%	0.36
$CaCO_3$, fine, 30%	0.32
Biodac, 18%	0.21
Biodac, 30%	0.17

TABLE 4.17 Mold shrinkage of filled LDPE. For materials—see Table 4.12. Data by Dr. Tatyana Samoylova, LDI Composites

Filler (% w/w)	Mold shrinkage (%)
Neat LDPE, virgin (control)	3.00
Neat LDPE, recycled	4.16
Talc, 30%	2.62
$CaCO_3$ coarse, 30%	4.22
$CaCO_3$, fine, 30%	3.37
Biodac, 18%	2.92
Biodac, 30%	2.39

Water absorption is low by neat LDPE and LDPE filled with talc and $CaCO_3$. Biodac®-filled LDPE, however, has significantly higher water absorption (Table 4.18).

After 7 days of underwater exposure of the filled LDPE, the neat LDPE and talc-filled LDPE retained sizes of their samples, $CaCO_3$-LDPE shrank by 0.05%, and Biodac®-filled LDPE shrank by 0.03%. After air-drying of the samples for 7 days, neat LDPE and talc-filled LDPE samples retained their dimensional sizes, while $CaCO_3$- and Biodac®-filled LDPE retained the same shrinkage of about 0.04%. There were no swell of the samples.

The above examples show some empirical data. In the absence of truly systematic studies in the area, it is rather risky to draw some general conclusions. However, even those scattered data provide some guidelines in further experimenting and justify certain anticipations in behavior of filled WPC products.

Silica (SiO_2)

About 19% of rice hulls by weight consists of silica. Rice hulls are used as a major filler (or among major fillers) in commercial WPCs, such as GeoDeck and Nexwood (though the second one is taken off the market). Those silicates are extracted by the plants from soil.

TABLE 4.18 Water absorption of filled LDPE. For materials—see Table 4.12. Data by Dr. Tatyana Samoylova, LDI Composites

Filler (% w/w)	Weight increase (%) of filled LDPE after underwater exposure		
	24 hrs	4 days	7 days
Neat LDPE, virgin (control)	0.04	0.04	0.04
Neat LDPE, recycled	0.04	0.04	0.04
Talc, 30%	0.02	0.02	0.02
$CaCO_3$ coarse, 30%	0.03	0.04	0.05
$CaCO_3$, fine, 30%	0.02	0.02	0.02
Biodac, 18%	0.03	0.05	0.15
Biodac, 30%	0.14	0.22	0.29

Besides the principal silicon oxide, common silica fillers include other oxides as minor inclusions or traces, such as Al_2O_3, Fe_2O_3, TiO_2, CaO, MgO, Na_2SO_4, and trace elements. Specific gravity (density) of silica is 1.9–2.2 g/cm^3, moisture content is typically from 0.1 to 7%, and can go up as high as 20%. Particle sizes of commercial silica grades, which are typically white powder, are from less than 1 to 40 µm, oil absorption between 20 and 330 g/100 g, specific surface area from 1 to 800 m^2/g.

In common use are three principal materials of silica: one natural (both crystalline and amorphous) and two so-called synthetic, products of thermal process (fumed silica grades) and wet process (precipitated silica). Particles of silica commonly have a spherical shape.

Kaolin Clay ($Al_2O_3 \cdot 2SiO_2 \cdot 2H_2O$)

Kaolin is used in LDPE in amounts of 20–45% (w/w) to increase abrasion resistance of the plastic.

Kaolin belongs to clay minerals, which consist of five groups: (1) kaolinite and halloysite, (2) illite, (3) montmorillonite and hectorite, (4) sepiolite and attapulgite, and (5) vermiculite. All of them are called clay fillers. All have a complex chemical composition, and generally consist of SiO_2 (50–60%), Al_2O_3 (25–35%), and small amounts of Fe_2O_3, TiO_2, CaO, MgO, K_2O, and Na_2O, all of them at about 1% or smaller. Kaolin clay, or China clay has a chemical composition of SiO_2 (40–60%), Al_2O_3 (25–45%), and small amounts of the same oxides, particularly Fe_2O_3, TiO_2, and K_2O.

Specific gravity (density) of clays is around 2.6 g/cm^3, Mohs hardness 2–2.5, oil absorption between 30–50 g/100 g. Moisture content is typically around 3%, but they can absorb to about 15% of moisture. Surface area varies typically between 8 and 70 m^2/g. Collectively "clay" means that the mixed aluminosilicate has small particles, typically between 0.2 and 7 µm.

Mica

Mica reportedly increases thermal properties and reinforces polyolefines.

Mica is a complex aluminosilicate, with a chemical composition of SiO_2 (40–50%), MgO (20–25%), Al_2O_3 (10–40%), K_2O (3–11%), and Fe_2O_3 (traces to 11%). Specific gravity (density) is 2.7 – 3.2 g/cm^3, Mohs hardness 2.5 – 4. Moisture content of mica is typically low (0.3 – 0.7%) and water absorption reaches 3–4%. Particle size of mica for different commercial brands varies between 4 and 70 µm. Oil absorption is between 65 and 72 g/100 g ([1], p. 112).

Because mica platelets are significantly larger than those in talc, and darker, they are much less in use as filler compared with talc. For some applications, though, a high aspect ratio of mica, often between 10 and 70, is attractive. However, the larger particle size of mica decreases impact resistance of filled polymers.

Mica group is a large one, with more than 30 minerals (muscovite, phlogopite, biotite are the most common among them). They are often found in limestones, dolomites, and other magnesium-rich rocks.

Mica is not in use—so far—as a mineral filler for WPCs. However, it was rather extensively studied in model WPC systems. Generally, it behaves similarly with calcium carbonate in some systems—decreases flexural modulus of 60-mesh pine wood flour (16.7% and 25% w/w)-filled oriented polypropylene from 10 to 20 to 30% loading level of mica (overall from 390,000–420,000 psi with no mica to 320,000 psi with 20 and 30% w/w of mica), and also decreases flex strength in the same system from 6500 psi (control) to 4000–5000 psi. As the author of the study [9] has noticed, comparisons with other phyllosilicates show mica to have the least benefits.

Wollastonite ($CaSiO_3$)

A general chemical composition of wollastonite is CaO (40–50%), SiO_2 (40–50%), and minor fractions of Fe_2O_3, Al_2O_3, MgO, MnO, and TiO_2. Its specific gravity (density) is around 2.9 g/cm^3, Mohs hardness is 4.5, specific surface area is low, between 0.4 and 4 m^2/g, and oil absorption between 20 and 50 g/100 g. The typical moisture content of wollastonite is low, less than 0.2%.

Wollastonite is the only white mineral which is fibrous. Its aspect ratio (the length to diameter ratio) is between 3 and 70, and typically between 3 and 20, with particle length of 8 – 650 μm.

A variety of particle size grades of wollastonite are supplied by multinational company NYCO which has production facilities in New York and Mexico. A high aspect ratio and low aspect ratio (powder) of a family of products are offered. The available median particle sizes for the products range from 2 to 40 μm and the surface area range is 0.5–4 m/g^2, which is very typical for wollastonites (see above).

Wollastonite is not yet used in commercial WPC materials, but it is currently under investigation to further enhance the properties of WPCs. The mineral is used worldwide in many plastic applications providing improvement in stiffness, impact, scratch resistance, lower thermal coefficient of expansion–contraction, and flame retardancy. The unique morphology of wollastonite and the variety size grades available can provide benefits that are not obtained by other minerals. In a number of cases, wollastonite has successfully replaced talc in plastic applications where further improvements in properties such as greater strength and improved scratch resistance were required.

Glass Fibers

There is no public information regarding the use of glass fiber in WPC deck boards. However, affordable glass fiber would certainly improve properties of composite materials.

Glass fibers have a general chemical composition of SiO_2 (50–60%), CaO (20–25%), Al_2O_3 (14–15%), and B_2O_3 (5–10%). Their specific gravity (density) is 2.5–2.7 g/cm^3, Mohs hardness 6–6.5. Glass fibers have a low moisture content and water absorption, generally below 0.5%. Aspect ratio of glass fibers is between low digits and high numbers, such as 1000. Fiber thickness is 10–20 μm, and fiber length from 50 μm (for milled grades) to 10–20 mm (for chopped grades).

Fly Ash

Fly ash is a powdery substance obtained from dust collectors of coal–electric utility power plants. Essentially, 60–90% of fly ash is glass. More specifically, fly ash generally consists of 30–60% of SiO_2, 10–20% of Al_2O_3, 5–10% of Fe_2O_3, 5–6% of MgO, and 2–45% of CaO.

Fly ash starts out as impurities in coal, mostly clay, shales, limestone, and dolomite, which ends up as ash, and fuse at high temperature becoming glass. Two U.S. classifications of fly ash are produced, Class C and Class F, according to the type of coal used. Class C fly ash, typically obtained from subbituminous and lignite coals, must have more than 50% total of silica, alumina, and iron oxide. Class F fly ash, typically obtained from bituminous and anthracite coals, has more than 70% of these oxides.

Fly ash is a major by-product of coal combustion and is produced in very large quantities throughout the world. In the United States alone, the annual volume of fly ash produced by power plants is reported to be in the range of 50 million tons per year. Of this total, between 10 and 20% is used in blended cements. Other uses of fly ash are as a filler material in concrete roadbases and subbases, structural fills, flowable fill, and grout [10], a filler in nylon- and polypropylene-based compositions [11], in lime bricks, blocks and tiles, in metallic matrix composites, utilizing, in particular, aluminum and aluminum alloys, growing media for ornamental flowers, and so on. However, a major part of produced fly ash, 70–75%, is being disposed of in landfills.

Some types of fly ash contain much of unburned carbon particles, making the fly ash "black," compared to a lighter colored fly ash, and eventually "white" fly ash, processed at higher temperatures and/or with longer residence times. The carbon, which is relatively soft and of low strength, tends to act as a lubricant between particles in a mix, such as a concrete mix, thereby reducing its mechanical properties. Hence, the regulations are promulgated by many states which limit the amount of carbon in fly ash used in the manufacture of concrete to less than 5% by weight, and probably less than 3% by weight, significantly below the 6–20% level frequently encountered in most coal combustion fly ash.

Besides, new combustion conditions increasingly being specified in order to minimize NO_x emissions in power plant stack gases result in increased carbon content in the fly ash produced under these new conditions. This further restricts the types and amounts of fly ash that can be utilized as a filler, and decreases commercial applications of coal combustion fly ash even more. Therefore, many methods have been developed to remove carbon particles from the fly ash and minimize the adverse effects of the carbon on characteristics of the filler materials. These methods include chemical combustion, gravitational, flotational, electrostatic, magnetic, and mechanical means, and combinations of these [12–19].

Generally, fly ash has a broad particle size distribution. Arbitrarily, there are five "modes" of the distribution, which are ultrafine (a median particle diameter of 0.05–0.2 μm), fine (0.3–1.0 μm), medium (10–25 μm), large (40–80 μm), and coarse (100–200 μm). Often, approximately 15% of particles are "fine," 60–70% "medium," and 20–30% "large." As a result of flame treatment, fly ash is a predominantly amorphous, noncrystalline material. As was mentioned earlier, most of fly ash is an aluminosilicate glass or a calcium aluminosilicate glass, with inclusions of lime (CaO), hematite (Fe_2O_3), ferrite spinel (Fe_3O_4), quartz (SiO_2), dicalcium silicate (Ca_2SiO_4), tricalcium aluminate ($Ca_3Al_2O_6$), and others [20]. Particle shape of fly ash is typically spherical, as a result of flame treatment, compared, for example, with calcium carbonate particles, which are highly irregular and blocky.

Fly ash is used as a filler in plastics such as polyethylene, polypropylene, PVC, nylons, ABS, and others. Because of its broad particle size distribution, fly ash can be used at high loadings, up to 75% by weight. Specific gravity of fly ash is typically between 2.1 and 3.1 g/cm^3, more often between 2.1 and 2.2 g/cm^3, Mohs hardness is about 3–5.5, specific surface area between 0.2 and 0.8 m^2/g, and oil absorption between 15 and 30 g/100 g. The last parameter determines the so-called "packing factor," which is approximately inversely proportional to oil absorption, and typically varies between 57 and 72%. Fly ash particles with a low packing factor (58–60%) resulted in very high viscosity of molten plastics at a loading of 70% by weight. Increase of the packing factor to 68% resulted in decrease of the viscosity by 2–3 times [20]. Moisture content of fly ash is typically between 2 and 20%.

Among earlier studies of fly ash as a filler for plastics, two U.S. Patents can be mentioned [21, 22]. In the first one, fly ash was mentioned in the claims as a filler for a thermosetting composition, in the second one, for a thermoplastic resin. No details were given, though, in examples of the patents.

A detailed study of deck boards filled with fly ash has been done with GeoDeck 2 × 6 hollow profiles (Table 4.19).

One can see that only at 30% of fly ash by weight (HDPE 34% w/w and 46% v/v), both flex strength and modulus (stiffness) are noticeably higher than that of control. However, weight of the boards increased from 1.71 lb/ft (control) to 1.94 lb/ft (30% w/w fly ash).

Table 4.19 shows a typical complex interplay between volume- and weight-fractions of a composite material when both are changed concurrently, particularly

TABLE 4.19 Flexural strength and flexural modulus of GeoDeck 2 × 6 hollow deck boards filled with fly ash (four-point load). The first three boards were made by keeping HDPE 39% (w/w) and replacing rice hulls with fly ash by weight. The second three boards were made by keeping HDPE 46% (v/v) and replacing rice hulls with fly ash by weight. Moment of inertia of the boards: control 0.736 in^4, fly ash filled 0.775 in.^{4} Control board (no fly ash) contained 39% HDPE by weight and 46% HDPE by volume. Data of the year of 2000 (early variants of GeoDeck)

Fly ash (% w/w)	Flexural strength (psi)	Flexural modulus (psi)
0	1911 ± 51	297,000 ± 25,000
10	2046 ± 126	297,000 ± 11,000
20	2247 ± 51	293,000 ± 18,000
10	1930 ± 129	260,000 ± 23,000
20	1911 ± 140	286,000 ± 47,000
30	2336 ± 51	383,000 ± 40,000

TABLE 4.20 Flexural strength and modulus of compositions of HDPE with different ratios of fly ash and wood flour, at a constant 40% total filler level [23]

Fly ash (% w/w)	Wood flour (% w/w)	Flexural strength (psi)	Flexural modulus (psi)
0 (neat HDPE)	0 (neat HDPE)	4580 ± 60	128,000 ± 4,000
40	0	7180 ± 90	305,000 ± 6,000
30	10	7920 ± 30	348,000 ± 13,000
20	20	8500 ± 70	389,000 ± 9,000
10	30	9060 ± 250c	439,000 ± 6,000
0	40	9240 ± 30	436,000 ± 4,000

when some components have low density and some have high, such as HDPE and fly ash in this particular case. Addition of a high weight fraction of a mineral causes an increase of a volume fraction of plastic. Hence, with 20% w/w of fly ash the volumetric amount of HDPE was higher than without fly ash at all (the top row in Table 4.19). As a result, flexural modulus went down. When the volumetric amount of HDPE was fixed at a certain level (46%), addition of fly ash caused a sharp increase in flexural modulus (the bottom row in Table 4.19).

In a combination of fly ash and wood flour in HDPE-filled composite, fly ash was making the material weaker and more flexible compared with only wood flour-filled HDPE (Table 4.20). However, fly ash makes the filled material stronger and stiffer compared with the neat plastic.

Measurements of tensile strength and modulus also confirm that wood flour makes more mechanically sound HDPE-based composite compared with fly ash as a filler (Table 4.21).

Table 4.22 shows impact resistance of the fly ash containing WPC boards (ASTM D 3763, high-speed puncture properties of plastics using load and displacement sensors), which was measured by the rate sensitivity of the material to impact (load versus deflection response). Test speed was of 3.3 m/s (10.8 ft/s).

TABLE 4.21 Tensile strength and modulus of compositions of HDPE with different ratios of fly ash and wood flour, at a constant 40% total filler level [23]

Fly ash (% w/w)	Wood flour (% w/w)	Tensile strength (psi)	Tensile modulus (psi)
0 (neat HDPE)	0 (neat HDPE)	3030 ± 30	107,000 ± 12,000
40	0	3640 ± 30	244,000 ± 9,000
30	10	3770 ± 30	287,000 ± 6,000
20	20	4120 ± 20	307,000 ± 3,000
10	30	4540 ± 30	348,000 ± 10,000
0	40	4960 ± 30	373,000 ± 10,000

TABLE 4.22 Impact resistance of GeoDeck 2 × 6 hollow deck boards filled with fly ash. Specimen size 102 × 102 mm. GRC Dynatup Model 8200 Series 930-V1.16, tup diameter 12.7 mm. The specimens did not exhibit any cracking. Average data of the five tests are shown

Composition (not all ingredients are shown)	Average peak load (lbs)	Average energy at peak loads (J)
39% HDPE, 28% rice hulls 28% Biodac no fly ash	411	5.0
39% HDPE, 57% rice hulls no fly ash	321	8.4
35% HDPE, 20% rice hulls 30% fly ash	468	16.7
50% HDPE, 43.5% fly ash	593	28.0

Generally, a replacement of rice hulls with fly ash led to a higher impact resistance (the two middle rows in Table 4.22). Biodac® effects impact resistance, however, in a complex way (the first two rows). The highest impact resistance (the last row) was due to the highest amount of HDPE and/or the highest amount of fly ash.

Wood flour was more beneficial compared to fly ash in terms of notched test of impact resistance of HDPE-filled materials and showed mixed data in terms of unnotched tests (Table 4.23).

Table 4.24 shows data on puncture resistance of fly-ash-filled GeoDeck boards. Again, fly ash increases mechanical resistance of the boards.

One can see from Table 4.24 that all fly-ash-filled boards are stronger to puncture compared with the control board. However, there were practically no systematic difference between fly ash boards. Apparently, a complex interplay between plastic

TABLE 4.23 Impact resistance of compositions of HDPE with different ratios of fly ash and wood flour, at a constant 40% total filler level [23]

Fly ash (% w/w)	Wood flour (% w/w)	Notched izod (J/m)	Unnotched izod (J/m)
0 (neat HDPE)	0 (neat HDPE)	54 ± 10	1000 ± 250
40	0	26 ± 1	325 ± 30
30	10	29 ± 1	185 ± 15
20	20	30 ± 1	180 ± 15
10	30	35 ± 5	165 ± 15
0	40	37 ± 1	180 ± 10

TABLE 4.24 Puncture resistance of GeoDeck 2 × 6 hollow deck boards filled with fly ash. The data show a load required to puncture a hollow board with a 0.87″ steel ball. The first three boards were made by keeping HDPE 39% (w/w) and replacing rice hulls with fly ash by weight. The second three boards were made by keeping HDPE 46% (v/v) and replacing rice hulls with fly ash by weight. Control board (no fly ash) contained 39% HDPE by weight and 46% HDPE by volume

Fly ash (% w/w)	Puncture load (lbs)
0	335
10	481
20	459
10	410
20	395
30	414

content (weight and volume-related amounts) and fly ash content largely compensated the effect of fly ash.

Fly ash significantly increased viscosity of the composite material (Table 4.25).

However, for boards in Table 4.19, because of a complex interplay between amount (per weight and per volume) of plastic, fly ash, and other ingredients, melt flow index does not change monotonously (Table 4.26).

A monotonous decrease of melt flow index in a series of HDPE-wood flour-fly ash composite materials was observed [23]. However, this decrease has resulted due to a shift from a neat HDPE to 40% fly-ash-filled HDPE to 40% wood-flour-filled HDPE. The data show that wood flour increased viscosity of the filled plastic much more compared with that of fly ash (Table 4.27).

TABLE 4.25 Melt flow index (190°C) of HDPE, GeoDeck and fly-ash-filled compositions. Data of 2000

Formulation	MFI, 2.16 kg	MFI, 10 kg
Neat HDPE	0.73 ± 0.01	13.2 ± 0.7
GeoDeck I	0.11 ± 0.02	1.78 ± 0.08
20% fly-ash-filled GeoDeck I	0.076 ± 0.013	1.24 ± 0.06
GeoDeck II	0.095 ± 0.008	1.58 ± 0.07
20% fly-ash-filled GeoDeck II	0.027 ± 0.001	0.61 ± 0.02
20% fly-ash-filled GeoDeck III	0.039 ± 0.003	0.90 ± 0.04

TABLE 4.26 Melt flow index (190°C) for compositions of GeoDeck hollow deck boards filled with fly ash. The first three boards were made by keeping HDPE 39% (w/w) and replacing rice hulls with fly ash by weight. The second three boards were made by keeping HDPE 46% (v/v) and replacing rice hulls with fly ash by weight. Control board (no fly ash) contained 39% HDPE by weight and 46% HDPE by volume. Data of the year of 2000

Fly ash (% w/w)	MFI under 2.16 kg	MFI under 10 kg
0	0.11	1.78
10	0.10	0.65
20	0.13	1.09
30	Poor run	Poor run
10	0.14	0.92
20	0.07	0.77
30	No flow	0.61

The above data show that fly ash was not beneficial for mechanical properties of the HDPE-wood flour-based composite. However, the flowability (MFI) of the molten composite was getting higher (viscosity was lower) when fly ash was replacing wood flour. Last but not the least, it was found that fly ash stabilized the HDPE–wood flour composition against heating. The onset of thermal decomposition for the 40% wood-flour-filled HDPE started at about 280°C, while that of 40% fly ash-filled HDPE started at approximately 490°C. The triple (HDPE-wood flour-fly ash) composite was between the two in this regard, showing a thermally more stable behavior than composites with wood flour alone [23].

Because fly ash is not a real, "active" flame retardant, plastics filled with it typically have practically the same ignition point and only slightly higher flame spread index compared to the neat plastic.

As a conclusion, fly ash shows mixed results with different properties of WPCs, and it may be successfully employed in some cases but not in others. Optimization is again the name of the game.

TABLE 4.27 Melt flow index (190°C) for compositions of HDPE with different ratios of fly ash and wood flour, at a constant 40% total filler level [23]

Fly ash (% w/w)	Wood flour (% w/w)	MFI under 2.16 kg, g/10 min
0 (neat HDPE)	0 (neat HDPE)	12.2
40	0	7.5
30	10	4.9
20	20	3.5
10	30	2.5
0	40	2.0

Carbon Black

Carbon black in amounts of 0.2–2.5% increases thermal and UV stability of LDPE and HDPE, and reportedly increases tensile strength. However, it is apparently not used in WPCs.

Carbon black consists of 95–99% of elementary carbon. Besides, it typically contains many trace elements, metals and nonmetals, such as sulfur. Specific gravity (density) of carbon black is typically 1.7–1.9 g/cm^3. It often contains between 0.1 and 2% of moisture. Particle size of commercial carbon black brands vary typically between 10 and 250 μm. Carbon black has a developed specific surface area, between 10 and 600 m^2/g.

Nanofillers and Nanocomposites

Rather recently, so-called nanoparticles were introduced into the field of additives for plastics, composites in general and WPCs in particular. Nanoparticles are employed in amounts typically below 10%, or more often below 5%, so they are not real "fillers" but rather additives. However, because even in such small concentrations they sometimes improve properties of materials, some call them "nanofillers."

Nanoparticles or nanofillers are collective terms for modified layered silicates (organoclay), graphite nanoflakes, carbon nanotubes, and a number of materials dispersed in the polymer matrix, when the particles' size is in order of nanometers (one thousands of micron), or tens of nanometers. A plastic filled with nanoparticles, typically in the range of 2–10% (w/w) is called a nanocomposite.

There are two basic types of nanocomposites, in which particles are intercalated or exfoliated. In an intercalated composite the nanodispersed filler still consists of ordered structures of smaller individual particles, packed into intercalated structures. Exfoliated particles are those dispersed into practically individual units, randomly distributed in the composite. Layered silicates, such as montmorillonite clays or organoclays, can be used in nanocomposites. Because clays are hydrophilic and polyolefines are hydrophobic, it is not easy to make a nanocomposite based on polyethylene or polypropylene because of their natural incompatibility.

When added to WPCs, layered nanoparticles typically do not improve flexural or tensile strength (though there were some reports on the beneficial effect of nanoparticles on flex strength), but significantly increase flexural modulus (stiffness). Sometimes they even increase water absorption by the WPC, making the final material worse in this regard [24].

Tables 4.28 and 4.29 show effect of a montmorillonite "nanoclay" employed as a masterbatch (PolyOne Nanoblend™ 1001) in polypropylene on polypropylene-based composite containing 50% (w/w) maple wood flour and on flexural strength and modulus of the WPC.

One can see that the coupling agent significantly increases flexural strength and insignificantly increases (or even decreases) flexural modulus of WPC however, the presence of the nanoclay uniformly decreases flex strength and produces almost no effect on flex modulus, particularly as show in Table 4.28. Table 4.29 gives mixed data regarding flexural modulus at low amount of nanoclay.

TABLE 4.28 Effect of a nanoclay on flexural strength and flexural modulus of polypropylene-based WPC containing 50% (w/w) maple wood flour (80-mesh) [24]. No coupling agent added

Nanoclay (% w/w)	Flexural strength (psi)	Flexural modulus (psi)
0	5945	565,500
2	5510	507,500
4	5220	493,000
10	4785	522,000

TABLE 4.29 The same as in Table 4.28, but in the presence of 4% of a coupling agent, polypropylene grafted with maleic anhydride [24]

Nanoclay (% w/w)	Flexural strength (psi)	Flexural modulus (psi)
0	9860	536,500
2	7250	436,500
4	6815	609,000
10	5800	638,000
20	—	667,000

Unlike WPC, neat polypropylene filled with the same nanoclay significantly increases its flexural strength and modulus (Table 4.30).

Clearly, the nanoclay conflicts with wood fiber in the polypropylene-based WPC, which results in the decrease of its both flex strength and modulus.

Nanoparticles can increase water absorption by WPCs. For example, after a month of underwater exposure of the WPC (50% polypropylene + nanoclay, w/w, 46% maple wood flour [80-mesh], and 4% coupling agent) no-nanoclay composite absorbed 2.7% of water, 4%-nanoclay composite absorbed 2.9% of water, and 10%-nanoclay composite absorbed 4.4% water [24]. It also makes nanoparticles undesirable in WPC products, unless some means are found to overcome this problem.

Weakening effect of nanoparticles was observed not only with wood-flour-filled polypropylene (Tables 4.28–4.30) but also with rice-hulls-filled polypropylene on both tensile and flexural properties (Tables 4.31 and 4.32).

TABLE 4.30 Effect of a nanoclay on flexural strength and flexural modulus of polypropylene [24]

Nanoclay (% w/w)	Flexural strength (psi)	Flexural modulus (psi)
0	5090	155,150
4	6310	234,900
10	6880	298,700
20	7290	394,400

TABLE 4.31 Effect of a nanoclay on tensile strength and tensile modulus of polypropylene, ExxonMobile Chemicals, PP1074E5, MFI 19, filled with rice hulls (1–2 mm length), with 5% of coupling agent Fusabond MB511D [25]

Nanoclay (% w/w)	Tensile strength (psi)	Tensile modulus (psi)
0	3770	138,000
1	3260	138,000
3	3190	142,000
5	2755	140,000

TABLE 4.32 Effect of a nanoclay on flexural strength and flexural modulus of polypropylene-based composite material (description of materials is given in Table 4.31) [25]

Nanoclay (% w/w)	Flexural strength (psi)	Flexural modulus (psi)
0	6820	312,000
1	6100	319,000
3	5800	326,000
5	5510	340,000

One can see that nanoclay consistently decreases tensile and flexural strength and modulus of the polypropylene-based composite material, and only slightly (less than 10%) increases tensile and flexural modulus of the composite.

It is not clear why nanocomposite particles weaken wood flour- and rice-hulls-filled polypropylene, but it seems to be a repetitive and reproducible phenomenon. Certainly, it can be hypothesized that the reason is the nonuniform dispersion of cellulose fillers and nanoclay particles, but it remains just a hypothesis.

CONCLUSIONS

Despite beneficial properties of mineral fillers, only few commercially available WPC deck boards include them along with cellulosic fiber, such as wood flour and/or rice hulls. As a result, many commercial deck boards can be installed at a span only less than 16 in. on center due to their high deflection under the code-prescribed load (low flexural modulus). This property of deck boards is analyzed in Chapter 7.

Apparently, the main reasons why WPC manufacturers do not use mineral fillers are product cost and wearing of the processing equipment. Many WPC manufacturers struggle to keep cost of materials as low as possible and try to minimize equipment maintenance and repair cost by all means.

Nanofiller mineral additives have not shown noticeable improvements in mechanical and other properties of WPC as yet, hence, they are not in commercial use in WPC materials.

Table 4.33 shows principal properties of mineral fillers which either are used or can be used readily in WPC.

TABLE 4.33 Some important properties of commercial mineral fillers used in plastics industry and/or in WPC materials (adapted from Ref. [1])

Material	Chemical formula	Specific gravity (g/cm^3)	Mohs hardness	Linear coefficient of thermal expansion ($\times 10^{-6}$ 1/F)	Median particle size (μm)	Specific surface area (m^2/g)	Oil absorption (g/100 g)
Calcium carbonate	$CaCO_3$	2.7–2.9	3–4	2–6	0.2–30	5–24	13–21
Talc	$Mg_3Si_4O_{10}(OH)_2$; SiO_2 45–65% MgO 25–30% CaO 0.5–13% Al_2O_3, Fe_2O_3	2.7–2.9	1–1.5	4	1.5–15	3–35	22–57
Biodac	Blend of cellulose fiber, $CaCO_3$ and kaolin clay	1/58			300		150
Silica	SiO_2	1.9–2.2	1[a]	0.3	1–40	1–800	20–330
Kaolin clay	$Al_2O_3 \cdot 2SiO_2 \cdot 2H_2O$ SiO_2 40–60% Al_2O_3 25–45% Fe_2O_3, TiO_2, CaO, MgO, K_2O, Na_2O	2.6	2–2.5		0.2–7	8–70	30–50
Mica	Complex aluminosilicate, SiO_2 40–50% MgO 20–25% Al_2O_3 10–40% K_2O 3–11% Fe_2O_3 0–11%	2.7–3.2	2.5–4	1–14	4–70		65–72

(Continued)

TABLE 4.33 *(Continued)*

Material	Chemical formula	Specific gravity (g/cm^3)	Mohs hardness	Linear coefficient of thermal expansion ($\times 10^{-6}$ 1/F)	Median particle size (μm)	Specific surface area (m^2/g)	Oil absorption, (g/100 g)
Wollastonite	$CaSiO_3$ CaO 40–50% SiO_2 40–50% Fe_2O_3, Al_2O_3, MgO, MnO, TiO_2	2.9	4.5	4	8–650	0.4–4	20–50
Fly ash	SiO_2 30–60% Al_2O_3 10–20% Fe_2O_3 5–10% MgO 5–6% CaO 2–45%	2.1–2.2	3–5.5		0.3–80	0.2–0.8	15–30
Glass fibers	SiO_2 50–60% CaO 20–25% Al_2O_3 14–15% B_2O_3 5–10%	2.5–2.7	6–6.5		50–20,000		
Carbon black	Carbon	1.7–1.9			10–250	10–600	

[a] Precipitated silica.

REFERENCES

1. G. Wypych. *Handbook of Fillers*, 2nd edition, ChemTec Publishing, Toronto, New York, 1999, p. 890.
2. B. English, N. Stark, and C. Clemons. Weight reduction: wood versus mineral fillers in polypropylene. In: *Fourth International Conference on Wood-Plastic Composites*, Forest Product Society, Madison, WI, May 12–14, 1997, pp. 237–244.
3. T. Dombrowski. Carbonate minerals in wood polymer composites (WPC). In: *WPCs 2006, The Changing Technology in WPCs. The First World Congress on Woodfiber/Plastics and Related Composites*, San Francisco, CA, April 2–4, 2006.
4. K. Kuck and H. Hansen. Mineral based concentrates for polypropylene injection molding: combining value and performance. In: *Fifteenth International Conference Additives 2006. Plastic Additives for Special Effects*, ECM, Plymouth, MI, Las Vegas, NV, January 30–February 1, 2006.
5. T. Dombrowski. Phyllosilixates in WPC applications: Talc and MinFlex™. In: *WPCs 2006, The Changing Technology in WPCs. The First World Congress on Woodfiber/Plastics and Related Composites*, San Francisco, CA, April 2–4, 2006.
6. O. Noel and R. Clark. The use of talc in wood-plastic composites. In: *Progress in Woodfibre-Plastic Composites*, Canadian Natural Composites Council; University of Toronto, Canada, Toronto, 2004.
7. O. Noel and R. Clark. Recent advances in the use of talc in wood-plastic composites. In: *The Global Outlook for Natural Fiber & Wood Composites*, Intertech, Portland, ME, New Orleans, LA, December 8–10, 2004.
8. O. Noel and R. Clark. Recent advances in talc-reinforced wood-plastic composites. In: *Intertech's 4th Conference of Natural Fiber & Wood Composites*, Intertech, Portland, ME, Orlando, FL, 2005.
9. T. Dombrowski. Mica in oriented polypropylene WPC. In: *WPCs 2006, The Changing Technology in WPCs. The First World Congress on Woodfiber/Plastics and Related Composites*, ECM, Plymouth, MI, San Francisco, CA, April 2–4, 2006.
10. U.S. Department of Transportation/Federal Highway Administration Report No. FHWA-SA-94–081, Fly ash facts for highway engineers, August 1995.
11. Electric Power Research Institute, Report CS-4765, September 1986.
12. U.S. Pat. No. 5,868,084 (1999). A. Bachik. Apparatus and process for carbon removal from fly ash.
13. U.S. Pat. No. 5,887,724 (1999). T.E. Weyand and C.J. Koshinski. Methods of treating bi-modal fly ash to remove carbon.
14. U.S. Pat. No. 6,068,131 (2000). R.W. Styron and J.-Y. Hwang. Method of removing carbon from fly ash.
15. U.S. Pat. No. 6,599,358 (2003). B.E. Boggs. Carbon scavenger fly ash pretreatment method.
16. U.S. Pat. No. 6,708,909 (2004). Y. Toda, S. Kojima, and T. Yamamoto. Separation device for unburned carbon in fly ash and separation method.
17. U.S. Pat. No. 6,730,161 (2004). V.I. Lakshmanan, R. Sridhar, N.P. Mailvaganam, and V.M. Malhotra. Treatment of fly ash.
18. U.S. Pat. No. 6,783,739 (2004). R.F. Altman. Fly ash treatment by in situ ozone generation.

19. U.S. Pat. No. 6,890,507 (2005). X. Chen, Y. Gao, R.H. Hurt, E.M. Suuberg, and A.K. Mehta. Ozone treatment of fly ash.
20. U.S. Pat. No. 6,916,863 (2005). R.T. Hemmings, R.L. Hill, and B.J. Cornelius. Filler comprising fly ash for use in polymer composites.
21. U.S. Pat. No. 4,058,406 (1977). D.A. Raponi. Cementitious composition.
22. U.S. Pat. No. 4,165,302 (1979). C.F. Armenti and J. V. De Juneas. Filled resin compositions containing atactic polypropylene.
23. J. J. Balatinecz, M. I. Khavkine, S. Law, and V. Kovac. Properties of polyolefin composites with blends of wood flour and coal ash. In: *Fifth International Conference on Woodfiber-Plastic Composites*, Forest Products Society, Madison, WI, May 26–27, 1999, pp. 235–240.
24. S.-K. Yeh, A. Al-Mulla, and R.K. Gupta. Influence of the coupling agent of polypropylene/clay nanocomposite-based wood-plastic composites. *ANTEC*, Society of Plastic Engineers, Brookfield, CT, 2005, pp. 1290–1294.
25. Y.H. Lee, T. Kuboki, C.B. Park, and M. Sain. Processing-property behaviors of PP/rice hull and PP/rice hull/nano-clay composites. In: *The Global Outlook for Natural Fiber & Wood Composites*, Intertech, Portland, ME, Lake Buena Vista, FL, 2005.

5

COMPOSITION OF WOOD–PLASTIC COMPOSITES: COUPLING AGENTS

INTRODUCTION

Among a wide range of additives in wood–plastic composites (WPCs), coupling agents stand out strongly. A few years ago they were called "crosslinking" agents. Now they are more often referred to as "coupling agents" or "compatibilizers" interchangeably. These terms are considered appropriate because coupling agents or compatibilizers improve the compatibility between cellulose fiber and polymer matrix, aid in fiber dispersion, hence, provide a better flowability of hot melts (though coupling agents generally increase melt viscosity), improve the melt elasticity and melt strength, and enhance mechanical properties of WPCs. Sometimes they are defined as "wetting wood fiber" agents. As it turned out, many of the coupling agents, or compatibilizers, do not form covalent bonds with wood fiber, or form them scarcely.

Some extend the terms "coupling agents" and "compatibilizers" to all bonding agents and surfactants, including lubricants because the latter also serve as dispersing agents and improve compatibility between cellulose fiber and polymer matrix. Some argue that compatibilizers, dispersing agents, and lubricants do not form strong adhesion at the fiber–plastic interface, which is supposed to be a distinct feature of coupling agents. In this chapter, we would not draw a clear line between these different and fuzzy areas, but rather consider those agents that are used in the industry of WPCs as "coupling agents."

It is very important to understand that just adding the "right" amount of a coupling agent will not necessarily improve properties of the WPC material. It is really important to highlight the significance of optimizing the manufacturing conditions, as well as the other additives in the formulation. This is yet to be fully understood, much less to be quantitatively predicted. Furthermore, it is very difficult to compare different sets of data from WPC boards made by different groups of people, using different production equipment, operated at different manufacturing speeds, with different coupling agents, and at their different levels. Despite this, some regularities in using coupling agents can now be identified. This is what this chapter is aimed at.

The principal function of coupling agents is to improve adhesion between cellulose fiber and plastic, and, therefore, to disperse the fiber in the resin as uniformly as possible, preferably by forming covalent bonds between fiber and plastic.

Why Such a Task?

WPCs, based on polyolefins, have one principal problem related to their integrity: It is commonly a blend of a hydrophobic plastic and hydrophilic wood (or hydrophilic cellulosic fiber). Hence, the adhesion between them is poor, the interface between plastic and wood filler is typically weak, and it fails to optimally transfer stress between the two phases, when loaded. Lubricants help to solve the problem; however, there are means to further improve bridging the interface by employing other mechanisms.

In order to implement those means as fully as possible, two issues should be solved: (1) distribute the filler in the plastic matrix as uniformly as possible, and (2) bridge the interface by coupling wood fiber and the plastic. Hence, the terms coupling agents, compatibilizers, crosslinking agents are in use. Throughout the chapter we will use the term coupling agents, and elaborate, when possible, a specific mechanism of "bridging" or "compatibilizing."

Studies into the matter have shown that a proper coupling agent should contain two functional domains: One domain capable to form entanglements, or segmental crystallization with the polymer matrix and the other capable to strongly interact with the filler—via covalent bond(s), ionic interactions, hydrogen bonds, and so on. These interactions commonly increase the interface adhesion between the filler and the matrix.

In order to illustrate, though semiquantitatively, the importance of the interface adhesion in composite materials, let us take a look at the so-called fracture energy, which is needed in order to disrupt, to break bonds across the interface between the substrate and the matrix. Bonds involving only physical van der Waals interactions yield fracture energy up to 0.1 J/m^2 approximately [1]. This is applicable to low- molecular-weight unentangled oligomers or adhesive bonds involving purely physical interactions. For covalent chemical bonds between adhering surfaces, fracture energy equals to about 1 J/m^2. This value is still unacceptably low for practical, long-term durable adhesion. For high- molecular-weight entangled polymers, the experimental values of fracture energy is 100–1000 J/m^2, that is, two to three orders of magnitude higher than that for covalent bonds [1].

The above is essentially a theoretical basis for coupling agents for WPC, consisting in an ideal case of two "domains," that is, a chemically reactive group capable to interact with cellulose and a polymeric chain capable to entangle with matrix polymers, as it was described above.

Several dozen coupling agents have been considered for using with WPC, among them silicates, titanates, organic acids, chlorotriazines, anhydrides, epoxides, isocyanates, acrylates, amides, imides, silanes, and polymeric compounds. Only few of them were introduced to the WPC industry by manufacturers and suppliers. Those coupling agents can be subdivided into (a) maleated polyolefins (such as maleic anhydride derivatized polyethylene and polypropylene), which bind with cellulose fiber either via hydrogen bonds, ionic interactions or (allegedly) covalent bond, (b) other bifunctional oligomers or polymers, which can interact with inorganic fillers via an ion-pair bond, (c) silanes, grafted onto polymers, which then (allegedly) covalently interact with hydroxyl groups of cellulose fiber, forming a Si—O—C linkage, (d) acrylic-modified polytetrafluoroethylene (PTFE), (e) chloroparafins, and (f) other compatibilizers, resulting in a better dispersion of fillers in the polymer matrix. "Allegedly" in this context refers to actual (industrial) WPC manufacturing conditions, at which formation of covalent bonds between coupling agents and wood fiber in the hot melt could be marginal. In specially designed laboratory conditions, particularly when wood fiber is treated with coupling agents separately from plastic matrix, and only then is added to the matrix, covalent bonds are undoubtedly formed at much higher degree.

This chapter considers coupling agents, available in the WPC industry, and their effects on mechanical properties of WPC, water absorption, and other related properties, as well as some details of interaction of coupling agents with WPC components, including both fillers and lubricants. As estimated, about five out of 30 manufacturers of WPC use coupling agents in their formulations, but only XTENDEX and E-Deck (both by Carney Timber Co., ON, Canada) have mentioned it in their ICC Report (NER-695, Legacy Report of 11/1/2004).

A BRIEF OVERVIEW OF THE CHAPTER

Coupling agents, when employed at proper conditions (appropriate temperature, minimal moisture content, and absence of compounds chemically blocking the coupling agents, such as metal-containing stearates in case of maleated polyolefins) can greatly improve mechanical properties of WPC. They can double flexural and tensile strength of WPC, increase flexural modulus (stiffness) by as much as 40%, double and triple the impact resistance (actual gain depends on a test method), decrease water absorption by two–four times (depending on duration of the immersion test), increase the density, hence, the resistance of the WPC to elements, including oxidation, microbial degradation, and so on.

A rather common assumption that coupling agents, such as maleated polyolefins or organosilanes, act via formation of covalent bonds with wood fiber is typically word of mouth rather than based on a solid experimental base. In fact, there were just a few attempts to identify covalent bonds between coupling agents and cellulose fiber, formed in the presence of molten polymer matrix in conditions, which were simulated industrial ones, and most of those attempts gave inconclusive and/or ambiguous results (see below).

Overall, effects of coupling agents on WPC at manufacturing conditions is poorly understood. It appears that the same coupling agents, such as maleated polyolefins, can either form covalent linkages with cellulose fiber, bringing about a significant improvement of properties of WPC or do not form covalent bonds practically at all, or form rather negligible amount of them, leading to marginal increases of WPC properties. Hence, coupling agents provide a great variety of effects on WPC in practice.

There is a general consensus among researchers in the area that coupling agents seem to improve the adhesion between the wood fiber and the resin, hence, an improvement in compatibility. How it happens, in terms of specific mechanisms, remains to be solved. It well might be that in different systems (different resin, fiber, coupling agent, temperature, and other conditions) a type of interaction effecting adhesion and compatibility is different, and covalent interactions might not be prevalent at all. In fact, an identification of the specific type of interaction between the coupling agent, the fiber, and the matrix is not the most important issue for a practitioner in the WPC area. If a coupling agents doubles and triples the strength and stiffness of a deckboard and a handrail, that is what matters. However, in order to understand how to achieve the effect of this scale and make it possible, we need to understand how coupling agents work and why they work that way. Hence, this chapter presents considerations of possible mechanisms of action of coupling agents.

It is not clear as yet how to maximize covalent bonds formation, except for removing moisture as much as possible from the formulation and replace metal-containing lubricants with nonmetal ones. The lubricant industry currently provides a good variety of nonmetal lubricants for WPC manufacturing. At any rate, effects of coupling agents on WPC are largely unpredictable, and optimizations of the system are typically empirical, basically a trial-and-error kind of experimentation. Generally, nonmetal lubricants cannot help coupling agents to form covalent bonds with wood fiber; they just do not conflict with coupling agents and do not remove the agent's reactive, functional groups from the reaction system.

It was observed that maleated polyolefins bring about a more significant effect on flexural modulus (up to 40%) of WPC materials manufactured industrially compared to those obtained in a laboratory, which is small scale. Based on this, it was conjectured that the coupling agents act more as dispersing agents for cellulose fiber in the resin. In small-scale laboratory tests, mixing is more intense and hence more dispersive; the dispersing effect of the coupling agents is minimal. In the industrial scale equipment, the dispersive mixing is not as effective, and the coupling agents "upgrade" the conditions more noticeably.

The most commonly employed coupling agents, such as maleated polylefins of the Polybond®, Integrate®, and Fusabond® series, show effects on WPC that are rather similar to each other. Those effects are largely determined by processing conditions for WPC and types of cellulose filler (lignified vs. bleached, for example), rather than by the coupling agents themselves.

Overall, only a handful of WPC manufacturers employ coupling agents in their industrial products. However, premium products, such as deck boards and railing systems certainly require good coupling agents, utilized at proper, optimized manufacturing conditions. This remains to be one of the most challenging, attractive, and priority areas of WPC research and development.

MALEATED POLYOLEFINS (POLYBOND, INTEGRATE, FUSABOND, EPOLENE, EXXELOR, OREVAC, LOTADER, SCONA, AND UNNAMED SERIES)

Maleated polyolefins are the most widely used coupling agents. They contain two functional domains: one, a polyolefin (typically high density polyethylene [HDPE] or polypropylene), which is able to form entanglements with the polymer matrix, to build into the matrix, and the second group, maleic anhydride, which is able to strongly interact with cellulose fiber at extrusion temperatures, covalently (apparently, not always), via hydrogen, or ionic bonds. Covalent interactions, when actually take place, occur with hydroxyl groups of cellulose and result in formation of a cellulosic ester and a free acid, still attached to the polymeric chain.

Maleated poleolefins are usually made by grafting maleic anhydride onto the polymer backbone (via radical reactions), resulting in 1–6% (w/w) of the covalently attached anhydride. There are some data (see below) that say that this covalently linked maleic anhydride can further covalently interact—at extrusion temperatures—with hydroxyl groups of cellulose, thereby strongly coupling the cellulose fiber to the plastic matrix. However, these data are often inconclusive and rather fall into a category of "alleged" mechanisms, "probable" reactions, or "likely" modes of interaction. Therefore, it would be fair to say that this alleged esterification probably creates a crosslinked cellulose-maleic half-ester structure, which is better dispersed in the polymeric matrix compared with the unmodified cellulose fiber.

It should be taken into account that maleated polyolefins can slowly react with air moisture during storage, and form free acid. As a result, chemical reactivity of the coupling agents decrease. Hence, care should be taken to keep maleated polyolefins dry, or heat them up before usage in order to regenerate the anhydride chemical structure.

Maleated polyolefins are usually used at 1–5% by weight in a WPC formulation, and their retail price is around $1.50/lb. Hence, they cost 1.5–7.5 ¢/lb of the formulation. This can be compared to the most expensive (in the final formulation) ingredient, a polymer matrix itself, which costs around 70 ¢/lb, and at 30–40% content (by weight), it costs 21–28 ¢/lb of the formulation. Therefore, coupling agents may add about 4–20% of a total cost of the materials in the formulation. Plastic often costs about 60–80% of the formulation.

Let us consider several of the most known maleated polyolefins as coupling agents in WPC. They are

- Polybond® series (Chemtura Corp., formerly Crompton and Great Lakes)
- Integrate™ series (Equistar Chemical)
- Fusabond® series (DuPont Industrial Polymers)
- Epolene series (Eastman Chemical)
- Exxelor series (Exxon Mobile Chemical)
- Orevac series (Arkema)
- Scona series (Kometra)
- Coesive series (Industrie Polieco—MPB, Italy)
- Licomont series (Clariant)

In addition to the list, maleated polymers as coupling agents are produced by Arkema as Lotader series. They are not made by reactive extrusion. These compounds are terpolymers of ethylene, maleic anhydride and methyl, ethyl, or butyl acrylates. Acrylates make the compound rubbery and, therefore, reportedly decrease flexural modulus.

Most of the above products (except Lotader series) are similar in kind, chemically, and differ by a type of plastic on which maleic anhydride is grafted on, by the amount of the grafted maleic anhydride, by the molecular weight, and hence, by the melt flow index (MFI).

Polybond series is represented by compounds shown in Table 5.1. Polybond 3039 differs from other maleated HDPE-based compounds by a smaller, 40-mesh size of their particles. The manufacturer recommends small-size micropellets of Polybond 3039 for applications where the ingredients in the WPC formation are preblended as powders prior to the addition to the extruder.

Integrate series is represented by HDPE- and linear low density polyethylene (LLDPE)-functionalized maleic anhydride (Table 5.2)

TABLE 5.1 Polybond® family of coupling agents

Polybond	Base polymer	Melt flow index (190°C/2.16 kg)	Maleic anhydride level (%)
3009	HDPE	5	1.0
3109	LLDPE	30	1.0
3029	HDPE	4	1.6
3039[a]	HDPE	4	1.6
3000	Polypropylene	400	1.2
3200[b]	Polypropylene	110	1.0
1001	Polypropylene	40 (230°C)	6.0

[a]A 40-mesh micropellet version of Polybond 3029.

[b]Number-average molecular weight of 12,300, weight-average molecular weight of 97,500 [2].

TABLE 5.2 Integrate™ family of coupling agents

Product, NE-	Base polymer	Melt flow index	Density, (g/cm^3)	Maleic anhydride level
556-004	HDPE	3.8	0.956	High
558-004	HDPE	3.9	0.958	Very high
433-003	LLDPE	2.7	0.933	High
534-003	LLDPE	2.6	0.934	Very high
542-013	LLDPE	13	0.943	Very high

Fusabond series, recommended by the manufacturer for WPC, consists of three maleic anhydride derivatized polyolefins, polyethylene (Fusabond MB-226D and Fusabond WPC-576D), and polypropylene (Fusabond MD-353D). The first and the last one are traditional grafted polyolefins, in which maleic anhydride covalently linked to the polymer chain. Fusabond WPC-576D is an ethylene anhydride copolymer, in which maleic anhydride is covalently incorporated into a polymer chain (Table 5.3)

As it is shown below, it appears that effects of lubricants on functional behavior of maleated polyolefins as coupling agents in WPC formulations often exceed functional differences between the coupling agents themselves.

The manufacturer claims that the new technology of incorporation of maleic anhydride into the polymer chain (such as in the case of WPC-576D) results in a higher level of reactive sites for chemical links with cellulose fiber, and hence, in a better performance of WPC-576D at lower additive concentrations compared to traditional

TABLE 5.3 Fusabond® family of coupling agents

Product	Base polymer	Melt flow index	Density (g/cm^3)	Maleic anhydride level
Recommended in WPC applications				
MB-226D	LLDPE	1.5	0.93	High
MD-353D	Polypropylene	450	N/A	Very High
WPC-576D	Ethylene copolymer	25	N/A	N/A
Other Fusabond® products				
MZ109D	Polypropylene	120	N/A	Low
M613-05	Polypropylene	120	N/A	Low
MD511D	Polypropylene	25	N/A	Low
MZ203D	Polypropylene	250	0.94	Low
MD353D	Polypropylene	450	N/A	Very high
MD411D	Polypropylene	290	0.90	Very high
MB100D	HDPE	1–2	0.96	High
MB265D	HDPE	12	0.95	High
MB439D	LLDPE	2.7	0.92	High
MB528D	LLDPE	6.8	0.92	Very high
MX110D	LLDPE	25	0.93	High

grafter derivatives [3,4]. The claim is hardly substantiated judging from data gathered in Tables 5.7 and 5.9. Though the data were obtained in various conditions, with various lubricants (or often without them) and with various grades of wood fiber, they illustrate that in general Fusabond, Polybond, and Integrate show similar effects on both strength and deflection of WPC materials. Flexural strength typically increases up to 30–110% with Polybonds, 30–120% with Integrates, 10–30% with "common" Fusabond MB-226D, and 10–20% with a new Fusabond WPC-576D.

When Fusabond WPC-576D and Integrate NE-542 were compared in the same WPC formulation (Table 5.9), they showed almost identical effects. Regarding Polybond and Integrate, data on tensile strength mirror those on flexural strength: The Polybond (at 2–2.5% w/w) increased tensile strength by 40–100% and Integrate by 60% (there were no data on Fusabond). Similarly, flexural modulus typically increased up to 10–25% with Polybond, 10–40% with Integrate, by 10% with "regular" Fusabond MB-226D and by 20–40% with the new Fusabond WPC-576D.

Du Pont also subdivides their Fusabond® products to Fusabonds P (based on polypropylene), E (polyethylene), C (EVA), A (acrylate polymers), and N (EP, EPDM, E-Octene). Some examples, besides shown in Table 5.3, are MC190D and MC250D (Fusabonds C), MG423D, ME556D and EB560D (Fusabonds A), and MF416D, MN493D, MN494D, and MO525D (Fusabonds N).

An alleged mode of action of maleated coupling agents, generally accepted for maleated polyolefins, is shown in Figure 5.1.

Figure 5.2 shows an actual effect of a coupling effect of a Fusabond®, in which Nylon granules are practically totally dispersed in the LLDPE matrix. The manufacturer claims that this effect is due to a covalent binding of Fusabond® to functional amino groups of Nylon, as shown in Figure 5.1.

Several types of maleic anhydride-grafted Epolene coupling agents are manufactured—PE-based Epolene G-2608 and PP-based Epolene E-43, G-3002, G-3003, and G-3015 (manufactured by Eastman Chemical Products). Some of their properties are shown in Table 5.4.

Exxelor series (Exxon Mobile Chemical) is represented by, for example, Exxelor PO1020 (graft level 1%, or "high") and PO1015 ("medium") (Table 5.5).

As with most of the other maleated polyolefins, Exxelor coupling agents increase the flexural modulus only slightly, that is, by 5–8% (at 1–2% of the coupling agent), and then it decreases at 2–3% of coupling agents. Impact resistance, however, is noticeably increased in the presence of either of the two Exxelors, from 7 kJ/m^2 without the coupling agents to 13.5–14 kJ/m^2 in the presence of 3% of either one (in 50% wood-fiber-filled polypropylene) in a notched Izod test, and from 4.7 kJ/m^2 without the coupling agents to 8.3–8.7 kJ/m^2, respectively, in an unnotched Charpy test [6].

Effect of Exxelor coupling agents slightly depended on the amount of wood fiber in a polypropylene-based composite. Exxelor PO1020 at concentrations of 1, 2, and 3% did not change flexural modulus at 40% wood fiber in the composite; at 50% of wood fiber load, its effect was within 7%, and at 60% its effect was within 15% between 0 and 1% of the coupling agent, and then stayed the same at 2 and 3% of the coupling agent.

Orevac series (Atofina; Arkema) is represented by Orevac CA 100 (graft level 1.0% MA, number average molecular weight of 8,822, weight average molecular

(a)
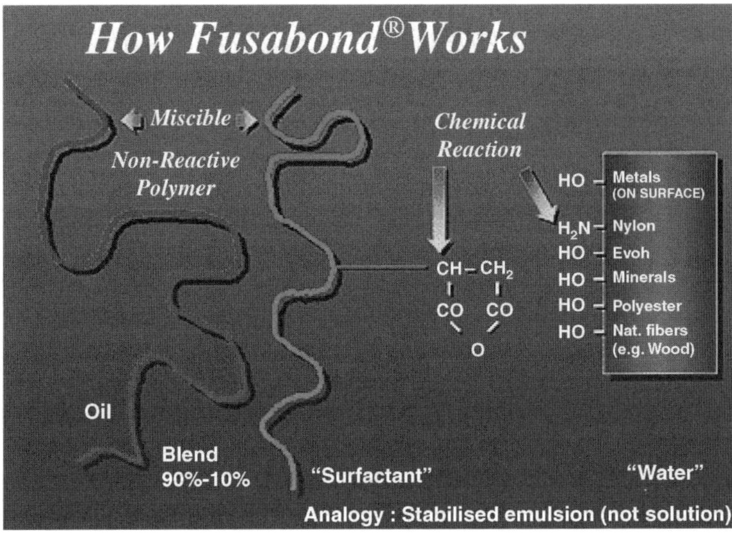

(b)
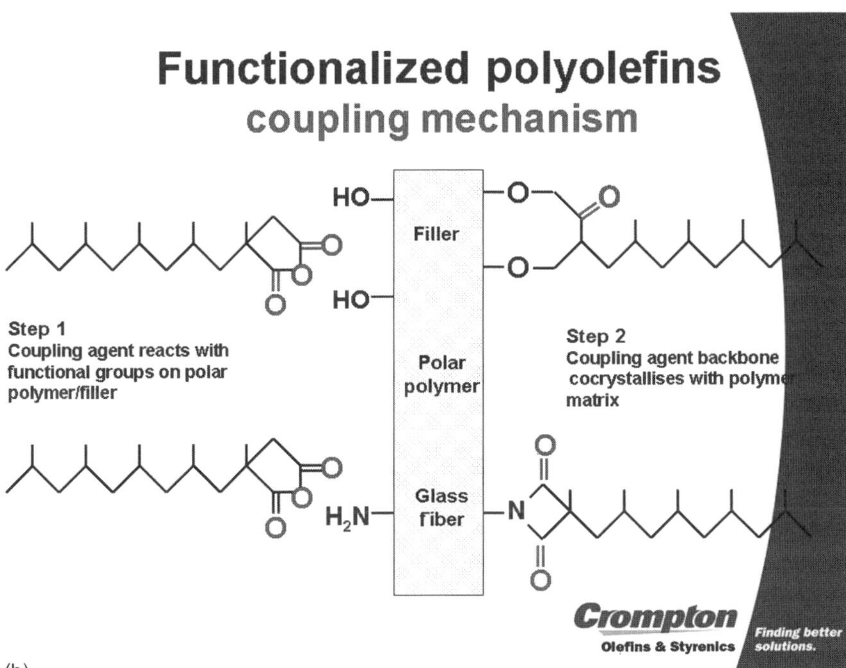

Figure 5.1 The charts suggest how a maleated polymer can noncovalently interact with a matrix resin, and covalently interact with chemical functional groups of a filler or other polymers in the blend (provided by Du Pont, courtesy David Dean [upper] and Crompton, courtesy Bill Sigworth [lower]).

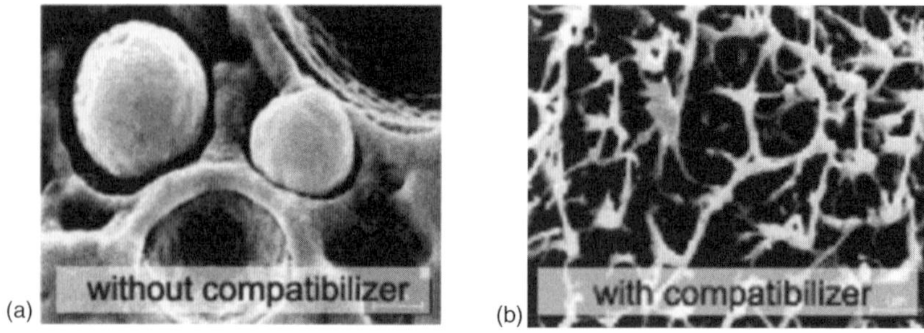

Figure 5.2 On the left—30% Nylon 6 with 70% LLDPE blend without a coupling agent. On the right—10% of a Fusabond® coupling agent was added to the blend (provided by Du Pont).

TABLE 5.4 Some properties of maleic anhydride and maleated polypropylenes (MAAP, Eastman Chemical Products)

Product	Acid number, mg/g	Density, g/cm^3	Weight average molecular weight	Number average molecular weight	Maleic anhydride[a]
Maleic anhydride	—	1.48	98	—	—
Epolene E-43	47	0.934	9,100	3,900	1.6
			20,171[b]	1,775[a]	3.7%[b]
Epolene G-3002	60	0.959	60,000	20,000	10.7
Epolene G-2608	8	—	65,000	—	1.5
Epolene G-3003	9	0.912	52,000	27,200	0.8%[b]
			84,400[b]	8,031[b]	

Acid number is the number of milligrams of KOH to neutralize 1 g of MAAP sample [5].
[a]Relative units, unless % indicated.
[b]Data listed in [2].

TABLE 5.5 Exxelor coupling agents (Exxon Mobile Chemical) [6]

Product	Base polymer, polypropylene	Melt flow index (230°C, 2.16 kg)	Maleic anhydride level
PO1015	Copolymer	150	Medium
PO1020	Homopolymer	430	High

TABLE 5.6 Some properties of maleated polypropylenes (products by Kometra and Clariant, Germany) [8]

Maleic anhydride content (% w/w)	MFI, 190°C, 2.16 kg (g/10 min)	Molecular weight	
		Weight average	Number average
0.5	80	167,000	54,900
0.9	109	153,000	59,400
1.3	30	160,000	60,400
1.4	1.9	295,000	50,300
1.5	97	151,000	58,200
2.0	67	140,000	44,100
3.1	>1000	34,700	4,810
7.0	>1000	24,300	4,840

weight of 94,328 [2], Orevac SM-7001 (0.4% MA), and Orevac 18307 maleated HDPE). Orevac CA 100, introduced into polypropylene-based WPC, did not show any effect on flexural and tensile modulus, increased flexural and tensile strength, slightly decreased impact strength (notched), and increased impact strength (unnotched) [7]. These data are listed and analyzed below.

Lotader series (Arkema) is represented by Lotader 3210 (recommended by the manufacturer for mineral fillers) and Lotader AX8900 (epoxy coupling agent).

Scona series (Kometra) is represented by Scona TPPP 8112 (graft level 1.3–2.0%) (Table 5.6).

Coesive series (Industrie Polieco—MPB, Italy) is represented by two LLDPE-based compounds (MFI of 1–2 g/10 min for both)—Coesive LL15 (0.36% w/w of maleic anhydride) and Coesive F30 (1% w/w of maleic anhydride).

Licomont series (Clariant, Germany) is represented by Licomont AR 504, that is a maleic anhydride-modified polypropylene, density of 0.89–0.93 g/cm^3, acid value of 37–45 mg KOH/g. Other Licomonts in the series are lubricants.

There are little data on industrial WPC applications of the last several series of coupling agents.

ORGANOSILANES (DOW CORNING Z-6020, MOMENTIVE A-172 AND OTHERS)

Silanes have found a rather wide application as coupling agents and compatibilizers in plastics filled with minerals [9,10]. A general formula for the silanes is as follows:

$$R—(CH_2)_n—SiX_3,$$

where R is the chemical group which possess an affinity for, and possibly reactivity with, the polymer matrix (amino, epoxy, vinyl, alkyl), X, a reactive (hydrolysable) chemical group which provides a covalent interaction with a filler, such as $—OCH_3$, $—OC_2H_5$, $CH_3COO—$, and *n*, typically 0 or 3, and up to 7 or higher.

An example of organosilane coupling agent is γ-glycidoxy-propyltrimethoxy-silane (Z-6040 Silane, manufactured by Dow Corning Corp., and A-187, manufactured by Momentive) [10].

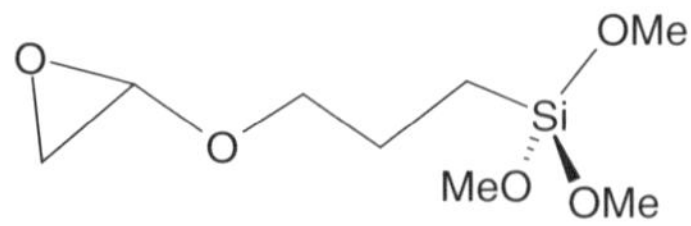

The way the application of this silane was described in this specific reference makes it dubious for an industrial process. Cellulose fiber was first dewaxed in an alcoholic solution for 24 h, and then was washed with distilled water and treated with the silane component (2% w/w) in an alcoholic solution for 24 h. The silane application was finished with a 4-h drying process in a vacuum at 75°C [10].

There are many different organofunctional groups available on organosilanes. Among the more commercially important are γ-methacryloxypropyltrimethoxy-silane (Dow Corning Z-6030 Silane and Momentive A-174), vinyl silanes (Dow Corning Z-6300 Silane and Momentive A-172), epoxy silanes (Dow Corning Z-6040 and Momentive A-186 and A-187), amino silanes (Dow Corning Z-6020 and Momentive A-1100), and alkyl silanes such as Dow Corning Z-6341 Silane (n-Octyl). Organosilanes that may find utility in WPCs include aminosilanes, long alkyl chain alkoxysilanes, vinyl alkoxysilanes, and their oligomers. This range of organic functionalities is quite useful for matching the silane to the particular polymer matrix of interest.

For optimum performance of the compound or composite, it is best to have an organosilane functionality that will react with the polymer matrix. For instance, an epoxy resin has the proper reactive groups to form covalent bonds with especially epoxy and amino silanes. For many thermoplastic polymers, this is made more difficult by the lack of reactive groups on the polymer. Here, the proper silane that can improve compatibility with the polymer will likely provide an improvement in physical properties.

Typical addition levels of silanes are 3–10% by weight, but efforts continue to work to decrease the required amount. However, as the cost of the organosilanes is in the range of $3.5–12.0/lb [10], at the current 3–10% use level, the silane would add from 10 cents to $1.20 per pounds of a WPC formulation. This is practically prohibitive, except for some very special applications, without further effort to enhance the performance and value of the organosilane solutions. If the organosilane technology can show more solid benefits in decreasing moisture uptake or improving durability and physical properties, they may be at value in applying this technology to WPCs, albeit at lower amounts and/or for a lower cost.

Vinyl trimethoxysilane grafted polyethylene and their analogous copolymers are commercially available from Borealis, Dow, TT Electronics (AEI), Crosspolimeri

(Italy). In early applications of organosilanes to WPCs, they were first grafted onto the wood fiber surface, and then mixed with plastic melts. This approach has insured the covalent binding of the coupling agents to cellulose and the subsequent entanglement with the polymer matrix. In the limited work with organosilanes, little benefit has been observed, and the cost may be prohibitive.

Vinyl functional silanes have been used for over 35 years to crosslink polyethylene. The most common commercial products using this crosslinking technology are crosslinked PE insulations for the global wire and cable market [11] and crosslinked HDPE pipes for potable water and natural gas. Some work recently has been done to combine this crosslinking technology with wood-filled polyethylene.

Crosslinking of wood fiber in WPC with organosilanes was exemplified by a reactive extrusion, that is, by pumping a combined solution of vinyltrimethoxy silane and dicumyl peroxide (12:1 w/w, total amount of the added silane solution of 2% w/w) into a vented extruder with a concurrent feed of wood flour and HDPE granules. The obtained granulated material was then extruded into profiles employing the same extruder, and the profiles were cured in a humid (100% RH) and hot (90°C) chamber [11]. As a result of the process, vinyl silanes were crosslinked with available hydroxyl groups of wood fiber in the presence of trace amounts of water. The crosslinking reaction proceeded in two steps—first, hydrolysis of the methoxyl groups in the silane to hydroxyl groups, and second, recombination of hydroxyl groups through a condensation step. As a result, the vinyltrimethoxy silane was grafted onto the polyethylene and wood flour [11]. Crosslinked polyethylene was insoluble in boiling *p*-xylene, while the noncrosslinked part was soluble. From this, a degree of crosslinking was determined, and it was 0% for the composite obtained without the coupling agent, 33% for the composite obtained in the presence of 2% of the silane, and 59% for the coupled and cured composite. At 4% of the coupling agent, the amount of crosslinked fraction in the composite was shown to be as high as 75–80% [11]. Mechanical properties of the composites are listed in Tables 5.7, 5.9 and 5.10.

METABLEN™ A3000 (ACRYLIC-MODIFIED POLYTETRAFLUOROETHYLENE, PTFE)

This coupling agent is manufactured by Mitsubishi Rayon America, subsidiary of Mitsubishi Rayon Co. Ltd. The company makes a series of Metablens, including impact modifiers (acrylic, silicone, and copolymer Metablens series C/E, W and S), processing aids (acrylic Metable series P), lubricants (acrylic Metablen series L), coupling agents (Acrylic-PTFE Metablen series A), and other modifiers.

Metablen A3000 is made in a small granular form (50 mesh size), and at 2–5% w/w, it reportedly improves dispersibility and flowability of hot melts and improves the impact resistance of WPC profiles (ratio of wood powder to polypropylene 80:20; data of Mitsubishi Rayon America). Examples of effect of Metablen A3000 on flexural strength and modulus of a WPC are shown in Tables 5.7 and 5.9.

OTHER COUPLING AGENTS

Dover Chemical Corporation produces resinous chlorinated paraffins under a name Chlorez®. All Chlorez® grades have a physical form of white powder (particles smaller than 50 mesh) with chlorine content around 70%. The manufacturer recommends Chlorez® as flame retardants (in a combination with antimony trioxide) and lately as a nonreactive coupling agent under a brand name Doverbond® (such as Doverbond DB 4300 or Doverbond 3000). The manufacturer claims that Chlorez® in the amount of 10% along with 3% of a lubricant in the WPC shows an effect of a coupling agent and increases the flexural strength and modulus of the product, as well as the UV and moisture resistance (private communication, Dover Chemical Corp.). DB 4300 lists at $1.50/lb [12,13].

Clariant produces plastic compatibilizers of Cesa-mix series, including Cesa-mix 8611 (for use in polyethylene and polypropylene) and Cesa-mix 8468 (for polypropylene-based WPC). However, they are not in active usage in the WPC as yet.

There are many coupling agents, compatibilizers, crosslinking agents and other dispersants described in the literature as experimental ones. They are not used industrially because of a number of reasons, mainly because of the lack of scale-up testing and the lack of financial considerations. Among them are, for example, *N*-vinylformamide-grafted polypropylene [14] and poly(*N*-acryloyl dopamine [15].

EFFECT OF COUPLING AGENTS ON MECHANICAL PROPERTIES OF WOOD–PLASTIC COMPOSITES: EXPERIMENTAL DATA

There is no doubt that maleated polymers as coupling agents often decrease water absorption by WPC, as well as increase flexural and tensile strength of WPC, sometimes up to 200% of the control (Table 5.9). However, they often do not increase or increase insignificantly the flexural and tensile modulus (Table 5.7), contrary to that as would have been expected if strong covalent bonds form between the wood filler and the polymer matrix (see below). In best and rather rare cases, this increase reaches 40% of the control (without the coupling agent), and it can be attributed to a dispersing effect of the coupling agents, whatever mechanism is involved in the dispersing action.

An alternative explanation of a different (in magnitude) effect of coupling agents on strength, on one hand, and modulus, on the other, is that strength is measured in high fiber stress conditions, technically at a break load, when the fiber–resin interface is significantly strained and stressed. At the same time, modulus is measured at low fiber stress conditions, which may be not fully translated to fiber–resin interface. In this case, effects of improved interfacial adhesion produced by coupling agents would be more noticeable in strength tests, such as flexural, tensile, and impact strengths, when nonelastic flow or fracture is dominating. This explanation does not go along, though, with many observations when flexural modulus of a WPC is affected more compared with its flexural strength under other factors. For example, freeze–thaw cycles decreased the flex strength of a wood (maple)-flour-filled HDPE

TABLE 5.7 Effect of coupling agents on flexural and tensile modulus of WPC

Type of coupling agents	Amount of coupling agents (%, w/w)	Tensile modulus (×1000 psi)	Flexural modulus (×1000 psi)
Thermomechanical pulp (30%) + polypropylene[a]			
None	0	355	500
Epolene E-43	3.0	357	493
Epolene G-3002	3.0	349	492
Bleached cellulose fiber (30%) + polypropylene[a]			
None	0	318	402
Epolene E-43	3.0	354	450
Epolene G-3002	3.0	338	413
Kenaf (50%) + homopolypropylene Solvay 1602[b,c]			
None	0	1349	—
Maleated polypropylene	3.0	1204	—
Kenaf (50%) + impact ethylene–propylene copolymer Amoco 3541[b,c]			
None	0	1,884	—
Maleated polypropylene	3.0	1,080	—
Kenaf (50%) + impact ethylene–propylene copolymer Amoco 3143[b,c]			
None	0	1,652	—
Maleated polypropylene	3.0	1,147	—
Wood flour (40%) + polypropylene[d]			
Neat polypropylene (control)	0	203	—
WPC (control)	0	348	—
Maleated polypropylene (Unite MP880, Aristech Chemical Corp.)	2.0	348	—
Wood flour (40%) + HDPE[e,f]			
None	0	467	470
Polybond 3009[g]	0.5	352	531
	1.0	388	508
	1.5	366	459
	2.5	366	502
	4.0	433	538
Wood flour (40%) + HDPE[h]			
None	0	—	232 ± 29
Vinyltriethoxy silane, not cured	2	—	276 ± 44
Vinyltriethoxy silane, cured	2	—	276 ± 29

(Continued)

TABLE 5.7 (*Continued*)

Type of coupling agents	Amount of coupling agents (%, w/w)	Tensile modulus (×1000 psi)	Flexural modulus (×1000 psi)
Wood flour (40%, 40–70 mesh) + HDPE[i,j]			
Neat HDPE (control)	0	—	160 ± 15
	4	—	116 ± 1
WPC (control)	0	—	435 ± 29
Vinyltrimethoxy silane	2	—	348 ± 15
	3	—	334 ± 15
	4	—	305 ± 29
	6	—	305 ± 15
Wood flour (55%, 40 mesh dried oak) + HDPE, no lubricant[k]			
None	0	—	330
Fusabond 226DE	2	—	430
Fusabond 100D	2	—	420
Wood flour (55%, 40 mesh dried oak) + polypropylene, no lubricant[l]			
None	0	—	530
Fusabond 353D	2	—	560
Fusabond 613-05	2	—	560
Wood flour (60%, 60 mesh pine) + HDPE + talc (5%) + a nonmetal lubricant (3%)[l]			
None	0	—	540
Fusabond EMB-226D	0.5	—	545
	1.0	—	530
	2.0	—	550
	3.0	—	590
Fusabond WPC-576D	0.5	—	550
	1.0	—	620
	2.0	—	710
	3.0	—	730
HDPE-based WPC, nonmetal lubricants[m]			
None	0	—	200
Fusabond WPC-576D	2.0	—	180
	2.0	—	240
	2.0	—	275
Integrate NE-542	2.0	—	245
	2.0	—	270
	2.0	—	280
Wood flour (50%, 80-mesh) + polypropylene[n]			
None	0	570	550
Polybond 3200	2.0	580	510
	4.0	620	520
	10	590	510

TABLE 5.7 (*Continued*)

Type of coupling agents	Amount of coupling agents (%, w/w)	Tensile modulus (×1000 psi)	Flexural modulus (×1000 psi)
HDPE-based WPC[m]			
None	0	—	240
Polybond 3029	1.5	—	300
	1.5	—	270
	1.5	—	290
HDPE-based WPC[o]			
None	0	—	245 ± 10
Silane Z-6341	1.0[l]	—	190 ± 10
PolySize 94	2.2	—	246 ± 6
Metablen A-3000	3.0	—	324 ± 16
Wood flour (60%, 60 mesh) + HDPE, no lubricants[p]			
None	0	—	520
Integrate NE 556-004	2.0	—	580
Integrate NE 558-004	2.0	—	610
Integrate NE 433-003	2.0	—	600
Integrate NE 534-003	2.0	—	590
Wood flour (60%, 60 mesh) + HDPE, a nonmetal lubricant[p]			
None	0	—	515
Integrate NE 556-004	2.0	—	560
Integrate NE 558-004	2.0	—	590
Integrate NE 433-003	2.0	—	550
Integrate NE 534-003	2.0	—	580
Oak fiber (50%, 40 mesh) + HDPE[q]			
None	0	—	550
Epolene G-2608	3.0	—	550
Maple wood fiber (50%, 40 mesh) + polypropylene[r]			
None	0	—	612
Exxelor 1015	1.0	—	645
	2.0	—	642
	3.0	—	638
Exxelor 1020	1.0	—	653
	2.0	—	660
	3.0	—	657

(*Continued*)

TABLE 5.7 (*Continued*)

Type of coupling agents	Amount of coupling agents (%, w/w)	Tensile modulus (×1000 psi)	Flexural modulus (×1000 psi)
Milled newsprint (40%) + polypropylene[s]			
Neat polypropylene	0	167 ± 6	160 ± 4
None	0	467 ± 6	512 ± 6
Orevac CA100	1.0	479 ± 3	512 ± 13
	2.0	476 ± 2	515 ± 10
	5.0	473 ± 4	512 ± 9
Wood flour (50%) + HDPE, a nonmetal lubricant (4%)[t]			
None	0	—	249
Polybond 3029	2	—	364 ± 11
Wood flour (60%) + HDPE, no lubricant[t]			
None	0	395	488
Polybond 3029	1	458	467
	2	492	539
	4	488	545
Wood flour (50%, fine particles) + polypropylene[u]			
Wood particles of smaller than 400 mesh size			
None	0	30.5	290
Polybond 3029	2	30.5	247
Wood particles of 200–300 mesh size			
None	0	31.9	290
Polybond 3029	2	34.8	319
Wood particles of 80–100 mesh size			
None	0	27.6	363
Polybond 3029	2	29.0	305
Wood flour (61%, pine) + HDPE, nonmetal lubricants[v]			
None (lub 1)	0	—	696 ± 30
Polybond 3029	3	—	727 ± 25
None (lub 2)	0	—	717 ± 33
Polybond 3029	3	—	773 ± 13

[a] Ref. [5]. [b] Ref. [17]. [c] Ref. [18]. [d] Ref. [19]. [e] Ref. [20]. [f] Ref. [21].

[g] Polybond 3109 gave lower values of flexural modulus of WPC. [h] Ref. [11]. [i] Ref. [22].

[j] Crystallinity was decreasing with increase of the coupling agent in the material, from 65 to 56% (in neat HDPE) and from 77 to 53% (in the WPC).

[k] Ref. [23]. [l] Ref. [4]

[m] Data by Ferro Corporation. Data obtained with different lubricants at the same amount of the coupling agent.

[n] Ref. [24].

[o] Data by Tatyana Samoylova (LDI Composites). (1) Cellulose-containing fillers were treated with 20% silane solution in isopropanol and vacuum-dried before compounding. (2) PolySize 94 (PolymerVentures Inc.). (3) No lubricants were added with Metablen A-3000.

[p] Ref. [25]. [q] Ref. [26]. [r] Ref. [6]. [s] Ref. [7].

[t] Compression molded samples. Bill Sigworth, personal communication.

[u] Ref. [27].

[v] Jonas Burke (Ferro), private communication.

by 21%, but flex modulus was reduced at the same conditions by 49% (after 15 cycles). In a pinewood-flour-filled HDPE the reduction effects were by 5 and 38%, respectively [16]. In other words, effects of factors on flex modulus as a measure of a relatively low fiber stress changes is not necessarily smaller compared to those in flex strength, when nonelastic flow or fracture is dominating.

Hence, the question why coupling agents typically effect strength more than deflection of WPC is still open.

Likewise, addition of maleated polypropylenes to 30% wood—70% PP composition led to increased tensile strength from 15% (at 1% of the coupling agent) to 40% (at 5% of the coupling agent), and then stayed the same to 13% of the coupling agent, while flexural modulus was fluctuating between 5 and 10% (above the control baseline) with 1–13% of the added maleated coupling agent [28].

Furthermore, some other properties of plastics in WPC, such as the glass transition temperature (T_g), either do not show changes in the presence of maleated polymers or even show some decrease, contrary to the expected if covalent bonds are formed (Table 5.8).

Table 5.8 shows that the introducing 7 and 14% of maleic anhydride into polystyrene noticeably increases the T_g of the plastic from 113 to 124 and further to 135°C. However, introducing of wood fiber to the maleated polymer causes almost no effect on the T_g in each case. These experiments cannot rule out the chemical bonding between the maleated matrix and the wood filler; however, they do not support it either.

One can reasonably argue that these data do not disprove covalent bond formation between the maleated polymer and wood fiber in the system, and might be even irrelevant. It is known that varying comonomer content in a copolymer (such as varying maleic anhydride in the styrene polymer backbone, as in maleated polystyrenes in Table 5.8) can lead to a change in the T_g of the resulting copolymer (as it was observed; see Table 5.8). This effect is driven by the change in molecular motion available to the entire copolymer chain. Even when wood fiber forms covalent bonds with anhydride groups, each "block" in copolymers, covalently bound with cellulose, can have their own distinct T_g, and no overall change in T_g can be observed in the SMA material.

This example shows how complicated and not self-evident can the observed effects and their interpretation be.

TABLE 5.8 Glass transition temperature for styrene-maleic anhydride (SMA) copolymers filled with aspen fiber

Maleated polystyrene, type	Amount of aspen fiber in WPC (% w/w)	Glass transition temperature (°C)
Polystyrene (control)	0	113
SMA-7	0	124
	40%	122
SMA-14	0	135
	40%	134

SMA-7 and SMA-14 contained 7 and 15% (w/w) of maleic anhydride, respectively [29].

An application of Fourier transform infrared spectroscopy (FTIR) to maleated polypropylene-treated wood fiber has indicated that esterification reaction between hydroxyl groups of lignified fiber (unbleached thermomechanical pulp, TPM) and the coupling agent does not occur [5]. The FTIR spectra did not indicate the presence of any distinct absorption bands near 1730 cm, which may be associated to other ester links besides those already present in wood fiber. However, tensile and flexural strength and impact resistance of the WPC in the presence of the maleated polypropylenes increased significantly (up to 280%) (Tables 5.9 and 5.10).

MECHANISMS OF CROSSLINKING, COUPLING AND/OR COMPATIBILIZING EFFECTS

It was logical to suggest that maleated polymers introduced into a polymer matrix filled with cellulose fiber form—at hot melt temperature—covalent ester links between the anhydride groups of the coupling agents and hydroxyl groups on the surface of wood fiber, as they do in model chemical systems. However, studies into the matter have presented no conclusive evidence of such covalent bonds in WPC, except maybe in some isolated cases.

Spectroscopic Studies

Probably, the principal reason why it is so difficult to find covalent ester bonds between coupling agents and wood fiber in WPC, employing spectroscopic observations, is that wood fiber itself contains plenty of hemicellulosic materials (besides cellulose and lignin), which in turn contain plenty of ether and ester bonds, creating a heavy background for spectroscopic measurements. It would have been more reasonable to use, for such studies, bleached cellulose or pure cellulose, such as of cotton fiber.

Indeed, unlike lignified wood fiber, bleached cellulose apparently formed ester bonds with maleated polypropylene, according to [5]. This can be attributed to an appearance of a weak shoulder at the 1645 cm absorption band, as well as the broad absorption band at 1722 cm in the digital subtraction spectra. Similar absorption bands (at 1730 cm and digital subtraction band at 1724 cm) appear after treatment of thermomechanical fiber with maleic anhydride, and appear near 1645 cm absorption band after treatment of bleached cellulose fiber with maleic anhydride. The absorption bands in the 1728–1721 cm region were assigned to ester links formed by the esterification reaction between hydroxyl groups of cellulose and anhydride groups of maleic anhydride.

Table 5.9 shows that the WPC filled with lignin-rich TMP in the presence of maleated coupling agents shows higher tensile and flexural strength compared to the bleached cellulose, despite the alleged formation of ester bonds with the coupling agent in the latter case. This implies that either ester bonds in fact are not formed between the coupling agents and the cellulose filler or other factors play more important role in the strength increase of the WPC. There is a general consensus among researchers in the area that coupling agents seem to improve the adhesion between

TABLE 5.9 Effect of coupling agents on tensile and flexural strength of WPC

Type of coupling agents	Amount of coupling agents (% w/w)	Tensile strength (psi)	Flexural strength (psi)
Thermomechanical pulp (30%) + polypropylene (70%)[a]			
None	0	5,020	9,510
Epolene E-43	3.0	6,400	11,340
Epolene G-3002	3.0	7,010	12,410
Bleached cellulose fiber (30%) + polypropylene (70%)[a]			
None	0	4,710	8,620
Epolene E-43	3.0	5,930	10,910
Epolene G-3002	3.0	6,270	11,090
Kenaf (50%) + homopolypropylene Solvay 1602 (50%)[b]			
None	0	4,830	—
Maleated polypropylene	3.0	9,540	—
Kenaf (50%) + impact ethylene–propylene copolymer Amoco 3541 (50%)[b]			
None	0	3,710	—
Maleated polypropylene	3.0	7,610	—
Kenaf (50%) + impact ethylene–propylene copolymer Amoco 3143 (50%)[b]			
None	0	3,380	—
Maleated polypropylene	3.0	7,270	—
Wood flour (40%) + HDPE[c,d]			
None	0	2,420	3,940
Polybond 3009[e]	0.5	2,780	4,830
	1.0	3,150	4,900
	1.5	3,420	5,410
	2.5	3,990	6,150
	4.0	4,420	6,990
Wood flour (50%) + polypropylene[c]			
None	0	3,180	5,260
Polybond 3200	2.5	4,550	7,500
	3.0	5,000	8,240
Wood flour (40%) + HDPE[f]			
None	0	—	2,800 ± 300
Vinyltriethoxy silane, not cured	2	—	5,200 ± 400
Vinyltriethoxy silane, cured	2	—	4,900 ± 300

(Continued)

TABLE 5.9 (*Continued*)

Type of coupling agents	Amount of coupling agents (% w/w)	Tensile strength (psi)	Flexural strength (psi)
Wood flour (40%, 40–70 mesh) + HDPE[g]			
Neat HDPE (control)	0	—	4,420 ± 130
	4	—	3,470 ± 100
WPC (control)	0	—	4,290 ± 250
Vinyltrimethoxy silane	2	—	6,920 ± 170
	3	—	7,080 ± 120
	4	—	6,660 ± 160
	6	—	6,260 ± 200
Wood flour (55%, 40 mesh dried oak) + HDPE, no lubricant[h]			
None	0	—	3,500
Fusabond 226DE	2	—	8,100
Fusabond 100D	2	—	8,000
Wood flour (55%, 40 mesh dried oak) + polypropylene, no lubricant[i]			
None	0	—	5,000
Fusabond 353D	2	—	9,000
Fusabond 613-05	2	—	8,000
Wood flour (60%, 60 mesh pine) + HDPE + talc (5%) + a nonmetal lubricant (3%)[j]			
None	0	—	3,100
Fusabond EMB-226D	0.5	—	3,100
	1.0	—	3,000
	2.0	—	3,500
	3.0	—	4,100
Fusabond WPC-576D	0.5	—	3,250
	1.0	—	3,550
	2.0	—	4,400
	3.0	—	5,450
HDPE-based WPC, nonmetal lubricants[k]			
None	0	—	2,790
Fusabond WPC-576D[h]	2.0	—	3,140
	2.0	—	3,510
	2.0	—	3,450
Integrate NP-542[h]	2.0	—	3,600
	2.0	—	3,670
	2.0	—	3,710
Wood flour (50%, 80-mesh) + polypropylene[l]			
None	0	3,190	6,090
Polybond 3200	2.0	4,710	8,260
	4.0	5,800	10,000
	10	6,160	10,450

TABLE 5.9 (*Continued*)

Type of coupling agents	Amount of coupling agents (% w/w)	Tensile strength (psi)	Flexural strength (psi)
HDPE-based WPC[m]			
None	0	—	3,130
Polybond 3029[h]	1.5	—	3,730
	1.5	—	3,610
	1.5	—	3,750
HDPE-based WPC[m]			
None	0	—	4,600 ± 100
Silane Z-6341	1.0	—	4,500 ± 100
PolySize 94	2.2	—	4,250 ± 250
Metablen A-3000	3.0	—	4,800 ± 100
Wood flour (60%, 60 mesh) + HDPE, no lubricants[n]			
None	0	2,100	4,000
Integrate NE 556-004	2.0	4,800	8,400
Integrate NE 558-004	2.0	4,900	8,900
Integrate NE 433-003	2.0	4,600	8,000
Integrate NE 534-003	2.0	4,600	8,600
Wood flour (60%, 60 mesh) + HDPE, 2% of a nonmetal lubricant[n]			
None	0	1,600	3,400
Integrate NE 556-004	2.0	4,000	7,500
Integrate NE 558-004	2.0	4,200	7,800
Integrate NE 433-003	2.0	3,400	6,600
Integrate NE 534-003	2.0	4,100	7,200
Oak fiber (50%, 40 mesh) + HDPE[o]			
None	0	3,500	5800
Epolene G-2608	3.0	6,000	7800
Milled newsprint (40%) + polypropylene[p]			
Neat polypropylene	0	4,410 ± 30	6,105 ± 130
None	0	4,680 ± 70	8,323 ± 160
Orevac CA100	1.0	5,930 ± 100	9,585 ± 250
	2.0	6,610 ± 100	10,300 ± 160
	3.0	7,530 ± 160	11,800 ± 150

(*Continued*)

TABLE 5.9 ***(Continued)***

Type of coupling agents	Amount of coupling agents (% w/w)	Tensile strength (psi)	Flexural strength (psi)
Wood flour (50%) + HDPE, a nonmetal lubricant (4%)[q]			
None	0	—	2,840
Polybond 3029	2	—	6,050 ± 300
Wood flour (60%) + HDPE, no lubricant[q]			
None	0	1,532	3,402
Polybond 3029	1	2,628	4,590
	2	3,279	6,726
	4	3,668	6,423
Wood flour (50%, fine particles) + polypropylene[r]			
Wood particles of smaller than 400 mesh size			
None	0	2,610	5,220
Polybond 3029	2	4,640	6,815
Wood particles of 200–300 mesh size			
None	0	2,900	5,365
Polybond 3029	2	5,075	7,250
Wood particles of 80-100 mesh size			
None	0	2,030	3,190
Polybond 3029	2	5,220	6,525
Wood flour (61%, pine) + HDPE, nonmetal lubricants[s]			
None (lub 1)	0	—	4,670 ± 90
Polybond 3029	3	—	9,100 ± 150
None (lub 2)	0	—	4,650 ± 85
Polybond 3029	3	—	8,820 ± 160

[a] Ref. [5]. [b] Ref. [17]. [c] Ref. [20]. [d] Ref. [21].

[e] Polybond 3109 gave lower values of flexural strength of WPC.

[f] Ref. [11]. [g] Ref. [22].

[h] Crystallinity was decreasing with increase of the coupling agent in the material, from 65–56% (in neat HDPE) and from 77–53% (in the WPC).

[i] Ref. [23]. [j] Ref. [4].

[k] Data by Ferro Corporation. Data obtained with different lubricants at the same amount of the coupling agent.

[l] Ref. [24].

[m] Data by Tatyana Samoylova (LDI Composites). (1) Cellulose-containing fillers were treated with 20% silane solution in isopropanol and vacuum-dried before compounding. (2) PolySize 94 (PolymerVentures Inc.). (3) No lubricants were added with Metablen A-3000.

[n] Ref. [25]. [o] Ref. [26]. [p] Ref. [7].

[q] Compression molded samples. Bill Sigworth, personal communication.

[r] Ref. [27].

[s] Jonas Burke (Ferro), private communication.

TABLE 5.10 Effect of coupling agents on impact resistance of WPC

Type of coupling agents	Amount of coupling agents (% w/w)	Izod, notched (J/m)	Izod, unnotched (J/m)
Kenaf (50%) + homopolypropylene Solvay 1602 (50%)[a]			
None	0	34	88
Maleated polypropylene	3.0	37	168
Kenaf (50%) + impact ethylene–propylene copolymer Amoco 3541 (50%)[a]			
None	0	41	76
Maleated polypropylene	3.0	74	212
Kenaf (50%) + impact ethylene–propylene copolymer Amoco 3143 (50%)[a]			
None	0	47	119
Maleated polypropylene	3.0	71	220
Wood flour (40%) + polypropylene (60%)[b]			
None	0	26.1	85.8
Maleated polypropylene (Unite MP880, Aristech Chemical Corp.)	2.0	23.5	86.1
Wood flour (40%) + HDPE[c]			
None	0	22	70
Polybond 3009[e]	0.5	21	84
	1.0	22	92
	1.5	21	115
	2.5	23	125
	4.0	30	155
Wood flour (40%) + polypropylene (60%)[d]			
Neat PP (control)	0	20.9	656
PP + 40% wood flour (control)	0	22.2	73
Maleated PP		21.2	78
Wood flour (40%) + HDPE[e]			
None	0	—	100
Vinyltriethoxy silane	2	—	220
Milled newsprint (40%) + polypropylene[f]			
Neat polypropylene	0	16 ± 2	730 ± 20
None	0	33 ± 2	105 ± 10
Orevac CA100	1.0	24 ± 1	115 ± 20
	2.0	24 ± 1	125 ± 15
	3.0	27 ± 2	190 ± 25
Wood flour (60%) + HDPE[g]			
None	0	0.7	—
Polybond 3029	1	0.7	—
	2	1.0	—
	4	1.3	—

[a] Ref. [17]. [b] Ref. [19]. [c] Ref. [21]. [d] Ref. [30]. [e] Ref. [11]. [f] Ref. [7].
[g] Reversed notch Izod, compression molded samples. Bill Sigworth, personal communication.

the wood fiber and the resin, hence, an improvement in compatibility. How it happens, in terms of specific mechanisms, remains to be solved. It well might be that in different systems (different resin, fiber, coupling agent, temperature, other conditions), a type of interaction affecting adhesion and compatibility is different, and covalent interactions might not be prevalent at all.

Near infrared spectroscopy (NIRS) of WPC material with maleated polyethylene and polypropylene showed "no obvious differences with coupling agent addition" [31]. The authors noted that using NIRS, it was "not possible to identify coupling agent in formulation."

FTIR microscopy was applied to study 10-μm thin slices from a polypropylene-based WPC with and without a maleated polypropylene [32]. Again, it was concluded that there was no distinguishable differences in the infrared spectra taken at the matrix of both composites, and there were no absorption bands present, either in the 1750–1700 cm^{-1} (assigned to ester groups) or at 1785 and 1715 cm^{-1} (characteristic absorption bands of maleated polypropylene). The authors concluded that while formation of ester links between hydroxyl groups in the wood fiber and anhydride groups of maleated polypropylene may have occurred, it cannot be claimed with accuracy because hydrogen bonds should also be considered.

A recent investigation into the possible ester bond formation between a maleated polypropylene and wood fiber when the reaction is performed in a molten polypropylene matrix came again to an ambiguous conclusion [28]. The authors have noticed that due to overlapping infrared absorption spectra for all the initial ingredients and (presumably) final products of the coupling reaction, the results are "not a clear proof of any kind," and "from this lack of evidence no statement can be made."

Rheological Studies

Coupling agents should influence the rheological behavior of hot melts of WPCs in a manner, opposite to that for lubricants and plastisizers. Generally, it is expected that coupling between wood fiber and a maleated polyolefin would increase the shear viscosity, while lubricants would decrease the viscosity. In reality, quantitative rheological description of coupling agents effects on molten WPC is complicated due to slip at the wall and entrance pressure drop. Besides, specific effects of coupling agents on rheology of a WPC depend on molecular weight of the polymer matrix, molecular weight distribution, and molecular characteristics of the coupling agent itself (Velichko Hristov, personal communication). At any rate, the rheological data obtained up to date were unable to provide information whether interaction between the coupling agent and wood fiber in the polymer matrix were covalent bonds, ionic interactions, hydrogen bonds, and so on.

Figures 5.3 and 5.4 provide examples illustrating the complexity of rheological behavior of WPC with and without coupling agents (Velichko Hristov, personal communications).

The upper panel of Figure 5.3 shows that Polybond 3109, having the lowest viscosity (MFI 30, see Table 5.1) among the three coupling agents and acting within the high-viscosity PE matrix (MFI 0.35), gives the best coupling and, hence, the highest

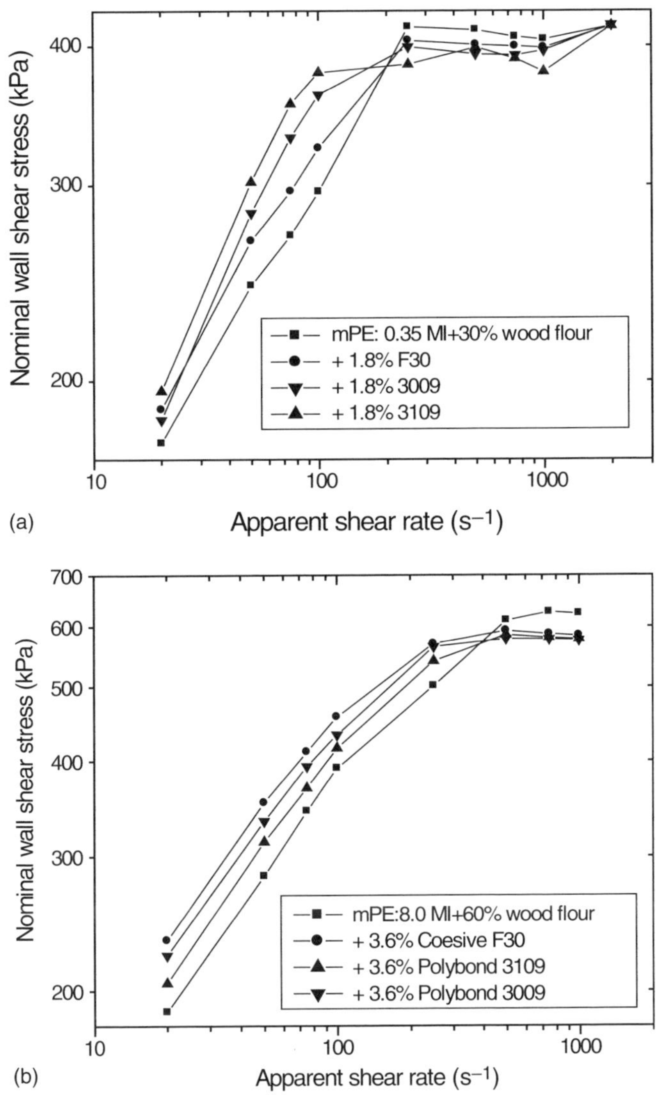

Figure 5.3 Effects of Polybond 3009, Polybond 3109, and Coesive F30 on flow curves of wood fiber-filled polyethylene-based composite. Left panel: MFI of polyethylene 0.35/10 min, 30% wood flour, coupling agents of 1.8% w/w. Right panel: MFI of polyethylene 8.0/10 min, 60% wood flour, coupling agents of 3.6% w/w (courtesy Velichko Hristov).

shear stress in the system. The higher viscosity Coesive (MFI 1–2) shows the lowest effect on shear stress among the three coupling agents.

The lower panel of Figure 5.3, with a highly filled PE (60% w/w), shows the opposite pattern. Here Polybond 3109, acting within the low-viscosity matrix (MFI 8.0), gives the worst coupling among the three agents. The higher viscosity Coesive shows the highest effect on shear stress, hence, the best coupling reaction.

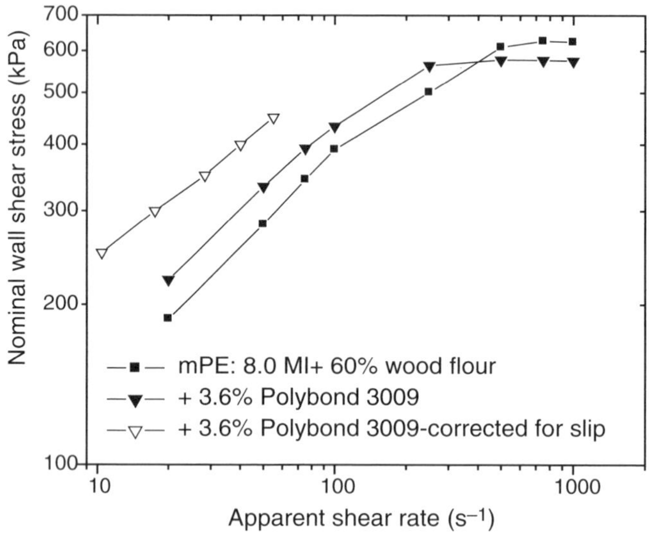

Figure 5.4 Effects of Polybond 3009 (3.6% w/w) on flow curves of wood fiber (60% w/w)-filled polyethylene-based composite. MFI of polyethylene is 8.0/10 min. The lower curve for Polybond 3009 is the same as that in Figure 5.3 (left-hand panel), the upper curve is the same but is corrected for wall slip (courtesy Velichko Hristov).

It is rather striking that, as it is seen from Figure 5.3, at high shear rates (above 200 s^{-1}) all three coupling agents produce shear stress lower than that of the control, uncoupled WPC. At the first thought, this is nonsense. However, it just reflects a complex interplay of various factors in the system, and the wall slip plays a significant role there. A correction for wall slip shows how important is the factor (Fig. 5.4).

If to add an observation that a WPC polyethylene matrix with a more narrow molecular weight distribution gives a higher shear stress (hence, better coupling) in the presence of maleated polyolefins compared to that for a broader molecular weight distribution PE as the matrix (Velichko Hristov, personal communication), it becomes clear that quantitative effects of coupling agents depend on many factors, far from to be understood and even less controlled in industrial manufacturing of WPCs with coupling agents. As it was said above, rheological studies provide plenty of information on flowability of WPC materials with and without coupling agents; however, this information has not been translated as yet into "mechanistic" descriptions on how coupling agents interact with cellulose fiber in filled plastics.

Kinetic Studies

The question regarding the mechanism(s) of the action of maleated coupling agents was approached via studying the crystallization kinetics in wood-fiber-filled polypropylene in the presence and the absence of a maleated polypropylene copolymer

(Honeywell A-C 950P) [33]. The introduction of cellulose fiber into a polypropylene melt provides a surface upon which crystals nucleate much faster compared to that in the bulk polymer. However, it turns out that the addition of the maleated polyolefin into the system leads to increased nucleation both on the surface of the wood and in the polymer matrix, as was shown by polarized microscopy. As a result, no differences were found in the kinetics of the crystal formation nucleated on the wood surface and in the bulk polymer. Besides, FTIR Spectroscopy of the interphase did not confirm covalent bonding of the wood with the maleated polypropylene because spectra of the two components overlapped, as it was described above on other similar examples. The authors [33] noted, in unison with many other investigators, that it is still unclear if improvements in mechanical strength of WPC with the addition of maleated polyolefins are the result of improved wood–plastic interaction (in general terms, including covalent, ionic, hydrogen bonds, etc., or all or some of the above), better dispersion of the wood fiber, or changes in the thermoplastic morphology, meaning (but not only) patterns of the polymer crystallization from hot melt.

Other Considerations

Clearly, if the coupling agents do not form covalent bonds with the wood fiber, but result in a noticeable increase in the WPC strength, there should be another mechanism for increase. Many authors use more careful language to describe an interaction of maleated polymers with wood fiber in WPC, such as “this may be due to the increase in compatibilization resulting from more contact between hydrophilic fiber and hydrophobic plastics” [34]. Or “the compatibilizer (Epolene G-3003) was used to minimize the incompatibility between the wood fibers and the polypropylene matrix. Utilizing Epolene G-3003 improved the fiber-matrix adhesion, resulting in a significant improvement in composite performance” [35]. Or “formation of an adhesive bridge between the wood fiber and the polymer matrix” [5]. Or “by greater wettability of wood fiber by the polymer matrix, and improved dispersion and orientation of wood fiber in the polymer matrix” [5]. Or “reduces the formation of agglomerates, and fibers are dispersed more uniformly” [5]. In any case, it can be concluded that the increase in the number of maleic anhydride groups on the coupling polymeric agent and a longer hydrocarbon chain of the coupling agent, which can cocrystallize more effectively with the polymer matrix, apparently lead to improved mechanical properties and lower water absorption of WPC.

The approach to covalently modifying cellulose fiber with a coupling agent, and only then mixing the modified fiber with a polymer matrix was originated in the 1960s [36–38]. It was used mainly in academic studies of nature of the bonding [39–43]. It was found, for instance, that when maleated polypropylenes covalently bound to wood fiber, grafted polymeric chains likely stretch away from the cellulose fiber surface, giving rise to a brushlike configuration in a polypropylene melt [41]. This provides a good interaction of fiber with the polymer matrix, and the interaction increases with increasing molecular weight of the grafted polypropylene. In turn, this improves adhesion between treated fibers and the plastic matrix (as was shown by peel testing), enhances stress transfer, and increases interphase thickness.

Some contributors to the field believe that maleated polymers do not form covalent bonds with hydroxyl groups of the filler but rather form ionic bonds via interactions with ionic groups on its surface. It is known, for example, that cellulose fiber contains many cationic and anionic groups on its surface.

The value of coupling agents in WPC is not only in their interactions, covalent, or otherwise with cellulose/wood fiber, but it is also in their interactions with inorganic fillers and flame retardants, such as aluminum trihydroxide (ATH), magnesium hydrohide, and calcium carbonate. The respective ionic interactions are claimed to be with maleated polymers, silanes, and other coupling agents grafted onto a polymer via double bonds or by crosslinking reactions using, for example, peroxide-cured rubbers [44].

However, adding of inorganic materials into a WPC formulation with a coupling agent can complicate an outcome. For example, addition of nanoclay to WPC (50% wood flour and 50% polypropylene) in the presence of Polybond 3200 has systematically decreased both tensile and flexural strength of the final material, for amounts of the nanoclay of 0, 2, 4, and 10%. Effect of tensile and flexural modulus was more complex. The authors concluded that the nanoparticles disrupt adhesion between wood flour and the polymer matrix [24].

Polypropylene-based maleated coupling agents can be used in HDPE-based WPCs, particularly if a small amount of polypropylene is added to the system. Table 5.11 compares two HDPE-based composites, containing rice hulls and Biodac® as cellulosic/mineral fillers, with Polybond 3009 (HDPE-based) and Polybond 3200 (polypropylene-based). As one can see, in all the cases both flexural strength and modulus are significantly increased.

TABLE 5.11 Effect of Polybonds 3009 (HDPE-based) and Polybond 3200 (polypropylene based) on flexural strength and flexural modulus of HDPE-based composites

HDPE (% w/w)	Rice hulls (% w/w)	Biodac (% w/w)	Polybond (% w/w)	Flexural strength (psi)	Flexural modulus (psi)
Polybond 3009 (no polypropylene in the system)					
35	4.5	60	0	2,130 ± 210	216,000 ± 2,000
35	4.0	60	0.5	2,090 ± 90	284,000 ± 26,000
35	3.5	60	1.0	2,345 ± 135	285,000 ± 21,000
35	1.5	60	3.0	2,720 ± 50	274,000 ± 11,000
35	0	59.5	5.0	2,850 ± 150	293,000 ± 16,000
Polybond 3200, polypropylene in the system 4.5% (except in the top row)					
25	24.5	50	0(PP = 0)	2,280 ± 120	240,000 ± 19,000
20	25	50	0	1,990 ± 50	229,000 ± 22,000
20	23	50	2.0	2,820 ± 130	290,000 ± 1,000
20	20	50	5.0	3,430 ± 160	337,000 ± 5,500

0.5% Struktol was used as a lubricant.

EFFECT OF COUPLING AGENTS ON WPC PROPERTIES: A SUMMARY

The following material is given and should be considered with a full understanding that scale of effects can vary due to multiple factors. The respective data have not been obtained, as a rule, under "optimum manufacturing conditions." The importance of this statement cannot be overestimated. For example, few folks appreciate an importance of the residence time in the extruder, which is directly related to the reaction time available for the anhydride groups in maleated polyolefins to react with cellulose. It is best to have this reaction time for at least one minute, or a little longer, if possible, to allow for the anhydride reaction to occur. If the time is too short for the reaction to proceed or the manufacturing conditions are not optimized, the coupling agent can still work as a compatibilizer, but much less effectively compared with that when covalent bonds are formed between the coupling agent and wood fiber.

What is an expected effect that a coupling agent can provide under the best scenario? In a very simplified case, when, say, 50% plastic–50% wood fiber are *ideally* blended into the WPC, and wood fiber is oriented along the flow, that is longitudinally, the flexural strength would be equal to a symmetrical superposition of the flexural strength of the matrix and the fiber. If we take flex strength for HDPE as 1400 psi (Chapter 2) and for wood fiber as 20,000 psi (Chapter 7), the ideal blend of the two would have flexural strength of 10,700 psi. In reality, it is of 1500–4400 psi for commercial wood–HDPE composite deck boards, up to 5000 psi for laboratory WPC, obtained at carefully controlled conditions, and up to 9000 psi, obtained in laboratory conditions and in the presence of coupling agents. At the lower end of this range is Trex boards, which, according to the manufacturer's data, has a flexural strength of 1423 psi. On the author's data, Trex boards have flex strength of 1900–2200 psi (Table 7.13). At the highest end of this range are wood-flour-filled HDPE (pine, 61–63%), flexural strength of which was 4670 ± 90 psi (without coupling agents) and 9100 ± 150 psi (in the presence of 3% Polybond 3029) (Jonas Burke, Ferro Corporation, private communication). Hence, in an ideal (and overly simplified) case, flexural strength can be increased—by introducing right coupling agent at right conditions and in the right amount—by about 240–700% for industrial WPC, and 230% for WPC, obtained in laboratory conditions. As in some cases some maleated polyolefins reportedly increased flexural strength by 100–200%, they did a good job.

Similarly, if we take flexural modulus for HDPE as 150,000 psi and for wood fiber as 1,500,000 psi (Chapters 2 and 7), the resulting flex modulus of a 50%–50% WPC would be 825,000 psi. In reality for industrial WPCs, exemplified again with Trex, it is 175,000 psi, which is about five times less. For best laboratory WPCs, flexural modulus is close to 700,000 psi in the absence of coupling agents (696,000 ± 30,000 psi and 717,000 ± 33,000 psi for the above-mentioned wood-flour-filled HDPE in the presence of two different lubricants) and slightly higher in the presence of coupling agents (727,000 ± 25,000 psi and 773,000 ± 13,000 psi, respectively, in the presence of 3% Polybond 3029 (Jonas Burke, Ferro Corporation, private communication). As a result, known maleated polyolefins improve flexural modulus of industrial WPC

TABLE 5.12 Effect of maleated polyolefins on flexural and tensile modulus of WPC (the data were assembled based on Tables 5.7 and 5.11)

Coupling agent	Increase of flexural modulus (%)	Increase of tensile modulus (%)
Polybonds	0–20	0–10
Integrates	10–40	N/A
Fusabonds	0–40	N/A
Epolenes	0–10	N/A
Exxelor	5–8	N/A
Orevac	1–2	2–3

by only 40% at best, and that of the best laboratory WPC by only a few percent (see Table 5.12) because there is not much room for improvement.

Apparently, some other factors also do not allow the flex modulus of commercial WPC deck boards to reach the state of the "ideal" blend. Such a factor can be, for instance, a porosity (void) in a typical WPC material. It probably would be a good idea to increase flexural modulus using a coupling agent, starting from a good, compacted material having a density (specific gravity) close to a maximum one. For example, Trex has an actual density of 0.91–0.95 g/cm^3, while its "theoretical" density, for a compacted material should be about 1.10 g/cm^3 (Chapter 6). In other words, coupling agents do not realize their potential when other factors hold low flexural modulus.

As the improvement of the flexural strength often does not match the improvement of the flexural modulus, one should make his priorities straight, regarding what particular property should be improved by adding a coupling agent, and how much. Coupling agents should be considered as a tool to further improve properties of an already very good board. It is not very productive to employ them when the board is clearly substandard in terms of density, water absorption, strength, and modulus. Other means should be considered first. After all, coupling agents are not cheap. At \$1.50/lb and 3% level, plus the same (at least) for a special, nonmetal lubricant, the material would cost 9 cents more per pound of the formulation. For many WPC deck boards, it means 30% higher material expenses.

Effect on Flexural and Tensile Modulus

The same coupling agents, both maleated polyolefins and silanes, affect flexural modulus of WPC quite differently, from decreasing through practically no effect, to increase up to 20–40% (Tables 5.7, 5.11, and 5.12).

Of the 32 cases of maleated polyolefins effects on WPC flexural modulus listed in Tables 5.7 and 5.11, in five cases there was a decrease in flex modulus (Polybond, Epolenes and a Fusabond), in 19 cases (Epolenes, Polybonds, Integrates, Fusabonds, Exxelors, Orevac) there was no change or an apparent increase within only 15%, that is within a common error margin (at some intermediate concentrations of coupling agents in their tested range, they decreased flex modulus, which also indicate that there were practically no overall effect), and in eight cases an increase in flex modulus (between 15 and 46%) was observed. These increases were caused by Fusabond WPC-576D at or above 2% w/w, Integrates, and Polybonds 3009, 3029 and 3200.

With respect to tensile modulus, of 11 cases in Table 5.7, five maleated polyolefins show decrease in the property (an Epolene, maleated polypropylenes and a Polybond), in five cases there was practically no effect (within 10% of the control), and in only one case an increase by 24% was observed (Polybond 3029).

Table 5.7 lists only three examples of effects of silanes on flexural modulus of WPC. In two cases a decrease of the property was observed, and in one case the increase was of 19 ± 20%, which was within the error margin.

It is reasonable to suggest that in those cases when there was a decrease in flexural or tensile modulus, or there were no any significant effects (increase within 10–15% of control), no appreciable amount of covalent bonds between the coupling agent and cellulose fibers was formed. The coupling agent was either functionally removed from the system by a metal-containing lubricant (which was not a case in most of data in Table 5.7) or covalent bonds were not formed for other reasons, such as unfavorable conditions, unfavorable mutual orientation of maleic anhydride and cellulose hydroxyl groups in hot melt, or other reasons. The coupling agents still could serve as compatibilizers, dispersants, lubricants, or so on, and hence, lead to a better adhesion between fiber and plastic and increase the flex and tensile strength and modulus of the resulting WPC. However, this effect would be due to a better dispersion of the cellulose filler and not because of the covalent crosslinking effect.

Effect on Flexural and Tensile Strength

A similar consideration (see above section) can be applied to flexural and tensile strength. The same coupling agents, both maleated polyolefins and silanes, affect flexural and tensile strength of WPC quite differently, from decreasing through practically no effect to increase up to 100–160% (Tables 5.9, 5.11, and 5.13).

Of 29 cases of maleated polyolefins effects on WPC flexural strength, listed in Table 5.9 and 5.11, in five cases the apparent increase was within 20%, which was within a common error margin, mainly for low concentrations of Polybonds and Fusabond (less than 2% w/w), and in 24 cases an increase of flex strength was observed in the range of 30–230% (Epolenes, Polybonds, Fusabonds, Integrates, and Orevac).

TABLE 5.13 Effect of maleated polyolefins on flexural and tensile strength of WPC (the data were assembled based on Table 5.9 and 5.11)

Coupling agent	Increase of flexural strength (%)	Increase of tensile strength (%)
Polybonds	30–110	50–150
Integrates	30–120	60–160
Fusabonds	10–230	N/A
Epolenes	20–30	30–70
Orevac	15–40	30–60

There was no correlation in the increase of flexural strength and flexural modulus in the presence of the same coupling agent. For example:

- A mediocre increase in flex modulus (12–17%) in the presence of Integrates gave concurrently the highest increase in flexural strength (100–123%).
- Polybond 3200 decreased, did not change, or slightly increased both flexural and tensile moduli, but resulted in 40–70% increase in flexural strength and 50–90% increase in tensile strength.
- Polybond 3029 only slightly increased flexural modulus (by 4–8%), but increased flexural strength of the same materials by 90–95%.
- Vinyltriethoxy silane decreased (by 40%) or practically did not increase flexural modulus, but gave 66–86% increase in flexural strength.
- Epolenes in most cases showed decrease or only slight increase of flexural and tensile modulus (Table 5.7), but increased the flexural and tensile strength by 20–40% (Table 5.9).
- Some maleated polypropylenes showed significant decrease (by 40–75%) of tensile modulus, but resulted in 100–150% increase in tensile strength.

These data show that increases in flexural and tensile strength are not necessarily the results of covalent bonding of coupling agents to wood fiber. Otherwise one should see a concurrent increase of both strength and modulus.

Effect on Water Absorption

Coupling agents often improve density of WPC. For example, with an increase in the amount of Fusabond® WPC-576D in an HDPE-based composite, containing 60% of wood flour w/w, from 0 to 3% w/w, density of the composite increases from 70.4 lb/ft^3 (1.13 g/cm^3) to 74.7 lb/ft^3 (1.20 g/cm^3), respectively [4]. This in turn leads to decrease of water absorption by the composite materials, which is almost invariably observed at the introduction of coupling agents (Table 5.14).

LUBRICANTS, COMPATIBLE AND NOT COMPATIBLE WITH COUPLING AGENT

As soon as the maleated coupling agents were introduced into the WPC, it was noticed that their effect often significantly depends on lubricants employed in the same system. The most striking was a conflicting effect between maleated polyolefins and metal stearate lubricants (Tables 5.15 and 5.16)

Table 5.15 shows that zinc stearate invariably decreases both strength and stiffness of WPC compared to the nonmetal lubricant. However, if without coupling agents this decrease was insignificant and within 1% for flex strength and modulus (certainly within error margin), in the presence of each of four of the coupling agents, this decrease reached 70–90% for flex strength and 10–20% for flex modulus, when

TABLE 5.14 Effect of coupling agents on water absorption of WPCs

Type of coupling agents	Amount of coupling agents (% w/w)	Immersion underwater (days)	Water absorption (%)
HDPE-based composite material (60% wood flour, 5% talc, 3% lubricant)[a]			
None	0	55	10
Fusabond® E MB-226D	0.5–1.0		9
Fusabond® WPC-576D	0.5–1.0		8
Polypropylene-based composite material (50% sisal)[b]			
None	0	5	11
Maleated polypropylene	1.0		6
	4.0		5
None	0	15	13
Maleated polypropylene	1.0		10
	4.0		8
HDPE-based composite material (40% wood flour)[c]			
None	0	30	10
Polybond 3009	1.0		6
	2.5		4
None	0	90	12
Polybond 3009	1.0		9
	2.5		6.5
HDPE-based composite material (rice hulls, bleached wood fiber and minerals)[d]			
None	0	1	2.4
Polybond 3029	1.5		1.4–1.8
None	0	4	3.6
Polybond 3029	1.5		2.2–2.7
None	0	7	4.2
Polybond 3029	1.5		2.7–3.3
HDPE-based composite material (60% wood flour, 40% HDPE)[e]			
None	0	50	18
Integrate NE534-003	2.0		4
None	0	225	20
Integrate NE534-003	2.0		7
Polypropylene-based composite material (40% maple fiber)[f]			
None	0	1	1.0
Exxelor PO1020	3		0.6
Polypropylene-based composite material (50% maple fiber)[f]			
None	0	1	2.2
Exxelor PO1020	3		1.2
Polypropylene-based composite material (60% maple fiber)[f]			
None	0	1	5.9
Exxelor PO1020	1		2.3
	3		1.9

TABLE 5.14 **(*Continued*)**

Type of coupling agents	Amount of coupling agents (% w/w)	Immersion underwater (days)	Water absorption (%)
Polypropylene-based composite material (40% wood fiber)[g]			
None(cascade mixer)	0	225	9.1
MAPP (high MW, low MA)	2		7.2
MAPP (low MW, high MA)	2		6.6
None (plastagglomeration)	0	225	8.4
MAPP (high MW, low MA)	2		6.8
MAPP (low MW, high MA)	2		6.0
None (twin-screw extruder)	0	225	7.6
MAPP (high MW, low MA)	2		5.2
MAPP (low MW, high MA)	2		5.1
Wood flour (61%, pine) + HDPE, nonmetal lubricants[h]			
None (lube 1)	0	14	22.6
Polybond 3029	3		10.8
None (lube 2)	0	14	22.2
Polybond 3029	3		9.3

[a] At less than 55 days water immersion time, the effect was proportionally lower [4].

[b] At less water immersion time (up to 9 days), the effect of the coupling agent was higher, and then gradually decreased to practically no difference after 30 days of immersion in water [18].

[c] At 4% of the coupling agent, water absorption slightly increased. The effect of the coupling agent practically did not depend on water immersion time period [20].

[d] Water absorption in the presence of coupling agents are shown in the presence of different lubricants. The lowest water absorption was with the same lubricants that provided the best increase in flexural strength and modulus, though the latter effects were only 19 and 23%, respectively (data by Ferro Corporation).

[e] Ref. [25].

[f] Ref. [6].

[g] Ref. [8].

[h] Jonas Burke (Ferro), private communication.

comparing two kinds of the lubricants, nonmetal and metal-containing one, and by 100–130% and 15–20%, respectively, when comparing no lubricant and metal stearate.

Nonmetal lubricants, however, were much less antagonistic, though they also decreased the mechanical properties of the WPC in the presence of the coupling agents by 10–20% for flexural strength and 2–9% for flex modulus.

Essentially, addition of 2% of zinc stearate to the WPC negated all gains in strength and stiffness obtained in the presence of coupling agents. In other words, both flexural strength and modulus were coming back to their values obtained without any of the coupling agents. Hence, zinc stearate kills maleated polymers in their

TABLE 5.15 Effect of ethylene bis stearamide (EBS)/zinc stearate on flexural strength and flexural modulus of a WPC (60% wood flour, 40% HDPE, integrate coupling agents)

Type of coupling agents	Type and amount of lubricants, (%)	Flexural strength (psi)	Flexural modulus (psi)
None	None	4000	520,000
	Nonmetal lube	3400	515,000
	EBS/zinc stearate	3400	510,000
Integrate NE556-004	None	8400	580,000
	Nonmetal lube	7500	560,000
	EBS/zinc stearate	4200	510,000
Integrate NE558-004	None	8900	610,000
	Nonmetal lube	7700	590,000
	EBS/zinc stearate	4100	520,000
Integrate NE433-003	None	8000	600,000
	Nonmetal lube	6600	550,000
	EBS/zinc stearate	3900	490,000
Integrate NE534-003	None	8600	590,000
	Nonmetal lube	7200	580,000
	EBS/zinc stearate	3800	480,000

All coupling agents and both lubricants were at 2% w/w [25].

effects on WPC. The same phenomena were observed in tensile strength and Izod impact strength of the materials [25].

Similar in kind data are shown in Table 5.16. A nonmetal lubricant (as a part of two different additive packages) noticeably (by 15–25%) increased both flexural strength and flexural modulus of the WPC in the absence of a coupling agent. The coupling

TABLE 5.16 Effect of metal-base (zinc stearate and SXT 2000) and nonmetallic lubricant (ST 3100) in formulated packages on flexural strength and flexural modulus of a WPC (61–63% wood flour, 34% HDPE, polybond 3029 as a coupling agent)

Amount of Polybond 3029	Type and amount of lubricants, (%)	Flexural strength, (psi)	Flexural modulus, (psi)
None	Zinc stearate	3,740 ± 150	575,000 ± 35,000
	SXT 2000/ RC-570	3,880 ± 100	611,000 ± 33,000
	SXT 3100/RC-544	4,670 ± 100	696,000 ± 30,000
	SXT 3100/RC-571	4,650 ± 85	717,000 ± 33,000
3%	Zinc stearate	3,350 ± 200	393,000 ± 19,000
	SXT 2000/ RC-570	3,600 ± 200	484,000 ± 34,000
	SXT 3100/RC-544	9,100 ± 150	727,000 ± 25,000
	SXT 3100/RC-571	8,820 ± 160	773,000 ± 13,000

All the lubricants and the coupling agent were at 3% w/w (Jonas Burke, Ferro Corporation, private communication.

agent, a maleated polyolefin, severely conflicted with zinc stearate (flexural modulus dropped by 46% and flexural strength—by 12%), but along with nonmetal lubricants almost doubled flexural strength and increased flexural modulus by 4–8%.

This antagonistic effect of zinc stearate on maleated coupling agents was explained by using infrared spectroscopy. It turned out that maleic acid, formed from grafted maleic anhydride in the course of the compounding in the presence of moisture, reacts with zinc stearate. This reaction is more thermodynamically favorable compared to the coupling of the maleated polymer to cellulose fiber (if the latter reaction takes place in any significant extent), and essentially removes the coupling agent from the resulting WPC [25]. Hence, mechanical (and other) properties of the WPC essentially come back to the material without coupling agents.

Another consequence of the suggested chemical interaction between maleated polymers and zinc stearate is that the WPC material starts to decompose at lower temperatures (at about 300°C compared to 350°C) compared to that compounded without the metal-containing lubricant [25]. However, knowing that metal-containing compounds serve as catalysts of oxidation of plastics (see Chapter 15), the earlier plastic degradation could have occurred without any coupling agents, just due to the presence of zinc stearate. A simple test in that case would be to increase an amount of an antioxidant in the system, and the earlier decomposition temperature point would have predictably returned back to normal. Hence, the above experiment [25] cannot be considered a proof of the chemical interaction of zinc stearate with the maleated polymer albeit does not contradict it.

The effect of conflicting of the maleated polyolefins with zinc stearate and other metal-containing stearates is commonly known in the industry. Therefore, a number of companies have developed nonmetal lubricants, such as Ferro Corporation' RC-553, RC-571, RC-572, RC-576, SXT 3100 (see Table 5.16; Ferro's SXT 2000 is a blend of metal stearates with other nonmetal lubricants [3]), Struktol' TPW-113 (Struktol's TPW 104 contains zinc stearate), Lonza' Glucolube WP-2200 (a new proprietary amide lubricant that contains no metal stearates [12]).

Nonmetal lubricants often do not improve properties of the resulting WPC containing maleated polyolefins (data in Table 5.16 provide a good exception regarding flexural strength), though they are almost invariably better than the same WPC obtained with metal-containing lubricants. For example, data in Tables 5.7 and 5.9 show that variation of nonmetal lubricants (each of 4.7% w/w) along with Polybond 3029 (1.5% w/w) resulted in a variation of flexural strength within 15–25% over a baseline (no coupling agent), and flexural modulus within 12–25%. With nonmetal lubricants (each of 5.0%) and Fusabond WPC-576D (2%) the variations were within 12–26% for flexural strength and from −11% to +38% for flexural modulus. When the Fusabond was replaced with Integrate NP-542 (2%), the variations were within 29–33% for flexural strength and within 20–40% for flexural modulus. One can see that nonmetal lubricants in these examples show more variations in flex modulus compared to flex strength. However, this can be only an apparent effect because flex modulus is typically measured with a higher error margin (higher variations) compared to that with flex strength.

At any rate, effects of lubricants with or without coupling agents in WPC are still largely unpredictable and mainly empirical. Nonmetal lubricants do not increase a chance for covalent bonds formation between coupling agents and wood fiber; they just do not result in a functional damage of maleated polyolefins compared to that by metal-containing lubricants.

REFERENCES

1. W.S. Gutowski. Physical model of interface and interphase performance in composite materials and bonded polymers. In: *Second International Wood and Natural Fibre Composites Symposium*, Kassel, Germany, June 28–29, 1999, pp. 6–1–6–11.
2. C.A. Correa, C.A. Razzino, and E. Hage. DMTA analysis of interface adhesion in wood-plastic composites. In: *Eighth International Conference on Woodfiber-Plastic Composites (and Other Natural Fibers)*, Madison, WI, 2005.
3. J. Markarian. Wood-plastic composites: current trends in materials and processing. In: *Plastics Additives & Compounding*, September–October 2005, pp. 20–26.
4. D. Dean. Influence of ethylene-anhydride copolymer coupling agents on the mechanical properties of HDPE based wood polymer composites. In: *Progress in Wood Fibre Plastic Composites 2006 International Conference*, Toronto, Canada, May 1–2, 2006.
5. M. Kazayawoko and J.J. Balatinecz. Adhesion mechanisms in woodfiber-polypropylene composites. In: *Fourth International Conference on Woodfiber-Plastic Composites*, Madison, WI, May 12–14, 1997, pp. 81–93.
6. J.-R. Schauder. Exxelor coupling agents for wood fibers reinforced composites. In: *Wood fibre Polymer Composites Symposium. Application and Trends*, Bordeaux, France, March 27–28, 2003.
7. M. Sain, S. Law, F. Suhara, and A. Bouiloux. Interface modification and stiffness correlation of natural fibre—polyolefin composite products. In: *Wood fibre Polymer Composites Symposium. Application and Trends*, Bordeaux, France, March 27–28, 2003.
8. A.K. Bledzki, V.E. Sperber, K. Specht, M.A. Letman and A. Viksne. Effect of defined waxes and coupling agents on moisture behavior of injection molded woodfiber-reinforced PP composites. In: *Eighth International Conference on Woodfiber-Plastic Composites (and Other Natural Fibers)*, Madison, WI, May 23–25, 2005.
9. J.D. Umpleby. Material choice for wood-plastic composites. In: *Wood-Plastic Composites—A Sustainable Future. International Conference*, Vienna, Austria, May 14–16, 2002.
10. J. Gassan and A.K. Bledzki. Dynamic-mechanical properties of natural fiber-reinforced plastics: the effect of coupling agents. In: *Fourth International Conference on Woodfiber-Plastic Composites,* Madison, WI, May 12–14, 1997, pp. 76–80.
11. M. Bengtsson, N.M. Stark, and K. Oksman. Crosslinked wood-thermoplastic composites—profile extrusion and properties. In: *Progress in Wood Fibre Plastic Composites 2006 International Conference,* Toronto, Canada, May 1–2, 2006.
12. L. Manolis Sherman. Wood-filled plastics—they need the right additives for strength, good looks & long life. *Plastics Technology*, July 2004.

13. L. Manolis Sherman. Close-up on technology—additives loads of new additives & colorants to debut at K 2004. *Plastics Technology*, September 2004.

14. C. Zhang, K. Li, and J. Simonsen. Improvement of interfacial adhesion between wood and polypropylene in wood-polypropylene composites. *J. Adhesion Sci. Technol.*, 2004, **18**(14), 1603–1612.

15. C. Zhang, K. Li, and J. Simonsen. A novel wood-binding domain of a wood-coupling agent: development and characterization. *J. Appl. Polym. Sci.*, 2003, **89**, 1078–1084.

16. Laurent M. Matuana. Resistance of HDPE/wood-flour composites to cyclic freezing and thawing exposures. In: *Eighth International Conference on Woodfiber-Plastic Composites*, Madison, WI, May 23–25, 2005.

17. D. Feng and A.R. Sanadi. Effect of compatibilizer on the structure-property relationships of kenaf fiber-polypropylene composites. In: *Fourth International Conference on Woodfiber-Plastic Composites*, Madison, WI, May 12–14, 1997, pp 157–160.

18. A.R. Sanadi and D.F. Caufield. Interphase effects on the mechanical and physical behavior of natural fiber composites. In: *Second International Wood and Natural Fibre Composites Symposium*, Kassel, Germany, June 28–29, 1999, pp. 9.1–9.18.

19. K. Oksman and C. Clemons. Effect of elastomers and coupling agent on impact performance of wood flour-filled polypropylene. In: *Fourth International Conference on Woodfiber-Plastic Composites*, Madison, WI, May 12–14, 1997, pp. 144–155.

20. B. Sigworth, L. Walp, R. Bacaloglu, and P. Kleinlauth. The role of additives in formulating WPC. In: *The Role of Additives in Formulating Wood-Plastic Composites, The Global Outlook for Natural Fiber & Wood Composites 2003*, New Orleans, LA, December 3–5.

21. B. Sigworth. Additives for wood-filled polyolefins. Coupling agents. In: *Progress in Woodfibre-Plastic Composites Conference 2002*, Toronto, Canada, May 23–24, 2002.

22. M. Bengtsson and K. Oksman. One-step process of silane crosslinked polyethylene/wood flour composites. In: *Eighth International Conference on Woodfiber-Plastic Composites*, Madison, WI, May 23–25, 2005.

23. D. Dean. Development of enhanced anhydride coupling agents for wood polymer composites. In: *WPC Conference 2004 "Realizing the Full Potential,"* Baltimore, MD, October 11–12.

24. S.-K. Yeh, A. Al-Mulla, and R.K. Gupta. Influence of the coupling agent on polypropylene/clay nanocomposite-based wood-plastic composites. In: *ANTEC 2005*, pp. 1290–1295.

25. M.G. Botros. Development of new generation coupling agents for wood-plastic composites. In: *The Global Outlook for Natural Fiber & Wood Composites 2003*, New Orleans, December 3–5, 2003.

26. T.J. Keener. New maleated polyethylene (MaPE) couplers. In: *Seventh International Conference on Woodfiber-Plastic Composites*, Madison, WI, May 19, 2003.

27. A.G. McDonald, L.W. Gallagher, S.T. Sundar, and M.P. Wolcott. The effect of wood surface modification and particle size on wood plastic composite performance. In: *Eighth International Conference on Woodfiber-Plastic Composites*, Madison, WI, May 23–25, 2005.

28. C. Burgstaller and W. Stadlbauer. Investigation of the effects of coupling agents in wood plastic composites. In: *Progress in Wood Fibre Plastic Composites 2006 International Conference,* Toronto, Canada, May 1–2, 2006.

29. J. Simonsen, R. Jacobson and R. Rowell. Wood fiber reinforcement of styrene-maleic anhydride copolymers. In: *Fourth International Conference on Woodfiber-Plastic Composites,* Madison, WI, May 12–14, 1997, pp. 215–220.
30. M.M. Stark and R.E. Rowlands. Effect of wood fiber characteristics on mechanical properties of wood/polypropylene composites. *Wood Fiber Sci.*, 2003, **35**(2), 167–174.
31. T.G. Rials, N.Labbe, S.S. Kelley, and D.J. Gardner. Near infrared spectroscopy assessment of wood-polyolefin extruded composites. In: *Eighth International Conference on Woodfiber-Plastic Composites (and Other Natural Fibers)*, Madison, WI, 2005.
32. V. Hristov and S. Vasileva. Dynamic mechanical and thermal properties of modified poly(propylene) wood fiber composites. *Macromol. Mater. Eng.*, 2003, **288**(10), 798–806.
33. D. Harper and M. Wolcott. Interaction between coupling agent and lubricants in wood-polypropylene composites. *Composites, Part A*, 2004, **35**, 385–394.
34. R.M. Rowell, S.E. Lange, and R.E. Jacobson. Effects of moisture on aspen-fiber/polypropylene composites. In: *Woodfibre-Plastic Composites Progress*, Toronto, Canada, May 23–24, 2002.
35. R. Kahraman, S. Abbasi, and B. Abu-Sharkh. Influence of Epolene G-3003 as a coupling agent on the mechanical behavior of palm fiber-polypropylene composites. *Int. J. Polym. Mater.*, 2005, **54**(6), 483–503.
36. D.J. Bridgeford, 1963, U.S. Pat. No. 3,083,118.
37. A.A. Gulina, R.M. Livshits, and Z.A. Rogovin. Synthesis of cellulose-polyacrylonitrile graft copolymers in the presence of the oxidation-reduction system cellulose Fe^{2+}–H_2O_2. II. Investigation of the effect of initiation conditions on the polymerization coefficient of polyacrylonitrile and the degree of conversion of cellulose. *Vysokomol. Soyed.*, 1965, **7**(9), 1529–1534.
38. U.S. Pat. No. 3,359,224 (1967). R.W. Faessinger and J.S. Conte.
39. H. Matsuda. Preparation and utilization of esterified woods bearing carboxyl groups. *Wood Sci. Technol.*, 1987, **21**(1), 75–88.
40. H. Matsuda. Preparation and properties of oligoesterified wood blocks based on anhydride and epoxide. *Wood Sci. Technol.*, 1992, **27**(1), 23–34.
41. J.M. Felix and P. Gatenholm. Formation of entanglements at brushlike interfaces in cellulose-polymer composites. *J. Appl. Polym. Sci.*, 1993, **50**(4), 699–708.
42. C.Q. Yang. Infrared spectroscopy studies of the effects of the catalyst on the ester cross-linking of cellulose by poly(carboxylic acids). *J. Appl. Polym. Sci.*, 1993, **50**(12), 2047–2053.
43. J.M. Felix, P. Gatenholm, and H.P. Schreiber. Controlled interactions in cellulose-polymer composites. 1: effect on mechanical properties. *Polym. Composites*, 1993, **14**(6), 449–457.
44. J. Schofield. New coupling agents enhance mechanical properties in filled polymers. In: *Plastics Additives & Compounding*, March/April 2005.

6

DENSITY (SPECIFIC GRAVITY) OF WOOD-PLASTIC COMPOSITES AND ITS EFFECT ON WPC PROPERTIES

INTRODUCTION

Importance of density (specific gravity) of wood-plastic composite (WPC) materials cannot be overestimated. By "density" we here mean not an absolute density of different WPCs, but a density of the same WPC material that can be lower compared to the highest possible density of the same WPC, determined by specific gravities of its ingredients.

Let us consider an example of, say, a Trex deck board. It consists of 50% w/w polyethylene (LDPE/LLDPE and/or HDPE) and 50% w/w of wood flour. Specific gravity of LDPE/LLDPE is 0.925 g/cm^3 and that of HDPE is 0.96 g/cm^3. Specific gravity of wood flour is 1.30 g/cm^3. These two ingredients define the density of the composite material at their "natural" compaction, which would be 1.08 g/cm^3 for LDPE/LLDPE-based Trex and 1.10 g/cm^3 for HDPE-based Trex (see the insert for calculations). Trex reported that the actual density is 0.91–0.95 g/cm^3 (Trex data). Hence, 14–21% of all Trex composite volume is taken by voids, porosity.

It is practically impossible to make industrial WPC boards without any porosity, hence, without any decrease in density compared to its "theoretical" value. Even traces of moisture in wood/cellulose fiber create steam at hot melt temperatures, hence, porosity. Plastic decomposition during processing produces volatile organic compound (VOC), hence, porosity. Wood extractives' decomposition produces VOC, hence, porosity. Wood fibers' lignin decomposition at plastic hot

Wood-Plastic Composites, by Anatole A. Klyosov

Calculation of density of Trex composite material. 100 g of the composite material contain 50 g of LDPE/LLDPE (density of 0.925 g/cm^3) or HDPE (density of 0.96 g/cm^3) and 50 g of wood flour (density of 1.30 g/cm^3). Each of these components takes the following volume: HDPE 50 g/0.96 g/cm^3 = 52.083 cm^3, (variant – LDPE/LLDPE 50 g/0.925 g/cm^3 = 54.054 cm^3), wood flour 50 g/1.30 g/cm^3 = 38.462 cm^3. Therefore, total volume of the 100 g of the composite will be 90.545 cm^3 (HDPE-based) or 92.516 cm^3 (LDPE-based). Hence, specific density of the composite is 100 g/90.545 cm^3 = 1.104 g/cm^3, or 100 g/92.516 cm^3 = 1.081 g/cm^3, respectively.

melt temperatures produce CO_2, hence, porosity. The faster the speed of the extrusion, the faster is the plastic decomposition, and higher the porosity. The very fact that any WPC deck board absorbs some water indicates the board porosity. Vented extruders are the best in terms of removing VOC, CO_2, and steam, decreasing porosity and increasing board density, and they bring the density close to the maximum one.

Calculation of density of more complex WPC materials, with the example of a three-component formulation. If we take 100 g of a composite material, containing, say, 50% w/w of HDPE (d = 0.96 g/cm^3), 30% of wood flour (d = 1.30 g/cm^3), and 20% of talc (d = 2.8 g/cm^3), each of these components take the following volume: HDPE 50 g/0.96 g/cm^3 = 52.083 cm^3, wood flour 30 g/1.30 g/cm^3 = 23.077 cm^3, talc 20 g/2.8 g/cm^3 = 7.143 cm^3.

Therefore, total volume of the 100 g of the composite will be 82.303 cm^3. Hence, specific density of the composite is 100 g/79.226 cm^3 = 1.22 g/cm^3.

The following example shows how even a small amount of VOC in the extruder can noticeably decrease density of the resulting WPC product. Suppose, only 0.25% of wood flour in WPC produces VOC due to decomposition at hot melt temperatures in the extruder. This is a really small figure, if to take into consideration that up to 30% of a lignified fiber can be converted to VOC as a result of heating, as shown in Chapter 3. If 400 lbs of lignified cellulose in the extruder release 1 lb of VOC (0.25% w/w), this amount would take 50 L as a gas volume (considering an average molecular weight of VOC about 200 Da (most of them are naphthalates, see Chapter 3) and that 1 mol of gas occupies 22.4 L at standard conditions; at hot melt temperatures the volume will be much higher, hence, the calculations are very conservative). At 50% w/w load of wood fiber into HDPE, 800 lbs of the composite material will be produced, with a specific gravity of 1.10 g/cm^3 and total volume of 330 L. Fifty liters of it, or about 15%, will be taken by gaseous VOC, and density will drop from 1.10 to 0.94 g/cm^3. If only 0.50% of wood flour is converted to VOC during the extrusion, density of the resulting WPC drops from 1.10 to 0.77 g/cm^3. This would be a great "foamed composite," except its pores would be open, irregular, and the material would be very weak and flexible.

Coupling agents often increase density of WPC. For example, with an increase of amounts of Fusabond® WPC-576D in an HDPE-based composite, containing 60% of wood flour w/w, from 0 to 3% w/w, density of the composite increases from 70.4 lb/ft^3 (1.13 g/cm^3) to 74.7 lb/ft^3 (1.20 g/cm^3), respectively [1]. This in turn leads to decrease of water absorption of the composite materials, which is almost invariably observed at the introduction of coupling agents.

The above data show that to keep density as high as possible for the given wood-plastic composition is very important for the quality of the WPC. However, density is still considered by many manufacturers of composite building materials as a factor determining a profile weight in terms of transportation expenses and convenience of handling during an installation of deck, and as a factor defining expenses for manufacturing and raw materials. However, density of a given composite material largely determines also its lifetime as shown in Chapter 15.

In this book we use terms "density" and "specific gravity" interchangeably. However, these two terms have a subtle but a principal difference. Density is measured in g/cm^3 (or, generally, in units of a ratio of weight to volume of the sample). Specific gravity is dimensionless because it is a ratio of weight to weight, that is, the weight of the sample to the weight of an equal volume of water at 4°C (39°F). At this temperature the density of water is 1.0 g/cm^3. Therefore, density and specific gravity have the same numeric value at 39°F. Specific gravity is also called "relative density."

Even in these two simple definitions of density and specific gravity there are some (insignificant for our purpose) assumptions. One is that instead of "weight" we should use the term "mass" because weight varies with the force of gravity and mass does not. However, for our purpose it does not matter. Second, we suppose to correct density and specific gravity data for temperature because at test temperatures water is commonly warmer than 39°F, and specific gravity will be slightly lower because the sample expands and its density slightly decreases. It also does not matter for our purpose because an error margin for density and specific gravity determinations is higher than those supposed corrections.

The water displacement test procedures (see below) gives specific gravity of the specimen because the procedure operates with a ratio of sample's weight in air to its weight in water. Hence, specific gravity is dimensionless. The float/sink test procedure (see below) gives density of the sample, measured in g/cm^3. Here is the difference: in the outer space, for example, the displacement procedure (weight to volume) would not work because it involves direct weighing of specimens, whereas density measuring procedures, including the float/sink procedure, would work the same way (well, if to solve some technical problems) as on the Earth.

In order to avoid comments to each figure of density or specific gravity, we will use both terms interchangeably and use units g/cm^3 with both of them.

TABLE 6.1 The effect of density (specific gravity) of GeoDeck composite pickets of the railing system on their flexural strength and modulus (stiffness)

Specific gravity (g/cm^3)	Flexural strength (psi)	Flexural modulus (psi)
1.03	1456	110,651
1.05	1650	130,500
1.08	1827	151,000
1.16	2270	192,500
1.24	2560	197,000

EFFECT OF DENSITY (SPECIFIC GRAVITY) OF WPC

Effect on Flexural Strength and Modulus

The decrease in density (increase in porosity) affects practically all important properties of WPC boards considered in this book. The lower the density, the lower the flexural strength (Table 6.1) and the flexural modulus (Tables 6.1 and 6.2). In Table 6.1, the density of 1.24 g/cm^3 is close to the maximum density of the composition.

Generally, there is a certain correlation between density, on the one hand, and flexural strength and modulus, on the other, for many other materials, and that correlation is not related to porosity. For example, there is a strong correlation ($R^2 = 0.984$) between density of all 38 polyethylene materials, listed in Table 7.49 of Chapter 7, including LDPE, LLDPE, HDPE, and their flexural modulus (Figure 6.1). Besides, mineral fillers in WPC materials increase density of the final product and also increase its flexural modulus. However, this chapter is mainly concerned about relationships between density and properties of WPC having the same formulation but produced at different regimes.

Effect on Oxidation and Degradation

The effect of density of WPC on its durability in terms of oxidative depolymerization of its plastic matrix is described in detail in Chapter 15. Briefly, porosity in WPC, which is directly related to the decrease of density (specific gravity) of the material, provides a chemically reactive area for oxygen. Oxygen flows into pores and attacks WPC "from inside," particularly at elevated temperatures. An increase in temperature by every 10°C accelerates the oxidative destruction of WPC by about

TABLE 6.2 The effect of density (specific gravity) of GeoDeck composite deck boards on their flexural modulus. Center point load, support span 14 inches

Specific gravity (g/cm^3)	Flexural modulus (psi)
1.07	182,840
1.10	215,040
1.12	261,225

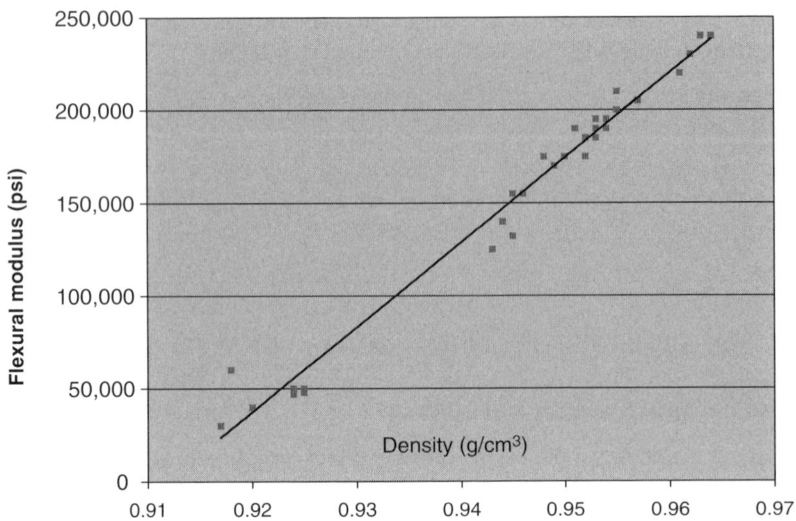

Figure 6.1 A correlation between polyethylene density and its flexural modulus. All materials are products of Chevron Phillips Chemical Company (see text). LDPE, LLDPE, and HDPE are shown with densities of 0.917–0.925, 0.918, and 0.943–0.964, respectively.

three times. On a hot sunny afternoon, when air temperature is, say, 90°F, deck surface is heated up to about 130–140°F. At 110°F in Phoenix, AZ, deck surface temperature reaches 160°F (70°C), and thermal oxidation of plastic in WPC accelerates by 3^5, that is, by 240 times. An additional increase of an available surface area for oxygen attack "from inside" as a result of porosity, which may reach many times, accelerates the oxidation dramatically (Table 6.3).

TABLE 6.3 Thermo-oxidation of GeoDeck composite boards in an airflow oven at 107°C (225°F) after 87 h, measured by residual load at failure (flex and shear strength) after boards conditioning. No antioxidants were added to the composite materials

Specific gravity (g/cm^3)	Residual load at failure compared to control (no heating) (%)
1.02	41
1.07	48
1.08	51
1.09	42
1.10	64
1.105	63
1.11	50
1.12	63
1.13	59
1.17	75

TABLE 6.4 The effect of board density on durability of GeoDeck composite boards with no added antioxidant in terms of half-life. Data are based on air-flow experiments

Specific gravity (g/cm^3)	Time to reach 50% reduction of board strength at 107°C (225°F), h
1.02	68
1.07	82
1.08	90
1.09	98
1.10	106
1.105	110
1.11	116
1.12	131
1.13	146
1.17	210

Table 6.3 shows an obvious trend—the higher the density, the higher the durability of deck boards. The durability in this particular case was measured as a decrease of a load at failure after exposure boards at the indicated high temperature for a certain time period.

The data can be presented in terms of half-life time of deck boards during the exposure (Table 6.4).

One can see that increase of board density slows down their deterioration by more than three times.

What is a mechanism of this reduction in strength? It is an oxidative degradation of the polymer matrix of WPC, as shown in Table 6.5.

These data can be shown in terms of HDPE chain length (Table 6.6).

One can see that a higher density board (the first row in Tables 6.5 and 6.6) showed only an insignificant drop in the integrity of its plastic. However, a lower density board (the last row) revealed catastrophic damage in terms of both polymer chain length and physical properties of the board. Clearly, a lower density board

TABLE 6.5 Average molecular weight (number-average, weight-average, and viscosity-average) of deck boards—freshly extruded and annealed at 200°F

Material	Average molecular weight of HDPE		
	Number-average	Weight-average	Viscosity-average
Composite board, d = 1.12 g/cm^3, not annealed	50,200	276,000	1,236,000
Composite board, d = 1.12 g/cm^3, annealed	41,700	257,900	1,009,000
Composite board, d = 1.09 g/cm^3, annealed	15,100	50,900	114,000

TABLE 6.6 Average chain length of polyethylene in deck boards—freshly extruded (upper line) or annealed as described in Table 6.5

	Average chain length of HDPE		
Material	Number-average	Weight-average	Viscosity-average
Composite board, $d = 1.12$ g/cm^3, not annealed	3600	19,700	88,000
Composite board, $d = 1.12$ g/cm^3, annealed	3000	18,400	72,000
Composite board, $d = 1.09$ g/cm^3, annealed	1000	3,600	8,100

provides more of its inner space for air oxygen to get in and to oxidize the material more rapidly. This phenomenon and its implications are discussed in more detail in Chapter 15, which shows that two powerful stabilizing factors that prolong lifetime of composite deckboards are board density and added antioxidants. These two factors are in a way functionally interchangeable, but taken together they work in synergism. Antioxidants block propagation of free radicals, and density controls amount of air oxygen flowing into pores of the composite matrix. High density of a composite material effectively blocks access of oxygen and slows down oxidative degradation.

Effect on Flammability, Ignition, Flame Spread

Clearly, porous WPC boards, having low density and having their pores filled with air oxygen, would maintain flame spread much more easily than the higher density boards.

The literature has apparently no data on the effect of density of WPC on flammability, but the issue is quite obvious. Table 6.7, showing ignition data with respect to several wood species, illustrates the concept.

One can see how much density of the materials effects ignition time. The lower the density of materials, the lower the ignition surface temperature.

TABLE 6.7 Flammability data for selected wood species [2]. Data are determined using ASTM E 906

		Ignition time (s) for heat flux	
Species	Density (g/cm^3)	18 kW/m^2	55 kW/m^2
Red oak	0.66	930	13
Southern pine	0.51	740	5
Redwood	0.31	741	3
Basswood	0.31	183	5

TABLE 6.8 Sidewise swelling of GeoDeck tongue-and-groove boards having different density (specific gravity) and being submerged to water for a total of 28 days. Swelling is expressed in absolute units of expansion

	Swelling, mills				
Density (g/cm^3)	24 h	5 days	7 days	14 days	28 days
1.125	2	8.5	9.3	19	32
1.15	1.5	4.5	5.3	13	23
1.17	0.75	8.0	5.5	13.5	24
1.21	0.75	6.3	5.8	13.5	23

Effect on Moisture Content and Water Absorption

Obviously, the higher the density, the lower the moisture content in WPC boards, the lower water absorption by the boards, and the less swell and buckling, the less are the microbial contamination and microbial degradation.

For the most dense GeoDeck deck boards (specific gravity 1.24–1.25 g/cm^3), moisture content is around 0.4–0.5% (brushed boards). For GeoDeck boards with much lower density (specific gravity 1.10 g/cm^3) moisture content is around 1.7%.

Generally, water absorption by composite materials depends on their porosity, amount of cellulose fiber, and their availability for incoming water. Because wood fiber in WPC is exposed into pores, it also increases water absorption by WPC.

Tables 6.8 and 6.9 show some experimental data for GeoDeck deck boards having the same formulation but different density as a result of different regimes of processing (speed and temperature of the extrusion) and different moisture content of the initial ingredients (rice hulls, first of all).

One can see that swelling is more pronounced with lower density boards than with high density boards, and the difference is even higher at a short-term water absorption.

Table 6.10 contains some data regarding water absorption by GeoDeck deck boards with different densities, obtained during a 5-year manufacturing period. Different

TABLE 6.9 Water absorption by GeoDeck tongue-and-groove boards having different density (specific gravity) and being submerged for a total of 28 days

	Water absorption (%)			
Density (g/cm^3)	24 h	7 days	14 days	28 days
1.125	1.77	4.31	5.96	8.43
1.15	1.52	3.34	4.68	6.55
1.17	1.47	3.19	4.41	6.13
1.21	1.49	2.94	3.98	5.44

TABLE 6.10 Water absorption by GeoDeck deck boards having different density (specific gravity) after 24 h submerging

Density (g/cm^3)	Water absorption (%)
1.04	4.3
1.06	3.8
1.07	2.9
1.08	2.9
1.09	2.7
1.10	2.5
1.10	2.04
1.115	1.85
1.12	1.7
1.15	1.8
1.16	1.7
1.18	1.5
1.21	1.1
1.25	1.02

densities of boards with the same formulation resulted from different manufacturing regimes such as manufacturing speed, temperature, and also amounts of antioxidants and moisture content in incoming cellulose fiber, as well as devolatilization of hot melt during the extrusion.

Clearly, the higher the density, the lower is the porosity, and lower the water absorption.

Effect on Microbial Contamination/Degradation

As it was noted earlier, pores in composite materials are typically open and form chains of pores, penetrating the whole matrix. Wood fiber is exposed into these pores. Hence, higher or lower degree of water absorption, depending on a lower or a higher WPC density. As a result, microbial contamination of the material in the matrix pores and voids, microbial degradation of wood particles (and in some cases, particles of minerals, which some microorganisms use as a food source), and in some acute cases, - microbial growth through the matrix of composite materials, take place.

Probability of such cases of microbial degradation is determined by accessibility of the composite matrix by microflora. This in turn is determined by a degree of porosity of the composite, density of the material (specific gravity), water absorption, content of minerals in the material (minerals often not used as food, but, conversely, play a role of a shield, blocking invasion of microbes into the matrix), and the presence of biocides, or antimicrobial agents.

Generally, the lower the density of deck boards, the higher is the likelihood of the microbial contamination and possibly microbial degradation.

TABLE 6.11 Effect of density (specific gravity) of GeoDeck composite pickets of the railing system on their shrinkage. Shrinkage was measured after exposure for 4 h in an oven at 200°F, 24 h after manufacturing. Data by Dr. Tatyana Samoylova, LDI Composites

Specific gravity (g/cm^3)	Shrinkage (%)
1.03	0.59
1.05	0.36
1.08	0.28
1.16	0.23
1.24	0.20

The Effect on Shrinkage

Studying of the shrinkage of GeoDeck deck boards, railing pickets, and so on has persistently indicated that the lower the density, the higher is the shrinkage. An example of this behavior is given in Table 6.11, which shows data obtained with GeoDeck composite pickets. The pickets were manufactured in the industrial extruder using vented and nonvented extruders, wet or dried pellets, and at various extrusion speed. By changing these conditions, pickets of various densities were manufactured.

The Effect on the Coefficient of Friction (The Slip Coefficient)

There are no data available on the effect of WPC density on the slip coefficient. However, it is known that polyethylene of lower density has a better traction than that of a higher density. In other words, HDPE is characterized by a low coefficients of friction, and the higher the density (specific gravity), the lower the static (and dynamic) coefficient of friction. For polyethylene density of 0.915 g/cm^3, coefficient of friction equals to 0.50; for 0.932 g/cm^3, it is equal to 0.30, and for 0.965 g/cm^3, it is equal to 0.10 [3].

The primary factors that control the coefficient of friction of polyethylene are its molecular characteristics, mainly its molecular weight and its distribution (number-, weight- and viscosity-average molecular weights), and a degree of crystallinity, that is, branching levels. This in turn affects molecular interactions between the polymer surface and any object in contact with it. Generally, coefficient of friction of polyethylene increases with the increase of molecular weight and branching levels, which also lead to the decrease of density (specific gravity).

Generally, LDPE has lower density and, hence has higher coefficient of friction than with HDPE-made composite deck boards.

GeoDeck composite deck boards often utilize HDPE with the density of 0.955 g/cm^3. Its coefficient of friction is about 0.15. However, when rice hulls and a granular blend of calcium carbonate/kaolin and delignified cellulose fiber are incorporated into the plastic matrix, the static coefficient of friction increased to 0.53, that is, to 350% compared to the initial HDPE.

It might appear that the easiest way to enhance friction of a WPC board is to change the initial plastic to that with a higher coefficient of friction, that is—in case of HDPE—to a low-density plastic and to a more "rubbery" HDPE. However, this might lead to problems in flowability of the composition in the extruder, to compromise its strength, and—more than that—its flexural modulus, that is, deflection, creep, and other properties of the final material. To change the plastic—it is always a trade-off and an optimization game. If the overall balance shows that the final material has acquired an appreciably higher coefficient of friction with other properties being more or less the same or within an accepted tolerance, if not even better, this can be called a success.

DENSITY OF CROSS-SECTIONAL AREAS OF HOLLOW PROFILES OF GEODECK WPC BOARDS

Test procedures for density (specific gravity) determinations are commonly too imprecise, that is, have too much of an error margin, to be employed for delicate studies of density distributions across WPCs. Those distribution charts would have been very useful for die design, for structural and functional analysis of WPC profiles, and for other research and development projects. However, by employing the sink/float procedure for density determinations, we have been able to obtain density distribution diagrams, some examples of which are shown in Figures 6.2–6.9. It should be noted that the maximum density of GeoDeck WPC profiles is 1.24–1.25 g/cm^3.

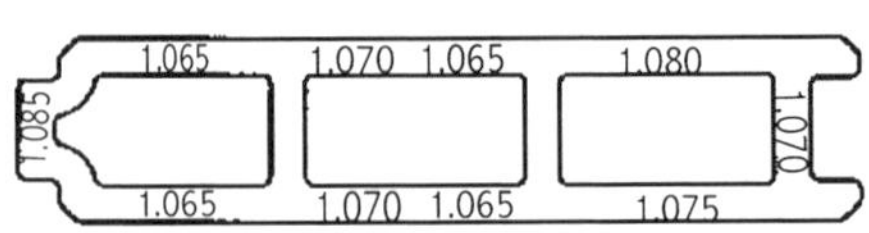

Figure 6.2 Density distribution in GeoDeck tongue and groove board of low overall density (d = 1.075 g/cm^3).

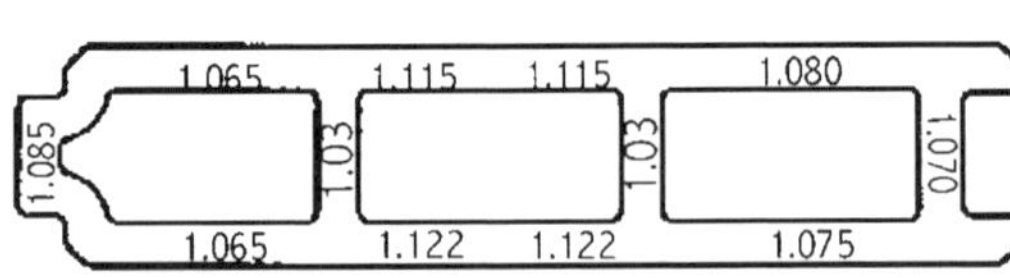

Figure 6.3 Density distribution in GeoDeck tongue and groove board of low overall density (d = 1.075 g/cm^3). Note the lowest densities were of the ribs.

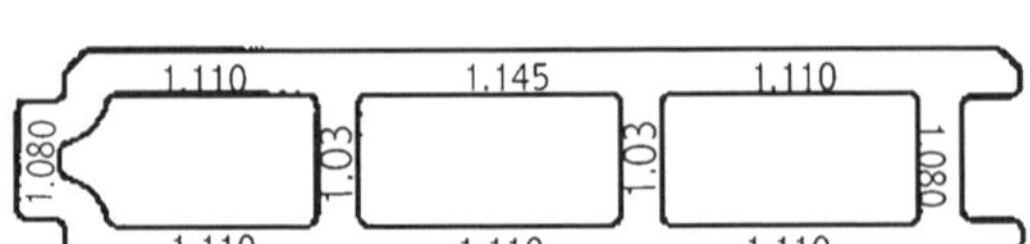

Figure 6.4 Density distribution in GeoDeck tongue and groove board of low overall density (d = 1.10 g/cm^3). Note the lowest densities were of the ribs.

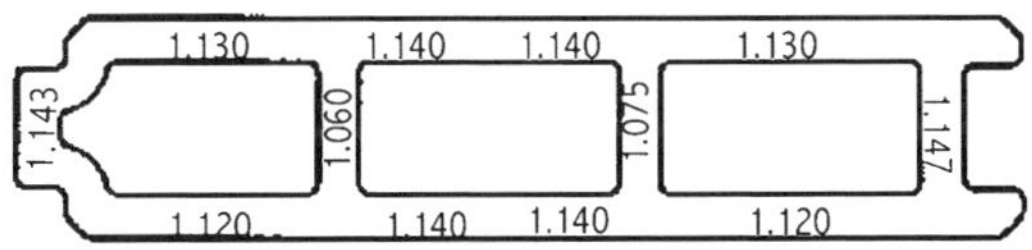

Figure 6.5 Density distribution in GeoDeck tongue and groove board of medium overall density (d = 1.12 g/cm^3). Note the lowest densities were of the ribs.

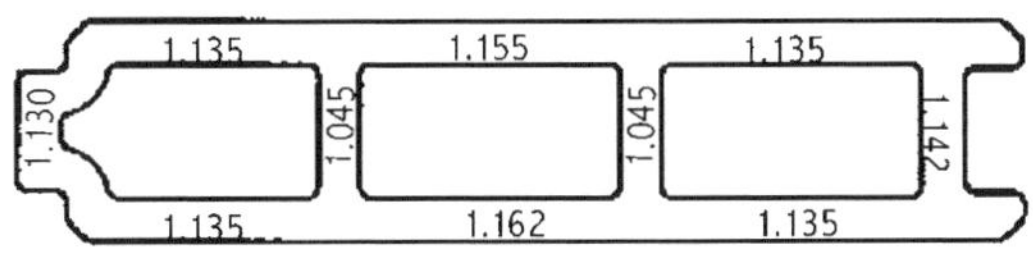

Figure 6.6 Density distribution in GeoDeck tongue and groove board of medium overall density (d = 1.135 g/cm^3). Note the lowest densities were of the ribs.

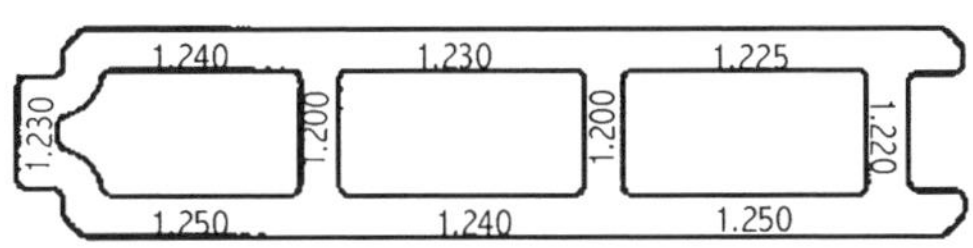

Figure 6.7 Density distribution in GeoDeck tongue and groove board of high overall density (d = 1.24 g/cm^3). Note the lowest densities were of the ribs. The board was manufactured using a vented extruder.

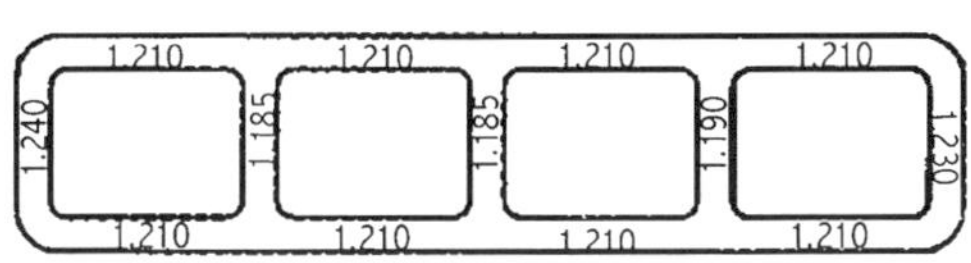

Figure 6.8 Density distribution in GeoDeck traditional board of high overall density (d = 1.21 g/cm^3). Note the lowest densities were of the ribs.

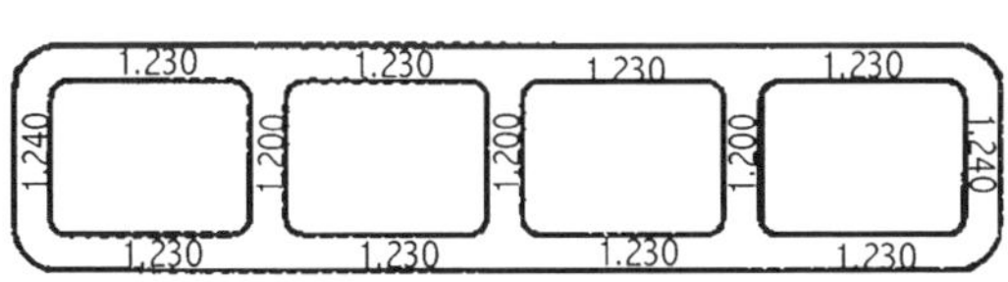

Figure 6.9 Density distribution in GeoDeck traditional board of high overall density (d = 1.22 g/cm^3). Note the lowest densities were of the ribs.

One can see that T&G (tongue and groove) profile shown in Figure 6.2 had much lower density than a maximum one. Overall porosity was accounted for 13% of the material volume. The main reason was that the profile shown in Figure 6.2 was obtained without added antioxidants. The material was rather quickly oxidized in an airflow oven, in a weathering box, and on actual decks, particularly in the South. The highest compaction of the material was in the tongue (d = 1.085 g/cm^3), but an overall distribution was not much different from the average overall density. Density of the ribs was not determined in this particular profile. However, when such measurements have been conducted, it was found that the lowest density was in the ribs (Fig. 6.3).

In another case a pattern of density distribution across the same, T&G GeoDeck profiles were quite different from those in Figure 6.2. The tongue and the groove had the same densities as those in Figure 6.2, however, the highest densities were in the flat panels of the profile. Top and bottom panels, whichever they were, had practically the same densities (differed by 0.005–0.007 g/cm^3, or 0.4–0.6% from each other), but the ribs had significantly lower densities, 1.03 g/cm^3, than the densities 1.115—1.122 g/cm^3 in the flat panels. Clearly, the hot melt flow was not uniform in density, and for some reason the density was the lowest in the rib area. As shown below, this makes the ribs the most vulnerable part of boards to oxidation and degradation compared to other segments of the board.

The profile shown in Figure 6.4 had a higher overall density of 1.10 g/cm^3, mainly because of a more uniform distribution of higher densities along the flat panels, except the relatively high density in the central part of the panel. Again, ribs had the lowest density of 1.03 g/cm^3.

An example of T&G profile with yet an higher overall density, 1.12 g/cm^3, is shown in Figure 6.5. The both flat panels reached density of 1.14 g/cm^3, the tongue and the groove had even higher densities of 1.143–1.147 g/cm^3, but the ribs had the lowest densities of 1.06–1.075 g/cm^3.

The extrusion of T&G board shown in Figure 6.6 has resulted in a board with relatively high densities of both flat panels (reaching 1.16 g/cm^3), tongue (1.13 g/cm^3) and groove (1.14 g/cm^3), but densities of the ribs again were the lowest ones, 1.045 g/cm^3. That is why despite a relatively high segmental densities of the board, its overall density was only 1.135 g/cm^3. The board, though, was resistant to oxidation and not crumbled compared to boards shown in Figures 6.2–6.5.

The same T&G board, made of the same composition as shown above, but produced on a vented extruder, showed much higher density (Fig. 6.7).

This board showed the density as high as it can get. Still, the ribs had the lowest density.

"Traditional," symmetrically shaped boards, such as those shown in Figure 6.8, typically demonstrate higher densities.

However, the ribs still showed slightly lower densities of 1.185–1.190 g/cm^3 compared to those of flat panels (1.21 g/cm^3) and edges of the board (1.23–1.24 g/cm^3).

Similarly, a "traditional" board shown in Figure 6.9 had very high densities of the flat panels (1.23 g/cm^3) and the edges (1.24 g/cm^3) still showed the lowest densities in the ribs (1.20 g/cm^3).

When GeoDeck boards, shown in Figures. 6.2–6.4, were placed into an airflow oven at 225°F, after several days of such an exposure ribs (but not flat panels or edges of the boards) turned brown. This "burn" color gradually disappears in the board's cross-section toward upper and lower panels of the board, that is toward the surface where density was the highest. A study of flexural and shear strength of the boards has shown that the dark ribs largely lost their strength. In other words, the darkening of the board cross-section served as a clearly visible pattern of density (specific gravity) distribution across the board. The darkening reflects the highest degree of "burning," that is, the oxidative degradation.

Segments of the board with lower density (ribs first of all) have higher porosity, provide the highest surface area for oxidation by air oxygen (which readily diffuses into pores), and lose their structural integrity with the fastest rate across the board. Therefore, in hollow profiles, such as in GeoDeck boards, the thermal degradation starts and proceeds from inside, from the ribs that have the lowest density (specific gravity).

DENSITIES AND WEIGHT OF SOME COMMERCIAL WOOD–PLASTIC DECK BOARDS

Table 6.12 lists density (specific gravity) of some commercial deck boards. Table 6.13 shows weights of some commercial deck boards.

TABLE 6.12 Density (specific gravity) of some commercial deck boards. Data were obtained using sink/float method (see below)

Brand	Density (g/cm^3)
Boardwalk	0.94
Trex (Saddle)	0.95
	0.91–0.95[a]
	0.92[b]
Monarch	1.06
Perfection	1.07
Fiberon (Buff Cedar)	1.075
	1.11[a]
Rhino Deck	1.08[a]
	0.92–0.96[a]
	1.13[b]
Evergrain (gray)	1.085
SmartDeck	1.10[b]
EverX	1.12[a]
Evergrain (Cedar)	1.14
CorrectDeck	1.15[a]
Nexwood	1.17[a]
	1.175–1.178[a]
Xtendex	1.17–1.21[a]
WeatherBest	1.20
UltraDeck	1.22
Timbertech	1.22
	1.23[a]
GeoDeck	1.24

[a] Manufacturer's data.

[b] Listed in http://nature.berkeley.edu/~fbeall/DeckTest/DeckbdMatls.htm

TABLE 6.13 Weight per lineal foot of some commercial deck boards. Data were obtained by weighing of actual boards. Board pictures are given in Figures 1.2–1.25

Brand	Board weight (lb/ft)
Pressure treated lumber (reference)	1.4 (dry)–1.7 (conditioned)
Carefree (USPL) (plastic lumber; reference)	1.19
EON (plastic lumber; reference)	1.39 (cedar)1.52 (redwood)
UltraDeck	1.48
Millenium	1.48
Procell	1.53
Xtendex	1.59
GeoDeck, Tongue and Groove	1.65
GeoDeck, Traditional	1.80
Nexwood	1.87
CrossTimbers	2.00
CorrectDeck	2.13–2.29
Timbertech	2.19–2.23
Fiberon	2.20
Perfection	2.21
ChoiceDek	2.22–2.32–2.38–2.45
Life Long	2.22
Boardwalk	2.23
EverX	2.30
Monarch	2.30
Evergrain (gray)	2.37
Rhino Deck	2.49
Evergrain (cedar)	2.54
WeatherBest	2.55
Trex (Winchester)	2.63
Trex (Saddle)	2.65
Trex (Madera)	2.68
SmartDeck	2.72
GeoDeck Heavy Duty	2.80

DETERMINATION OF DENSITY OF WOOD–PLASTIC COMPOSITES USING A SINK/FLOAT METHOD

A quick and reliable method for density determination of WPC materials is a sink/float procedure. It is readily applicable in a laboratory as well as at a plant. LDI Composites routinely uses this procedure for regular QC of GeoDeck deck boards at manufacturing sites.

The current method employs a series of glycerol–water mixed solutions. Glycerol and water are fully compatible liquids. Neat glycerol has the density of 1.25 g/cm^3, water 1.0 g/cm^3. Hence, density of WPC materials (in this particular case) in the range of 1.0–1.25 g/cm^3 can be measured, practically with accuracy up to 0.002 g/cm^3. When a specimen (a full cross-section of a deck board or other profile) is carefully immersed into a glycerol–water system with a certain density, it will either sink, or float, or "hang" in the solvent being in equilibrium with the system in terms of their densities. When the specimen sinks, its density is higher than that of the liquid. When the specimen floats, its density is lower than that of the liquid. When the specimen is in equilibrium, its density is equal to that of the liquid.

At density difference between the two, the solvent and specimen, of 0.01 g/cm^3 (for example, when density of a liquid is 1.11 g/cm^3 and density of the sample is 1.12 g/cm^3), velocity of specimen sinking is rather high. At 0.002 g/cm^3 density difference between the two (such as 1.118 and 1.120), velocity of specimen sinking is slow but quite noticeable. After some experience with the procedure, density differences of 0.002 g/cm^3 in two specimens are easily noticeable. However, in most cases such accuracy is not required.

There are three parts in the floating/sinking samples procedure. First, it is a preparation of standard solutions with known densities. Glycerol's density is $d = 1.25$ g/cm^3. If it is diluted with water 50:50, the density of the resulting solution is, naturally, 1.125 g/cm^3. A general formula for densities in glycerol-water systems is

$$d = 1.25x + (1-x),$$

where
x = fraction of glycerol ($x = 1.0$ for neat glycerol, $x = 0$ for pure water), *v*/*v*.

Examples
at $x = 1$, $d = 1.25$; at $x = 0$, $d = 1.00$; at $x = 0.5$, $d = 1.125$; at $x = 0.75$ (750 mL glycerol + 250 mL water), $d = 1.1875$; at $x = 0.25$ (250 mL glycerol + 750 mL water), $d = 1.0625$.

It is recommended to use a 1-L measuring cylinder and dilute glycerol with water to 1 L of a resulting solution. A complete mixing takes vigorous shaking, so the resulting solution should be left for about an hour to get rid of microbubbles of air in the liquid.

The second part is to measure an actual density of the resulting solution. Typically it is somewhat different from the calculated number. Take 50-mL measuring cylinder, carefully weigh it, fill with 50 mL of the resulting solution, carefully weigh it, and take a difference (weight of the liquid). Divide weight of the 50-mL liquid (with accuracy to a milligram) by 50 (mL), and you get a density of the resulting liquid with a reasonable accuracy. Of course, other methods for density determinations of liquids can be used, for example, employing a hydrometer, many of which have a density scale of a 0.0005 g/cm^3 accuracy.

The third part is to measure density of a sample. Shake the sample well or/and gently knock on a hard surface to shake off small particles and dust (to avoid bringing dirt to the liquid system), carefully (avoid air bubbles) immerse into the liquid, punch/drag up and down in the liquid to shake off air bubbles, using an appropriate hook, and see whether the sample floats, sinks, or hangs in the middle of the solution. If it sinks—go up along the battery of liquids, until it hangs. If it floats—go down along the battery. Soon one will be able to "see" densities judging from speed of the sample coming up or down. Of course, the latter "method" is applicable only when the density of a liquid is around the specimen's density, hence, the sample is moving up or down rather slowly.

ASTM TESTS RECOMMENDED FOR DETERMINATION OF THE SPECIFIC GRAVITY (DENSITY)

ASTM D 6111"Standard Test Method for Bulk Density and Specific Gravity of Plastic Lumber and Shapes by Displacement"

The test method covers the determination of the bulk density and specific gravity of materials in their "as manufactured" form. Therefore, this test method evaluates a "product," not an inherent "material property."

Note of the author: If the material has voids, an increased degree of porosity, and so on, it would result in a decreased "specific gravity," or density. This would be the product property, not a "material" property per se. The same material, obtained without voids, for instance, in a vented extruder, will have higher density or specific gravity.

It is important that in this method a sample should represent the whole cross-section of the product.

Note of the author: The above material ("density of cross-sectional areas of hollow profiles of GeoDeck boards") shows how significantly nonhomogeneous a product can be in terms of its density at different segments of its cross-section.

Essentially, the test consists of weighing the specimen in air, then immersing it into water in a special cage equipped with a sinker (in case if the specimen is lighter than water), determining a weight of the specimen in water upon immersion, and calculating its bulk specific gravity.

Specific gravity of the sample is calculated as follows:

$$d = a/(a-c)$$

where a = weight of the specimen in air, c = weight of the specimen in water, c is calculated as $(b-w)$, where b = overall weight of the completely immersed specimen, cage, sinker, and partially immersed wire, holding the cage and the sinker, and b = overall weight of the completely immersed cage, sinker, and partially immersed wire.

Notes of the author:

(a) If a specimen has the same specific gravity as water ($d = 1.0$), its weight in water will be zero ($c = 0$). Indeed, from the above equation $d = a/a = 1.0$.

(b) If a specimen has the specific gravity equal to 2.0, its weight in water equals to 1/2 of its air weight ($c = a/2$). Indeed, from the above equation at $c = a/2$, $d = 2.0$.

(c) If a specimen is lighter than water, c has a negative value. In these cases d is less than 1.0. For example, at $d = 0.5$, $c = -a$.

Notes of the author: Density for samples can be obtained by multiplying their specific gravities by the coefficient equal to water density at a given temperature (0.9991 at 15°C, 0.9982 at 20°C, 0.9975 at 23°C, 0.9957 at 30°C, and so forth). The coefficient can be found in editions such as *Handbook of Chemistry and Physics, CRC press* (cont. editions). As water densities are measured in g/cm^3, densities have the same dimension. However, in order to make these calculations meaningful, precision of specific gravity determination should be no less than 0.09–0.4% for the above cases. Realistically, error margins in density and specific gravity determinations are much higher. The ASTM lists results of a round-robin test for specific gravity determination for two samples of plastic lumber materials conducted in six laboratories, with five 28-g specimens for each test. Test results obtained within one laboratory varied between 1.7 and 4.5%; test results obtained by different laboratories varied between 2.0 and 5.2%.

Bulk density in g/cm^3 can be converted to lb/ft^3 multiplying by 62.43. For example, density of 2.00 g/cm^3 corresponds to 124.86 lb/ft^3.

The ASTM test method lists requirements to test specimens, water, to a balance and precision of its measurements, to cage, wire, sinker, and so on. The method recommends to conduct the test at water temperature of 23 ± 2°C.

ASTM D 792 "Standard Test Method for Density and Specific Gravity (Relative Density) of Plastics by Displacement"

This test method is a prototype for the above procedure, and methodologically is the same thing, except it aims at solid plastics and not at plastic lumber. Two test methods are described, method A—for testing solid plastics in water and method B—for testing solid plastics in liquid other than water. It is recommended to use a one-piece specimen of 1–50 g by weight in air and of a volume not less than 1 cm^3. The test method recommends to conduct the test at water temperature of 23 ± 2°C.

The test method is illustrated with data in methods A and B for several plastics, conducted in six laboratories, and each test result was based on two individual determinations. The test method does not disclose "liquids other than water" that were used in method B. Precision of the procedure is usually fair. Some data are shown in Table 6.14.

Note of the author: Data in Table 6.14 are modified compared to that of ASTM D 792, with respect to the main figures and their standard deviations. It does not make much sense to present data in a way chosen by the said ASTM procedure, for example, for polypropylene as 0.9007 ± 0.00196, for polyvinyl chloride as 1.3396 ± 0.00243, and so on. First, standard deviations are typically determined with a precision hardly better than 30%, particularly using just a few experimental points as in the above example of the ASTM. The ASTM precision in the

TABLE 6.14 Specific gravity of some plastics

	Specific gravity	
Material	Average, within-laboratory standard deviations	Average, between-laboratory standard deviations
Polypropylene (in water)	0.901 ± 0.002	0.901 ± 0.003
Polypropylene (in a nonwater liquid)	0.902 ± 0.001	0.902 ± 0.002
Polyvinyl chloride (in water)	1.340 ± 0.002	1.340 ± 0.006
LDPE (in a nonwater liquid)	0.9215 ± 0.0011	0.9215 ± 0.0020
HDPE (in a nonwater liquid)	0.968 ± 0.001	0.968 ± 0.002

above figures pretends that standard deviations are determined with better than 0.01% accuracy. Second, the above (given in the ASTM) figures pretend that the standard deviation is determined with a higher precision (to the fifth digit) than the principal figure (to the fourth digit). Third, there is not much sense to indicate decimals in the principal figure when deviations are larger than those decimals. This provides a misleading statement about a high accuracy of measurements. For example, one cannot measure lengths down to one-thousands of an inch using a ruler calibrated by inches, or calculate an average from a few measurements, each of them in whole inches, with a "precision" of one-thousand of an inch.

ASTM D 1505 "Standard Test Method for Density of Plastics by the Density-Gradient Technique"

This test method gives a significantly better accuracy than the above procedures and is designed to yield accurate results better than 0.05%.

Note of the author: This accuracy would translate to an error margin of a density, for example, of 1.159 ± 0.001 g/cm^3

The ASTM method is based on using liquid systems to fill density-gradient columns, or "tubes." Examples of such liquid systems are shown in Table 6.15.

A binary solution of choice (400–600 mL) should be prepared by mixing the liquid ingredients (see Table 6.15) and adding both the liquids to the mixture, such that the resulting density should be approximately equal to the desired lowest density. The solution should be calibrated with glass floats, requirements to which and a calibration procedure are given in the test method. Glass floats can be purchased from American Density Materials (3826 Springhill Rd., Staunton, VA, (540)887-1217). In the range, for example, 1.00 and 1.25 g/cm^3, they cost $86 a piece, and one might need to have five glass floats to cover the range (1.05, 1.10, 1.15, 1.20, and 1.25 g/cm^3).

TABLE 6.15 Liquid systems for density-gradient tubes

System	Density range (g/cm^3)
Methanol–benzyl alcohol	0.80–0.92
Isopropanol–water	0.79–1.00
Isopropanol–diethylene glycol	0.79–1.11
Ethanol–carbon tetrachloride	0.79–1.59
Toluene–carbon tetrachloride	0.87–1.59
Water–sodium bromide	1.00–1.41
Water–calcium nitrate	1.00–1.60
Carbon tetrachloride–trimethylene dibromide	1.60–1.99
Trimethylene dibromide–ethylene bromide	1.99–2.18
Ethylene bromide–bromoform	2.18–2.89

Once a liquid system is prepared, a test specimen should be gently placed into the tube. The test procedure recommends to place three specimens into the tube with a liquid system. After the liquid in the tube and the specimens reached equilibrium, which often requires 10 min or more, the operator reads the level to which test specimens sink, in comparison with the height of each glass float, and averages the three values. The densities of the samples may be determined graphically or by calculation from the levels. The formula for numerical calculations is given in the test method.

Precision of the procedure is good, and it is in the same range that the method was designed for (see above). Tables 6.16–6.17 show data obtained in a number of laboratories (round robin test) for molded polyethylene samples of nominal densities of 0.92–0.96 g/cm^3.

One can see that the accuracy of the procedure is of 0.01–0.03% for within-laboratory tests and 0.09–0.11% for between-laboratory tests.

TABLE 6.16 Polyethylene density determined according to ASTM D 1505 for four molded samples in 22 laboratories

	Density (g/cm^3)	
Polyethylene	Average, within-laboratory standard deviations	Average, between-laboratory standard deviations
1	0.9196 ± 0.0003	0.920 ± 0.001
2	0.9319 ± 0.0001	0.9319 ± 0.0008
3	0.9527 ± 0.0003	0.953 ± 0.001
4	0.9623 ± 0.0006	0.962 ± 0.001

TABLE 6.17 Polyethylene density determined according to ASTM D 1505 for seven samples in 7 to 11 laboratories. Each laboratory obtained six test results for each material

	Density (g/cm^3)	
Polyethylene	Average, within-laboratory standard deviations	Average, between-laboratory standard deviations
1	0.9139 ± 0.0003	0.9139 ± 0.0009
2	0.9177 ± 0.0002	0.9177 ± 0.0008
3	0.9220 ± 0.0003	0.9220 ± 0.0007
4	0.9356 ± 0.0004	0.936 ± 0.001
5	0.9528 ± 0.0005	0.953 ± 0.001
6	0.962 ± 0.001	0.962 ± 0.001
7	0.9633 ± 0.0004	0.963 ± 0.001

ASTM D 1622 "Standard Test Method for Apparent Density of Rigid Cellular Plastics"

Note of the author: This test method is mentioned here as an example of a different approach to determining apparent density, that is, by direct measurement of three dimensions of a specimen and calculation of the specimen volume, and then by a simple division of the specimen weight (W) in air by the specimen volume (V), $d = W/V$.

This test procedure is applicable to foamed materials. Terminology of the test standard includes an "apparent core density" and an "apparent overall density." The first term is applicable to cellular plastics that have their forming skin removed. The second term is applicable to cellular plastics that have their skin intact. According to the ASTM procedure, the term "density" with respect to foamed plastics shall be interpreted accordingly.

The precision of the test method is lower than that of the above ASTM procedures (Table 6.18).

TABLE 6.18 Round-robin test data involving four materials tested in five laboratories. Each test result was the average of five individual determinations. The nature of the materials was not disclosed in the ASTM D 1622

	Density (kg/m^3) (= 0.001 g/cm^3)	
Material	Average, within-laboratory standard deviations	Average, between-laboratory standard deviations
1	37.51 ± 0.42	37.51 ± 0.56
2	49.63 ± 0.30	49.63 ± 0.46
3	26.03 ± 0.14	26.03 ± 0.66
4	20.79 ± 0.59	20.79 ± 1.11

Note of the author: Again, as have been indicated not once before, data in Table 6.18 are misleading in a sense that they pretend that accuracy of the determinations is much higher than it is in reality. It does not make much sense to indicate density as precise as 20.79, while the error margin is ± 0.59. In other words, the density listed as accurate to the second decimal, while even the first decimal is within error margin. Here is how Table 6.18 can be presented in a more correct way (see Table 6.19).

TABLE 6.19 Round-robin test data involving four materials tested in five laboratories. Each test result was the average of five individual determinations. The Table was modified compared to that in ASTM D 1622 with respect to figures and their standard deviations

	Density (kg/m^3) (= 0.001 g/cm^3)	
Material	Average, within-laboratory standard deviations	Average, between-laboratory standard deviations
1	37.5 ± 0.4	37.5 ± 0.6
2	49.6 ± 0.3	49.6 ± 0.5
3	26.0 ± 0.1	26.0 ± 0.7
4	20.8 ± 0.6	21 ± 1

One can see that the accuracy of the procedure is between 1 and 3% for within-laboratory tests and between 1 and 5% for between-laboratory tests.

ASTM D 1895 "Standard Test Methods for Apparent Density, Bulk Factor, and Pourability of Plastic Materials"

These test methods cover the measurements of apparent density of pellets, granules, powders, flakes, and so on. They are not directly related to density (specific gravity) determinations of WPC profiles, however, similar procedures are used for measuring bulk densities of WPC pellets. There are three test methods given in the ASTM procedure (test methods A, B, and C), and all of them result in pouring of about 100–500 g sample of granules, powders, and so on into a measuring cap or a measuring cylinder, recording volume, occupied by the material, and weight of the material, and calculating the apparent density as weight divided by volume. The test method lists results of the measurement of apparent density of eight different commercial plastics conducted in six laboratories, each test was based on three individual determinations. Within-laboratory standard deviations were of 0.1–0.8%, and between-laboratory standard deviations were of 2–5%.

There are a number of ways how to keep density of wood-plastic composites as high as possible, and every situation presents a balance between performance and cost. Drying the ingredients, particularly cellulose fiber, slowing down the extrusion speed, lowering the compounding and extrusion temperature, decreasing attrition of components during the processing, employing vented extruders, introducing antioxidants and coupling agents—these are the most appropriate and realistic actions.

REFERENCES

1. D. Dean. Influence of ethylene-anhydride copolymer coupling agents on the mechanical properties of HDPE based wood polymer composites. In: *Proceedings of Progress in Wood Fibre Plastic Composites 2006 International Conference*, Toronto, Canada, May 1–2, 2006.
2. R.H. White and M.A. Dietenberger. *Fire Safety. Wood Handbook*, Forest Products Society, Madison, WI, 1999, Chapter 17, p. 17.7.
3. A.J. Peacock. *Handbook of Polyethylene: Structures, Properties, and Applications*, Marcel Dekker, New York, 2000, p. 203.

7

FLEXURAL STRENGTH (MOR) AND FLEXURAL MODULUS (MOE) OF COMPOSITE MATERIALS AND PROFILES

INTRODUCTION

Mechanical properties of composite deck boards and components of railing systems—properties such as flexural, compressive, shear, tensile, impact, and creep, and their flexural strength and stiffness in particular are among the most important characteristics. Flexural strength and stiffness are of a prime concern of ICC-ES acceptance criteria, such as AC-174 "Acceptance criteria for deck board span ratings and guardrail systems (guards and handrails)," not only "as is" at ambient temperature, but also at different temperatures and before and after weathering.

At the same time, the methods for determination of flex strength and flex modulus of materials in general and composite materials in particular are rather tricky and may lead to gross deviations from correct values. Though, "correct values" also depend on the basic definitions, experimental setups, and interpretations of experimental data. Viscoelastic behavior of plastics and plastic-based composite materials as well as a certain nonuniformity of the matrix of composite materials also add more complications to measurements and interpretations of experimental values of flex strength and modulus.

Basic Definitions and Equations

In this chapter we will mainly consider simple beam bending. Composite deck boards in terms of their flexural properties are simple beams. A whole deck board,

Wood-Plastic Composites, by Anatole A. Klyosov

or its "as manufactured" pieces (without altering surfaces beyond cutting to length), or cut rectangular pieces are all simple beams. However, in the first two cases, using whole board or its cut piece, one determines "product property"; in the last case (cut piece with altered machined surface) one determines "material property." These two kinds of properties can provide significantly different values of strength and modulus of composite shapes.

In flexural testing, the deformation of the specimen is measured by the deflection at the center of the specimen.

When a load is placed on a specimen, stress and strain result. Stress is the internal resistance to the load as the applied force. Strain is the amount of deformation caused by this stress, such as deflection in bending, contraction in compression, and elongation in tension.

ASTM D 790-03, which we will consider below, refers to "maximum surface stress," whereas the earlier version, ASTM D 790-97, refers to maximum fiber stress and says that "the maximum stress in the outer fibers occurs at the midspan." Because maximum fiber stress is often used in strain and stress literature, the term "outer fibers" needs an explanation. It has nothing to do with the actual fiber in the material. The term outer fibers refers to the material near the specimen surfaces, where the maximum strains occur when the specimen is loaded at, say, midspan, as described in ASTM 790.

When the specimen is loaded and bent, the side of the specimen in contact with the loading nose is put into compressive strain mode, while the opposite, bottom side, is put into a tensile, stretching strain mode. This results in a gradient of strains across the depth of the specimen, ranging from maximum compressive strain at the top surface, at the loading nose, gradually changing to zero strain, and therefore zero stress, at the neutral axis (a half-depth in rectangular specimens), and then gradually changing to a maximum tensile strain at the bottom surface. Hence, the term outer fibers refers in this case to the material next to the loading nose and at the midspan at the opposite surface. The neutral axis is right in between, at the half-depth of rectangular samples, such as in many deck boards.

The three dimensions of the specimen under the load in a flexural test are defined as follows:

- Length, the longer dimension perpendicular to the direction of the force application.
- Width, the shorter dimension perpendicular to the direction of the force application.
- Depth, the dimension in line with the direction of the force application.

The common term "thickness" is not in a typical use in flexural tests. However, we will use all three common terms—depth, height, and thickness.

Strain and stress in a specimen are largely defined by the three dimensions of the specimen, by location of the load application, and by the specimen's profile, that is, a configuration of its cross section. Solid composite deck boards are typically represented by just one, flat, rectangular section. Hollow boards, having ribs, can be described by a number of sections, both filled and empty. Overall, the total section can be broken into rectangular elements, and such an operation provides a basis for strain and stress calculations. Those rectangular elements are measured by the area and expressed in square inches.

Flexural strength

for center-point load:

$$S = PLH/8 \times I$$

for third-point load:

$$S = PLH/12 \times I$$

for quarter-point load:

$$S = PLH/16 \times I$$

For uniformly distributed load

$$S = WbHL^2/16 \times 144 \times I$$

Moment of inertia

for a square profile:

$$I = B^4/12$$

for a rectangular profile:

$$I = BH^3/12$$

for a hollow square profile:

$$I = (B^4 - b^4)/12$$

for a hollow rectangular profile:

$$I = (BH^3 - bh^3)/12$$

for a hollow rectangular profile with one rib:

$$I = (BH^3 - bh^3 + wh^3)/12$$

Flexural modulus

for center-point load:

$$E = PL^3/48 \times DI$$

for third-point load:

$$E = PL^3/56.5 \times DI$$

for quarter-point load:

$$E = PL^3/70 \times DI$$

For uniformly distributed load:

$$E = 5WbL^4/384 \times 144 \times DI$$

where
S = flexural strength, in psi;
I = moment of inertia, in in.^{4};
P = ultimate load, in lb (for flexural strength) or a load (for flexural modulus);
D = deflection, in in., at the load, P;
L = support span, in in.;
B = specimen overall width, in in.;
b = specimen's hollow part width, in in.;
H = specimen overall depth, in in.;
h = specimen's hollow part depth, in in.; and
W = uniformly distributed load, lb/ft^2.

Deflection

for center-point load:

$$D = PL^3/48 \times EI$$

for third-point load:

$$D = PL^3/56.5 \times EI$$

for quarter-point load:

$$D = PL^3/70 \times EI$$

For uniformly distributed load:

$$\mathrm{D} = 5\mathrm{WbL}^4/384 \times 144 \times \mathrm{EI}$$

where
D = deflection, in in., at the load, P;
S = flexural strength, in psi;
I = moment of inertia, in in.^{4};
E = flexural modulus, in psi;
P = ultimate load, in lb (for flexural strength) or a load (for flexural modulus);
L = support span, in in.;
B = specimen overall width, in in.;
b = specimen's hollow part width, in in.;
H = specimen overall depth, in in.;
h = specimen's hollow part depth, in in.; and
W = uniformly distributed load, in lb/ft.^{2}

Moment of Inertia

One of the most important parameters of the property of the section is its moment of inertia. It is a measure of the strength and stiffness of a board (generally, of a beam). The moment of inertia is a rigidity factor, and it is defined by the section's geometry and size. The moment of inertia is expressed in inches raised to the fourth power (in.^{4}).

There are tabulated moments of inertia for typical sections, such as rectangular, triangle, circular, oval, and so on, solid and hollow, with different locations of neutral axis or base lines for each one. Because we will consider the simplest cases, that is, solid rectangular and hollow rectangular composite deck boards, which can be broken down into smaller elements, the respective formulas will be rather simple.

For a solid rectangular board in a flexural mode (that is, about its neutral axis), the moment of inertia is

$$I = bh^3/12, \qquad (7.1)$$

where I = moment of inertia, b = board width, and h = board depth (height).

For a standard board with b = 5.5 in. and h = 1.25 in., the moment of inertia is equal to

$$I = 0.895 \text{ in.}^4.$$

This is, for example, the moment of inertia for Trex boards.

For hollow boards with ribs, the moment of inertia is calculated by elements, adding solid and subtraction hollow sections, each one—if rectangular—is determined by the above formula (7.1).

Let us consider GeoDeck, Traditional (hollow) board. It has standard overall dimensions, that is, B = 5.5 in. and H = 1.27 in. Here capital letters denote the overall dimensions, small letters denote sections/elements. Besides, GeoDeck has three ribs, each 0.2 in. thick, and its wall thickness is 0.25 in. in the top, bottom, and side panels (Fig. 7.1).

Moment of inertia of this profile is defined by the following formula:

$$I = (BH^3 - bh^3 + 3wh^3)/12. \qquad (7.2)$$

Here, from the moment of inertia for the overall section $BH^3/12$, the moment of inertia of the "hollow section" (including the ribs) $bh^3/12$ was subtracted, where b = the width of the hollow section, 5.5″ − 0.25″ − 0.25″ = 5.0 in., and h = the depth of

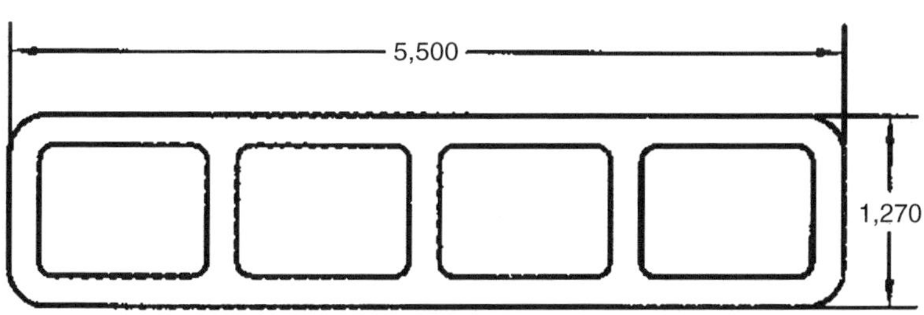

Figure 7.1 GeoDeck Traditional composite board, cross section.

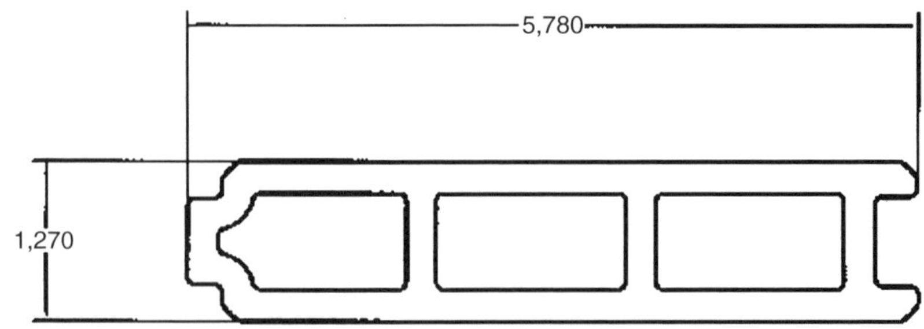

Figure 7.2 GeoDeck Tongue and Groove composite board, cross section.

the rib, 1.27″ − 0.25″ − 0.25″ = 0.77 in.. Finally, three moments of inertia, for each of the ribs, were added, that is, $3wh^3/12$, where w = rib thickness, 0.20 in. That is, for GeoDeck traditional board with said dimensions

$$I = 0.771 \text{ in.}^4.$$

That is, a direct comparison of two moments of inertia, that of solid Trex board ($I = 0.895$ in.^{4}) and hollow GeoDeck board ($I = 0.771$ in.^{4}), immediately shows that if they were made from the same material and tested in the same conditions and at the same span, a transition from the solid profile to the hollow profile would result in 0.895/0.771 = 1.16 times weaker and 1.16 times more flexible board. However, a gain in weight of the material will be much more significant, namely, 1.94 times.

This simple calculation illustrates a whole concept of hollow profiles. In the above example, a loss of 16% of the mechanical properties gives almost a double economy in the material. What is more important is a matter of judgment of the manufacturer (or a commissioner) in every particular case.

Another example, for a bit more complicated profile, is of GeoDeck tongue and groove board (Fig. 7.2).

Here tongue and groove are not structural elements in terms of flex strength or modulus, because they do not provide support for a load when a single board is tested. Hence, the relevant dimensions in this case are as follows:
$B = 5.125$ in., $H = 1.27$ in., $b = 4.6875$ in., $h = 0.75$ in., and $w = 0.2$ in.

The formula for the moment of inertia is

$$I = (BH^3 - bh^3 + 2wh^2)/12 \tag{7.3}$$

and for the GeoDeck tongue and groove board with said dimensions

$$I = 0.724 \text{ in.}^4.$$

However, on a deck, when tongue and groove are locked and become structural elements, the moment of inertia is the same as that for the GeoDeck traditional board (see above, $I = 0.771$ in.^{4}). That is, tongue and groove provide 6% of board's rigidity.

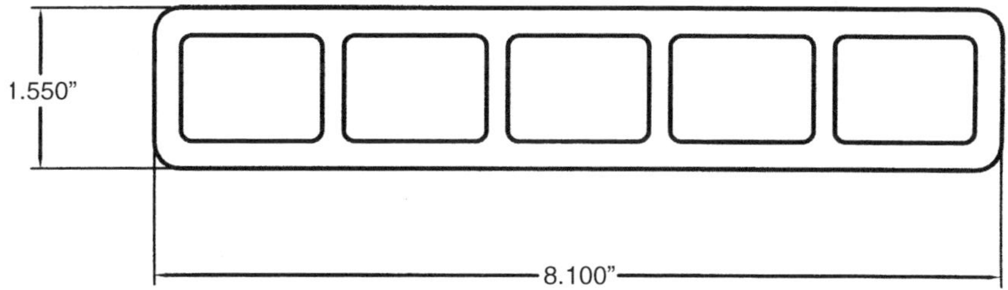

Figure 7.3 GeoDeck Heavy Duty composite board, cross section.

Finally, for GeoDeck Heavy Duty board (Fig. 7.3), which is a wider and a deeper profile, the moment of inertia is equal to

$$I = (BH^3 - bh^3 + 4wh^2)/12, \qquad (7.4)$$

where $B = 8.10$ in., $H = 1.55$ in., $b = 7.60$ in., $h = 1.05$ in., and $w = 0.20$ in.

and for GeoDeck Heavy Duty board with said dimensions

$$I = 1.858 \text{ in.}^4$$

As it will be shown later in this chapter, flexural strength and modulus, and the ultimate load (load at break, or load at failure) are all proportionally dependent on the moment of inertia. Thus, it can be concluded right away that a break load of a GeoDeck board, whatever its length would be, for Heavy Duty board will be 240% (140% higher) of that for Traditional board or Tongue and Groove board (on a deck).

Bending Moment

When a load is placed on a partially supported board, it sets up bending moment (M) along the board's length. This in turn causes a stress in the cross section of the board.

The bending stress is zero at the neutral axis (see above), increases linearly toward the outer surface of the board, and reaches maximum at the outer surface. This causes the board to bend, to deflect.

The bending stress at any point of the board is determined by the following formula:

$$\sigma = Mc/I, \qquad (7.5)$$

where σ is the bending stress, in psi; M is the bending moment, in in. lb; and c is the distance from neutral axis to the point at which stress is measured or calculated.

For rectangular deck boards, solid or hollow, the distance c is typically equal to $d/2$, that is, half-a-depth of the board. Hence,

$$\sigma = Mh/2I. \qquad (7.6)$$

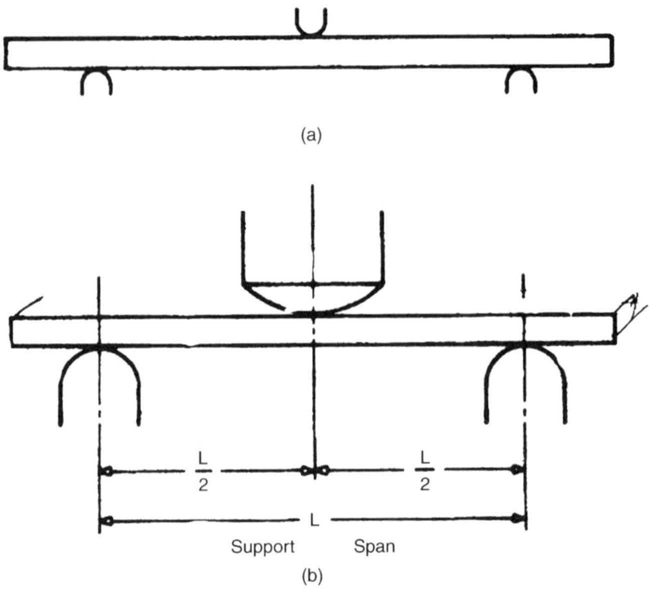

Figure 7.4 Three-point loading, allowable range of loading nose and support span. (a) Minimum radius 1/8 inch (3.2 mm), (b) maximum radius supports = 1.6 times specimen depth; maximum radius loading nose = 4 times the specimen depth. Source of the figure—ASTM D 790-03, Copyright ASTM International. Reprinted with permission.

The bending moment is determined from standard beam diagrams for different types of beam, such as fixed end beam (cantilever), guided end, simply supported end, partial distributed load, and their various combinations. We will consider only maximum bending moment, which results in a board failure, and only for four types of load application to the board resting on two supports with the distance (span) L between them:

- Concentrated loading at midspan, or 3-point (3-pt) loading (see Fig. 7.4).
- 1/3-pt loading (4-pt load), at which the board is loaded at two points (by means of two loading noses), each an equal distance from the adjacent support point. The distance between the loading noses (i.e., the load span) is one third of the support span (see Figs. 7.5 and 7.6).
- 1/4-pt loading (4-pt load), at which the board is loaded at two points (by means of two loading noses), each an equal distance from the adjacent support span. The distance between the loading noses is one half of the support span. That is, each of the two loads is applied at one fourth of the span from the respective end (see Fig. 7.7).
- Uniform loading.

The most severe stress for the given load is at 3-pt loading. Maximum moment at center (at midspan) is

$$M = PL/4, \tag{7.7}$$

where P = load, and L = span.

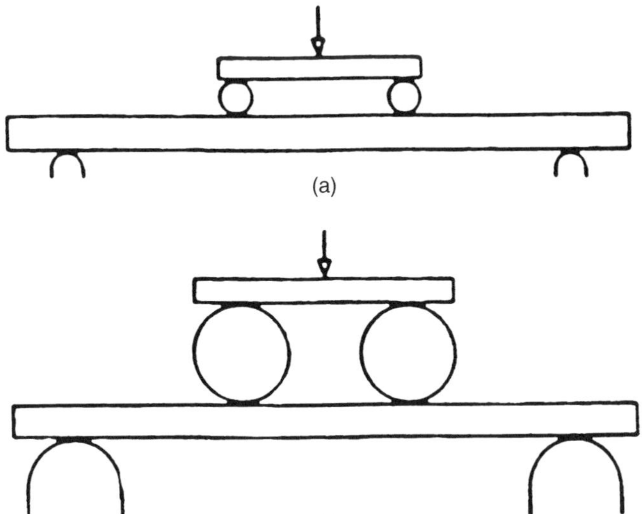

Figure 7.5 Four-point loading. The distance between the loading noses is one-third of the support span (see the loading diagram in Fig. 7.6). (a) Minimum radius = 1/2 inch (12.7 mm), (b) maximum radius supports and loading nose = 1.5 times the specimen depth. Source of the figure—ASTM D 6109-03, Copyright ASTM International. Reprinted with permission.

For 1/3-pt loading (center),

$$M = PL/6. \tag{7.8}$$

For 1/4-pt loading (center),

$$M = PL/8. \tag{7.9}$$

For uniformly distributed load,

$$M = WbL^2/8, \tag{7.10}$$

where W = a uniformly distributed load, in psi; and b = width of the specimen, in in.

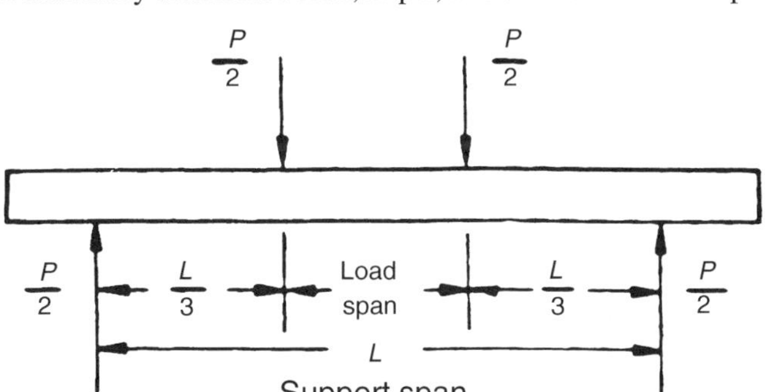

Figure 7.6 Four-point loading diagram. Source of the figure—ASTM D 6109-03, Copyright ASTM International. Reprinted with permission.

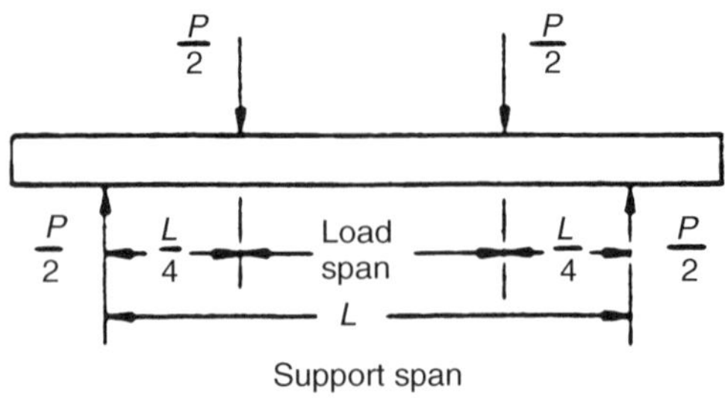

Figure 7.7 Quarter point loading diagram. Source of the figure—ASTM D 6272-02, Copyright ASTM International. Reprinted with permission.

Introducing the maximum moments to Eq. (7.6), we will obtain the bending stress for

- 3-pt load (center)

$$\sigma = PLh/8 \times I, \tag{7.11}$$

  where σ is the bending stress, in psi;
- 1/3-pt loading (center)

$$\sigma = PLh/12 \times I; \tag{7.12}$$

- 1/4-pt loading (center)

$$\sigma = PLh/16 \times I; \tag{7.13}$$

- Uniformly distributed load

$$\sigma = WbhL^2/16 \times 144 \times I \tag{7.14}$$

(the coefficient of 144 appeared due to transition from square inches to square feet). The above formulas will be employed in the next section for the determination of flexural strength and flexural modulus of composite deck boards.

ASTM RECOMMENDATIONS

ASTM D 790, "Standard Test Methods for Flexural Properties of Unreinforced and Reinforced Plastics and Electrical Insulating Materials"

Note of the author: This procedure was one of the most widely used of the mechanical property tests for wood–plastic composite (WPC) materials, as the specimen geometry is very simple as well as the testing equipment. However, it became more and more obvious that material properties, tested by this method, were not necessarily corresponding to the product properties, such as composite deck board.

Composite products are often not precisely uniform in their composition; sometimes they contain voids, their surface (particularly unbrushed) often contains more plastic compared to the "bulk" material, and so on. To make data obtained more realistic, ASTM D 6109 was developed (see below).

The ASTM procedure describes procedures for determining flexural properties of materials in the form of rectangular bars. The ASTM emphasizes that it is applicable to only those materials that break or fail in the outer surface within the 5% strain limit at midspan. The method employs a 3-pt loading system (concentrated load) applied at midspan of a simply supported beam (Fig. 7.4).

Note of the author: The maximum strain in the outer surface at concentrated load is equal to

$$r = 6Dh/L^2, \tag{7.15}$$

where r = maximum strain; D = midspan deflection at break, in in.; L = support span; and h = depth of the beam, in in.

Therefore, for a typical composite deck board (h = 1.25 in.) tested at 16-in. span (L), with the deflection at a break of 1 in., maximum strain is

$$r = 6 \times 1 \text{ in.} \times 1.25 \text{ in.}/256 \text{ in.}^2 = 0.0293,$$

and this value of the maximum strain (2.93%) is within the norm of ASTM D 790. The same is valid for thinner composite deck boards, such as 1 in., 15/16 in., and 13/16 in. (1.0″, 0.9375″, and 0.8125″, respectively). These are all commercially available composite deck boards. For them at 16-in. span and 1-in. deflection at break, the value of the maximum strain is in the range of 2.34–1.90%.

The limit can be exceeded at a shorter span and/or a larger deflection at break, such as at 12-in. span:

- for a 1.25-in. thick board and 1-in. deflection at break (maximum strain of 5.2%), or
- for a 13/16-in. thick board and 1.5-in. deflection at break (maximum strain of 5.1%).

ASTM D 790 is subdivided into two procedures, procedure A and procedure B. The only difference between them is in the crosshead speed. Procedure A was designed for materials that break at comparatively small deflections, and the crosshead motion should be such as to provide the rate of straining of the outer surface of 0.01 min^{-1}. It can be used for measurement of flexural strength and modulus. Procedure B was designed for materials that break at comparatively large deflections, and the rate of straining of the outer surface should be 0.10 min^{-1}, that is, 10 times as fast for the same thickness of the specimen. It can be used for measuring flexural strength only. Procedure B is normally employed if the specimen does not rupture within the 5% strain limit when procedure A is used.

The ASTM provides a formula for calculation of the rate of crosshead motion:

$$R = ZL^2/6h, \tag{7.16}$$

where R = rate of crosshead motion, in in./min; L = support span, in in.; Z = rate of straining of the outer surface, in min^{-1} (0.01 in procedure A and 0.10 in procedure B); and h = depth of beam, in in.

For example, for a composite deck board of 1.25 in. in depth and at 16-in. support span, the rate of crosshead motion should be (procedure A)

$$R = 0.01 \times 256/6 \times 1.25 = 0.34 \text{ in./min.}$$

For procedure B, it should be 3.4 in./min.

The ASTM procedure indicates that a support span-to-depth ratio should be 16:1. Decreasing the ratio much below 16:1 would move the test from a flexural mode to a shear mode and would effectively increase the apparent flexural strength of the material to very high values. The procedure does not recommend to use the ratio below 14:1. However, for some highly anisotropic composites, the procedure recommends to avoid shear effects as much as possible, particularly when flexural modulus data are required, and increase the span-to-depth ratio to 20:1, 32:1, 40:1, and even to 60:1.

The procedure specifies a testing machine, loading noses and supports (see Fig. 7.4), micrometers for measuring the width and depth of the test specimen, and conditioning of the test specimens. At least five specimens for each sample should be tested according to the procedure. It should be pointed that according to AC 174, flexural testing of minimum 15 specimens is recommended.

The readout of the procedure is the load and the respective deflection until failure. The load at which the specimen fails is called the ultimate load, or load at failure.

The stress in the outer surface occurring at the midspan is calculated using the formula (see eq. (7.11) above)

$$\sigma = PLh/8 \times I, \tag{7.17}$$

σ is the bending stress, in psi.

For a solid rectangular specimen, for which the moment of inertia $I = bh^3/12$,

$$\sigma = 3PL/2bh^2, \tag{7.18}$$

where σ = stress in the outer surface at midspan, in psi; P = load at a given point on the load–deflection curve, in lb; L = support span, in in.; b = width of beam tested, in in.; and h = depth of beam tested, in in.

For the maximum bending stress, which occurs at the ultimate load (point of rupture),

$$S = 3PL/2bh^2, \tag{7.19}$$

where S = the maximum bending stress, or flexural strength, in psi; and P = the ultimate load, or load at failure.

The above formula is provided in ASTM D 790.

In a more general form, Eq. (7.19) can be represented as

$$S = PLH/8 \times I. \tag{7.20}$$

In many test reports and specifications, the flexural strength is called MOR, or modulus of rupture.

ASTM D 790 also describes determination of the modulus of elasticity (MOE), or flexural modulus, which is the ratio of stress to corresponding strain. It is calculated by drawing a tangent to the steepest initial straight-line portion of the load–deflection curve, which is essentially a load at which the specimen deflects by 1 in.

The ratio of stress (Eq. (7.11)) to strain (Eq. (7.15)) for a solid rectangular beam is equal to

$$E = PL^3/4bh^3D \tag{7.21}$$

where E = modulus of elasticity (flexural modulus) in bending, in psi; P = the load at the steepest initial straight-line portion of the load–deflection curve, in lb; L = support span, in in.; b = width of beam tested, in in.; h = depth of the beam tested, in. in.; and D = deflection at the load P.

This formula can be slightly simplified as

$$E = L^3m/4bh^3, \tag{7.22}$$

where m = slope of the tangent to the initial straight-line portion of the load–deflection curve, in lb/in.

Equation (7.22) is provided in ASTM D 790.

In a more general case, for other configuration of profiles, solid or hollow, Eq. (7.22) can be rewritten as

$$E = PL^3/48 \times I \times D$$

or

$$E = L^3m/48 \times I \tag{7.23}$$

where I = moment of inertia, in in.^{4}; and m = slope of the tangent to the initial straight-line portion of the load–deflection curve, in lb/in.

This equation is applicable for the concentrated (3-pt) load.

Precision of flexural modulus measurements is usually good. ASTM D 790-97 lists examples for six different plastics tested in six different laboratories. For within-laboratory tests, standard deviations of the average were 0.8 ± 0.6% of the principal figure, and for between-laboratory tests, standard deviations of the average were 1.2 ± 0.8%.

ASTM D 6109, "Standard Test Methods for Flexural Properties of Unreinforced and Reinforced Plastic Lumbers"

As it was mentioned above, ASTM D 6109 was developed to take into account some nonuniformity of material. Therefore, the procedure covers the determination of flexural properties of profiles with rectangular or square cross sections, using "as manufactured" pieces without any altering or machining of surfaces beyond cutting to length. Hence, this procedure evaluates the flexural properties of profiles as a "product" and not as a "material." When both are tested, material often shows better mechanical properties compared to the profile.

Note of the author: The method aims at testing manufactured products composed of more than 50% (w/w) of plastic. However, it is accepted as the principal method in AC 174 for composite decking and railing without any restrictions regarding plastic content.

The principal difference between ASTM D 790 and D 6109, besides testing a material and a product (see above), is that ASTM D 790 employs a concentrated load (3-pt) test, whereas in ASTM D 6109 the beam is loaded at two points (by means of two loading noses), each an equal distance from the adjacent support point. The distance between the loading noses (i.e., the load span) is one third of the support (Figs. 7.5 and 7.6).

Unlike ASTM D 790, in which the specimen should fail (ruptured) within 5% maximum outer surface strain, in ASTM D 6109 this should occur before a maximum outer surface strain of 3% is reached.

Note of the author: The maximum strain in the outer surface at 1/3 span loading is equal to

$$r = 4.71 \times Dh/L^2, \tag{7.24}$$

where r = maximum strain; D = midspan deflection at break, in in.; L = support span; and h = depth of the beam, in in.

It is about 27% lower compared to the strain for 3-pt load (concentrated load), see Eq. (7.15). For a typical composite deck board (h = 1.25 in.) tested at 16-in. span (L), with the deflection at a break of 1 in., maximum strain is

$$r = 4.71 \times 1\text{ in.} \times 1.25\text{ in.}/256\text{ in.}^2 = 0.023,$$

and this value of the maximum strain (2.3%) is within the norm of ASTM D 6109. The same is valid for thinner composite deck boards, such as 1 in., 15/16 in., and 13/16 in. (1.0″, 0.9375″, and 0.8125″, respectively). These are all commercially available composite deck boards. For them at 16-in. span and 1-in. deflection at break, the value of the maximum strain is in the range of 1.49–1.84%.

The limit can be exceeded at a shorter span and/or a larger deflection at break, such as at 12-in. span:

- for a 1.25-in. thick board and 1-in. deflection at break (maximum strain of 4.1%), or
- for a 13/16-in. thick board and 1.5-in. deflection at break (maximum strain of 4.0%).

ASTM D 6109 is subdivided into two procedures, procedure A and procedure B. The difference between them is that procedure A was designed to test products in the flat, or "plank," position, whereas procedure B tests products in the edgewise, or "joist," position. Hence, different requirements for the rate of straining of the outer surface and, respectively, for the crosshead motion. For procedure A the rate of straining of the outer surface should be 0.01 min^{-1}; for procedure B it should be in the range of 0.002–0.003 min^{-1}.

The ASTM provides a formula for calculation of the rate of crosshead motion:

$$R = 0.185 \times ZL^2/h, \tag{7.25}$$

where R = rate of crosshead motion, in in./min; L = support span, in in.; Z = rate of straining of the outer surface, in min^{-1} (0.01 in procedure A and 0.002–0.003 in procedure B); and h = depth of beam, in in.

For example, for a composite deck board of 1.25 in. in depth and at 16-in. support span, the rate of crosshead motion should be (procedure A)

$$R = 0.185 \times 0.01 \times 256/1.25 = 0.38 \text{ in./min.}$$

For procedure B, it should be 0.076–0.11 in./min.

The ASTM procedure indicates that a support span-to-depth ratio should be 16:1. Decreasing the ratio much below 16:1 would move the test from a flexural mode to a shear mode and would effectively increase the apparent flexural strength of the material to very high values. The procedure does not recommend to use the ratio below 14:1 and above 20:1.

Note of the author: ASTM D 6109 does not mention the effect of some anisotropic composites on flexural modulus (and flexural strength, but in a lesser degree) at a reduced span-to-depth ratio (less than 16:1, and particularly less than 10:1). Examples are given below, in the next section. In order to avoid shear effects as much as possible, particularly when flexural modulus data are required, an increased span-to-depth ratio (equal or higher than 16:1) should be considered.

The procedure specifies a testing machine, loading noses and supports (see Figs. 7.5 and 7.6), and conditioning of the test specimens. At least five specimens for each sample should be tested according to the procedure. It should be pointed that according to AC 174 "Acceptance criteria for deck board span ratings and guardrail systems (guards and handrails)," flexural testing of minimum 15 specimens is recommended.

The readout of the procedure is the load and the respective deflection until failure.

The stress in the outer surface occurring at the midspan is calculated using the formula (see Eq. (7.12))

$$\sigma = PLh/12 \times I,$$

where σ = bending stress in outer surface throughout load span, in psi; P = load at a given point on load–deflection curve, in lb; L = support span, in in.; and I = moment of inertia.

In a more general form, Eq. (7.12) can be represented as

$$S = PLh/12 \times I, \tag{7.26}$$

where S is the maximum bending stress, of flexural strength, in psi.

For a solid rectangular specimen, for which the moment of inertia $I = bh^3/12$,

$$\sigma = PL/bh^2, \tag{7.27}$$

where b is the width of beam tested, in in.; and h = depth of beam tested, in in.

For the maximum bending stress, which occurs at the ultimate load (point of rupture),

$$S = PL/bh^2, \tag{7.28}$$

where S = the maximum bending stress, or flexural strength, in psi; and P = the ultimate load, or load at failure.

The above formula is provided in ASTM D 6109.

ASTM D 6109 also describes determination of the modulus of elasticity, which is the ratio of stress to corresponding strain. It is calculated by drawing a tangent to the steepest initial straight-line portion of the load–deflection curve, which is essentially a load at which the specimen deflects by 1 in.

The ratio of stress (Eq. (7.28)) to strain (Eq. (7.24)) for a solid rectangular beam is equal to

$$E = PL^3/4.71 \times bh^3D, \tag{7.29}$$

where E = modulus of elasticity (flexural modulus) in bending, in psi; P = the load at the steepest initial straight-line portion of the load–deflection curve, in lb; L = support span, in in; b = width of beam tested, in in.; h = depth of the beam tested, in. in; and D = deflection at the load P.

This formula can be slightly simplified as

$$E = L^3m/4.71bh^3, \tag{7.30}$$

where m = slope of the tangent to the initial straight-line portion of the load–deflection curve, or the load resulting in 1-in. deflection at midspan, in lb/in.

Eq. (7.30) in a slightly modified form is provided in ASTM D 6109.

In a more general case, for other configuration of profiles, solid or hollow, Eq. (7.29) can be rewritten as

$$E = L^3m/56.5 \times I, \tag{7.31}$$

where I = moment of inertia, in in.^{4}

This equation is applicable for the 4-pt load (1/3-span loading).

ASTM D 6272, "Standard Test Methods for Flexural Properties of Unreinforced and Reinforced Plastics and Electrical Insulating Materials by Four-Point Bending"

ASTM D 6272 was developed for the same purpose as ASTM D 790 (see above) but for two variants of four-point bending, that is, one third of support span, and one half of support span. These are also called third-point load and quarter-point load, respectively (Figs. 7.5 and 7.6, and 7.7, respectively).

Unlike ASTM D 6109, which also employs third-point load (Figs. 7.5 and 7.6), ASTM D 6262 procedure does not use "as manufactured pieces" (though it is not excluded), but rectangular solid bars cut or molded directly. Both third-point load and quarter-point load according to ASTM D 6272 are applied to a simply supported beam.

According to the procedure, a beam of rectangular cross section rests on two supports and is loaded at two points by means of two loading noses, each an equal distance from the adjacent support point. The distance between the loading noses (the load span) is either one third (Figs. 7.5 and 7.6) or one half (Fig. 7.7) of the support span.

The ASTM procedure indicates that a support span-to-depth ratio should be 16:1. Decreasing the ratio much below 16:1 would move the test from a flexural mode to a shear mode and would effectively increase the apparent flexural strength of the material to very high values. The procedure does not recommend to use the ratio below 14:1 and above 20:1. However, for some anisotropic composites, the recommended support-to-depth ratios are 32:1, 40:1, and even 60:1, particularly when flexural modulus data are required.

The same as in ASTM D 790, the specimen should fail (ruptured) within 5% maximum outer surface strain. As it was described above (see ASTM D 6109), the maximum strain (at midspan) in the outer surface at 1/3 span loading is equal to (see Eq. (7.24))

$$r = 4.71 \times Dh/L^2,$$

where r = maximum strain; D = midspan deflection at break, in in.; L = support span; and h = depth of the beam, in in.

whereas for quarter-point loading, the maximum strain (at midspan) in the outer surface is equal to

$$r = 4.36 \times Dh/L^2 \tag{7.32}$$

or 8% less (38% less compared to center-point load, or 3-pt load). This difference comes from a difference between the location of the maximum bending moment and maximum axial fiber stress. In 4-pt bending (both third-point load and quarter-point load), the maximum axial fiber stress is uniformly distributed between the loading noses. In center-point load, the maximum axial fiber is located immediately under the loading nose.

Note of the author: That is why center-point loading is the most severe one among the three modes of loading, and ultimate load (break-point load) for center-point load is supposed to be two times lower compared to quarter-point load, and 1.5 times lower compared to third-point load, when the same materials and profiles are tested. In reality, there might be some slight deviations from these coefficients (1.5 and 2.0). If deviations are noticeable, there should be some reasons: anisotropicity of materials, too short support span (i.e., a noticeable effect of shearing compared to flexural), nonhomogeneity of the material, some deviations of the specimen from flat position, and so forth.

Overall, ASTM D 6272 is similar to ASTM D 790 and ASTM D 6109. It describes procedure A and procedure B in the same terms as those in ASTM D 790. The only differences in ASTM D 6272 concern equations describing crosshead motion, strain and stress specifically for quarter-point load. If, such as in ASTM D 6109, the rate of crosshead motion for one-third load is (see Eq. (7.25))

$$R = 0.185 \times ZL^2/h,$$

where R = rate of crosshead motion, in in./min; L = support span, in in.; Z = rate of straining of the outer surface, in min^{-1} (0.01 in procedure A and 0.05 in procedure B); and h = depth of beam, in in. then the rate of crosshead motion for quarter-point load is

$$R = 0.167 \times ZL^2/h. \tag{7.33}$$

The procedure specifies a testing machine, loading noses and supports, and conditioning of the test specimens. At least five specimens for each sample should be tested according to the procedure. It should be pointed that according to AC 174, flexural testing of minimum 15 specimens is recommended.

The readout of the procedure is the load and the respective deflection until failure.

The stress in the outer surface occurring at the midspan is calculated using the formula

$$\sigma = PLh/12 \times I$$

for one-third load (see Eq. (7.12)),
and

$$\sigma = PLh/16 \times I$$

for quarter-point load (see Eq. (7.13)),
where σ = bending stress in outer surface throughout load span, in psi; P = load at a given point on load-deflection curve, in lb; L = support span, in in.; and I = moment of inertia.

For a solid rectangular specimen, for which the moment of inertia $I = bh^3/12$, Eq. (7.12) for one-third load transforms to Eq. (7.27)

$$\sigma = PL/bh^2$$

and Eq. (7.13) for quarter-point load transforms to

$$\sigma = 3PL/4bh^2, \tag{7.34}$$

where b = width of beam tested, in in.; and h = depth of beam tested, in in.

For the maximum bending stress, which occurs at the ultimate load, that is, at the moment of break (point of rupture), for third-point load (see Eq. (7.28))

$$S = PL/bh^2$$

and for quarter-point load

$$S = 3PL/4bh^2, \tag{7.35}$$

where S = the maximum bending stress, or flexural strength, in psi; and P = the ultimate load, or load at failure.

The above equations are provided in ASTM D 6272.

ASTM D 6272 also describes determination of the modulus of elasticity, which is the ratio of stress to corresponding strain. It is calculated by drawing a tangent to the steepest initial straight-line portion of the load–deflection curve, which is essentially a load at which the specimen deflects by 1 in.

The ratio of stress to strain for a solid rectangular beam for third-point load is equal to (see Eq. (7.29))

$$E = PL^3/4.71 \times bh^3D$$

and for quarter-point load

$$E = PL^3/5.81 \times bh^3D, \tag{7.36}$$

where E = modulus of elasticity (flexural modulus) in bending, in psi; P = the load at the steepest initial straight-line portion of the load–deflection curve, in lb; L = support span, in in.; b = width of beam tested, in in.; h = depth of the beam tested, in in.; and D = deflection at the load P.

This formula can be slightly simplified for third-point load as (see Eq. (7.30))

$$E = L^3m/4.71bh^3$$

and for quarter-point load

$$E = L^3m/5.83bh^3, \tag{7.37}$$

where m = slope of the tangent to the initial straight-line portion of the load–deflection curve, or the load resulting in 1-in. deflection at midspan, in lb/in.

The last two equations in a slightly modified form are provided in ASTM D 6272.

In a more general case, for other configuration of profiles, solid or hollow, Eq. (7.30) and (7.37) can be rewritten as

$$E = L^3m/56.5 \times I \tag{7.38}$$

for third-point load, and

$$E = L^3m/70 \times I \tag{7.39}$$

for quarter-point load,
where I = moment of inertia, in in.^{4}.

FLEXURAL STRENGTH OF COMPOSITE DECK BOARDS

English Units and SI Units

In this chapter and throughout the book, we will use English units for measurements, that is, the load in pounds and the specimen dimensions in inches. Hence, the units of stress and strength (maximum stress) are in pounds per square inch, that is, psi.

Many people and many ASTM procedures use SI system, which is not the same as the metric system. The SI units are derived from the metric system units. However, for example, the SI system does not allow the use of the centimeter as a unit. It uses meters.

Here is a short table with conversions of some English units to SI units and back (Table 7.1).

TABLE 7.1 Conversion factors between English units and SI units

1 in.	0.0254 m
1 in.^{2}	645 mm^2
1 lbf	4.448 N
1 psi	6,895 Pa
1000 psi	6.895 MPa
100,000 psi	689.5 MPa
1 m	39.370 in.
1 N	0.2248 lbf
1 kN	224.8 lbf
1 Pa (1 N/m^2)	1.4504×10^{-4} psi
1 MPa (10^6 N/m^2)	145.0 psi
1 GPa = 1000 MPa (10^9 N/m^2)	145,000 psi

Center Point Load, or Concentrated Load (3-pt Load)

First, let us consider how typical composite deck boards would satisfy the ASTM D 790 criterion on the 5% strain limit at midspan in the outer surface of boards (Table 7.2).

TABLE 7.2 Limiting values for deflection at midspan for 3-pt load for GeoDeck boards (1.25-in. depth)

Support span (in.)	Deflection at 5% (limiting) strain at break (in.)	Actual deflection at break (in.)	Actual values of maximum strain in the outer surface (%)
14	1.31	0.425 ± 0.005 (Traditional)	1.6% (Traditional)
		0.41 ± 0.01 (Tongue and Groove)	1.6% (Tongue and Groove)
		0.37 ± 0.01 (Heavy Duty)	1.8% (Heavy Duty)
20	2.67	0.79 ± 0.03 (Traditional)	1.5% (Traditional)
		0.70 ± 0.04 (Tongue and Groove)	1.3% (Tongue and Groove)
22	3.23	0.69 ± 0.01 (Heavy Duty)	1.3% (Heavy Duty)

As one can see, for GeoDeck composite deck boards, maximum strain in the outer surface between 14″ and 22″ support span is in the range of 1.3–1.8%. This is way below the limiting 5% value. Generally, composite deck boards easily satisfy this limiting criterion.

One more example, with a Trex sample (Winchester), GeoDeck composite material, and pressure-treated lumber (PTL), conducted in 1999 according to ASTM D 790 (a year is indicated here and in some examples below just to indicate that the tested boards and the respective results may not relate to the current production). Table 7.3 shows these data.

Another data for PTL are shown in Table 7.4.

Generally, these data show a typical difference in flexural strength between wood and WPC materials. In Table 7.3, this difference is almost five times in favor of wood; overall this difference is between 2 and 8 times. Despite those differences, maximum strain in the outer surface (i.e., strain at break) was in all the cases

TABLE 7.3 Flexural strength test results (center-point load) with a Trex and GeoDeck samples and PTL

Material	Width (in.)	Depth (in.)	Deflection at break (in.)	Maximum strain in the outer surface (%)	Ultimate load (lb)	Flex strength (psi)
Trex	1.04	1.19	0.50 ± 0.02	1.8 ± 0.1%	154	2197
PTL	1.06	1.01	0.53 ± 0.06	1.6 ± 0.2%	550	10,683
GeoDeck[a]	5.5	1.27	0.52	2.0%	836	2368

Support span 14″. Average data are shown.

[a] Crosshead speed 0.257 in./min, radii of supports 0.197 in., and radius of the loading nose 0.197 in.

TABLE 7.4 Flexural strength test results (center-point load) for a PTL sample

Width (in.)	Depth (in.)	Ultimate load (lb)	Flex strength (psi)
0.95	0.46	374	16,744

Support span 6″. Center-point load.

TABLE 7.5 Flexural strength test results (center-point load) with a GeoDeck samples

Material	Width (in.)	Depth (in.)	Deflection at break (in.)	Maximum strain in the outer surface (%)	Ultimate load (lb)	Flex strength (psi)
Composition 1	0.377	0.329	0.058 ± 0.012	2.9 ± 0.6%	39.4	2897
Composition 2	0.340	0.305	0.070 ± 0.010	3.2 ± 0.5%	30.6	2902
Composition 3	0.304	0.302	0.100 ± 0.020	4.5 ± 0.9%	27.3	2954
Composition 4 (long fiber)	0.287	0.285	0.090 ± 0.004	3.8 ± 0.2%	32.9	4234

Average data are shown.
Support span 2″

TABLE 7.6 Flexural strength test results (third-point load) for Trex composite deck board (material, machined surface)

Width (in.)	Depth (in.)	Ultimate load (lb)	Flex strength (psi)
5.0	1.172	787 ± 13	2292 ± 38

Support span 20″. Average data are shown.

below 5% (see also Table 7.5 for GeoDeck). Another Trex flexural data, measured in October 1999, are shown in Table 7.6.

One can see that two different test data for Trex boards (Tables 7.3 and 7.6) obtained with samples of two different sizes (with 5.0″ and 1.04″), different support spans (20″ and 14″), and using two different load applications (third-point load and center-point load, respectively) gave flex strength values that differed by only 4% (2292 and 2197 lb).

Next year, in 2000, Kadant Composites has started to make its composite material, properties of which are shown in Table 7.7.

Support span should not—in theory—change flexural strength of profiles. However, in reality it does change results, to a certain extent (Table 7.8).

One can see that in all nine pairs of data, a longer support span gives slightly higher values of flexural strength for respective profiles. This is apparently a result of anisotropic nature of the extruded profiles of composite materials. The effect is not very significant for flex strength, typically around 6%. However, for flex modulus, as will be shown later, this effect is much higher.

TABLE 7.7 Flexural strength test results (center-point load) with GeoDeck composite deck board (Traditional)

Width (in.)	Depth (in.)	Ultimate load (lb)	Flex strength (psi)
5.48	1.26	906	2782

Support span 16″. Average data are shown. Moment of inertia = 0.823 in.^{4}

TABLE 7.8 Effect of support span on apparent flexural strength of composite deck boards

Profile	Moment of Inertia (in.^{4})	Support span (in.)	Ultimate load (lb)	Flexural strength (psi)
Board 1,	0.733	20	578	2462
composition A		14	851	2539
Board 1,	0.733	20	498	2121
composition B		14	801	2390
Board 2,	0.688	20	481	2186
composition A		14	748	2380
Board 2,	0.688	20	418	1880
composition B		14	614	1954
Board 3,	1.779	22	1002	2354
composition A		14	1664	2487
Board 4,	0.873	20	596	2219
composition A		14	971	2530
Picket,	0.696	28	183	1741
composition A		14	404	1841
Picket,	0.696	28	191	1817
composition B		14	451	2056
Handrail,	1.14	32	350	2149
composition A		14	848	2279

Center-point load.

Third-Point Load (4-pt Load, or 1/3-Span Load)

Third-point load (Fig. 7.8) is less aggressive load in terms of the maximum strain in the outer surface of the profile under the load compared to center-point load. As it was discussed earlier, the maximum strain in the outer surface at 1/3-span load

Figure 7.8 Flexural third-point load test in progress.

TABLE 7.9 Limiting values for deflection at 4-pt load (1/3-span load) for GeoDeck boards (1.25-in. depth)

Support span (in.)	Deflection at 3% (limiting) strain at break (in.)	Actual deflection at break (in.)	Actual values of maximum strain in the outer surface (%)
20	2.04	1.13 ± 0.09 (Traditional) 0.92 ± 0.05 (Tongue and Groove)	1.7 ± 0.1% (Traditional) 1.4 ± 0.1% (Tongue and Groove)
22	2.47	0.97 ± 0.10 (Heavy Duty)	1.5 ± 0.2% (Heavy Duty)

TABLE 7.10 Limiting values for deflection at midspan for 4-pt load (third-point load) for commercial composite deck board and actual data

Board	Depth (in.)	Deflection at 3% (limiting) strain at break (in.)	Actual deflection at break (in.)	Actual values of maximum strain in the outer surface (%)
Trex	1–1/4	2.04	1.57 ± 0.16	2.31 ± 0.24
Fiberon	31/32	2.63	1.67 ± 0.11	1.90 ± 0.12
GeoDeck	1–1/4	2.04	1.13 ± 0.09	1.66 ± 0.13
Choice Deck	1–1/4	2.04	1.03 ± 0.20	1.52 ± 0.29
UltraDeck	1–3/32	2.33	0.96 ± 0.05	1.24 ± 0.06

Support span 20 in. in all the cases.

(Fig. 7.6) is about 27% lower compared to the strain for center-point load. Hence, third-point load is closer to a uniformly distributed load than a load at midspan. Tables 7.9 and 7.10 show the actual values of maximum strain in the outer surface (i.e., strain at break). Despite those differences, maximum strain in the outer surface (i.e., strain at break) was in all the cases well below 3%.

It was mentioned earlier that if the support span-to-depth ratio is noticeably less than 16:1, this would move the test from a flexural mode to a shear mode and would effectively increase the apparent flexural strength of the material to very high values. This effect is illustrated in Table 7.11

The second value of flexural strength, at a higher span-to-depth ratio (average 16.1), is 44% higher compared to that at a longer support span (average 4.3).

Flexural Strength of Composite Deck Boards

The following values, listed in Tables 7.12 and 7.13, were determined in December of 2000 and July 2003, respectively.

It should be noted that a shorter support span (such as 14 in., as in Table 7.13) and center-point load give somewhat higher figures for flexural strength compared to a longer span (20 in.) and four-point load (see, e.g., Table 7.12).

TABLE 7.11 Flexural strength for GeoDeck composite material (cut from a deck board) at two different span-to-depth ratios for third-point load

Width (in.)	Depth (in.)	Span (in.)	Span-to-depth ratio	Ultimate load (lb)	Flex strength (psi)
1.91	0.239	3.82	16.0	68	2381
1.91	0.227	3.82	16.8	71	2756
1.91	0.246	3.82	15.5	83	2743
				Average 2630 ± 210 psi	
1.32	0.238	0.988	4.15	294	3897
1.37	0.218	0.988	4.53	255	3870
1.30	0.227	0.988	4.35	243	3584
				Average 3780 ± 170 psi	

A comparison of Tables 7.3, 7.5–7.8, and 7.11–7.13 allows us to outline some pattern regarding flexural strength of composite deck boards, as materials and the whole profiles. Here are the main features of this pattern:

- Flexural strength of the *material* is typically higher than that of the *profile.*
  Example with Trex:
  – for the material, S = 2197 psi (Table 7.3, 3-pt load) and 2292 psi (Table 7.6, 4-pt load, machined sample);
  – for the profile, S = 1625 psi (4-pt load).
  Example with GeoDeck:
  – for the material, S = 2897 psi (Table 7.5, 3-pt load) and 2630 psi (Table 7.11, 4-pt load);
  – for the profile, S = 2782 psi (Table 7.7, 3-pt load), 2462 psi (Table 7.8, 3-pt load), and 2319 psi (Table 7.12, 4-pt load).
- Flexural strength appears to be higher when tested using 3-pt load compared with 4-pt load.
  Examples with GeoDeck, material:

TABLE 7.12 Flexural strength values for actual composite deck boards, determined using ASTM D 6109

Board	Depth (in.)	Profile	Moment of inertia (in.^{4})	Load at failure (lb)	Flexural strength (psi)
Trex (Winchester)	1–1/4	Solid	0.895	698 ± 88	1625 ± 205
Choice Deck	1–1/4	Solid	0.832	781 ± 53	1956 ± 133
Fiberon	31/32	Solid	0.407	874 ± 38	3467 ± 151
GeoDeck	1–1/4	Hollow	0.733	816 ± 24	2319 ± 68
UltraDeck	1–3/32	Hollow	0.460	843 ± 17	3341 ± 67

Support span 20 in. in all the cases. Third-point load.

TABLE 7.13 Flexural strength values for actual composite deck boards (and PTL as a reference), determined using ASTM D 790

Board	Depth in.	Profile, plastic	Moment of inertia, in.^{4}	Load at failure, lb	Flexural strength, psi
PTL	1.0	Solid	0.458	5211	19,911
Ponderosa pine					9,400 (literature data)
CorrectDeck (brown)	15/16	Solid, PP	0.339	1687	6,202
CorrectDeck (yellow)	7/8	Solid, PP	0.276	1424	6,001
CorrectDeck (red)	15/16	Solid, PP	0.373	1587	5,303
Fiberon(Buff Cedar)	15/16	Solid, PE	0.360	817 ± 11	3721 ± 48[a]
Rhino Deck(one-side brushed)	1.0	Solid, PE	0.450	913 ± 62	3551 ± 243[b]
Rhino Deck(embossed)	1.0	Solid, PE	0.446	877 ± 54	3440 ± 213[b]
WeatherBest	15/16	Solid, PE	0.390	790 ± 24	3368 ± 102[c]
Nexwood	1–1/4	Hollow, PE	0.663	931 ± 18	3072 ± 59[d]
Perfection	15/16	Solid, PE	0.347	642 ± 10	2995 ± 47[e]
Evergrain (Epoch)	15/16	Solid, PE	0.393	690 ± 13	2919 ± 55
Boardwalk	1–1/16	Solid, PVC	0.564	821 ± 12	2726 ± 40
GeoDeck	1–1/4	Hollow, PE	0.734	909 ± 35	2709 ± 104
Integrated Composite Technology	1–1/2	Hollow, PE	1.11	1088 ± 8	2573 ± 19
ChoiceDeck (gray)	1.27	Solid, w/ grooves, PE	0.923	1018 ± 17	2451 ± 41
ChoiceDeck (yellow)	1.18	Soild, w/ grooves, PE	0.732	832 ± 10	2347 ± 28
Trex (Saddle)	1–1/4	Solid, PE	0.895	900 ± 25	2200 ± 61[f]
TimberTech	1–1/2	Engineered, PE	1.193	904 ± 12	1989 ± 26[g]
EverX	1–1/16	Hollow, PE	0.524	551 ± 8	1957 ± 28[h]
Trex (Madeira)	1–1/4	Solid, PE	0.895	769 ± 15	1879 ± 37[f]

Support span 14 in. in all the cases. Center-point load. Data by Dr. Tatyana Samoylova and the author, unless indicated. Footnotes to the table show flexural strength data obtained by manufacturers and listed in the respective technical data sheets.

Data published by the manufacturer.

[a] Fiberon, 2814 psi.

[b] Rhino Deck, 3160 psi.

[c] WeatherBest, 3069 psi (5/4″ × 6″); 1925 psi (2″ × 6″).

[d] Nexwood, 2346 psi (board), 3098 psi (material); 2710 psi (Tongue and Groove), 3240 psi (REB).

[e] Perfection, 2814 psi.

[f] Trex, 1423 psi; 1465 psi (reported to CCMC, Canada, as 10.1 MPa [CCMC 13125-R]).

[g] TimberTech, 3260 psi.

[h] EverX, 2650 psi.

Other data by manufacturers:

Carefree, 2050 psi;

Timberlast (Kroy Building Products), 2800 psi;

USPL, 1700 psi;

XTENDEX (Composite Building Products International), 3000 psi (board), 3420 psi (material);

Durawood (Eaglebrook Products), 1762 psi;

Comptrusion Corp., 2108 psi;

TriMax (filled with fiberglass), 2900 psi;

PlasTEAK, 1320 psi (plastic decking lumber);

LifeLong Decking (Brite Manufacturing), 2510 psi (solid), 1930 psi (hollow).

– 3-pt load, $S = 2897$ psi (Table 7.5);
– 4-pt load, $S = 2630$ psi (Table 7.11).
Examples with GeoDeck, proflile:
– 3-pt load, $S = 2782$ psi (Table 7.7) and 2462 psi (Table 7.8);
– 4-pt load, $S = 2319$ psi (Table 7.12).

- Flexural strength appears to be higher when tested at span-to-depth ratio lower than 16:1 and particularly when lower than 10:1.
  Examples are given in Tables 7.8 and 7.11.

Hence, direct comparisons of flexural strength data published on different composite deck boards should be viewed with the respective reservations. Data on flexural modulus, as it is shown below, are subjected to these factors even much greater.

Flexural Strength of Materials Versus Profiles

In 2004–2005 American Architectural Manufacturers Association (AAMA) had arranged an independent testing of commercial composite boards provided by six manufacturers. All boards were tested in the same conditions and using the same equipment and the same operator, and data are shown in Table 7.14.

Figures in Table 7.14 do not reflect a shape of the composite board, size of profile (solid or hollow), but just the formulation—type and amount of plastic, type and amount of fillers, coupling agents, other additives, and so on.

It is interesting that one of the strongest materials (B) in Table 7.14 turned out to be one of the weakest deck profile (or the weakest one, considering the error margins). It can be a result of an inconsistency of the final product across the profile (voids, areas of different density).

TABLE 7.14 Flexural strength of materials and profiles of six commercial composite boards

	Flexural strength (psi)	
Manufacturer (code)	Material	Profile
A	3300 ± 290	4400 ± 570
B	3100 ± 310	2855 ± 50
C	2900 ± 180	3100 ± 100
D	2900 ± 100	3040 ± 40
E	2200 ± 180	3140 ± 80
F	2100 ± 210	2700 ± 260

The boards were tested according to ASTM D 790 (materials; center-point load) and ASTM D 6109 (profiles; third-point load). The specimens were 5″ by ½″ by ¼″ (materials) and "as is" (profiles). The support span was 4.0″ (materials) and 16″ (profiles), and the crosshead speed was 0.110 in./min (materials). Five samples (materials) and 15 samples (profiles) of each product were tested. Boards were provided by six different commercial manufacturers. The order of materials in the table is from higher to lower flex strength values of materials and does not indicate the particular manufacturer.

TABLE 7.15 Flexural strength for a series of different profiles made from the same composite material

Profile	Support span (in.)	Moment of inertia (in.^{4})	Ultimate load (lb)	Flexural strength (psi)
6 × 6 Post	60	47.04	2917 ± 163	1860 ± 104
4 × 4 Support rail	56	8.828	1104 ± 38	2334 ± 80
4 × 4 Post	52	8.828	929 ± 26	1824 ± 51
2 × 8 Deck Board	22	1.779	1371 ± 70	2147 ± 110
Banister rail	32	1.140	473 ± 29	1937 ± 119
Soundwall Board	20	0.873	862 ± 22	2139 ± 54
Traditional Board	20	0.733	816 ± 24	2317 ± 68
2 × 2 Picket	28″	0.696	281 ± 9	1708 ± 55
Tongue and Groove Board	20″	0.688	669 ± 27	2026 ± 82
			Average 2032 ± 219 psi	

Third-point load was used from all the measurements.

Flexural Strength for the Same Material but for Different Profiles

Table 7.15 shows an example when the same composite material (GeoDeck) was extruded into different profiles, and flexural strength was determined for each of them.

The table shows that the flexural strength values vary for different profiles in a rather narrow range of ± 11%. There is not any correlation of flex strength with a shape, span, moment of inertia, ultimate load, and so forth. Apparently, the variations result from deviations of the profile from their "theoretical" behavior when stressed and strained.

Comparison of Center-Point Load and Third-Point Load

As it was shown above, center-point loading is the most severe among the three modes of loading—center-point, third-point, and quarter-point loading. Theoretically, ultimate load (break-point load) for center-point load should be 1.5 times lower compared to third-point load, when the same materials and profiles are tested. In reality, there might be some slight deviations from this coefficient due to anisotropicity of materials, too short support span (i.e., a noticeable effect of shearing compared to flexural), nonflat position of a specimen, and so forth.

Some examples of this effect are given in Table 7.16.

Table 7.16 shows that for most cases the ratio between ultimate loads for center-point load and one-third point load is not exactly 1.5, as should follow from Eq. (7.20) and (7.26). It is around 1.50 only for square cross-sectional pickets. For all flat profiles (including an almost oval-shaped handrail profile), the ratio is equal to 1.38 ± 0.04 (an average for seven profiles). Apparently, this deviation from the theoretical figure of 1.5 is a reflection of anisotropic properties of the tested composite materials.

TABLE 7.16 Comparison of center-point load and third-point load data for GeoDeck composite deck boards

Profile	Support span (in.)	Type of the load	Ultimate load (lb)	Ratio between 4-pt and 3-pt load ultimate loads	Flexural strength (psi)
Board 1,	20	Center	578	1.41	2462
composition A		One-Third	816		2317
Board 1,	20	Center	498	1.35	2121
composition B		One-Third	673		1911
Board 2,	20	Center	481	1.39	2186
composition A		One-Third	669		2027
Board 2,	20	Center	418	1.32	1880
composition B		One-Third	550		1667
Board 3,	22	Center	1002	1.37	2354
composition A		One-Third	1371		2147
Board 4,	20	Center	596	1.45	2219
composition A		One-Third	862		2139
Picket,	28	Center	183	1.54	1741
composition A		One-Third	281		1708
Picket,	28	Center	191	1.50	1817
composition B		One-Third	287		1744
Handrail,	32	Center	350	1.35	2149
composition A		One-Third	473		1937

The data were obtained using the same batch of boards on the same day, using the same machine.

This results in lower values for flexural strength determined by using 4-pt load compared to 3-pt load (Table 7.16). On average, the difference is $9 \pm 3\%$ in favor of center-point load compared to third-point load in terms of flexural strength test values.

Quarter-Point Load (4-pt Load, 1/4-point Load)

Quarter-point loading is defined as loading the beam at two points (by means of two loading noses), each at an equal distance from the adjacent support point. The distance between the loading noses (i.e., the load span) is one half of the support. In other words, the loading noses are located at one fourth of the span from each end of the beam.

Quarter-point loading was prescribed to test deck boards and railing systems in AC 174 (an older version, effective May 1, 2002) by the following language: "Quarter-point loading shall be used for this (flexural) test. When a stair tread performance rating is desired, both the quarter-point load test and a center-point load test shall be used." In the subsequent September 2004, June 2005, and June 2006 editions, the recommendation regarding quarter-point load was removed, and third-point load was recommended instead, by referring to ASTM D 7032 and D 6109.

A common reason for using quarter-point loading is that it is more "gentle" compared with both concentrated load (3-pt load) and 1/3-span loading, as it is obvious from equations for their respective bending moments. For 3-pt loading, maximum bending moment at midspan is

$$M = PL/4,$$

for 1/3-span loading (four-point loading) it is

$$M = PL/6,$$

and for quarter-point loading it is

$$M = PL/8.$$

That is, the last bending moment is the lowest one among the three modes at the same load and the same support span. It is considered that quarter-point loading is close to uniform loading in this regard.

The stress in the outer surface occurring at the midspan for quarter-point loading is calculated using the formula (see Eq. (7.13))

$$\sigma = PLh/16 \times I,$$

where σ is the bending stress in outer surface throughout load span, in psi; P = load at a given point on load–deflection curve, in lb; L = support span, in in.; and I = moment of inertia.

For the maximum bending stress, which occurs at the ultimate load (point of rupture),

$$S = PLh/16 \times I, \qquad (7.40)$$

where S = the maximum bending stress, or flexural strength, in psi; and P = the ultimate load, or load at failure.

Examples of quarter-point loading and the respective flexural strength of two GeoDeck composite profiles are given in Table 7.17.

The flex strength values in Table 7.17 differ by 27%. Apparently, such a difference is a result of a different span-to-depth ratio in these two tests. The ratio was 12.8 for Traditional board and only 10.4 for Heavy Duty board, which was a significant

TABLE 7.17 Flexural strength values for GeoDeck Traditional and Heavy Duty—Commercial composite deck boards, determined using quarter-point load

Board	Moment of Inertia (in.^{4})	Ultimate load (lb)	Flexural strength (psi)
Traditional	0.709	1521 ± 66	2682 ± 116
Heavy Duty	1.779	2433 ± 55	2106 ± 47

In both the cases the support span is 16 in. Data of August, 2002.

deviation from ASTM D 6109 and ASTM D 790 requirements calling for the ratio being preferably 16 or above, and not below 14 (see above). There might have been some other reasons for a lower flex strength for the second board—not exactly a flat position during the testing, for example.

Uniformly Distributed Load

Uniformly distributed load is not tested typically at testing facilities because of some technical difficulties. Commonly, tests are conducted using one of those 3-pt or 4-pt loads, described above, and the uniform load is calculated using standard equations.

For a uniformly distributed load on a straight beam (elastically stressed) with its left and right ends simply supported [1], the maximum fiber stress, or flexural strength, can be expressed by the following formula:

$$S = WbhL^2/16 \times 144 \times I \tag{7.41}$$

where W = uniformly distributed load, in lb/ft^2; b = width of the beam, in in.; h = depth of the beam, in in.; L = support span, in in.; and I = moment of inertia, in in.^{4}

Coefficient 144 just reflects a transition from square inches to square feet.

Code Section 1606.1 of the *BOCA National Building Code/1999* requires the minimum uniformly distributed live load to be 100 lb/ft^2 for main floors, exterior balconies, and other structural systems. In order to evaluate the related properties of composite decking components, the following are calculations of stress imposed on said components at their flexural failure under superimposed uniformly distributed load.

It was shown above that the flexural stress at center-point load is described by the formula (7.20)

$$S = PLH/8 \times I.$$

At the moment of break, load P equals to the ultimate load, or load at failure, and S equals to the flexural strength value. In order to reach the same maximum stress in the outer surface of the board, the uniformly distributed load (W) should be as follows:

$$WbhL^2/16 \times 144 \times I = PLH/8 \times I;$$

that is, for a center-point load (P)

$$W = 288 \times P/bL \tag{7.42}$$

or, similarly, for third-point load

$$WbhL^2/16 \times 144 \times I = PLH/12 \times I,$$

$$W = 192 \times P/bL. \tag{7.43}$$

TABLE 7.18 Flexural strength of GeoDeck Traditional deck boards at different temperatures

Temperature (°F)	Ultimate load (lb)	Flexural strength (psi)
−10	1120 ± 18	3580 ± 60
74	641 ± 4	2050 ± 10
120	402 ± 7	1280 ± 20

Center-point load. Support span 15 in. Loading nose radius 0.25 in., support nose radii 0.50 in., and test speed 0.290 in./min.

Or, finally, for quarter-point load

$$WbhL^2/16 \times 144 \times I = PLH/16 \times I,$$

$$W = 144 \times P/bL. \tag{7.44}$$

One can see that quarter-point load (in psi) is indeed equivalent to uniformly distributed load (in psf) with respect to stress in the outer surface. That is why quarter-point load is often employed as a proxy for uniformly distributed load.

Effect of Temperature on Flexural Strength of Composite Materials

Generally, HDPE-based composite deck boards reduce their flexural strength by about 30–60% when temperature changes from ambient to 120–130°F. Three examples with GeoDeck boards are given in Tables 7.18–7.20, and several more cases are given in Table 7.21.

An average difference between the values of flexural strength at 70°F and 125°F in this particular case is 1.47 ± 0.18.

TABLE 7.19 Flexural strength of GeoDeck Tongue and Groove deck boards at different temperatures

Temperature (°F)	Ultimate load (lb)	Flexural strength (psi)
−20	1403 ± 28	3080 ± 60
74	870 ± 27	1910 ± 60
130	673 ± 17	1480 ± 40

Quarter-point load. Support span 20 in.

TABLE 7.20 Flexural strength of GeoDeck Tongue and Groove deck boards at different temperatures

Temperature (°F)	Ultimate load (lb)	Flexural strength (psi)
−20	1002 ± 62	3530 ± 210
74	705 ± 19	2490 ± 60
125	494 ± 25	1760 ± 90

Third-point load with a support span of 24 in. (data of July 2006).

TABLE 7.21 Flexural strength of profiles of six commercial composite boards

Manufacturer (code)	Flexural strength (psi) 70°C	125°F
A	4400 ± 570	3500 ± 200
B	3140 ± 80	2320 ± 120
C	3100 ± 100	2310 ± 90
D	3040 ± 40	1750 ± 80
E	2855 ± 50	1830 ± 100
F	2700 ± 260	1700 ± 160

The boards were tested according to ASTM D 6109 (profiles; third-point load). The support span was 16″. Fifteen samples of each product was tested. Boards were provided by the manufacturers for testing to AAMA. The order of materials in the table is from higher to lower flex strength values and does not indicate the particular manufacturer.

Effect of Commercial HDPE Materials on Flexural Strength of Composite Deck Boards

As it was said in this book and not once, commercial HDPE provide a whole spectrum of polymers in terms of their molecular weight distribution even at the same values of specific gravity and melt flow index. They all have different mechanical properties and different flowability, which, of course, affects their performance, and flexural strength of the respective product among different properties.

Table 7.22 shows a series of flexural strength values obtained for the same composite deck boards (the same composition and profile), but made from different HDPE (and LLDPE in one case).

TABLE 7.22 Flexural strength values for Traditional, and Tongue and Grove GeoDeck deck boards, made using different HDPE (and LDPE in one case) commercial materials

Deck board profile	Manufacturer or a trade name, abbreviated	Ultimate load, lb	Flexural strength, psi
Traditional	P (Sheet and profile extrusion grade)	880 ± 12	2430 ± 33
	P(Blow molding grade)	847 ± 17	2375 ± 48
	P (LLDPE)	770 ± 20	2160 ± 56
Tongue and Grove	P (ethene–butene copolymer)	1067 ± 5	3227 ± 15
	P (Sheet and profile extrusion grade)	1015 ± 8	3117 ± 25
	C	857 ± 12	2674 ± 37
	S	855 ± 2	2667 ± 6
	C	872 ± 15	2657 ± 46
	E	885 ± 17	2609 ± 50
	A (Sheet and profile extrusion grade)	871 ± 12	2540 ± 35
	T	812 ± 14	2512 ± 43
	T	778 ± 16	2407 ± 50
	D	681 ± 2	2125 ± 6

Names of manufacturers or trade names of HDPE/LLDPE are abbreviated to the first letter. Center-point load, with the support span of 14 in. for all the cases. Both ultimate load and flexural strength values are given, because dimensions of the boards were slightly different in some cases (apparently, reflecting different flowability of the materials).

TABLE 7.23 Effect of density (specific gravity) of GeoDeck composite pickets of the railing system on their flexural strength

Specific gravity (g/cm^3)	Flexural strength (psi)
1.03	1456
1.05	1650
1.08	1827
1.16	2270
1.24	2560

Obviously, flexural strength for composite deck boards made with different HDPE materials differs quite appreciably.

One more example shows flexural strength of a Tongue and Grove composite board of the same composition, except 8% (w/w) of HDPE was replaced with 8% (w/w) of LDPE. In this case, flexural strength decreased from 2850 ± 119 to 2698 ± 40 psi.

Effect of Density (Specific Gravity) of Composite Materials on Flexural Strength

When WPC materials are manufactured with the same formulation but at a different speed, using vented or nonvented extruders, different moisture content of cellulosic filler, and so on, the resulting profiles often have different density (specific gravity). As a rule, the higher the specific gravity, the higher the flexural strength (Table 7.23).

However, in some cases, particularly in a rather narrow range of specific gravity, the effect is not that consistent (Table 7.24)

Flexural Strength of Neat HDPE and Other Plastics, and Comparisons with that for WPCs

Neat HDPE (and LDPE, LLDPE) are typically not reported in terms of their flexural strength, because they normally do not break at conditions of ASTM D 790. Some data indicate that HDPE shows flex strength around 1400 psi. Filling HDPE with wood fiber, rice hulls, and other plant fiber material increases flexural strength of resulting composites—to about 1600–2200 psi for Trex composites and higher, for other composite materials, up to about 3000 psi. Further increase of flex strength of

TABLE 7.24 Effect of density (specific gravity) of GeoDeck composite deck boards on their flexural strength

Specific gravity (g/cm^3)	Moment of inertia of the boards (in.^{4})	Flexural strength (psi)
1.07	0.766	2251
1.10	0.751	2168
1.12	0.755	2397

Center-point load, support span 14 in.

TABLE 7.25 Flexural strength of HDPE–rice hulls composite deck boards

HDPE content (%, (w/w))	Deflection at failure (in.)	Load at failure (lb)	Flexural strength (psi)
37%	1.35 ± 0.17	673 ± 18	1911 ± 51
50%	>2.2 (did not break)	>800–900 lb (did not break)	>2400

Support span 20 in. Third-point load.

wood–HDPE composites, to about 3800 psi, is typically achieved by using coupling and crosslinking agents.

Flexural strength of neat polypropylene can reach 6000–7000 psi (and commercial wood–PP composites have flex strength in the range of 5300–6200 psi), of PVC—to 10,000–16,000 psi, and Nylon—to 14,000–16,500 psi (type 6, cast) and 8000 psi (6/10 Nylon). It is interesting, that a PVC-based solid composite board has flex strength of only about 2700 psi (Table 7.13).

Effect of Plastic Content on Flexural Strength of Composite Materials

When composite deck boards having a different plastic content are tested for flexural strength, data obtained are often interpreted for face value. For example, that increase in the plastic content leads to the increase in flexural strength.

In some cases it is observed indeed (Table 7.25).

However, increase in plastic content is often accompanied with change in other important factors, more directly related to flexural strength, such as change in the filler content (balancing the change of plastic content), density (specific gravity) of the boards, and so on. This effect is illustrated in Table 7.26.

A Deck Board Used as a Stair Tread

A deck board used as a stair tread has to withstand a much more severe flexural test compared to a regular deck board. This test is described in AC 174, and ASTM D 7032. Principally, the acceptance criteria aim at determining a span rating indicating the ability of stair tread to comply with the building code. For example, a deck span rating is typically 12/100 or 16/100, that is, the deck board is recognized by the code when installed on deck joists spaced maximum 12 or 16 in. on center, respectively, and supports the load of 100 lb/ft^2 multiplied by a safety factor of 2.5.

TABLE 7.26 Effect of HDPE content on flexural strength of GeoDeck deck boards

HDPE content (%, (w/w))	Specific gravity of the board (g/cm^3)	Flexural strength (psi)
35	1.16	2524 ± 204
37	1.21	2804 ± 40
39	1.22	2940 ± 10
41	1.24	3000 ± 10

Support span was 14 in. in all the cases.

According to AC 174, the performance of deck boards used as stair treads shall meet the following structural requirements, besides specified requirements to deck boards (referred to in ASTM D 7032):

- A minimum concentrated load of 750 lb (applied over a 2 × 2 in. area at midspan),
- A minimum concentrated load of 300 lb (average of 15 specimens) at midspan at the deflection at 1/180th of the test span,

or

- The maximum deflection of 0.125″ at 300-lb -concentrated load at midspan, plus adjustments for end use as stated in Section 5.4 of ASTM D 7032 (Creep-Recovery Test).

These requirements to deck boards used as stair treads are much more severe compared to requirements to regular deck board, among them an ability to hold a uniformly distributed load of 100 lb/ft^2 multiplied by a safety factor of 2.5, hence, 250 lb/ft^2.

In terms of strain in the board, a concentrated load of 750 lb (P) at midspan of a board of 5.5″ width (b) and 16″ support span (L) is equivalent to the uniformly distributed load (W) described by Eq. (7.42)

$$W = 288 \times P/bL,$$

that is, 288 × 750/5.5 × 16 = 2455 lb/ft^2. This load is 10 times higher than that required for performance of a regular board on deck.

Let us consider these criteria on a comparative basis in more detail.

As it was indicated above, to apply a uniformly distributed load testing is a technically difficult task; therefore, testing is usually done using three- or four-point load. As it was shown above, Eq. (7.42)–(7.44) describe equivalency between these loads in terms of their equal flexural stress at the outer surface:
For center-point load (see Eq. (7.42)),

$$W = 288 \times P/bL,$$

where P is a center-point load,
for third-point load (P is the respective total load, see Eq. (7.43))

$$W = 192 \times P/bL,$$

for quarter-point load (P is the respective total load, see eq. (7.44))

$$W = 144 \times P/bL.$$

Therefore, if load at failure for uniformly distributed load is 250 lb/ft^2 for a deck board of 5–1/2 in. wide and 16 and 20 in. long, the same stress will be reached for values, shown in Table 7.27.

All these values are much less than actual loads at failure for these boards, equal to 906 lb (center-point load, 16″ span, Table 7.7), 669–874 lb (third-point load, 20″ span, Tables 7.6 and 7.12), and 481–596 lb (center-point load, 20″ span, Table 7.8), that is, between 5 and 12 times lower than the actual flexural strength values. In

TABLE 7.27 Load at failure (ultimate load) equivalent to 250 lb/ft^2 for a uniformly distributed load for a deck board of 5–1/2 width

Load	Load in lb, equivalent to 250 lb/ft^2 at the uniformly distributed load at support span	
	16 in. at center	20 in. at center
Center-point load	76.4	95.5
Third-point load	114.6	143.2
Quarter-point load	152.8	191.0

other words, composite deck boards show 500–1200% of the flexural strength compared with that required by the building code.

Not so for the requirements for stair tread.

Let us consider GeoDeck deck boards used for stair tread at 16 in. at center. The code requires that a stair tread should hold a concentrated load of 750 lb, or about 10 times higher than the load in Table 7.27, that is, 10 times than the deck code requirement including 2.5 times safety factor (100 lb/ft^2 $\times$ 2.5).

An actual test with GeoDeck at 16 in. at center showed that of 906 lb, 21% above the stair tread code requirement.

The code also requires that a stair tread should deflect no more than span/180 under the concentrated load of 300 lb (on average of 15 specimens). A test showed that at a stair tread span of 16 in., the deflection at 16/180 = 0.0889 in., is reached at 301 $\pm$ 20 lb, just one pound above the requirement, on average.

These figures allow to calculate flexural modules of the GeoDeck deck board, using Eq. (7.45)

$$E = PL^3/48 \times DI, \tag{7.45}$$

where E = flexural modules (see the next section); P = load at the given deflection D, in lb; L = support span; I = moment of inertia (0.8233 in.^{4} in this case); and D = deflection at the given load P, in in.

At P = 301 lb, L = 16 in., I = 0.8233 in.^{4}, and D = 0.0889 in., flexural modulus is equal to 350,934 psi.

This is a rather high figure for flexural modulus for composite deck boards (see the next section). Hence, for many composite deck boards stair tread, in order to satisfy AC 174, support span for stair tread should be a step down from 16 in., that is, 12 in.

However, not all composite deck boards would satisfy even this low support span requirement. Trex deck board, for example, has flexural modulus equal to 175,000 psi (listed on Trex technical data sheet). Using Eq. (7.45), one can calculate that at L = 12 in., $I = bh^3/12$ = 0.895 (b = 5–1/2″, h = 1–1/4″), load at the deflection of L/180 = 0.0667″ would be equal to 290 lb. In order to satisfy the code for 12-in. stair tread, either the board should be thicker (such as in another Trex decking board brand, 1.50 in. thick), or flex modulus should be a little higher, at least of 181,000 psi.

Table 7.28 shows maximum spacing of constructions supporting commercial deck boards accepted as stair treads by ICC-ES and published in manufacturers ICC-ES reports.

TABLE 7.28 Maximum spacing of constructions supporting commercial deck boards accepted as stair treads by ICC-ES and published in manufacturer's ICC-ES reports

Brand name (composition, w/w)	Manufacturer	ICC-ES Report No.	Board dimension (in.)	Recommended span (o.c.) (in.)
CorrectDeck	Correct Building Products, LLC	ESR-1341 6/1/2005	5/4 × 6	16
GeoDeck	LDI Composites	ESR-1369 6/1/2006	1.25 × 5.5	16
Deck Lok (100% PVC)	Royal Crown Limited	ESR-1051 6/1/2005	1.5 × 5.9	16
Brock Deck (100% PVC)	Royal Crown Ltd.	NER-705 (Legacy) 5/1/2005	1.5 × 6	16
Genova (100% PVC)	Genova Products, Inc.	ESR-1904 6/1/2005	1.5 × 5.5	16
Presidio (100% PVC)	Westech Building Products, Inc.	NER-710 (Legacy) 7/1/2004	1.5 × 6 (nominal) 1.5 × 5.68 (actual)	16
XTENDEX,E-Deck	Carney Timber Company	NER-695 (Legacy) 11/1/2004	5/4 × 5–3/8 1.5 × 5.5	12 (5/4 board) 16 (2 × 6 board)
ChoiceDek, Dreamworks, LifeCycle, MoistureShield, A.E.R.T.	Advanced Environmental Recycling Technologies; Weyerhauser	NER-596 2/1/2006	5/4 × 6	12
Life Long®	Brite Manufacturing, Inc.	ESR-1279 6/1/2005	1.25 × 6 (nominal)	12
TimberTech®	TimberTech Limited	ESR-1400 6/1/2005	1.0 × 5.43 (actual)	12
Oasis	Alcoa Home Exteriors, Inc.	ESR-1425 6/1/2005	5/4 × 6	12
Cross Timbers	Elk Composite Building Products, Inc.	ESR-1590 6/1/2005	5/4 × 6 (nominal) 1.0 × 5.5 (actual)	12
UltraDeck	Midwest Manufacturing Extrusion	ESR-1674 11/1/2004	15/16 × 5–3/16 1.125 × 5–3/16	12
Millenium	Millenium Decking	ESR-1603 6/1/2006	1.0 × 6.6	12
Teck Deck (100% PVC)	Outdoor Technologies Inc.	21-26 (Legacy) 2/1/2004	1.5 × 5.625	12

TABLE 7.28 (*Continued*)

CertainTeed PVC Deck Plank (100% PVC)	CertainTeed Corp.	NER-605 4/1/2005	1.5 × 5.5	12
Perma–Deck (recycled 100% plastic)	Cascades Re-Plast	Legacy 21-91 5/1/2005	1.19 × 5.5 1.44 × 5.44	12
Liberty™ (100% PVC)	Outdoor Technologies Inc.	Legacy 22-39 5/1/2005	1.5 × 5.5	12
EON (100% polystyrene)	CPI Plastics Group Ltd	ESR-1300 3/1/2005	1.25 × 5.5	12
Trex® (50% wood fiber, 50% polyethylene)	Trex Company, Inc.	ESR-1190 6/1/2005	1.25 × 5.5 1.5 × 5.5 1.5 × 7.25	10.5 12 12
Boardwalk	CertainTeed Corp.	NER-576 3/1/2004	5/4 × 6	N/A
			2 × 4	8
			2 × 6	12
WeatherBest	Lousiana Pacific Corp.	ESR-1088 6/1/2005	15/16 × 5–7/16	9
EverX, Latitudes, Veranda	UFP Ventures II, Inc.	ESR-1573 6/1/2005	1.0 × 5.4 0.9 × 5.2	8
Premier	Composatron Manufacturing Inc.	ESR-1481 6/1/2005	1.0 × 5.44	8
Monarch (55% wood fiber, 40% HDPE)	Green Tree Composites, LLC	ESR-1084 2/1/2005	1.0 × 5.5	Not rated
Rhino Deck 50% wood fiber, 50% thermoplastic)	Master Mark Plastic Products	ESR-1461 6/1/2005	1.0 × 5.38	N/A
C-Clip Deck (PVC)	Kroy Building Products, Inc.	Legacy 21-90	2 × 8	N/A
Endura Board	Crawford Industries, LLC	ESR-1890 6/1/2005	5/4 × 6 (nominal) 1.0 × 5.5 (actual)	Outside the scope of the report
Fiberon, Perfection, Veranda	Fiber Composites, LLC	22-41 (Legacy) 10/1/2004	0.95 × 5.30 (actual)	N/A
Epoch/Evergrain	Epoch Composite Products, Inc	ESR-1625 6/1/2005 NER-630 (Legacy) 4/1/2006	1 × 6 2 × 6	N/A
Carefree	U.S. Plastic Lumber	97-63.01 12/1/04	1.5 × 5.5	N/A

Plastic lumber and WPC decks are also shown in this table.

FLEXURAL MODULUS OF COMPOSITE DECK BOARDS

Flexural strength of the material or the profile is basically, in a simplified description, a force applied per square inch at the sample failure, normalized with the support span and the specimen cross-sectional configuration. In the same manner, flexural modulus is a force applied per square inch and causing a certain deflection of the sample, normalized with the support span and the specimen cross-sectional configuration.

The principal readout of a flex strength test is a load at failure, which linearly depends on a support span, that is, on a support span in the first power. An increase of the support span leads to a proportional decrease of the load at failure for the same specimen.

The principal readout of a flex modulus is a load causing the given deflection, or, alternatively, a deflection caused by the given load. Both of them are much more sensitive to a support span compared to that of a load at failure in a flex strength test. The load/deflection ratio depends on a support span in the third power.

For wood–cellulose fiber composite materials, flex strength is measured typically in thousands of psi (pounds per square inch), whereas flex modulus is measured in hundreds of thousands of psi.

Flexural modulus is typically much more sensitive to fillers, support span, moisture, temperature, and other factors, compared to flexural strength. Also, flexural modulus rather than flexural strength is often a controlling factor in the applicability of composite decking and railing systems for stair tread and decking at certain support span. Many composite deck boards cannot satisfy the building code when using at 16 in. on center for stair tread and 24 in. on center for decks. Examples are given below in this chapter.

Center-Point Load, or Concentrated Load (3-pt Load)

ASTM procedures for measuring flexural modulus, as well as the criteria to satisfy the ASTM D 790 requirements on the limiting values of deflection, that is, 5% strain limit at midspan in the outer surface of boards, were described in the preceding section of this chapter (see Tables 7.2, 7.3, and 7.5).

Flexural modulus for center-point load is given by the following formula:

$$E = PL^3/48 \times DI, \tag{7.46}$$

where D = deflection, in in., at the load P, in lb; P = load, in lb; L = support span, in in.; and I = moment of inertia, in in.^{4}.

Table 7.29 shows data obtained for a Trex sample (Winchester), GeoDeck composite material, and PTL, conducted in 1999 according to ASTM D 790.

Another data for Trex and PTL are shown in Table 7.30.

Generally, these data show a typical difference in flexural modulus between wood and WPC materials. In Table 7.29, this difference is between 3.5 and 6 times in favor of wood; overall, this difference is typically between 2 and 10 times.

Another Trex flexural data, measured in December 2000, but using third-point load, are shown in Table 7.31.

TABLE 7.29 Flexural modulus test results (center-point load) with a Trex and GeoDeck samples and PTL

Material	Width (in.)	Depth (in.)	Flex modulus (psi)
Trex	1.04	1.19	214,000 ± 8,000
PTL	1.06	1.01	1,244,000 ± 63,000
GeoDeck[a]	5.5	1.27	358,000 ± 16,000

Support span 14″. Average data are shown.
[a] Crosshead speed 0.257 in./min, radii of supports 0.197 in., and radius of the loading nose 0.197 in.

One can see that three different test data for Trex boards (Tables 7.29–7.31) obtained with samples of three different sizes (with 1.04″, 1.16″ and 5.5″), different support spans (6″, 14″, and 20″), and using two different load applications (third-point load and center-point load) gave flex modulus values that differed by only 12% from the average (191,200 ± 23,750 psi). This might be explained that span-to-depth ratio, to which flex modulus is very sensitive, was not too much different in the three tests, that is, 12, 15, and 16.

Support span should not—in theory—change flexural modulus of profiles. However, in reality it does change results rather significantly (Table 7.32)

One can see that in all nine pairs of data, a longer support span gives significantly higher values of flexural modulus for respective profiles. This is apparently a result of anisotropic nature of the extruded profiles of composite materials. Although for flex strength this effect was around 6% (see the preceding section), for flex modulus it is much higher, about 28% on average.

Third-Point Load (4-pt Load, or 1/3-Span Load)

Limiting values for deflection at 4-pt load (1/3-span load) were described above, and it was shown that the maximum strain in the outer surface (i.e., strain at break) when composite deck boards are tested is typically below 3%, as it is prescribed by ASTM D 6109 (see Tables 7.9 and 7.10).

TABLE 7.30 Flexural modulus test results (center-point load) for a Trex and PTL sample

Material	Width (in.)	Depth (in.)	Flex modulus (psi)
Trex	1.16	0.41	166,600 ± 9,000
PTL	0.95	0.46	1,025,000 ± 56,000

Support span 6″. Center-point load.

TABLE 7.31 Flexural modulus test results (third-point load) for Trex composite deck board

Width (in.)	Depth (in.)	Flex modulus (psi)
5.5	1.25	193,000 ± 19,000

Support span 20″. Average data are shown.

TABLE 7.32 Effect of support span on apparent flexural modulus of composite deck boards

Profile	Moment of Inertia (in.^{4})	Support span (in.)	Deflection at failure (in.)	Flexural modulus (psi)
Board 1, composition A	0.733	20	0.79	371,000
		14	0.425	315,000
Board 1, composition B	0.733	20	0.93	276,000
		14	0.675	252,000
Board 2, composition A	0.688	20	0.70	373,000
		14	0.41	318,000
Board 2, composition B	0.688	20	1.29	226,000
		14	0.66	222,000
Board 3, composition A	1.779	22	0.69	357,000
		14	0.37	271,000
Board 4, composition A	0.873	20	0.84	353,000
		14	0.475	221,000
Picket, composition A	0.696	28	1.18	267,000
		14	0.38	209,000
Picket, composition B	0.696	28	1.58	299,000
		14	0.525	268,000
Handrail, composition A	1.14	32	1.51	353,000
		14	0.46	218,000

Center-point load.

It was mentioned earlier that if the support span-to-depth ratio is noticeably less than 16:1, this would move the test from a flexural mode to a shear mode and would effectively decrease the apparent flexural modulus of the material to much lower values. Data in Table 27 show, though, that for this effect to be quite noticeable, the span-to-depth ratio should not necessarily be much lower than 16:1. Even a move of the ratio from 16:1 to 11.2:1 results in 28% (on average) decrease of flexural modulus, such as from 300,000 to 234,000 psi, as an example.

Flexural Modulus of Composite Deck Boards

The following values, listed in Tables 7.33 and 7.34, were determined in December of 2000 and July 2003, respectively.

It should be noted that a shorter support span (such as 14 in., as in Table 7.34) and center-point load give, as a rule, lower figures for flexural modulus compared to a longer span (20 in.) and four-point load (see, e.g., Table 7.33). The difference between the two sets of data reaches 28–35%.

A comparison of Tables 7.33 and 7.34 (and Table 7.37 below) allows us to outline some patterns regarding flexural modulus of composite deck boards:

- Flexural modulus appears to be higher when tested using 4-pt load compared with 3-pt load.

TABLE 7.33 Flexural modulus values for actual composite deck boards, determined using ASTM D 6109

Board	Depth (in.)	Profile	Moment of inertia (in.^{4})	Deflection at failure (in.)	Flexural modulus (psi)
Choice Deck	1–1/4	Solid	0.832	1.03 ± 0.20	180,000 ± 20,000
Trex (Winchester)	1–1/4	Solid	0.895	1.57 ± 0.16	193,000 ± 19,000
Fiberon	31/32	Solid	0.407	1.67 ± 0.11	564,000 ± 74,000
GeoDeck	1–1/4	Hollow	0.733	1.18 ± 0.05	374,000 ± 39,000
UltraDeck	1–3/32	Hollow	0.460	0.96 ± 0.05	502,000 ± 27,000

Support span 20 in. in all the cases. Third-point load.

- Examples with GeoDeck:
  4-pt load, $E = 374{,}000$ psi,
  3-pt load, $E = 289{,}000$ psi.
- Examples with Fiberon:
  4-pt load, $E = 564{,}000$ psi,
  3-pt load, $E = 419{,}000$ psi.

- Flexural modulus appears to be lower when tested at span-to-depth ratio lower than 16:1

  Examples are given in Table 7.32.

Hence, direct comparisons of flexural modulus data published on different composite deck boards should be viewed with the respective reservations.

Flexural Modulus of Materials Versus Profiles

In 2004–2005 AAMA had arranged an independent testing of commercial composite boards provided by six manufacturers. All boards were tested in the same conditions and using the same equipment and the same operator, and data are shown in Table 7.35.

Figures in Table 7.35 does not reflect a shape of the composite board, size of profile (solid or hollow), but just the formulation—type and amount of plastic, type and amount of fillers, coupling agents, other additives, and so on. It should be noted here that only one tested board had the lowest both flex strength and flex modulus from the whole series of boards. The strongest board was not the stiffest one, and generally, there was no any correlation between flex strength and modulus of these boards.

One can see that for upper five profiles, flex modulus is close to each other, practically within the error margin of the test. Only the last material profile (F) is the most flexible one, and it is the weakest one (Table 7.35).

Flexural Modulus for the Same Material but for Different Profiles: Solid and Hollow Deck Boards

Table 7.36 shows an example when the same composite material (GeoDeck) was extruded into different profiles, and flexural modulus was determined for each of them.

TABLE 7.34 Flexural modulus values for actual composite deck boards (and PTL as a reference), determined using ASTM D 790

Board	Depth (in.)	Profile	Moment of inertia (in.^{4})	Flexural modulus (psi)
Ponderosa pine				1,290,000 (literature data)
PTL	1.0	Solid	0.458	1,248,180
CorrectDeck (yellow)	7/8	Solid	0.276	717,300 ± 9,300[a]
CorrectDeck (brown)	15/16	Solid	0.339	650,500 ± 8,500[a]
CorrectDeck (red)	15/16	Solid	0.373	531,400 ± 11,200[a]
WeatherBest	15/16	Solid	0.390	526,000 ± 23,000[b]
Evergrain (Epoch)	15/16	Solid	0.393	433,000 ± 15,000
Perfection	15/16	Solid	0.347	430,000 ± 9,000[c]
Fiberon(Buff Cedar)	15/16	Solid	0.360	419,000 ± 3,000[d]
Rhino Deck(one-side brushed)	1.0	Solid	0.450	368,000 ± 86,000
Rhino Deck(embossed)	1.0	Solid	0.446	314,000 ± 12,000
Integrated Composite Technology	1–1/2	Hollow	1.11	298,000 ± 3000
GeoDeck	1–1/4	Hollow	0.734	289,000 ± 12,000
Nexwood	1–1/4	Hollow	0.663	284,000 ± 3000[e]
EverX	1–1/16	Hollow	0.524	252,300 ± 1900[f]
ChoiceDeck (gray)	1.27	Solid, w/ grooves	0.923	219,900 ± 8500
Trex (Saddle)	1–1/4	Solid	0.895	176,500 ± 1100[g]
ChoiceDeck (yellow)	1.18	Solid, w/ grooves	0.732	175,700 ± 3200
Boardwalk	1–1/16	Solid	0.564	175,000 ± 6000[h]
TimberTech	1–1/2	Engineered	1.193	156,000 ± 8000[i]
Trex (Madera)	1–1/4	Solid	0.895	151,000 ± 19,000[g]
Carefree (neat HDPE)	1–1/5	Solid	0.658	93,400 ± 8000[j]

Support span 14 in. in all the cases. Center-point load. Data by Dr. Tatyana Samoylova and the author, unless indicated. Footnotes to the table show flexural modulus data obtained by manufacturers and listed in the respective technical data sheets.

Data published by the manufacturer:

[a] CorrectDeck (polypropylene-based), 850,000 psi.

[b] WeatherBest, 541,000 psi (5/4″ × 6″); 338,000 psi (2″ × 6″).

[c] Perfection, 561,000 psi.

[d] Fiberon, 561,000 psi.

[e] Nexwood, 310,000 psi (board), 496,000 psi (material); 331,588 psi (Tongue and Groove); 445,000 psi (REB).

[f] EverX, 420,000 psi.

[g] Trex, 175,000 psi; 224,800 psi (reported to CCMC, Canada, as 1550 MPa [CCMC 13125-R]).

[h] Boardwalk, 231,000 psi.

[i] TimberTech, 621,000 psi.

[j] Carefree, 110,000 psi.

Other data by manufacturers:

Timberlast (Kroy Building Products), 540,000 psi.

USPL, 220,000 psi (board); 290,000 psi (material).

XTENDEX (Composite Building Products International), 396,000 psi (board), 340,000 psi (material).

Durawood (Eaglebrook Products), 104,156 psi.

Comptrusion Corp., 504,600 psi.

TriMax (filled with fiberglass), 325,000 psi (board); 420,000 psi (material).

PlasTEAK, 200,000 psi and 238,000 psi (plastic decking lumber).

LifeLong Decking (Brite Manufacturing), 400,000 psi (solid); 391,000 psi (hollow).

TABLE 7.35 Flexural modulus of materials and profiles of six commercial composite boards

	Flexural modulus (psi)	
Manufacturer (code)	Material	Profile
A	470,000 ± 60,000	531,000 ± 27,000
B	420,000 ± 20,000	551,000 ± 31,000
C	390,000 ± 70,000	624,000 ± 91,000
D	390,000 ± 10,000	512,000 ± 35,000
E	290,000 ± 40,000	565,000 ± 67,000
F	260,000 ± 20,000	398,000 ± 55,000

The boards were tested according to ASTM D 790 (materials; center-point load) and ASTM D 6109 (profiles; third-point load). The specimens were 5″ by 1/2″ by 1/4″ (materials) and "as is" (profiles). The support span was 4.0″ (materials) and 16″ (profiles), and the crosshead speed was 0.110″ min (materials). Five samples (materials) and 15 samples (profiles) of each product were tested. Boards were provided by six commercial manufacturers to AAMA for tests. The order of materials in the table is from higher to lower flex strength values of materials and does not indicate the particular manufacturer.

One can see that flexural modulus values vary for different profiles in the range of ±12%, that is, approximately the same range as that for flexural strength (11%, see above). There is no any correlation with a shape, span, moment of inertia, deflection at failure, and so on. Apparently, the variations result from deviations of the profile from their "theoretical" behavior when stressed and strained.

When plotted, flexural strength values appear to show a linear correlation with flexural modulus (Fig. 7.9). It shows that in a wide range of the mechanical properties, fillers and coupling agents generally increase both strength and stiffness of composite materials. Upon transition from HDPE to polypropylene (three upper left points on the graph), both strength and stiffness of the composite materials are increased.

TABLE 7.36 Flexural modulus for a series of different profiles made from the same composite material

Profile	Support span (in.)	Moment of inertia (in.^{4})	Deflection at failure (in.)	Flexural modulus (psi)
6 × 6 Post	60	47.04	1.80 ± 0.09	277,000 ± 17,000
4 × 4 Support rail	56	8.828	2.73 ± 0.26	378,000 ± 29,000
4 × 4 Post	52	8.828	1.66 ± 0.12	328,000 ± 17,000
2 × 8 Deck Board	22	1.779	0.97 ± 0.10	367,000 ± 35,000
Banister rail	32	1.140	1.91 ± 0.12	361,000 ± 59,000
Soundwall Board	20	0.873	1.05 ± 0.08	281,000 ± 21,000
Traditional Board	20	0.733	1.18 ± 0.05	374,000 ± 39,000
2 × 2 Picket	28	0.696	1.57 ± 0.11	323,000 ± 78,000
Tongue and Groove Board	20	0.688	0.92 ± 0.05	389,000 ± 65,000
			Average 342,000 ± 42,000 psi	

Third-point load was used from all the measurements.

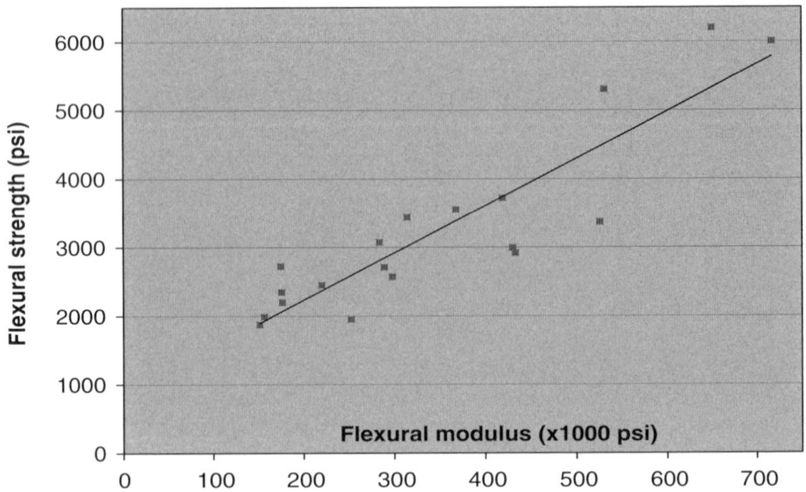

Figure 7.9 Flexural strength and flexural modulus for WPC deck boards listed in Tables 7.13 and 7.34.

Comparison of Center-Point Load and Third-Point Load

Flexural modulus for third-point load is given by the formula

$$E = PL^3/56.5 \times DI, \tag{7.47}$$

where D = deflection, inch, at the load P, in lbs; P = load, in lbs; L = support span, in in.; and I = moment of inertia, in in.^{4}.

Comparison of Eq. (7.47) with Eq. (7.46) shows that at the same support span and the load, the same profile (with the same moment of inertia) should deflect by 18% less under third (4-pt)-point load compared to concentrated, center-point load at midspan.

However, the above equations—theoretically—should give the same values for flexural modulus, despite center-point or third-point load was employed at the test. In reality, though, the data show some deviation from this rule. A series of examples is given in Table 7.37.

Table 7.37 shows lower values for flexural modulus determined by using 3-pt load compared to 4-pt load. On average, the difference is 9 ± 7% in favor of third-point load compared to center-point load in terms of flexural modulus test values. The difference is very close to that (9 ± 3%) for flexural strength values (see above).

Quarter-Point Load (4-pt Load, 1/4-Point Load)

Flexural modulus for quarter-point load is given by the formula

$$E = PL^3/70 \times DI,$$

where D = deflection, in in., at the load P, in lb; P = load, in lb; L = support span, in in.; and I = moment of inertia, in in.^{4}.

TABLE 7.37 Comparison of center-point load and third-point load data for GeoDeck composite deck boards

Profile	Support span (in.)	Type of the load	Deflection at failure (in.)	Ratio between 4-pt and 3-pt load deflection at failure	Flexural modulus (psi)
Board 1,	20	Center	0.79	1.49	371,000
composition A		One-Third	1.18		374,000
Board 1,	20	Center	0.93	1.45	276,000
composition B		One-Third	1.35		297,000
Board 2,	20	Center	0.70	1.31	373,000
composition A		One-Third	0.92		389,000
Board 2,	20	Center	1.29	1.26	226,000
composition B		One-Third	1.63		266,000
Board 3,	22	Center	0.69	1.41	357,000
composition A		One-Third	0.97		367,000
Board 4,	20	Center	0.84	1.25	253,000
composition A		One-Third	1.05		281,000
Picket,	28	Center	1.18	1.33	267,000
composition A		One-Third	1.57		323,000
Picket,	28	Center	1.58	1.25	299,000
composition B		One-Third	1.98		332,000
Handrail,	32	Center	1.51	1.26	353,000
composition A		One-Third	1.91		361,000

The data were obtained using the same batch of boards on the same day, using the same machine.

As one can see, quarter-point load is the most gentle of all (center-point, third-point, and quarter-point) types of loading. At the same load and support span, and for the same configuration (same moment of inertia), deflection of the profile is 46% lower (a ratio of 70/48) compared to center-point load, and 24% lower (a ratio of 70/56.5) compared to third-point load.

Examples of quarter-point loading and the respective flexural modulus of two GeoDeck composite profiles are given in Table 7.38.

The flex modulus values in Table 7.38 differ by 28%. Because the respective flexural strength values (Table 7.17) differed by 27%, it is tempting to suggest that this difference can be accounted to the respective deviation in one of the two values of the moment of inertia, which could easily happen due to a complex profile. It might

TABLE 7.38 Flexural modulus values for GeoDeck Traditional and Heavy Duty – Commercial composite deck boards, determined using quarter-point load

Board	Moment of Inertia (in.^{4})	Flexural modulus (psi)
Traditional	0.709	261,000 ± 20,000
Heavy Duty	1.779	204,000 ± 17,000

In the both cases the support span is 16 in. Data of August 2002.

have been, as it was suggested earlier, a result of a different span-to-depth ratio in these two tests. The ratio was 12.8 for Traditional board and only 10.4 for Heavy Duty board, which was a significant deviation from ASTM D 6109 and ASTM D 790 requirements calling for the ratio being preferably 16 or above, and not below 14 (see above). There might have been some other reasons for a lower flex strength for the second board—not exactly a flat position during the testing, for example.

Uniformly Distributed Load

Flexural modulus for uniformly distributed load on a straight beam (elastically stressed) with its left and right ends simply supported is given by the formula

$$E = 5WbL^4/384 \times 144 \times DI,$$

where W = uniformly distributed load, in lb/ft^2; I = moment of inertia, in in.^{4},; L = support span, in in.; b = specimen overall width, in in.; and D = deflection at the uniformly distributed load W.

Coefficient 144 in the formula reflects a transition from square inches to square feet.

Snow on a Deck

As it was described above, a deck should satisfy a building code requirement of the uniformly distributed load of 250 lb/ft^2. (100 lb/ft^2 of the requirements itself, with a factor of safety of 2.5). In some snowy areas there is a more stringent requirement for a deck of the uniformly distributed load of 400 lb/ft^2. Let us consider a joist support span rating for this latter case with an example of GeoDeck.

Strength At ambient (74°F) flexural strength of GeoDeck is 2490 psi and flexural modulus is 323,000 psi. At—20°F the figures are equal to 3530 and 564,000 psi, respectively (Tables 7.20 and 7.41). Because we consider a snow load, let us take a modest 32°F. Below this temperature GeoDeck is stronger and stiffer. A simple linear extrapolation (there is no reason to seriously doubt this assumption) shows that at 32°F flexural strength would be 2955 psi and flexural modulus 430,593 psi.

An alternative two sets of data (Tables 7.18, 7.19 and 7.39, 7.40) give us flexural strength and modulus for GeoDeck at ambient temperature, at −10 and at −20°F. Linear extrapolation of these numbers (flex strength and modulus separately, of course) to 32°F gives us flexural strength values of 2430 and 2820 psi, and flexural modulus of 445,000 and 360,000 psi, respectively (from two separate sets of data).

From Eq. (7.41) we get that the uniformly distributed load at failure is equal to

$$W = 16 \times 144 \times S \times I/bhL^2,$$

where W = uniformly distributed load, lb/ft^2; S = flexural strength, in psi; b = width of the board, in in.; h = depth of the board, in in.; L = support span, in in.; and I = moment of inertia, in in.^{4}.

Snow Load on a "Traditional" GeoDeck Hollow Board (Width 5.5 in., Depth 1.25 in., Moment of Inertia 0.784 in.^{4}, and Support Span of 16 in.) The ultimate uniformly distributed load at 32°F at support span of 16 in. for GeoDeck boards with flexural strength of 2955 psi (see above) equals to 3033 lb/ft^2. This more than seven times exceeds the snow load of 400 lb/ft^2. For the alternative flexure strength of values of 2430 and 2820 psi at 32°F, the ultimate uniformly distributed load equals to 2494 lb/ft^2 and 2894 lb, that is, again more that six and seven times, respectively, exceeds the snow load of 400 lb/ft^2. One can see that any number for flex strength of GeoDeck gives quite a safety margin (6–7 times) over the required load resistance of 400 lb/ft^2.

Snow Load on a "Heavy Duty" GeoDeck Hollow Board (Width 8.10 in., Depth 1.55 in., Moment of Inertia 1.858 in.^{4}, and Support Span of 24 in.) Using the extrapolated to 32°F flexural strength values of 2955, 2430 and 2820 psi (see above), one can obtain the ultimate uniformly distributed load at support span of 24 in. for the board equal to 1749 lb/ft^2, 1438 lb/ft^2, and 1669 lb/ft^2, that is, 3.6–4.4 times higher than the required 400 lb/ft^2 potential snow load.

Deflection Let us now consider deflection of a deck under such a snow load. For uniformly distributed load, the following equation for deflection at midspan is applicable:

$$D = 5WbL^4/384 \times 144 \times E \times I,$$

where D = deflection in midspan, in in.; and E = flexural modulus, in psi.

Snow Load on a "traditional" GeoDeck Hollow Board (Width 5.5 in., Depth 1.25 in., Moment of Inertia 0.784 in.^{4}, and Support Span of 16 in.) At 32°F under the snow load of 400 lb/ft^2. and at the flexural modulus of 431,000 psi, 445,000 psi, or 360,000 psi (the first was determined at 24-in. span under third-point load; second was determined at 20-in. span under quarter-point load, and the third was determined at 15-in. span under center-point load, which is less accurate), the deflections would be equal to 0.039, 0.037, and 0.046 in., respectively. For a support span of 16 in. (on center between joists), an allowable deflection is 16/180 = 0.089 in. This means that deflection under 400 lb of snow load is well within the allowable limit for all alternative experimental data.

Snow Load on a "heavy duty" GeoDeck Hollow Board (Width 8.10 in., Depth 1.55 in., Moment of Inertia 1.858 in.^{4}, and Support Span of 24 in.) Using the extrapolated to 32°F flexural modulus values of 431,000, 445,000, and 360,000 psi (see above), one can obtain the deflection of the heavy duty board at support span of 24 in. for the board equal to 0.12, 0.12, and 0.145 in., respectively. The allowed deflection is 24/180 = 0.133 in., which is good for the first two flex modulus values, and 9% off-limit for the third value of flex modulus (360,000 psi). However, in this case of snow load, it is irrelevant, because the sense of the allowable deflection is to avoid tripping when someone is walking on the deck. This is hardly applicable when deck is covered with 400 lb/ft^2 of snow.

TABLE 7.39 Flexural modulus of GeoDeck Traditional deck boards (I = 0.734 in.^{4}) at different temperatures

Temperature (°F)	Flexural modulus (psi)
−10	485,000 ± 18,000
74	242,000 ± 5000
120	109,000 ± 6000

Center-point load. Support span 15 in. Loading nose radius 0.25 in., support nose radii 0.50 in., and crosshead speed 0.290 in./min.

At–20°F the boards are about 20–30% stronger and 30–40% stiffer compared to that at 32°F. Hence, they are even more resistant to snow load compared to that at milder temperatures.

Effect of Temperature on Flexural Modulus of Composite Materials

Generally, HDPE-based composite deck boards reduce their flexural modulus by about 1.7–2.2 times when temperature changes from ambient to 120–130°F. This is much more compared to the respective change of flexural strength of HDPE-based composites, which is about 30–60% over the same temperature range. Three examples with GeoDeck boards are given in Tables 7.39–7.41, with an average change in said temperature range of 2.1 ± 0.3.

Table 7.42 shows data for several more commercial WPC boards in the temperature range of 70–125°F. The data are split into two groups: for one, flex modulus in said range has changed by 1.8 ± 0.1 times, that is, in the same range as that above for HDPE-based boards (within an error margin), and for another group the ratio was 1.3 ± 0.1 times (boards C and D). Clearly, the boards were based not on HDPE. They have distinctly higher flexural modulus at 125°F compared with HDPE-based boards.

TABLE 7.40 Flexural modulus of GeoDeck Tongue and Groove deck boards at different temperatures

Temperature (°F)	Flexural modulus (psi)
−20	623,000 ± 52,000
74	304,000 ± 20,000
130	137,500 ± 5600

Quarter-point load. Support span 20 in.

TABLE 7.41 Flexural modulus of GeoDeck Tongue and Groove deck boards at different temperatures

Temperature (°F)	Flexural modulus (psi)
−20	564,000 ± 21,000
74	323,000 ± 6000
125	187,000 ± 15,000

Third-point loading with a span of 24 in. (data of July 2006).

TABLE 7.42 Flexural modulus of profiles of six commercial composite boards

Manufacturer (code)	Flexural modulus (psi)	
	70°F	125°F
A	531,000 ± 27,000	282,000 ± 35,000
B	551,000 ± 31,000	298,000 ± 24,000
C	624,000 ± 91,000	474,000 ± 48,000
D	512,000 ± 35,000	391,000 ± 102,000
E	565,000 ± 67,000	339,000 ± 31,000
F	398,000 ± 55,000	235,000 ± 27,000

The boards were tested according to ASTM D 6109 (profiles; third-point load). The support span was 16″. Fifteen samples (profiles) of each product were tested. Boards were provided by six manufacturing companies to AAMA for testing. The order of materials in the table is from higher to lower flex strength values and does not indicate the particular manufacturer.

Effect of Commercial HDPE on Flexural Modulus of Composite Deck Boards

Commercial HDPE materials provide a whole spectrum of polymers in terms of their molecular weight distribution even at the same values of specific gravity and melt flow index. They all have different mechanical properties and different flowability, which, of course, affects their performance and flexural modulus of the respective product among different properties.

Table 7.43 shows a series of flexural modulus values obtained for the same composite deck boards (the same composition and profile) but made from different HDPE (and LLDPE).

TABLE 7.43 Flexural modulus values for Traditional and Tongue and Grove GeoDeck deck boards, made using different HDPE (and LLDPE) commercial materials

Deck board profile	Manufacturer or a trade name (abbreviated)	Flexural modulus (psi)
Traditional	P (Sheet and profile extrusion grade)	169,000 ± 22,000
	P (Blow molding grade)	172,000 ± 17,000
	P (LLDPE)	166,400 ± 6500
Tongue and Grove	P (ethene–butene copolymer)	254,500 ± 3200
	P (Sheet and profile extrusion grade)	217,000 ± 13,000
	C	226,000 ± 14,000
	S	204,700 ± 5,800
	C	205,000 ± 7000
	E	194,500 ± 4800
	A (Sheet and profile extrusion grade)	177,000 ± 21,000
	T	224,000 ± 8000
	T	179,000 ± 9000
	D	179,900 ± 3700

Names of manufacturers or trade names of HDPE/LLDPE are abbreviated to the first letter. Center-point load, with the support span of 14 in. for all the cases.

TABLE 7.44 Effect of density (specific gravity) of GeoDeck composite pickets of the railing system on their flexural modulus

Specific gravity (g/cm^3)	Flexural modulus (psi)
1.03	110,651
1.05	130,500
1.08	151,000
1.16	192,500
1.24	197,000

Obviously, flexural modulus values for composite deck boards made with different HDPE materials differ quite appreciably, in the range of 166,000–254,000 psi.

One more example shows flexural modulus of a Tongue and Grove composite board of the same composition, except that 8% (w/w) of HDPE was replaced with 8% (w/w) of LDPE. In this case flexural modulus decreased from 284,000 ± 2,000 to 239,000 ± 17,000 psi.

Effect of Density (Specific Gravity) on Flexural Modulus

When WPC materials are manufactured with the same formulation but at a different speed, using vented or nonvented extruders, different moisture content of cellulosic filler, and so on, the resulting profiles often have different density (specific gravity). As a rule, the higher the specific gravity, the higher the flexural modulus (Table 7.44).

Principally, the data are similar in their trend as those for neat polyethylene, shown in Figure 7.9.

Flexural strength values for the boards shown in Table 7.45 did not vary much, and the change was not consistent (see Table 7.24); however, flexural modulus trend with density was very clear.

Effect of Plastic Content on Flexural Modulus of Composite Materials

It is often assumed that increase of the plastic content in composite deck boards would decrease their flexural modulus. However, it is not always so. Increase in plastic content is often accompanied by changes in other important factors, more directly related to flexural modulus. Such factors are, for example, the filler content

TABLE 7.45 Effect of density (specific gravity) of GeoDeck composite deck boards on their flexural modulus

Specific gravity (g/cm^3)	Flexural modulus (psi)
1.07	182,840
1.10	215,040
1.12	261,225

Center-point load, support span 14 in.

TABLE 7.46 Effect of HDPE content on flexural modulus of GeoDeck deck boards

HDPE content (%, (w/w))	Specific gravity of the board (g/cm³)	Flexural modulus (psi)
35	1.16	164,000 ± 1000
37	1.21	155,600 ± 2300
39	1.22	158,250 ± 1000
41	1.24	155,500 ± 11,000

Support span was 14 in. in all the cases.

(balancing the change of the plastic content), density (specific gravity) of the boards, and so on. Besides, a change of the filler content is often accompanied by a change in moisture content in the formulation, hence, change in specific gravity of the final formulation. This effect is illustrated in Table 7.46.

As one can see, the increase in HDPE content does not result in any noticeable systematic change in flexural modulus of the material. All changes, if any, are within error margin of the obtained values of flex modulus.

Somewhat different data are shown in Table 7.47. Here three composite boards were compared; each one was a blend of HDPE and rice hulls. The overall depth of the boards were 1.25″ (37% HDPE) and 1.28″ (50% HDPE); cross-sectional areas were 3.425 in.^{2} (37% HDPE), and 4.019 in.^{2} and 3.859 in.^{2} (50% HDPE).

Here a significant increase in HDPE content lead to a significant decrease in flexural modulus of the boards. The 50% HDPE boards were slightly stronger compared to the 37% HDPE board (flex strength was 2542, 3.101 and 2420 psi, respectively) but significantly more flexible (Table 7.47). Because the 50% HDPE boards had a little thicker walls compared to those of the 37% HDPE board, along with a noticeable change in their density, there were too many poorly controllable factors to consider. Nevertheless, the flex modulus values were calculated considering the respective moments of inertia, which included the respective dimensional parameters. As expected, such a large increase in the HDPE content led to a large decrease of the flex modulus.

The same boards were tested using the support span of 20 in., compared to that of 14 in. in Table 7.47. As expected, their flexural strength values were significantly higher (Table 7.48).

Again, a significant increase in HDPE content led to a significant decrease in flexural modulus.

TABLE 7.47 Flexural modulus of three HDPE–rice hulls composite deck boards

HDPE content (%, (w/w))	Specific gravity of the board (g/cm³)	Weight of the board (lb/ft)	Flexural modulus (psi)
37%	1.12	1.66	252,000 ± 8000
50%	1.02	1.78	169,800 ± 4000
50%	0.97	1.63	174,200 ± 3000

Support span 14 in. Center-point load.

TABLE 7.48 Flexural modulus of HDPE–rice hulls composite deck boards

HDPE content (%, (w/w))	Deflection under 100 lb (in.)	Load at 1 in. deflection (lb)	Flexural modulus (psi)
37%	0.052 ± 0.006	1669 ± 146	322,000 ± 28,000
50%	0.065 ± 0.010	1202 ± 92	204,000 ± 16,000

Support span 20 in. Third-point load.

FLEXURAL MODULUS OF NEAT HDPE AND OTHER PLASTICS AND COMPARISONS WITH THAT FOR WPCs

As it was mentioned above, flexural strength data for neat polyethylene are typically not reported in the literature, because the respective samples normally do not break at conditions of ASTM D 790. They just bend. Therefore, data on flexural modulus of neat polyethylene are plentiful in the literature. They are often listed in specifications of plastics.

Table 7.49 contains such data. As examples, we consider polyethylene materials manufactured by Chevron Phillips Chemical Company (Woodlands, TX).

As one can see, LDPE in this list is the most flexible, with flex modulus between 30,000 and 50,000 psi; LLDPE (one sample) has flex modulus of 60,000 psi, and HDPE has flex modulus in the range of 125,000–240,000 psi.

There is a strong correlation ($R^2 = 0.984$) between density of all 38 polyethylene materials from Table 7.49, including LDPE, LLDPE, HDPE, and their flexural modulus (Fig. 7.10).

Filling with wood fiber, rice hulls, and other plant fiber material increases flexural modulus of resulting composites—to about 151,000–176,000 psi for Trex composites, 220,000–300,000 psi for other composite materials, including those containing minerals and not very efficient crosslinking (and/or coupling) agents, and 300,000–530,000 psi for HDPE-based composites, containing efficient crosslinking agents (see Table 7.34). Further increase of flex modulus of WPCs, to about 720,000 psi, is typically achieved by using polypropylene.

Flexural modulus of neat polypropylene is similar with that of HDPE and is in the range of 170,000–250,000 psi. Flex modulus for high-impact polypropylene is of 100,000–200,000 psi. It is interesting that flex modulus for wood fiber—polypropylene composites (CorrectDeck, Table 7.34) is much higher, in the range of 530,000–730,000 psi.

Flexural modulus for Nylon is in the range of 140,000–410,000 psi for different types of Nylon. For Nylon 6/10 it is in the lower range, 160,000–280,00 psi; for Nylon 6 and 6/6 flex modulus reaches 390,000 and 410,000 psi, respectively. For rigid PVC flex modulus is in the range of 380,000–540,000 psi [2]. For PVC-based composite (Boardwalk, Table 7.34), flex modulus equals to 175,000 psi, and it is one of the lowest on the market. Thus, for the PVC-based composite (Table 7.34), both flex strength and flex modulus are much lower than those for neat PVC, that is, one half to one third for flex modulus, and one fourth to one sixth for flex strength.

For AZEK cellular PVC board (width 5.51″, depth 0.745″, and moment of inertia 0.19 in.^{4}), flexural modulus equals to 165,480 psi. Though it is significantly lower compared

TABLE 7.49 Flexural modulus data for polyethylene materials (Marlex®) supplied by Chevron Phillips Chemical Company

Polyethylene, type (density, g/cm^3)	Marlex® Index	Flexural modulus (psi)
LDPE(0.917–0.925)	1003	30,000
	1007	30,000
	1009	30,000
	1122	40,000
	1400	47,000
	1412	48,000
	KN226	48,000
	1409	50,000
	4538	50,000
	5104	50,000
LLDPE (0.918)	7104	60,000
HDPE (0.943–0.964)	HMN TR-942	125,000
	HMN TR-945	132,000
	HHM TR-480X	140,000
	C513UV	155,000
	9005	155,000
	9503H	155,000
	HHM 4903	170,000
	HXM 50100	175,000
	9506H	175,000
	9018	175,000
	9006	185,000
	9012	185,000
	HHM 5202-02BN	190,000
	C514	190,000
	C516	190,000
	C590	200,000
	HXB TR-512	195,000
	9004	195,000
	9512H	195,000
	9514H	205,000
	HHM 5502BZ	210,000
	K605	220,000
	K608	220,000
	9708	230,000
	EHM 6007	240,000
	K606	240,000
	9402	240,000

Span-to-depth ratio was 16:1, crosshead speed 0.5 in./min.

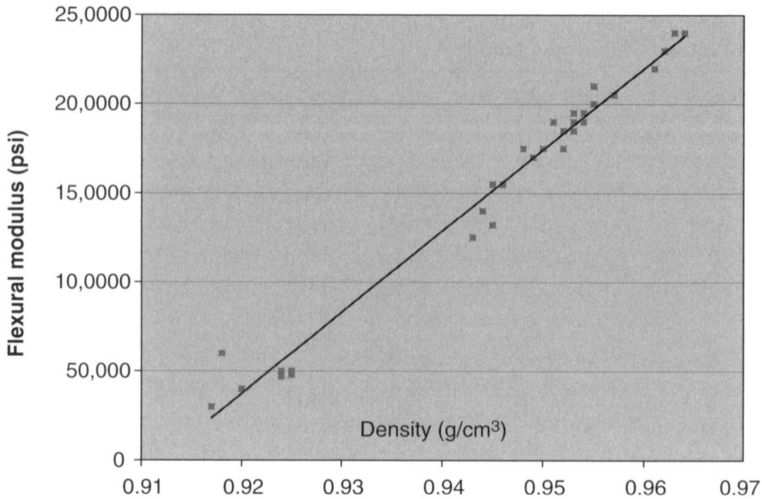

Figure 7.10 A correlation between polyethylene density and its flexural modulus. All materials are products of Chevron Phillips Chemical Company (see Table 7.49). LDPE, LLDPE, and HDPE are shown with densities of 0.917-0.925, 0.918, and 0.943-0.964, respectively.

to that for neat PVC (see above), it is still not too bad for a material with specific gravity of only 0.54 g/cm^3. Due to its low flex modulus (high flexibility), the board with a support span of 14″ did not break. At 201 lb center-point load, the board deflected to 0.393″.

A DECK BOARD USED AS A STAIR TREAD: A CRITICAL ROLE OF FLEXURAL MODULUS

Acceptance criteria for deck board span ratings and guardrail systems (AC 174) prescribe that the concentrated (center-point) load at the deflection at 1/180 of the span shall be recorded. According to the AC and 2000 International Building Code, Section 1607.1, this load should be minimum 300 lb (on average of 15 specimens). As an alternative requirement, the maximum deflection under 300 lb concentrated load should be 0.125 in. (Section 4.1.1 of AC 174).

Let us consider two examples of stair tread made of

- solid 5.5″ × 1.25″ board, such as Trex (I = 0.895 in.^{4}, E = 193,000 psi); and
- hollow 5.5″ × 1.25 board, such as GeoDeck Traditional (I = 0.733 in.^{4}, 374,000 psi).

At a span of 16 in. on center, deflection of stair tread under 300 lbs will be defined by the equation given above (7.46):

$$D = PL^3/48 \times EI,$$

where D = deflection, in in.; P = 300 lbs, center-point load; L = span, 16 in.; E = flexural modulus; and I = moment of inertia.

For the solid board (see above) at a span of 16 in., the allowed deflection is 16″/180 = 0.089″. In this particular case of the solid board, the deflection would be equal to 0.148″. Too much. The span would not pass.

Even for a span of 12 in., with the allowed deflection of 12″/180 = 0.067″, the deflection for this solid board under concentrated load of 300 lbs would be 0.063″, that is, only slightly less than the allowed one.

That is why Trex recommends its composite boards only for a 12-in. span, and only for thick 2 × 6- and 2 × 8-in. boards, when used as a stair tread. For a standard 5/4 × 6-in. board, Trex recommends only a maximum center-to-center spacing of 10.5 in. (ICC-ES Report ESR-1190, issued June 1, 2005).

For the hollow GeoDeck board (see above) at a span of 16 in. with the allowed deflection of 0.089″, the calculated deflection would be equal to 0.093″. Close, but not enough. To pass with these numbers for flex modulus (374,000 psi) and moment of inertia (0.733 in.^{4}), a span should be of 15–3/4″. In reality, the hollow GeoDeck Traditional board passed the requirement for 16-in. span for stair tread (L/180) at 301 lb concentrated load (average of 15 tests), a pound above the required 300 lb.

Calculations show that for a stair tread with a 24-in. span (allowed deflection 0.133″), a composite board should have a $E \times I$ value of 548,016 lb × in.^{2}. This would be applicable to a hollow GeoDeck Heavy Duty composite deck board (8.1″ × 1.55, I = 1.858 in.^{4}, E = 374,000 psi, and $E \times I$ = 694,892 lb × in.^{2}). For a solid board of a standard dimension (5.5″ × 1.25″, I = 0.895 in.^{4}), flexural modulus should be at least 776,416 psi, and composite deck boards of such stiffness are not available as yet on the current market (see Table 7.34), except those made of wood. For thin solid board, such as 5.5″ × 15/16″ (I = 0.378 in.^{4}), flexural modulus applicable for stair tread with 24″ span should be at least 2,054,000 psi, which is much higher than that for typical wood (Table 7.34).

Table 7.28 above shows the recommended support span values (on center) for commercial composite and plastic lumber deck boards used as stair treads.

DEFLECTION OF COMPOSITE MATERIALS: CASE STUDIES

We have considered the above deflection of deck boards that serve as stair treads. Obviously, the same principle can be applied in the calculations of deflection of composite profiles in other cases: deflection of soundwalls under force of wind (in this case an equation for uniformly distributed load should be applied), deflection of an animal farm fence under weight of an animal leaning toward a board, deflection of a deck board under a hot tub installed on the deck, deflection of a handrail under a cantilever force, and so on.

Let us consider some examples.

Deflection and Bending Moment of a Soundwall Under Windloads

Soundwall is a noise barrier, built on an existing highway or freeway (or as part of a new highway project) to shield residences from the road noise. Typically, soundwalls are constructed next to residences where noise level is or above the 66-dB threshold and must be able to achieve at least a 5-dB noise reduction. The noise threshold is established by Federal guidelines. Sixty six decibels is an effective steady noise equivalent to fluctuating traffic noise over an hour. Traditionally, soundwalls are 8–16 ft of height, depending on specific design needs, and constructed as decorative concrete blocks.

There were several attempts to construct and test soundwalls made of WPC materials. Although noise barrier properties of those were fair and good, composite soundwalls typically had two major problems: large deflection of the wall and high cantilever load on the vertical posts.

Here is one example. A wall section consisting of 14 tongue-and-groove 8-ft boards (width 5–3/16″ of each) on top of each other has a dimension of 6 ft (high) by 8 ft (long). It is supported by one 6-ft (above the ground) post. At wind pressure of 25 lb/ft^2 (see below), the bending moment at the cantilever section is equal to

$$M_{\text{wind}} = WLH^2/2,$$

where W = uniformly distributed wind pressure, H = height of the wall, and L = length of the wall section.

For the calculations, a factor of 144 should be introduced to the denominator, if a windload is measured in lb/ft^2. Hence, the bending moment at W = 25 lb/ft^2 equals to

$$M_{\text{wind}} = 25\ \text{lb/ft}^2 \times 96\ \text{in.} \times 72^2/2 \times 144 = 43{,}200\ \text{in. lb.}$$

This bending moment is concentrated on a 6-ft post, holding the 6 × 8 ft wall, and can be simulated in a testing lab.

This bending moment can be translated onto a single post, resting on two supports (3-pt load). In this case, the bending moment is equal to

$$M_{\text{test}} = PL/4,$$

where P = center-point load, in lb, and L = distance between two supports (support span).

Because the two bending moments should equal to each other,

$$PL/4 = 43{,}200\ \text{in. lb.}$$

The center-point loads at different span values should be set as

$$P = 2400\ \text{lb at } 72''\ \text{span},$$
$$P = 2880\ \text{lb at } 60''\ \text{span, and}$$
$$P = 3840\ \text{lb at } 45''\ \text{span}.$$

For 1/3-pt load (4-pt load), the bending moment is equal to

$$M_{\text{test}} = PL/6,$$

and for a 1/3-pt load test,

$$PL/6 = 43{,}200 \text{ in. lb},$$

and

$$P = 3600 \text{ lb at } 72'' \text{ span},$$
$$P = 4320 \text{ lb at } 60'' \text{ span, and}$$
$$P = 5760 \text{ lb at } 45'' \text{ span.}$$

Conversely, from an ultimate load placed on a post, it can be determined that how much high wind pressure the vertical post can hold (Table 7.50). Bending moment for all three cases in Table 7.50 was $M = PL/6$.

Deflection of the soundwall can be determined using the following equation for uniformly distributed load:

$$D = 5WbL^4/384 \times 144 \times EI,$$

where W = uniformly distributed load, in lb/ft^2; b = width of the soundwall (96 in. in this case); L = height of the soundwall (72 in. in this case); E = modulus of elasticity of the board, 389,000 psi in this case; and I = moment of inertia of the soundwall, that is, 14×0.865 in.^{4} = 12.11 in.^{4} in this case (14 boards in a 6-ft high soundwall).

For (ultimate) wind pressure of 19.4 lb/ft^2 (Table 7.50), deflection of the soundwall made from the composite materials would be 0.96 in. For a higher wind pressure, the soundwall post will snap at its base.

To simplify the calculations, there is no need to use dimensions of the whole soundwall and a combined moment of inertia for all boards of the soundwall. We can consider deflection of only one board, because total wind pressure (per the total soundwall area), total soundwall area, and moment of inertia of the boards will increase in parallel with the wall height. Because moment of inertia of one soundwall tongue-and-groove board is 0.865 in.^{4} (see above), deflection of one board would be

TABLE 7.50 1/3-pt ultimate load onto 60″-support span for soundwall 6″ × 6″ posts, made from three different formulations

Soundwall boards composition (w/w)	Ultimate load (P) (lb)	Ultimate wind pressure (lb/ft^2)
37% HDPE, 30% rice hulls, 30% Biodac, 2% coupling agent	3350 ± 240 lb	19.4
45% HDPE, 53% rice hulls	3330 ± 140 lb	19.3
33% HDPE, 65% rice hulls	3000 ± 100 lb	17.4

The ultimate wind pressure was calculated for 6 × 8 ft soundwall. Bending moment was calculated using the formula $M = PL/6$, in which P = ultimate load in the table, $L = 96''$. Hence, $W = P/172.8$ lb/ft^2.

TABLE 7.51 Wind design pressure values calculated for soundwalls of height between 10 and 20 ft

Basic wind speed (mph)	Wind design pressure (lb/ft^2)
Soundwalls between 10 and 16 ft high	
70	8.9
80	11.7
90	14.8
100	18.2
Soundwalls of 20 ft high	
70	9.7
80	12.7
90	16.0
100	19.8

determined by the above formula, where b = width of one board and $I = 0.865$ in.^{4}. Deflection of one board would be the same as the whole wall, that is, 0.96 in.

To design WPC soundwalls capable of holding against a more intense wind (see Table 7.51), wood, steel, and aluminum post inserts were tested. Results of some of these tests are shown in Table 7.52.

Before that—what is, actually, a wind pressure of 19.4 lb/ft^2? Is it much or little? What wind speed it corresponds to? What are the code design values?

Wind design pressure, as specified in Code Section 1609.7 of the *BOCA National Building Code/1999*, should be calculated as a combination of the windward and leeward wall design pressures. The wind design pressure values in Table 7.51 were calculated for soundwalls of height between 10 and 16 ft, and for 20 ft, for exposure category B (urban and suburban areas).

Hence, for soundwalls up to 16 ft high, the capability to withstand win pressure of 19.4 lb/ft^2 is within the code.

TABLE 7.52 Testing of a two-panel soundwall system (18 eight-ft long tongue-and-groove boards composed each panel) inserted in a tongue-and-groove manner into three 6″ × 6″ posts, each 12-ft long; one middle post and two side posts

"Height" of the post/insert exposed "above ground"	Total load	Deflection of the middle post at its end	Uniformly distributed load, equivalent of constant wind pressure
8 ft	56 sandbags = 3920 lb	5″	28.8 lb/ft^2. (no failure)
	4200 lb	5–3/4″	30.9 (no failure)
8 ft	4760	8″	35.0 (no failure)
6 ft	2100 lb/66 ft^2 (one panel)	n.d.	31.8 (no failure)
6 ft	2730 lb/one panel	n.d.	41.4 (left post failed)
6 ft	2930 lb/one panel (200 lb added on the central post)	n.d.	44.4 (central post failed)

Total panel area was 136 ft^2. 12-gauge 3″ × 5″ rectangular steel tubing (0.108 in. cold roll steel) was inserted into each post. Sandbags, 70 lb each, were loaded uniformly—as much as possible. The assembly (dead load) weighed 640 lb; it is not included to the table below.

For a comparison, for category C (open flat terrain areas) the allowable design wind pressure is 29.7 lb/ft^2 at basic wind speed of 100 mph.

Now, let us consider a soundwall of 15 ft (180″) high and a basic wind speed of 100 mph in category C (see above) conditions. Again, there is no need to consider the whole soundwall with a combined moment of inertia for all 15-ft long boards. We can consider deflection of only one board, as it was described above. Because moment of inertia of one soundwall tongue-and-groove board is 0.865 in.^{4} (see above), deflection of one board would be equal to

$$D = 5WbL^4/384 \times 144 \times EI,$$

where W = uniformly distributed load, in lb/ft^2 (29.7 lb/ft^2 in this case); b = width of a soundwall board (5.5 in. in this case); L = length of the soundwall (180 in. in this case); E = modulus of elasticity of the board, 389,000 psi in this case; and I = moment of inertia of the soundwall board (0.865 in.^{4} in this case).

Thus, deflection of the soundwall will be equal to 46 in. (!) Clearly, soundwalls of this height or length (15 ft) cannot be used in areas of Category C to withstand wind of 100 mph speed, unless their configuration and/or material properties are greatly improved.

Table 7.53 shows a reinforcing effect of 12-gauge steel inserts of various configurations.

TABLE 7.53 Continuation of Table 7.52, but inserts were of various configurations into each of the three 6″ × 6″ post

Metal insert	Total load	Deflection of the middle post at its end	Ultimate uniformly distributed load (equivalent of constant wind ultimate pressure)
Z-shape	35 sandbags = 2450 lb/51 ft^2. (one panel)	7″ at failure	48 lb/ft^2 Metal insert bent.
C-shape	2030 lb/51 ft^2	10″ at failure	39.8
U-shape	1820 lb/51 ft^2	12″, and continued to 14″ without any additional load	35.7
2 × L-shape (forming a tight rectangular tubing)	2940 lb/51 ft^2	n.d.	57.6
2 × C-shape (similar with a tight I-beam)	2940 lb/51 ft^2	n.d.	57.6 (no failure)
2 × C-shape (similar with a tight I-beam)	4200 lb/136 ft^2 (whole double panel area)	n.d.	30.9 (failure)

All other conditions and measurements were the same as those in Table 7.52, except the height of the wall/insert was 6 ft "above the ground." Total two-panel area was 136 ft^2. The one-panel area was 8.5 ft (width) by 6-ft (height), that is, 51 ft^2.

TABLE 7.54 Ultimate load test results for twelve 6″ × 6″ wood fiber–HDPE composite posts of 90″–138″ inches long

Post length (in.)	Moment arm (in.)	Metal insert (gauge)	Load at failure (lb)
	I-beam		
90	66	18	850
90	66	16	1060
90	66	14	1410
90	66	12	2520
	Tube 3″ × 5″		
90	66	12	2340
	I-beam		
114	78	18	750
114	78	16	1040
114	78	14	1370
114	78	12	2430
	Tube 3″ × 5″		
114	78	12	2380
138	100	12	1880
138	100	12	1700

The lower part of a post was clamped in the 14-in. jig. The force application was 4″ below the end of the metal insert. The metal insert was flush with the fixed (lower) end of the post and ended 6″ below the top of the post (with 90″ post) or 18″ below the top of the post (with 114″ and 138″ long posts).

One can see that 12-gauge metal inserts can practically double the ultimate uniformly distributed load.

Further tests with a steel *I*-beam of various thicknesses (12–18 gauge) and 12 gauge 3″ × 5″ steel tube gave results shown in Table 7.54.

Finally, Table 7.55 shows maximum, or critical, windloads exemplified with a WPC soundwall system 8 ft long and 6–12 ft high.

TABLE 7.55 Maximum/critical windloads for the GeoDeck 8-ft long soundwall system with 6″ × 6″ post steel inserts

	Estimated windloads (lb/ft^2) that the GeoDeck 8-ft long soundwall sections with 6″ × 6″ posts (with inserts indicated) and the indicated wall heights can withstand			
Metal insert	6-ft	8-ft	10-ft	12-ft
"I"-beam, 18 gauge	33	19	12	8
"I"-beam, 16 gauge	44	25	16	11
"I"-beam, 14 gauge	58	33	21	15
"I"-beam, 12 gauge	103	58	37	26
Tube 3″ × 5″, 12 gauge	101	57	36	25

Critical soundwall heights are shown in the table.

Deflection of a Fence Board

Let us consider a fence around a section of an animal farm. The fence is constructed as posts connected by composite deck boards. Animals can lean against a board, deflecting it up to a certain extent depending on the force applied, length of the board, and its flexural modulus.

Table 7.56 shows to which extent fence boards can deflect for a number of the following situations.

TABLE 7.56 Deflection of solid and hollow composite deck boards exemplified with Trex and GeoDeck boards, respectively

Board	Dimension (in.)	Board length (ft)	Force applied (lb)	Deflection (in.)
Trex, Solid	5.5 × 1.25	6	100	4.5″
			200	(9″) (broken at 129 lb at 5.8″ deflection)
			400	(18″) (broken at 129 lb at 5.8″ deflection)
		8	100	(10.7″) (broken at 97 lb at 10.4″ deflection)
			200	(21″) (broken at 97 lb at 10.4″ deflection)
			400	(43″) (broken at 97 lb at 10.4″ deflection)
GeoDeck, Hollow	5.5 × 1.25	6	100	2.8″
			200	(5.7″) (broken at 151 lb at 4.3″ deflection)
			400	(11.3″) (broken at 151 lb at 4.3″ deflection)
		8	100	6.7″
			200	(13.4″) (broken at 113 lb at 7.6″ deflection)
			400	(26.9″) (broken at 113 lb at 7.6″ deflection)
	8.1 × 1.55	6	100	1.1″
			200	2.2″
			400	(4.5″) (broken at 309 lb at 3.5″ deflection)
		8	100	2.7″
			200	5.3″
			400	(10.6″) (broken at 232 lb at 6.2″ deflection)

Center-point load at midspan. Break load at certain pound-force was calculated taking into account flexural strength of the materials (ASTM D 6109) as 1625 (Trex) and 2319 psi (GeoDeck), determined at the longest support span available (20 in.), see Table 7.12.

Three kinds of composite boards are used, with their respective moment of inertia (I) and flexural modulus (E):

a. Solid 5.5″ × 1.25″ board, such as Trex ($I = 0.895$ in.^{4}, $E = 193{,}000$ psi)
b. Hollow 5.5″ × 1.25 board, such as GeoDeck Traditional ($I = 0.733$ in.^{4}, 374,000 psi)
c. Hollow 8.1″ × 1.55 board, such as GeoDeck Heavy Duty ($I = 1.858$ in.^{4}, 374,000 psi).

Flexural modulus values are those determined for the longest support span for which data are available (20 in.), see Tables 7.25, 7.31, and 7.33.

Two board lengths are used: 6 and 8 ft.

Three values of forces applied to a board are considered: 100, 200, and 400 lb.

In these cases we will apply a modified Eq. (7.46) for deflection at center-point load (by converting inches to feet):

$$D = 36 \times PL^3/EI,$$

where D = deflection, in in.; P = load, in lb; L = length of the board, in ft; E = flexural modulus, in psi; and I = moment of inertia, in in.^{4}.

The data in Table 7.56 and corresponding calculations show that only heavy duty composite deck boards (with moments of inertia higher than 4.0 in.^{4} at flex strength of about 2000 psi) or those having flexural strength above 5500 psi (with moment of inertia of about 1.0 in.^{4}) can withstand center point force of 300 lb at board length 8 ft or higher. For example, an 8-ft solid board 2.1-in. thick and 5.5-in. wide ($I = 4.24$ in.^{4}) and with $S = 2000$ psi can withstand a midspan force of 337 lb. These calculations are performed using the modified Eq. (7.20)

$$P = 8SI/LH,$$

where S = flexural strength, in psi; I = moment of inertia, in in.^{4}; L = support span (length of the board in this case), in in.; and H = depth of the board, in in.

Similarly, an 8-ft solid board 1.25-in. thick and 5.5-in. wide ($I = 0.895$ in.^{4}) and with $S = 5500$ psi can withstand a midspan force of 328 lb.

It seems that an ordinary composite deck board (flexural strength of 2000–3000 psi, see Table 7.13) of a standard size (5.5 in. by 1–1.5 in.) with moment of inertia of 0.5–1.5 in.^{4} cannot be securely used for the application described in this section.

Deflection of WPC Joists WPC materials are not used as joists for decks. These composites are not there as yet. They are not considered as *structural* materials, in a sense that they do not match wood regarding wood flexural strength and flex modulus.

Let's, however, consider Trex-based 1.5″ × 8″ solid composite board as a joist. Moment of inertia of such a board would be

$$I = bh^3/12,$$

and for $b = 1.5$ and $h = 8.0$″, moment of inertia is 64.0 in.^{4}.

At flexural modulus of 193,000 psi (see above), deflection of a 12-ft board

$$D = 36 \times PL^3/EI$$

under 300 lb (center-point load) is 1.51″,
where D = deflection, in in.; P = load, in lb; L = length of the board, in ft; E = flexural modulus, in psi; and I = moment of inertia, in in.^{4}.

Because $L/180$ = 0.8 in., the deflection by 1.51″ is higher than allowed (0.8 in.). Thus, 12-ft Trex board cannot be used as a joist. Only at L = 8.7 ft (allowed deflection is 104.4″/180 = 0.58″) the deflection at 300 lb (center-point load) is 0.576″, that is, within the allowed deflection. However, 8.7″ is too short for a joist of a deck in most of cases.

For a hollow 1.55″ × 8.1″ GeoDeck Heavy Duty deck board as a joist, moment of inertia is 30.25 in.^{4}. At flexural modulus of 374,000 psi, deflection of a 12-ft board under 300 lb (center-point load) is 1.65″. Only at L = 8.3 ft (allowed deflection is 99.6″/180 = 0.55″), the deflection at 300 lb (center-point load) is 0.55″, that is, allowed deflection. However, 8.3″ is too short for a joist of a deck in most of cases.

This can be compared with a joist made of PTL. For a solid 1.5″ × 8″ (moment of inertia 64.0 in.^{4}) board as a joist, with flexural modulus of 1,248,180 psi (Table 7.34), deflection of a 12-ft board under 300 lb (center-point load) is 0.23″. Allowed deflection is 144″/180 = 0.80″.

For a 16-ft board (PTL or an equivalent wood board), at allowed deflection of 1.067″, deflection under 300-lb load (center-point load) is 0.55″.

For a 20-ft PTL board at allowed deflection of 1.33″, deflection under 300-lb load (center-point load) is 1.08″, which is still within the norm.

These examples show how far conventional composite boards are from structural materials like wood for applications such as deck joists.

Deflection of a Deck Under a Hot Tub

A typical hot tub for, say, five persons has dimensions of 93″ × 78″, that is, takes 50.375 ft^2. Total weight of the hot tub consists of its own weight (1100 lb), water (350 gal, or 2900 lb), and people (5 × 200 lb = 1000 lb), for a rather heavy scenario, total 5000 lb. Therefore, the hot tub produces a uniformly distributed load of about 100 lb/ft^2. This is close to a highest load of a hot tub on a deck.

100 lb/ft^2 is a typical code requirement (such as by ICC-ES) for uniformly distributed live loads, including those for decks. This corresponds to a rather low center-point load, which can be calculated using formula (7.42), given above:

$$W = 288 \times P/bL,$$

where P = a center-point load, in lb; B = board width, in in.; and L = board length, in in.

For a 100 lb/ft^2 (0.694 psi), a corresponding center-point load, resulting in the same outer fiber stress, equals to only 31 lb. This load is 20–25 times lower than a break load at midspan for a typical composite deck board. Hence, deflection of a deck under a hot tub would be minimal, particularly because a hot tub sits on 5–6 joists.

The main (potential) problem with a hot tub on a composite deck is not a deck deflection or a flexural strength of deck boards but a (potential) creep of a plastic-based composite material, particularly in the South. This can lead to a partial "sinking" of a hot tub into the deck. Of course, joists would eventually stop the "sinking" even in a worst case scenario; however, this might leave a permanent dent on the deck of a size of the hot tub base. Warranty would not cover the damage.

As far as the author knows, there have been no cases of sinking of a hot tub into GeoDeck deck from the weight and heat, although it could and probably would happen. The best way to go would be to place the hot tub directly on top of the joists and build the deck around it.

Another concern regarding placing a hot tub on a composite deck is a possibility of an accidentally breaking a board by impact, directed—by a not careful unloading of the hot tub—between joists.

Deflection of a Hollow Deck Board Filled with Hot Water

A deck was installed at 22.5° angle on joist mounted at 16 in. between their centers. A report came that after several hot summer days, with air temperature of 95+ degrees; boards were sagging in the middle of the span of the deck. Normally, it has not happened with similar boards even at warmer air temperature. Spare boards found in the garage were tested for deflection (flexural modulus), and it was within norm.

An inspection of the deck showed that the hollow boards were filled with water. It was reasonable to suggest that the excessive deflection of the boards was caused by a combination of three factors: weight of the water-filled boards, high temperature, and a span slightly higher than normal.

Verification of the suggestion:

(1) *Span*: The distance between joists is 16 in. on center. At 22.5° angle the span will be equal to $\sqrt{16^2 + 8^2} = 17.9$ in.

(2) *Weight*: The hollow board weight 1.8 lb per lineal foot. In a common situation, at ambient temperature and a span of 16 in. on center, a deflection of the boards themselves under the uniform load (dead load) is

$$D = 5WbL^4/384 \times 144 \times EI,$$

where D = deflection, in in., at midspan under the uniformly distributed load; I = moment of inertia (0.8 in.^{4} in this particular case); E = flexural modulus, in psi (360,000 psi in this particular case); L = support span, in in. (16″ in this particular case); b = specimen overall width, in in. (5.5″ in this particular case); and W = uniformly distributed load (3.93 lb/ft^2 in this particular case).

Thus, deflection of a deck board at midspan in the common situation is

$$D = 5 \times 3.93 \text{ lb/ft}^2 \times 5.5 \text{ in} \times 65{,}536 \text{ in.}^4/384 \times 144 \times 360{,}000 \text{ lb/psi} \times 0.8 \text{ in.}^4 = 0.00044''$$

All factors being equal, but at 17.9 in. on center, deflection at midspan will be

$$D = 0.00070''.$$

Weight of the hollow boards, filled with water, is almost twice as much, namely 7.5 lb/ft^2. Hence, deflection at midspan will be

$$D = 0.0013''.$$

(3) *High temperature*: At 125°F flexural modulus of the boards decreased by 73% compared to that at 70°F, and becomes 208,000 psi. That means the temperature coefficient of decreasing flex modulus is 1.105/10°F. At 100°F temperature of the a deck surface in the South is 50° higher than the air temperature, that is, 150°F for the deck surface in our particular case. Taking into account the temperature coefficient, flexural modulus of the boards at 150°F will be 162,000 psi. Hence, deflection of the boards filled with hot water at midspan at 17.9-in. span will be

$$D = 0.0029''$$

This is 660% higher deflection than that at ambient temperature and corresponds to an apparent increase of the span from 16 to 26 in. No wonder, there was an excessive deflection of boards, which as 150°F on the deck surface caused creep and sagging.

Deflection and Creep of Composite Deck Boards

When composite deck boards are loaded for an extended time period, they develop creep. This property of plastic-based materials is quite understandable, because plastic flows. The real issue is not to prevent creep but to minimize it. Three obvious ways to minimize creep are (a) to change the span, (b) to lower the weight/load, and (c) to reinforce the composition.

Deflection of the loaded board can be predicted provided that the load and its location on the board is known, as well as the span, the moment of inertia, and the flexural modulus of the board, and assuming that the load and the deflection are within the linear relationship between each other. If the load is outside of this relationship (higher), a deflection would be higher than calculated using formulas

$$D = PL^3/48 \times EI$$

for a center-point load and

$$D = PL^3/56.5 \times EI$$

for a third-point load,
where D = deflection, P = load, L = support span, E = flexural modulus, and I = moment of inertia.

Let us consider all the three ways (see the first paragraph of this section), using practical examples.

Example 1. A Trex board, width 5.5 in., depth 1.25 in., moment of inertia 0.895 in.^{4}, flexural modulus 175,000 psi (the manufacturer's data), support span 22 in., and a third-point load of 100 lb.

An expected deflection would be

$$D = 100 \text{ lb} \times 10{,}648 \text{ in.}^3/56.5 \times 175{,}000 \text{ lb/in}^2 \times 0.895 \text{ in.}^4 = 0.12 \text{ in.}$$

In reality, the immediate deflection was of 0.095 in., that is, 26% lower than predicted, not higher. This can be explained by an inaccurate value of the flexural modulus reported by the manufacturer. Indeed, flex modulus for this particular Trex product, calculated from the above experiment, was 221,000 psi, not 175,000 psi. Other figures for Trex board flexural modulus were of 193,000 ± 19,000 psi (Table 7.30), 214,000 ± 8,000 psi and 224,800 psi (Table 7.34, footnote).

The board under the load of 100 lb was left for 24 h, and dynamics of the additional deflection (sinking) was shown in Table 7.57:

TABLE 7.57 Deflection dynamics under a 100 lb load (a third-point load), a 5.5 in. × 1.25 in. Trex board, support span 22 in.

Time elapsed after placing the third-point load of 100 lb on a 22-in.-span Trex board	The board's deflection (in.)
Instant (within 1 min)	0.095
2 min	0.104
5 min	0.111
10 min	0.117
13 min	0.120
15 min	0.122
20 min	0.125
30 min	0.130
47 min	0.136
1 h	0.140
1 h 16 min	0.144
1 h 30 min	0.147
2 h 30 min	0.155
4 h	0.165
7 h 30 min	0.175
24 h	0.196

After 24 h, the load was removed, and dynamics of the board recovery was as follows:

TABLE 7.58 Dynamics of the Trex board recovery after the 100-lb load was placed for 24 h and then removed (see Table 7.57).

Time elapsed after removing of 100-lb load from the 22-in.-span Trex board	The board residual deflection (in.)
2 min	0.092
4 min	0.087
6 min	0.083
13 min	0.077
25 min	0.071
45 min	0.065
1 h	0.062
1 h 34 min	0.057
2 h	0.055
3 h	0.050
5 h 30 min	0.045
7 h 30 min	0.041
24 h	0.034

One can see that 0.095″ was an instant deflection, 0.101″ was an additional "sinking" deflection (106% of the instant deflection), and not recovered after 24 h was 0.034″. The recovery after 24 h was 83% [(0.196–0.034)/0.196] of the total deflection (instant and "sinking").

Example 2. An UltraDeck hollow board, width 5.22 in., depth 1.09 in., moment of inertia 0.46 in.^{4}, flexural modulus 502,000 psi (Table 7.33), support span 22 in., and a third-point load of 100 lb.

An expected deflection would be

$$D = 100 \text{ lb} \times 10{,}648 \text{ in.}^3/56.5 \times 502{,}000 \text{ lb/in.}^2 \times 0.46 \text{ in.}^4 = 0.082 \text{ in.}$$

In reality, the immediate deflection was of 0.145 in., that is, much higher than predicted. This can be explained as indicated above, that at the 100-lb load, a load-deflection curve was well outside of its linear portion.

The board under the load of 100 lb was left for 24 h, and dynamics of the additional deflection (sinking) was shown in the table 7.59.

After 24 h, the load was removed, and dynamics of the board recovery was as shown in Table 7.60.

One can see that 0.145″ was an instant deflection, 0.062″ was an additional "sinking" deflection (43% of the instant deflection), and not recovered after 24 h was 0.026″. The recovery after 24 h was 87% [(0.207–0.026)/0.207] of the total deflection (instant and sinking, combined).

Therefore, UltraDeck, having much higher flexural modulus and much lower moment of inertia (hollow board) than those for Trex, showed much higher

TABLE 7.59

Time elapsed after placing the third-point load of 100 lb on a 22-in.-span UltraDeck board	The board's deflection (in.)
Instant (within 1 min)	0.145
2 min	0.156
3 min	0.159
4 min	0.161
5 min	0.162
7 min	0.164
10 min	0.167
20 min	0.171
33 min	0.175
45 min	0.177
1 h 05 min	0.180
2 h 06 min	0.185
4 h 40 min	0.192
6 h 40 min	0.195
8 h 40 min	0.197
24 h	0.207

initial (instant) deflection than that of Trex (0.145″ vs. 0.095″) but better recovery (87% vs. 83%), hence, lower creep.

TABLE 7.60

Time elapsed after removing of 100-lb load from the 22-in.-span UltraDeck board	The board residual deflection (in.)
2 min	0.070
3 min	0.068
4 min	0.066
5 min	0.065
10 min	0.060
20 min	0.057
45 min	0.052
1 h 05 min	0.0500
1 h 30 min	0.048
2 h	0.046
3 h 50 min	0.042
8 h 10 min	0.031
24 h	0.026

Now, let us see how a high content of a mineral filler may affect creep in some particular examples (Examples 3 and 4).

Example 3. A rice-hulls-filled HDPE (39% w/w, 46% v/v) hollow board, width 5.5 in., depth 1.25 in., moment of inertia 0.736 in.^{4}, flexural modulus 297,000 psi, support span 22 in., and a third-point load of 100 lb.

An expected deflection would be

$$D = 100 \text{ lb} \times 10{,}648 \text{ in.}^3/56.5 \times 297{,}000 \text{ lb/in.}^2 \times 0.736 \text{ in.}^4 = 0.086 \text{ in.}$$

In reality, the immediate deflection was of 0.115 in., that is, 34% higher than predicted. This can be explained as indicated above, that at the 100-lb load, a load-deflection curve was outside of its linear portion.

The board under the load of 100 lb was left for 24 h, and dynamics of the additional deflection (sinking) was shown in the following Table:

TABLE 7.61

Time elapsed after placing the third-point load of 100 lb on a 22-in.-span board	The board's deflection (in.)
Instant (within 1 min)	0.115
2 min	0.122
6 min	0.130
8 min	0.132
10 min	0.134
15 min	0.137
20 min	0.140
30 min	0.144
45 min	0.147
1 h	0.151
1 h 30 min	0.155
2 h 10 min	0.161
4 h 20 min	0.171
6 h	0.177
8 h	0.186
24 h	0.209

After 24 hr, the load was removed, and dynamics of the board recovery was as shown in Table 7.62.

One can see that 0.115″ was an instant deflection, 0.094″ was an additional "sinking" deflection (82% of the instant deflection), and not recovered after 24 h was 0.024″. The recovery after 24 h was 89% [(0.209–0.024)/0.209] of the total deflection (instant and sinking, combined).

Thus, the last board, having both flexural modulus and moment of inertia, as well as the initial (instant) deflection in between those of Trex and UltraDek, showed a better recovery than either of them (89 vs. 83, and 87%), hence, lower creep.

Example 4. A rice hulls and fly ash (30% w/w) filled HDPE (34% w/w, 46% v/v) hollow board, width 5.5 in., depth 1.25 in., moment of inertia 0.775 in.^{4}, flexural modulus 383,000 psi, support span 22 in., and a third-point load of 100 lb.

An expected deflection would be

$$D = 100 \text{ lb} \times 10{,}648 \text{ in.}^3/56.5 \times 383{,}000 \text{ lb/in.}^2 \times 0.775 \text{ in.}^4 = 0.063 \text{ in.}$$

TABLE 7.62

Time elapsed after removing of 100-lb load from the 22-in.-span board	The board residual deflection (in.)
2 min	0.084
4 min	0.077
6 min	0.074
10 min	0.069
15 min	0.065
30 min	0.059
50 min	0.053
1 h 35 min	0.047
2 h 45 min	0.043
5 h 30 min	0.036
7 h 45 min	0.033
8 h 40 min	0.031
24 h	0.024

In reality, the immediate deflection was of 0.115 in., that is, much higher than predicted. This can be explained as indicated above, that at the 100-lb load, a load-deflection curve was well outside of its linear portion.

The board under the load of 100 lb was left for 24 h, and dynamics of the additional deflection (sinking) was shown in the following table.

TABLE 7.63

Time elapsed after placing the third-point load of 100-lb on a 22-in.-span board filled with 30% fly ash (w/w)	The board's deflection (in.)
Instant (within 1 min)	0.115
2 min	0.126
4 min	0.132
7 min	0.136
10 min	0.140
15 min	0.143
30 min	0.149
50 min	0.155
1 h 05 min	0.158
1 h 40 min	0.162
2 h 30 min	0.165
4 h	0.174
5 h 15 min	0.178
6 h 50 min	0.183
8 h	0.186
24 h	0.204

After 24 h, the load was removed, and dynamics of the board recovery was as follows:

TABLE 7.64

Time elapsed after removing of 100-lb load from the 22-in.-span board filled with 30% fly ash (w/w)	The board residual deflection (in.)
1 min	0.082
2 min	0.077
3 min	0.074
4 min	0.071
6 min	0.068
10 min	0.063
20 min	0.057
40 min	0.050
1 h	0.045
2 h	0.037
3 h	0.033
5 h 45 min	0.026
8 h 45 min	0.021
24 h	0.016

One can see that 0.115″ was an instant deflection, 0.089″ was an additional "sinking" deflection (77% of the instant deflection), and not recovered after 24 h was 0.016″. The recovery after 24 h was 92% [(0.204–0.016)/0.204] of the total deflection (instant and sinking, combined).

Thus, the last board, filled with rice hulls and fly ash (30% w/w), showed the best recovery, hence, lower creep.

Now, let us consider a uniformly distributed load at different board support spans.

Example 5. A rice-hulls-filled HDPE hollow deck board, width 5.5 in., depth 1.25 in., moment of inertia 0.784 in.^{4}, and flexural modulus 252,000 psi.

An expected deflection (D) for the board at 16″ span under a uniformly distributed load of 100 lb/ft^2 is as follows:

$$D = 5WbL^4/384 \times 144 \times EI,$$

where $W = 100$ lb/ft^2, $B = 5.5$ in., $L = 16$ in., $E = 252{,}000$ psi, and $I = 0.784$ in.^{4}.

and a resulting deflection is 0.0165 in. (close to 1/64 in.).

If the span is 14 in., the deflection will be 0.00967 in., that is, less than 1/96 in. Hence, this is an observation: Reducing the span from 16 in. to 14 in. reduced the deflection under a uniformly distributed load by more than 50%.

The same deflection at 14-in. span will be reached at 33 lb at 3-pt load (midspan load). Indeed, for this type of load

$$P = 48 \times EID/L^3,$$

where P = a midspan load, L = span (14 in.), and D = deflection (0.00967 in.).
and the load at these conditions is 33.4 lb.

Hence, one more observation: At these conditions a uniformly distributed load of 100 lb/ft^2. corresponds to a midspan load of only one third of its value.

The above deflection is 14/0.00967 = 1448 times less compared to the span. The most stringent requirement for boards by building codes is 360 times less compared to the span, that is, 14/360 = 0.0389 in. for the 14-in. span.

If we consider a design load as the load causing this deflection of L/360, it would be equal to

$$P = 48 \times EI \times 0.0389/14^3 = 134.4 \text{ lb}.$$

A direct experiment showed that the deflection of 0.0389 in. at 14-in. span of the board was reached at 133 lb (a difference of 1% from the calculated value).

The further test on deflection and creep has been done with a 2.0 × design load, by placing the 270-lb load at the midspan (span of 14 in.) of the board described above.

Example 6. A rice-hulls-filled HDPE hollow deck board, width 5.5 in., depth 1.25 in., moment of inertia 0.784 in.^{4}, and flexural modulus 252,000 psi.

Under a midspan load of 270 lb and a span of 14 in., the instant deflection was 0.096 in. It was more than twice compared with that at the design load, because the deflection was outside of the linearity zone (deflection vs. load). The load was left on the board for 24 h, then removed. The immediate (1 min) board recovery was to 0.051″ from the initial (zero load) base. The following board recovery is shown in Table 7.65.

The board has recovered by 76% from the instant deflection.

As soon as the last reading was done, the same board was reloaded with 2.5 × design load, that is, with 336 lb. The immediate deflection went from 0.023 (see the table above, bottom line) to 0.1455 in., that is, by an additional 0.1225 in. The load was left on the board for 24 h. After its removal, the recovery went as shown in table 7.66.

One can see that after the first recovery of 76% (from the instant deflection), the recovery after reloading with 2.5 × design load for 24 h was 69% total, or 82% from the first residual recovery.

Example 7. The board had the same profile as that in Example 6, but it was a little stiffer, and its flexural modulus was 297,000 psi. The span compared to Example 6 was increased from 14 to 24 in.

TABLE 7.65

Time elapsed after removing of 270-lb load from the midspan of a 14-in.-span board	The board residual deflection (in.)
1 min	0.051
2 min	0.047
4 min	0.045
7 min	0.044
12 min	0.041
15 min	0.040
40 min	0.036
1 h 20 min	0.033
3 h 30 min	0.031
5 h 30 min	0.0275
7 h	0.026
22 h	0.023
24 h	0.023

The design load was calculated in the same manner as in Example 6, that is, a load placed at midspan and causing $L/360$ ($24/360 = 0.0667$ in.) deflection. Such a load would be equal to

$$P = 48 \times EI \times 0.0667/24^3 = 53.9 \text{ lb.}$$

TABLE 7.66

Time elapsed after removing of 336-lb load from the midspan of a 14-in.-span board	The board residual deflection (in.)
1 min	0.089
2 min	0.085
3 min	0.0825
4 min	0.081
5 min	0.080
8 min	0.076
10 min	0.075
36 min	0.066
46 min	0.0655
1 h	0.061
1 h 45 min	0.0595
2 h 10 min	0.057
3 h 40 min	0.055
6 h 10 min	0.052
8 h 30 min	0.050
24 h	0.045

In terms of stress of the board, this 3-pt load is close to the uniformly distributed load of 100 lb/ft^2 (the latter would cause a deflection of 0.0709 in. at the same conditions).

A direct experiment showed that the deflection of 0.0667 in. at 24-in. span of the board was reached at 54 lb (practically, no difference from the calculated value).

The further test on deflection and creep has been done with a 2.0 × design load, by placing the 108-lb load at the midspan (span of 24 in.) of the board.

The instant deflection was 0.160 in. It was more than twice compared with that at the design load, because the deflection was again outside of the linearity zone (deflection vs. load), as it was described above. The load was left on the board for 24 h, then removed. The immediate (1 min) board recovery was to 0.084″ from the initial (zero load) base. In other words, the board recovered by 0.076″, or 48% of the initial instant deflection. The following board recovery is shown below.

TABLE 7.67

Time elapsed after removing of 108-lb load from the midspan of a 24-in.-span board	The board residual deflection (in.)	The board recovery %
1 min	0.084	48
2 min	0.081	49
3 min	0.079	51
4 min	0.078	51
6 min	0.076	52.5
8 min	0.075	53
10 min	0.0735	54
12 min	0.072	55
15 min	0.071	56
20 min	0.070	56
30 min	0.0685	57
40 min	0.0675	58
1 h	0.066	59
1 h 20 min	0.065	59
15 h	0.0425	73
16 h	0.04175	74
18 h	0.041	74
24 h	0.040	75

The board has recovered by 75% from the instant deflection.

Similar to Example 6, as soon as the last reading was done, the same board was reloaded with 2.5 × design load, that is, with 135 lb. The immediate deflection went from 0.040 (see the table above, bottom line) to 0.287 in., that is, by an additional 0.247 in. The load was left on the board for 24 h. After its removal, the recovery went as shown in table 7.68.

One can see that after the first recovery of 75% from the instant deflection, the recovery after reloading with 2.5 × design load for 24 h was 71% total (from zero-load point, or from the total deflection of 0.287 in.), or 82% from the second deflection (0.247 in.).

TABLE 7.68

Time elapsed after removing of 135-lb load from the midspan of a 24-in.-span board	The board residual deflection (in.)	The board recovery with respect to the initial, zero-load point (%)
1 min	0.149	48
2 min	0.144	50
3 min	0.141	51
4 min	0.1385	52
5 min	0.137	52
6 min	0.1355	53
8 min	0.133	54
10 min	0.131	54
16 min	0.1265	56
30 min	0.120	58
45 min	0.116	60
1 h	0.113	61
1 h 30 min	0.110	62
17 h	0.086	70
19 h	0.084	71
24 h	0.08325	71

This means that for the boards at 14-in. span and 24-in. span, the residual creep (after 24-h load) was about the same, namely 75–76% from under 2 × design load and 69–71% from under 2.5 × design load (82% recovery from the instant deflection for the both boards if we consider only the second creep).

Example 8. The board had the same profile as that in Examples 6 and 7, but it was loaded not only with rice hulls, but also with Biodac®, and its flexural modulus was 389,000 psi. The span was kept as that in Example 7, that is, 24 in.

The design load was calculated in the same manner as in Examples 6 and 7, that is, a load placed at midspan and causing $L/360$ (24/360 = 0.0667 in.) deflection. Such a load would be equal to

$$P = 48 \times EI \times 0.0667/24^3 = 70.6 \text{ lb}.$$

In terms of stress of the board, this 3-pt load is close to the uniformly distributed load of 100 lb/ft^2 (the latter would cause a deflection of 0.0709 in. at the same conditions).

A direct experiment showed that the deflection of 0.0667 in. at 24-in. span of the board was reached at 71 lb (practically, no difference from the calculated value).

The further test on deflection and creep has been done with a 2.0 × design load, by placing the 142-lb load at the midspan (span of 24 in.) of the board.

The instant deflection was 0.153 in. It was more than twice compared with that at the design load, because the deflection was again outside of the linearity zone (deflection vs. load), as it was described above. The load was left on the board for 24 h, then removed. The immediate (within 30 s) board recovery was to 0.057″ from the initial (zero load) base. In other words, the board recovered by 0.096″, or 63% of the initial instant deflection. The following board recovery is shown below:

TABLE 7.69

Time elapsed after removing of 142-lb load from the midspan of a 24-in.-span board	The board residual deflection (in.)	The board recovery (%)
1 min	0.0520	66
2 min	0.0490	68
3 min	0.0476	69
4 min	0.0463	70
5 min	0.0455	70
6 min	0.0446	71
7 min	0.0432	72
8 min	0.0426	72
10 min	0.0422	72
23 min	0.0378	75
32 min	0.0355	77
40 min	0.0345	77
51 min	0.0334	78
1 h 45 min	0.0290	81
3 h 10 min	0.0270	82
5 h	0.0256	83
6 h 15 min	0.0246	84
7 h	0.0240	84
8 h	0.0234	85
24 h	0.0170	89

The board has recovered by 89% from the initial, instant deflection.

Similar to Examples 6 and 7, as soon as the last reading was done, the same board was reloaded with 2.5 × design load, that is, with 177 lb. The immediate deflection went from 0.017 (see the table above, bottom line) to 0.203 in., that is, by an additional 0.186 in. The load was left on the board for 24 h. After its removal, the recovery went as shown in table 7.70.

One can see that after the first recovery of 89%, the recovery after reloading with 2.5 × design load for 24 h was 92% total (from zero-load point, or from the total deflection of 0.203 in.), or practically 100% from the second deflection (0.186 in.).

This means that after the second deflection, under the 2.5 × design load, the board is fully recovered to the level of the first deflection, under 2 × design load. This experiment shows a power of Biodac® as a mineral–cellulose filler for HDPE-based composite boards.

GUARDRAIL SYSTEMS

Handrail assemblies shall be designed to resist a load of 50 lb per linear foot applied in any direction at the top (2000 International Building Code, Section 1607.7.1).

TABLE 7.70

Time elapsed after removing of 177-lb load from the midspan of a 24-in.-span board	The board residual deflection (in.)	The board recovery with respect to the initial, zero-load point (%)
1 min	0.072	65
2 min	0.067	67
3 min	0.064	68
4 min	0.0615	70
5 min	0.0595	71
6 min	0.058	71
8 min	0.055	73
10 min	0.0535	74
20 min	0.0475	77
33 min	0.043	79
1 h	0.038	81
1 h 50 min	0.03225	84
3 h 15 min	0.02925	86
4 h 35 min	0.025	88
6 h 25 min	0.02225	89
7 h 45 min	0.021	90
24 h	0.016	92

This means that a 6-ft railing should withstand a 300-lb load, and a 8-ft railing should withstand a 400-lb load applied at the top at midspan in a cantilever fashion, or horizontal direction, parallel to the ground, that is, typically the most damaging application. However, this is not all. AC 174 requires to multiply this load by (a) *a safety factor* $\times$ 2.5 that brings the required load to 750 lb for a 6-ft railing system and to 1000 lb for an 8-ft railing; and (b) *a temperature factor*, which is approximately 1.5 for HDPE; this brings the required load to 1125 lb for a 6-ft railing and to 1500 lb for an 8-ft railing system.

Few WPC railing systems could satisfy these code conditions. In this situation many manufacturers use an exception in the code (Exception 1, Section 1607.7.1, 2000 International Building Code), which allows to employ only the single, concentrated load of 200 lb for railing systems designed for one- and two-family dwelling, despite the length of the railing system. In fact, with a required safety factor ($\times$2.5), this concentrated load shall be 500 lb. The temperature factor (see above) in this case is not required. This center-point load is—theoretically—equivalent to a 1000 lb quarter-point load as a proxy for a uniformly distributed load.

Figure 7.11 shows a uniform load test (actually, quarter-point loading test) of a railing system.

Figure 7.12 shows a failure of a handrail in a composite railing system under these conditions, that is, under concentrated 500-lb load. Despite a metal insert, the system failed. The test requires that after 500-lb concentrated load is reached, it should be continuously applied for 60 s. The rail shown in Figure 7.12 held the 500-lb load and failed at 56 s.

Figure 7.11 A uniform loading test (actually, a quarter-point loading test) of a railing system.

Figure 7.12 A 500 lb concentrated load test on a wood-HDPE composite railing system with an aluminum tube reinforced handrail.

For in-fill test (a normal load horizontally applied over 1 ft^2 area, including openings and spaces between components, see Fig. 7.13), the respective components of the railing system (such as pickets or spindles) should withstand a load of 50 lb pounds (2000 International Building Code, Section 1607.7.1.2). However, again, AC 174 requires including of the safety and temperature factors that makes for HDPE-based railing system $50 \times 2.5 \times 1.5 = 187.5$ lb as a principal requirement for infill test.

The following are descriptions of the tests performed with GeoDeck railing system according to those requirements.

(a) *In-fill load test*: As it was described above, the test specimen, which was a row of spindles (pickets), must be capable of satisfactorily resisting a load of 187.5 lb applied over a 1 ft^2 area normal to the in-fill. The load was applied at a position on the in-fill that represented the "worst case" loading and deflection scenarios. The system passed the requirement with no failure, no evidence of disengagement of any components (posts, spindles, handrail, etc.), and no visible cracks in any component.

(b) *Uniform load test*: For a 6-ft railing system, the top rail of the system (guard and handrail) was subjected to two separate tests in which a maximum quarter-point load (as a proxy for uniformly distributed load) of 1125 lb was applied vertically and horizontally. When no reinforcing elements were introduced (such as metal inserts into the handrail, arrogances), the handrail failed at only 478 lb applied horizontally, in a cantilever mode.

Figure 7.13 In-fill load test set-up.

(c) *Concentrated load test*: If the above two tests were satisfactory, two separate concentrated tests were to be conducted, of 750 lb each, being applied at the top rail and at the top of a single post. The load should be continuously applied normal to the top rail at the maximum railing system height. When the applied load reaches 200 lb, the deflection at the point of loading should be recorded. The load of 750 lb in this case is a product of 500 lb basic test load multiplied by a temperature factor of 1.5. In the described case the test was not performed, because the preceding test (uniform load test) failed.

The temperature factor 1.5 for HDPE-based composition was determined experimentally by measuring composite bottom rail and spindle flexural strength and flexural modulus at ambient temperature (70°F) and 130°F. The respective flex modulus values for bottom rails were 360,330 ± 33,600 psi and 174,820 ± 32,600 psi; therefore, "high temperature loss" was 185,510 ± 66,200 psi, or 51 ± 9%. Similarly, for spindles the respective ambient- and high-temperature flexural modulus were 266,000 ± 33,900 psi and 140,200 ± 16,700 psi, and the "high temperature loss" was 125,800 ± 50,600 psi, or 47 ± 19%.

Another composite railing by a different manufacturer was made of about one third of wood flour and two thirds of ABS. The material behaved quite differently with respect to temperature. Flex strength for bottom rails at 70°F and 130°F were 5200 ± 200 psi and 4400 ± 200 psi, respectively. This means that, "high temperature loss" was 800 ± 400 psi, or 15% only. For the balusters the figures were 6000 ± 100 psi and 4800 ± 80 psi, respectively, and "high temperature loss" was 1200 psi, or 20%. Flex modulus values for bottom rails were 560,000 ± 40,000 psi and 500,000 ± 70,000 psi at 70°F and 130°F, respectively; hence, the "high temperature loss" was 11%, and for balusters the respective figures were 460,000 ± 20,000 psi and 390,000 ± 30,000 psi; hence, the "high temperature loss" was 15%. The highest figure, 20%, was taken for the temperature adjustment, unlike 50% for HDPE-based railing system (see above).

As a result, for the ABS-based railing system, the required load values were set as follows: (a) For in-fill test 50 lb × 2.5 (safety factor) × 1.2 (temperature factor) = 150 lb on ft^2. (cf. 187.5 lb for the HDPE-based railing system, see above); (b) for uniform load test 50 lb × 6 ft (length of the railing system) × 2.5 × 1.2 = 900 lb (cf. 1125 lb for the HDPE-based railing system above), and (c) for concentrated load test (at top of the rail) 500 lb × 1.2 = 600 lb (cf. 750 lb for the HDPE-based railing system above). The ABS-based railing system met all the three requirements.

In June of 2005 and June 2006, AC174 was partially modified. It is now based on a newly introduced ASTM D 7032-05. Regarding guardrail system (guard and handrail), the ASTM says (Section 6.2) that for in-fill load test a railing system should resist a load of 125 lbf applied over a 1 ft^2 area normal to in-fill (Section 6.2.2). The figure of 125 lbf came from 50-lbf load multiplied by the safety factor of 2.5. This requirement is essentially the same as described above.

Likewise, for uniform load test according to ASTM D 7032 (and 2000 IBC, Section 1607.7.1), the top rail of the guard system shall resist a uniform load of 125 lb/ft applied vertically and horizontally. The procedure indicates that quarter-point loading shall be deemed to be equivalent to uniform loading. Again, the figure of 125 lb/ft came from 50 lb per lineal foot (2000 IBC, Section 1607.7.1) multiplied by the safety factor of 2.5. This requirement is also essentially the same as described above.

Finally, according to new AC174, a concentrated 500-lb load shall be applied to the guardrail system at the maximum guardrail height (ASTM D 7032, Sections 6.2.4 and 6.3). This figure came from a single concentrated load of 200 lb (2000 INC, Section 1607.7.1.1), multiplied by the safety factor of 2.5.

However, new AC174 (effective July 1, 2006, Section 5.1.1) says: "When compliance with SBC code requirements is desired, the following test loads in ASTM D 7032 shall be increased. The stated in-fill load in Section 6.2.2 of ASTM D 7032 shall be increased to 400 lbf (1.78 KN), plus any adjustments required" (temperature and moisture, UV light, and freeze–thaw effects). This means that, the load requirements were increased by at least 3.2 times.

Besides, as it was stated above, ASTM D 7032 says (Section 6.2.3) that the top rail of the tested guard system shall resist a uniform load (actually, a quarter-point loading) of 125 lb/ft, applied vertically and horizontally. The new AC174, however, states that when compliance with SBC code requirements is desired, the stated uniform load in Section 6.2.3 of ASTM D 7032 shall be increased to 235 lb/ft (applied in an outward direction at an angle of 58° from horizontal, plus any adjustments [temperature and moisture, UV light, and freeze–thaw effects]).

Composite (and PVC) Railing Systems for Which ICC-ES Reports were Issued Until November 2006

Because the acceptance criteria AC 174 do not concern materials from which railing systems are made, whether they be wood, concrete or WPCs, composite materials appeared to be in a difficult situation. Nonstructural (as yet) materials should in fact serve as structural, such as in railing systems. That is why most of manufacturers of composite (or plastic) railing systems use reinforcing components, such as wood posts covered with composite (or plastic) sleeves, and metal (steel or aluminum) inserts in support (top and bottom) rails.

Table 7.71 lists composite (and plastic) railing systems for which the respective ICC-ES reports were issued until November 2006.

Twenty two railing systems are listed in Table 7.71. Thirteen of them can hold a concentrated load of 200 lb at top rail midspan, that is, an old BOCA requirements, inferior compared to the current ICC AC 174 requirement. For another nine railing systems, a load requirement either is not a part of their ICC-ES Report or it is not specified. For two railing systems there are indications that they have passed the necessary requirements; however, specific numbers are not included.

Clearly, WPC-based railing systems are generally—with only a few exceptions—cannot be considered as yet as sound structural products.

TABLE 7.71 Composite (and plastic) railing systems listed with ICC-ES legacy or regular reports

Brand name	Manufacturer	Material	ICC-ES report (issue date)	Comments on metal inserts and passed loads
Trex Designer Handrail	Trex Company Inc.	50–50 wood–polyethylene	ESR-1190 (6/01/05)	Concentrated load of 200 lb to the top rail
TimberTech Ornamental Railing System	TimberTech Ltd.	Wood flour and HDPE	ESR-1400 (6/01/05)	The post has not been evaluated according to the code and is not a part of the ICC report.
Epoch/Evergrain	Epoch Composite Products, Inc.	50% wood fibers and 50% "plastic," compression molding	ESR-1625 (6/01/05)	Posts are not part of the ICC report
Epoch/Evergrain	Epoch Composite Products, Inc.	50% wood fibers and 50% "plastic," compression molding	Legacy NER-630 (4/01/06)	200 lb concentrated load at midpoint top rail of 6-ft span. 50 plf horizontally and 100 plf vertically
GeoDeck	LDI Composites Co.	40% HDPE, 28% rice hulls, 28% Biodac	Legacy 21–71 (9/01/06)	200 lb concentrated load at midpoint top rail (BOCA requirement)
Weatherbest Premium Railing	Louisiana Pacific Corporation	60% wood flour, 28% HDPE, 12% other ingredients (incl. phenolic resin)	ESR-1088 (6/01/05)	All solid members, except post sleeve. Length 6 ft maximum. 4″ × 4″ wood inserted into post sleeve
Boardwalk guardrail system	Certainteed Corporation	55–65% PVC, 35–45% hardwood fiber	Legacy NER-576 (3/01/04)	ICC-ES does not allow to use as handrails. Maximum post spacing is 4–6 ft on center. 50 plf horizontally to the top of the rail. Concentrated 200 lb at midspan of the top rail. Horizontally 50 plf horizontally + 100 plf vertically.
ChoiceDek Guardrail (Premium, Dreamworks, LifeCycle, MoistureShield, etc.)	A.E.R.T.	"Wood–polymer composite"	Legacy NER-596 (2/01/06)	Not specified

TABLE 7.71 (*Continued*)

UltraDeck	Midwest Manufacturing Extrusion	60% wood filler and 35% HDPE, 5% talc	ESR-1674 (11/01/04)	A top-rail steel insert. No details given in the report on the load
Oasis Composite Rail System	Alcoa Home Exteriors, Inc.	55% wood flour and 45% virgin HDPE	ESR-1425 (6/01/05)	Post sleeves with 4″ × 4″ wood post insert. 6 ft length. "Load specified in Table 16-B of the UBC (Uniform Building Code)." It is a load of 200 lb.
XTENDEX	Carney Timber Company	60% rice husks and 40% HDPE, coupling agents	Legacy NER-695 (11/01/04)	Steel tube reinforced 5″ × 5″ post and top rail. Post sleeve over wood post. Railing system 5′–4″ wide. Load of 50 lb per lineal foot horizontally and 100 plf vertically, concentrated load of 200 lb in any direction on the top rail
Carefree	U.S. Plastic Lumber	HDPE, no fillers	Legacy 97-63.01 (12/01/04)	200 lb concentrated load at midpoint top rail (BOCA requirement)
Cross Timber Guardrail system	Elk Composite Building Products, Inc.	Wood flour and polypropylene	ESR-1590 (6/01/05)	The use as a handrail is not included to the ICC-ES report.
Dream Rail	Thermal Industries	PVC	Legacy 97-55 (6/1/05)	Aluminum or wood top and bottom rail inserts, steel rail support straps. 200 lb concentrated load at midpoint top rail (BOCA requirement)
Teck Deck (Boston, Heritage, Bellevue, VinylGard, Bristol, Heatland, etc.)	Outdoor Technologies	Modified PVC, no fillers	Legacy 21-26 (2/01/06)	Wood, steel or aluminum post inserts. 200 lb concentrated load at midpoint top rail (BOCA requirement)

(*Continue*)

TABLE 7.71 *(Continued)*

Brand name	Manufacturer	Material	ICC-ES report (issue date)	Comments on metal inserts and passed loads
Fiberon, Fiberail, other names	Fiber Composites	PVC	Legacy 22-41 (10/01/04)	4″ × 4″ wood inserted into posts. Concentrated 200-lb load to the top rail; max 6 and 8 ft length (for two different posts)
Genova Vinyl Railing System	Genova Products	PVC	ESR-1904 (6/01/05)	Top and bottom rails with aluminum insert (p-channels)
Sheerline PVC Railing	L.B. Plastics, Inc.	PVC	Legacy NER-571 (3/01/06)	PVC post sleeve, wood post. Aluminum "U" shaped reinforced top rail (for one railing product of two). Simultaneous 50 plf horizontal and 100 plf vertical loading
CertainTeed PVC Railing Systems: Cambridge and Oxford	Certainteed Corporation	PVC	Legacy NER-605 (4/01/05)	U-shaped aluminum top and bottom rail reinforcing; 1–5/8″ galvanized steel tube posts reinforcing. Load of 50 lb per lineal foot, concentrated load of 200 lb in any direction on the top rail
Yardcrafters Vinyl Railing systems	GSW Building Products	PVC	ESR-1738 (6/01/2005)	G90 galvanized, U-shaped steel insert. Load not specified.
Vinyl Guardrail Systems (Sentinel and Reliant)	Westech Building Products, Inc.	PVC	Legacy NER-710 (7/01/04)	Reinforced top rail, aluminum post insert or PVC sleeve over 4″ × 4″ wood post. 200 lb concentrated load at midpoint top rail (BOCA requirement)
Triple Crown Fence	Royal Crown Ltd.	PVC	Legacy NER-705 (5/01/05)	Galvanized steel top and bottom rail inserts. 200 lb concentrated load at midpoint top rail (BOCA requirement)

All information in the table taken from the respective reports.

COMBINED FLEXURAL AND SHEAR STRENGTH: A "SHOTGUN" TEST

As it was noted above, flexural tests with WPCs should be conducted, as a rule, with a support span-to-depth ratio (of the profile) not less than 16:1. The less the ratio, the more the contribution of the shear effects. As a result, an apparent "flexural strength" can be increased, decreased, or stay approximately the same, but an apparent "flexural modulus" progressively decreases, if standard formulas for their calculation are employed. For example, when a support span was decreased from 20 to 14″, flex strength for GeoDeck increased on average by 6% (Table 7.8) and flex modulus decreased by 28% (Table 7.32). When a support span decreased from 16 to 4.3″, flex strength increased by as much as 44% (Table 7.11). In the last case it had a significant component of shear strength.

However, these tests at a small ratio of span:depth can serve as a quick test for strength of the material (or profile), without specifying which particular "strength" is measured. An example is shown in Figure 7.14. Two profiles of two different WPCs were compared in terms of deflection versus load. This was not a test for material; this was a test for overall strength and deflection of the given profile. If we calculate (mistakenly) "flexural strength" and "flexural modulus" from the data, for GeoDeck, for example, these figures would gave been equal to 1913 and 57,000 psi, respectively. Hence, flexural strength in this particular case decreased by about a third, but flex modulus decreased from 380,000 to 57,000 psi, that is, by almost seven times.

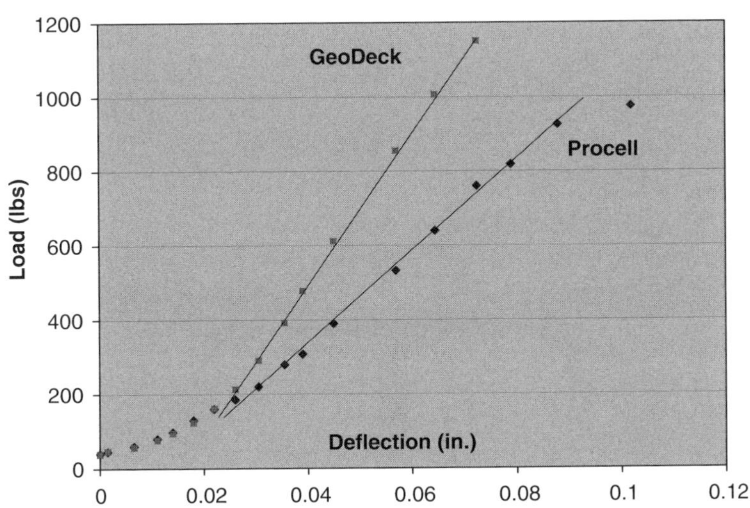

Figure 7.14 Flexural/shear strength of GeoDeck hollow board and Procell solid board. Three-point load, span of 4.75 inch, board thickness 1-1/4″ (GeoDeck) and 15/16″ (Procell), moment of inertia 0.80 in^4 (GeoDeck), 0.38 in^4 (Procell). Break load of 3093 lbs (GeoDeck) and 3135 lbs (Procell), that is, 1.4% difference.

MATHEMATICAL MODELING OF WPCs AND THE REAL WORLD

Mathematical modeling for this section was done by Professor Alexander Chemenda and his team (University of Nice, France), whom I am greatly indebted.

This mathematical exercise was aimed at optimizing a shape of WPC board shown in Figure 7.1. The principal goal was to minimize the cross-sectional area in such a way that the product would by reduced by at least 10% by weight, at the same time retaining (or improving) its mechanical properties exemplified by flexural strength and modulus, and impact resistance.

Distribution of stress under a midspan load (823 N/m, i.e., 25.9 lb/width of the profile in this case) is shown in Figure 7.15. One can see when the maximum stress is developing.

When the board is hit by the falling ball between the ribs, as shown in Figure 7.16, it undergoes deformation. Red-colored contours in the figure correspond to tensile deformations going beyond the elasticity limit and resulting in tensile fractures. One can see that the fracture is more developed along the more internal rib.

In order to minimize the cross-sectional area of the above board, and, hence, the board weight, while preserving (or improving) mechanical properties of the board, an extra rib was introduced and the channels were changed into and arch-like shape (Fig. 7.17, the lower profile).

The modified board had 11% lower mass (calculated). Mathematical modeling (see Fig. 7.18) showed that the modified board had about the same flexural strength and 9% higher flexural modulus compared to the conventional board (the upper profile in Fig. 7.17).

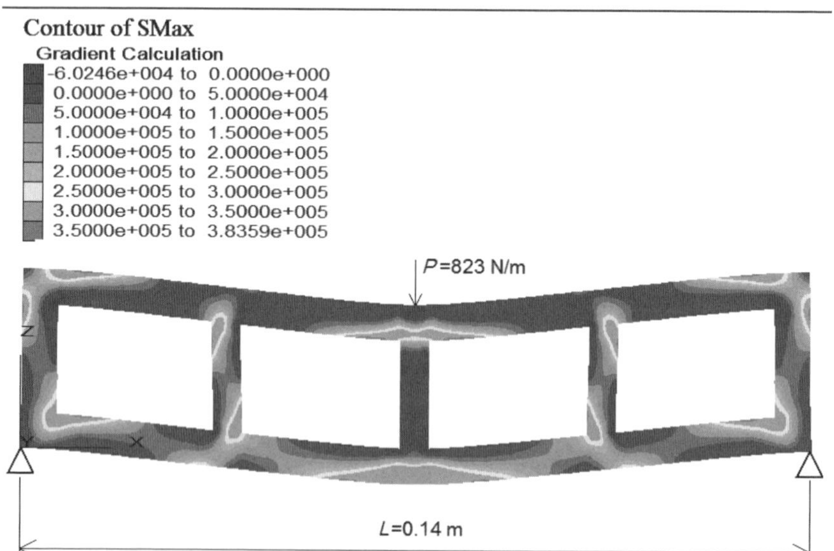

Figure 7.15 Distribution of the maximum stress in a WPC deck profile under a load of 25.9 lb/width of the board (width is equal to 5.5 in.). The calculations are done for flexural strength of the material equal to 2796 psi, and flexural modulus equal to 250,762 psi.

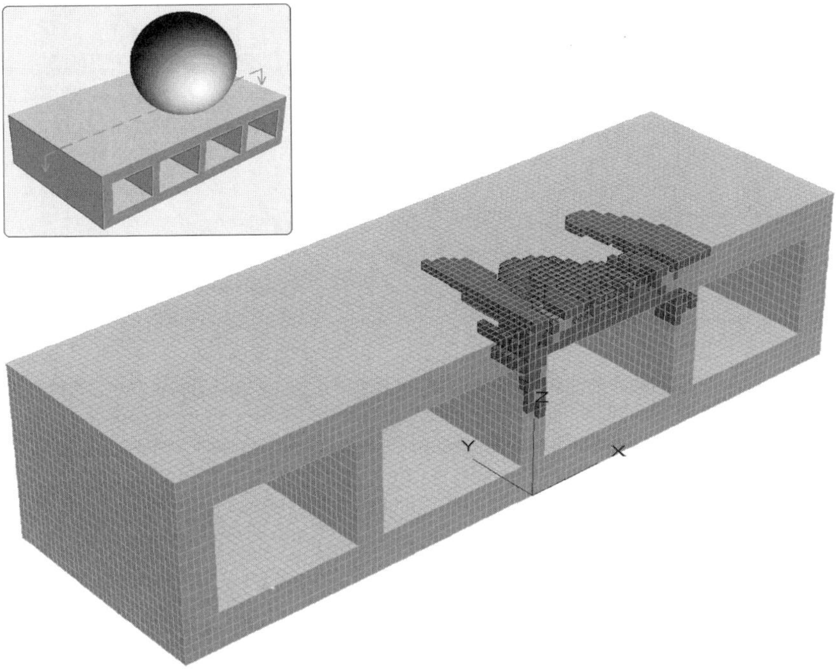

Figure 7.16 Tensile deformation of the WPC board under falling weight. Red-colored contours correspond to tensile deformations going beyond the elasticity limit and resulting in tensile fractures. See color insert.

The following figures illustrate a numerical experiment on determining impact resistance of the two profiles: conventional and modified. A steel ball (diameter of 76.20 mm = 3 in.) was (mathematically) dropped from 36-in. weight at the board surface at different locations. Predictably, the most vulnerable areas were those between ribs. Figures 7.19 and 7.20 show vertical displacement of the material within both boards at the moment when the ball reached its maximum depth. At this moment the velocity vector of the ball changes sign, and the ball starts to move upward.

Figures 7.19 and 7.20 show that the calculated vertical displacements of the material is practically the same for the both boards and equal to 0.18 cm (close to 1/16 in.). The dent diameter in both cases was 2.3 cm (0.9 in.). These figures correspond to actual experimental figures with actual boards. Figure 7.21 also shows that the modified board undergoes smaller damage under the impact compared to the conventional board.

One can see in Figure 7.21 that the tensile damage zones are more developed in the conventional composite board (left-hand side) and went deeper as fractures along the rib. The modeling data also show that the modified board undergoes smaller damage under the impact compared to the conventional board.

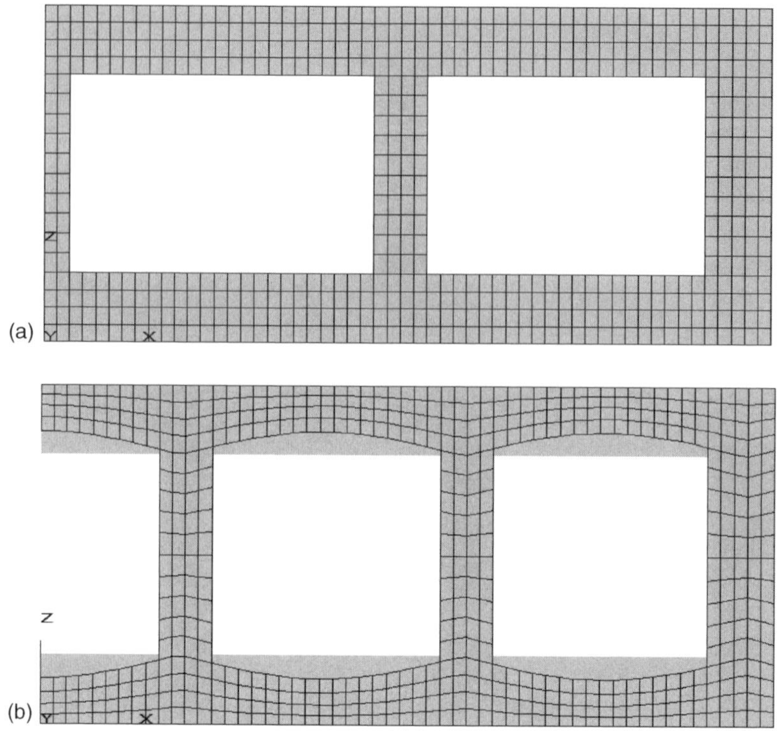

Figure 7.17 A conventional (top) and the modified board (bottom). Only half-board profiles are shown; the whole profile can be presented by mirror reflection of the half-board.

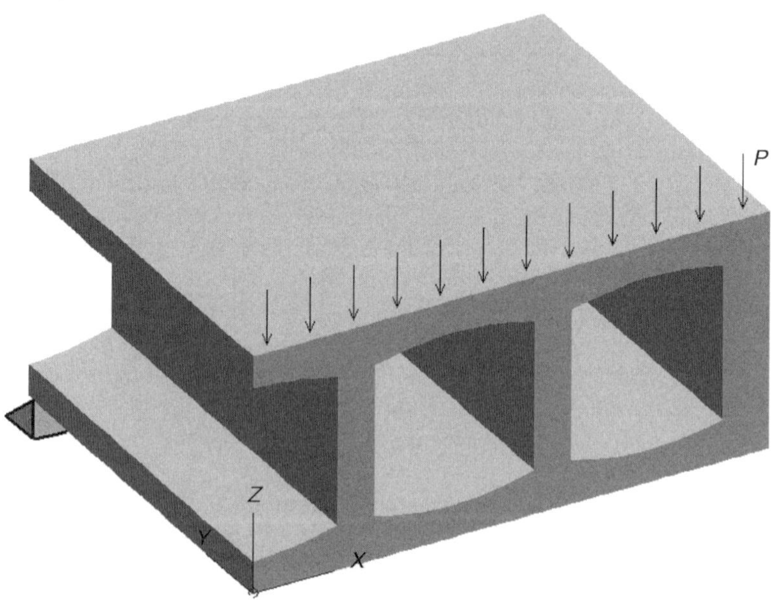

Figure 7.18 A template for the flexural numerical experiments.

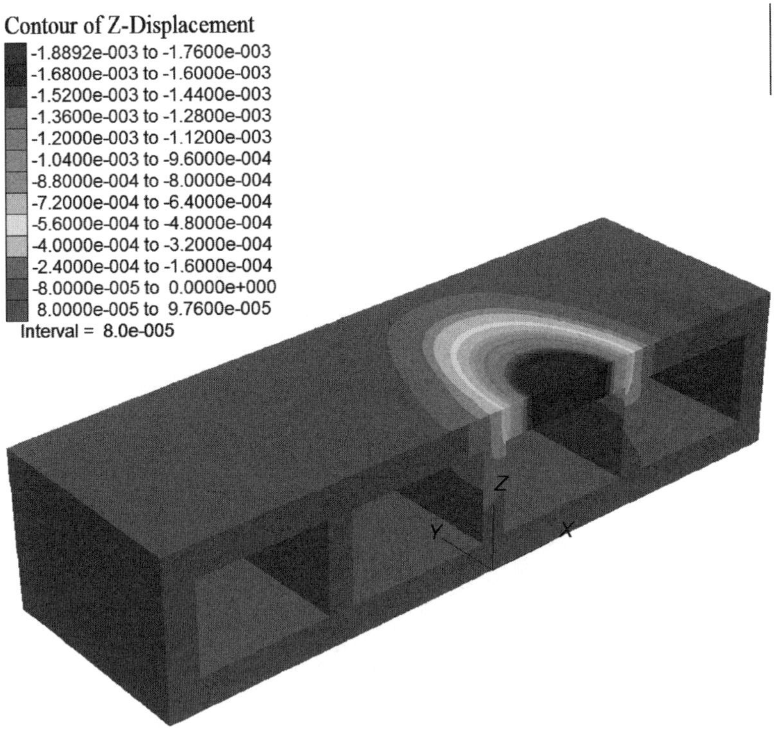

Figure 7.19 Vertical displacement of the material within the conventional composite board caused by an impact of the 3-in. (diameter) steel ball dropped from 36-in. height at the surface between ribs. The displacement and the deformation are showed at the moment when the ball reached its deepest position. At this moment, only the area around the impact location is deformed; the deformation front has not yet reached the peripheral parts of the board.

Verification of the Mathematical Model with Actual Conventional and Modified Composite Boards

A die was made, and the modified board (Figs. 7.17, 7.18, and 7.20) was extruded using the same HDPE filled with rice hulls and Biodac® composite formulation.

Weight Indeed, the modified board was 14% lighter compared with the conventional board:

Unbrushed boards

- Conventional—1.88 ± 0.01 lb/ft
- Modified—1.65 ± 0.01 lb/ft

Brushed boards

- Conventional—1.84 ± 0.01 lb/ft
- Modified—1.62 ± 0.01 lb/ft

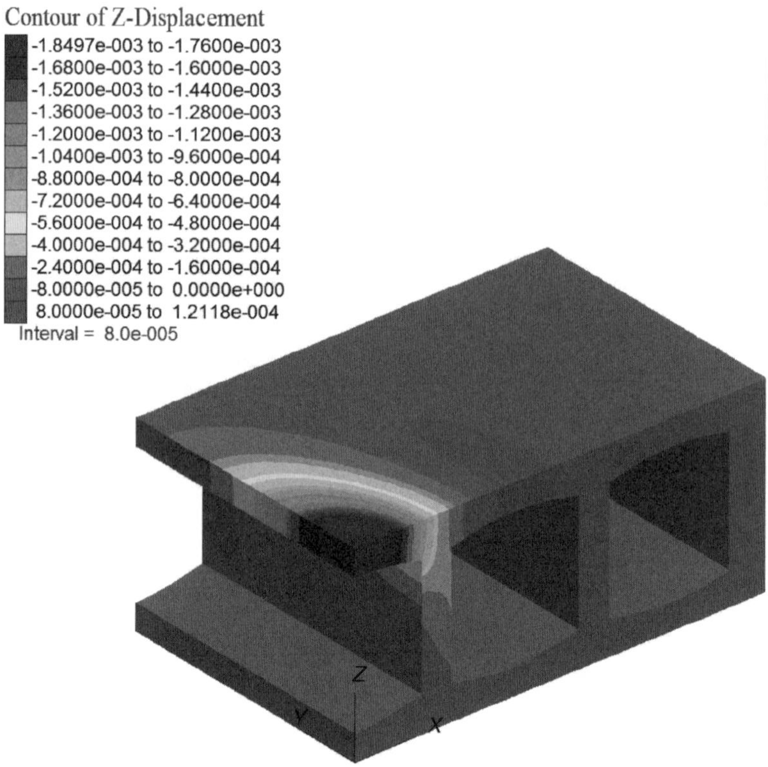

Figure 7.20 Vertical displacement of the material within the modified composite board caused by an impact of the 3-in. (diameter) steel ball dropped from 36-in. height at the surface between ribs. The displacement and the deformation are showed at the moment when the ball reached its deepest position. Only a half of the profile is shown.

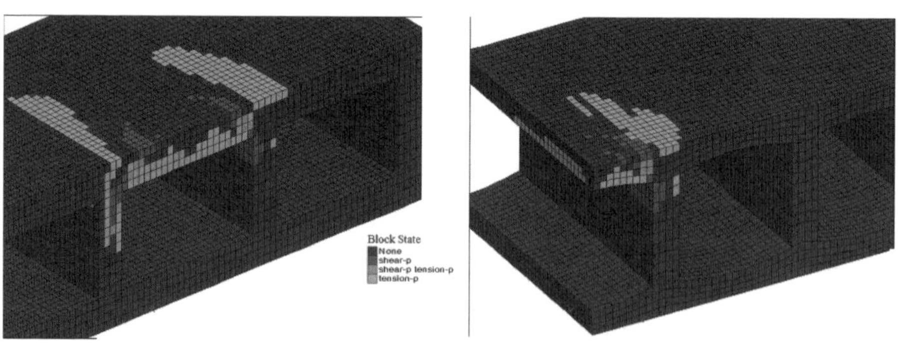

Figure 7.21 Tensile damage zones (light-blue color) as the result of the impact of a 3-in. steel ball falling from a height of 36 in. (see Figs 7.19 and 7.20). A conventional composite board is shown in the left-hand side and the modified board on the right-hand side. See color insert.

However, mechanical properties of the modified board were practically the same or even lower.

Flexural Strength (ultimate load, 14-in. span, center-point load)
Unbrushed boards

- Conventional—1,039 ± 53 lb
- Modified—923 ± 32 lb (13% lower)

Brushed boards

- Conventional—1043 ± 33 lb
- Modified—888 ± 18 lb (17% lower)

Flexural Modulus (as deflection under specified load, 14-in. span, center-point load)
Unbrushed boards

- Conventional—0.054 in. under 210 lb
- Modified—0.053 in. under 210 lb

Brushed boards

- Conventional—0.041 in. under 180 lb
- Modified—0.042 in. under 150 in.
- Conventional—0.023 in. under 110 lb
- Modified—0.021 in. under 90 lb

One can see that deflection of the modified composite board is the same or slightly higher compared to that of the conventional board.

Impact Resistance Impact resistance in terms of dents and cracks caused by a 16-lb load falling from 2, 3, and 4 ft height was practically the same for both conventional and modified composite board.

An obvious conclusion is that although mathematical modeling has shown a clear benefit of the modified composite hollow board over the conventional one in flexural modulus and impact resistance, a direct experiment showed that these properties are about the same for both the boards; Furthermore, the mathematical modeling showed that flexural strength should be the same for both the boards, a direct experiment showed that the modified board is 13–17% weaker, that is, practically parallel with the weight reduction of 14%.

That is essentially why that experiment with the mathematical modeling was conducted in the first place. In the numerical experiment, the composite material was assumed to be uniform and isotropic. In reality, it is nonuniform and anisotropic. Hollow board, by virtue of its manufacturing, bear so-called spider lines, making the board actually welded from its fragments. Those spider lines present the stress concentrators and are weak areas in the boards. Boards often break along spider lines.

Therefore, mathematical/numerical modeling of mechanical properties of composite hollow deck boards should take into account those important structural features of the boards.

REFERENCES

1. W.C. Young. *Roark's Formulas for Stress and Strain*, 6th edition, McGraw-Hill, Inc., New York, 1989, p. 101.
2. J.F. Shackelford and W. Alexander, *CRC Material Science and Engineering Handbook*, 3rd edition, CRC Press, Boca Raton, FL, 2001, pp. 794–802.

8

COMPRESSIVE AND TENSILE STRENGTH AND MODULUS OF COMPOSITE PROFILES

INTRODUCTION

Compressive and tensile strength and modulus of wood–plastic composite (WPC) materials are not among standard parameters required for composite characterizations. They are not included in ICC-ES acceptance criteria or other requirements. Hence, we will consider them here very briefly.

In a tensile test, the specimen ("dogbone") is gripped at each end and pulled apart. The "dogbone" shape of a specimen is chosen to minimize the likelihood to break the sample at or near the grips (jaws). The detailed shape (curvature) of a specimen is described in ASTM D 638, "Test method for tensile properties of plastics."

In short, a typical readout in tensile tests is elongation of the specimen, which is a stretch, or extension of the specimen. The elongation is typically expressed in percent of the stretch related to gage length (the distance between two reference points within the center section of the "dogbone." In other words, percent of elongation equals to 100× tensile strain.

In a compressive test, the specimen, often machined as a right prism whose length is twice its width, is squeezed (compressed).

Wood-Plastic Composites, by Anatole A. Klyosov

BASIC DEFINITIONS AND EQUATIONS

For a tensile test or a compressive test, the cross-sectional area (width × thickness) of the specimen should be determined. Stress is the ratio of the load (P) to the cross-sectional area (A):

$$\text{Stress } (\sigma) = P/A \text{ (psi)}.$$

Ultimate tensile strength is the tensile stress at the specimen rupture.
For rectangular specimens, maximum tensile strength equals to

$$T = P/bd,$$

where T is the maximum tensile strength, in psi; P is the ultimate load, in lb; b is the width of the specimen, in in.; and, d is the depth of the specimen, in in..

Strain in a tensile test is the ratio of elongation to gage length (see above), typically expressed in percent values.

$$\text{Strain } (\varepsilon) = \Delta L/L \times 100\%.$$

Ultimate tensile percent elongation (at break) is the elongation at the specimen rupture.

Tensile or compressive modulus is the ratio of stress to strain at any point along the initial straight portion of the stress–strain (load–deformation) curve:

$$E = \sigma/\varepsilon \text{ (psi)}.$$

Tensile modulus of elasticity equals to

$$E = PL/\Delta Lbd,$$

where P is the arbitrary load on the load–deformation plot, lb; L is the length of the specimen, in.; ΔL is the elongation of the specimen under the load P, in.; b is the width (base) of the specimen, in.; and d is the depth of the specimen, in..

ASTM RECOMMENDATIONS

ASTM D 638, "Standard Test Methods for Tensile Properties of Plastics"

The ASTM procedure describes procedures for determining tensile properties of materials in the form of standard dumbbell-shaped test specimens, with thickness between 1.0 mm (0.04 in.) and 14 mm (0.55 in.). The ASTM emphasizes, where directly comparable results are desired, that all samples should be of equal thickness.

The ASTM procedure describes in detail a typical testing machine, consisting of fixed member, movable member, grips (fixed or self-aligning grips), drive mechanism, load indicator, and crosshead extension indicator. The test procedure also describes suitable instruments (extensometers) that shall be used for determining the distance between two designated points within the gage length of the test specimen as the specimen is stretched.

Because tested materials can have different thicknesses of their samples and can be designated, for example, as rigid and semirigid plastics or reinforced composites, test specimens are subdivided into five types (I through V) according to recommended specimen dimensions. These dimensions are described in detail in the ASTM procedure. Test specimens can be prepared by machining, cutting, or molding.

For isotropic materials, at least five specimens for each sample are required to be tested. For anisotropic materials, 10 specimens should be tested, five normal to and five parallel with the principle axis of anisotropy, for each sample.

Speed of testing shall be from 1 mm (0.05 in.)/min to 500 mm (20 in.)/min, depending on specimen type and degree of rigidity of the material, so that rupture occurs within 30 s to 5 min of testing time.

Calculations of tensile strength and tensile modulus of elasticity are described in the text above.

Tensile strength can be measured at yield or at break. In plastics, when a curve of stress versus strain is determined, a continued increase in deformation without an increase in load is observed rather often. In other words, it is the first point on the stress–strain curve at which an increase in strain occurs without an increase in stress. The point at which a curve first shows such a behavior is defined as the yield point. Yield point is also defined as the first point on the stress–strain curve when it shows zero slope. At this point the tensile strength at yield is measured. Otherwise, if the specimen ruptures at a continuous increase of load, the tensile strength at break is measured and reported.

Precision of flexural modulus measurements is usually fair. ASTM D 638-03 lists examples for several different plastics, including polypropylene, tested by eight laboratories using the Type I specimen, of nominal 0.125-in. thickness. Each test result was based on five individual determinations. Each laboratory obtained two test results for each material. For polyethylene, tensile modulus of elasticity was 210,000$\pm$8900 psi for within-laboratory tests and 210,000$\pm$71,000 psi for between-laboratory tests. Tensile strength at yield for polypropylene was 3630$\pm$22 psi for within-laboratory tests and 3630$\pm$161 psi for between-laboratory tests. Elongation at yield was 8.79$\pm$0.45 and 8.8$\pm$5.9% for within- and between-laboratory tests, respectively.

Tensile strength at break for polypropylene was 2970$\pm$1540 psi for within-laboratory tests and 2970$\pm$1650 psi for between-laboratory tests. Elongation at break was 293$\pm$51 and 293$\pm$119% for within- and between-laboratory tests, respectively.

The ASTM procedure notes that tensile strength and elongation at break values for unreinforced polypropylene are generally highly variable due to inconsistencies in necking or “drawing” of the center section of the test specimen. Therefore, the

TABLE 8.1 Tensile strength at yield for polyethylene plastics, obtained by ten different laboratories (ASTM D 638)

		Tensile strength at yield (psi)		
			Deviation	
Material	Test speed (in./min)	Average	Within-laboratory tests	Between-laboratory tests
LDPE	20	1,544	52	64
LDPE	20	1,894	53	61
LLDPE	20	1,879	74	100
LLDPE	20	1,791	49	76
LLDPE	20	2,900	56	88
LLDPE	20	1,730	64	96
HDPE	2	4,101	196	372
HDPE	2	3,523	176	478

ASTM procedure recommends employing of tensile strength and elongation at yield as more reproducible tests, related in most cases to the practical usefulness.

For polyethylene plastics, the ASTM procedure lists the following values for tensile strength at yield, obtained by 10 different laboratories (Table 8.1).

Tensile yield elongation, obtained by eight laboratories (Table 8.2).

TABLE 8.2 Tensile yield elongation for polyethylene plastics, obtained by eight different laboratories (ASTM D 638)

		Tensile yield elongation (%)		
			Deviation	
Material	Test speed (in./min)	Average	Within-laboratory tests	Between-laboratory tests
LDPE	20	17.0	1.3	3.2
LDPE	20	14.6	1.0	2.4
LLDPE	20	15.7	1.4	2.9
LLDPE	20	16.6	1.6	3.3
LLDPE	20	11.7	1.3	2.9
LLDPE	20	15.2	1.3	2.6
HDPE	2	9.3	1.4	2.8
HDPE	2	9.6	1.2	2.8

Note of the author: In the ASTM D 638-03 procedure, these data are given in a different format, such as (for the last line) $9.63 \pm 1.23\%$ (for within-laboratory tests). Such a notation does not have much sense, and it is misleading regarding an "apparent" precision. First, when the average data deviates by more than one unit ($\pm 1.23\%$), it has little sense to give the principal figure as 9.63%, that is, pretending that the measurement is done with a precision to a third digit, when even the first one

TABLE 8.3 Tensile strength at break for polyethylene plastics, obtained by nine different laboratories (ASTM D 638)

	Tensile strength at break (psi)		
		Deviation	
Material	Average	Within-laboratory tests	Between-laboratory tests
LDPE	1,592	52	75
LDPE	1,750	67	103
LLDPE	4,379	127	219
LLDPE	2,840	79	144
LLDPE	1,679	34	47
LLDPE	2,660	119	166

is questionable. Second, deviations are rarely determined with a precision better than about 30%. That is, to give 1.23, it is to send a misleading message that the deviation is determined with precision better than 1%. At best, the data in the table should be given as 9.6 ± 1.2%. More adequate would be even 10 ± 1%.

Tensile strength at break, obtained by nine laboratories (test speed of 20 in./min), is shown in Table 8.3.

Tensile break elongation, obtained by nine laboratories (test speed of 20 in./min), is shown in Table 8.4.

TABLE 8.4 Tensile break elongation for polyethylene plastics, obtained by nine different laboratories (ASTM D 638)

	Tensile break elongation (%)		
		Deviation	
Material	Average	Within-laboratory tests	Between-laboratory tests
LDPE	567	32	60
LDPE	569	62	89
LLDPE	890	26	114
LLDPE	64	7	12
LLDPE	803	26	104
LLDPE	782	42	97

ASTM D 5083, "Test Methods for Tensile Properties of Reinforced Thermosetting Plastics Using Straight-Sided Specimens"

This particular ASTM procedure is not directly relevant to wood–thermoplastic composites, and therefore it is only briefly described here. The procedure is recommended for the determination of the tensile properties of thermosetting reinforced plastics. Experience with this test method has been limited to glass-reinforced thermosets. The principal difference with ASTM D 638, using dogbone-shaped specimen, is that ASTM D 5083 uses test specimens of uniform nominal width.

The test method can be used for testing materials with thickness up to 0.55 in. (14 mm). The apparatus recommended for ASTM D 5083 procedure is similar in kind to that in ASTM D 638. Test specimens shall be in the form of a rectangular prism, with the preferred overall length more than 10 in., width of 1.0 in., and thickness between 0.08 in. (2 mm) and 0.55 in. (14 mm). The recommended standard speed of testing shall be 0.2 in. (5 mm)/min for stress testing and 0.08 in. (2 mm)/min for tensile modulus determinations.

Tensile strength is calculated, as usually, by dividing the maximum load by the original minimum cross-sectional area of the specimen, and is reported in psi or in Pascal (pound-force per square inch) as "tensile strength at yield" or "tensile strength at break," whichever term is applicable. Tensile modulus of elasticity is calculated from the initial linear portion of the load–extension curve, by dividing the difference in stress on any segment on this straight line by the corresponding difference in strain, and taking into account the average cross-sectional area of the test specimens.

The procedure is illustrated in the description of the ASTM method using test results for six materials (realistically round average figures for tensile strength at break and tensile modulus of elasticity are given here):

- a bulk molding compound (6100 psi; 2,040,000 psi)
- a sheet molding compound (9700 psi; 1,740,000 psi)
- a vinylester/glass fiber mat reinforced (12,900 psi; 1,303,000 psi)
- a urethane resin/glass fiber mat reinforced (16,500 psi; 1,130,000 psi)
- a polyester resin/glass fiber mat reinforced (17,800 psi; 1,629,000 psi)
- a pultruded ladder rail (82,000 psi; 4,360,000 psi),

ASTM D 695, "Standard Test Method for Compressive Properties of Rigid Plastics"

Compressive properties include compressive strength, modulus of elasticity, yield stress, and deformation beyond yield point. The ASTM procedure covers determinations of all of them. In all cases, tested specimens are loaded in compression at relatively low uniform rates of straining or loading. Compressive yield point is the first point on the stress–strain curve at which an increase in strain occurs without an increase in stress. In other words, it is the load under which the specimen starts to move continuously without an increase in the load. Also, many plastic materials will continue to deform in compression until a flat disk is produced, without breaking of the specimen. In those cases the compressive stress (nominal) increases steadily in the process, without failure of the material. Compressive strength typically has no meaning in such cases.

The ASTM test describes in detail the apparatus for measuring compressive properties, preparation of test specimens, conditioning, and procedures. The standard test specimen is recommended to be in the form of a right cylinder or prism whose length is twice its width or diameter. Preferred specimens are 0.5″ × 0.5″ × 1.0″ (prism) or 0.5″ × 1.0″ (cylinder) in size. Though, for different types of compressive behavior, the ASTM recommends different sizes of the specimens.

When testing material is of anisotropic nature (or may be suspected to be of that nature), such as WPC materials, duplicate sets of test specimens are recommended to be prepared having their long axis parallel with and normal to the suspected direction of anisotropy. This is often called "direction of length" and "direction of thickness." In wood specimens it is called "parallel to grain" and "perpendicular to grain." At least five specimens shall be tested for each sample (five for isotropic materials and 10 for anisotropic materials). The standard speed of testing is typically 0.05 in./min (1.3 mm/min), unless the material reaches the yield point. In that case it might be desirable to slightly increase the speed, as described in the ASTM D 695.

The compressive strength is calculated by dividing the maximum compressive load by the original minimum cross-sectional area of the specimen. The compressive yield strength is calculated in the same manner, but instead of compressive load at break, the compressive load at the yield point is used. The compressive modulus of elasticity is calculated in the usual manner, by dividing the compressive stress taken as a point on the initial linear portion of the load-deformation curve by the corresponding strain.

The ASTM gives several examples of compressive strength and modulus of plastics. For polystyrene, for example, compressive strength, determined in six different laboratories, is equal to 15,370±200 psi (within-laboratory measurements) and 15,400±500 psi (between-laboratory measurements), and compressive modulus of elasticity is equal to 563±10 psi (within-laboratory measurements) and 560±100 psi (between-laboratory measurements).

ASTM D 6108, "Standard Test Methods for Compressive Properties of Unreinforced and Reinforced Plastic Lumbers"

This ASTM method is an adaptation of ASTM D 695 to "as manufactured" plastic/composite lumber and shapes (=whole profiles), unlike small cut specimen in ASTM D 695. In ASTM D 6108 the entire cross section (cut as specified) is loaded in compression. Hence, ASTM D 6108 describes test method for evaluating compressive properties of a product, whereas ASTM D 695 describes test method for evaluating compressive properties of a material.

The reason for the adapted ASTM procedure is that plastic lumber (and WPC products) is generally nonuniform through the cross section, and small cut specimens sometimes give large variations in compressive (and other) properties.

Both methods are essentially the same. The standard test specimen is recommended to be in the form of the "as manufactured" profile whose height is twice its minimum width or diameter. Five specimens shall be tested for each sample. The standard speed of testing/loading shall result in a strain rate of 3%, that is, 0.03 in./in./min. At this speed a typical compression test on plastic/composite lumber is expected to last 1–5 min. All other conditions, tests, and calculations are essentially the same.

Precision of compressive strength and modulus measurements is usually fair. ASTM D 6108-03 lists examples of two plastic lumber shapes, tested by five laboratories. For compressive strength the profiles give values of 1600±40 and 1920±150 psi, respectively (within-laboratory testing), with standard deviations of 2.5 and 7.8% for within-laboratory testing, and 14 and 22% for between-laboratory testing. For compressive modulus of elasticity, the profiles give values of 82,400±2900 and 100,800±7300 psi, respectively (within-laboratory testing), with standard deviations of 3.5 and 7.2% for within-laboratory testing, and 25 and 32% for between-laboratory testing.

Tensile, compressive, flexural rearrangements of a sample morphology result in a dimensional change to the sample in response to an applied external force. The nature of the response and its intensity can be correlated with morphological and molecular characteristics of the sample. Two of the most important mechanical properties are stress and strain of materials and profiles, developed under a series of loads. The ultimate stress of the materials is often expressed as strength and the initial (transient but sustained) strain as a function of load is expressed as modulus of elasticity. This is related to both tensile and compressive properties.

TENSILE STRENGTH OF COMPOSITE MATERIALS

Table 8.5 shows tensile strength of several plastics. Polyethylene is the weakest among them in terms of tensile strength, and its tensile strength increases with density. However, tensile strength of nonrigid PVC can be even weaker than that of LDPE. Polypropylene follows rigid PVA and Nylon as the strongest plastics in this category.

Table 8.6 shows tensile strength of wood and some plastic lumber and wood fiber–plastic composite materials. Pressure-treated lumber (PTL) is described in the

TABLE 8.5 Tensile strength of plastics [1]

Plastic	Tensile strength (psi)
Polyethylene	
Low density (0.910–0.925)	1,400–2,500
Medium density (0.926–0.940)	2,000–2,400
High density (0.941–0.965)	2,900–4,400
Polypropylene	4,500–6,000
PVC	
Nonrigid	1,000–3,500
Rigid	5,500–8,000
Nylon	
Type 6	9,500–12,500
Type 12	7,100–8,500
Type 6/6	11,200–11,800
Type 6/10	7,100–8,500

TABLE 8.6 Tensile strength values for actual composite deck boards, determined using ASTM D 198 or D 638, parallel to the length (or as indicated). Data are reported by the manufacturers or by the author.[a] Data on wood are given in [2] or by the author's measurements[a]

Board	ASTM	Tensile strength (psi)
		Material
Pressure-treated lumber	D 198	22,200 (parallel to grain), 1,400 (perpendicular to grain)
Pressure-treated lumber	D 638	6,410±480[a]
Ponderosa pine	N/A	420 (perpendicular to grain)
PlasTEAK	D 638	3,055
Nexwood	D 638	2,840 (material), 2,230 (profile)
TimberTech	D 638	1,810
GeoDeck	D 638	1,660±33[a,b]
Trex	D 638	1,370±80[a]
Strandex (Comptrusion Corp.)	D 198	1,204
EverX	D 638	1,200
Timberlast	D 638	1,200
Trex	D 198	854
USPL Composite	D 198	780

[a] The author's measurements.

[b] Sample dimensions, 0.500 in.×0.234 in. (average); cross-head speed, 0.2 in./min, 50 mm gage length; maximum tensile strength 1660±33 psi (panel, length direction), 1610±28 psi (rib, length direction); elongation at break, 1.7±0.1% (panel), 1.5±0.1% (rib).

literature as the strongest one (22,200 psi), having a huge gap in tensile strength compared to that of plastics. However, our own data obtained with commercially available PTL showed more modest values (6410 psi), though still much higher than that of HDPE-based composite materials (780–3000 psi). However, because two different ASTM (D 638 and D198) were employed for measurements of tensile strength, data in Table 8.6 cannot be compared directly across the table.

Tensile strength of HDPE generally decreases when filled to become HDPE-based composite materials. Tensile strength moves from 2900–4400 psi (neat HDPE) to 800–2800 psi in composite materials.

Likewise, tensile strength of polypropylene decreases when filled with 40-mesh ponderosa pinewood flour (0–60% w/w) and continues to decrease when filled with 70–85% of Lignocel C 300 (wood fiber, particle size 200–300 μm, granular size 3–8 mm). It goes down from 4060 psi for neat polypropylene to 2900 psi for 60% wood-flour-filled PP to 1300 psi for 85% Lignocel-filled PP [3].

Moisture significantly decreases the tensile strength of polypropylene-based WPC with high wood flour content. For example, for 80%-substituted polypropylene, the tensile strength decreases from 2400 psi (zero moisture content) to 860 psi (25%

moisture, w/w). For 85%-substituted polypropylene, the tensile strength decreases from 1250 psi (zero moisture content) to 290 psi (30% moisture content) [3].

COMPRESSIVE STRENGTH OF COMPOSITE MATERIALS: EXAMPLES

TABLE 8.7 Compressive strength of plastics [4]

Plastic	Compressive strength (psi)
Polypropylene	5,500–6,500
PVC, rigid	10,000–11,000
Nylon	
Type 6	9,700
Type 6/6	4,900
Type 6/10	3,000

TABLE 8.8 Compressive strength (data of September–November 1999) of PTL, Trex, and GeoDeck (discontinued composition and current composition, respectively)

Material	ASTM	Dimension	Ultimate load (lb)	Compressive strength (psi)
Pressure-treated lumber	D 695	1.0″ × 1.0″ × 2.0″	5,830 ± 170	5,830 ± 170
Trex	D 695	1.0″ × 1.0″ × 2.0″	2,928 ± 50	2,930 ± 50
Trex	D 6108	5.0″ × 1.18″ × 2.36″	16,350 ± 550	2,771 ± 94
GeoDeck	D 695	0.39″ × 0.39″ × 0.78″	495 ± 14	3,250 ± 90
GeoDeck	D 6108	Base area 3,425 in.^{2}	11,970 ± 1,020	3,495 ± 300

Tests were conducted in longitudinal direction.

TABLE 8.9 Effect of composition of GeoDeck 6 in. × 3 6 in. soundwall post on compression strength

Composition	Compressive strength (psi)
42% HDPE, 43% rice hulls, 10% Biodac, 2% coupling agent, 3% additives	2,710 ± 200
33% HDPE, 6% rice hulls, 55% Biodac, 6% additives	2,780 ± 80
33% HDPE, 65% rice hulls, 2% additives	3,340 ± 140
45% HDPE, 53% rice hulls, 2% additives	3,380 ± 285
37% HDPE, 30% rice hulls, 30% Biodac, 2% coupling agent, 1% additives	3,910 ± 300

One-foot long samples were tested (ASTM D 6108).

TABLE 8.10 Effect of mineral fillers (Biodac) on rice-hulls-filled HDPE-based composite materials

Profile	Ratio between rice hulls and Biodac in the formulation (%)		Board base area (in.^{2})	Ultimate compressive load (lb)	Compressive strength (psi)
	Rice hulls	Biodac			
Traditional board (5.5 in. × 1.25 in.), hollow	100	0	3.425	11,380±470	3,220±140
	50	50		11,970±1,020	3,495±300
Tongue-and-groove board	100	0	3.500	9,890±540	2,830±150
	50	50		12,080±240	3,450±70
Picket	100	0	1.723	5,540±160	3,215±93
	50	50		5,500±60	3,192±35

Deck boards (depth of 1.25 in.) of two profiles and 2 × 2 pickets were tested. ASTM D 6108, sample height of 2.5 in. (boards) or 4.0 in. (pickets) (2 × depth).

TENSILE MODULUS OF ELASTICITY OF COMPOSITE MATERIALS

For the majority of plastics (within the same group) the tensile modulus of elasticity increases approximately linearly with the degree of crystallinity [5]. The data for linear and branched polymers follow the same approximate relationship. It is not clear whether the same molecular principles is applicable to WPCs (Table 8.13), but their tensile modulus of elasticity vary between different brands quite significantly.

Tensile modulus of polypropylene increases when filled with 40-mesh ponderosa pinewood flour (0–60% w/w) and continues to be at the same level or further

TABLE 8.11 Compressive strength of GeoDeck flat panel and a rib (see Figure 7.1)

Sample	Load direction	Maximum compression strength (psi)	Yield stress (psi)	Yield strain (%)
Panel	Length	5,430±205	3,420±36	7.9±0.7
	Thickness	Did not fail	3,540±68	11±2
Rib	Length	6,270±510	3,180±36	9.4±1
	Thickness	6,500±2,060	No zero-slope yield	N/A

Sample dimensions, panel: (a) length direction 0.250″ × 0.250″ × 0.500″, (b) thickness direction 0.250″ × 0.250″ × 0.250″. Cross-head speed: 0.05 in./min. Sample dimensions, rib: 0.225″ × 0.250″ × 0.500″. Cross-head speed 0.05 in./min.

TABLE 8.12 Compressive strength values for actual composite deck boards, determined using ASTM procedures as indicated. Data are given for parallel[a] and perpendicular[b] directions of compression. Data are reported by the manufacturers. Data on wood are given in [3] or by the author's measurements[c]

Board	ASTM	Compressive strength (psi)
Pressure-treated lumber	D 198	9,500[a]
		2,100[b]
Pressure treated lumber[c]	D 695	5,830 ± 170[a]
Ponderosa pine	D 198	5,320[b]
		580[b]
GeoDeck (rib)	D 695	6,270 ± 510[a]
GeoDeck (panel)	D 695	5,430 ± 205[a]
GeoDeck	D 6108	3,495 ± 300[a]
XTENDEX	D 6108	4,400
Maxituf (plastic lumber)	D 695 (yield)	3,900
Rhino Deck	D 198	3,419[a]
	D 143	2,163[b]
CorrectDeck	D 6108	3,280[a]
TimberTech	D 695	3,031[a]
Fiberon	D 198	1,736[a]
	D 143	2,718[b]
Perfection	D 198	1,736[a]
	D 143	2,718[b]
Boardwalk	D 198	2,448
Strandex (Comptrusion Corp.)	D 198	2,428[a]
		2,441[b]
EverX	D 143	2,350[a]
		2,350[b]
Timberlast	D 143	2,350[a]
		2,350[b]
USPL Composite	D 198	2,300[a]
		2,300[b]
PlasTEAK	N/A	2,080[a]
		1,110[b]
Nexwood	D 6108	1,970[a]
Trex	D 198	1,806[a]
	D 143	1,944[b]
Carefree	D 6108	1,740[a]

increases when filled with 70–85% of Lignocel C 300 (wood fiber, particle size 200–300 μm, granular size 3–8 mm). It goes up from 189,000 psi for neat polypropylene to 670,000 psi for 60% wood-flour-filled PP to 840,000 psi for 80% Lignocel-filled PP [3].

TABLE 8.13 Tensile modulus values for actual composite deck boards, determined in parallel to grain/to the length direction or as indicated. Data are reported by the manufacturers or by the author.[a] Data on wood are given in [3] or determined by the author[a]

Board	ASTM	Tensile modulus (psi)
Pressure-treated lumber	N/A	2,300,000 (parallel to grain), 166,000 (perpendicular to grain)
Pressure-treated lumber[a]	D 638	1,340,000±240,000
TimberTech	D 638	879,000
EverX	N/A	450,000
GeoDeck (panel)[a,b,c]	D 638	421,000±8,100
GeoDeck (rib)[a,b,d]	D 638	403,000±5,300
PlasTEAK	D 638	150,000

[a] Determined by the author.

[b] Sample dimensions 0.500 in. × 0.223 in. (average), gage length 50 mm, cross-head speed 0.2 in./min.

[c] Elongation at break 1.7±0.1% (GeoDeck panel).

[d] Elongation at break 1.5±0.1% (GeoDeck rib).

Moisture significantly decreases the tensile modulus of polypropylene-based WPC with high wood flour content. For example, for 80%-substituted polypropylene, the tensile modulus decreases from 710,000 psi (zero moisture content) to 150,000 psi (25% moisture content, w/w). For 85%-substituted polypropylene, the tensile modulus decreases from 540,000 psi (zero moisture content) to 90,000 psi (30% moisture content) [3].

COMPRESSIVE MODULUS OF COMPOSITE MATERIALS

As for tensile modulus of plastics (Table 8.14), the compressive modulus of composite deck boards (Table 8.15) follows the same trends as the elastic modulus with respect to a sample's degree of crystallinity [6].

TABLE 8.14 Tensile modulus of plastics

Plastic	Tensile modulus (psi)
HDPE	170,000
Polypropylene	190,000
Nylon	400,000
PVC	410,000

TABLE 8.15 Compressive modulus values for actual composite deck boards, parallel[a] or perpendicular[b] to the length. Data are reported by the manufacturers or by the author.[c] Data on wood are given in [3] or determined by the author[c]

Board	ASTM	Compressive modulus (psi)
Pressure-treated lumber	N/A	2,300,000[a], 162,000[b]
Pressure-treated lumber[c]	D 695	401,000±44,000[a,c]
EverX	D 4761	550,000[a]
Timberlast	D 4761	550,000[a]
GeoDeck board (rib)[c,d]	D 695	179,000±46,200[a,e]
		127,000±12,900[b]
GeoDeck board (panel)[c,d]	D 695	144,000±31,400[a,f]
		114,000±30,100[b,g]
Trex[c]	D 695	139,700±4,400[a,c]

[a] Parallel to the length or length direction.

[b] Perpendicular to grain or thickness direction.

[c] Determined by the author.

[d] Sample dimensions 0.250″ × 0.250″ × 0.500″ (length direction) and 0.250″ × 0.250″ × 0.250″ (thickness direction), cross-head speed 0.05 in./min.

[e] Yield strain 9.4±1.0% (GeoDeck rib, length direction).

[f] Yield strain 7.9±0.7% (GeoDeck panel, length direction).

[g] Yield strain 11±2% (GeoDeck panel, thickness direction).

REFERENCES

1. J.F. Shackelford and W. Alexander. *CRC Material Science and Engineering Handbook*, 3rd edition, CRC Press, Boca Raton, FL, 2001, p. 600–606.
2. M.A. Svoboda and R.W. Lang. Property profiles and structure-property-relationships of polypropylene-wood-composites with high wood content. In: *Wood-Plastic Composites, A Sustainable Future. The Proceedings of International Conference,* Applied Market Information, Bristol, UK, Vienna, Austria, May 14–16, 2002.
3. R.M. Rowell, Specialty treatments. In: *Wood Handbook,* Forest Products Society, Madison, WI, 1999, p. 19–7.
4. J.F. Shackelford and W. Alexander. *CRC Material Science and Engineering Handbook*, 3rd edition, CRC Press, Boca Raton, FL, 2001, p. 679–680.
5. A.J. Peacock. *Handbook of Polyethylene: Structure, Properties, and Applications,* Marcel Dekker, New York, 2000, p. 133.
6. A.J. Peacock. *Handbook of Polyethylene: Structure, Properties, and Applications*, Marcel Dekker, New York, 2000, p. 146.

9

LINEAR SHRINKAGE OF EXTRUDED WOOD–PLASTIC COMPOSITES

INTRODUCTION

There are two principally different types of temperature-driven dimensional changes in plastics and plastic-based composites: linear shrinkage (which we will further call shrinkage) and expansion–contraction. Shrinkage is both the *irreversible* process and the result of the process. Expansion–contraction is a *reversible* process.

In this chapter I will describe the shrinkage—its origin, magnitude, manifestation, and implications, including deck failures and warranty claims because of excessive shrinkage on decks. Failure in this context means not a physical collapse, of course, but a rejection of the product (deck) by a customer (deck owner). I will also describe preventive measures that help to eliminate the shrinkage in the field. By definition, this shrinkage is a postmanufacture shrinkage.

ORIGIN OF SHRINKAGE

Let us consider polyethylene. Its long molecules, consisting of tens and hundreds of thousands of methylene groups (—CH_2—), cannot exist as long stretched polymeric chains, which might be called a "primary structure." This would be energetically unfavorable. In reality, in order to decrease its energy and to increase entropy of the system, that is to acquire the most disoriented state, in full accordance with the second law of thermodynamics, polyethylene molecules take a maximally economical

Wood-Plastic Composites, by Anatole A. Klyosov

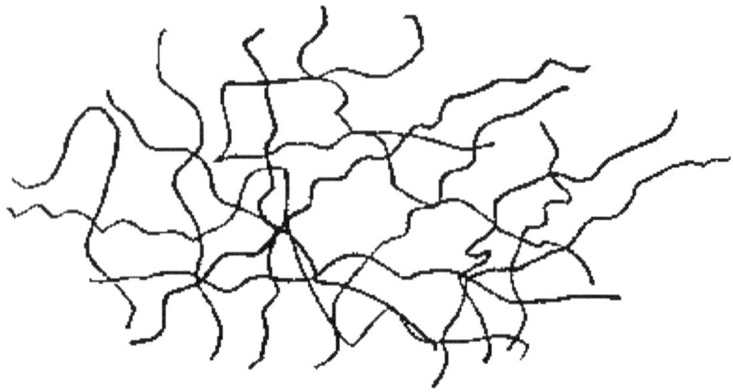

Figure 9.1 A typical presentation of an amorphous polymer structure. In reality, polymeric chains make many more interactions with each other, forming more like a "bean" structure, rather than being tightly packed.

configuration. Often this is represented as a "random" structure (Fig. 9.1), but in reality it is not so.

This configuration includes as many internal interactions as possible ("secondary structure"), and as a result it is randomly coiled, or folded into a shape resembling a bean. This might be called a "tertiary structure."

In a solid state and in a hot melt, being still or moving slowly, polyethylene molecules preserve this "bean" shape. Besides, in the solid state HDPE is a semicrystalline polymer, that is it contains both well-packed, crystalline areas and amorphous regions (Fig. 9.2).

In a hot melt, of course, crystals are molten. However, the faster the hot melt flows, the more elongated (aligned) the polyethylene molecules become in the direction of

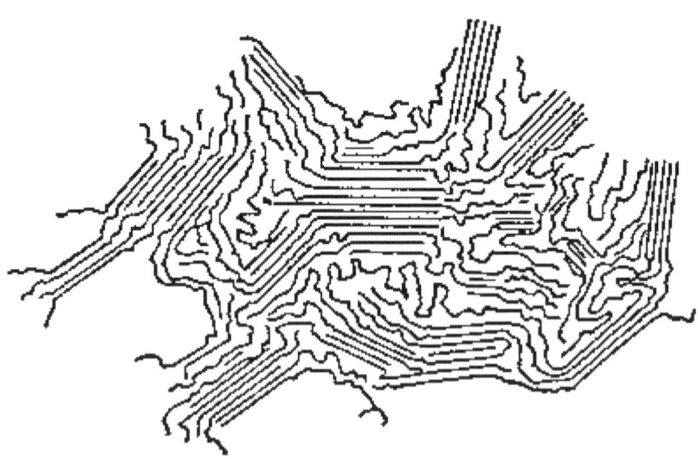

Figure 9.2 A typical presentation of the solid state HDPE [1]. Both well-packed, crystalline areas, and amorphous regions are seen.

the flow, and the more they are oriented in the direction of the flow. This reminds of an orientation of wood logs in a water basin. On a lake surface, or on a surface of a slow river, there is not any particular orientation of wood logs. Their orientation is practically random. However, in a fast and narrow stream the wood logs are oriented along the stream. As soon as the stream stops, such as in the next wide lake, the logs gradually reorient themselves in a random manner.

This is what happens when polyethylene molecules come through a die at the end of an extrusion process. They are oriented along the flow, and the degree of their alignment is highly dependent on the rate of deformation. The higher the extrusion speed, the higher the deformation, particularly when plastic comes through the die. After the die, it takes some time for polymer molecules to get reoriented to a random manner again, and to recrystallize. But the manufacturers often do not give the polymer that chance. If the temperature and, hence, mobility of polymer molecules is maintained for several hours or even several days, this time would be enough for the polymer to come back to their thermodynamically "settled" random and crystalline orientation. This in turn would result—after cooling and solidification—in dimensionally stable (in terms of irreversible shrinkage) shape of polyethylene. Of course, subsequent heating of the shape would lead to its dimensional expansion because of temperature-driven increased fluctuations, oscillation, and frequencies of polymer molecules, and a follow-up cooling would lead to a reversible contraction of the shape as a result of decreased fluctuations of polymer molecules. These expansions–contractions can be repeated unlimitedly. We will consider them in Chapter 10.

However, in reality a profile coming from the die is forcefully cooled, often using a water bath. Polymer molecules oriented along the flow are "frozen" in the thermodynamically unfavorable state and cannot move noticeably until temperature is up again. This is related to polymer molecules in both a neat plastic and plastic-based composites. In such a "frozen" state composite deck boards are often placed into a warehouse and eventually sold to an end user, a customer.

Once such "frozen" boards are installed on a deck, and as soon as the sun is up and boards are heated, polymer molecules start to move, steadily reaching their thermodynamically "settled," favorable final "bean" shape. The board becomes shorter compared to its preceding, "frozen" state.

In fact, we consider here two factors, a thermodynamic and a kinetic one. Thermodynamics provides a driving force to molecules of plastic to reorient themselves to the most energetically favorable configuration of the molecules. Generally, it does not depend on temperature. However, a kinetic factor, which determines the speed with which the molecules change their configuration to the most favorable, is largely—in fact, exponentially—dependent on temperature. "Frozen" molecules in this regard are molecules that move very slowly. It will take years for them to reach that favorable, energetically minimized configuration. Under a hot, direct sun it might take only weeks, and even days. When a deck surface temperature increases from 60° F in the morning to 130° F in the afternoon, polymer molecules move about 100 times faster. For such processes temperature coefficient is close to 2 per 10° F, that is in the velocity of the configurational change doubles per each 10° F increase temperature. Therefore, from the 60 to 130° temperature range polymer molecules move 2^7 times faster, that is 128 times faster.

SIZE OF SHRINKAGE

The overall shrinkage reflects a summarily transition of polymer molecules from the initial, "hot," elongated, right-out-of-the-die average length to the final, folded, "bean" shape. For HDPE, this overall shrinkage accounts for approximately 3% of the length, depending on average molecular weight of HDPE, its degree of crystallinity, die configuration, flow speed, and so on. Generally, shrinkage of semicrystalline polymers is between 1 and 4%, compared to 0.3–0.7% for amorphous polymers. The formation of crystallites results in better packing of the polymer chains and, therefore, increases the shrinkage. However, most of this shrinkage takes place before a board cools down right-off-the-die. For solid composite boards, because of their mass, the cooling takes a long time, up to 2 days (despite their forceful cooling online), during which the full shrinkage is practically accomplished, often to 99% and more of their overall value. That is why solid boards usually do not show any noticeable postmanufacture shrinkage (within 0.02–0.05% of their length). This corresponds to a postmanufacture shrinkage of 1/32 in. –1/8 in. for a 20-foot board or 1/48 in. –1/16 in. for a 12-foot board. Typically, this is not an issue.

Hollow composite deck boards are cooled much faster, and the residual, postmanufacture shrinkage is noticeably higher compared to solid boards. In hollow boards as much as 15% of the overall shrinkage is still "stored," waiting for temperature to go up, such as on a deck, under the direct sun. Residual, postmanufacture shrinkage in hollow boards reaches 0.3–0.5%, that is around 1 in. for a 20-foot board. Because boards normally shrink equally from the both sides, this leaves a 1-in. gap between board butts (Figs. 9.3 and 9.4). Usually it takes a few months for the gap to develop, but in the South or during a heat wave anywhere in the United States a gap can form after a few weeks or even days. Figure 9.5 shows a smaller

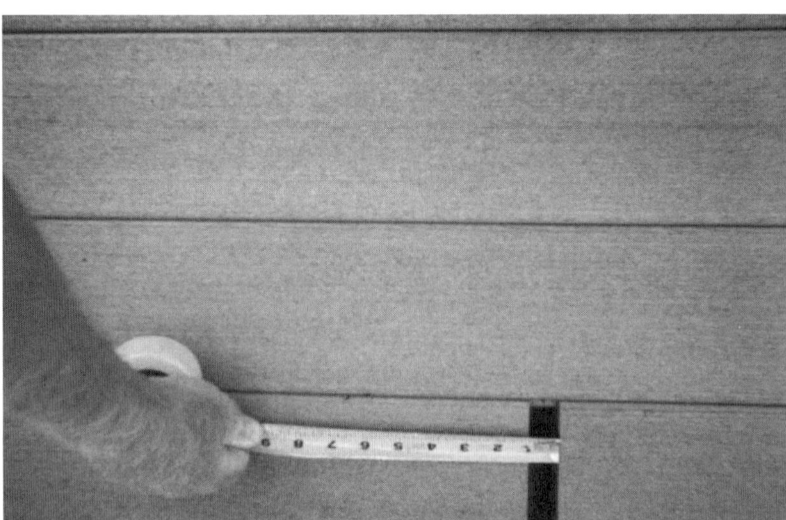

Figure 9.3 A 1-in. gap due to shrinkage of composite deck boards on a deck.

Figure 9.4 A 1-in. shrinkage gap.

gap, about 3/8-in. Figures 9.8 through 9.38 provide some examples of shrinkage on real decks.

EFFECT OF DENSITY (SPECIFIC GRAVITY) OF WPC ON ITS SHRINKAGE

Studying of shrinkage of GeoDeck deck boards, railing pickets, and so on have persistently indicated that the lower the density, the higher the shrinkage. An example

Figure 9.5 A 3/8-in. shrinkage gap.

TABLE 9.1 Effect of density (specific gravity) of GeoDeck composite pickets of the railing system on their shrinkage

Specific gravity (g/cm3)	Shrinkage (%)
1.03	0.59
1.05	0.36
1.08	0.28
1.16	0.23
1.24	0.20

Shrinkage was measured after exposure in an oven at 200° F for 4 h, 24 h after manufacturing. Data by Dr. Tatyana Samoylova, LDI Composites.

of this behavior is given in Table 9.1, which shows data obtained with GeoDeck composite pickets. The pickets were manufactured in the industrial extruder using vented and nonvented extruders, wet or dried pellets, and at various extrusion speed. By changing these conditions, pickets of various density were obtained.

EFFECT OF EXTRUSION REGIME ON SHRINKAGE

Typically, the higher the extrusion speed, the higher the shrinkage. This is clearly understandable in the light of what was described above about the origin of shrinkage. Besides, slower extrusion often leads to a higher density, which in turn results in a smaller shrinkage. On top of it, a vented extruder typically results in a higher density of the product (by removing volatile organic compounds, steam, etc.) that also gives smaller shrinkage. This can be exemplified with GeoDeck deck boards manufactured in two different regimes: one, with a relatively large throughput (600 lb/h) using nonvented extruder, and another, with a lower throughput (409 lb/h) using vented extruder. The first set of composite deck boards had specific gravity of 1.08 g/cm3, and their shrinkage was of 0.45% (after exposure for 4 h at 200° F); the second set of boards had specific gravity of 1.24 g/cm3 and shrinkage of 0.14%. The last regime has also provided less intensive cooling, and the product was pushed through the die rather than pulled out of it.

ANNEALING OF COMPOSITE BOARDS

A preventive measure against the postmanufacture shrinkage is annealing, or heating up boards in a kiln, or in a specially designed heating chamber. Under laboratory conditions it takes 4 h for heating a piece of board in an oven at 190° F to have the shrinkage completely done. Alternatively, it takes only a few minutes for composite boards to be completely annealed if they were immersed in boiling water (Table 9.2).

Such an effective annealing of the boards in boiling water was due to the fact that the boards were hollow, and water accessed the material from both outside

TABLE 9.2 Shrinkage of GeoDeck composite hollow deck boards (unbrushed) in boiling water

Time for exposure of boards to boiling water (min)	Shrinkage measured after 24 h of cooling and conditioning (%)	Moisture of the boards after 24 h	Moisture of the boards after 48 h
0 (control)	0	—	—
0 (control)	0.36 (after 4 h heating in an oven)	—	—
1	0.32	0.13	0.11
2	0.40	0.16	0.11
5	0.48	0.23	0.21
10	0.52	0.39	0.34

Shrinkage was measured 24 h after the boards were immersed into boiling water and pulled out. Water had an access to the material from both outside and inside of the boards.

and inside. When the ends of the boards were sealed with end cups, annealing was significantly less efficient (Table 9.3).

Moisture absorption of all boards described in Table 9.3 and boiled for 1–5 min, measured after 24 h, was 0.096±0.003%.

After all experimentations, an air heating annealing chamber was built at the GeoDeck plant. It takes one load of 140 bundles of boards, 96 boards in a bundle, and total 13,440 boards of more than 215,000 lineal feet. Obviously, it takes much more time to heat all this mass of boards, compared to only 4 h for a few board pieces in a laboratory oven. Generally, a heating time in a large annealing chamber at a plant depends on the mass of boards loaded into the chamber. For the amount of boards indicated above, it takes 24 h to have the annealing accomplished, of which about 20 h goes for heating up and cooling down such a mass of the material.

TABLE 9.3 Shrinkage of GeoDeck composite hollow deck boards (unbrushed) in boiling water

Time for exposure of boards to boiling water (min)	Shrinkage measured after 24 h of cooling and conditioning (%)
0 (control)	0
0 (control)	0.36 (after 4 h heating in an oven)
1	0.18
2	0.25
3	0.30
4	0.34
5	0.37

Ends of all boards were sealed with end cups. Water had an access to the material only from outside of the boards. Shrinkage was measured 24 h after the boards were immersed into boiling water and pulled out.

A typical observation shows that, for example, a GeoDeck board is cut online off the extruder while still warm, at the length of the board 243 in. (20′3″), and left on the floor until next day, becomes shorter by 1 in. (0.41% of the "initial" length, that is a part of the manufacture shrinkage). Placing this 242 in.-board in the annealing chamber at 190° F for 24 h makes it an inch shorter again (another 0.41% shrinkage of the manufacture shrinkage). Without the annealing, this 0.41% would be a post-manufacture shrinkage, with a good likelihood that it would happen on a deck after some exposure. Now, if this annealed 241-in. board is placed in a heating chamber (or a laboratory oven) again, the observed shrinkage is less than 0.010-in., that is less than 0.05%.

If to take 3% as an overall shrinkage of the material right off the die to a complete stop, then approximately 2% of it takes place online, before the board is cut, about 0.5% in the course of cooling of the board after it was cut, and about 0.5% during the annealing (or on a deck, if no annealing). This is related primarily to hollow composite HDPE-based boards.

WARRANTY CLAIMS: GEODECK COMPOSITE DECKBOARDS

Kadant Composites has experienced the shrinkage problem with hollow composite deck boards manufactured from the beginning of 2001 until the summer of 2003. By the beginning of 2003, after receiving the first nine warranty claims regarding shrinkage (the first claim was filed on September 5, 2002), the problem was realized and identified; in the summer of 2003 all production was treated using a third party kilns, and by the beginning of 2004 the plant was operating on its own, with specially designed annealing chamber. By October 2006, Kadant Composites has received 721 warranty claims regarding shrinkage of its composite boards; practically all of them were made before the summer of 2003. There were practically no warranty claims regarding shrinkage of decks made of boards manufactured after the summer of 2003.

Let us consider the warranty claims as common laboratory experimental data points. There are three characteristic events on the time frame: (a) time when the boards were manufactured, (b) time when the deck was installed, and (c) time when the warranty claim was filed. In only few cases, time when the boards were manufactured was known and recorded in warranty claims. Time period between boards manufacturing and the deck installation varies greatly, from several months to several years. In fact, in some warranty claims it was indicated that the boards were installed 2–3 years after they were purchased. Thus, a warranty claim filed lately could be related to boards made several years back. There was one warranty claim, filed in June 2006, regarding shrinking boards on the deck, installed in January 2006. The claimant has indicated that the boards for the deck were purchased back in the beginning of 2004. They were apparently manufactured in 2003.

Figure 9.6 shows a profile of shrinkage warranty claims dated according to their installation time. All the boards, related to claims, that could be identified

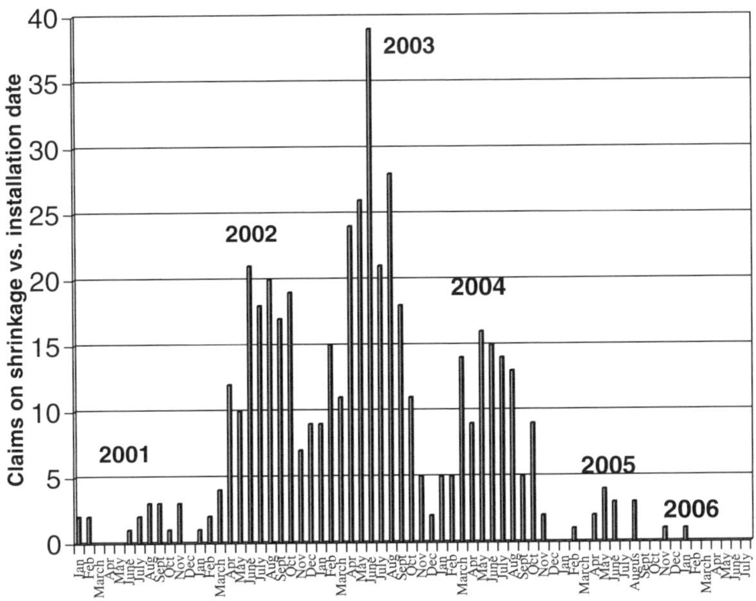

Figure 9.6 GeoDeck shrinkage warranty claims with respect to installation dates, 488 cases total with recorded installation dates.

according to their manufacturing date were made in 2001–2003. Obviously, it takes time for manufactured boards to get on a deck. This time span depends on the duration of storage of the product in the plant warehouse, sales, delivery, inventory of a distributor, delivery to the customer, and installation. Sometimes it takes several years to deplete the inventory in a lumberyard, particularly with a typical "first come, last sold" situation on store shelves. Then, it takes some time for shrinkage to happen. For a direct sunlight in the South it might take a few weeks or even days; for a covered deck, or in the North, it might take several months to several years or longer. Finally, it takes time for an owner to realize the problem and to file a warranty claim. As data show, an average time between the installation of a deck and filing the shrinkage warranty claim is equal to 13 ± 5 months.

Figure 9.6 shows the phase out of warranty claims regarding shrinkage. Clearly, the peak of claims was related to decks installed in 2003 (209 claims), then in 2002 (140 claims), and in 2004 (107 claims); total 456 claims, or 93% of total, *for which installation dates were identified*. There are only 14 cases reported in 2005 and one case reported in 2006, and all of them were apparently related to boards made in or before 2003.

The problem with shrinkage of composite deck boards is over at Kadant Composites. However, the experience is certainly worth considering by other manufacturers and researches in the field.

TABLE 9.4 GeoDeck decks. Number of warranty claims regarding shrinkage in 2002–2006

State	Number of claims
CT	80
NY	56
MD	44
NC, MO	34 each
IL, WI, KS	32 each
PA	30
VA, NJ	28 each
CO, MA	26 each
OH	22
AZ	20
IN, TN	19 each
OR	18
KY, MI, MN	17 each
SC	15
WA	11
GA	9
NM	8
CA, TX	7 each
IA, NE	6 each
DE	4
FL, VT	3 each
AR, LA, RI, NV	2 each
OK, WY, NH	1 each
Total	721

As Kadant has received warranty claims regarding shrinkage of GeoDeck boards from all over the United States, it is of interest to examine the effect of various climatic conditions on the shrinkage. In this case we will analyze not the degree of shrinkage, but the reason for it, prompting the owner to complain. The analysis gave an interesting result—unlike oxidation and crumbling of the same boards (Chapter 15), which progressively increased from North to South, the shrinking did not show any clear pattern in regard to geography.

Table 9.4 lists number of warranty claims on shrinking of GeoDeck boards (made in 2001–2003) by the States. It should be noted that amount of boards produced in 2001 was too small compared to 2002–2003, so it can be neglected; no warranty claims on shrinkage was filed in 2001, and only eight of them were filed in 2002.

By themselves, these data do not tell us much. For instance, number of claims in New England and New York states greatly exceed those in the southern states, and some Great Lakes states produced more claims than Arizona and Florida (the latter has only three claims in all), but may be there are much more decks installed in the North in the first place? California has less crumbled decks than Connecticut, but maybe there are more decks in Connecticut to begin with?

TABLE 9.5 Number of warranty claims due to composite board shrinkage (in 2003–2006), and number of claims per $100,000 of sales ($R$) in 2003 in the states indicated

States	Number of warranty claims (shrinkage)	R
NY, MA, VT, NH, CT, ME, RI	168	6.0
MN, WI, IL, IN, MI, OH, ND, SD	139	2.3
PA, MD, VA, NJ	134	5.0
KY, TN, NC, SC, GA, LA, MS	96	5.7
CA, NV, CO, WA, OR, WY	65	7.3
AR, MO, IA	42	5.4
KS, OK, NE	39	4.2
AZ	20	3.9
TX, NM	15	2.9
FL	3	0.9
Total	721	4±2

Total amount of sales in 2003 was $17,058,000 (data by Principia Partners, see Table 1.3).

In order to normalize these data with respect to quantities of decks in the regions, the volume of sales in the regions was considered. The continental United States territory was subdivided in this analysis into 10 regions, which essentially reflect distributorships of GeoDeck in the United States, the data are shown in Table 9.5, and the dynamics of shrinkage by the regions is given in Figure 9.7. For subsequent analysis an average cost of material (GeoDeck) per one deck was taken as $3500. This was done in order to convert sales by the regions into amount of GeoDeck decks built in the region.

Table 9.5 shows that the ratio of shrinkage claims across the United States is approximately proportional to sales in the regions, in an average amount of 4±2 warranty claims per $100,000 sales in the region. Considering that an average deck is about $3500 worth of composite material at the sales price level (Table 9.5), shrinkage warranty claims amount on average 14±7 per 100 decks across the United States. In other words, sales can indeed be used as a normalized factor in calculations of probability of decks failure in terms of shrinkage (as well as the oxidative-degradation crumbling, as shown in Chapter 15).

It certainly attracts attention, as shown in Table 9.5 and Figure 9.7, that the Southern states such as Florida, Texas, New Mexico, and Arizona produced low rate of shrinkage, less than that in New England and the Atlantic states. A low rate of shrinkage in the Great Lakes area is understandable, but why is it higher than that in Florida and close to those in Texas and New Mexico?

A relatively low amount of shrinkage warranty claims in the South can be explained by a high speed of shrinkage in hot climate. The boards shrunk even before they got on a deck. They practically completed their shrinkage in the course of their storage in a lumberyard and/or on a job site, before being installed. In the North a likelihood of this fast shrinkage is much less, and board shrunk on the deck, prompting a complaint by the deck owner.

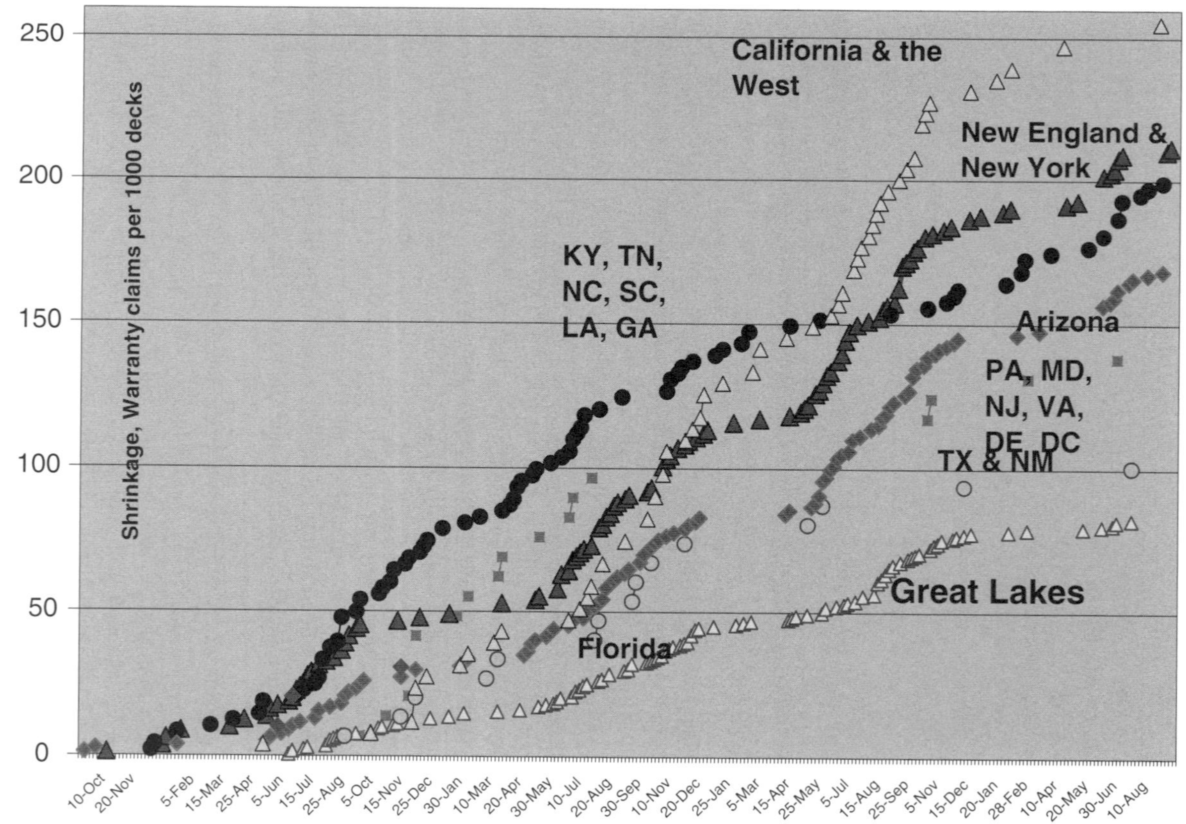

Figure 9.7 Dynamics of GeoDeck shrinkage warranty claims by months from September 2002 to October 2006, shown (vertical axis) as amount of decks subjected to formal complains per 1000 of GeoDeck decks in the area.

EXAMPLES OF COMPOSITE BOARDS SHRINKAGE ON A DECK

Figure 9.8

Figure 9.9

Figure 9.10

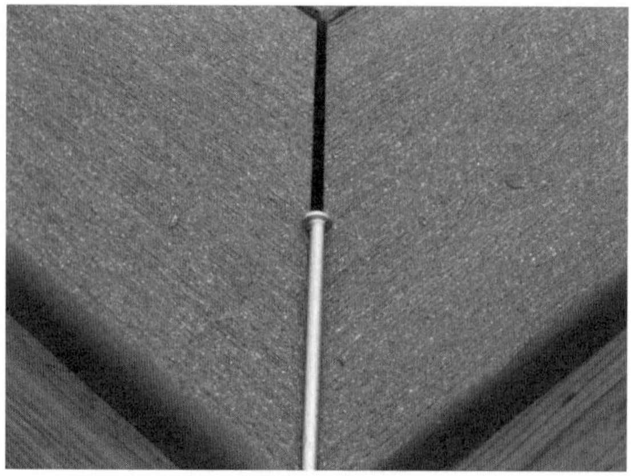

Figure 9.11

Figure 9.12

Figure 9.13

Figure 9.14

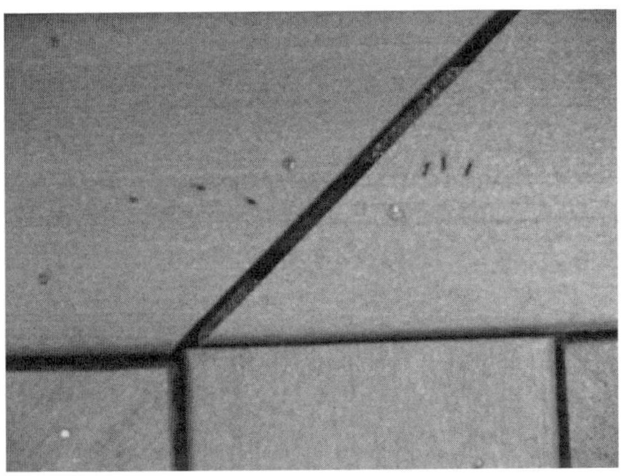

Figure 9.15

Figure 9.16

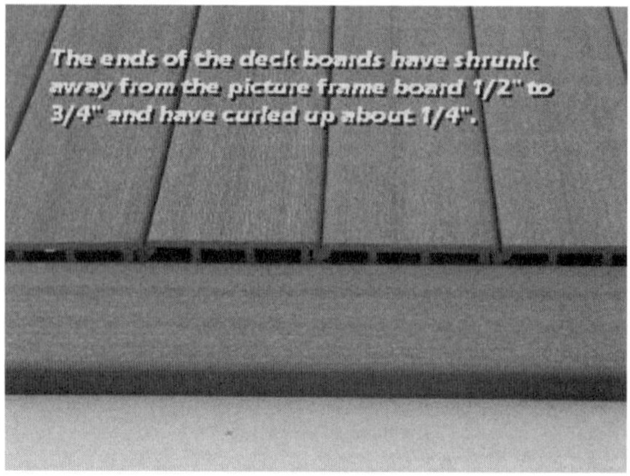

Figure 9.17

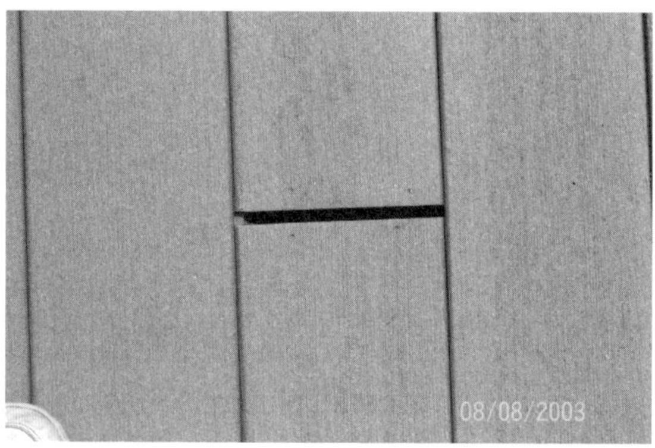

Figure 9.18

Figure 9.19

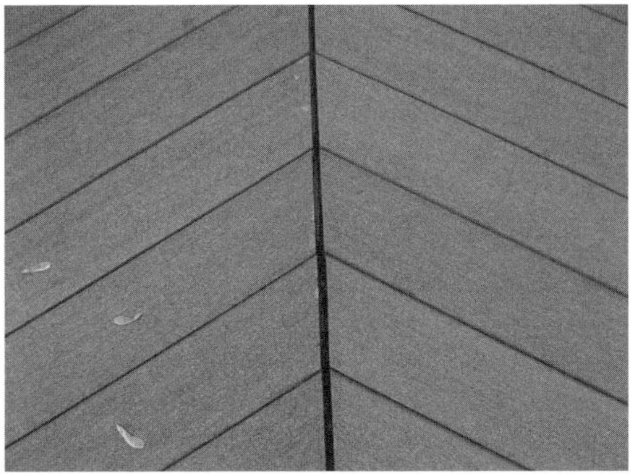

Figure 9.20

Figure 9.21

Figure 9.22

Figure 9.23

Figure 9.24

Figure 9.25

Figure 9.26

Figure 9.27

Figure 9.28

Figure 9.29

Figure 9.30

Figure 9.31

Figure 9.32

Figure 9.33

Figure 9.34

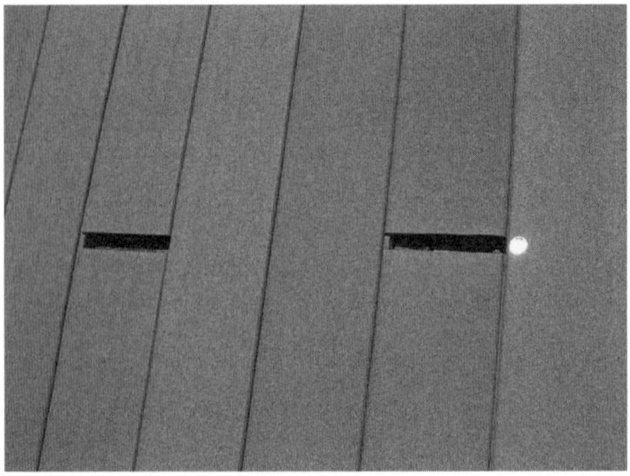

Figure 9.35

Figure 9.36

Figure 9.37

Figure 9.38

REFERENCES

1. W.M.D. Bryant, R.H.H. Pierce Jr., C.R. Lindegren, and R. Roberts. Nucleation and growth of crystallites in high polymers. Formation of spherulites. *J. Polym. Sci.* 1955, **16**(82), 131–142.

10

TEMPERATURE DRIVEN EXPANSION–CONTRACTION OF COMPOSITE DECK BOARDS: LINEAR COEFFICIENT OF THERMAL EXPANSION–CONTRACTION

INTRODUCTION

As it was described in the preceding chapter, there are two principally different types of temperature-driven dimensional changes in plastics and plastic-based composites: linear shrinkage and linear expansion–contraction. Shrinkage is an irreversible process and the result of the processing. Expansion–contraction is a *reversible* process, and it should not depend on the processing, though there is not enough data to conclude it decisively.

Expansion–contraction is a universal phenomenon and is observed with all solid objects, liquids, and gases, that is, with everything composed of atoms and molecules. Atoms and molecules oscillate at any temperature above the absolute zero. The higher the temperature, the higher the amplitude of the oscillation. This in turn increases the effective volume of objects, hence, their dimensions in all directions. The higher the temperature, the higher the expansion. The lower the temperature, the higher the contraction. Composite deck boards become wider, thicker, and longer during a hot day, and narrower, thinner, and shorter during a cool night. These expansions–contractions are repeated unlimitedly, every day and every night.

Physical objects can be isotropic or anisotropic by their nature. Isotropic objects expand and contract equally in all directions. Anisotropic objects expand and contract differently in length, width, and/or depth. A piece of wood is anisotropic in terms of expansion–contraction as a result of orientation of cellulose fiber. That is why wood is characterized by three linear coefficients of expansion–contraction, of

Wood-Plastic Composites, by Anatole A. Klyosov

Figure 1.9 Nexwood.

Figure 1.10 Rhino Deck.

Figure 1.11 SmartDeck.

Figure 1.16 Ultradeck.

Figure 1.22 Life long.

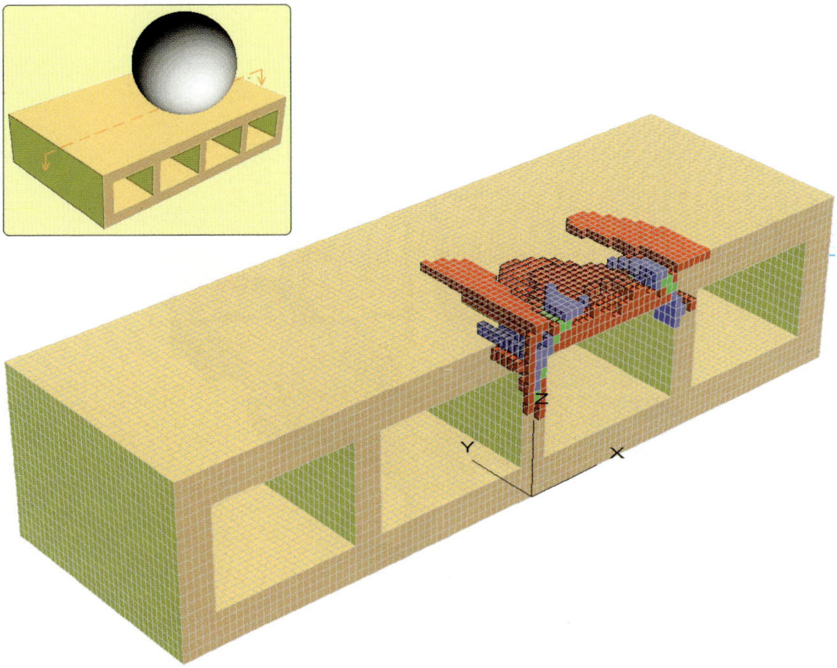

Figure 7.16 Tensile deformation of the WPC board under falling weight. Red-colored contours correspond to tensile deformations going beyond the elasticity limit and resulting in tensile fractures.

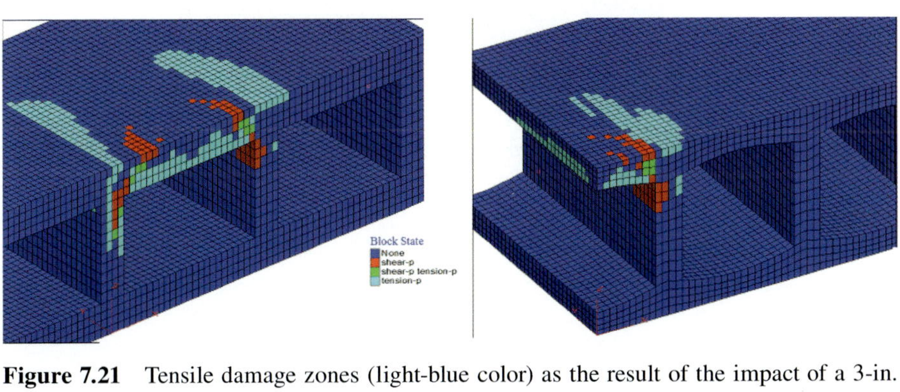

Figure 7.21 Tensile damage zones (light-blue color) as the result of the impact of a 3-in. steel ball falling from a height of 36 in. (see Figs 7.18 and 7.19). A conventional composite board is shown in the left-hand side and the modified board on the right-hand side.

Figure 16.2 The Hunter Lab color space in a cube form. It is based on the Opponent-Colors Theory. The theory assumes that the human eye perceive color as pairs of opposites, namely, light-dark, red-green, yellow-blue (with HunterLab permission).

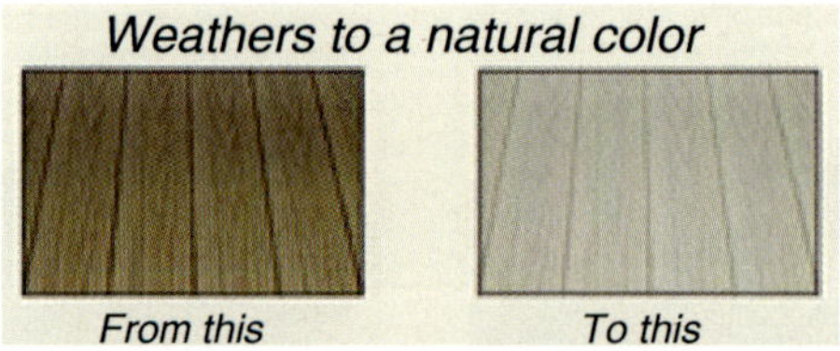

Figure 16.3 Fading of UltraDeck wood-plastic composite deck board, according to the manufacturer's data (© Midwest Manufacturing Extrusion).

which two—measured across the grain (radial and tangential)—appear to be equal to each other, and are about 5–10 times higher than that measured along the grain. This will be discussed in more detail at the end of this chapter.

LINEAR COEFFICIENT OF EXPANSION–CONTRACTION

Linear coefficient of expansion–contraction (k) can be determined by the following formula:

$$k = \Delta L/(L \times \Delta T)$$

where

ΔL is the expansion or contraction (in inches, centimeters, or other appropriate units) of the given material over the temperature range of ΔT, and L is the "initial" length (width, depth) of the material in the beginning of the temperature range, the upper or the lower end.

It does not matter in which direction temperature changes in the given temperature range (ΔT) — increases or decreases—provided that the object is in equilibrium with temperature across the object, that is, the temperature does not change too fast. For heating or cooling, the coefficient of expansion–contraction should remain the same in the given range of temperature.

The above formula shows that the dimension of the coefficient of expansion–contraction is 1°F or °C, whichever was used for temperature measurements.

The above formula can be used for calculations of expansion–contraction of, say, composite deck boards on a deck, however, with many reservations. Let us consider these reservations in some detail. This would represent one more examples of how carefully laboratory data should be applied to the real world.

As a rule, coefficients of linear expansion–contraction are calculated based on measurements of length of samples *freely placed*, that is, in completely unrestrained conditions, under carefully controlled temperature. Freely placed here means that they are under unrestricted temperature-driven movement. For example, for GeoDeck boards the temperature coefficient is equal to 3.58×10^{-5} 1/°F. This followed from an experimental fact that a 2-in. long GeoDeck sample initially placed at −30°F became 12.17 mil (0.01217−in.) longer at 140°F.

$$k = 0.01217/2 \times 170 = 3.58 \times 10^{-5} \text{ 1/°F}$$

This experiment was conducted in a fused silica quartz dilatometer, as described in ASTM D 696 and ASTM E 228 (see below).

This value is reasonably well reproducible in "the real world" conditions, even if the GeoDeck boards are not "completely unrestrained" and the measurements are not quite precise. For example, 20-ft long GeoDeck boards were used to make a lower deck on a boat, and a butt-to-butt gap was chosen to be 5/16″. The deck was assembled at about 60°F. The boat owner has noticed that the gaps were closed due to temperature-driven board expansion when the temperature reached 90°. On air cooling the boards contracted, and at about 60° the gaps were again at 5/16″ (0.3125″). Hence, the coefficient of linear temperature expansion–contraction was approximately

$$k = 0.3125/240 \times 30 = 4.3 \times 10^{-5}\ 1/°\text{F}$$

Considering that the measurements were conducted with apparently not a small error margin, and the boards were restrained (nails), the above figure fits pretty well with the professional test results.

It should be noticed that for the above example at temperatures higher than 90°, the boards would be seriously restrained by their physical butt-to-butt contact. The further expansion would create stress in the boards, and eventually the boards could buckle. A probability of the buckling would depend mainly (but not only) on two parameters—an actual linear expansion and compressive modulus of the boards. However, as buckling of that kind practically has never happened with GeoDeck boards on thousands of decks, even in Arizona where surface temperature of boards reach 160°F, it seems that the mechanical properties of the boards can hold at those kinds and levels of stress.

SOME RESERVATIONS IN APPLICABILITY OF COEFFICIENTS OF EXPANSION–CONTRACTION

Now, let us calculate a longitudinal expansion, that is, an increase of the length of a 16-ft GeoDeck board on a deck at the *air temperature* increase from 60°F in the morning to 90°F in the afternoon. The situation would be very different for a board in a shade, compared to that exposed to direct sunlight. Surface temperature of the board in a shade would follow the air temperature, that is, ΔT in the above example would be 30°F. Using the above formula one can find that the increase of the length of the board, ΔL, will be equal to $kL\Delta T$, that is, $\times\ 3.58 \times 10^{-5}\ \text{F}^{-1} \times 192\ \text{in.} \times 30°\text{F} = 0.21$ in., or close to one-quarter of an inch. This is in the shade or on a covered deck. This board, or a 16-ft long deck made of these boards *not fastened to the frame* (that is completely unrestrained, see above) would experience a cyclical swing by about 1/4″ in length from 60° in the night to 90° in the afternoon and back.

Not so for a board exposed to sunlight. As it is described in Chapter 15, surface temperature of a deck under direct sunlight exceeds the air temperature by 40° in the North, and 50° in the South. That is, temperature change of the deck surface in the above example would be from 60° in the morning to 140° (in the South), with the temperature swing of 80°, not 30°. Hence, the linear expansion–contraction for the same deck board, freely placed under direct sunlight in the South, with all other conditions equal, would be 80/30, or almost three times larger compared with that in the shade. More precisely, it will be $3.58 \times 10^{-5}\ \text{F}^{-1} \times 192\ \text{in.} \times 80°\text{F} = 0.55$ in., or more than half-an-inch. This is not a little difference for a day-by-night-by-day expansion–contraction of a deck.

In reality, it is not exactly so. And here comes one more reservation regarding comparing laboratory (ASTM) data with the real world. In reality the deck is fastened, and this creates forces counteracting against the forces that cause the linear expansion–contraction. The stronger the fasteners, the higher these counteracting forces. At some point these counteracting forces can actually "overpower" expansion–contraction of composite boards. This can be reached using proper screws as fasteners.

As a result of it, 16- and 20-ft GeoDeck boards, which—when placed freely–expand–contract by 1/2–3/4 in. under direct sun in the South (compared to their length on cool nights), but on a deck fastened by screws they typically show linear expansion–contraction of only 1/16 in., and no more than 3/32 in., that is, about eight times less. This is practically negligible movement.

ASTM TESTS RECOMMENDED FOR DETERMINATION OF THE LINEAR COEFFICIENT OF THERMAL EXPANSION–CONTRACTION

ASTM D 696 "Standard Test Method for Coefficient of Linear Thermal Expansion of Plastics Between −30°C and 30°C with a Vitreous Silica Dilatometer"

The test method covers determination of the coefficient of linear thermal expansion for plastic materials, using a vitreous silica push-rod dilatometer. Vitreous silica is also known as fused silica quartz. Dilatometer is an "expansion-meter," as dilation is expansion of materials when they are heated. The test method recommends that ASTM E 228 shall be used for temperature range other than −30 to 30°C, because ASTM E 228 operates in the temperature range of −30 to 140°F (−34°C to +60°C).

The temperature range of −30 to 30°C (−22 to 86°F), recommended in ASTM D 696, is chosen to be a convenient range for plastics, taking into account temperature sensitivity of some plastics, and is the range that covers the temperatures in which plastics are most commonly used.

Note of the Author: This, of course, is not true for composite deckboards, which often experience temperatures on deck surfaces in the range of 100–130°F in the North, and up to 150–170°F in the South.

Sample size according to the test procedure shall be between 50 and 125 mm (between 2″ and 5″). The shorter the sample, the lower the precision of the measurement. The longer the sample, the temperature of the sample may become difficult to control.

The central element of the dilatometer is the vitreous silica tube. Vitreous silica is quartz having the coefficient of thermal expansion 100 times less than that of most plastics. The sample is placed into the tube with one end of the sample against a spring loaded push-rod, slide into a cylindrical furnace, which is placed in the −30°C controlled environment, such as liquid bath. The ASTM test describes details of the test. As the sample contracts, the motion is transmitted and detected by a transducer mounted on the tube. In other words, the transducer measures the amount that the sample and push-rod contract compared to the tube.

The temperature of the sample is measured, for example, with a chromel–alumel thermocouple and electronic device reference. The microvolt signal along with the output of the LVDT (linear variable differential transformer) signal conditioner are measured with a computer-based data acquisition system. The temperature and change in sample length relative to the ambient temperature are printed and written to file. Typically, a 1mV noise signal limits the resolution to 0.1 μm (4×10^{-6} in., or four-thousands of a mil). The ASTM recommends the accuracy of the measurements not to be less than ±1 μm, or 4×10^{-5} in., for any length change.

After the measurements are done at −30°C (−22°F), the ASTM procedure prescribes to change the transfer the dilatometer to the +30°C (86°F) bath, and conduct the measurements as described above. Both the measurements should be repeated and finally conducted at room temperature. The coefficient of linear thermal expansion over the temperature range is calculated using the formula given above. Average coefficient is referenced to room temperature.

Precision of the coefficient of linear thermal expansion is usually fair. ASTM D 696-98 lists an example with nine plastics, tested in five different laboratories (a round robin test). The data for three plastics are shown in Table 10.1.

It should be noted that, surprisingly, and probably mistakenly, ASTM D 696-98 does not specify units for coefficients, 1/°F or 1/°C in the above table. What makes things even worse, these figures do not correspond to coefficients of thermal expansion to those in either temperature scale units (see Table 10.2). Therefore, the table above can be used for evaluations of standard deviations only.

Note of the Author: Data in Table 10.1 are modified compared to ASTM D 696, with respect to the main figures and their standard deviations. It does not make much sense to present data in a way chosen by the ASTM, for example, for polypropylene as 158.2 ± 3.38, for polyethylene as 63.0 ± 0.454, and for Nylon as 130.7 ± 2.83 (all $\times 10^{-6}$, as given in ASTM D 696-98). First, standard deviations are typically determined with a precision hardly exceeding 30%. The ASTM precision in the above figures pretends that standard deviations are determined with better than 0.1% accuracy. Second, there is no much sense to indicate decimals when deviations are larger than a whole unit. This provides a misleading statement about the high accuracy of measurements. One cannot measure lengths down to one-thousands of an inch using a ruler calibrated by inches or calculate an average from a few measurements with a "precision" of one-thousand of an inch.

TABLE 10.1 Coefficient of linear thermal expansion, 10^{-5}, listed in ASTM D 696-98

Material	Average, within laboratory standard deviations	Average, between laboratory standard deviations
Polyethylene	6.30 ± 0.05	6.3 ± 0.2
Polypropylene	15.8 ± 0.3	15.8 ± 1.2
Nylon 66	13.1 ± 0.3	13.1 ± 0.8

TABLE 10.2 Linear coefficients of thermal expansion–contraction of plastics, determined in accord with the procedure of ASTM D 696

Material	Notes	$\times 10^{-5}$ 1/°F
Polyethylene, molded or extruded	Low density (0.910–0.925), MFI 0.3 – 26	8.9–11
	Medium density (0.926–0.940), MFI 1 – 20	8.3–16.7
	High density (0.941–0.965), MFI 0.2 – 15	8.3–16.7
Polypropylene	—	3.8–5.8
PVC	—	2.8–3.3
Polyester	Thermoplastic	5.3
Nylon 6	General purpose	4.8
Nylon 6/6	General purpose, molded or extruded	1.7
Nylon 6/10	General purpose	1.5

Note of the author: Table 10.2 provides linear coefficients of thermal expansion of a number of polymers, taken from Ref. [1]. These values generally correspond to those often quoted in the literature, unlike those listed in ASTM D 696-98.

ASTM D 6341 "Standard Test Method for Determination of the Linear Coefficient of Thermal Expansion of Plastics Lumber and Plastic Lumber Shapes Between −30 and 140°F (−34.4 and 60°C)"

This test method is developed specifically for plastic-based building materials. It is principally different from ASTM D 696 (see above) in two major aspects: (a) It is applicable not to "cut samples" of the material but to whole cross-sectioned members, in "as manufactured" form, and (b) measurements of the length at temperatures are conducted manually, using a caliper.

As it is explained in the ASTM, this test method evaluates the properties of plastic-based profiles as a product and not a material property test method.

Other important features of the ASTM procedure are that the standard test specimen is a minimum of 12 in. in length, and the measurements are conducted at three temperatures, −30°F, 73°F, and 140°F (−34°C, 23°C, and 60°C), no more than 1 min after removal from the temperature chamber.

ASTM E 228 "Standard Test Method for Linear Thermal Expansion of Solid Materials with a Vitreous Silica Dilatometer", (Withdrawn)

Note of the author: This ASTM was withdrawn in May 2005. However, because it was rather often used in the literature, the scope and procedure will be briefly described here.

The test method was different from ASTM D 696 only by the recommended temperature range, that is, from −180 to 900°C (−292 to +1652°F). The method was recommended for metals, ceramics, glasses, rocks and minerals, plastics, woods, and other reinforced matrix composites. It was generally applicable to materials having linear expansion coefficients above 0.5×10^{-5} 1/C (0.3×10^{-5} 1/F). The recommended size of the specimen was at least 1″ long and between 0.2″ and 0.4″ in diameter. All test procedures and the equipment were the same as those in ASTM D 696.

LINEAR COEFFICIENT OF THERMAL EXPANSION–CONTRACTION FOR WOOD–PLASTIC COMPOSITES. EFFECT OF FILLERS AND COUPLING AGENTS

Mineral fillers, such as calcium carbonate, have linear coefficients of thermal expansion (CTE) about 20 times lower than those of plastics and about 10 times lower than those of wood-plastic composites (WPCs). CTE of minerals are close to CTE of wood (along grain), which in turn has CTE of $0.17 - 0.25 \times 10^{-5}$ 1/F ($0.31 - 0.45 \times 10^{-5}$ 1/C). CTE of wood will be discussed below in more detail.

The effect of fillers on the linear coefficient of thermal expansion–contraction is far from being understood. It appears that the largest effect is caused by the degree of anisotropy of the filler and the filler orientation in the flow (and in the final product). It also appears that characteristics of HDPE (density, degree of crystallinity) might be very important. For example, of the three composite materials from Table 10.3—Old Geodeck, Nexwood, and UltraDeck—the first two have similar composition (HDPE and rice hulls), and the third one contains HDPE and wood flour. The CTE for the first two are similar to each other and higher, that is, 3.5 and 3.2 compared to the third one, UltraDeck, with 2.1 ($\times 10^{-5}$ 1/F). Extrusion conditions (speed, pressure, temperature profile) might also be important for the CTE values.

CTE values for HDPE in textbooks and in the literature vary a great deal. Some sources show the same figure that was determined by us and listed in Table 10.3,

TABLE 10.3 Effect of short and long fiber (wheat straw) on the coefficient of expansion–contraction of a polypropylene-based composite in the −40 to +25°C temperature range [3]

	Coefficient of thermal expansion–contraction, 1/C ($\times 10^{-5}$), between −40 and 25°C	
Filler (%)	Thickness	Width
None, Neat polypropylene (control)	7.37	8.00
	Long fiber	
30	7.26	6.83
50	7.00	5.89
	Short fiber	
30	6.71	6.93
50	6.39	6.29

that is, 8.6×10^{-5} 1/F. Some sources give the following CTE values: for low-density polyethylene, $5.6 - 12.2 \times 10^{-5}$ 1/F; for medium density, $7.8 - 8.9 \times 10^{-5}$ 1/F; for high-density polyethylene, $6.1 - 7.2 \times 10^{-5}$ 1/F (all ASTM D 696). These values can be compared with the coefficients given in Table 10.3, that is, 8.9–11 for low-density HDPE, 8.3–16.7 for medium density, and 8.3–16.7 for high density (all $\times 10^{-5}$ 1/F). Clearly, some other factors affect the CTE for HDPE, rather than just the polymer density.

The amount of the filler certainly affects the value of CTE. Apparently, with all other things equal, CTE is inversely proportional to a filler amount, and the introducing of about 50% filler (by volume) decreases CTE by about 50%. For Fasalex (Table 10.3), containing only 15–20% of plastic, CTE is about 5–6 times lower than that of HDPE.

Figure 10.1 illustrates the observation that the lower the amount of cellulose fiber in plastic, the lower is the CTE.

However, the key words in the above statement are "with all other things equal." When one changes the amount of the same filler in a WPC, a linear relationship between the amount and the CTE is often observed, such as in Figure 10.1. However, when different fillers are used, a resulting pattern is generally unpredictable. Even with the same filler introduced in different amounts, results are rather puzzling, as shown in Table 10.3.

As one can see, both long and short fibers (wheat straw) change CTE (in the thickness direction) very little, by only 5 and 15%, respectively, after substitution of 50% of plastic with filler. In width direction change of CTE was larger, by 36 and 27%, respectively.

The same pattern was observed in a different temperature range (Table 10.4).

Here the long fiber did not change CTE at all, or even increased it (in the thickness direction), but reduced by 50% in the width direction after substitution of 50% of plastic with filler. With short fiber, changes of CTE were mixed depthwise and reduced up to 33% widthwise. Overall, the data are all over the place. As the authors did nor provide standard deviations or other error margins, meaning of these data is not clear.

TABLE 10.4 Effect of short and long fiber (wheat straw) on the coefficient of expansion–contraction of a polypropylene-based composite in the 25–100°C temperature range [3]

Filler (%)	Coefficient of thermal expansion–contraction, 1/C ($\times 10^{-5}$), between 25 and 100°C: Thickness	Width
None, Neat polypropylene (control)	15.2	16.1
	Long fiber	
30	17.6	13.1
50	15.5	10.9
	Short fiber	
30	15.9	13.6
50	12.8	12.1

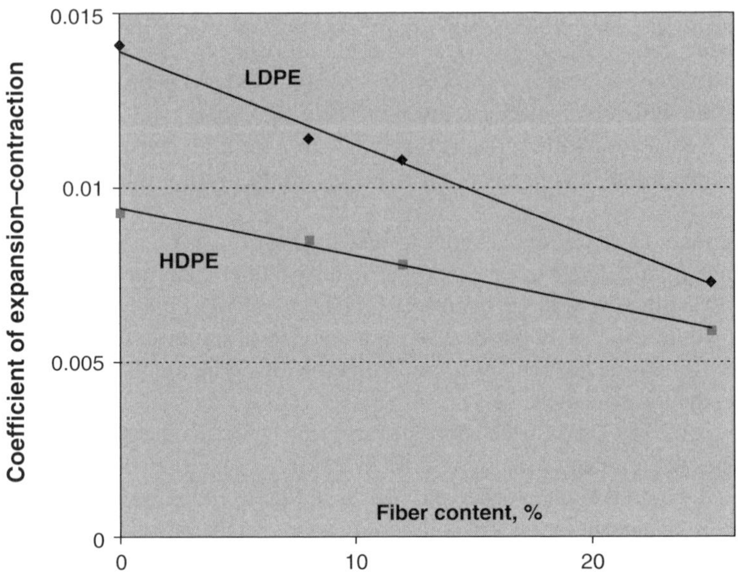

Figure 10.1 Effect of cellulose fiber content in LDPE and HDPE on the coefficient of linear expansion–contraction of the composite material (expressed in mm/°C, and determined in the temperature range of –18°C to 49°C) [2].

One can see from Figure 10.1 that in the range from 0 to 25% of cellulose fiber, the coefficient of linear expansion–contraction decreases practically linearly with the fiber content.

Table 10.5 shows a collection of data on coefficients of expansion–contraction for WPCs, including commercially available ones, and for some neat plastic lumber boards, as references.

Most of the composites listed in Table 10.5 have the widthwise CTE higher compared with that for the lengthwise CTE (except that in neat HDPE and HDPE filled with a mineral filler, fly ash). Apparently this is a result of the anisotropy of the fillers, wood fiber, or rice hulls.

Example: A case study

GeoDeck tongue and groove boards were used for a low deck construction on a boat. Twenty-ft boards were mounted butt-to-butt with a 5/16 in. gap between them at a temperature of 65°F. It was noticed that at 95°F the gap was closed. Back to 65° the gap was 5/16 in. again.

Hence, the coefficient of temperature expansion–contraction was 0.3125″ (5/16″)/240″ $\times$ 30°F $=$ 4.3 $\times$ 10^{-5} 1/°F. This corresponds approximately to the coefficient of expansion–contraction of GeoDeck in Table 10.5.

In 2004–2005, AAMA (American Architectural Manufacturers Association) had arranged an independent testing of commercial composite boards provided by six

TABLE 10.5 Linear coefficient of thermal expansion–contraction for WPCs, determined in accord with ASTM D 6341. Data were obtained by Dr. Tatyana Samoylova, LDI Composites

	Coefficient of thermal expansion, $\times 10^{-5}$ 1/F	
Composite deck board	Lengthwise	Widthwise
Neat HDPE board (as a reference)	8.6	7.1
HDPE willed with fly ash	4.2	4.0
GeoDeck (40% HDPE)	4.0	5.3
GeoDeck (36% HDPE)	3.6	5.1
Same with Biodac replaced with talc (experimental sample)	3.4	N/A
Same with Biodac replaced with wollastonite (experimental sample)	2.25	N/A
Old GeoDeck(HDPE and rice hulls)(not a commercial product)	3.5	4.6
Same with lower amount of rice hulls	3.0	N/A
GeoDeck recipe (40% HDPE) with rice hulls replaced with saw dust (not a commercial product)	2.7	5.1
Same, with 36% HDPE	2.2	5.0
Nexwood	3.2	5.1
	2.7[a]	4.9[a]
Boardwalk	2.3	1.2
Evergrain (Epoch)	2.7	6.6
U.S.A. Inteque Resources (polypropylene-based board)	2.6	2.9
Trex	2.2	7.0
	1.6–1.9[a]	3.5–4.3[a]
Monarch	2.2	5.2
Timbertech	2.1	5.5
	1.6[a]	3.7[a]
UltraDeck	2.1	5.0
Rhino Deck	2.0	5.0
	1.9[a]	
Weatherbest	1.7	4.9
Fiberon	2.0	—
Xtendex	2.2[a]	—
EverX	1.6[a]	—
CorrectDeck	1.5[a]	—
Timberlast	1.6[a]	—
USPL (50% wood fibers, 50% HDPE)	4.5	—
Fasalex (European WPC floring, 70–75% wood, 15–20% plastic)	0.8	3.6
Plastic lumber boards, as references		
Ecobord[a] (fiberglass-filled)	4.1	—
TriMax (fiberglass-filled)	3.4	—
PlasTeak[a]	6.4	—

[a]Manufacturer's data.

TABLE 10.6 Coefficient of linear thermal expansion of materials of six commercial composite boards. The boards were tested according to ASTM D 696. The specimens were 1/4″ by 1/2″ by 3″, in the longitudinal direction of the deck board. Two samples of each product were tested. The order of materials in the table is from higher to lower value of the coefficient and does not indicate the particular manufacturer

	Coefficient of thermal expansion, $\times 10^{-5}$, lengthwise	
Manufacturer (code)	1/°C	1/°F
A	4.14	2.30
B	4.02	2.23
C	3.76	2.09
D	3.74	2.08
E	3.55	1.97
F	2.74	1.52

manufacturers. All boards were tested in the same conditions, using the same testing equipment and the same testing operator, and the data are given in Table 10.6.

These values of coefficient of CTE practically correspond to the following lengthwise expansion–contraction of 20-ft composite deck boards when temperature changes between 10 and 130°F (Table 10.7).

TABLE 10.7

Manufacturer (code)	Lengthwise expansion–contraction of 20-ft composite deckboard, in.
A	0.66
B	0.64
C	0.60
D	0.60
E	0.57
F	0.44

One can see that the change in length of a 20-ft board is approximately between half an inch and three-thirds of an inch.

It is of interest that rice hulls as a filler results in the higher values of CTE compared with that for wood flour (Table 10.5), 3.6 and 2.7 ($\times 10^{-5}$ 1/°F), respectively. The plastic (HDPE) here was the same, hence, this effect must have resulted from the cellulose fiber and/or accompanying components (lignin, silicates). This effect was studied in more detail, and the data are given in Table 10.8.

Indeed, wood flour resulted consistently in a lower CTE compared to rice hulls in the longitudinal direction, but not to the transverse direction. The effect (lengthwise) was as much as 56 ± 11% in favor of wood flour. Densities of rice hulls and wood flour can hardly explain such a difference, though cannot be completely dismissed.

Again, replacing 10% of HDPE (from 40 to 36%, Table 10.8) with a filler decreased CTE by about the same 10%, taking into account an error margin.

TABLE 10.8 The linear coefficient of thermal expansion–contraction for extruded WPCs, filled with rice hulls (20–80 mesh size) or 40 mesh sawdust, determined in accordance with ASTM D 6341. Biodac (granular porous filler having 50% cellulose fiber and 50% mineral, see Chapter 4) was 28% in all the cases. MFI for the HDPE was 0.3. Data by Dr. Tatyana Samoylova, LDI Composites

Formulation (composite's density)	Coefficient of thermal expansion, $\times 10^{-5}$ 1/°F	
	Lengthwise	Widthwise
HDPE 40%, rice hulls 32% (1.26 g/cm^3)	4.0 ± 0.1	5.3 ± 0.1
HDPE 40%, saw dust 32% (1.24 g/cm^3)	2.7 ± 0.1	5.3 ± 0.4
HDPE 36%, rice hulls 36% (1.25 g/cm^3)	3.6 ± 0.2	4.6 ± 0.3
HDPE 36%, saw dust 36% (1.22 g/cm^3)	2.2 ± 0.2	4.8 ± 0.4

Table 10.9 lists data regarding CTE of LDPE and HDPE, filled with cellulose fiber.

As one can see, CTE for neat LDPE is about 50% higher than that for neat HDPE. With the increase of the filler, this difference is narrowing down, but still amounts to about 20% (Table 10.10).

One can see that the coupling agent does not change the CTE in a longitudinal direction, and reduces it by approximately 30% in width and thickness directions. However, the extruded (presumably) samples are largely anisotropic, and their CTE is higher in the transverse directions by 190–250% (width) and by 290–380% (thickness).

Linear coefficient of thermal expansion–contraction for wood is significantly lower compared to those for plastics and WPCs. It appears to be independent of species (hardwoods and softwoods) and specific gravity, and equal (along the grain) to 0.17 –0.25 × 10^{-5} 1/°F (0.31–0.45 × 10^{-5} 1/°C). CTE measured across the grain (radial and tangential) is proportional to wood specific gravity and is about 5–10 times higher than those along the grain, that is, about 1 to 2 × 10^{-5} 1/F. These values for wood are practically independent of temperature over the temperature range

TABLE 10.9 The linear coefficient of thermal expansion–contraction for LDPE and HDPE filled with cellulose fiber (wastepaper). The published article [2] indicated only the absolute expansion values (in mm) per 1°C, without providing the size (length) of the samples. It was assumed that all the samples were of 60 mm in length, and the data were recalculated to CTE (1/°F)

Material	Filler, cellulose fiber, (%)	Coefficient of thermal expansion, × 10^{-5} 1/°F
HDPE	0	8.6
	8	7.9
	12	7.2
	25	5.5
LDPE	0	13.1
	8	10.6
	12	10.0
	25	6.8

TABLE 10.10 The linear coefficient of thermal expansion–contraction for HDPE-based WPC. The published article [4] did not provide details on the formulation of the composite material. MAPP — maleic anhydride derivative of polypropylene (see Chapter 5).

Material	Dimension	Coefficient of thermal expansion, $\times 10^{-5}$ 1/°F
HDPE-based WPC	L	1.41 ± 0.12
	W	3.51 ± 0.20
	T	5.39 ± 0.22
Same, with a coupling agent (MAPP)	L	1.43 ± 0.10
	W	2.75 ± 0.19
	T	4.13 ± 0.31

of −60 to 130°F [5]. These data are related to dry wood, because for wet wood, expansion–contraction and shrinkage are mixed up.

Dry wood again, as with the shrinkage (see the preceding chapter), is a superior material compared to plastic and WPCs with respect to thermal expansion–contraction. The trouble is that in the real world wood is almost never dry, hence, subject to shrinkage, microbial degradation, fading, and so on.

REFERENCES

1. J.F. Shackelford and W. Alexander, *CRC Material Science and Engineering Handbook*, 3rd edition, CRC Press, Boca Raton, FL, 2001, pp. 510–513.
2. B.W. english and R.H. Falk. Factors that affect the application of woodfiber-plastic composites. In: *Proceedings of the Forest Products Society No. 7293*, Forest Products Society, Madison, WI, 1996. pp. 189–194.
3. D.A. Johnson, D.A. Johnson, J.L. Urich, R.M. Rowell, R. Jacobson, and D.F. Caufield. Weathering characteristics of fiber-polymer composites. In: *Fifth International Conference on Woofiber-Plastic Composites*, Forest Products Society, Madison, WI, May 26–27, 1999.
4. D.J. Gardner and N.M. Stark. Understanding the durability of wood-plastic composites. In: *The Global Outlook for Natural Fiber and Wood Composites*, Intertech, Portland, ME, New Orleans, LA, December 8–10, 2004.
5. W. Simpson and A. TenWolde. Physical properties and moisture relations of wood. In: *Wood Handbook*, Forest Products Society, Madison, WI, 1999, Chapter 3, pp. 3–21.

11

SLIP RESISTANCE AND COEFFICIENT OF FRICTION OF COMPOSITE DECK BOARDS

INTRODUCTION

Slip resistance is quantitatively characterized by the static coefficient of friction. It is also—informally–called traction. A sufficient slip resistance is important with respect to human locomotion safety. It is of a particular concern for elderly people, as prescribed in Americans with Disabilities Act (ADA), which set a quantitative limit to the static coefficient of friction, now retracted (see below).

Wooden decks are rather slip resistant, particularly when wet. This would be illustrated and explained below in this chapter. Wood–plastic composites (WPCs) are typically less slip resistant compared to pressure-treated lumber. Obviously, a composite deck should not compromise human safety in that regard. This chapter considers quantitative data with respect to slip resistance of wood and composite decks, the ways, how these data were obtained, and means how to make composite decks less slippery.

Definitions

Generally, *friction* is the resistance to motion of two moving objects or surfaces that touch. This definition can be reversed as follows: Friction is the force between surfaces in contact that resists their relative moving, or sliding motion.

Wood-Plastic Composites, by Anatole A. Klyosov

"Moving" here is not necessarily physically moving, but also having a tendency to move under action of a certain force having a component directed parallel to the plane of contact.

Friction between solid surfaces is similar in kind with viscosity in rheology (see Chapter 17). Viscosity is a measure of fluid friction. Viscosity describes a fluid's internal resistance to flow, that is, to moving.

The frictional force is a function of (1) the force pressing the surfaces together and (2) the coefficient of friction between the materials.

Friction is not necessarily higher when the surface is rough. In fact, roughness is a minor factor affecting friction. Rough surfaces results in abrasion, not in friction. Friction is often higher between smooth surfaces (see glass-to-glass coefficient of friction in Table 11.1). It is known that two glass plates placed together are hard to slide against each other. Molecular forces between two polished objects can be very high. Hence, high friction forces between them.

Coefficient of friction is the ratio of the above-mentioned component of force (parallel to the plane of contact) required to overcome friction, to the vertical component of the object weight (or, generally, normal force applied through the object).

Clearly, there are two principal factors that determine friction, or slip resistance: nature of the surface and weight of the object. Friction is directly proportional to the normal force.

The coefficient of friction is an empirical measurement, that is, it has to be measured experimentally and cannot be found through theoretical calculations.

Static coefficient of friction is the coefficient of friction of a body that just overcomes the resistance to slipping, provided that the vertical component that results in a contact pressure is not less than 1 psi (6.9 kPa) and not more than 13 psi (90 kPa) applied uniformly over the area of mutual contact.

Dynamic coefficient of friction is the coefficient of friction of a body in motion, once slipping has begun and the body continues sliding at a relatively constant

TABLE 11.1 Static coefficient of friction for selected surfaces (different sources, including [1])

Contact surfaces	Static coefficient of friction
Rubber–rubber	1.16
Glass–glass	0.94
	0.9–1.0
Steel–steel	0.74
Car tire–asphalt	0.72
Car tire–grass	0.35
Wood–brick	0.60
Leather–wood (dry)	0.3–0.4
Wood–felt	0.29
Wood–wood (dry)	0.28
	0.25–0.50
Wood–wood (wet)	0.2
Bone joints	0.01

velocity of not less than 1.5 in./s and not more than 6 in./s, and the vertical component that results in a contact pressure is not less than 1 psi and not more than 13 psi applied uniformly over the area of mutual contact.

The coefficient of static friction is generally greater than the coefficient of dynamic friction on the same surface.

Slip resistance is the property of surface that prevents slipping. In case of composite deck board, this property either reflects a nature of the matrix (a blend of plastic and fillers), or it could be designed by special means (brushing, embossing, introducing slip modifiers).

Explanations and Some Examples

Friction is independent of the surface area, of speed (except when the objects are resting), and of temperature. This is in theory. However, in the real world, the contact area does matter. Furthermore, in the case of polymers, in general, and composite materials, in particular, the actual contact area is difficult to determine due to deformations of the plastic. In reality, applied force, test temperature, sliding rate, and duration of the test, all are important.

The concept of friction was recognized more than 300 years ago by Guillaume Amontons (1663–1705). In his article "De la résistance causée dans les machines," published in 1699, Amontons first established that there existed a proportional relationship between friction force and the mutual pressure (or force) between the bodies in contact. That is why to obtain the coefficient of friction, we divide friction force by normal force.

Hence, if it takes 5 lb of horizontal force to move a 10 lb block resting on a floor, the static coefficient of friction is 0.5.

In a simplified experiment, the static coefficient of friction is the tangent of the angle from the vertical at which slipping begins to occur. If a shoe placed on a deck board starts moving at 45° angle, the static coefficient of friction for this particular case is 1.0. If a shoe stars moving at 27° angle, static coefficient of friction is 0.51.

Floor or deck surfaces, to be considered slip-resistant, should have a static coefficient of friction of 0.5 or greater. This was also a code requirement of ADA (Americans with Disability Act).

In order to provide some reference values, Table 11.1 shows static coefficient of friction for some materials.

In 1990, the Americans with Disabilities Act made recommendations for the static coefficient of friction to be not less than 0.5. These recommendations have since been removed. ADA has retracted the recommendations because there were numerous ways to measure the coefficient of friction for different materials, and no single test device or procedure has been identified in the ADA recommendations.

SLIP RESISTANCE OF PLASTICS

High-density polyethylene (HDPE) is characterized by a low coefficients of friction, and the higher the density (specific gravity), the lower the static (and dynamic) coefficient of friction. For polyethylene density 0.915 g/cm^3, coefficient of friction equals to 0.50, for 0.932 g/cm^3 to 0.30, and for 0.965 g/cm^3 it is equal to 0.10 [2].

TABLE 11.2 Static coefficient of friction (ASTM D 2047 and ASTM C1028) or slip index (ASTM F 1679) for plastics and plastic lumber. Data provided by the manufacturers (marketing materials, ICC-ES reports and listed on manufacturer's Web sites) or taken from Ref. [2], p. 205. Data for Nylon and polyester are taken from Ref. [3]

Material or a trade name	Manufacturer or source	ASTM	Static coefficient of friction
HDPE	[2]	N/A	0.50 (density 0.915 g/cm^3) 0.30 (d = 0.932 g/cm^3) 0.10 (d = 0.965 g/cm^3)
HDPE–nylon rope	Timothy W. Manning www.mra.org/ManningPaper.pdf	C 1028[a]	0.13–0.25 (dry)
			0.50 (wood–nylon rope as a reference)
HDPE	[2]	N/A	0.23
LDPE		N/A	0.50
Nylon 6,6		N/A	0.36
Polystyrene		N/A	0.38
Nylon 6		N/A	0.39
PVC		N/A	0.50
Polypropylene (isotactic)		N/A	0.67
Nylon 6	[3]	N/A	0.32 (dynamic)
Nylon 6.6		N/A	0.04–0.13
Polyester		D 1894	0.16–0.17
Teck Deck(PVC)	Outdoor Technologies Inc.	C 1028[a]	0.71 (dry) 0.71 (wet)
Carefree(HDPE)	Carefree Building Products	F 1679	0.52 (dry) 0.41 (wet)
Liberty(PVC)	Outdoor Technologies Inc.	C 1028[a]	0.78 (dry) 0.58 (wet)
Dream Deck(PVC)	Thermal Industries, Inc.	D 2047	0.6 (dry)
CertainTeed PVC Deck(PVC)	CertainTeed Corp.	F 1679	0.70 (dry) 0.65 (wet)
Sheerline(PVC)	L.B. Plastics, Inc.	C 1028[a]	1.01 (dry) 0.78 (wet)
C-Clip Vinyl Deck	Kroy Building Products, Inc.	N/A	0.64 (dry) 0.47 (wet)

[a]ASTM C 1028-96 "Standard test method for determining the static coefficient of friction of ceramic tile and other like surfaces by the horizontal dynamometer pull-meter method" was withdrawn in December 2004, with no replacement. The method covered the measurement of static coefficient of friction of ceramic surfaces under both wet and dry conditions while utilizing Neolite heel assemblies.

The primary factors that control the coefficient of friction of HDPE are its molecular characteristics, mainly its molecular weight and its distribution (number-, weight-, and viscosity-average molecular weights), and a degree of crystallinity, that is, branching levels. This in turn effects molecular interactions between the polymer surface and an object in contact with it. Generally, coefficient of friction of polyethylene increases with the increase of molecular weight and branching levels, which also leads to decrease in density (specific gravity). As a result, the coefficient of friction of HDPE has a bell-shape form, going through a maximum. This relationship is a consequence of the viscoelastic nature of polyethylene [2]. The uphill rise of the coefficient of friction of polyethylene is determined by its viscous component, the decrease is determined by its elastic component.

Data on the coefficient of friction for different plastics are collected in Table 11.2.

SLIP RESISTANCE OF WOOD DECKS

Generally, wood decks are rather slip resistant, particularly when wet. Tables 11.1, 11.3, and 11.4 list some quantitative data. According to Table 11.1, leather shoes on wood deck show the static coefficient of friction of 0.3–0.4 (depending on type of wood) when dry; no data were available in the cited source regarding wet surface.

However, when tested using ASTM F 1679 (the Variable Incidence Tribometer procedure), CCA pressure treated southern yellow pine showed slip index of 0.92 (dry wood) and 0.88 (wet wood) [ICC-ES Legacy Report NER-682], which is generally higher than slip resistance of WPC materials (Table 11.3).

Similar data, shown in the same Table 11.3, were obtained using a different ASTM D2394 (the movable sliding unit procedure using a leather patch) that showed the static coefficient of friction for wood of 0.72 (dry, along the grain and transverse) and 0.86–0.89 (wet, along the grain, and transverse, respectively).

The above data are supported, in kind, by other data in Table 11.3, which systematically compares slip resistant of different composite and plastic deck boards along with pressure-treated lumber deck board.

Overall, it appears that pressure-treated lumber is less slippery than composite deck boards, and wet wood is less slippery than dry wood.

SLIP RESISTANCE OF WOOD–PLASTIC COMPOSITE DECKS

To date, no in-depth studies have been done to compare the slip resistance or static coefficient of friction between wood and WPCs. ASTM test methods exist for various materials, but the results can vary greatly, depending on factors such as the type of shoe sole material used in the test (leather vs. rubber) and whether the test sample is wet and/or dry [4]. This was written 5 years ago, but the situation is still the same.

Table 11.3 contains some information gathered from different sources.

TABLE 11.3 Static coefficient of friction (ASTM D 2047) or slip index (ASTM F 1679) for wood and WPCs. Data provided by the manufacturers (marketing materials, ICC-ES reports, and listed on manufacturer's Web sites)

WPC, trade name	Manufacturer	ASTM	Static coefficient of friction
Wood (SYP)	NER-682	F 1679	0.92 (dry) 0.88 (wet)
Wood	NER-695	D 2394	0.72 (dry, along grain) 0.89 (wet, along grain) 0.72 (dry, transverse) 0.86 (wet, transverse)
TREX	Trex Company, Inc.	D 2047 F 1679	0.53–0.55 (dry) 0.59–0.70 (dry) 0.70–0.75 (wet)
EverX (Veranda)	UFP Ventures II, Inc.	F 1679	0.55 (wet) 0.55 (dry) 0.40–0.65
Timberlast (Strandex technology)	Kroy Building Products, Inc.	F 1679	0.40–0.65
Lakeshore	Georgia-Pacific Corporation	F 1679	0.55 (wet) 0.55 (dry)
Premier	Composatron Mfg. Inc.	F 1679	0.55 (wet) 0.55 (dry)
Boardwalk	CertainTeed Corp.	D 2047 F 1679	0.52–0.54 (dry) 0.87–0.88 (wet) 0.70 (dry) 0.64 (wet)
GeoDeck	Kadant Composites, Inc.	D 2047	0.50–0.53 (dry) 0.85–0.90 (wet)
Evergrain	Epoch Composite Products, Inc.	D 2047	0.88 (dry) 0.96 (wet)
Nexwood	Composite Technology Resources, Ltd.	D 2394 C 1028	0.32 (along grain) 0.48 (against grain) 0.58 (dry) 0.64 (wet)
Choice Dek	Weyerhauser/AERT Inc.	D 2394	0.62 (dry) 0.46 (wet)
XTENDEX, Fine Grain	Composite Building Products International, Inc. Carney Timber Company	D 2394	0.36 (dry) 0.41 (wet)(along grain) 0.40 (dry) 0.55 (wet) (transverse)

TABLE 11.3 (*Continued*)

WPC, trade name	Manufacturer	ASTM	Static coefficient of friction
XTENDEX, Wood Grain			0.30 (dry) 0.38 (wet) (along grain) 0.38 (dry) 0.50 (wet) (transverse)
CorrectDeck	Correct Building Products, LLC	C 1028	0.77–0.80 (dry) 0.72–0.77 (wet)
Rhino Deck	Master Mark Plastics Products	D 2394	0.22 (dry) 0.57 (wet)
TimberTech	TimberTech Ltd	D 2047	0.55 (dry) 0.88 (wet)
Fiberon	Fiber Composites, LLC	D 2394	0.52 (dry) 0.54 (wet)

In 2004–2005, AAMA (American Architectural Manufacturers Association) had arranged an independent testing of commercial composite boards provided by six manufacturers. All boards were tested in the same conditions and using the same equipment and the same operator, and data are shown in Table 11.4.

As one can see, there is no correlation between the results obtained by these two ASTM procedures.

TABLE 11.4 Static coefficient of friction (ASTM D 2394) and Slip Resistance (ASTM F 1679) of the materials of six commercial composite boards. The specimens were cut from production boards to a length of 9″. Three samples of each product were tested in ASTM D 2394 and three samples in ASTM F 1679. The tests were performed on a dry surface on each sample at 0°, 90°, 180°, and 270° with 0° and 180° being in the longitudinal direction with respect to the board length. The results in the table are the least average values (for three samples) from the four orientations. The order of materials in the table is from a higher to lower value of the static coefficient of friction (ASTM D 2394) and does not indicate the particular manufacturer

Manufacturer (code)	Static coefficient of friction (ASTM D 2394)	Slip resistance ASTM F 1679
A	0.65	0.54
B	0.59	0.51
C	0.56	0.81
D	0.54	0.60
E	0.49	0.50
F	0.47	0.62

ASTM TESTS RECOMMENDED FOR DETERMINING STATIC COEFFICIENT OF FRICTION

ASTM D 2047 "Standard Test Method for Static Coefficient of Friction of Polish-Coated Floor Surfaces as Measured by the James Machine"

The ASTM procedure primarily describes measuring the slip resistance of surface to shoe material tested, such as a shoe sole, in terms of static coefficient of friction. The test employs the laboratory-only device named James machine which measures friction between a tested surface and a square pad of leather 3 × 3 in. by 0.25 in. thick. The James machine applies a known constant vertical force to the test pad, and then applies an increasing lateral force until a slip occurs. The test table of the machine moves forward uniformly at a rate of 1 in./s until the piece of leather mounted on a "shoe" slips and the vertical column of the machine drops. The principal readout of the test procedure is a recording on the chart graph marked with coefficient of friction lines. Hence, the machine provides results directly in values of static coefficient of friction. The test is conducted on three panels of the materials in four mutually perpendicular directions, giving total twelve readings for each material.

Note of the author: The above ASTM procedure indicated (1. Scope) that the test method is not intended for use on wet surfaces. This follows from the definition of the static coefficient of friction according to which the two surfaces should be in direct contact with each other, and nothing, including water, should be between them, on the interface. Therefore, by NBS definition (US National Bureau of Standards, now National Institute of Standards and Technology), all static friction meters must be used on dry, clean surfaces. Formally, static coefficient of friction cannot be determined on a wet deck. That is why the ASTM standard F 1679, described below, and applicable to wet surfaces, does not mention *coefficient of friction*, but only *slip resistance* and *slip index*.

Note of the author: The ASTM test does not specify specimen sample. However, standard size composite boards are often too narrow for the James machine. It is recommended that boards were cut to approximately 11″ long and two pieces were glued together to make up samples that were 9″ × 11″.

ASTM F 1679 "Standard Test Method for Using a Variable Incidence Tribometer (VIT)"

Tribometry is a scientific field of measuring, interpretation, and studying friction. The ASTM procedure aims at determining the slip resistance of test surfaces to shoe material tested, such as a shoe sole, or related material (test feet) in either the laboratory or field under dry, wet, or contaminated conditions. The ASTM test does not mention "static coefficient of friction" (see above). It is described recording of the slip index directly from the protractor scale.

Unlike James machine (see above), the English XL tribometer does not rely on gravity, but is powered by a small carbon dioxide cartridge at a set pressure (see

Figure 11.1 The English ***XL*** Variable Incidence Tribometer (VIT) or the slipmeter (weight 3.85 lb).

Fig. 11.1). The test foot is applied against the test surface, the actuating (pneumatic) cylinder causes swinging (slippage) motion of the test foot, and the instrument displays the slip index on its protractor scale. It ensures consistent operation by the application of uniform force for each test, and permits reliable metering of inclined surfaces such as ramps. The application of vertical and horizontal forces is simultaneous, thus avoiding residence -time and permitting reliable measurement of wet surfaces.

ASTM D 2394 "Standard Method for Simulated Service Testing of Wood and Wood-Base Finish Flooring"

This ASTM test method is not mentioned in acceptance criteria for WPCs, but it was employed in characterization of some of them (see Table 11.3 for Nexwood, Choice Dek, XTENDEX, Rhino Deck, and Fiberon, and Table 11.4). Therefore, it would be briefly described in this chapter.

The ASTM procedure is used for determining both static and dynamic (sliding) coefficient of friction. The procedure employs the testing machine preferably with autographic equipment for recording load and head travel. The testing machine is of the low capacity type, as the loads obtained are usually less than 25 lb. Least reading should be approximately 0.1 lb. According to the ASTM procedure, friction is determined in the most slippery direction, that is, along the grain of wood.

The apparatus employed in the ASTM procedure is an assembly consisting of a weighed sliding unit, a cleated bed for holding the specimen, a movable unit attached to the weighing mechanism, a chain or a cable, and a pulley. The sliding unit is a 25-lb piece of metal with a piece of plywood and a piece of shoe sole leather (4 × 4½″) glued to it. Hence, the sliding unit has the smooth face of the leather in contact with the specimen. The static coefficient of friction is determined by dividing the force (load) required to move the sliding unit in contact with the specimen from a stationary position by the mass of the sliding unit (25 lb). The dynamic coefficient of friction is determined in the same way, except that the average force that is measured is required to maintain movement at a rate of separation of the heads of the testing machine of 2 in./min.

Typically, the plot of force (load) versus head travel is a series of pulsations after the initial high force of static friction. These peaks and valleys of sliding friction should be averaged to get the mean values of dynamic friction.

Note of the author regarding ASTM tests employing tribometers: Overall, there are six ASTM test methods and, hence, six tribometers, aimed at determining slip resistance. As only two of them, ASTM D 2047 and ASTM F 1679, are recommended for WPCs (see ASTM D 7031 and ASTM D 7032), and one more, ASTM D 2394, is described above; we will mention the other three procedures only briefly.

ASTM F 609 "Standard test method for using a horizontal pull slipmeter (HPS)": The basic principle of the HPS, a dragsled class of slipmeter, is the pulling of a footwear or surrogate material against a walkway surface under a fixed load at a constant velocity. The HPS consists of a 10-lb weight onto which a slip index meter is attached. This component is attached to a nylon string and pulled by a capstan-headed motor. The ASTM method is approved only for dry testing.

ASTM F 1678 "Standard test method for using a portable articulated strut slip tester": The method is similar in principle with ASTM D 2047 (the James machine); however, it employs a portable device and can test actual floors or decks. Besides, it uses a graduated rod that provides a direct reading from the device. Some calculations are required to convert this to a slip-resistance measurement. In the literature, the method is sometimes called Mark I.

ASTM F 1677 "Standard test method for using a portable inclineable articulated strut slip tester": The method, which is also called Mark II, is approved for dry and wet testing. It is also based on gravity, as the James machine and the Mark I tests, but can avoid stick-slip effects on wet surfaces by eliminating the residence time (or time delay) between the application of the vertical and horizontal forces. Like the Mark I, it is a portable device. It uses a 10-lb weight on an inclineable frame with a test foot suspended just above the walkway surface. Each time the angle is set to a more horizontal position, the weight is released, until a slip occurs. The slip-resistance reading can be taken directly from the instrument.

SLIP RESISTANCE USING AN INCLINED-PLANE METHOD

As it is practically impossible to compare slip resistance or static coefficients of friction of various WPC materials using the same experimental approach (Table 11.3), I have undertaken—while preparing this book for the publication—a simplified comparison using the standard incline-plane method. This method is a standard experiment in mechanical physics and involves tilting a platform to the point where movement of a material first occurs. The tangent of the angle of the plane at the point of movement is equal to the static coefficient of friction.

For this study, the method was performed using actual composite deck boards. Each test was performed using the weight of 11.01 lb (5.0 kg) placed in a shoe having a leather sole. This resulted in a pressure application of about 1 psi. Measurements were repeated a minimum of twenty times per material pair at various board angles. Results are given in Table 11.5. All tests were performed in the most slippery direction, that is, along the grain and along the board.

One can see that the values of static coefficient of friction are spread in a rather wide range. The three lowest values in the table are essentially for neat HDPE because the surface for unbrushed GeoDeck deck boards (not a commercial product) is practically a thin film of the plastic. Brushing increases the

TABLE 11.5 Static coefficient of friction for plastic or WPC deck boards, measured using a simplified inclined plane method. All measurements were done in the most slippery directions, that is, along the grain. Date for dry and wet pressure treated lumber are given for a comparison

Deck board	Static coefficient of friction
GeoDeck, Driftwood, unbrushed(not a commercial product)	0.28
GeoDeck, Cedar, unbrushed (not a commercial product)	0.28
Carefree (plastic lumber, HDPE)	0.30
Perfection	0.33
GeoDeck, Cedar, brushed	0.33 (dry) 0.41 (wet)
GeoDeck, Driftwood, brushed	0.36
Nexwood, gray, unbrushed side	0.36
Weatherbest, not embossed (back) side	0.36
Choice Dek	0.37
Nexwood, gray, brushed side	0.38
Old GeoDeck (HDPE & rice hulls)(not a commercial product after 2002)	0.39
Integrated Composite Technology	0.39
International Paper (experimental, not a commercial product)	0.40
Weatherbest, gray, embossed	0.41
Nexwood, pink, brushed side	0.43
GeoDeck, Mahogany, brushed	0.43
Evergrain, Cape Cod Grey, embossed	0.43
Fiberon	0.45
Evergrain, Cedar, embossed	0.46
Trex, Winchester	0.48
Trex, Saddle	0.48
Trex, Madera	0.48
Evergrain, Grey, embossed	0.48
Azek (cellular RVC)	0.50
Pressure-treated lumber (wood)	0.54 (dry) 0.66 (wet)

coefficient by 0.05–0.15 units, depending on a quality/wearing of brushes. In case of Nexwood, brushing did not bring much of a friction (0.36 and 0.38, respectively). With the Weatherbest product, embossing brought only 0.05 units (0.36 and 0.41, respectively).

Perfection is a licensee of Fiberon, and a difference between the coefficients of friction of the respective materials is surprising. It appears that they use different brands of plastic (HDPE), such as HDPE of different density. If Fiberon uses HDPE of a lower density while Perfection uses a higher density HDPE (see below, next section in this chapter), it can explain the difference between them in Table 11.5.

Trex products show consistently high value of the static coefficient of friction, which is the highest end of all composite materials. Only Evergrain embossed board, produced by compression molding (not extrusion "along the grain"), matches Trex boards. It might be the result of low-density polyethylene (LDPE), rather than HDPE, which Trex has used for their product. Generally, LDPE has lower density and, hence, higher coefficient of friction compared with HDPE-made composite deck boards.

Pressure-treated lumber has a significantly higher static coefficient of friction compared with WPCs. This is a remarkable feature of wood, along with it unsurpassable strength and stiffness compared to WPC materials. If not a relatively low durability of wood, its sensitivity to biodegradation, and an increasing scarcity of wood of a good quality, appropriate for deck materials, composite materials would hardly be competitive to real wood.

It is of interest that counter intuitively wet wood, along with wet composite materials, shows higher friction compared to dry materials. Surface of wet deck boards is less slippery than that of dry deck boards. This phenomenon is a result of "sticktion" or stick-slip effect of wet surfaces, which in turn originates in a capillary nature of wood and extruded composite materials.

Both wood and WPC are porous materials. They absorb water. They are heavily penetrated by capillary. In fact, it is capillary/pores that absorb water. The higher friction of these materials when wet is the result of water being squeezing out of the pores/capillary between two contacting surfaces—during a slip test or between a foot and the deck surface, creating a temporary bond between these surfaces. When tested using an inclined plane, as described in Table 11.5, the weight-loaded leather-sole shoe moves much slower on the wetted surface compared with a fast sliding on the respective dry surface.

Strictly speaking, coefficients of friction obtained on wet surfaces cannot be called "static coefficient of friction," as by accepted definition there should be nothing between the tested surfaces that might effect the friction, including water. However, as we aim at realistic slip behavior of deck surfaces, which are directly related to safety issues, in this context there is no real difference how to call the obtained values—static coefficients of friction or slip coefficients. In any case we operate with quantitative measurements of slip resistance directly applicable to realistic situations.

EFFECTS OF FORMULATION OF COMPOSITE DECK BOARDS ON SLIP RESISTANCE: SLIP ENHANCERS

A term "slip modifier" is often used in extrusion of plastics and composite materials where slip-stick of the hot melt at the die is typically observed at too slow or too fast melt flow (see Chapter 17). To avoid confusion, we will use in this chapter the term "slip enhancer" or "friction enhancer." Friction enhancer is a material that increases friction between two surfaces so as to attain a specific friction level in a controlled manner.

Fillers in composite materials, such as wood fiber and (if any) mineral fillers, are natural friction enhancers, though, strictly speaking, they typically increase abrasion, not friction. It is not important, though, for this consideration is aimed at safety, and it is not an academic discussion about differences between molecular interactions and abrasiveness. As it was described above, neat HDPE is characterized by a low value of coefficients of friction, and the higher the density (specific gravity), the lower the static coefficient of friction. For HDPE density between 0.915 and 0.965 g/cm^3, coefficient of friction drops from 0.50 to 0.10. However, for most of the commercial composite deck boards static, coefficient of friction is above 0.50 (see Table 11.3).

For instance, GeoDeck composite deck boards often utilize HDPE with density of 0.955 g/cm^3. Its coefficient of friction is about 0.15. However, when rice hulls and a granular blend of calcium carbonate/kaolin and delignified cellulose fiber is incorporated into the plastic matrix, the static coefficient of friction increases to 0.53, that is, to 350% compared to the initial HDPE. This is nevertheless still close to the value of 0.50 that is considered to be safe.

It might appear that the easiest way to enhance friction of a WPC board is to change the initial plastic to that with a higher coefficient of friction, that is—in case of HDPE—to a low density plastic or to a more "rubbery" HDPE. However, this might lead to problems in flowability of the composition in the extruder, to compromise its strength and—more than that—its flexural modulus, that is, deflection, creep, and other properties of the final material. To change the plastic is always a trade-off and an optimization game. If the overall balance shows that the final material has acquired an appreciably higher coefficient of friction with other properties being more or less the same or within an accepted tolerance, if not even better, this can be called a success.

As rubber has a much better coefficient of friction compared with polyolefins, it might be helpful to add rubber powder or small particles into, say, HDPE-based composite matrix. Coarse grades of calcium carbonate could serve the same purpose (this would be again a certain trade-off in properties of the final composite material). Additional benefits can be obtained if the same additive/filler enhances both friction and impact resistance (a rubber might be a good candidate in this case).

Brushing and embossing would almost invariably increase the roughness of the surface and, hence, the slip resistance. Again, this is not directly related to "true" friction, that is, resistance to molecular interactions between two contacting surfaces, either moving against each other or intended to move. However, this is covered by a general term "slip resistance" that is under consideration in this chapter.

REFERENCES

1. R.A. Serway. *Physics for Scientists and Engineers*, 4th edition, Harcorut Brace College Publishers, Orlando, FL, 1995, p. 126.
2. A.J. Peacock. *Handbook of Polyethylene: Structures, Properties, and Applications*, Marcel Dekker, New York, 2000, p. 203.
3. J.F. Shackelford and W. Alexander (Eds.), *CRC Materials Science and Engineering Handbook*, 3rd edition, CRC Press, Boca Raton, FL, 2001, pp. 736, 1651.
4. P. Krishnaswamy and R. Lampo. From waste plastics to markets for plastic-lumber bridges. *ASTM Standardization News*, American Society for Testing Materials, West Conshohocken, PA, December 2001.

12

WATER ABSORPTION BY COMPOSITE MATERIALS AND RELATED EFFECTS

INTRODUCTION

One of the most remarkable features of composite materials is a relatively low water absorption compared to that of wood. This is one of the few properties due to which composites are clearly better than wood lumber. And "better" here is directly related to dimensional stability and a better durability of the material, first of all its resistance to microbial degradation.

When a piece of wood is immersed in water, after 24 h its weight increases by about 25%. In other words, wood absorbs about 25% of water by weight for the said time period. After a much longer exposure time, wood can double its weight (100% water absorption). In trees, moisture content can range from about 30% to more than 200% by weight [1]. For wood–plastic composite (WPC) materials, these figures typically are 0.7–2% after 24 h, 1–5% after a week, and up to 18–22% after several months.

Clearly, the belief that cellulose fibers or wood particles in composites are encapsulated with plastic is not completely valid. Particularly, it is not valid when composites contain a significant amount of cellulose, such as above 40%. Commercial composites often contain 50–65% cellulose. There are enough cellulose particles in these composites to form extended and contacting chains, along which water penetrates into the bulk of the materials. Hence, water absorption on a singular and double-digit percentage scale (after weeks- and months-long exposure), by weight.

Wood-Plastic Composites, by Anatole A. Klyosov

As it will be shown below, water absorption mainly occurs at the outer layers of composite materials and progressively decreases into the bulk of the matrix. Therefore, numerical figures of "water absorption" are related to only total weight increase of the material and have nothing to do with a distribution of water across the material. From this viewpoint, hollow profiles absorb almost a double (well, less than that, depending on the configuration of the hollow profile) amount of water compared to the same material in the form of a solid profile.

A relatively high water absorption by composite materials leads to a higher weight of wet profiles, possible decrease in their strength and increase in their deflection, swelling and a resulting pressure on neighboring structures, which can result in buckling, warping, higher chance of their microbial inhabitation and the following degradation, freeze-and-thaw induced deterioration of mechanical properties of materials, among others. These effects will be considered in this chapter.

"Near-Surface" Versus "Into the Bulk" Distribution of Absorbed Water in Composite Materials

As the phenomenon of a sharp gradient of water distribution from the near-surface layers into the bulk of composites turned out to be very important for microbial inhabitation, this issue was considered in a chapter of this book, describing microbial degradation of composites. It is known that the longer the composite materials are immersed into water, the higher is the water absorption. For example, for composite deck boards at ambient temperature, water absorption after 24 h is typically between 0.7 and 2% by weight, after 7 days between 1 and 5%, after 100 days about 13–22%. These values depend on temperature, and the higher the temperature, the higher the water absorption [2]. However, water absorption by the top layer of a composite board (1 mm in depth, 50:50 mix of wood:plastic) after 24 h is in excess of 15% [3]. Other data show [2] that water absorption by the 5-mm top layer of Trex deck board in the temperature range from 5 to 25°C is (after 22 days of immersion in water) 45 and 60%, respectively. In the 5–10-mm (from the surface) layer, water absorption is about 3%, and at 10–15-mm layer it is about 1%.

Another example of moisture content in a WPC at different distances from the surface is given in Table 12.1.

Water absorption by composite materials depends on their porosity, amount of cellulose fiber, and their availability to incoming water. Composite materials are typically porous, and their degree of porosity is determined by moisture of the incoming raw materials and processing conditions (primarily, local overheating) that translate to a density (specific gravity) of the resulting profile. The higher the moisture content in the initial formulation, the higher the amount of VOC formed during the processing, the higher the porosity, the lower the density (specific gravity) of the material, the higher the water absorption.

Pores in composite materials are typically open, and form chains of pores, penetrating the whole matrix. Wood fiber is exposed to these pores. Hence, higher

TABLE 12.1 Moisture content in a WPC board (0.5 in. thickness, that is, 12.7 mm) at different distances from the surface

Distance from the top surface		
(mm)	(in.)	Moisture content (%)[a]
0.5	0.02	30
2.8	0.11	18
5.0	0.20(~3/16″)	5
7.2	0.28	5
9.6	0.38	12
12	0.47	26

The board (51% of 20-mesh pine wood flour, 45% HDPE, 1% talc) was exposed outdoors for an undisclosed period of time [4].

[a]Microbial decay initiation point is at about 25% of moisture.

or lower degree of water absorption. Water penetrates inside the composite matrix very slowly. In wood this penetration is much faster and can reach very high moisture level rather deep into the matrix. For example, the outer 5-mm layer of wood can have as much as 75% of water content after 190 days of immersion, whereas the 15–18-mm layer can contain 65% of water [2]. In Trex material these figures are 50 and 1%, respectively. After 180–215 days of immersion of Trex and Strandex materials in water, the overall water absorption is 16% (and increasing) for Trex and 11% (and increasing) for Strandex. Those are solid composite samples, 4″ in length [2]. The author of this book has studied pressure development as a result of swelling of GeoDeck hollow deck board immersed in water (see below). After 75 days, pressure was still climbing up, albeit slowly.

EFFECT OF MINERAL FILLERS ON WATER ABSORPTION

Mineral fillers generally do not absorb water (or absorb very little of it), so, naturally, they decrease the water absorption. For example, a composite deck board consisting of 65% rice hulls and 35% HDPE (minis minor additives) absorbed 3.0 ± 0.1% water for 24 h, whereas the same composition except 32% of rice hulls was replaced with Biodac® (a 50–50% blend of delignified cellulose and minerals [calcium carbonate and kaolin/clay]) absorbed in the same conditions 1.8 ± 0.1% water. Moisture content in these two composite boards was 0.64 ± 0.04 and 0.39 ± 0.02%, respectively.

Another example employed some experimental deck boards containing 65% rice hulls (11% silicates by weight), 28% Biodac® and 28% rice hulls (19% minerals), and 70 and 79% of fly ash (Table 12.2).

There are plenty of similar data in the literature, illustrating that the more the minerals, the lower the water absorption, and the longer the underwater exposure, the higher the water absorption.

TABLE 12.2 Water absorption by some composite deck boards

Time period of submerging in water (days)	Amount of minerals (w/w) in HDPE-based composites			
	11% minerals	19% minerals	70% minerals	79% minerals
	Water absorption (%)			
1	2.5	1.4	0.24	0.20
14	7.8	4.9	0.47	0.42
38	13.0	8.1	0.66	0.53
100	20.7	13.6	1.02	0.83

The test was performed according to ASTM D 1037.

SWELLING (DIMENSIONAL INSTABILITY), PRESSURE DEVELOPMENT, AND BUCKLING

It is well known that wood undergoes changes in dimension with changes in moisture content. Increase in moisture content causes swelling, and decrease in moisture content causes shrinkage. The following equation can be used for evaluating these dimensional changes. For wood this equation is considered satisfactory between 6 and 14% of moisture content.

$$\Delta L = LC\Delta M$$

where

ΔL = change in dimension (swelling),
L = initial dimension (tangential direction or radial direction) at certain moisture content,
ΔM = change in moisture content, and
C = coefficient of dimensional change in tangential (C_T) or radial (C_R) direction.

Coefficient C numerically is a percent-dimensional change ($\Delta L/L$) at 1% change in moisture content in the material. Several examples of the values of these two coefficients are given in Table 12.3.

Table 12.3 (and other available data) shows that the coefficients for dimensional change of wood (per 1% moisture content) are in the range of 0.2–0.4% (tangential) and 0.1–0.2% (radial). Due to morphology of a wood log, the radial direction includes both width and thickness.

In WPC materials a third coefficient appears, that is, different dimensional change in thickness (depth) and width. The last two parameters are separate values, by virtue of the very mode of extrusion. For example, a GeoDeck board (low density, see Tables 12.4–12.6) after submerging for a month absorbed 11.9% water and increased its length by 0.64%, width by 2.25%, and thickness by 4.8%. Therefore, the respective coefficients C_T (tangential or lengthwise), C_W (widthwise), and C_D (depthwise) in this case are equal to (per 1% of water absorption) 0.054, 0.19, and 0.40%. One can see that the lengthwise dimensional change coefficient for this composite material is

TABLE 12.3 Coefficients for dimensional change of wood within the range of moisture content of 6–14% [5]

	Dimensional change coefficients (%)	
Species	C_T (tangential direction)	C_R (radial direction)
Cedar, yellow	0.208	0.095
Pine, ponderosa	0.216	0.133
Pine, western white	0.259	0.141
Maple, red	0.289	0.137
Red oak, commercial	0.369	0.158
Mahogany	0.238	0.172

significantly lower (4–6 times lower) than that of wood, the widthwise coefficient for this composite material is of the same order of magnitude and only slightly higher than that for wood, and the depthwise coefficient is significantly (2–4 times) higher than that of wood.

As it was mentioned above, coefficients for dimensional change of wood are considered reliable when measured in the range of moisture content of 6–14%. For composite materials a respective range is not established as yet due to little experimental data available. As WPCs absorb water much slower compared to that of wood, it appears that water absorption data (moisture content and swelling) for the first 24 h cannot provide reliable data for coefficients of dimensional change. Let us consider some experimental data for GeoDeck deck boards (Table 12.4) having the same formulation but different density as a result of different regimes of processing (speed and temperature of the extrusion) and different moisture content of the initial ingredients (rice hulls, first of all).

Table 12.4 shows that swelling is more pronounced with low-density boards compared with high-density boards, and the difference is even higher at a short-term water absorption. Table 12.5 shows the data, but as a percentage of swelling to the initial width of the boards, and Table 12.6 shows the apparent coefficients of dimensional change, C_W.

TABLE 12.4 Sidewise swelling of GeoDeck tongue-and-groove boards having different density (specific gravity) and being submerged to water for a total of 28 days

		Swelling (mills)				
Density (g/cm^3)	Initial width (in.)	24 h	5 days	7 days	14 days	28 days
1.125	5.692	2	8.5	9.3	19	32
1.15	5.726	1.5	4.5	5.3	13	23
1.17	5.746	0.75	8.0	5.5	13.5	24
1.21	5.752	0.75	6.3	5.8	13.5	23

Swelling is expressed in absolute units of expansion.

TABLE 12.5 Sidewise swelling of GeoDeck tongue-and-groove boards having different density (specific gravity) and being submerged to water for a total of 28 days

	Swelling, in percent of the initial board width			
Density (g/cm^3)	24 h	7 days	14 days	28 days
1.125	0.0351	0.163	0.334	0.562
1.15	0.0262	0.0917	0.227	0.402
1.17	0.0131	0.0957	0.235	0.418
1.21	0.0130	0.0999	0.235	0.400

Table 12.6 shows that after 24 h of submersion underwater, an apparent C_W value is the smallest one, particularly for higher density composites. After 7 days in water, an apparent C_W is 2–4 times higher than that after 24 h and is almost the same across the range of densities ($C_W = 0.032 \pm 0.005\%$). After 14 days $C_W = 0.054 \pm 0.04\%$, and after 28 days of underwater submerging, $C_W = 0.067 \pm 0.05\%$. It is about two times lower than the average radial coefficient of dimensional change for wood (C_R, Table 12.3) equal to $0.14 \pm 0.03\%$. As one can see, C_W practically does not depend on the density of composite materials (exemplified by GeoDeck deck boards).

More examples are given in Table 12.7 for long-term water submerging of composite deck boards. If the data are averaged across different composites in Table 12.7 (11 data points), $C_T = 0.037 \pm 0.016\%$, $C_W = 0.13 \pm 0.05\%$, and $C_D = 0.32 \pm 0.08\%$. For wood (Table 12.3), $C_T = 0.26 \pm 0.06\%$ and $C_R = 0.14 \pm 0.03\%$.

This again shows that in a lengthwise direction composite materials swell much less than wood (seven time less on average), in widthwise direction wood and composite deck boards are just about the same, and in the depthwise direction composites swell more than wood. However, a depthwise swelling usually is not a concern on decks, except some special cases.

More data on a short-term water absorption (24 h of submerging) for composite materials show that C_T (lengthwise) and C_W (widthwise) apparent coefficients

TABLE 12.6 Apparent coefficients (C_W) of widthwise dimensional change for GeoDeck tongue-and-groove boards having different density (specific gravity) and being submerged to water for a total of 28 days

	Apparent coefficients (C_W) of widthwise dimensional change (%)			
Density	24 h	7 days	14 days	28 days
1.125	0.0198	0.0378	0.0560	0.0667
1.15	0.0172	0.0275	0.0485	0.0614
1.17	0.0089	0.0300	0.0533	0.0682
1.21	0.0087	0.0340	0.0590	0.0735

For data on water absorption of these boards, see Table 12.12.

TABLE 12.7 Dimensional change coefficients, %

Board	Water absorption (%)	Length increase (%)(C_T)	Width increase (%)(C_W)	Depth increase (%)(C_D)
GeoDeck old	11.9(30 days)	0.64(0.054%)	2.25(0.19%)	4.8(0.40%)
Same	21.8(110 days)	1.1(0.050%)	2.0(0.09%)	4.0(0.18%)
Same	17.1(100 days)	0.86(0.050%)	1.8(0.11%)	4.4(0.26%)
GeoDeck newmedium density(d = 1.16 g/cm^3)	13.6(100 days)	0.5(0.037%)	1.8(0.13%)	2.5(0.18%)
Same	7.65(117 h)	0.15(0.020%)	0.52(0.068%)	2.7(0.35%)
Same	9.4(45 days)	0.50(0.053%)	0.96(0.10%)	3.5(0.37%)
GeoDeck, density 1.12 g/cm^3	8.7(45 days)	0.29(0.033%)	0.84(0.097%)	3.0(0.34%)
GeoDeck, density 1.16 g/cm^3	8.7(45 days)	0.43(0.049%)	0.90(0.10%)	3.1(0.36%)
Trex	14.3(110 days)	0.17(0.012%)	2.3(0.16%)	5.3(0.37%)
TimberTech	13.0(100 days)	0.50(0.038%)	2.3(0.18%)	4.3(0.33%)
UltraDeck	16.8(100 days)	0.20(0.012%)	3.6(0.21%)	6.3(0.38%)

C_T : tangential (lengthwise), C_W : widthwise, and C_D : depthwise coefficients.

are much lower (0.016–0.018 and 0.03–0.04%, respectively) compared to those at a long-term water absorption (average C_T = 0.037% and C_W = 0.13%, see above). Therefore, they are not included in Table 12.7.

When a material swells, it produces pressure toward any object resisting the moisture-driven expansion. That is why deck boards on a deck are normally installed with a certain gapping, such as 1/4–3/8-in. between boards (Fig. 12.1–12.3).

Figure 12.1 A typical gapping of deck boards on a deck.

Figure 12.2 A typical gapping of deck boards on a deck.

If gapping is too narrow and cannot accommodate the swelling, this situation leads to buckling, distortions, cracking, and other types of failure (Fig. 12.4–12.9)

Clearly, for such distortions and mechanical failure of composite deck boards, a significant pressure/load should have been developed. It was determined that the

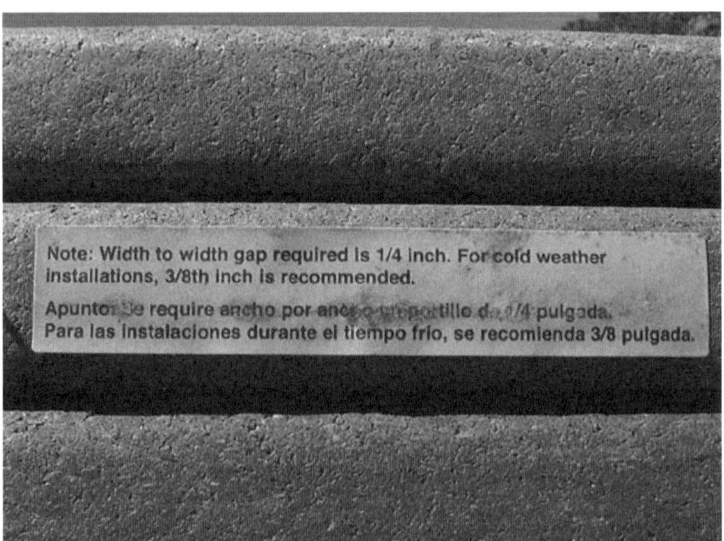

Figure 12.3 An instruction for board gapping (Trex boards).

Figure 12.4 Distortions of WPC deck boards due to water absorption.

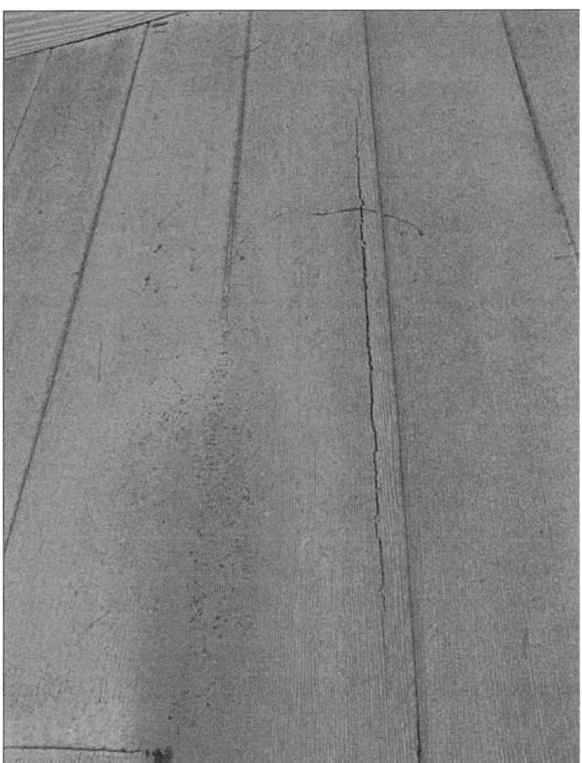

Figure 12.5 Moisture-induced distortion and mechanical failure of WPC deck boards.

Figure 12.6 A crack in a composite deck board as a result of buckling due to water absorption and insufficient gapping.

pressure can reach thousands of pounds. The experiment was conducted with a GeoDeck hollow deck board, 13.5 in. long, 896 g (close to 2 lb) weight, immersed in water at room temperature. A pressure being developed as a result of swelling in the widthwise direction was measured using a load cell.

At the same time, the length of the boards was measured. After 24 h the board length was increased by 0.05″ (0.37%), after a week by 0.10″ (0.74%), and after a

Figure 12.7 Moisture-induced dimensional instability of a WPC material.

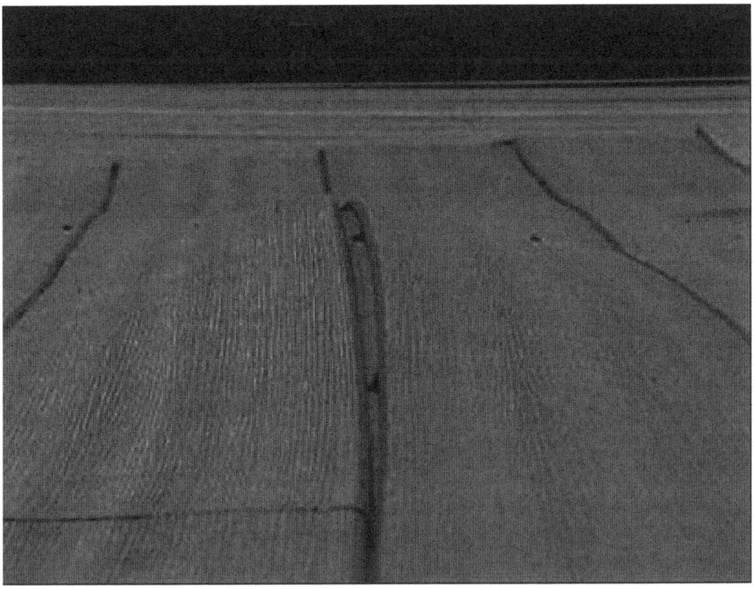

Figure 12.8 Moisture-induced dimensional instability of a WPC material.

month by 0.20″ (1.5%). After a month-and-a-half the board width was increased by 0.25″ (1.85%), and the further increase, if any, could not be detected within 0.005″ error margin until the end of the experiment after two-and-a-half months from the beginning. Meanwhile, pressure continued to increase until the end of the

Figure 12.9 Moisture-induced dimensional instability of a WPC material.

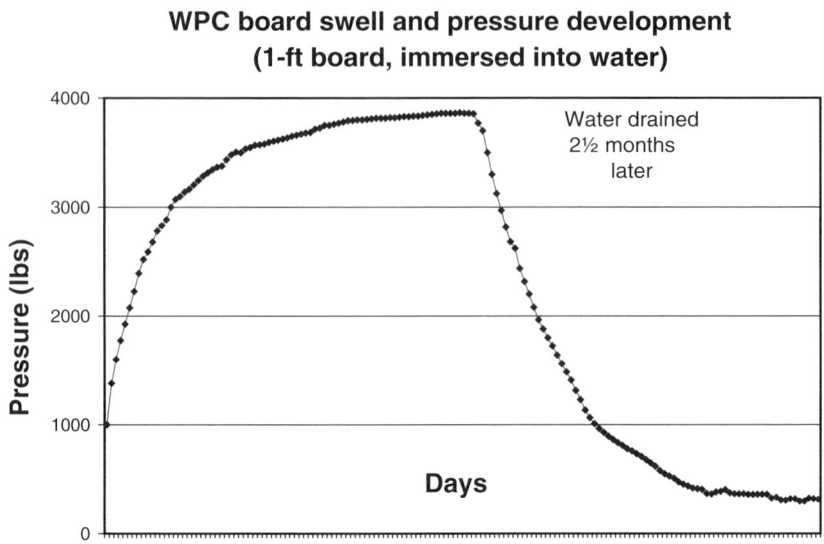

Figure 12.10 Swelling and pressure development by a composite deck board (13.5-in. long and 5.5-in. wide) immersed in water, for $2\frac{1}{2}$ months. The initial pressure, caused by holding in clamps, was 1000 lb. After 80 days, in water was drained and the board was getting dry with the resulting pressure release for the following $2\frac{1}{2}$ months.

experiment, though slowly, and reached almost 3000 lb (Fig. 12.10). It should be taken into account that such a large swelling of the board in longitudinal direction (1.85%) was apparently because a swell in the sidewise direction was blocked. Hence, such a high pressure in the sidewise direction.

Such a pressure developed by the board can explain distortions and mechanical failure of deck boards in case of a tight side-by-side contact with swelling neighboring boards.

After 80 days water was drained, and the board was getting dry for the following 2½ months. Pressure quickly dropped to the initial value (produced by holding clamps) and continued to decrease until it reached a value about 700 lb below the initial value (Fig. 12.10). It apparently reflects a compression of the board by the pressure developed against holding clamps.

The data show how long it takes to fill the composite board—its pores and capillary—with water. In this experiment it took 80 days and the process was not complete by the end of that period. The data also show how long it takes to drain water from pores and capillary of the composite board. In this experiment it took more than 2 months. When the board surface appeared to be completely dry, the bulk of the board was still filled with water.

When the same experiment was conducted, but with water filled only in the inner, hollow part of a GeoDeck board, the pattern was principally the same (Fig. 12.11). After 22 days, pressure increased from 400 lb (caused by holding clamps) to 1600 lb, that is, increased by 1200 lb. For the same time period, a

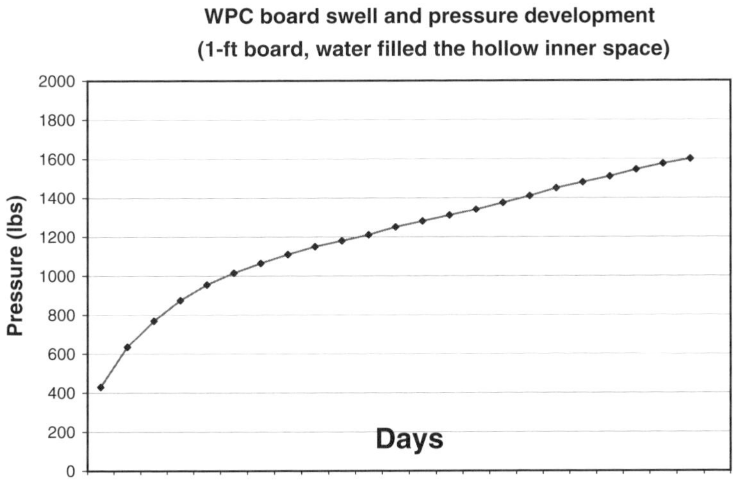

Figure 12.11 Swelling and pressure development by a composite deck board (13.5-in. long and 5.5-in. wide) immersed in water, for 22 days. The initial pressure, caused by holding clamps, was 400 lb.

similar board, completely submerged in water, increased its pressure by 2300 lb (Fig. 12.10).

In summary, water gets into composite materials through channels, pores, and capillary. Some of them are readily available, and they are filled with water first, for an hour or two. Water absorption for the first 24 h consists only a little fraction of a full capacity of a composite material to absorb water. Some pores and capillary are poorly available, and they are gradually and slowly filled with water for several months, even when the board is fully submerged in water. Buckling is not caused by water absorbed in the first 24 h. It is very unlikely that buckling would be caused by plain rain, when water dries fast. Buckling is caused by a long contact with water. In that case pressure as a result of swelling can reach several thousand pounds, which would indeed cause buckling in a case of a tight contact between the boards. Buckling typically results from an improper installation of a composite deck—causing a prolonged contact with water (from outside of from inside of deckboards, such as for hollow boards), lack of proper gapping, and so forth.

After 1 day, GeoDeck board develop pressure of 250–300 lb (expansion by 0.75–2 mills, see Table 12.4), after 7 days 1400 lb (expansion by 6–9 mills), after 28 days 2500 lb (expansion by 23–32 mills).

For low-density boards (d = 1.06), after 24 h expansion is by 6–7 mills. After 5 days, low-density boards (d = 1.06) expand by 27–39 mills. Medium-density boards (d = 1.09 and 1.10) expand by less than 3 mills after 24 h and by 15–16 mills after 5 days. Hence, buckling is more likely to happen after a shorter contact with water for low-density boards.

SHORT- AND LONG-TERM WATER ABSORPTION

The above experiment shows once more that there is a short-term and a long-term water absorption. When a composite material is immersed in water, during the first hours and days, water fills the most accessible pores and capillary, and penetrates the most available cellulose particles and fibers, commonly in the top layers of the board. This pattern of water absorption is tested according to ASTM D 570 and ASTM D 1037 (see below), during the first 24 h. Water absorption during the following days is never linear with time, it is more like the pattern shown in Figure 12.11. In fact, it is practically never linear with time even during the first 24 h. With each composite material water absorption has its own dynamics, determined by the material morphology, and 24 h on this time scale was chosen for ASTM 1037 quite arbitrarily, just as a means for data to appear in a standard format. It is very likely that some composite materials absorb water faster than others during the first 24 h, and then slower during the following time span. That is why it is useful to determine water absorption after several time periods (see Table 12.8).

Draining of water from a bulk material occurs very slowly compared to that from surface and near-surface layers. It takes place by a slow diffusion, not as a result of air movement, as at the surface area.

To compare a short- and long-term water absorption, an experiment was conducted, in which several deck boards were submerged in water for a period from 24 h to 110 days. The experiment was performed in 2001, and two old (later discontinued) GeoDeck boards were chosen, one containing only rice hulls (65%) and HDPE (35% minus minor additives), another containing rice hulls and Biodac® in approximately equal amounts (by weight). The second board had a very low density (specific gravity), of 1.09 g/cm^3, compared with the current (2006) boards of 1.19–1.24 g/cm^3.

Data such as listed in Table 12.8 can be plotted against the square root of exposure time, as shown in Figure 12.12. This approach to data analysis is based on the

TABLE 12.8 Water absorption by some deck boards

	Water absorption (%)			
Time period of submerging in water (days)	Pressure treated lumber	Trex	GeoDeck, rice hulls as a filler	GeoDeck, rice hulls and Biodac, low density (1.09 g/cm^3)
1	23.6	3.04	4.59	3.11
4	32.2	4.59	7.28	5.26
11	40.1	6.58	10.6	7.44
14	42.6	7.20	11.8	8.19
39	44.3	9.94	16.5	11.6
49	48.5	10.8	18.7	13.2
110	48.6	14.3	21.8	17.1

The test was performed according to ASTM D 1037.

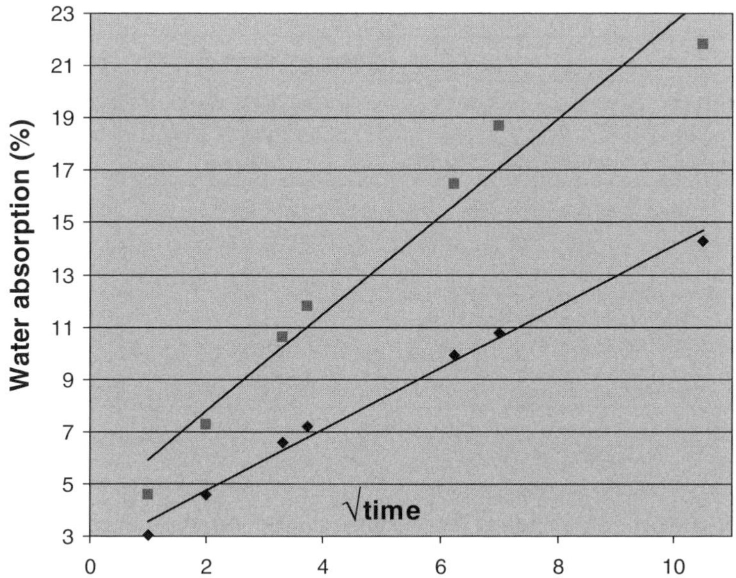

Figure 12.12 An illustration of a typical Fickian diffusion behavior of water absorption by two WPC deck boards, a solid Trex board (lower plot), and a hollow GeoDeck board filled with only rice hulls (see Table 12.8). The horizontal axis—a square root from underwater exposure time.

so-called Fickian diffusion and utilizes Fick's second law of diffusion. According to that law, the concentration gradient is the driving force for diffusion, and the amount of component (water in this particular case) diffused is a function of time. More specifically, Fick's law predicts that the mass of water absorbed increases linearly with square root of time and then gradually slows until it reaches equilibrium plateau. The slope of the plot is defined by the diffusion coefficient (hence, by porosity), sample thickness, and other factors.

In fact, the above ("Fickian") consideration with respect to WPC presents only a rough approximation. In the Fick's law the material is considered thin, with only one-directional description of the mass transfer. In cases such as described in Table 12.8 and Figure 12.12, a profile has a considerable thickness and water can penetrate into the tested material in several directions. This calls for at least a correction for the "edge effect" and for maximum moisture content in the material. For a longer underwater exposure, absorbed water attains a saturation, and the straight line as shown in Figure 12.12 asymptotically approaches the maximum absorption level. Hence, plots such as shown in Figure 12.12 have a limited range of validity.

Another similar experiment, such as shown in Table 12.8, was conducted with a different set of boards (GeoDeck, having only rice hulls as a filler, was later discontinued), but following a different time schedule (Table 12.9).

TABLE 12.9 Water absorption by some deck boards

Time period of submerging in water (days)	Water absorption (%)			
	TimberTech	UltraDeck	GeoDeck, rice hulls as a filler	GeoDeck, rice hulls and Biodac (density 1.16 g/cm^3)
1	1.5	1.9	2.5	1.4
4	2.8	3.5	3.7	2.3
7	3.5	4.6	5.0	3.2
14	4.6	6.6	7.8	4.9
28	6.5	9.9	11.0	6.9
38	7.5	11.8	13.0	8.1
100	13.0	16.8	20.7	13.6

The test was performed according to ASTM D 1037.

Water absorption depends very significantly on the shape of composite profiles, and its dependence on brushed or unbrushed profiles is under study. Boards described in Tables 12.8 and 12.9 were brushed. For a comparison, unbrushed GeoDeck pickets (2″ × 2″ nominal cross-sectional size, for a composite railing system) showed the following figures for a short- and long-terms water absorption:

24 h 0.76%
7 days 1.4%
116 days 5.4%.

Obviously, these figures are much lower compared with data in Tables 12.8 and 12.9, though GeoDeck profiles were made from the same material (HDPE, rice hulls and Biodac®).

Pickets of the same size were also made from a polypropylene–Nylon blend (carpet scrap)—and subjected to water submerging for a short- and long-term test. Despite a generally recognized hygroscopicity of Nylon, the pickets absorb only a little water (Table 12.10)

The data in Table 12.10 show that polypropylene in the blend noticeably reduces water absorption by Nylon. Biodac® obviously absorbs water, along with Nylon, and 37.5% of polypropylene cannot stop it.

In yet another experiment a 1.7 ft-long GeoDeck hollow board (3.0 lb by weight), with wall thickness of 0.25 in, was sealed on one end, placed vertically, and completely filled with water. The goal of the experiment was to examine if after a long exposure, water penetrates through the board's walls. In other words, if after a prolonged exposure the board would be wet outside. Water in the sealed board was kept flush with the upper end by adding fresh water. After 4 months (120 days) the experiment was ended because no water or any wetness appeared on the outer wall

TABLE 12.10 Water absorption by pickets (2″ × 2″ nominal cross-sectional size) made from different materials

Time period of submerging in water (days)	Composition of pickets			
	75% polypropylene–25% carpet scrap	50% polypropylene–50% carpet scrap	37.5% polypropylene–12.5% carpet scrap–50% Biodac	GeoDeck
	Water absorption (%)			
1	0.24	1.3	3.0	0.76
2	0.27	1.4	—	—
7	0.30	1.4	—	1.4
116	0.58	2.2	11.6	5.4

of the board. Water was poured out, the board was placed upside down for 10 min (as prescribed ASTM D 1037 for similar, but standard tests), and weighed. The amount of the absorbed water was 15.8% by weight. This figure is close to the respective figures in Tables 12.8 and 12.9.

ASTM RECOMMENDATIONS

Most of the ASTM recommendations applicable for WPCs have been developed either for plastics (ASTM Committee D20) or for wood (ASTM Committee D7). ASTM procedures developed specifically for WPCs have started to appear only lately. This situation is applicable for majority of WPC properties—slip, flammability, microbial degradation, oxidative degradation, among others, but test methods for some of them are essentially the same, regardless of wood or plastic; for some they are quite different. Recommended test methods for water absorption are quite different for wood or plastics. The principal ASTM procedures for both of them are given below.

Applicability of a particular procedure for WPCs are defined by the respective Acceptance Criteria (AC 174), developed and modified by ICC-ES.

ASTM D 570, "Standard Test Methods for Water Absorption of Plastics"

The ASTM procedure covers the determination of the relative rate of absorption of water by plastic when immersed. Ideally, diffusion of water (or liquid in general) into plastics for long-time immersion procedure follows the square root of immersion time, with the initial slope of the graph proportional to the diffusion constant of water in the plastic. The plateau region with little or no change in weight represents the saturation water content of the plastic. The ASTM recommends that the saturation level is reached when the increase in weight per 2-week period averages less than 1% of the total increase in weight or 5 mg, whichever is greater.

The ASTM notices that the rate of water absorption may be widely different through each edge and surface. It may be greater through cut edges than through molded surface. Hence, quantitative correlation of water absorption with the surface area should be limited to similarly shaped specimens and to closely related materials. Also, for materials with widely varied density, relation between water-absorption values on a volume as well as weight basis may need to be considered.

The ASTM procedure specifies in detail the dimensions of specimens for molded plastics (in the form of a disk), plastics (in the form of a square), sheets, rods, and tubes. For example, the test specimen for molded plastics shall be in the form of a disk 2 in. in diameter and 1/8 in. in thickness. Specimens should be preferably dried for 24 h at 50°C. (122°F).

The procedures for water absorption for immersed specimens are subdivided into short-term (2 h and 24 h) immersion and long-term immersion. All tests should be done in distilled water at ambient temperature (22–24°C or 72–75°F). Two-hour immersion is recommended for samples having a relatively high rate of water absorption and for thin specimens. At the end of the immersion time, the specimens should be removed from the water one at a time, all surface water wiped off, and weighed. For long-term immersion, the specimens should be removed, weighed, and placed back in water at-24 h time point, 7 days, and every 2 weeks, until the test is terminated or the specimens reach the water saturation level (see above).

Note of the author: Distilled (deaerated) water is recommended to minimize the problem of air bubbles clinging to the submerged specimen. Air bubbles shield the specimen from contact with water and reduce water absorption.

Besides, the ASTM method includes boiling water immersion tests (30-min and 2-h tests) and 2-day immersion test at 50°C. At the end of the immersion time the specimens should be removed from boiling water, placed in distilled water maintained at room temperature for cooling, and after 15 min removed, wiped off, and weighed.

Percentage increase in weight during immersion in all the cases is calculated by dividing the weight gain by the initial (dry or conditioned) weight:

$$\text{Increase in weight (\%)} = (\text{wet weight} - \text{conditioned weight})/\text{conditioned weight} \times 100$$

ASTM D 1037, "Standard Test Method for Evaluating Properties of Wood-Based Fiber and Particle Panel Materials"

The ASTM procedure covers an extended set of tests of wood-based materials, among them (Sections 100–111) tests for water absorption and thickness swelling, and for moisture content. The procedure describes two water absorption methods: Method A, for 2 plus 22 h submersion period, and Method B, for single continuous

24-h submersion period. Thus, Method A provides information on the short-term and longer term water absorption and thickness swelling performance. This method is commonly not in use with WPCs.

The ASTM methods recommend test specimens of 12 by 12 or 6 by 6 in. in size. After conditioning as described in the procedure, the specimen shall be weighed and the width, length, and thickness measured. The specimen shall be submerged horizontally or vertically under 1 in. of distilled water at 20°C (68°F). After 24 h of submersion (Method B) the specimen shall be suspended for 10 min to drain, then remaining water wiped off, the specimen weighed, and the thickness measured. The procedure notes that specimens placed in water vertically will absorb considerably more water then those placed horizontally.

Notes of the author: The last statement of the ASTM procedure is arguable. Table 12.11 shows data obtained with three commercial WPC boards specimens (12 in. long and 5.5 in. wide) after 24 h submersion in water, according to ASTM D-1037 [6]. One can see that water absorption in that case was practically the same whether the samples were placed into water vertically or horizontally.

After submersion, weighing, and measurements, the specimen shall be dried in an oven at 103°C (217°F) until approximately constant weight is attained. Water absorption and moisture content shall be calculated with respect to conditioned and dry weight of the specimen, respectively, such as

$$WA = 100[(W_2 - W_1)/W_1]$$

where
WA = water absorption, %
W_1 = weight of the conditioned specimen, g
W_2 = weight of the specimen after 24 h of underwater submersion, g.

$$MC = 100[(W_1 - W_0)/W_0]$$

where
MC = moisture content, %
W_1 = weight of the conditioned specimen, g
W_0 = weight of the specimen when oven-dry, g.

The thickness swelling shall be expressed as a percentage of the original thickness.

TABLE 12.11 Water absorption by three WPC boards (5.5 in. × 12 in.) of the same composition, placed underwater vertically and horizontally for 24 h.

	Water absorption after 24 h (%)	
WPC board	Horizontal position	Vertical position
1	3.5	3.6
2	1.8	1.8
3	0.4	0.4

ASTM D 2842 "Test Methods for Water Absorption of Rigid Cellular Plastics"

The ASTM procedure covers the determination of the water absorption of cellular plastics (lighter than water) by measuring the change in buoyant force resulting from immersion. It can be applicable to WPC, having density less than 1.0, in its methodological part of underwater weighing of specimens.

The procedure employs an underwater weighing jig and describes two configurations of such jigs. The recommended size of a test specimen is 6″ × 6″ × 3″. Two procedures are described. Procedure A considers weight of a dry specimen and its underwater weight after a 4-day submerging period. Procedure B considers weight of a dry specimen, initial weight of the submerged specimen, and weight of the submerged specimen after a 4-day submerging period. Water absorption (%) by the specimen volume is calculated and reported, along with "water absorbed per unit of surface area," apparent specimen volume, surface area, "true specimen volume," and "volume of surface cells."

The ASTM method describes as an example polyisocyanurate (Procedure A) and extruded polystyrene (Procedure B), both samples 3 in. thick, in terms of the average water absorption as volume percent. For polyisocyanurate it is 2.06 ± 0.14% (within-laboratory tests) and 2.1 ± 0.5% (between-laboratory tests). For extruded polystyrene it is 0.17 ± 0.04% (within laboratory tests) and 0.17 ± 0.08% (between-laboratory tests).

ASTM D 6662 "Standard Specification for Polyolefin-Based Plastic Lumber Decking Boards"

Section 6.3.3 "Hygrothermal Cycling" of the ASTM describes a test procedure for freeze–thaw evaluation of deck boards. The procedure consists of three cycles: each one consisting of a specimen weighing and total submerging underwater for 24 h, wiping off, and weighing again within 20 min. If a weight gain in more than 1% of the initial weight of the specimen, the specimen shall be resoaked until the weight change after 24 h submerging time is less than 1% compared to the preceding submerging. In terms of the procedure, such specimens have reached water-absorption equilibrium. After reaching this equilibrium, the specimens shall be frozen to −20°F (−29°C) for 24 h, and then returned to room temperature. This process comprises one hygrothermal cycle. It should be repeated two more times, for a total of three cycles. Then the specimens shall be returned to room temperature and tested as described in ASTM 6109, for flexural strength and flexural modulus. Any obvious physical changes in the specimens should be reported. Flex strength and modulus should retain 90% of the original values.

ASTM D 7032 "Standard Specification for Establishing Performance Ratings for Wood–Plastic Composite Deck Boards and Guardrail Systems (Guards or Handrails)"

Section 4.7 "Freeze–Thaw Resistance Test" of ASTM D 7032 essentially repeats the above description from ASTM D 6662, with a couple of changes. First, not 15, but five specimens are required for the procedure of D 7032. Second, after completion of each hygrothermal cycle, when the specimens are returned to room temperature, they should be conditioned at room temperature for 24 h (ASTM D 7032). This conditioning was not specified in ASTM 6662, so it was not clear how soon the specimens should be immersed in water after first and second hygrothermal cycles.

EFFECT OF CELLULOSE CONTENT IN COMPOSITE MATERIALS ON WATER ABSORPTION

Obviously, the higher the cellulose content in a WPC material, the higher is the water absorption, unless each cellulose particle is fully encapsulated in the plastic and inaccessible to water. Indeed, water absorption typically shows an almost linear increase with the increase in cellulose fraction in the composite. For example, with cellulose content in LDPE-based composite material of 10, 20, 30, 40, and 50%, the "equilibrium" (as authors of the study believed) water absorption after 20–30 days was equal to 1, 4, 6, 9, and 12% (w/w), respectively [7].

Another typical study [8] showed that in wood (aspen)–plastic (polypropylene) composite the increase in wood fraction—0, 30, 40, 50, and 60%—led to the increase in water absorption as 0, 2, 5, 7, 9, and 11%, respectively, after 10 weeks of submersion under water. Moisture content at 90% relative humidity had increased in the same direction.

A similar study was conducted using wood–LDPE composite material with wood content of 20, 30, 40, 50, 60, 70, and 80%. Water absorption followed a practically linear dependence, from 5 to 40% (w/w) after a certain time period [9].

It is a quite a common knowledge in WPCs that the higher the cellulose content in the composite, the higher the water absorption.

EFFECT OF BOARD DENSITY (SPECIFIC GRAVITY) ON WATER ABSORPTION

Earlier it was shown (Tables 12.4–12.6) that the lower the density, the higher the swelling of composite deck boards being submerged underwater. Table 12.12 shows short- and long-term water absorption for the same boards.

TABLE 12.12 Water absorption by GeoDeck tongue-and-groove boards having different density (specific gravity) and being submerged for a total of 28 days

	Water absorption (%)			
Density	24 h	7 days	14 days	28 days
1.125	1.77	4.31	5.96	8.43
1.15	1.52	3.34	4.68	6.55
1.17	1.47	3.19	4.41	6.13
1.21	1.49	2.94	3.98	5.44

TABLE 12.13 Water absorption by GeoDeck deck boards having different density (specific gravity) after 24 h of submerging

Density (g/cm^3)	Water absorption (%)
1.04	4.3
1.06	3.8
1.07	2.9[a]
1.08	2.9[b]
1.09	2.7[c]
1.10	2.5
1.10	2.04[d]
1.115	1.85[e]
1.12	1.7[f]
1.15	1.8
1.16	1.7
1.18	1.5
1.21	1.1[g]
1.25	1.02[h]

[a]After 5 days 7.65%, after 7 days 9.1%.

[b]After 5 days water absorption 6.4%.

[c]After 5 days water absorption 7.6%, after 7 days 8.3%.

[d]After 7 days 5.1%.

[e]After 2 days 2.9%.

[f]After 45 days 8.7%.

[g]After 7 days 2.2%.

[h]After 7 days 2.2%.

Table 12.13 contains some assorted data regarding water absorption by GeoDeck deck boards with different densities, obtained during a 5-year manufacturing period. Different densities of boards with the same formulation resulted from different manufacturing regimes, such as manufacturing speed, temperature, and also amounts of antioxidants and moisture content in incoming cellulose fiber, as well as devolatilization of hot melt during the extrusion.(Fig. 12.13)

Clearly, the higher the density, the lower the porosity and the lower the water absorption.

Typically, the lower the plastic content, the higher the water absorption at the same formulation, except that the plastic is substituted with cellulose fiber (Table 12.14).

It is not surprising that the increase in plastic in the composite leads to lower water absorption; however, it is counterintuitive that the increase in plastic content leads to a higher specific gravity of the material. HDPE is the lightest component in the composite (d = 0.96 g/cm^3), the other two principal ingredients are rice hulls and Biodac®, with specific gravities of 1.50 and 1.58 g/cm^3. However, in this particular case increase in the fillers content led to an increase in the porosity of the composite; hence, the lower density and the higher water absorption.

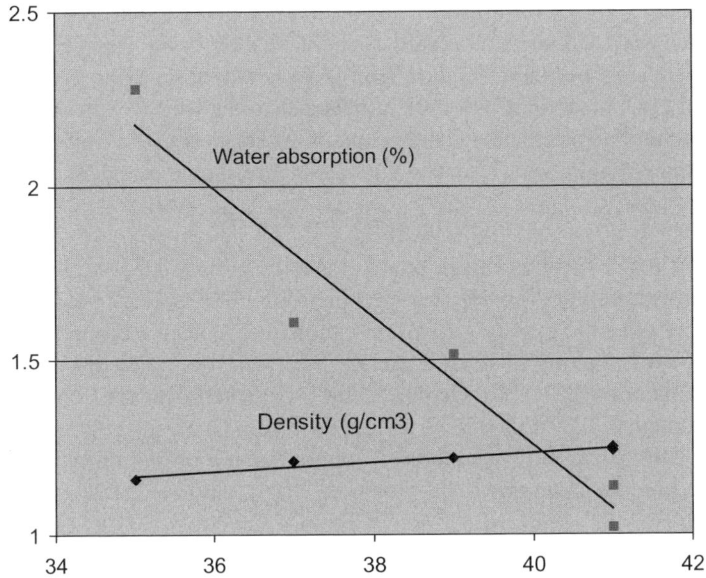

Figure 12.13 Effect of HDPE content (horizontal axis) in WPC material on density (specific gravity) of the composite and its water absorption (data in Table 12.13).

MOISTURE CONTENT OF WOOD AND WOOD–PLASTIC COMPOSITES

As it was indicated above, moisture content in live trees can range from about 30% to more than 200% by weight [1]. WPC materials contain much less moisture, even in very humid conditions. Here, moisture content is related to moisture absorbed from the air. Typically, moisture content in composites fluctuates seasonally, increasing in summer time and decreasing in winter. Moisture content depends on density of the composite material: the lower the density, the higher the moisture content. Sometimes, in the real life conditions, moisture content is related to uncontrolled amount of water getting to the composite material, both from the air and from occasional rains.

TABLE 12.14 Water absorption by GeoDeck deck boards with different amount of HDPE

Amount of HDPE (%, w/w)	Specific gravity (g/cm^3)	Water absorption (%)
41	1.25	1.02 ± 0.01%
41	1.24	1.14 ± 0.01%
39	1.22	1.52 ± 0.08%
37	1.21	1.61 ± 0.04%
35	1.16	2.28 ± 0.35%

The data are given for 24 h of underwater submerging.

One of the examples of moisture content may be that I found in pressure-treated lumber purchased at a local lumberyard. A segment of the board weighed 1191.5 g, and after 24 h of drying the board weighed 916.9 g. Weight loss was 274.6 g, or 23.05% (decrease from the "wet weight") or 29.9% (decrease to the "dry weight"). As ASTM D 1037 recommends to calculate moisture content with respect to the "initial dry weight," or "conditioned weight," moisture content in this case should be calculated as

$$274.6\ \text{g}/916.9\ \text{g} = 29.9\%$$

For the most dense GeoDeck deck boards (specific gravity 1.24–1.25 g/cm^3), moisture content is around 0.4–0.5% (brushed boards). For GeoDeck boards with much lower density (specific gravity 1.10 g/cm^3), moisture content is around 1.7%. Earlier GeoDeck boards, having 35% HDPE and 65% rice hulls (later discontinued), had moisture content of 0.6–0.7%. GeoDeck fines, collected near the compounder, had moisture content of 0.87 ± 0.07%. Brushing dust from GeoDeck boards had moisture content of 2.06 ± 0.13%. These figures greatly depend on humidity and are given here just as random examples.

EFFECT OF WATER ABSORPTION ON FLEXURAL STRENGTH AND MODULUS

For many WPCs an effect of water absorption on their flexural properties is not very significant. Let us consider, for example, data on GeoDeck hollow deck board (tongue and groove) at maximum water absorption, that is, up to its saturation level (data obtained by PFS Corporation, Madison, WI). For a relatively dry composite board (moisture content 0.63 ± 0.04%), the ultimate load was 870 ± 27 lb (support span 16″), flexural strength was 1909 ± 59 psi, and flexural modulus was 304,000 ± 20,000 psi. After the board was saturated with water, the ultimate load was 928 ± 16 lb, flexural strength was 2037 ± 35 psi, and flexural modulus was 296,000 ± 27,000 psi. This means, its flexural strength slightly increased (by 6.7%), and flexural modulus was within error margin with its initial value (or 2.6% lower, comparing the average figures).

In another set of similar experiments, conducted 3 years later, for support span of 24″, ultimate load was 705 ± 19 lb, flexural strength was 2491 ± 64 psi, and flexural modulus was 323,000 ± 6000 psi. After the board was saturated with water, ultimate load was 714 ± 45 lb, flexural strength was 2542 ± 158 psi, and flexural modulus was 321,000 ± 17,000 psi. This means, water saturation of GeoDeck practically did not affect its flexural strength and modulus.

In yet another example for a different commercial solid WPC board, the ultimate load at ambient conditions was 1521 ± 28 lb (moisture content 0.34 ± 0.04%) at support span of 16″, flexural strength was 3,692 ± 55 psi, and flexural modulus was 430,000 ± 17,000 psi. After the board was saturated with water, the ultimate load was 1616 ± 26 lb, flexural strength was 3789 ± 81 psi, and flexural modulus was 448,000 ± 15,000 psi. This means, its flexural strength and modulus for water-saturated board again slightly increased, by 2.6 and 4.2%, respectively.

FREEZE–THAW RESISTANCE

As with the effect of water absorption on flexural strength, effects of a few repeated freeze–thaw on composite deck boards are also not very significant. For example, three-cycle freeze–thaw (hygrothermal) test on GeoGeck composite deck boards, conducted by PFS Corporation (Madison, WI), showed an increase in flexural strength from 1909 ± 59 to 2033 ± 35 psi, or by 6.5%.

Effect of Board Density on Freeze–Thaw Resistance—A Case Study

In another example, the effect of freeze–thaw was compared with GeoDeck boards manufactured in 2003 (board density 1.12 g/cm^3) and 2004 (board density 1.18 g/cm^3). After a saturation of the boards with water, the first cycle of freeze–thaw and the subsequent 24-h submerging underwater, the boards gained weight (from the initial) of 3.79% (density 1.12 g/cm^3) and 3.74% (density 1.19 g/cm^3). After the additional cycle of freeze–thaw and the subsequent submerging of the boards for 24 h, the boards gained weight (from the initial) of 4.03 and 3.95%, respectively. It was noticed that at drying in an oven at 75°C the boards with density of 1.12 g/cm^3 were releasing water much faster compared with the boards with density of 1.18 g/cm^3. After all three cycles of freeze–thaw, the ultimate load (break load) for the boards changed not very significantly, namely, to 93.7% (boards with density 1.12 g/cm^3) and to 96.8% (boards with density 1.18 g/cm^3). For 2003 boards, (density 1.12 g/cm^3) flexural strength (which was calculated including small variations in moments of inertia of the boards) changed from 2330 psi (control) to 2253 psi (after freeze–thaw), that is, by 3.3%. For 2004 boards (density 1.18 g/cm^3) flexural strength changed from 2579 psi (control) to 2497 psi (after freeze–thaw), that is, by 3.2%.

These changes are highlighted in Table 12.15.

Therefore, board density (in this range, between 1.12 and 1.18 g/cm^3) did not play any noticeable role in resistance to triple freeze–thaw. The resistance itself was very good, with a change in flexural strength only about 3% from the initial value.

TABLE 12.15 Effect of three-cycle freeze–thaw treatment of GeoDeck boards

		Weight gain, after submerging to water (%)		Ultimate load compared to control samples (not freeze–thawed) (%)	Flexural strength		
Year of making	Density (g/cm^3)	Two cycles	Three cycles		Control (psi)	After three cycles (psi)	Change (%)
2003	1.12	3.79	4.03	93.7	2330	2253	3.3
2004	1.18	3.74	3.95	96.8	2579	2497	3.2

Data are average of five samples for each batch.

Effect of Board Density and Weathering on Freeze–Thaw Resistance—A Case Study

In yet another example, the effect of freeze–thaw was compared with GeoDeck boards manufactured in 2003 (they did not have an added antioxidants, board density 1.12 g/cm^3) and then weathered in a chamber for 500 h, and with these manufactured in 2004 (containing antioxidants, board density 1.18 g/cm^3) and then weathered for 500 h and for 6000 h. All three sets of the boards were compared with the respective control.

The boards were placed in water, and the boards gained weight as given below (Table 12.16).

TABLE 12.16 Water absorption by GeoDeck boards made in 2003 (no antioxidants added) and 2004 (antioxidants added) prior to a freeze–thaw study

	Water absorption (%)	
Deck boards, data of manufacturing and density	After 6 days	After 9 days
2003, 1.12 g/cm^3		
Control (not weathered)	3.81 ± 0.23%	4.19 ± 0.34%
Weathered for 500 h	4.60 ± 0.45%	5.08 ± 0.53
2004, 1.18 g/cm^3		
Control (not weathered)	3.51 ± 0.24%	3.94 ± 0.28%
Weathered for 500 h	3.73 ± 0.33	4.20 ± 0.31
Weathered for 6000 h (there was no control for these boards)	2.82 ± 0.22	3.08 ± 0.16

As between 6 and 9 days of underwater submerging, the weight gain was less than 1% for all boards, all of them were considered as having reached the equilibrium in water in terms of readiness for freeze–thaw test. The freeze–thaw test was conducted as described above. Data are shown in Table 12.17.

One can see that weathered or nonweathered boards, with density 1.12 or 1.18–1.21 g/cm^3, all of them did not play any noticeable role in triple freeze–thaw. The resistance was very good, with a change in flexural strength not more than 6%

TABLE 12.17 Effect of three-cycle freeze–thaw treatment of GeoDeck boards

Year of making (duration of weathering in the box)	Density (g/cm^3)	Ultimate load compared to control samples (not freeze–thawed) (%)
2003(nonweathered board)	1.12	97
2003(weathered for 500 h)		97
2004(nonweathered boards)	1.18	94
2004(weathered for 500 h)		100
2004(weathered for 6000 h)	1.21	105

Data are the average of four samples for each batch.

TABLE 12.18 Effect of 15 consecutive freeze–thaw cycles on mechanical properties of a WPC board (44% HDPE, 50% wood flour, 6% lubricant) [10]

Freeze–thaw cycles	Flexural strength (psi)	Flexural modulus (psi)
	Maplewood flour as a filler	
0	3002	236,350
15	2364	120,350
	Pinewood flour as a filler	
0	3582	320,450
15	3408	200,100

(nonweathered board) and typically less compared to the initial flexural strength of boards.

Effect of Multiple Freeze–Thaw Cycles

More significant effects of multiple freeze–thaw cycles on flexural strength and flexural modulus of WPC boards were observed, when the boards did not contain mineral fillers or had some other compositional or/and structural features not yet fully understood. Some examples are given in Table 12.18.

One can see that 15 freeze–thaw cycles decreased flex strength of a wood (maple) flour-filled HDPE by 21% and flex modulus by 49%. In a pinewood flour-filled HDPE the reduction effects were 5 and 38%, respectively.

It can be conjectured why pine-derived fiber as a filler produced much stronger (by 19%) and much stiffer (by 36%) board compared to that filled with maple-derived filler, and why the maple flour-filled WPC became so week and flexible after the freeze–thaw cycles compared to the pine- flour-filled board; however, we do not have clear, direct, examined, and verified answers.

COMPARISON OF WATER ABSORPTION OF SOME COMPOSITE DECK BOARDS AVAILABLE IN THE MARKET

Table 12.19 shows that water absorption by commercial WPC deck boards differs rather significantly. The lowest water absorption was shown by a compression-molded deck board, Evergrain/Epoch (0.3% after 24 h and only 1% after 7 days underwater submerging). The rest of the products showed a variation mainly between 0.7 and 1.9% (24 h underwater), and one product absorbed an exceptionally high amount of water. It is not identified in the table. The product had noticeably high porosity and contained many voids. In the beginning of underwater submerging it was heavier than water, but at the end it was practically floating.

In 2004–2005, AAMA (American Architectural Manufacturers Association) had arranged an independent testing of commercial composite boards provided by

TABLE 12.19 Water absorption values for actual composite deck boards, determined by the author using ASTM D 1037 for 24 h and 7 days of submerging periods

	Water absorption (%)	
Board	24 h submersion	7 day submersion
Pressure- treated lumber	24[a]	36
WPC	16[b]	21
Trex, Saddle	1.9	4.8
Trex, Madeira	1.9	4.5
Trex, Winchester	1.8[c]	4.5
Perfection	1.6	4.3
ChoiceDeck	1.7	3.9
EverX	1.7[d]	3.7
TimberTech, red	1.3[e]	3.9
TimberTech, gray	1.0[e]	3.4
Rhino Deck, red and gray	1.05[f]	2.9
GeoDeck	1.02	2.2
Fiberon	1.0	2.2
Nexwood, red and gray	0.9[g]	2.2
Dreamworks	0.85	2.2
Evergrain, cedar	0.8	1.9
Nexwood, Sierra Brown	0.7[g]	1.9
WeatherBest	0.7[h]	1.7
Evergrain, gray(compression molded)	0.3	1.0
Others	See footnote[i]	

Data reported by the manufacturers are given in footnotes. Not all boards are the current ones; however, they were purchased on the market. WPC is a commercial board that showed such an inferior performance that I did not want to disclose the manufacturer.

[a] Ponderosa pine 17.2% (from some assorted data).

[b] Abraded 4.61%, unabraded 1.30%.

[c] Sanded surface 4.3%, unsanded surface 1.7%.

[d] 0.7%.

[e] <1.0%.

[f] Moisture content 0.65%; water absorption was not reported.

[g] <1.0%.

[h] from 0.49% to 2.35%.

[i] USPL 1.66%; Correct Deck 0.8%; Carefree (100% HDPE) less than 0.1% after 11 weeks.

manufacturers. All boards were tested in the same conditions and using the same equipment and the same operator, and data are shown in Table 12.20.

One can see that water absorption data in Table 12.20 are significantly higher compared with some figures reported in Table 12.19. This emphasizes again that in order to compare the respective figures, the data are preferably to be obtained at the same conditions.

TABLE 12.20 Water absorption (ASTM D 570) by materials of six commercial composite boards

Manufacturer (code)	Water absorption after 7 days of immersion
A	4.1 ± 0.2
B	4.75 ± 0.02
C	6.5 ± 0.3
D	7.2 ± 0.8
E	9.7 ± 0.1
F	13.9 ± 0.2

Two 3″ by 1/2″ by 1/4″ specimens of each board was tested after 7 days of underwater submersion. Prior to immersion, specimens were conditioned by oven drying at 105°C and then cooled in a desiccator. Boards were provided by manufacturers for testing. The order of materials in the table is from the lowest to the highest value of water absorption and does not indicate the particular manufacturer.

REFERENCES

1. Forest Products Laboratory. *Wood Handbook*, 1999, pp. 3–5.
2. W. Wang and J.J. Morrell. Water sorption characteristics of two wood-plastic composites. *Forest Prod. J.* 2004, **54**(12), pp. 209–212.
3. M. Manning. Borates as biocidal additives for WPC. In: *The Global Outlook for Natural Fiber and Wood Composites 2003*, New Orleans, LA, December 3–5, 2003.
4. M. Gnatowski. Water absorption by wood-plastic composites in exterior exposure. *8th International Conference on Woodfiber-Plastic Composites (and Other Natural Fibers)*, Madison, WI, May 23–25, 2005.
5. Forest Products Laboratory. *Wood Handbook*, 1999, pp. 12–16 to 12–17.
6. M. Manning. Creating value in WPC products with anti-microbials and stain resistant additives. In: *WPC Conference 2004*, Baltimore, MD, Principia Partners, Cleveland, OH, October 11–12, 2004.
7. J. Kajaks and S. Reihmane. Thermal and water sorption properties of polyethylene and linen yarn production waste composites. In: *2nd International Wood and Natural Fibre Composites Symposium*, Universitat Gesamthochschule Kassel, Institut fur Werkstofftechnik, Kassel, Germany, 1999, pp. 39/1-39/5.
8. R.M. Rowell, S.E. Lange, R.E. Jacobson. Effects of moisture on aspen-fiber/polypropylene composites. In: *Progress in Woodfibre-Plastic Composites, Conference Proceedings*, Canadian Natural Composites Council, University of Toronto, Toronto, Canada, 2002.
9. F. Yehia, S. Law, M.T. Kortschot. Water absorption in wood fiber reinforced polyethylene. In: *Progress in Woodfibre-Plastic Composites, Conference Proceedings*, Canadian Natural Composites Council, University of Toronto, Toronto, Canada, 2002.
10. L.M. Matuana. Resistance of HDPE/wood-flour composites to cyclic freezing and thawing exposures. In: *8th International Conference on Woodfiber-Plastic Composites*, Forest Products Society, Madison, WI, May 23–25, 2005.

13

MICROBIAL DEGRADATION OF WOOD–PLASTIC COMPOSITE MATERIALS AND "BLACK SPOTS" ON THE SURFACE: MOLD RESISTANCE

INTRODUCTION

Microbial Effects on Wood–Plastic Composites

There are three "levels" of microbial effects on wood–plastic composites. First, when mold forms colored spots on the surface of deck boards, but it does not degrade the material, which remains structurally sound. In this case mold feeds itself by dust, airborne particles such as pollen, and so on. Second, when mold and other fungi insignificantly consume some ingredients of the composite formulation, using them as nutrients, vitamins, or so on. Third, when fungi specifically and rapidly attack wood/cellulose fiber in the composites, which in turn causes diminishing of mechanical properties of the materials. Those are called decay fungi or cellulolytic fungi.

Many fungi are classified as mildew, or mold, or wood-staining fungi (all are synonyms) rather than wood-degrading fungi because they primarily just discolor wood. Mildew fungi generally do not cause a reduction of cellulose or composite materials in strength. However, as both mildew and decay fungi grow in the same conditions, the presence of molds can be taken as a signal of a potential decay. Besides, many molds cause allergies, and hence, are hazardous for health.

Wood-Plastic Composites, by Anatole A. Klyosov

Mold and Spores

Mold is a type of fungus that can weaken building materials including composite deck boards, make composites discolored or cover them with black, green, red, and other colored spots that can be very difficult to remove, and can make people sick with allergies, asthma, or respiratory problems, particularly in those who are already poor in health and/or have a pulmonary condition. Mold is one of the champions of survival on the Earth. Mold spores exist in astronomical numbers virtually everywhere around us. Air currents carry spores on decks, as well as onto all other surfaces.

About 250 thousand mold spores would fit on the head of a pin. In order for mold spores to get awakened and to grow, they require moisture, warm temperature, oxygen, and a food source. On an outdoor deck, all these factors are very difficult to control. Mold can use dust and organic dirt as food source, and oxygen is around. Within hours of coming into contact with water, spores wake up and sprout germ tubes, which in turn develop a network of hyphae within days, sometimes hours. Hyphae extract nutrients from dust, dirt, wood particles, pollen, among others., and form thousands of new spores. Air currents pick them up and deliver elsewhere. Mold spores are ubiquities. They are fact of life on earth. In North America alone, there are more than 100,000 species of mold.

Moisture and Ventilation: Critical Moisture Content

It is assumed, based on many observations, that cellulose having moisture content of 19% or lower will generally not support the growth of mold. Mold spores enter a kind of hibernation, able to maintain it sometimes for decades. They are practically invisible but still can make some people get allergiy sneeze, cough, and can make them feel sick. Composite deck boards rarely have moisture of their cellulose fiber above 19%, except in the very top and thin layer after a wood rain, but that layer dries fast and spores do not have time to wake up and sprout germ tubes. They dry out and go into hibernation state again.

However, in humid, moist areas, particularly with inadequate deck ventilation (to help keep the product relatively dry), for instance, when deck boards are installed too close to the ground, or when a deck forms a "box" isolated underneath, or when deck boards are wet and covered, moisture content in composite deckboards can exceed 20–25% and retain as such for a long time. With a lack of proper biocides, mildewcide, or other antimicrobial agents in the formulation of the composite material, mold can be rampant. That is why installation instructions for many composite decks direct to make a deck at least 12 in., and preferably 24 in. from the grade or rooftop, or provide a wider space between boards (such as 3/16″ or even 1/4″). Some installation instruction say that failure to adhere to proper ventilation may void the warranty.

Initially, WPC building materials were considered as being completely resistant to microbial degradation because wood fibers are completely immersed into plastic. It turned out not to be true. WPC materials are typically porous, and a degree of porosity is determined by the moisture of incoming raw materials (cellulose fiber, first and foremost) and processing conditions (primarily, local overheating) that translates to a density (specific gravity) of the resulting profile. Pores in composite materials are typically open, and form chains of pores, penetrating the whole matrix. Wood fiber is exposed into these pores. Hence, there is a higher or lower degree of water absorption. As a result, there is microbial contamination of the material into the matrix pores and voids, microbial degradation of wood particles (and on some cases particles of minerals, which some microorganisms use as a food source), and in some acute cases, microbial growth through the matrix of composite materials.

Probability of such cases of microbial degradation is determined by the accessibility of the composite matrix by microflora. This in turn is determined by a degree of porosity of the composite, density of the material (specific gravity), water absorption, content of minerals in the material (minerals often play a role of a shield, blocking invasion of microbes into the matrix), and the presence of biocides or antimicrobial agents.

It seems that water absorption serves as one of the key parameters in microbial growth in WPC materials. When a whole cross section of a composite deck board is tested by immersion in water, water absorption after 24 hr is typically between 1 and 3% by weight, after 7 days between 3 and 10%, after 20 days about 8 to 15% (see Chapter 12). These values depend on temperature, and the lower the the temperature, the lower the water absorption [1]. However, water absorption by the top layer of a composite board (1 mm in depth, 50:50 mix of wood:plastic) was in excess of 15% after 24 hr [2]. On other data, water absorption by the 5-mm top layer of Trex deck board in the temperature range from 5 to 25°C was 45 and 60%, respectively. This level of moisture content is well in excess of that necessary to support fungal decay. In fact, authors [1] noticed that when the 25°C trial was run for 30 days using Trex samples, a thick microbial film was developed on the surface of the material. This is a rather common observation in the course of long-term water absorption studies.

Generally, there is a clear correlation between an overall moisture content in WPC and its susceptibility to microbial degradation.

Wood Decay Fungi

Wood decay fungi, such as *Trichoderma* sp., after they attach themselves to the surface, secrete a whole family of cellulolytic enzymes. The enzymes, acting in concert, hydrolyze cellulose fiber eventually to glucose, the principal building sugar of cellulose chains, and the glucose is consumed by the fungus. The fungus grows, proliferates into pores and other voids of the composite material, and makes more enzymes to attack cellulose, until a significant part of the cellulose (theoretically all cellulose in the composite) is digested and utilized by the fungus. This obviously diminishes the mechanical properties of the composite material. Besides, the surface of the composite material can be largely covered by the fungi, or even produce actual mushrooms grown on the cellulose (Fig. 13.1).

Figure 13.1 Dime-sized mushrooms growing on a WPC board. Source of the photo: U.S. Borax; Plastic Technology, September 2005.

Biocides and "Mold Resistance"

Some WPC materials contain biocides to inhibit microbial growth, while others do not. However, it is not always easy to tell whether those biocides in chosen amounts in the given composite material are actually effective.

Biocides are often called—depending on their assumed application—as fungicides, mildewcides, antibacterials, or preservatives. Biocides essentially suppress the growth of different microorganisms, but in a different extent, depending on mechanisms of action of biocides and life features of microorganisms.

The whole area of "mold resistance" in composite materials is rather ill-defined. Typically, it is not clear, what "resistance" is under consideration: resistance to *microbial degradation* of the composite material (weight loss, loss of the integrity of the material) or resistance to *mold growing at the surface*, without penetration into the matrix, without violation of the structural integrity of the material.

In other words, microbial degradation of WPC means deterioration of the material, while mold growing at the surface does not mean actual deterioration but brings about a merely visual factor, though aesthetically repulsive. The composite material in the last case is still intact, as well as the original material. A plain pressure-washing would remove mold from the surface of a composite material.

There is a wide variety of "biocides" recommended (by manufacturers) for application in WPCs. Some of those biocides are totally ineffective in the real world; some are effective but only toward certain microbial species, not necessarily troublemakers for the particular composite in a given geographical region. Some of them will leach out of the composite rather soon. Some of them, on the contrary, are barely water-soluble and are "locked-in" in the composite making "biocides" practically isolated from the sites of a potential microbial attack.

What makes things even worse is that the standard ASTM tests for microbial resistance conducted against "standard" white and brown rot fungi or other microorganisms typically show that WPCs are excellent in terms of "microbial resistance." Composites never fail in these tests. However, as it was implied above, these standard tests typically do not reflect "the real world." As soon as one blends plastic and cellulose fiber, one can expect almost guaranteed "no decay" status of the composite material according the ASTM standards.

Let us consider the current situation with biocides for wood and WPCs and some science behind it.

PRESERVATIVES FOR WOOD LUMBER

Wood used in construction of decks, docks and buildings, utility poles, railroad ties, and bridge ties is typically treated with a chemical preservative to make it resistant against microbial degradation. Among those chemical preservatives, the most widely used were chromated copper arsenate (CCA), ammoniacal copper quat (ACQ), pentachlorophenol, and creosote. CCA was the most prevalent preservative due to its low cost, and because it provided a dry and paintable surface after its application and left the wood relatively leach-resistant [3].

Traditionally, solid sawn lumber products are dipped or pressure treated with solutions of preservative chemicals.

CCA

CCA had been used to pressure treat lumber since the 1940s. Since the 1970s, the majority of the wood used in outdoor residential settings has been CCA-treated wood.

However, as a result of growing concerns regarding leaching of arsenic salt around the home and in children's play areas, the Environmental Protection Agency (EPA) announced on February 12, 2002 a voluntary decision by the industry to move away from the consumer use of CCA-treated lumber by December 31, 2003, in favor of new alternative wood preservatives. According to the EPA, these alternatives of CCA are alkaline copper quarternary (ACQ), copper boron azole (CBA), cyproconazole, and propioconazole. Besides, as an alternative to pressure-treated wood, EPA has suggested to use WPC materials.

Currently, EPA classifies CCA as a restricted use product, for use only by certified pesticide applicators. Effective from December 31, 2003, pressure-treated wood containing CCA is no longer being produced for use in most residential uses, with certain exceptions. In fact, the EPA documents says "On or after May 16, 2003, any sale, distribution, or use of existing CCA stock by the registrants is prohibited. (…) It is illegal to treat wood intended for consumer use with CCA."

ACQ

ACQ is a water-based preservative, composed of a combination of copper oxide, quat as didecyldimethylammonium chloride, and either ammoniacal (ACQ type B) or amine (ACQ type D). ACQ does not contain arsenic or chromium and its application also provides a clean paintable surface.

PCP (The U.S. EPA Data)

Pentachlorophenol (PCP) was one of the most widely used biocides in the United States prior to regulatory actions to cancel and restrict certain nonwood preservative uses of PCP in 1987. It now has no registered residential uses. Its commercial uses include utility poles, fences, shingles, walkways, building components, piers, docks and porches, and flooring and laminated beams. Additionally, there are agricultural uses (which are sometimes referred to as "outdoor residential"), that is, wood protection treatment to buildings/products and fencerows/hedgerows. Prior to 1987, PCP was registered for use as a herbicide, defoliant, mossicide, and as a disinfectant, but now all these uses are cancelled. All nonpressure and nonthermal treatment uses (i.e., spray uses) are or soon will be deleted from the registrants' labels. Spray uses for these products were also deleted, effective from December 31, 2004. This action leaves only pressure and thermal treatments of PCP.

In Canada, PCP is used primarily to treat wood poles, piles, bridge timbers, exterior laminated timbers, bridge decking, and fence posts.

In the United States eight products containing PCP are currently registered with EPA, among them Dura-treet 40 wood preserver and Penta 5 sure-treat wood preserver.

Creosote (The U.S. EDA Data)

Creosote is a wood preservative used for commercial purposes only; it has no registered residential uses.

Creosote is obtained from high temperature distillation of coal tar (itself a mixture of hundreds of organic substances). Over 100 components in creosote have been identified. It is used as a fungicide, insecticide, miticide, and sporicide to protect wood and is applied by pressure methods to wood products, primarily utility poles and railroad ties. This treated wood is intended for exterior/outdoor uses only. Its commercial uses include railroad ties (70%), utility poles (15–20%), and other miscellaneous commercial uses (10–15%).

Creosote penetrates deeply into and remains in the pressure-treated wood for a long time. Exposure to creosote may present certain hazards. Creosote is a possible human carcinogen and has no registered residential uses. Therefore, precautions should be taken both when handling the treated wood and in determining where to use the treated wood. It should be noted that such exposure usually occurs only when one comes in contact with railroad ties and/or utility poles.

Unlike wood, composite materials do not need to be dipped or pressure-treated for preservatives, or biocides, to be easily incorporated into their matrix. Biocides as pellets or powder are mixed with other ingredients, compounded, and extruded into shapes.

MICROORGANISMS ACTIVE IN DEGRADATION AND STAINING OF COMPOSITE MATERIALS

There are three groups of microorganisms that can stain, deface, and degrade wood, plastic, and WPC materials. They are

- fungi
- bacteria
- algae.

Fungi are often called molds, and those which are particularly active in degradation of lignin are called wood-rot fungi (and mushrooms), or brown-rot and white-rot fungi. Fungi are aerobic microorganisms, that is, they need oxygen for their growth. They are particularly active from about 86°F (30°C) to about 104°F (40°C). They need water for their activity, and are active at 85% humidity and higher. They prefer slightly acidic conditions, and are typically disseminated by spores.

Bacteria often cause staining of surface by their metabolic products. They are reproduced by division, and only lack of available food can keep them away from fast and unlimited multiplications. They often need more water for their growth and metabolism. A number of bacterial species are able to live at elevated temperatures (thermophilic bacteria).

Algae are different from fungi and represent photosynthetic microorganisms. They need sunlight and prefer neutral pH.

Molds

The most common fungi (molds) that attack WPC materials, are listed below. They can be specific to different ingredients of composites, which they consume as food source. The damage to composites caused by microbial action can result from either consumption of these ingredients and, thereby, reducing mechanical properties of the product, causing discoloration, or production of extracellular metabolites, pigments, which stain the surface of the material with colored spots varying from green to red to black (Fig. 13.2). In the latter case, the microbes not necessarily physically damage the composite or cause its degradation, but they spoil the product and make it look rather ugly.

The abbreviation sp. below indicates the fundamental biological classification, a subdivision of a genus. Organisms that belong to the same species show persistent differences from members of allied species. When mate, species can generally interbred only among themselves.

Figure 13.2 Mold stains on a WPC.

For example, species of some fungi (molds) are

Aspergillus sp.
Penicillium sp.
Paecilomyces sp.
Trichoderma sp.
Chaetomium sp.
Gliocladium sp.
Aureobasidium sp.
Brown-rot fungi
White-rot fungi
Some mushrooms (e.g., *Lentinus lepideus*).

Their brief descriptions are given below. More detailed description can be found, for example, on a Web site of Environmental Microbiology Laboratory, Inc. [4].

• *Aspergillus* sp. are ubiquitous microorganisms. There are approximately 200 species of *Aspergillus*, among them are *Aspergillus fumigatus*, *A. flavus*, *A. niger*, *A. versicolor*, *A. oryzae*, *A. terreus*, and so on. They are found in soil, decaying plant debris, among others, and are disseminated by dry spores, carried typically by wind.

Aspergillus sp. are common allergens (hay fever, asthma, pneumonitis, etc.). Water requirement on a substrate depends on species, and typically varies between 71 and 94%. This is often described as Aw index, that is, "available water" index. Aw refers to a range of minimum water requirements, and is expressed as Aw = ERH/100 (air), where ERH is equilibrium relative humidity. For *Aspergillus* species Aw = 0.71–0.94.

Under magnification, *Aspergillus* fungi look as shown in Figure 13.3.

One of the *Aspergillus* fungi, *Aspergillus niger*, is recommended by ASTM G 21 "Standard practice for the determining resistance of synthetic polymeric materials to fungi" for the determination of the effect of fungi on the properties of synthetic polymeric materials.

- *Penicillium* sp. are ubiquitous microorganisms. There are approximately 200 species of *Penicillium*. They are found in soil, decaying plant debris, fruit rot, house dust, wallpaper, among others, and are disseminated by dry spores and carried typically by wind or insects. Free spores are indistinguishable from *Aspergillus*, and are commonly referred to as "spores typical of *Penicillium/Aspergillus*." *Aspergillus* sp., *Penicillium* sp. are common allergens (hay fever, asthma, pneumonitis, etc.). For *Penicillium* species, Aw = 0.78–0.86. Colonies are usually in shades of blue, green, and white.

One of the *Penicillium* fungi, *Penicillium pinophilum*, is recommended by ASTM G 21 (see above) for the determination of the effect of fungi on the properties of synthetic polymeric materials.

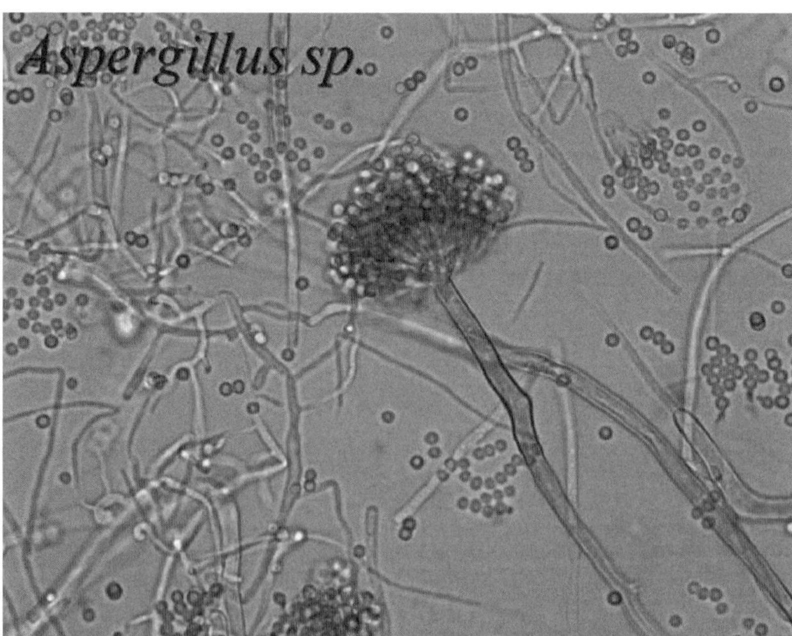

Figure 13.3 *Aspergillus* sp. (© Environmental Microbiology Laboratory).

• *Paecilomyces* sp. are closely related to *Penicillium*; free spores are similar to *Penicillium/Aspergillus. Paecilomyces* sp. are ubiquitous microorganisms. There are 9 – 30 known species of *Paecilomyces.* They are found in soil, decaying plant material, legumes, jute fibers, PVC, timber, leather, among others, and are disseminated by dry spores and carried typically by wind. *Aspergillus, Penicillium* and *Paecilomyces* sp. are common allergens (hay fever, asthma, pneumonitis, etc.). For *Paecilomyces* sp, Aw = 0.79–0.84. Colonies are represented by ocher and lilac pigments and do not produce blue or green colors.

• *Trichoderma* sp. are ubiquitous microorganisms. There are approximately 20 species of *Trichoderma*, among them are *T. viride, T. harzianum,* and so forth. They are found in soil, decaying wood, citrus fruit, tomatoes, paper, textile, among others, and are disseminated by wet spores and carried typically by rain, water splash, insects, wind—when spores dry out. *Trichoderma* sp. are strongly cellulolytic fungi. They release a complex of cellulolytic enzymes which efficiently hydrolyze cellulose fiber. As many other fungi, *Trichoderma* sp. are common allergens (hay fever, asthma, pneumonitis, etc.). Colonies are usually blue–green and yellow–green. *Trichoderma viride* produces a distinctive coconut odor.

• *Chaetomium* sp. are ubiquitous microorganisms. There are approximately 80 species of *Chaetomium.* They are found in soil, seeds, cellulosic materials such as wood and straw, sheetrock paper, among others. (see Fig. 13.4). *Chaetomium* sp. are strong cellulolytic fungi (see above). Spores are formed inside fruiting bodies, and come out at opening and spread by wind, insects, water splash. As many other fungi, *Chaetomium* sp. are common allergens (hay fever, asthma).

Figure 13.4 *Chaetomium* sp. (© Environmental Microbiology Laboratory).

One of the *Chaetomium fungi*, *Chaetomium globosum*, is recommended by ASTM G 21 (see above) for the determination of the effect of fungi on the properties of synthetic polymeric materials.

• *Gliocladium* sp. are common and are most closely related to *Penicillium* and *Paecilomyces*. They are widespread in decaying vegetation and in soil.

• *Gliocladium* spores are formed in sticky masses and are not easily disseminated by air currents. Spores do not have distinctive morphology and are usually categorized as "other colorless." Allergenicity of *Gliocladium* sp. has not been well studied.

One of the *Gliocladium fungi*, *Gliocladium virens* is recommended by ASTM G 21 (see above) for the determination of the effect of fungi on the properties of synthetic polymeric materials.

• *Aureobasidium* sp. are ubiquitous microorganisms. There are approximately 15 species of *Aureobasidium*. They are found in soil, forest soils, fresh water, fruit, wood, bathrooms and kitchens, on shower curtain, tile graft, textiles, among others. (see Fig. 13.5). Wet spores are disseminated by wind when spores dry out. Many other fungi, *Aureobasidium* sp. are common allergens (hay fever, asthma, pneumonitis, etc.). Colonies are usually of the shades of cream to pink, becoming dark brown with age.

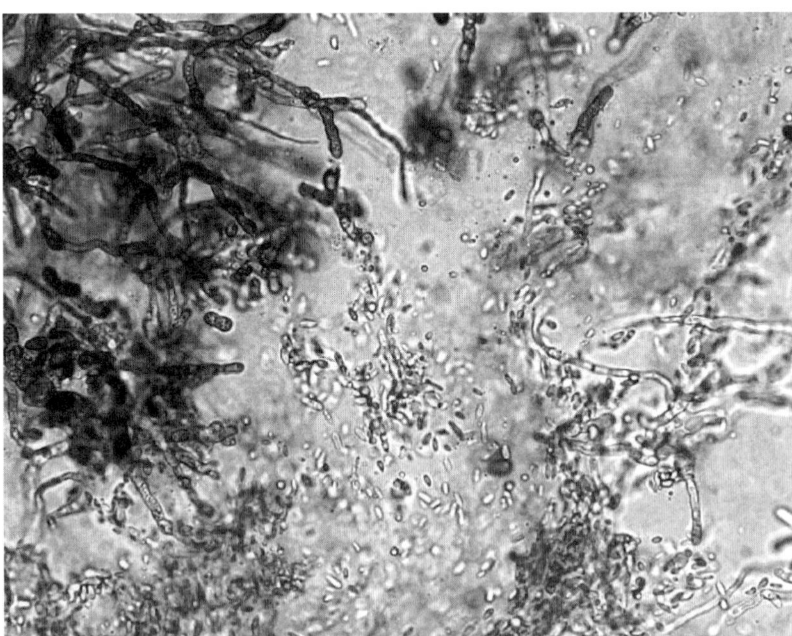

Figure 13.5 *Aureobasidium* sp. (© Environmental Microbiology Laboratory).

One of the *Aureobasidium* fungi, *Aureobasidium pullulans*, is recommended by ASTM G 21 (see above) for the determination of the effect of fungi on the properties of synthetic polymeric materials.

• Brown-rot fungi belong to basidiomycetes, which consist of the so-called macrofungi, including mushrooms. It is one of the major classes of fungal organisms. As follows from their name, brown-rot fungi cause brown rot, or dry rot of wood. Among these fungi are *Poria placenta* (particularly tolerant to copper and zinc compounds), *Poria incrassate*, *Serpula lacrimans*, *Gloephyllum trabeum* (particularly tolerant to phenolic and arsenic compounds), and many others. These fungi can digest the cellulose components of wood. A small amount of decay markedly alters the strength of the wood. If the macroscopic fungus fruiting body is collected, it may be identified as a polypore (bracket fungus). The polypores belong to a limited group of fungi capable of attacking wood and using it for food. The wood also shrinks and becomes some shade of brown. The natural habitat of the polypores is wood, that is slash in forests. No information is available regarding health effects or toxicity of these fungi. Allergenicity has not been studied.

Two of these fungi, *Gloephyllum trabeum* and *Poria placenta,* are recommended by ASTM D 1413 "Standard test method for wood preservatives by laboratory soil-block cultures" for the determination of the effect of fungi on the properties of wood and WPC materials.

• White-rot fungi also belong to polypores (see above), such as *Trametes versicolor* and *Coriolus versicolor.* The ultimate result of the wood decay under their action is that all components (cellulose and lignin) are removed in differing proportions at different rates, and the decayed wood is light-colored. The strength factor is more slowly altered, and frequently the decayed wood is still usable in the early stages of decay. No information is available regarding health effects or toxicity of these fungi. Allergenicity has not been studied.

One of these fungi, *Coriolus versicolor,* is recommended by ASTM D 1413 (see above) for the determination of the effect of fungi on the properties of wood and WPC materials

• There are many mushrooms that can grow on wood. Typically, they are not used for quantitative testing, except of *Lentinus lepideus* (Fig. 13.6), which is recommended by ASTM D 1413 (see above) for the determination of the effect of fungi on the properties of wood, as it is particularly tolerant to creosote. It apparently does not provide a particular interest in testing WPC materials.

There are several fungi that were identified on WPC materials (deck boards) and do not belong to the above list. They are exemplified by *Alternaria alternata* and *Papulospora* sp.

Alternaria alternata is an airborne fungus. It causes the so-called brown spot, one of the most destructive leafspot diseases on tobacco, peppers, tomatoes, and other plants. Fungi *Alternaria* (*Alternaria* species include around 50 known fungi) are commonly

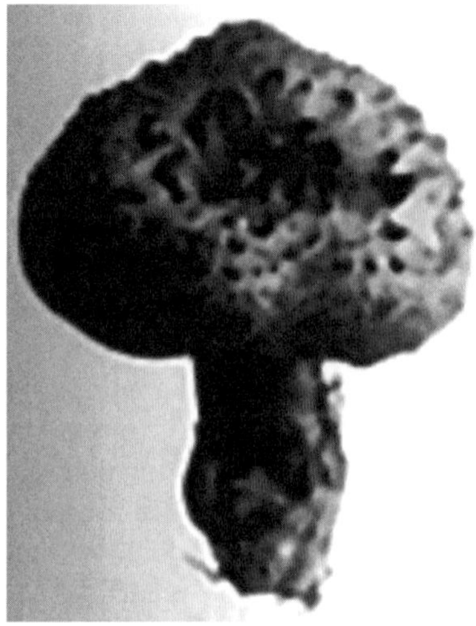

Figure 13.6 *Lentinus lepideus* (image provided by the Illinois Mycological Association, with permission).

isolated from plants, soil, food, and indoor air environment. Recently *Alternaria alternata* was found on a composite deck in Wisconsin (see Case study 3 below). It typically forms melanin-like pigmented circular spots, ranging from 1/4 to 1 1/4 in. in diameter. Initially, these spots have colored halos around them, which are made of thousands of tiny spores. Each of the spores requires a certain level of moisture and temperature to germinate on the surface. Poor ventilation helps the fungus to produce brown hyphae, multiply, and to make more brown spots on a deck. Ventilation reduces moisture content, spores cannot germinate, and new spots cannot develop. *Alternaria alternata* produces toxic metabolites that cause asthma and may develop pulmonary emphysema.

Papulospora sp. is a fungus found in soil, decaying plants, manure, and paper. This fungus was recently found on a composite deck in Wisconsin (see Case study 3 above).

There are many other typical types of mold that can be found on wood and composite materials, such as *Epicoccum* sp., *Fusarium* sp., *Geotrichum* sp., *Stachybotrys chartarum*, *Ulocladium* sp., and others.

Besides mold (fungi), bacteria can also invade wood and composite materials. Among those that were found in WPC are *Pseudomonas aeruginosa* and *Streptoverticillium recticulum*, producing red pigment. Bacteria are more seldom than fungi to occupy composite materials because typically bacteria need more water for their growth compared to fungi. As WPC absorb much less water compared to wood at the same conditions, amount of water even at the surface of WPC is often not enough for bacterial growth.

Black Mold

Black mold is of a particular concern regarding WPC materials due to its visibility (Fig. 13.7).

Figure 13.7 Black mold as black dots on the handrail surface.

Often black mold results from a combined action of two fungi, *Gonatobotryum* sp. and *Ceratocystis*. The first one, black mold, is parasitic on *Ceratocystis*, and so are found where *Ceratocystis* is found, most particularly lumber. *Ceratocystis* sp. is a ubiquitous mold (see Fig. 13.8). There are 56 known species of it. They are disseminated by wet spores and by insects. Most homes built with lumber have areas of growth of both *Ceratocystis* and *Gonatobotryum* on wood framing inside walls. Virtually all lumberyards have some percentage of boards with areas of this black mold growth. No information is available regarding health effects or toxicity of the

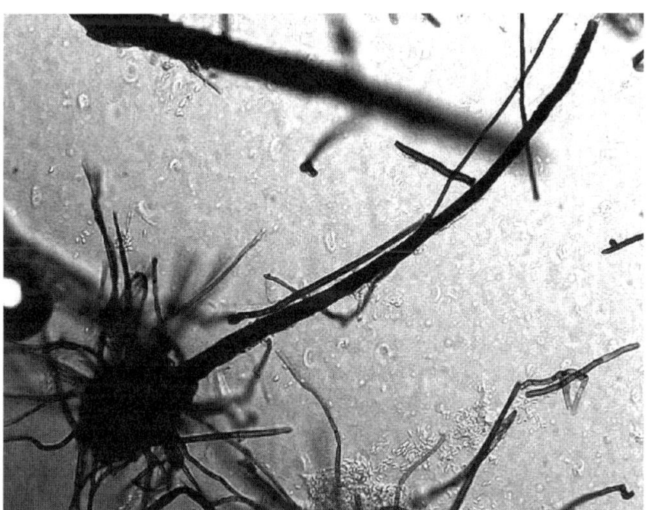

Figure 13.8 *Ceratocystis* sp. (© Environmental Microbiology Laboratory).

mold. Allergenicity has not been studied. Other natural habitats include soil and rotten wood. They are also parasitic on certain other plants.

Black Algae

The black stains that one can see on composite deck boards and/or railing systems are actually the darkened, dead cells of green algae *Gloeocapsa magma*, not mildew. This algae use calcium carbonate as their initial food source, and hence, they might appear when calcium carbonate is used as a filler in the composite material. Besides calcium carbonate (not necessarily their food source), algae use sunlight for an energy source and water. Algae grow rapidly under high moisture conditions and can displace mildew when the material is wet most of the time. Algae do not directly damage the surface of composite materials because they do not contain necessary amounts of cellulolytic enzymes and, hence, do not attack cellulose fiber. However, black spots that they create are highly visible and ugly (see Figs. 13.9 – 13.11).

Figure 13.9 Black mold and algae on a wood-plastic composite railing post.

Figure 13.10 Black mold and algae on railing posts.

Besides, algae, as many other fungi showing as wet spots on surface of composite materials, make the color fading. Apparently, they consume metal-based colorants on the surface. For example, when algae were removed from the surface of GeoDeck board, stored in humid place without ventilation, the surface under algae spots faded from 56.4 ± 0.7 units (on the Hunter scale) to 59.4 ± 0.2 units of lightness. This fading significantly exceeded that under direct sunlight (0.4 units after 1000 h in a weathering box), that is normally less than 1–2 units for outdoor weathering at the above conditions.

Case Study 1: Staining with a Microbial Pigment

Composite handrails were packed at the manufacturing plant by wrapping of them into corrugated paper. When—a few summer months later—a lamberyard personnel unwrapped the package, they saw that the paper was significantly decomposed by mold, producing black-colored pigment. The pigment stained the handrails

Figure 13.11 Black mold and algae on a post.

(Fig. 13.12). There was no degradation of the rails, but the stains looked very unappealing. Microbiological tests showed that the microbes belong to black mold fungi *Gonatobotryum* sp. and *Epicoccum* sp. (see above, in Black mold section).

Case Study 2: Deck as a Mold Incubator

The owner (Milwaukee, WI) built an elevated—about 5 ft high—composite deck completely enclosed with the same composite boards, thereby creating a box perfectly insulated from any air movement. There were no windows in the box. The top of the box, that is, the deck itself, was pretty warm on a warm day, and rainwaters dripping into the box created a warm and very humid environment. The "ceiling" of the box, that is, the bottom side of the deck, was warm to touch from the inside of the box. After several months, the owner noticed small dark spots on the "ceiling" of the box, when she looked from the inside of her house through a small door, normally closed. There were no spots on the sidewalls of the box.

Figure 13.12 Black-colored metabolic products of *Gonatobotryum* sp. and *Epicoccum* sp. on the surface of composite handrails.

A microbiologist that visited the place called it "a perfect incubator" for mildew.

A microbiological analysis has identified two different types of fungi in the samples: *Alternaria alternata* and *Papulospora sp.* (description is given below).

The owner was advised to make a side window in the "box." After two windows were cut, on opposite sides of the deck, mold disappeared rather soon.

Case Study 3: Black Mold due to Low Density of a Composite Material and High Moisture

In the August of 2004, the owner noticed black spots on a composite railing post and balusters. This was the only reported case (see Figs. 13.9 – 13.11) out of more than 30 thousand decks by the same brand installed. The affected deck was in Maryland.

Observations of the affected materials under the light microscope has shown that it is a microbial contamination, and the black spots are located on calcium carbonate particles (fillers) and rice hulls fibers. In some cases, the black substance covered calcium carbonate or rice hulls particles completely.

A simple dry wire brush was performed with a follow-up wetting the surface with water to clean the majority of dust from the profiles. After several passes with a wire brush and a wet towel, the surface was completely clean of black spots.

Further investigation has shown that the affected composite profiles did not contain antioxidants and, hence, had low density (specific gravity), such as 1.08 g/cm^3 rather than typical 1.16–1.20 g/cm^3. As it is discussed in Chapter 15, lack of antioxidants lead to plastic degradation during processing, which in turn results in VOC formation, increase of porosity, and low density of the composite. This porous material absorbed much more water, namely 3–4% w/w compared to the standard amount of 1.0% and provided more likely source for mold growth, particularly in the upper layer of the profile.

MICROBIAL INFESTATION OF WOOD–PLASTIC COMPOSITE MATERIALS

Requirements for Microbial Growth on Wood and Wood–Plastic Composites

It is generally recognized that there are a number of requirements for microbial growth on wood and WPCs:

- Digestible substrate (cellulose fiber, plasticisers, and other "consumable" additives, as well as low molecular weight fragments of plastic, with molecular weight lower than 500. They can form as a result of thermal- and photooxidation of plastics (see Chapter 15))
- Chemical growth factors (carbohydrates, metals, minerals, and other nutrients)
- Water (preferably 20–30% moisture content)
- Oxygen (for aerobic microorganisms)
- Favorable temperature (preferably 15–40°C)
- Favorable pH range (slightly acid).

These factors are largely self-explanatory. Low-molecular fragments are more accessible to microbial systems of fungi (which are often secreted by fungi and have to diffuse into the matrix they are attached to). The higher the temperature, the more active the enzymes, unless the temperature is too high (temperature maximum for their activity is often between 40 and 45°. C, that is, 100–110°F). Cellulolytic enzymes have pH optimum for their activity between 4 and 6, that is, in a slightly acidic area.

Two factors slow down, or even eliminate, altogether some microbial growth in WPC:

- Minerals as fillers in a relatively high amount (20–60%);
- Slightly alkaline pH (7.5–9). This can be a result of calcium carbonate, for example, used as a filler.

The reason is that the minerals serve as a shield physically protecting cellulose fibers from cellulolytic enzymes diffusing into the matrix. At pH 7.5–9, the enzymes have 10–1000 times lower activity, depending on the types of the enzymes, pH, and the dissociation constants of catalytically active groups in the enzyme active center.

Here are a number of features related to the microbial degradation or staining of WPCs:

- Microorganisms attach themselves to most surfaces and then multiply;
- Microorganisms stain the WPC surface with their metabolic products;
- Microorganisms consume primarily cellulose filler;
- Nutrients for microorganisms are available in the surrounding media;
- Water absorption by WPC as a factor in microbial degradation;
- There is no meaningful correlation between laboratory data on microbial degradation of WPC and the real world.

The latter is because laboratory data are typically obtained with specific and well-characterized microorganisms, while the real world provides all kinds of microorganisms, not necessarily including those employed in the laboratory. Laboratory data demonstrate general regularities of microbial degradation, such as effect of temperature, nutrients, pH, effect of antimicrobial agents, and so on. Each one of these factors can be very different in the real world.

SENSITIVITY AND RESISTANCE OF COMPOSITE MATERIALS TO MICROBIAL DEGRADATION: EXAMPLES

- The higher the content of wood, the faster the microbial degradation;
- The higher the moisture content, the faster the microbial degradation;
- The higher the amount of biocides, the slower the microbial degradation;
- Generally poor correlation between accelerated laboratory test data and the natural outdoor exposure.

Mold resistance of composite materials can be considered from different viewpoints:

(a) Based on standard laboratory microbial tests (such as ASTM D1413, D2017, or G21);

(b) Based on observations in the real world, that is, microbial effects on actual decks installed in various parts of the country;

(c) Based on knowledge accumulated from the first two items.

An example of the first approach was given in Ref. [2]. The AWPA (the American Wood-Preservers' Association) Soil Block test method has been employed. Samples were polyethylene-wood composite materials with wood content ranging from 50 to 70% (w/w). Weight losses as a result of microbial degradation were up to 18% when treated for 16 weeks with white-rot organism, *Trametes versicolor*, and up to 7% when treated for 12 weeks with brown-rot organism, *Gloeophyllum trabeum*, for composite samples with various wood content. White-rot fungi were more active due to the nature of wood in the composite, which was predominantly hardwood. Considering that the wood content was 50%, this yield a loss for the wood component of 36 and 14%, respectively, for a said time period. Light and scanning electron microscopy showed that in all of the samples that exhibited weight losses, the degradation was most severe on the surface that was in direct contact with the decay organism at the start of the test. The data were updated in a later communication [5] according to which weight loss of some WPC boards after 16 weeks of exposure to decay fungi was greater than 20%. This equated to a loss of the wood component of more than 40%.

The above findings are rather typical and often reproduced with various microorganisms and composite materials. It is well known and rather obvious that the more wood in the composite, the higher is the microbial degradation, with other conditions equal. For example, composites containing a 70/30 wood–HDPE blend was more susceptible to fungal attack, whereas two different 50/50 wood–HDPE composites experienced little or no attack in the same experimental conditions [6]. The authors, using scanning electron microscopy, found the presence of voids between the wood and plastic in the composites not exposed to fungi, whereas the same composites exposed to fungi revealed that the fungi had thoroughly colonized wood particles contacting with voids, and that fungal hyphae were also prevalent in the voids deeper in the composite, particularly with 70/30 wood–HDPE material.

Similarly, it was shown that wood–polypropylene composites with wood (20 mesh ponderosa pine particles) content 30, 40, 50, 60, and 70% lost weight under action of brown-rot fungus, *Gloeophyllum trabeum*, for 12 weeks in the following order: 0.2, 2,4, 3.6, 4.4, and 9.7%, respectively. Control pine sample lost 17.2% of its weight. In the presence of 3% of zinc borate (w/w), there was no weight loss [7].

In a recent example of similar studies very thin samples of composite boards, having as high as 60% wood flour and 40% of polypropylene, smeared with potato dextreose agar or malt extract agar (microbial growth media) have been employed. Those samples were inoculated with different types of fungi (brown-rot fungi, *Gloeophyllum trabeum* and *Postia placenta*, and white-rot fungus, *Trametes versicolor*) and were kept moist for 12 weeks' incubation. Concurrently, similar samples were employed in the soil block test (ASTM D 1413). Sample boards exposed to agar showed as much as 40–50% weight loss (at 70–90% moisture content), and those exposed to soil lost 40–45% weight (at 70–80% moisture content). In the both cases, there was practically linear correlation between moisture content and weight loss [8].

Studying microbial degradation of 50% WPCs by brown-rot and white-rot fungi (*G. trabeum* and *T. versicolor*, respectively), it was found that the decay was more pronounced as the wood particle size increased [9]. The authors found that decay in the pine-based composites was more sensitive to particle size than that in the maple-based composites. It is noticeable from the above examples that for the same soil block test, ASTM 1413 or similar to it, the weight loss figures are quite different, and vary from fractions of a percent to quite high numbers. Indeed, as it was shown, extruded composite material samples are much more sensitive to soil block test compared to compression-molded and injection-molded sample. After 12 weeks, the respective samples lost 6, 2, and 3% of their weight, and the same, but boiled, samples lost 23, 2.5, and 0.5% of their weight, respectively [10].

Even lubricants can contribute to mold growth. Zinc stearate (6%) noticeably increased mold susceptibility of the WPC, a blend of HDPE (24–28% w/w) and maple wood flour (70% w/w), and a clear dose response was observed. It was noticed that the higher lubricant content, the more was the mold growth [11].

It was described in Chapter 5 that coupling agents decrease water absorption by a WPC material. It can be conjectured that coupling agents can, therefore, decrease susceptibility of the composite to microbial degradation. Indeed, it was confirmed that low levels of coupling agents (such as 0.5% w/w) can decrease microbial degradation of a composite material (DuPont data).

Zinc borate (Borogard® ZB, U.S. Borax Inc.) in amounts of 0.5, 1 and 2% (w/w) in the composite materials practically stopped the microbial degradation in the latter, when exposed in the AWPA laboratory soil block procedure. In all the cases, weight loss values were less than 1.1%. Thus, the data clearly demonstrated the significant improvement in protection against fungal decay in the presence of the said amount of zinc borate. It was noticed, however, that the use of zinc borate did not completely control surface mold on WPC decks exposed in outdoor field tests, but improved the visual appearance of the composite boards [2].

Other data from independent sources indicate that zinc borate in less than 1% amount is not enough for outdoor protection of some composite materials, particularly in warm and humid environment. In these situations, zinc borate is needed in WPC between 2 and 5% (see Table 13.3) [11].

It is known that some composite deckboards (and other components of composite decking systems) show rather significant microbial degradation. This has never been observed with GeoDeck. There are about 30,000 GeoDeck decks built in the United States, and there is not a single known case of a microbial degradation of GeoDeck. There was not a single warranty claim in this regard.

The main reason, because GeoDeck contains mineral particles as part of its composition. About one-third of GeoDeck by weight is mineral fillers (calcium carbonate and kaolin/clay). Minerals form a natural shield against microorganisms and do not allow them to get into the board matrix, that is, a blend of plastic–fiber minerals. When we tested biological degradation of wood fiber compared to the mineral–wood fiber mix (that we actually have as a filler), the biodegradation (using cellulose enzymes as an agent) was about 50 times slower in case of the fiber–mineral mix. This shows how effective are minerals in slowing down biodegradation. Most of compositions in the deckboard business (and Trex among them) do not contain minerals.

Another very important reason is that GeoDeck has a low water absorption index, compared to many other composite materials. For example, pressure-treated lumber absorbs 24% of water being immersed into water for 24 h. Trex absorbs 4.3% water in the same conditions (sanded surface) and 1.7% (unsanded surface), according to their own published data. The latter figure was confirmed in our studies (1.8%). GeoDeck in the same conditions absorbs only 1.0% of water (brushed surface). The less the water in the material, the less is the chance for microorganisms to get in and proliferate. Microbes need water for their life cycle.

Overall, GeoDeck has a very good degree of mold resistance. We have not studied composite deckboards available in the market on a comparative basis with respect to mold resistance; however, the above considerations allows us to believe that GeoDeck is one of the most mold resistant materials in the market. Besides, we know that GeoDeck is the most fade resistant among all known deckboards in the market (we have done the study), partly based on the same reasons described above.

ASTM TESTS FOR MICROBIAL GROWTH AND DEGRADATION OF WOOD-PLASTIC COMPOSITES

ASTM D 1413 "Standard Test Method for Wood Preservatives by Laboratory Soil-Block Cultures"

Note of the author: This procedure was originally aimed at microbial degradation of wood, shaped in a form of milled cubes or blocks. The blocks were placed on top of soil that was pretreated, sterilized, and inoculated with certain fungi. Hence, it is knows as "soil-block cultures." Typically, wood is degraded quite noticeably in the test conditions (as examples, see Tables 13.1 and 13.2). WPC materials often (if not always) show "no weight loss" rank when tested in accord with ASTM D 1413 (Tables 13.1 and 13.2).

TABLE 13.1 Percent weight loss after exposure to brown-rot fungus *(Gloeophyllum trabeum)* for 12 weeks in accord with ASTM D 1413

Material exposed to fungus	Weight before the exposure (g)	Weight after the exposure (g)	Weight loss (%)
Southern yellow pine	4.66	3.06	34.3
	4.73	3.39	28.3
	4.66	3.12	33.0
	4.53	3.35	26.0
	4.61	3.22	30.2
	Average		30 ± 3%
Same, treated with CCA	3.43	3.13	8.75
	3.35	3.03	9.55
	3.55	3.27	7.89
	3.67	3.29	10.35
	3.50	3.20	8.57
	Average		9 ± 1%
GeoDeck, composite) material(deck board)	2.68	2.65	1.12
	2.66	2.66	0
	2.62	2.62	0
	2.62	2.62	0
	2.61	2.60	0.38
	Average		0.3 ± 0.5%

TABLE 13.2 Percent weight loss after exposure to white-rot fungus *(Trametes versicolor)* for 12 weeks in accord with ASTM D 1413

Material exposed to fungus	Weight before the exposure (g)	Weight after the exposure (g)	Weight loss (%)
Sweetgum sapwood	5.87	2.40	59.1
	5.17	1.01	80.5
	4.88	2.08	57.4
	5.21	1.00	80.8
	4.28	1.03	75.9
	Average		71 ± 12%
Same, treated with CCA	3.63	3.33	8.26
	3.63	3.31	8.82
	3.66	3.36	8.20
	3.77	3.44	8.75
	3.65	3.37	7.67
	Average		8.3 ± 0.5%
GeoDeck, composite material(deck board)	2.86	2.86	0
	2.88	2.88	0
	2.91	2.91	0
	2.88	2.88	0
	2.90	2.90	0
	Average		0%

In fact, the original procedure was directed no so much at just determining the extent of wood degradation, but by comparing various wood preservatives or the antimicrobial agents with each other. Quantitatively, the principal outcome of the ASTM was the determination of a minimum amount (a threshold concentration) of a biocide that is effective in preventing decay of selected species of wood by selected fungi. The biocide (or a preservative) is dissolved in water or a suitable organic solvent and applied to the tested wood block by means of its impregnation at well-defined and standard conditions.

The ASTM procedure describes in detail the conditioning of wood samples, dissolution of preservatives, impregnation, preparation of soil, bottles, sterilization procedures, inoculation of wood samples with one or more strains of wood-destroying fungi, incubation of samples and duration of the test, and so on. The principal readout of the test procedure is a loss of weight of the wood sample after its exposure to microbial cultures.

According to the test, wood blocks (or composite blocks, in our case) should be cubes milled to 0.75 in. (19 mm).

For softwood sapwood samples, the following fungi are recommended by the ASTM procedure:

- *Gloeophyllum trabeum* (brown-rot fungus, particularly tolerant to phenolic and arsenic compounds),

- *Lentinus lepideus* (particularly tolerant to creosote),
- *Poria placenta* (particularly tolerant to copper and zinc compounds).

For hardwood sapwood samples, the following fungi are recommended:

- The three above fungi,
- *Coriolus versicolor* (a white-rot fungus).

Generally, the ASTM test aims at comparing the same wood samples, uniformly impregnated with solutions of wood preservatives (biocides) in appropriate gradient concentrations so as to leave in the conditioned blocks of wood, after removal of the solvent, a range of retentions running from below to above the anticipated threshold. The lowest retention shall be low enough to permit fungus attack and consequent decay and definite weight loss for the various test fungi employed.

When applied to composite materials, the ASTM test aims at determining a degree of microbial attack, in terms of weight loss of the composite material, compared to wood samples—typically untreated with preservatives and treated with CCA. Often five samples of each kind are studied with each of the test fungi, typically brown- and white-rot fungi. Duration of the microbial treatment at the controlled conditions, described in the ASTM test, is for 12 weeks.

Examples: Wood In the above tests, weight loss under the attack of the brown-rot fungus (*Gloeophyllum trabeum*) on southern yellow pine (SYP) was 30% (ASTM 1413), while when a similar test—using the same wood specie and the same microorganism—was performed according to AWPA E10, weight loss was 37.6%, that is, close enough to the above figure. In the presence of 0.58% zinc borate, weight loss was 0.9% [12].

Untreated aspen showed weight loss of 24.5% under action of the brown-rot fungus, and 53.2% under action of a more aggressive white-rot fungus (*Trametes versicolor*). In the second case, in the presence of incorporated zinc borate (0.58%) weight loss was 8.3%.

Untreated birch showed 64.6% weight loss under the action of the white-rot fungus.

Examples: Wood–Plastic Composites A few results of such a testing can be found in actual specifications/descriptions, available in the literature:

- Trex "rating no decay"
- Boardwalk "no decay"
- Nexwood "$<3\%$ (high resistance)"; 1.4–1.8 (scale 1 = excellent; 4 = very bad)
- TimberTech "No decay"
- EverX "equivalent or superior to preservative treated lumber"

- Timberlast™ "equivalent or superior to preservative treated lumber"
- GeoDeck "no decay" (see Tables 13.1 and 13.2)

Notes of the author: Most of the composite deck board materials show good fungus resistance when tested in accord with the ASTM D 1413 procedure, against "standard" white- and brown-rot fungi. In fact, I have never seen any ASTM D 1413 test data that would have shown a problem with any tested commercial composite deck board.

A much higher microbial resistance of WPC materials compared to untreated wood or wood treated with common preservatives can be explained by encapsulation of cellulose fiber in WPC with plastic. Though the encapsulation is not complete, as it was described above in this chapter, it nevertheless decreases the microbial attack to such an extent that the conditions of ASTM D 1413 do not allow the microbial degradation to be noticeable. In the real world, however, with a vast variety of cellulose-degrading microorganisms and much longer exposure time, microbial degradation of WPC sometimes becomes alarming. "Customer concern" Web sites contain dozens of descriptions of mold on WPC materials.

One desperate composite deck user wrote from Kentucky in February 2005: "I am writing to make sure as many people know about the problems I and others are having with (that composite material—AK). I built a deck in 2003 that gets full sun all day long. This past summer it developed mold spots all over it. ... When I contacted the manufacturer they said that mold was an act of god and that they weren't responsible."

In conclusion, though ASTM D 1413 is generally useful for a comparison of WPC materials with standard wood samples, it does not predict actual resistance of composite materials with respect to microbial degradation in the real world.

Notes of the author: A number of testing of wood and WPC sensitivity to microbial attack has been done using the American Wood-Preservers' Association Standard E10 (AWPA E10, see above for an example), "Standard method of testing wood preservatives by laboratory soil-block cultures." Though the Standard is more detailed compared to ASTM D 1413, it essentially describes the same procedure. For determining the thresholds of new preservative formulations, the procedure requires to test a minimum of three species of both brown-rot and white-rot fungi on the same sample of treated wood. According to the E10 standard, test blocks should be cubes milled to 0.55 in. (14 mm) or 0.75 in. (19 mm). Recommended test fungi are *Gloeophyllum trabeum* (= *Lenzites trabea*) and *Poria placenta* (brown-rot), and *Coriolis versicolor* (= *Trametes versicolor*) and *Irpex lacteus* (= *Polyporus tulipiferae*) (white-rot). Besides these, E10 lists six more brown-rot fungi (*Lentinus lepideus, Coniophora puteana, Serpula lacrimans, Polyporus palustris, Poria incrassate*, and *Poria xantha*) and three white-rot fungi (*Pleurotus ostreatus, Xylobous frustulatus*, and *Trametes lilacino-gilva*) as additional microbial species for testing. Recommended duration of the test is 8 weeks for 0.55-in. blocks and 12 weeks for the 0.75-in. blocks. Calculations of weight-loss are the same as in ASTM D 1413.

ASTM D 2017 "Standard Method of Accelerated Laboratory Test of Natural Decay Resistance of Woods" (Discontinued)

The standard method was withdrawn in 2002. The main reason to describe it in this chapter is that this method was referenced in ASTM D 7031 ("Standard guide for evaluating mechanical and physical properties of wood–plastic composite products"), issued in 2004, and in ASTM D 7032 ("Standard specification for establishing performance ratings for wood-plastic composite deck boards and guardrail systems (guards or handrails)"), issued in 2005.

ASTM D 7031-04 says (5.21.1): "Resistance to fungal decay shall be determined in accordance with accepted methods. Test Method D 2017 (or its alternate Test Method D 1413) is commonly used for this purpose." It also says (Note 11): "This is an accelerated laboratory decay test. Results are subjective and comparisons between tests and materials should be used with caution. However, mean specimen weight losses greater than 5% or significantly greater than controls should be cause for concern."

Statements of the above paragraph are essentially repeated in ASTM 7032-05 (4.8.1): "Resistance to fungal decay shall be determined in accordance with Test Method D 2017 or D 1413." Note 4 in ASTM D 7032 repeats the Note 11 in ASTM D 7031.

The procedure of ASTM D 2017 is very similar to that of ASTM 1413, except for some technicalities. For example, in D 2017 wood samples should be sawed into blocks of 25 × 25 × 9 mm in size, while in D 1413 they should be cubes milled to 19 × 19 × 19 mm in size. Test fungi are the same in both ASTM procedures except *L. lepideus* was eliminated in D 2017. Duration of the test (microbial exposure period) is 12 weeks in D 1413, and a minimum of 8 weeks with subsequent weekly intervals for additional samples, until 60% weight loss or 16 weeks duration, whichever comes first. ASTM D 2017 classifies "Class of Resistance to a Specified Test Fungus" as follows:

• 0–10% weight loss	Highly resistant
• 11–24%	Resistant
• 25–44%	Moderately resistant
• 45% or above	Slightly resistant or nonresistant

The ASTM gives examples in each of these categories:

• Highly resistant or resistant	Redwood, western red cedar, white oak
• Moderately resistant	Douglas-fir, western larch
• Slightly resistant or nonresistant	Hemlocks, true firs, spruces, beech, birches

ASTM E 2180 "Standard Test Method for Determining the Activity of Incorporated Antimicrobial Agent(s) in Polymeric or Hydrophobic Materials"

This ASTM procedure is not in the mainstream of testing or studying of composite materials. Code documents do not refer it with respect to WPC building materials.

However, there certainly is a gap in our understanding of antimicrobial protection of composites, toward both microbial decay and microbial staining of the surface, even without microbial degradation of the matrix. Microbial staining makes products aesthetically unappealing and hazardous for health, particularly for people suffering from allergy.

The ASTM procedure determines an ability of microbes to stick to the surface of polymeric materials (containing incorporated biocides) and continue to inhabit it. The same procedure can be applicable to composite materials. Clearly, if the composite contains biocide, which diffuses to the surface, it would diminish microbial colonies on the surface. If, on the contrary, biocide is firmly "immobilized" within the matrix (e.g., if it is practically water-insoluble and does not move to the surface), it would not prevent microbial staining. Obviously, a biocide should not be too water-soluble and promptly leached out of the material, and it should not be practically irreversibly bound within the matrix. Actually, if the biocide quickly leaves the matrix, such as wet deck boards, the ASTM procedure would also be helpful in determining this fact.

ASTM E 2180 aims at comparing counts of microbial colonies at the surface in a control (untreated) sample and a sample made in the presence of biocides and incorporated into the matrix. A sample can be either freshly produced, or washed with water over a certain time period, or taken from the field.

According to the procedure, a thin layer of the inoculated agar slurry (0.5–1.0 mL) is pipetted onto the test (matrix containing biocides) and control sample (no biocides). After 24 h (or other specified time, up to 96 h, for more resistant microorganisms), surviving microorganisms are recovered via elution of the agar slurry inoculum from the test substrate, serial dilutions are made, poured or spread on agar plates, incubated for 48 h, and microbial colonies are counted and recorded. Three bacteria are recommended by the ASTM procedure: gram-positive *Staphylococcus aureus* and gram-negative *Pseudomonas aeruginosa* or *Klebsiella pneumoniae.*

ASTM G 21 "Standard Practice for Determining Resistance of Synthetic Polymeric Materials to Fungi"

This practice covers the determination of the effect of fungi on the properties of synthetic polymeric materials. Technically, any composite material, shape and profile can be tested in accordance with this practice. Any particular property of plastics can be chosen as a readout, such as changes in optical, mechanical, and electrical properties.

Five test fungi in preparing the cultures are recommended as follows:

- *Aspergillus niger*
- *Penicillium pinophilum*
- *Chaetomium globosum*
- *Gliocladium virens*
- *Aureobasidium pullulans*

As a number of other organisms may be of specific interest for certain polymeric materials, other pure cultures of organisms may be used in accordance with the ASTM standard practice.

Test specimens may be a 50 × 50 mm (2″ × 2″) piece, a 50-mm (2″) diameter piece, or a piece at least 76-mm (3 in.) long cut from the material to be tested.

The standard length of the test is 28 days of incubation (or less, for samples exhibiting a growth rating of two or more). If the test is for visible effects only, the ASTM standard practice gives a list of rating as follows (regarding observed growth on specimens):

- None — Rating 0
- Traces of growth (less than 10%) — Rating 1
- Light growth (10–30%) — Rating 2
- Medium growth (30–60%) — Rating 3
- Heavy growth (60% to complete coverage) — Rating 4

EFFECTS OF FORMULATION ON SENSITIVITY AND RESISTANCE OF WOOD–PLASTIC COMPOSITES TO MICROBIAL DEGRADATION

It was described above that the minerals in WPC materials serve as a shield preventing microbial cellulolytic enzymes to get to the cellulose fiber. Higher density of composite profiles (hence, low porosity) makes lower water absorption and decreases the chances for microbial degradation of the composite materials. Higher density can be achieved using antioxidants (that prevent formation of volatiles during the processing of hot melt) and/or employing vented extruders. Another way to increase density of products is to slow down extrusion rate; however, this is typically not an economical way of operation. Certainly, resistance of WPC to microbial degradation can be increased by the introduction of biocides, and it will be discussed later in this chapter.

At the lack of proper ventilation, microbes can occupy practically any surface, WPC or not (Fig. 13.13). Several flat pieces of HDPE on my backyard were covered with green algae, namely the backsides, not accessible by the direct sunlight. No wonder, WPC can carry spots of mold when placed in a warm and wet environment, particularly with no draft. Mold grows even indoors, on shower bath plastic curtains. Even when no cellulose fibers are available for microbes (such as with flat pieces of neat HDPE), they consume pollen and other airborne particles. Brushed WPC boards always contain cellulose-derived dust on their surface.

BIOCIDES USED (ACTUALLY OR UNDER CONSIDERATION) IN WOOD–PLASTIC COMPOSITES

Zinc Borate (e.g., Borogard [U.S. Borax], Fiberguard [Royce International])

Zinc borate is the most popular additive to wood–fiber composite materials, primarily because of its relatively low cost. A typical cost of zinc borate is around

Figure 13.13 Mold on a composite deck board.

\$1.30–1.50 per pound. As the realistic amount of zinc borate in composite materials to be effective (though, not necessarily universally effective) is 2–3% by weight, the overall expenses for this additive are 2.6–4.5 cents per pound of a composite material, or 10–15% of a total cost. This is not a low figure. However, it may be much less than that for other biocides, offered on the current market by their manufacturers.

Borates (such as boric acid, borax, disodium octaborate tetrahydrate) have been used as broad-spectrum wood preservatives for over 50 years. Zinc borate has a shorter history as wood preservative, but it is well accepted in that capacity. There is a belief that, generally, zinc borate is more effective against fungal decay compared to mold inhibition, though systematic studies, using different composite materials, are apparently lacking.

Zinc borate is barely soluble in water. Therefore, it is much more resistant to leaching compared to many other biocides.

Zinc borate has a complex chemistry of its dehydration. An overall chemical equation can be presented as follows:

$$2ZnO{\bullet}3B_2O_3{\bullet}3.5H_2O \rightarrow 4ZnO{\bullet}B_2O_3{\bullet}H_2O \rightarrow 2ZnO{\bullet}3B_2O_3$$

Thermogravimetric analysis of various forms of zinc borate shows that regular zinc borate (such as Borogard ZB®, U.S. Borax Inc., Valencia, CA) does not dehydrate until approximately 290°C (554°F). An active release of water starts at about 330°C. The forming $4ZnO{\bullet}B_2O_3{\bullet}H_2O$ further dehydrates at about 490°C, resulting in an anhydrous form $2ZnO{\bullet}3B_2O_3$ (a private communication from Steven Kulbieda, U.S. Borax Inc.). Thus, zinc borate does not decompose at compounding and extrusion of polyolefin-based WPC.

FiberGuard-ZB (Zinc Borate, Royce International) is recommended by the manufacturer to be combined with wood particles before mixing with the plastic. According to the manufacturer, its loading must be at least 0.5% and not exceed 10% (Technical Data Sheet). Its mean particle size is typically in the range of 10 μm, density 2.8 g/cm^3, and its solubility is less than 0.28% by weight.

There were reports that zinc borate can be an effective preservative against microbial degradation of composite materials in amounts of 1% and less. At 3%, it was reported to protect against surface mold problems. In some cases, 1.8–2.0% of zinc borate was shown to protect WPC test decks during 12–14 month exposure [5]. There are some data that WPC samples containing zinc borate and exposed in the field for more than 3 years have exhibited "minimal depletion" from the surface [5]. However, systematic studies on zinc borate preventing surface mold are apparently not known.

Other data on zinc borate efficiency against molds are less optimistic. For example, 1% of zinc borate practically did not effect mold development on a 60-mesh maple wood flour (70% w/w)-filled HDPE (Table 13.3). Only 3 and 5% of zinc borate were effective (Table 13.3).

It appears that zinc borate is effective in tests conducted in accord with ASTM D 1413 and the similar ones, such as AWPA E10 (American Wood Preservers' Association. 1999. E10–91; Standard Method of Testing Wood Preservatives by Laboratory Soil-Block Cultures. AWPA, Cranbury, TX, pp. 397–407). That is, against well-defined cultures such as brown-rot fungus, *Gloeophyllum trabeum*. Efficiency of zinc borate incorporated in composite materials in the real world is much less known. For example, while zinc borate (Borogard ZB®) was found to completely suppress the weight loss of a polypropylene–wood composite material by *G. trabeum* (60% PP, 37% ponderosa pine particles of 20 mesh size, 3% Borogard ZB®, compression molding in laboratory conditions), it was apparently

TABLE 13.3 Effect of zinc borate (in the WPC composition) on mold development on a 60-mesh maple wood flour (70% w/w)-filled HDPE [11]

	Mold rating(1, 2—hardly noticeable; 3, 4—slightly noticeable; 5–7—very noticeable; 8,9—severe coverage; 10—completely covered by dense mold fungi)			
Week	Zn Borate content in the WPC, % (w/w)			
	0	1.0	3.0	5.0
1	1	0	0	0
2	4	4	0	0
3	6	6	1	0
4	7	8	1	1
5	8	8	1	1
6	9	9	1	1
7	10	10	1	1
8	10	10	1	1

ineffective in a ground-contact field test in Hawaii [7]. Furthermore, it was found that zinc borate is effective in laboratory conditions in 1% concentration, while in the field conditions it was ineffective in 1, 3, and 5% concentrations. It left unclear what was the main reason of such inefficiency of zinc borate—lack of inhibition of given (local) microorganisms, premature leaching (a rapid leaching of zinc borate was found from the stakes above the ground, at the ground line, and below the ground line, with boron depletion between 50 and 75% from the initial amount), or both.

Regarding an efficiency of zinc borate in the soil block test, it was shown that with brown-rot fungi, *G. trabeum*, acting on extruded HDPE composites containing 50% pine wood floor, zinc borate effectively suppressed weight loss as follows: 47% suppression with 0.2% zinc borate, 53% suppression with 0.5% zinc borate, and 72% suppression with 1% of zinc borate [10].

Zinc borate does not completely suppress black mold fungi *Gonatobotryum* sp. and *Epicoccum* sp. Both fungi excrete dark-colored (black) metabolic products. Both were isolated from the surface of a GeoDeck composite handrail (see Fig. 13.10, Case 1). When cultured on agar plates, the first fungus in standard conditions had spread a dark spot of 25–30 mm in diameter, the second fungus spread to 12 mm. In the presence of 0.5% of zinc borate, the spots were 8–9 and 4–7 mm, respectively, but the second fungus had started to produce a brown color, apparently as a response to zinc borate, and spread it to 30 mm. With 1.0 and 1.5% of zinc borate, the spots were still 8–9 mm with the first fungus (i.e., there was no change from 0.5 to 1.5% of zinc borate), while the second fungus still formed a black spot of 3–5 mm in diameter, and a brown spot of 10–15 mm. Higher amounts of zinc borate gave the same results. Therefore, zinc borate partially inhibits, but does not stop the growth of the above black molds.

Introducing zinc borate (in amounts up to 3% w/w) to a WPC formulation does not effect the flexural strength and slightly increases flexural modulus (stiffness) of the materials (Table 13.4), which is understandable, as ZnB is an inorganic material. Besides, zinc borate in this amount does not effect the water absorption, as well as the postmanufactured shrinkage and the coefficient of expansion–contraction (Table 13.5).

Table 13.5 shows that water absorption by a WPC board in the presence of 3% of zinc borate is about the same as that without ZnB after 24 h, and can be slightly

TABLE 13.4 Effect of zinc borate on flexural strength and modulus of GeoDeck boards

Zn Borate (%)	Flex strength (psi)	Flex modulus (psi)
0	3200 ± 40	266,000 ± 4000
1.0	3320 ± 10	272,000 ± 7000
2.0	3350 ± 70	292,000 ± 1000
3.0	3320 ± 20	296,000 ± 45,000

TABLE 13.5 Effect of zinc borate on some properties of GeoDeck boards

Zn Borate (%)	Shrinkage (%)(±0.2)	Water absorption (%) 24 hrs(±0.05)	Water absorption (%) 7 days(±0.05)	Coefficient of expansion–contraction, $\times 10^{-5}$ 1/F(±0.2)
0	0.29	0.84	1.97	3.9
1.0	0.30	0.79	2.14	3.9
2.0	0.30	0.70	1.90	3.9
3.0	0.28	0.81	2.22	3.7

higher than that without ZnB after 7 days. After 2-week water soak, an HDPE-based WPC with ZnB can absorb noticeably more water compared with that without zinc borate [13]. Testing of seven WPC materials with different cellulosic fibers (rice, pine, maple, kenaf, jute, hemp, coconut) with and without ZnB for 2 weeks underwater showed that in four WPC materials water absorption increased in the presence of ZnB, in two there was no change, and in one sample water absorption was slightly lower in the presence of ZnB. The largest effect (in absolute %) was with pine wood flour-filled HDPE, in which after 2 weeks, water absorption without ZnB was 7%, and in the presence of ZnB it was 9.5%. With rice hulls, it was 2.2% and 4.0%, respectively [13].

Barium Metaborate, Busan

Commercial forms of barium metaborate monohydrate, such as Busan® 11-M1 (Buckman Larorатories, Memphis, TN), have been used for over 40 years in coating applications. Among its useful properties, such as protection against corrosion, tannin staining, UV light, as well as being fire retardant, barium metaborate apparently protects against microorganisms. Its chemical formula is $BaB_2O_4{\bullet}H_2O$. Barium metaborate monohydrate is a white powder.

There are no data available to me on barium metaborate application in WPC materials. Though it might be considered for this purpose, its required effective level might reach 5–10%. This can be affordable only if besides its antimicrobial effect, barium metaborate would add more value to the composite material, due to its other useful properties.

Folpet, Fungitrol 11, Intercide TMP (carboximide)

Folpet is a protective leaf fungicide (this and the following references on Folpet—if not indicated—were taken from EXTOXNET, Extension Toxicology Network, Oregon State University). Its chemical name is N-[(Trichloromethyl)thio]phthalimide, and the chemical class is carboximide. It is of a relatively low molecular weight (297). Besides Folpet, it is known under trade names Cosan T, Faltan, Folnit, Folpel, Folpan, Folpex, Flatan, Fungitrol 11, Intercide TMP, Orthoraltan 50, Orthophaltan, Phthaltan, Phaltan, Sanfol, Spolacid, Trifol, Vinicoil, and Thiophal. Its mixed formulation include captafol, cymoxanil, dinocap, mancozeb, matalaxyl, and a mix of Folpet with aluminum phosethyl.

Folpet appears as colorless crystals. It is sold as powders and dusts. Generally, Folpet is available as an industrial biocide (Federal Register: August 25, 2004).

Folpet is practically insoluble in water (0.1 mg in 100 mL). Its solubility in isopropanol is 1.25 mg/100 mL (at 20°C), in benzene 2.2 mg/100 mL. Folpet melts at 177–180°C (351–356°F). Basic manufacturer of Folpet is Zeneca Ag Products, Wilmington, DE.

In the presence of water, Folpet slowly hydrolyzes at low temperatures, and rapidly at high temperatures.

Folpet inhibits cell division in many microorganisms. As a fungicide, it is used to control cherry leaf spot, rose mildew, rose black spot, and apple scab and is also used on berries, flowers, fruits, and vegetables for seed treatment.

As a fungicide, Folpet is used in paints and plastics for treatment of internal and external structural surfaces of buildings.

The acute oral toxicity of Folpet for rats was not reached at 10 g/kg, for mice it was 2440 mg/kg. It does cause skin irritation in rabbits; the acute dermal toxicity for rabbits was not reached at 5 g/kg according to one source, but according to another its dermal LC_{50} of 22.6 g/kg was reported for rabbits. Folpet is considered slightly toxic by ingestion. It is not considered as an eye irritant to rabbits. Acute inhalation exposure to Folpet may cause irritation of the mucous membranes. Inhalation of dust or spray mists and contact with the eyes can also result in local irritation.

Chronic toxicity, reproductive effects, and teratogenic, mutagenic, and carcinogenic effects of Folpet are described in the above reference (Federal Register). The half-life of Folpet in human blood is about 1 min; Folpet degrades rapidly to phthalimide and ultimately to phthalic acid and ammonia.

Folpet (as Fungitrol 11, minimum 88% of the active ingredient) was reported to be effective against a variety of microorganisms, including "pink-staining" microbes.

Table 13.6 shows the activity of Folpet for a number of microorganisms, both bacteria and fungi.

One can see that the principal ingredient of Fungitrol 11 (minimum 88% of Folpet) is a very potent antimicrobial agent, active in amounts below 0.1%, and sometimes below 0.01% (100 ppm), on both bacteria and fungi *in vitro*. Recommended amounts of Folpet in plastics and WPCs is between 0.25 and 0.75%.

Another study was performed on the effect of Fungitrol 11 on PVC films, 0.015 mc thickness placed on agar plates in Petri dishes. Each agar surface was previously inoculated or spray inoculated with mixtures of microorganisms. The test systems were covered and incubated for three weeks at 30°C for the mixed fungi (*Aspergillus niger, Penicillium islandicum, Chaetonium globosum, Trichoderma sp.*) and 1–2 days for the mixed bacteria (*Bacillus subtilis, Staphylococcus aureus, Brevibacterium sp., Pseudomonas aeruginosa*). At the end of the incubation periods, the samples were monitored for zones of inhibition and pink-staining.

TABLE 13.6 The minimal inhibitory concentration of Fungitrol 11 for a number of microorganisms in laboratory media [14]

Microorganisms	The minimal inhibitory concentration, (ppm) (1000 ppm = 0.01%)
Bacteria	
Bacillus subtilis	25
Brevibacterium sp.	25
Staphylococcus aureus	50
Aerobacter aerogenes	75
Proteus vulgaris	100
Streptomyces rubrireticuli	100
Pseudomonas aeruginosa	1000
Fungi	
Lentinus tigrinus	35
Penicillium funiculosum	50
Penicillium islandicum	50
Penicillium glaucum	50
Saccharomyces cerevisiae	50
Torula rubra	50
Torula utilis	50
Aspergillus flavus	75
Coniophora puteana	75
Aureobasidium pullulans	75
Aspergillus niger	100
Chaetomium globosum	100
Candida albicans	100
Polyporus versicolor	100
Rhodotorula mucilaginosa	100
Sclerophoma pityophila	150
Trichoderma sp.	500
Candida krusei	500
Alternaria alternata	750
Algae	20

The data (Table 13.7) show what might have been generally expected: leaching for 72 h does not effect the performance of Folpet because it is practically water-insoluble, and such a short time of immersion in water had not extracted the biocide; accelerated weathering (UV light and water) makes leaching (and maybe hydrolysis) of Folpet more pronounced, and the amount of the active biocide in the matrix is decreased; the higher the biocide concentration, the higher is its antimicrobial effect.

Biocides for WPCs should be fairly thermostable, in order to withstand compounding and extrusion temperatures. Table 13.8 shows a thermostability and efficiency of a number of biocides after they were exposed in a mix with molten

TABLE 13.7 Activity of Fungitrol 11 for the mixed bacteria and fungi (see the text) in agar plate tests

		Fungitrol 11		
Microbial system	Control	0.25%	0.50%	0.75%
Mixed bacteria				
As is	0	+	+	++
Leached for 72 h (in water)	0	+	+	++
Weathered for 100 h	0	+	+	+
Weathered for 200 h	0	0	+	+
Weathered for 300 h	0	0	+	+
Mixed fungi				
As is	0	+	++	+++
Leached for 72 h (in water)	0	+	++	+++
Weathered for 100 h	0	+	+	++
Weathered for 200 h	0	+	+	++
Weathered for 300 h	0	+	+	++
Pink staining test				
As is	0	+	++	+++
Leached for 72 h (in water)	0	+	++	+++
Weathered for 100 h	0	+	+	+
Weathered for 200 h	0	0	+	+
Weathered for 300 h	0	0	+	+

Accelerated weathering was performed in the regime of 102 min UV light/18 min UV light + water spray. Rating: 0—no effect; + slight; ++ moderate; +++ strong antimicrobial effect. The data were provided by ISP (International Specialty Products), Lombard, IL; Sherman Oaks, CA; Wayne, NJ.

plastic in a WPC for 10 min, and then tested *in vitro* with a mixture of fungal spores. One can see that concentration-wise Folpet was a third biocide in the efficience after Bethoguard (no fungal defacement in the presence of 0.25% or less of the biocide after four weeks) and Zn Pyrithione (no fungal defacement in the presence of 0.25% of the biocide). Folpet showed no fungal defacement in the presence of 0.50% of the biocide after 4 weeks. Chlorothalonil and IPBC were less efficient, and OBPA, DCOIT, and Zinc Borate were inferior compared with the above compounds (Table 13.8).

TABLE 13.8 Activity of various biocides after their thermal exposure for 10 min at 220°C and a subsequent incubation of WPC test plates (1.5 × 1.5 × 0.5 cm) in an agar plate inoculated with a mixture of fungal spores according to ASTM G21-96

Biocide	Concentration, % (w/w)	Weeks of exposure in the agar plate test 1	2	3	4
Control (no biocides)	0	0	8	9	9
Folpet	0.25	0	1	2	8
	0.50	0	0	0	0
Zn Borate (Boroguard ZB)	0.70	1	3	4	6
	1.00	1	2	4	4
	2.00	1	2	3	3
Zn Pyrithione (Zn Omadine)	0.05	0	6	10	10
	0.10	0	1	3	6
	0.25	0	0	0	0
IPBC (Fungitrol 400)	0.50	0	1	3	10
	0.75	0	0	0	1
	1.00	0	0	0	0
Chlorothalonil (Nuocide 960)	0.25	0	1	3	10
	0.50	0	0	0	0
OBPA (Vinizene BP 5-5)	0.60	0	0	2	7
	0.80	0	0	0	7
	1.00	0	0	0	5
DCOIT (Vinizene 4000)	0.50	0	1	6	9
	2.00	0	0	1	1
	3.50	0	0	0	0
Bethoguard (Bethoxazin)	0.25	0	0	0	0
	0.50	0	0	0	0

Number on the right-hand side show degrees of fungal defacement of the WPC samples after 1, 2, 3, and 4 weeks of fungal exposure [14].

Some commercial biocides, recommended for WPCs, contain about 30% of the active ingredient and cost around $4.00 per pound. As recommended amounts of active Folpet in plastics and WPCs are between 0.25 and 0.75% (on some other, more optimistic estimates, between 0.2 and 0.3%), it takes between 0.75 and 2.3% (more optimistically, between 0.6 and 0.9%) of the commercial Folpet-based biocide to be added into the formulation. It would cost between 3 and 9 cents per pound of the final formulation (more optimistically, between 2.4 and 3.6 cents per pound of the composite). In many cases, this expense is on a high side in composite materials, because it amounts to 15–50% of all raw material cost (optimistically, 10–15%). Unless the microbial degradation problem for the composite material is really noticeable, manufacturers can hardly afford such an expense.

A close analog of Folpet is Captan, which also belongs to carboximides. Its chemical name is *N*-trichloromethylthio-3*a*,-4,7,7*a*-tetrahydrophthalimide. It is of a relatively low molecular weight (301). Besides Captan, it is known under trade names Fungitrol C, Merpan, Orthocide-406, and Vancide 89.

Fungitrol C is sold by International Specialty Products (Wayne, New Jersey). It is thermostable up to 200°C, and the recommended application dose is 0.25–0.75% (w/w).

Chlorothalonil (tetrachloroisophthalonitrile), Nuocide 960

Similar to the compounds in the above section is Chlorothalonil, or tetrachloroisophthalonitrile, a member of the chloronitrile chemical family, and also a pesticide. It is an aromatic halogen compound, with a molecular weight of 266. An example of its efficiency is shown in Tables 13.8 and 13.9.

OBPA, Intercide ABF (10,10′-Oxybisphenoxyarsine), Vinizene BP 5–5

OBPA is a common abbreviation for 10,10′ oxybisphenoxyarsine, or 10,10′-oxydiphenoxyarsine. Its melting point is 184–185°C. OBPA is practically insoluble in water (5 ppm, or 0.0005%, or 0.5 mg per 100 mL at 20°C). Its LD_{50} in rats equals to 35–50 mg/kg, that is, its toxicity is rather high.

Akcros Chemicals (in the United States—Akcros Chemicals America, New Brunswick, NJ) makes a few brands of fungicidal products based on OBPA as the active ingredient, all under a name Intercide ABF, which use PVC or PVC/PVA copolymer as carriers, among other liquid and solid carriers. Activity of one of those products, Intercide ABF-2-DIDP, in two different tests is shown in Table 13.10. Other set of data is shown in Table 13.8.

The data in Table 13.10 show a high level of performance of the biocide against all the test fungi. *A. niger* was more resistant in the Disk Test, but equally well suppressed in the NSA Test.

TABLE 13.9 Effect of chlorothalonil (in the WPC composition) on mold development on a 60-mesh maple wood flour (70% w/w) -filled HDPE [11]

	Mold rating (1, 2—hardly noticeable; 3, 4—slightly noticeable; 5–7—very noticeable; 8,9—severe coverage; 10—completely covered by dense mold fungi)			
Week	Chlorothalonil content in the WPC, % (w/w)			
	0	0.5	1.0	1.5
1	1	0	0	0
2	4	0	0	1
3	6	1	0	1
4	7	3	0	1
5	8	6	1	2
6	9	7	2	2
7	10	7	3	2
8	10	8	3	3

TABLE 13.10 Activity of an OBPA-based biocide (Intercide ABF-2-DIDP) versus single-test fungi, NSA, and Disk tests (plastisol folmulation)

	Zone of inhibited germination (mm)				
Test	*Penicillium funiculosum*	*Paecilomyces variotti*	*Aspergillus niger*	*Trichoderma longibrachiatum*	*Chaetomium globosum*
Control (no biocide)	0	0	0	0	0
NSA Test	>32	29	>32	>32	30
Disk Test	15	12	7	11	13

Concentration of the active ingredient 500 ppm (0.05%), or 2.5% per biocide with the carrier.
Note 1: Intercide ABF-2-DIDP consists of a 2% solution of OBPA in DIDP (a plasticiser).
Note 2: NSA Test and Disk Test are internal Akcros methods, both performed with mixed fungal spores. Data were provided by Akcros Chemicals.

Minimum inhibition concentration (ppm of the active ingredient in PVC film) against a number of bacteria (*Citrobacter freundii, Escherichia coli, Proteus vulgaris, Staphylococcus aureus, Salmonella choleraesuis, Enterococcus faecalis, Pseudomonas aeruginosa, Klebsiella pneumoniae*) varied between (≤200 and 400–600 ppm (≤0.02–0.06%). It should be noted here that these bacteria are practically not relevant to WPCs, but rather to plasic-made biomedical devices. This example just shows a range of active antibacterial concentrations of the biocide.

It should also be noted that biocides made for applications in biomedical area are typically much more expensive than those designed for use in WPCs. A cost of the former often reaches as much as $50/lb, that is, 40–50 times higher compared to the WPC biocides. Being applied at 1 or 2.5% by weight at $50/lb, biocides (active ingredients plus the carriers) would add 50 cents or $1.25 per pound of the composite material, which often exceeds two to six times the total cost of a formulation in composite materials. 1 or 2.5% of a biocide by weight is not much, because the active ingredient typically constitutes a very small fraction in a commercial biocide, with carriers, fillers, and stabilizers that constitute the rest.

For example, the above described biocide Intercide ABF-2-DIDP contains 2% of the active ingredient; that is, 500 ppm (0.05%) of the latter contains in 2.5% of the fungicide (active ingredient plus the carrier) in a composite formulation. At 2.5% w/w, even $2/lb cost for a biocide, or 5 cents per pound of the composite, would make the cost rather prohibitive.

However, according to the published material by Akcros Chemicals "Intercide® ABF" (Product Bulletin INT1/E2), Intercide ABF-2-DIDP is recommended to be added to PVC formulations at 2.5% of the total formulation.

A similar biocide, Intercide ABF-5-DIDP in the amount of 0.05 and 0.1% (w/w) of the active principle (OBPA), that is, 1.0 and 2.0% (w/w) of the commercial biocide, was tested in a woodfiber–PVC composite (data obtained courtesy of Akcros Chemicals). The unprotected composite (no biocides) did not prevent an active fungal growth in the testing system. The same was observed with the composite containing 1% of the biocide (0.05% of the active principle), that is, there were no zone of inhibited germination around the sample. At 2% of the biocide in the composite (1000 ppm or 0.1% of the active principle), the zone of the inhibited germination extended to 10–30 mm. The authors noticed that wood–PVC composition required more biocides compared to the neat PVC for protection of the materials.

IPBC, Polyphase®, Troy®, Intercide IBF (2-iodo-2-propynyl-*n*-butylcarbamate, 3-iodo-2-propynyl-*n*-butylcarbamate)

IPBC is a common abbreviation for 3-iodo-2-propynyl-*n*-butylcarbamate. It was initially developed and mostly used as a pesticide. IPBC is an off-white crystalline solid, slightly soluble in water (156 ppm, or 0.0156% at 20°C), but soluble in most organic solvents and alcohols. Its melting point is 68°C. LD_{50} in rats (acute oral effects) equals to 1056 and 1795 mg/kg (females and males, respectively), LD_{50} in rabbits (acute skin effects) is higher than 2000 mg/kg. IPBC causes a slight irritant effect on rabbit skin.

Troy Corporation (Newark, NJ) sells IPBC under the trade names Polyphase® P100 and Troy® EX685. Polyphase® P100 (amount of the active ingredient 97%) is recommended to be used in WPCs from 0.3 to 1% per total weight of a composite. Troy® EX685 (amount of the above ingredient 5–10%, plus 30–60% of zinc-containing inorganic and organic compounds) decomposes at 230°C (446 °F). The last compound is partly soluble in water (data received from Troy Corporation).

Akcros Chemicals (in the United States—Akcros Chemicals America, New Brunswick, NJ) makes a few brands of fungicidal products based on IPBC as the active ingredient, all under a name Intercide IBF, which use PVC or PVC/PVA copolymer as carriers, among other liquid and solid carriers. Activity of one of those products, Intercide IBF-8-DIDP, in two concentrations and in two different tests is shown in Table 13.11.

The data in Table 13.11 show a certain level of performance of the biocide against all the test fungi, though its level is somewhat below that for OBPA (Table 13.10).

A chemical isomer of IPBC is Fungitrol 400, 3-iodo-2-propynyl butylcarbamate. It is used as a mildewcide in many applications, including lumber treatment and plastic coating. Fungitrol 400 is stable up to 180°C, and its recommended application dose is 0.25 to 1.0% w/w.

OIT, DCOIT, Octhilinone, Micro-Chek, Intercide OBF (2-*n*-Octyl-4-isothiazolin-3-one)

OIT is a common abbreviation for 2-*n*-octyl-4-isothiazolin-3-one, or octhilinone. This is a liquid compound, boils at very low pressure (0.01 mm/Hg) at 120°C. Its LD_{50} in rats equals to: oral—457 mg/kg, inhalation (LC_{50}) 1.5 mg/kg, 4 h. Dermal LD_{50} (rabbit)—160 mg/kg. Another chemical variant of it is DCOIT, or Dichlorooctyl-4-isothiazoline.

TABLE 13.11 Activity of an IBPC-based biocide (Intercide IBF-8-DIDP) versus single-test fungi, NSA, and Disk tests (plastisol folmulation)

		Zone of inhibited germination (mm)				
Test	Conc. (%)	*Penicillium funiculosum*	*Paecilomyces variotti*	*Aspergillus niger*	*Trichoderma longibrachiatum*	*Chaetomium globosum*
Control (no biocide)	—	0	0	0	0	0
NSA Test	0.1	5	8	5	4	7
	0.2	26	13	>32	8	15
Disk Test	0.1	9	5	7	1	4
	0.2	14	11	13	4	4

In the "concentration" column, concentration of active ingredient is shown. Data were provided by Akcros Chemicals.

Ferro Corporation (Cleveland, OH) makes OIT under the brand name Micro-Chek, for applications as a mildewcide for PVC, polyurethane, and other polymer compositions. Micro-Chek is a 4% solution of the active ingredient in several different carriers. It is recommended to be used between 1 and 2.5% of the total formulation weight [15].

Akcros Chemicals (in the United States—Akcros Chemicals America, New Brunswick, NJ) makes a few brands of fungicidal products based on OIT as the active ingredient, all under a name Intercide OBF. Activity of one of those products, Intercide OBF-8-DIDP, in two concentrations and in two different tests is shown in Table 13.12.

The data in Table 13.12 show a certain level of performance of the biocide against all the test fungi, though its level is somewhat below that for OBPA (Table 13.10)

Zinc Pyrithione, Zinc Omadine, Intercide ZNP, Zinc Derivative of Mercaptopyridine 1-oxide

Zinc pyrithione, or zinc bis-(2-pyridinethiol-1-oxide) is a zinc derivative of 2-mercaptopyridine 1-oxide (in another nomenclature), or zinc omadine. It serves as an ingredient in *Head & Shoulders* shampoo. For oral administration, its LD_{50} in rats equals to 92–266 mg/kg, in mice 160–1000 mg/kg, and in dogs 600 mg/kg.

Akcros Chemicals makes Intercide ZNP, containing zinc omadine. In fact, a range of products are available, tailored to meet the requirements of customers in terms of active ingredient concentration and carrier material.

Minimum inhibition concentration (ppm of the active ingredient in PVC film) against a number of bacteria (*Citrobacter freundii, Escherichia coli, Proteus vulgaris, Staphylococcus aureus, Salmonella choleraesuis, Enterococcus faecalis, Pseudomonas aeruginosa, Klebsiella pneumoniae*) varied between ≤750 and 1000–2000 ppm (≤0.075–0.2%). Again, this example just shows a range of active

TABLE 13.12 Activity of an OIT-based biocide (Intercide OBF-8-DIDP) versus single-test fungi, NSA, and Disk tests (plastisol folmulation)

Test	Conc. (%)	Zone of inhibited germination (mm) *Penicillium funiculosum*	*Paecilomyces variotti*	*Aspergillus niger*	*Trichoderma longibrachiatum*	*Chaetomium globosum*
Control (no biocide)	—	0	0	0	0	0
NSA Test	0.1	22	16	4	7	11
	0.2	25	21	22	14	17
Disk Test	0.1	8	4	2	0	4
	0.2	9	10	5	3	7

In the "concentration" column, concentration of active ingredient is shown. Data were provided by Akcros Chemicals.

antibacterial concentrations of the biocide and illustrates that zinc omadine is about 2–5 times less efficient as an antibacterial agent compared with OBPA (see above).

Thiabendazole, Irgaguard F3000, 2-(4-Thiazolyl)-1*H*-benzimidazole, 4-(2-Benzimidazolyl)thiazole, Thiabendazole, MK-360, TBZ

Ciba (Tarrytown, NJ) that sells this well-known fungicide under a commercial name "Irgaguard® F3000" said that it is a broadly effective fungicide for wood–polyolefin and wood-PVC composites. This fungicide is conventionally used for spoilage control of citrus fruit, for treatment and preventing of Dutch elm disease in trees, and for control of fungal diseases of seed potatoes. A list of fungi, inhibited by this compound, contains 130 names [16].

BIOCIDES: ACCELERATED LABORATORY DATA AND THE REAL WORLD

It was recognized long ago that conflicting data may be obtained from accelerated laboratory tests and long-term outdoor exposure when determining the fungal resistance (or susceptibility). Furthermore, when data obtained from accelerated Petri dish tests are assumed to be valid in long-term exposures in the real world, conclusions and long-term decisions can be completely opposite and erroneous.

There are two main reasons to use laboratory tests such as described in the ASTM procedures described above. One, to compare the rate of microbial attack on selected wood, on one hand, and the composite material, on the other, by the selected fungus (or fungi, or bacteria). Another, to determine an antimicrobial efficiency of a chosen biocide against the selected mold species, and to find its "threshold" concentration. There is no scientific justification to extrapolate the observed data to a long-term exposure in the field. First, the exposure site contains a different set of microorganisms, in different amounts. Second, amount of the biocide incorporated into the composite material will diminish with time due to leaching, decomposition under UV light, chemical conversions, among others.

Every time when a comparative test is conducted in a laboratory and in the real world using the same composite material and the same biocide(s), data obtained provide very valuable information. Unfortunately, results of a very few tests of this kind are available.

An interesting and important study was performed recently by KibbeChem, Inc. and International Specialty Products [14]. Two sets of exposure of a wood–talc–plastic composite (35% HDPE, 45% 200 mesh white oak flour, 14.5% talc, 5.5% Structol TPW104) have been conducted, one was a chamber testing, according to ASTM D 3273, another test was performed outdoors, in the south Ohio (Figs. 13.14 through 13.19). Test panels of 5 × 15 cm were exposed on test fences for 2 years. It should be noted that none of the biocides tested were performed in the heavily stressed southern Florida Everglades exposure. All biocides failed to protect the boards at all concentrations tested in about 6 months [14]. In South Ohio and New Jersey Folpet at about 0.3% (w/w) offers protection (Fig. 13.14). Folpet also did not produce any significant effect on flexural properties of WPCs based on HDPE, polypropylene, and PVC (Tables 13.13–13.15).

The comparative data obtained in South Ohio are shown in Table 13.16, and another set of data obtained in New Jersey is shown in Table 13.17. In a few instances

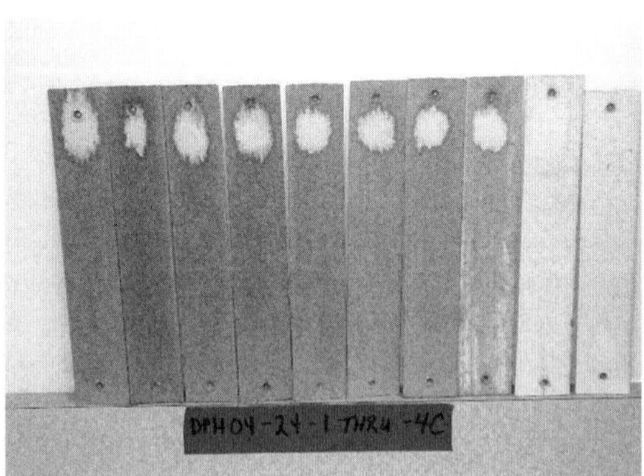

Figure 13.14 An outdoor (south Ohio) biocide two-year study using WPC (35% HDPE, 45% wood flour, 14.5% talc). Panels: 1 and 2 — control (no biocides); 3 and 4 — tributyltinoxide (TBTO) 0.05% and 0.1%; 5, 6, and 7 — zinc omadine 0.1%, 0.2%, and 0.3%; 8, 9, and 10 — folpet 0.1%, 0.2% and 0.3%. Sodium hypochlorite solution (bleach) was applied to the top of the panels as a diagnostic indicator test for mold pigmentation [14].

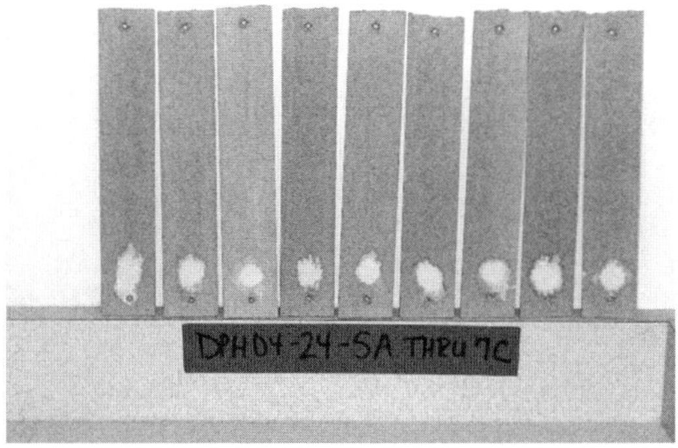

Figure 13.15 Conditions—as in Figure 13.14. Panels: 1, 2, and 3—Chlorothalonil 0.1, 0.2, and 0.3%; 4, 5, and 6—Bethoguard 0.05, 0.1 and 0.2%; 7, 8 and 9—Barium Metaborate 1, 2, and 3%.

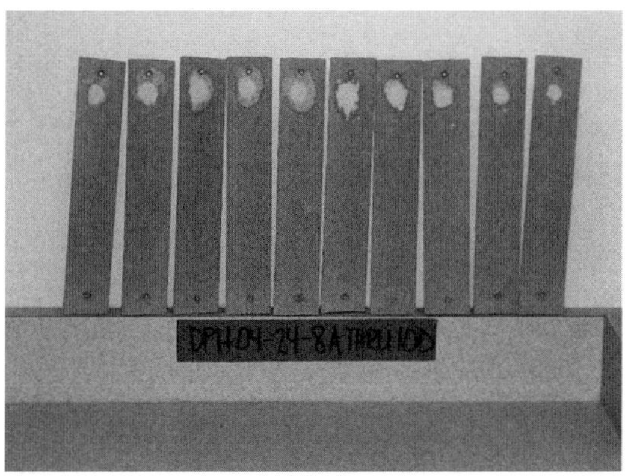

Figure 13.16 Conditions—as in Figure 13.14. Panels: 1, 2, and 3—Copper thiocyanate 0.1, 0.2, and 0.3%; 4, 5, and 6—Zinc caprylate 0.1, 0.2, and 0.3%; 7, 8, 9, and 10—"Biocide-1" (undisclosed) 0.5, 1, 1.5, and 2%.

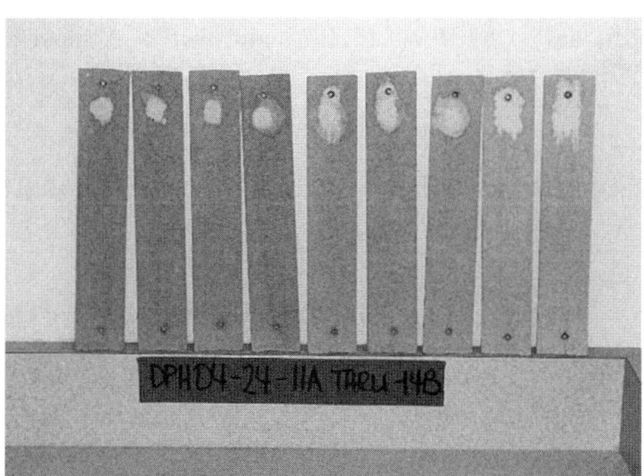

Figure 13.17 Conditions—as in Figure 13.14. Panels: 1, 2, and 3—"Biocide-2" (undisclosed) 0.5, 1, and 2%; 4 and 5—"Biocide-3" (undisclosed) 3 and 4%; 6 and 7—"Biocide-4" (undisclosed) 3% and 4%; 8 and 9—"Biocide-5" (undisclosed) 3 and 4%.

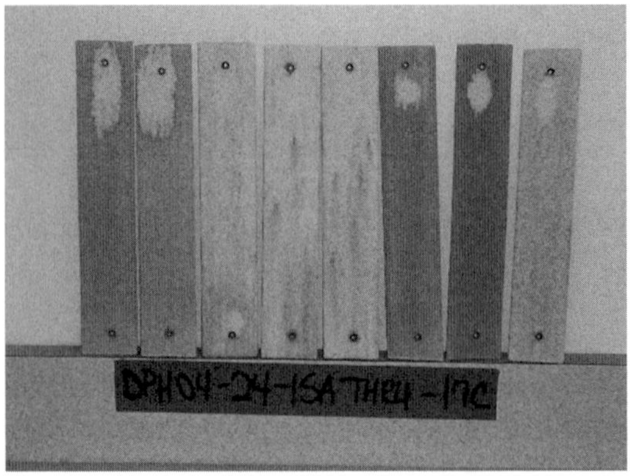

Figure 13.18 Conditions—as in Figure 13.14. Panels: 1 and 2—“Biocide-6” (undisclosed) 3 and 4%; 3, 4, and 5—Fungitrol 400 (IPBC) 0.1, 0.2, and 0.3%; 6, 7, and 8—“Biocide-7” (undisclosed) 2, 3, and 5%.

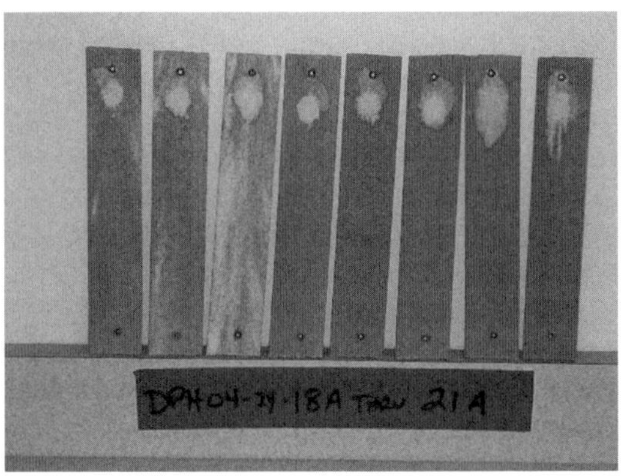

Figure 13.19 Conditions—as in Figure 13.14. Panels: 1, 2, and 3—“Biocide-8” (undisclosed) 0.5, 0.75, and 1%; 4, 5, and 6—OBPA 0.25, 0.75, and 1%; 7—Copper hydroxide 0.5%; 8—Doverbond 3000 5.5%.

TABLE 13.13 Effect of Folpet on flexural properties of HDPE-based wood-plastic composite [14]

Folpet, concentration in WPC boards (%), (w/w)	Flexural strength (psi)	Flexural modulus (psi)
0	1,876 ± 24	340,000 ± 60,000
0.3	1,834 ± 33	390,000 ± 9,000
0.5	1,716 ± 18	420,000 ± 80,000

Tests were conducted in accordance with ASTM D 6109. Standard deviations are rounded compared with those given in [14], to make the principal figures and standard deviations consistent with each other.

TABLE 13.14 Effect of Folpet on flexural properties of polypropylene-based WPC [14]

Folpet, concentration in WPC boards (%), (w/w)	Flexural strength (psi)	Flexural modulus (psi)
0	2,382 ± 40	470,000 ± 24,000
0.3	1,962 ± 40	390,000 ± 20,000
0.5	2,037 ± 32	400,000 ± 26,000

Conditions as in Table 13.13.

of a biocide activity, there was some correlation between the two tests ("artificial" and "natural"). For instance for Fungitrol 400, between 0.1 and 0.3% (w/w) of the compound, an excellent biocidal activity was observed both in a chamber and in an outdoor test in South Ohio, but not in New Jersey, with a different natural set of microorganisms. However, in most cases there was no correlation, neither in South Ohio nor in New Jersey.

For example, in South Ohio, Folpet showed excellent data in the natural environment, and practically no activity in the ASTM D 3273 test; copper thiocyanate and zinc caprylate, on the contrary, showed excellent "chamber" activity, but no activity at all in the natural setup. OBPA at 1% w/w also showed excellent activity in the chamber, but no activity outdoors.

In the New Jersey test there was also very poor correlation between biocide protection in the environmental chamber and outdoors. Many biocides showed excellent indoors protection, such as OBPA, copper hydroxide, chlorothalonil, bethoguard, but none of them showed fairly noticeable protective activity in the natural conditions (Table 13.17). Only two biocides, IPBC and Folpet, have shown what can be considered as a fair, but still not very good correlation of their biocidal activity in the environmental chamber and outdoors.

The lack of correlation between natural and laboratory exposure has a simple explanation. Many biocides are rather specific to the fungi, employed in the ASTM tests (see above), but they are not active toward many other natural microorganism, causing degradation of WPC materials. Other biocides possess a broad spectrum of toxicity toward variable microorganisms. Those are evidently preferred in composite formulations.

TABLE 13.15 Effect of Folpet on flexural properties of PVC-based WPC [14]

Folpet, concentration in WPC boards (%), (w/w)	Flexural strength (psi)	Flexural modulus (psi)
0	5,620 ± 140	850,000 ± 36,000
0.3	5,670 ± 50	810,000 ± 26,000
0.5	5,500 ± 120	880,000 ± 16,000

Conditions as in Table 13.13.

TABLE 13.16 Comparison of biocides in an indoor chamber study (ASTM D 3273) and in outdoor study (south Ohio, natural environment)

Biocide	Amount (weight %)	Performance: Indoor (ASTM D 3273)	Performance: Outdoor (natural environment)
Control	—	0	0
TBTO (tributyltinoxide)	0.05	0	0
	0.1	0	0
Zinc Omadine	0.1	+	0
	0.2	+	0
	0.3	+	0
Rohm & Haas SB-1-PR (OBPA)	0.25	0	0
	0.75	+	0
	1	+++	0
Bethoguard	0.05	0	0
	0.1	+	0
	0.2	+	0
Barium metaborate	1.0	0	0
	2.0	0	0
	3.0	0	0
Copper thiocyanate	0.1	+	0
	0.2	+++	0
	0.3	+++	0
Zinc caprylate	0.1	+	0
	0.2	++	0
	0.3	+++	0
Copper hydroxide	0.5	+	+
Doverbond 3000	5.5	0	+
Folpet	0.1	0	++
	0.2	0	+++
	0.3	0	+++
Fungitrol 400(IPBC)	0.1	+++	+++
	0.2	+++	+++
	0.3	+++	+++

Rating: 0—no noticeable effect; +—fair, ++—good; +++—great. The biocides are listed from the least efficient to the most efficient (Folpet) in an outdoor test.

TABLE 13.17 Comparison of biocides in an indoor environmental chamber study (ASTM D 3273) and in outdoor study (Piscataway, NJ, natural environment)

Biocide	Amount (weight %)	Performance: Indoor (ASTM D 3273) after 4 months	Performance: Outdoor (natural environment) after 3 months
Control	—	5	10
TBTO (Tributyltinoxide)	0.025	4	10
	0.1	5	10
Zinc Omadine	0.1	4	10
	0.3	4	7
Copper Thiocyanate	0.1	9	10
	0.3	7	10
Zn Caprylate	0.1	9	10
	0.3	6	10
OBPA	0.25	3	10
	0.50	1	10
Copper Hydroxide	0.5	2	10
Chlorothalonil	0.2	2	9
	0.3	0	7
Barium Methaborate	1.0	10	9
	3.0	5	9
Bethoguard	0.1	1	9
	0.2	0	9
Thiabendazole	2.0	2	8
	3.0	2	7
DCOIT	0.50	1	6
	0.75	1	4
IPBC (Fungitrol 400)	0.1	0	3
	0.3	0	2
Folpet	0.1	4	1
	0.3	3	0

Rating on a scale from 0 to 10, where 10 corresponds to complete coverage by surface microbial growth [14]. The biocides are listed from the least efficient to the most efficient (Folpet) in an outdoor test.

REFERENCES

1. W. Wang and J.J. Morrell. Water sorption characteristics of two wood-plastic composites. *Forest Prod. J.*, 2004, **54**(12), 209–212.
2. M. Manning. Borates as biocidal additives for WPC. In: *The Global Outlook for Natural Fiber & Wood Composites 2003*, New Orleans, LA, December 3–5, 2003.

3. U.S. Pat. No. 6,495,134. B.L. Illman, V.W. Yang, L.A. Ferge.
4. http://www.emlab.com/app/fungi/Fungi.po
5. M. Manning. Creating value in WPC products with antimicrobials and stain resistant additives. *WPC Conference 2004*, Principia Partners, Cleveland, OH, Baltimore, MD, October 11–12, 2004.
6. M. Mankowski and J.J. Morrell, *Wood Fiber Sci.*, 2000, **32**(3), 340–345.
7. S.A. Verhey and P.E. Laks. Strength loss following fungal attack on wood fiber/thermoplastic composites. In: *Progress in Woodfibre–Plastic Composites, Conference Proceedings*, Toronto, Canada, May 23–24, 2002.
8. J.A. Silva Guzman, J.J. Morrell, and B.L. Gartner. Comparison of agar (MEA and DPA) and soil block tests for assessing decay of wood plastic composites. In: *Eighth International Conference on Woodfiber Composites*, Madison, WI, May 23–25, 2005.
9. S.A. Verhey and P.E. Laks, Wood particle size affects the decay resistance of woodfiber/thermoplastic composites. *Forest Prod. J.*, 2002, **52**, 78–81.
10. C. Clemons and R. Ibach. Fungal exposure of wood filled polyethylene composites: laboratory evaluation. In: *Progress in Woodfibre–Plastic Composites, Conference Proceedings*, Canadian Natural Composites Council, University of Toronto, Toronto, Canada, May 23–24, 2002.
11. P.E. Laks, J.K. Vehring, S.A. Verhey, and D.L. Richter. Effect of manufacturing variables on mold susceptibility of wood-plastic composites. In: *Eighth International Conference on Woodfiber-Plastic Composites (and Other Natural Fibers)*. Forest Products Society, Madison, WI, May 23–25, 2005.
12. U.S. Pat. No. 6,368,529. J.D. Lloyd, M.J. Manning, and F.M. Ascherl. Lignocellulosic composite.
13. R.E. Ibach. Moisture and fungal evaluations for natural durability of compression molded fiber-plastic composites. In: *Progress in Wood-Fibre Plastic Composites 2006 International Conference*, Toronto, Canada, May 1–2, 2006.
14. W. Woods. Efficacy of a unique fungicide in wood filled plastic. In: *Conference Proceedings of Durability in Wood Plastic & Natural Fiber Composites 2006*, Intertech, Portland, ME, San Antonio, TX, December 4–5, 2006.
15. E.W. Flick (Ed.), *Plastic Additives*, Vol. 1, Noyes Publications William Andrew Publishing, Norwich, New York, 2001, p. 107.
16. G. Capocci, U. Stadler, and M. Reyes. Anti-staining and anti-fungal additives for woodfiber-plastic composites. In: *Conference Proceedings of the Global Outlook for Natural Fiber & Wood Composites 2004*, Intertech, Portland, ME, New Orleans, LA, December 8–10, 2004.

14

FLAMMABILITY AND FIRE RATING OF WOOD–PLASTIC COMPOSITES

INTRODUCTION

Acceptance criteria for deck boards and components of railing systems (ICC-ES' AC 174, effective July 2006) are based on ASTM D 7032 and specify the flame spread rating to be not greater than 200 when tested in accordance with ASTM E 84.

This chapter explains the meaning of the above statements. It describes flammability and smoke/toxic gases evolution at burning of wood compared to wood–plastic composite (WPC) materials and products of different compositions and profiles. It also explains flammability and fire ratings and indexes as quantitative measures for fire hazard and fire safety, and fire performance characteristics in general of wood and composites.

The building codes or fire codes regarding material requirements are based on three basic characteristics of materials: combustibility, flame spread, and fire endurance. Wood and most thermoplastic-based composites are combustible materials. For regulatory purposes, materials are classified according to their flame spread index (FSI). As it will be described in more detail below in the section "ASTM recommendations," for determining of flame spread index, materials are tested according to ASTM E 84 in a form of a 24-ft long and 20-in. wide assembled panel. This panel completes the top of the 25-ft long tunnel furnace. Hence, ASTM E 84 test is often called "the 25-ft tunnel test." The panel is ignited at one end and burned under specified forced draft conditions. The flame front position spread along the panel is recorded as a function of time. An FSI is calculated from these data using

Wood-Plastic Composites, by Anatole A. Klyosov

a prescribed formula. In other words, a FSI is a measure of the overall rate of flame spreading in the direction of a specified airflow. As set points, FSI for inorganic reinforced cement board surface is set as 0, and for select grade oak surface is arbitrarily established as 100 under the specified conditions (ASTM E 84, Section 9.2).

There are four basic categories, or classes, for FSI: Class A, with FSI between 0 and 25; Class B, with FSI between 26 and 75; Class C, with FSI between 76 and 200; and below Class C, with FSI above 200 (unclassified materials). Classes A, B, and C sometimes are called Classes I, II, and III.

In the same "25-ft tunnel test" the so-called smoke development index (SDI) can also be determined. The smoke measurement is based on the percentage of retardation of light passing through the tunnel exhaust stream and detected by a photocell, and then data obtained are converted to the SDI, with red oak flooring set at 100. AC 174 does not specify any particular SDI as the code requirement, but the industry generally considers SDI above 450 as hazardous and not acceptable, particularly for interior finish.

These are introductory data necessary for the following considerations, and they will be discussed in more detail later in this chapter.

FLAMMABILITY OF WOOD

The FSI for ordinary wood species is typically between 100 and 200, for some special cases it is as low as 60–70 (Table 14.1) To burn, wood should be exposed to heat

TABLE 14.1 ASTM E 84 flame spread indexes for 19-mm-thick solid lumber [1]

Species	Flame spread index	Smoke developed index
Yellow poplar	170–185	—
Southern pine	130–195	—
Ponderosa pine	105–230 (Average 154)	—
Eastern white pine	120–215	122
Cypress	145–150	—
Red pine	142	229
Sweetgum	140–155	—
Walnut	130–140	—
Cottonwood	115	—
Birch	105–110	—
Maple	104	—
Oak (red, white)	100	100
Spruce, Sitka	74–100	74
Douglas-fir	70–100	—
Yellow-cedar	78	90
Redwood	70	—
Fir, Pacific silver	69	58
Spruce, eastern	65	—

an air/oxygen. Burning occurs in stages: thermal decomposition with evolution of volatiles and heat release, ignition of flammable volatiles, combustion as flames, flame spread with evolution of smoke and toxic gases, and charring. Each stage can proceed differently depending on wood density, morphology, and composition (amount of lignin, etc.), the rate of heating, the temperatures, the moisture content, and the extent of air ventilation.

Generally, the same stages take place in burning of WPCs. However, composites provide much more variations in their chemical composition, density, the nature of plastics, plastic content, amount of fillers, and so on.

IGNITION OF COMPOSITE MATERIALS

Composite materials, as well as plastics, do not ignite, *per se*. Ignition happens when the flammable material reaches a certain temperature in atmosphere that contains sufficient amount of oxygen. Ignition can be piloted, that is, in the presence of a flame (or another ignition source), or unpiloted, such as in a furnace with temperature at or above the ignition point.

If to place a piece (or pellets, sheet, film, etc.) of HDPE into an oven at ambient temperature, close the oven door, program temperature at a rather high level (say, 600°C), and let it rise, at some point one can hear a pop inside the oven. Opening the door would reveal the sample in flames. To get ignited, the sample went through stages, such as melting the plastic, partial thermal decomposition resulting in evolution of volatiles, including flammable ones, and at some temperature the point of ignition of those volatiles and the sample itself will be reached. For polyethylene this temperature is about 360–367°C (680–693°F) [2], depending on a size of the sample.

Ignition point of filled HDPE, or HDPE-based composites, would be about the same provided that the material does not contain active flame retardants. Indeed, filled or not filled, HDPE in the material would still melt and emit flammable VOC (volatile organic compounds), which would ignite at a certain temperature point, close to that one for neat HDPE. A difference would depend mainly on the amount of HDPE in the material and the size of the sample.

For example, at the same conditions and for 3-g samples of HDPE and GeoDeck, ignition temperature was 806 and 815°F (435°C), respectively. Experiments with GeoDeck have shown that at 430°C ignition did not occur for more than 10 min; at 435°C the specimen ignited at 6:30 min, at 440°C it ignited at 5:45 min, and at 450°C the ignition occured at 4:00 min.

HDPE filled with a flame-retardant ATH (aluminum trihydrate) showed the ignition temperature of 445°C (833°F). The 10°C delay was caused by ATH, which releases water vapor at about 180–240°C (360–460°F), which that in turn cools down the material.

HDPE-based WPC materials all have rather similar ignition points. When tested at the same conditions, they showed almost the same self-ignition temperatures (SIT):

TABLE 14.2 Flammability data for selected wood species [3]

Species	Density (g/cm^3)	Ignition time (s) for heat flux 18 (kW/m^2)	55 (kW/m^2)
Red oak	0.66	930	13
Southern pine	0.51	740	5
Redwood	0.31	741	3
Basswood	0.31	183	5

Data are determined using ASTM E 906.

TimberTech 405°C,
Trex and TekDek 410°C,
GeoDeck 435°C.

GeoDeck has shown a slightly higher ignition point because it contains the highest amounts of inorganic fillers, more than 20%. In these conditions, pressure-treated lumber showed SIT at 430–435°C (806–815°C).

A sample surface is ignited by the flow of energy, or heat flux, from a heat source. Table 14.2 shows some values of heat flux and the respective ignition time for several wood species.

One can see how much the increase of the imposed heat flux decreases time to ignition, and how much the density of the materials affects ignition time. The lower the density of materials, the lower the ignition surface temperature. Moisture content in the material is also very important.

When the material contains some flame retardants, for example, ATH, its ignition point often shifts up. ATH, as it will be explained below in more detail, releases water at a certain temperature level. Water cools down the material and increases an apparent ignition time; it also reduces the heat produced by the burning material and therefore quenches flames.

Table 14.3 shows data on ignition temperatures of commercial WPCs, reported by the manufacturers.

FLAME SPREAD INDEXES AND FIRE RATING OF COMPOSITE MATERIALS

Essentially, FSI indicates how rapidly the fuel, emanating from the burning material, travels along the testing panel and reaches the ignition temperature under the heating by the flame front and the controlled heating, and the controlled air draft. The FSI greatly depends on the material's thermal conductivity, heat capacitance, thickness of the panel, shape (solid or hollow), amount of inorganic fillers (if any), flame-retardants (if any), and so on.

Some plastic lumber or WPC deck boards are not suitable to test according to ASTM E 84. As the fire exposure in the ASTM test is on the underside of a horizontal

TABLE 14.3 Ignition temperatures (FIT or SIT) for plastic-based commercial composite deck boards, reported by manufacturers and determined according to ASTM D 1929

Deck board	Manufacturer	Principal ingredients	FIT or SIT	Ignition temperature, °C (°F)
EverX	Universal Forest Products	HDPE 50%, wood flour 50%	SIT	436 (817)
			FIT	355 (671)
USPL	U.S. Plastic Lumber	HDPE, wood flour	SIT	387 (729)
			FIT	381 (718)
E-Z Deck	Pultronex Corporation	Polyester, glass fiber	FIT, SIT	>343 (> 650)
Ecoboard (nonreinforced)	Trelleborg Engineered Products, Inc.	100% polyethylene	FIT	350 (660)
Ecoboard (reinforced)		HDPE 70%, LDPE 10%, fiberglass 20%	FIT	350 (660)
Trex	Trex Company	Polyethylene 50%, wood flour 50%	SIT	395 (743)
			FIT	370 (698)
GeoDeck	LDI Composites	HDPE 40%, rice hulls 28%, Biodac 28%	SIT	435 (815)
Nexwood	Nexwood Industries	40% HDPE, 60% rice hulls	SIT	450 (842)
ChoiceDek	Advanced Environmental Recycling Technologies	50% HDPE, 50% wood fiber	SIT	394 (741)
			FIT	387 (729)
Boardwalk	CertainTeed Corporation	PVC, wood flour	SIT	345 (653)
			FIT	361 (682)

testing panel, when the material melts and drips or is not self-supporting, the test is commonly terminated.

There are three fire ratings on flame spread, based on the numerical indexes calculated from test data, according to ASTM E 84:

Class A: Flame spread index 0–25

Class B: Flame spread index 26–75

Class C: Flame spread index 76–200

FSIs for commercial WPC materials and products published by manufacturers (in ICC–ES reports and elsewhere) are given in Table 14.4.

Table 14.5 shows available data on burning rate of some composite materials determined according to ASTM D 635.

TABLE 14.4 Flame spread indexes for commercial WPCs determined according to ASTM E 84

Deck boards (solid or hollow)	Profile	Manufacturer	Principal ingredients	Flame spread index
USPL	Solid	U.S. Plastic Lumber	HDPE, wood flour	76
Millenium	Hollow	Millenium Decking	PVC, wood flour	60
Nexwood	Hollow	Nexwood Industries	HDPE, rice hulls	65
Life Long	Solid, channeled and hollow	Brite Manufacturing	Polyethylene, wood flour	<200
GeoDeck	Hollow	LDI Composites	HDPE, rice hull, Biodac	100 ± 3
Trex	Solid	Trex Company	Polyethylene, wood flour	80, between 75 and 200,120[a]
Monarch	Solid	Green Tree Composites	HDPE, wood flour (saw dust)	130
Boardwalk	Solid	CertainTeed Corporation	PVC, wood flour	25
Epoch (Evergrain)	Solid	Epoch Composite Products	Polyethylene, wood flour (compression molding)	Between 76 and 200
Rhino Deck	Solid	Master Mark Plastic Products	Thermoplastic, wood fiber	169
Timber Tech	Hollow	TimberTech	HDPE, wood fibers	75
EverX	Hollow	Universal Forest Products Ventures II	HDPE, wood fiber	46
WeatherBest	Solid	Louisiana-Pacific Corporation	HDPE, wood flour, 12% talc + phenol-formaldehyde resin	<200
E-Z Deck	Solid	Pultronex Corporation	Polyester, glass fiber	80
Presidio	Hollow	Westech Building Products, Inc.	100% PVC	No more than 75
XTENDEX	Hollow	Carney Timber Company	HDPE, rice husks	104
ChoiceDek	Channeled	Advanced Environmental Recycling Technologies, Inc.	HDPE, wood fiber	100

For a comparison, some other neat or filled plastics are shown.
[a] CAN/ULC-S102.2-M88

TABLE 14.5 Burning rate of WPCs according to ASTM D 635

Composite material	Burning rate (in./min)
Trex	0.71[a]
Rhino Deck	0.69–0.86[b]
Nexwood	0.68
GeoDeck	0.29[c]
E-Z Deck	0.28

[a]As an example: burning time 248 ± 13 s, burning distance 75 mm, burning rate 18.1 ± 1.0 mm/min, or 0.71 in/min.

[b]Density of smoke (ASTM D 2843-93) 5.5% per UBC 26-5.

[c]As an example: burning time 609 ± 112 s, burning distance 75 mm, burning rate 7.4 ± 1.4 mm/min, or 0.29 in./min.

EFFECT OF MINERAL FILLERS ON FLAMMABILITY

Many inert fillers, such as calcium carbonate, talc, clay, cellulose fiber, glass fiber, and so on, can slow down flame spread—by just "removing food" for flame propagation, or slow heat generation, favor charring, and so on—but they do not significantly change the ignition point. They do not act as "active" flame retardants, which typically produce some counter-flame or counter-ignition matters, such as water or inflammable gases. They act rather by diluting the fuel in the solid (plastic) phase. Calcium carbonate evolves inert gases (carbon dioxide) at about 900°C, which is too high in order to serve as a flame retardant.

For example, when ASTM D 635 procedure was used, a "basic" composite material (42% HDPE, 57% rice hulls, and 1% lubricant) burnt for 3′48″. For a comparison, a neat HDPE sample burnt for 3′30″ (±12″). With 10% fly ash added to the "basic" composition, the resulting composite burnt for 4′05″. With 25% fly ash, for 4′42″. With 40% fly ash, for 8′09′ (in some experiments, flame self-extinguished after about 5 min).

When compared with active flame retardants, such as decabrom (see below), the addition of 10% decabrom resulted in a delay of burning time of the sample to 5′50″, with 15, 20, or 30% decabrom the samples did not burn at all. The sample with 10% decabrom and 2.5% antimony oxide did not burn as well. Understandably, the sample with 12% decabrom and 3.5% antimony oxide did not burn too.

SMOKE AND TOXIC GASES, AND SMOKE DEVELOPMENT INDEX

Smoke and toxic gases associated with fire represent one of the most important problems, particularly when plastic-containing materials are burning. Smoke contains gases, solid particles, droplets of liquids, including water and molten plastic. Smoke is harmful to health, obscures vision, causes choking and sometimes death. This explains a general concern, which general public expresses regarding halogen-containing flame retardants, even when tests consistently show that some of them are harmless.

TABLE 14.6 Smoke developed indexes for commercial WPCs, determined according to ASTM E 84

Deck boards(solid or hollow)	Profile	Manufacturer	Principal ingredients	Smoke development index
GeoDeck	Hollow	LDI Composites	HDPE, rice hull, Biodac	166 ± 7
Rhino Deck	Solid	Master Mark Plastics		245
Nexwood	Hollow	Nexwood Industries	HDPE, rice hulls	340
ChoiceDek	Channeled	Advanced Environmental Recycling Technologies, Inc.	HDPE, wood fiber	360
Millenium	Hollow	Millenium Decking	PVC, wood flour	440
Timber Tech	Hollow	Timbertech Ltd.	HDPE, minerals, wood fiber	<450
Boardwalk	Solid	CertainTeed	PVC, wood flour	>450
Epoch (Evergrain)	Solid	Epoch Composite Products	Polyethylene, wood flour (compression molding)	>450
Trex	Solid	Trex Company	Polyethylene, wood flour	285 <450 >500[a]
XTENDEX	Hollow	Carney Timber Company	HDPE, rice husks	1597

[a] CAN/ULC-S102.2-M88.

Generally, but not necessarily, the higher the FSI, the higher the smoke developed index (SDI) (see Table 14.1). For wood the SDI is typically between 60 and 230. WPCs also often exceed these values (see Table 14.6).

Virgin (rigid) poly(vinyl chloride) contains about 57% chlorine and shows low flammability. PVC-based WPCs have FSI between 25 and 60. When burns, PVC pyrolyzes to form HCl and volatile organic (aromatic) compounds, which evolve large amounts of smoke.

FLAME RETARDANTS FOR PLASTICS AND COMPOSITE MATERIALS

Flame retardants for plastics and WPCs are completely different from those of wood materials. Wood is typically impregnated with solutions of flame retardants, commonly salts, such as monoammonium and diammonium phosphate, ammonium sulfate, zinc chloride, sodium tetraborate, boric acid, and guanylurea phosphate. In plastics and WPCs, however, flame retardants are added as solids directly into the formulation. Hence, flame retardants for plastics and WPCs should be temperature-resistant, in order not to be decomposed during processing.

PVC-based composites, unlike many other thermoplastic-based composites, are "natural" fire-resistant materials (see Chapter 2). Their FSI is in the range of 25–60 (Table 14.4), in Class A–B range. However, PVC also often employs flame retardants, mainly ATH, for further decrease in its flammability.

Generally, all groups of plastics utilize flame retardants when needed for their performance and specific application. Below is a short chart showing flame retardants that are used the most in the respective plastics.

- PE and PP — ATH (almost ten times more than the second FR, chlorinated compounds)
- PVC — ATH (twice as much as compared with the second FR, organophosphorus compounds, nonhalogenated)
- ABS — Brominated compounds and antimony oxides

Consumption of magnesium hydroxide compared to ATH as flame retardants is lower 15–20 times by volume and about 10 times lower by a dollar value [4].

Very few, if any, of commercial WPCs employ added flame retardants to become Class A or Class B deck boards. A rather common belief was that it does not make much sense: What is good in a fireproof deck if the house is burnt but the deck stays? However, because of massive brush fires, particularly in Southern California, Arizona, Colorado, New Mexico, and Oregon, a number of legislations are considering state laws requiring fire-resistant decks. For example, the state of California recently approved a new construction building code for new construction that calls for improved fire resistance from many building products including WPC decking. The new legislation will be effective starting January 2008 [5]. If the California regulations are adopted by some other states, Classes A and B WPC decks, containing effective flame retardants, might soon become a commodity in the market.

Alternatively (or, rather, along with) the opportunity for PVC-based products will be boosted.

Flame retardants for plastic and WPCs can be subdivided into the following principal groups according to their physical action:

- Water-releasing and cooling, creating a "heat sink";
- Halogen (gas) forming and flame poisoning, or flame choking;
- Char forming, formation of a protective layer.

As it was indicated above, many inert fillers, such as calcium carbonate, talc, clay, glass fiber, and so on, serve as "passive" flame retardants, by just "removing food" for flame propagation, or slow heat generation, favor charring, and so on, but they do not significantly change the ignition point.

Zinc borates, such as $4ZnO \times B_2O_3 \times H_2O$, $2ZnO \times 3B_2O_3 \times 3.5H_2O$, were shown to act as synergists with ATH and $Mg(OH)_2$ as flame retardants and smoke suppressants [6]. Their pyrolysis leads to a formation of a protective vitreous barrier, as low melting glasses can do. An example of zinc borate as a commercial flame retardant, recommended for plastics, is Firebrake® ZB by Borax (U.S. Borax Inc., Valencia, CA). This material releases its bound water at temperatures exceeding 290°C (554°F). At this temperature, ATH already releases about 10% of its water, if temperature increases by 10°C/min (see below). Firebrake® is claimed by its manufacturer to be flame retardant at 3–25% in halogen-containing plastics, or at higher levels with halogen-free plastics. It is typically used along with ATH or magnesium hydroxide. Median particle size of three main brands of Firebrake® are as follows: Firebrake® ZB, 7–9 μm, Firebrake® ZB-Fine, 3 μm, and Firebrake® ZB-XF, 2 μm (top particle size 12 μm). As Firebrake® has a low water solubility (less than 0.28%), it washes out only a little from composite materials used outdoors.

The way that flame retardants work is better understood considering the way how materials burn. As temperature of solid materials increases above certain value, they decompose via pyrolysis and release flammable gases. These gases burn with oxygen in the air, and the flame propagates, or spreads. Organic molecules decompose via free radicals to give free carbon that can react with oxygen to give CO and CO_2.

Active flame retardants inhibit or suppress combustion by several mechanisms, as briefly mentioned above:

- By releasing water (ATH, i.e., alumina trihydrate, $Al(OH)_3$; magnesium hydroxide $Mg(OH)_2$) that acts as heat sinks and prevents oxygen to get to flammable compounds, or by forming of a protective layer and by dilution and coating. ATH alone accounts for about 20% of all flame retardants used in plastics.
- By forming nonflammable gases to poison flame, to shield flammable materials from oxygen, insulating them, as halogenated flame retardants do. Brominated flame retardants remove free radicals in the gas phase and as a result prevent or slow down the burning process by reducing heat generation and by producing flammable gases. Brominated flame retardants account for about 32% of flame retardants, plus 6% goes for chlorinated flame retardants. Halogenated flame retardants are subdivided into halogenated organic and halogenated organophosphate esters.
- By acting as char formers, as phosphorous flame retardants do. They are also subdivided into nonhalogenated organophosphate esters, ammonium polyphosphate, and others. When heated, they produce a solid form of phosphoric acid that in turn chars the material and shields it from releasing of flammable gases feeding flames. Phosphorous flame retardants account for about 20% of flame retardants in the industry (mainly not with polyolefins). Boron compounds also work as char formers [2].

Some of the inorganic compounds, such as antimony trioxide (Sb_2O_3), or boron-based compounds, such as zinc borate, function as synergists rather than directly as flame retardants but enhance the effectiveness of the latter. Antimony trioxide is used mainly with halogenated flame retardants.

Flame Retardants in Plastics

Flame retardants are used very little in WPC materials compared to large scale application in plastics, such as in electrical applications (TBBA in epoxy laminated printed circuit boards, existing there as brominated epoxy polymers, or BEOs) or carpets. In epoxy resins it is a reactive chemical, existing as part of the polymer chain, TBBPA (tetrabromobisphenol-A) [7]. Another example—polybromostyrene flame retardants for use in Nylon 6 and Nylon 66 applications for connectors and high-temperature polyamide applications (PDBS 80, BrPS, BrPS-1, Firemaster® PBS-64HW, and Firemaster® CP-44HF, the latter is a copolymer of di- and tri-bromostyrene with glycidyl methacrylate).

Flame retardants that are often used in polycarbonate/ABS plastics (such as in computer industry) include nonhalogen triaryl phosphates, such as RDP [resorcinol bis (diphenyl phosphate)] and BDP [bisphenol A bis(diphenyl phosphate)]. As BDP has lower phosphorus content compared to that of RDP, more of it should be used to match the flammability performance of RDP. In one particular study using polycarbonate–ABS alloy, 9% of RDP or 12.3% of BDP was employed and showed equal to each other and excellent flame retardant properties [8].

Restrictions or Prohibitions of Some Brominated Flame Retardants

Use of some of the flame retardants, particularly brominated ones (PBDEs, or polybrominated diphenyl esters), was prohibited by European Union Risk Assessment program. Among them were pentabromodiphenyl ether (pentaBDE), which was used primarily in polyurethane foam, and octabromodiphenyl ether (octaBDE), used mainly in electrical and electronic equipment and automobiles (both were prohibited by EU in August, 2004). In the United States, production of pentaBDE was ceased in 2004 and its manufacture and import into the United States is prohibited. Octa-BDE was scheduled to be prohibited in 2006 in a number of states in the United States [9]. Both pentaBDE and octaBDE are linked to fetal development and thyroid problems; however, the concentrations of PBDEs in human blood serum and breast milk have been doubling every 2–5 years, according to the Centers for Disease Control & Prevention (CDC).

Decabromodiphenyl ether (decaBDE) was recognized as safe, with "no need for risk reduction measures" both in the United States and Europe [9], though CDC lists it as a possible human carcinogen. Currently, under the investigation by EU are the following flame retardants: TBBPA (tetrabromobisphenol-A), HBCD (hexabromocyclododecane), TCEP [tris(2-chloroethyl)ethyl)phosphate], TCPP [tris(2-chloropropyl)phosphate], TDCP [tris(2-chloro-1-(chloromethyl)ethyl) phosphate], and V-6 [2,2-bis(chloromethyl)trimethylene bis (bis(2-chloroethyl)phosphate)]. The main application for TCEP, TCPP, TDCP, and V-6 is polyurethane foam.

It looks like, however, that the "safe status" of decaBDE might be soon reconsidered. Recent researches have shown that microorganisms present in North American and European soil break down relatively stable decaBDE into octaBDE and then pentaBDE, both toxic. These microorganisms were identified as *Sulfurospirillum* and *Dehalococcoides* [10].

Chlorine-Containing Flame Retardants

Chlorine-containing flame retardants can be divided into three groups: aliphatic, alicyclic (cycloaliphatic), and aromatic. Their thermostability is increased in this order, but flame retardant efficiency is decreased in the same order. This reciprocal tendency is common among flame retardants. Clearly, the higher the thermal stability, the higher the temperature at which the flame retardant becomes chemically active and functional as a flame retardant.

A common disadvantage of chlorine-containing flame retardants is that they have to be added in quantities, which in turn decrease mechanical properties of the polymer materials. The same situation in terms of large amount that should be added into the base material holds for mineral flame retardants as well (ATH, $Mg(OH)_2$); however, minerals typically improve both flexural modulus (stiffness) and flexural strength of composites.

Aliphatic chlorine flame retardants are represented by chloroparaffins, with chlorine content between 40 and 70%. They typically have a poor thermal resistance, as their dechlorination often starts at 180°C (356°F); hence, their application is restricted by polyethylenes and PVC.

Dover Chemical Corporation produces resinous chlorinated paraffins under a name Chlorez® and advocates their usage as flame-retardant additives in plastics. All Chlorez® grades have a physical form of white powder (particles smaller than 50 mesh) with chlorine content around 70%. Chlorez® is not recommended to be processed above 180°C (356°F). Indeed, an attempt to compound one of these materials with a WPC at 190°C (374°F) has resulted in a rather violent decomposition of the material. The melt turned dark, and some foaming occurred with an accompanying strong smell.

On the contrary, chloroaromatic compounds, such as decachlorodiphenyl, are too stable thermally to be widely used as flame retardants.

An example of a successful application of Chlorez® in making of an HDPE-based WPC at 325°F (die lip at 340°F) was provided by Dover Chemical Corporation (Thomas Kelley). The base WPC formulation was 40-mesh pine flour (52–55% w/w), and MFI of HDPE was 0.4. A profile was extruded using a 35-mm conical twin-screw extruder from Cincinnati Milacron at WMEL, Washington State University Chlorez® was employed at 10–15% w/w along with 2.5–5% of Sb_2O_3 (antimony trioxide, or ATO) as a synergist. For comparison, Decabrom (10% w/w) along with 3% of Sb_2O_3 was also evaluated. According to the company data, the composite deck boards loaded both Chlorez®/ATO and Decabrom/ATO had a regular shape; however, Chlorez®-loaded boards showed better fire-resistant properties compared to Decabrom-loaded boards.

An example of cycloaliphatic chlorine flame retardants is hexachlorocyclopentadiene (Dechlorane Plus®, or CFR). It is typically used in polyester resins in which hexachlorocyclopentadiene is converted into functional derivatives by maleic anhydride to give the "net anhydride." It is the most stable chlorinated flame retardant(CFR). CFR is commonly used in synergism with antimony oxide, zinc borate, zinc oxide, and iron oxides. In Nylon 66, CFR is often used in concentrations of 8–25%, along with 1–10% of Sb_2O_3, zinc borate, or iron oxide [11].

ATH (Aluminum Trihydrate) and MDH (Magnesium Hydroxide)

A typical example of "active" flame retardants is represented by ATH, that is, aluminum trihydrate. When ATH is heated with a heating rate of 10°C/min, it starts releasing water at about 225°C (437°F). At 300°C, ATH releases 12% water by weight. At 334°C (633°F), the rate of water reaches its maximum, with about 28% of water release, and the process slows down, reaching 35% of water release at 900°C. Water release reduces the heat from the oxidized plastic and quenches flames.

$$2Al(OH)_3 \rightarrow Al_2O_3 + 3H_2O$$

This is an endothermic decomposition, with $\Delta H = 298$ kJ/mol (71.2 kcal/mol).

Generally, ATH is considered to be thermally stable at around 180–200°C (360–400°F), and—with some reservations—until 216°C (420°F). Some data indicate the decomposition of ATH in the temperature range of 180–240°C (360–460°F). ATH is usually cream-colored, free-flowing powder with specific gravity of 2.42 g/cm^3 loose bulk density of 70 lb/ft^3 (1.12 g/cm^3), and the so-called packed bulk density of 55 lb/ft^3 (1.36 g/cm^3). A typical commercial grade of ATH (for example, product of Alcan Chemicals, Cleveland, OH), contains up to 15% of powder with mesh size of 100 and larger particles, 67–87% with mesh size 200–200, and between 1 and 12% of mesh size 325.

Magnesium hydroxide converts to magnesium oxide and water above 300–330°C (570–630°F), and according to other data in the range of 330–460°C (630–860°F), which at any rate is in a significantly higher temperature range compared with that of aluminum trihydrate. Besides, this is more endothermic reaction, with a higher enthalpy of decomposition, hence, more efficient as a "heat sink."

$$Mg(OH)_2 \rightarrow MgO + H_2O$$

This is also an endothermic decomposition, with $\Delta H = 380$ kJ/mol (90.8 kcal/mol).

Obviously, ATH cannot be used in thermoplastics other than polyethylene and PVC, and in the respective plastic-based composite materials. A common disadvantage of ATH and MDH (magnesium hydroxide) is that in order to provide a sufficient level of flame retardancy, they have to be used in large amounts, such as 50–65% and not below 40%.

As it could have been expected, the decomposition of ATH to Al_2O_3 during the heating of the polymer resulted in an increase of the ignition time [12].

ATH Dehydration: A Quantitative Approach

Ashing of a sample of plastic (or a plastic-based composition) containing ATH can determine how much the ATH was dehydrated during the processing.

Though aluminum trihydrate, $Al_2O_3 \bullet 3H_2O$, or $[Al(OH)_3]_2$, theoretically contains three water molecule, in reality it typically contains a slightly reduced amount of water. A direct analysis of eight commercial ATH samples resulted in only 2.86 water molecules on average per Al_2O_3 in ATH.

This was determined as follows. Eight ATH samples were studied:

1. RIH-30, 10 mc, Riverland Ind. Inc.
2. Haltex
3. Huber, Surface modified HYMODTH SL, 36SL
4. Huber, Surface modified HYMODTH SP, 36SP
5. Huber SB-36
6. Huber SB-336
7. Huber Mold^{TM} A120
8. Alcoa Inc., 10 mc.

Theoretical molecular weight of $Al_2O_3 \bullet 3H_2O$ is 156.

Molecular weight of Al_2O_3 (ash) is 102.

Therefore, ash content in ATH, having 100% of bound water, is 102/156 = 65.38%.

If a whole water molecule–on average—is lost as a result of a temperature treatment of some other effects, the resulting aluminum dihydrate $Al_2O_3 \bullet 2H_2O$ will produce 102/138 = 73.91% of ash by weight. If two water molecule are lost, the resulting aluminum monohydrate $Al_2O_3 \bullet H_2O$ will produce 85.0% of ash. Obviously, if all three water molecules are lost, resulting in pure aluminum oxide Al_2O_3, after ashing of it all 100% of ash will be observed.

At more gentle temperature treatment, only a fraction of water—on average—can be lost. In other words, some ATH molecules will lose a water molecule, and some ATH molecules will not. Below is a brief chart showing how ash content of respective ATH samples would look like (Table 14.7).

TABLE 14.7 An amount of ash (per cent) after temperature treatment (at 525°C for 24 h) of Aluminum Trihydrate (ATH) until constant weight. Theoretical data.

Quantity of water molecules present on average per ATH molecule	Ash content (%)
3.0	65.38
2.9	66.15
2.8	66.93
2.7	67.73
2.6	68.55
2.5	69.39

TABLE 14.8 Determination of water content in various commercial ATH samples from their ashing (at 525°C for 24 h) data. The manufacturers names are listed on p. 474.

ATH sample (see list above)	Ashed sample weight (g)	Ash weight (g)	Ash, % of the initial sample weight	Average quantity of water molecules per ATH
1	1.514	1.011	66.78	2.82
		1.004	66.31	2.88
2	1.379	0.927	67.22	2.77
		0.920	66.72	2.83
3	1.889	1.254	66.38	2.87
		1.244	65.85	2.94
4	1.765	1.183	67.03	2.79
		1.175	66.57	2.85
5	2.102	1.406	66.89	2.81
		1.397	66.46	2.86
6	2.081	1.390	66.79	2.82
		1.382	66.41	2.87
7	1.246	0.838	67.26	2.76
		0.832	66.77	2.82
8	1.414	0.945	66.83	2.81
		0.940	66.46	2.86
		Average	66.67±0.36	**2.86±0.04**

When all eight ATH samples listed above were ashed at 525°C for 24 h (until constant weight), results looked as given in Table 14.8.

Thus, all eight ATH sample had a very close to each other amount of bound water, namely 2.86, rather than the theoretical amount of 3.0 water molecules.

The same approach can be applied to plastics, filled with ATH (or any other filler). A compounded mix of 60% ATH and 40% HDPE was prepared and a sample of the filled plastic was ashed. Theoretically, the amount of ash should be 39.23% by weight. This figure can be obtained by taking 65.38% ash content in ATH and taking 60% of it (because the ATH was diluted by 40% of HDPE). If to introduce the correction that there were only 2.86 water molecules per ATH (see above), the amount of ash should be 40.0%.

Any figure for a weight percentage of the ashed HDPE/ATH higher than 40.0% would indicate that the ATH in the material was partially (or completely) decomposed compared with the initial ATH.

An actual experiment showed ash content in four HDPE (40%)—ATH (60%) samples, prepared with ATH samples 2, 3, 7 and, 8 above. Ash contents were 40.16, 40.13, 39.78, and 40.19%, respectively. These data indicate that there was practically no decomposition of ATH in the course of compounding with ATH (less than 0.1 water molecule per ATH molecule was lost). The compounding was done using the Brabender mixing head.

When more fillers, such as fly ash, zinc borate, and a colorant, were added to the formulation and compounded with HDPE and ATH, ATH lost 0.6 water molecule on average, as it was shown by ashing. Repeat with the same formulation but with a gentler mixing resulted in a loss of only 0.3 water molecule per ATH molecule.

Flame Retardants with Wood–Plastic Composites

Decabrom and antimony oxide were tested as flame retardants with HDPE-based composite materials. Tests were conducted according to ASTM D 635 (plank size of 125 × 13 × 7 mm, 75 mm of length flame spreading). Decabrom in amounts of 12 to 20–30% prevented the composite from burning, with or without added antimony oxide (2.5–3.5%). 10% decabrom did not prevent burning, but in the presence of 2.5–10% of antimony oxide it made flame consistently die. Antimony oxide by itself in the amount of 20% did not affect noticeably the burning, as well as 3% Vitrolite.

Nanoparticles as Flame Retardants

Lately, flame retarding effects of nanoclay particles on flammability of WPCs, particularly in the presence of coupling agents, have attracted attention. Typically, conventional flame retardants are used at filling levels of 40–60% (w/w) and even higher. Nanofillers can reportedly avoid this disadvantage of traditional flame retardants. Nanoparticles or nanofillers are collective terms for modified layered silicates (organoclay) or carbon nanotubes dispersed in the polymer matrix, when the particles' size is in order of nanometers, or tens of nanometers. A plastic filled with nanoparticles typically in the range of 2–10% (w/w) is called a nanocomposite.

There are two basic types of nanocomposites in which particles are intercalated or exfoliated. In an intercalated composite, the nanodispersed filler still consists of ordered structures of smaller individual particles packed into intercalated structures. Exfoliated particles are those dispersed into practically individual units, randomly distributed in the composite. Layered silicates, such as montmorillonite clays or organoclays, can be used in nanocomposites. As clays are hydrophilic and polyolefins are hydrophobic, it is not easy to make nanocomposite based on polyethylene or polypropylene because of their natural incompatibility.

Both intercalated and exfoliated nanocomposites, containing 3–5% of nanoparticles (w/w), reportedly show better or comparable flame resistance compared with plastics filled up to 30–50% with traditional flame retardants. Another way to increase flame retardancy is to combine ATH or magnesium hydroxide with organoclays. It was reported that organoclays and some classical flame retardants, such as brominated compounds, showed a synergism between them [13].

In some cases only small amount of nanoparticles, such as 5% w/w, was claimed to be necessary to significantly reduce flammability of WPC. According to Ref. [14], a clay particle of a size of 8 μm consists of about 3000 platelets, which can be exfoliated into particles of 200 nm in length and 1 nm in thickness and intercalated into packets of these platelets of the same length of 200 nm, and thickness of 30 nm.

TABLE 14.9 Heat release in $Mg(OH)_2$ filled polypropylene [15]

Amount of $Mg(OH)_2$ in polypropylene (%, w/w)	$Mg(OH)_2$ particle size	Burning heat release (J/g-K)
0 (neat PP)	—	1183
65	5 μm	310
65	1 μm	290
65	Nanoparticles	261
65	32.5% of 5-mc + 32.5% of nanoparticles	273
65	32.5% of 1-mc + 32.5% of nanoparticles	253

According to the authors, burning rate (ASTM D 635) of the WPC in the presence of the nanoclay as a part of the composition decreased from the initial (no nanoclay) 29 to 13 mm/min. In another experiment, burning rate decreased from the initial 42 to 32 mm/min in the presence of 5% (w/w) of nanoclay, and from 32 to 26 mm/min in the presence of 0.5% of the nanoclay.

Generally, nanomaterials as flame retardants do not have commercial applications. Data obtained are commonly recognized as preliminary, and they are described here just as preliminary as well. One more example of such data is a study of heat release and char formation (in per cent units) at burning of polypropylene filled with magnesium hydroxide (5 μm, 1 μm particles, and nanoparticles) (Table 14.9).

Thus, addition of the nanoparticles to a regular size $Mg(OH)_2$ led to a significant decrease of heat release capacity of filled polypropylene.

Regarding char formation, pure polypropylene showed a residual char of 0.5%. In the presence of 65% of $Mg(OH)_2$ (regular or nano), a residual oxide or char was 51.3–54.8%, whereas an expected (theoretical) amount of char was 42.8–44.9%. The authors concluded that the formation of noncombustible carbon char from polypropylene was higher in the presence of the magnesium oxide [15].

ASTM RECOMMENDATIONS

Note 1: ASTM's policy is not to use descriptive terms such as "nonflammable," "flame retardant," "self-extinguishing," "*non*-burning," and similar. According to ASTM, results of any of fire test methods must be described in numbers, such as "flame spread index of 75," or "flame spread index below 200," or "a burning rate of 0.72 in./min," or "a burning distance of 75 mm."

Note 2: Numerical fire test results obtained using different ASTM procedure generally cannot be comparable and/or cannot be translated to expected results of, say, method ASTM E 84, using some empirical coefficients. There are too many noncontrollable factors involved in small and large-scale burning.

ASTM D 635 "Standard Test Method for Rate of Burning and/or Extent and Time of Burning of Plastics in a Horizontal Position"

The ASTM fire test procedure describes a small-scale laboratory determination of the relative linear rate of burning of plastics and plastic-based composites. Specimens can be in the form of bars, molded or cut from sheets, plates, or panels and tested in the horizontal position. The results of the test are intended to serve as a preliminary indication to their flammability.

In summary, a bar specimen of 125 mm (5 in.) long, 13 mm (0.5-in.) wide, and about 3 mm (0.12 in.) thick (but not more than 13 mm [0.5 in.] thick) is supported horizontally at one end, and the free end is exposed to a specified gas flame for 30 s. If the specimen burns to the 100 mm (3.94 in.) mark from the ignited end, time of burning of 75 mm (3 in.) and average burning rate (in in./min or mm/min) are reported. If the specimen does not burn 100 mm, time and extent of burning are measured and reported along with a pattern of burning and/or flame self-extinguishing. There should be no forced or induced draft allowed during the test.

Precision of the test procedure is usually fair. ASTM D 635–03 lists examples for several different plastics, among them are polyethylene, ABS (acrylonitrile-butadiene-styrene), and acrylic, tested by eleven laboratories, using three specimens for each material (Table 14.10).

For polyethylene, average rate of linear burning was 15.2 ± 0.7 mm/min for within-laboratory tests and 15.2 ± 1.3 mm/min for between-laboratory tests. For ABS, it was 27.9 ± 2.1 mm/min for within-laboratory tests and 27.9 ± 4.1 mm/min for between-laboratory tests. For acrylic, it was 29.7 ± 1.7 mm/min and 29.7 ± 2.2 for within- and between-laboratory tests, respectively.

Note: ASTM E 1321 "Standard test method for determining material ignition and flame spread properties" is a different test in which a vertically oriented sample is employed. This test is not in a general use in plastic and WPC industry.

ASTM D 1929 "Standard Test Method for Determining Ignition Temperature of Plastics"

The ASTM fire test procedure describes a laboratory determination of the flash ignition temperature (FIT) and spontaneous ignition temperature, or self-ignition temperature (SIT) of plastics and composites using a hot-air furnace. FIT is the minimum temperature at which sufficient flammable gases are emitted to ignite by a

TABLE 14.10 Average linear burning rate

Plastic	Specimen thickness, mm (in)	Average rate of linear burning, mm/min (in/min)
Polyethylene	3.0 (0.118)	15.2 (6.0)
Acrylonitrile-butadiene-styrene	3.2 (0.126)	27.9 (11.0)
Acrylic	3.0 (0.118)	29.7 (11.7)

TABLE 14.11 Flash ignition temperature (FIT)

Plastic	Physical form	Average FIT, °C (°F)	Repeatability (°C)	Reproducibility (°C)
Phenol–formaldehyde resin	Solid bar	430(806)	9	117
Polyamide 6	Granules	413(775)	8	38
Polystyrene	Granules	378(712)	10	27
Polyurethane foam	25 mm(1 in.) thickness	349(660)	12	66
PVC film	0.15 mm(6 mills) thickness	327(621)	11	45

pilot flame. SIT is the minimum temperature at which ignition occurs in the absence of any additional flame ignition (pilot) source.

Test values serve to rank materials according to ignition susceptibility under the actual use conditions. The procedure notes that specimens containing high levels of inorganic fillers are difficult to evaluate; also, that the same material tested in different forms may give different results.

The ASTM procedure describes in detail the hot-air ignition furnace with accessories, consisting primarily of an electrical heating unit, air source, specimen holder, thermocouples, pilot flame, and timing device.

A specimen may be in the form of pellets, powder, sheet, film, plastic cellular, or composite materials, with a specimen weight of 3.0 ± 0.2 g. The ASTM procedure describes in detail cutting or folding sheet or film materials and conditioning test specimens.

The temperature of 400°C shall be used when no prior knowledge of the probable flash ignition temperature range is available. Other starting temperatures may be selected when information of the materials is available. The only principal difference between FIT and SIT procedures is that the second one is conducted without the pilot flame. At the end of the first 10 min of the heating, depending on whether ignition has or has not occurred, temperature shall be lowered or raised, and the test shall be repeated with a fresh specimen. The lowest air temperature at which a flash or self-ignition is observed during the 10-min period is the flash- or selfignition temperature.

Precision of FIT and SIT measurements is usually fair. ASTM D 1929–96 lists examples for several different plastics; among them are PVC, polystyrene, nylon, polyurethane, and phenol–formaldehyde resin (see Tables 14.11 and 14.12), tested by seven laboratories, using three replicates of each material. "Repeatability" in this case is the difference (in°C) between two averages, each one determined from three specimens of identical test material, using the same apparatus by the same analyst within a short time interval.

"Reproducibility" in this case is the difference (in°C) between two averages, each one determined from three specimens of identical test material, found by two operators working in different laboratories.

TABLE 14.12 Self-ignition temperature (SIT)

Plastic	Physical form	Average SIT, °C(°F)	Repeatability, (°C)	Reproducibility, (°C)
Phenol–formaldehyde resin	Solid bar	482(900)	14	103
Polystyrene	Granules	458(856)	12	59
Polyamide 6	Granules	439(822)	31	56
PVC film	0.15 mm(6 mills) thickness	438(820)	13	64
Polyurethane foam	25 mm(1 in.) thickness	370(698)	11	61

One can see that self-ignition temperature is higher than flash ignition, which is understandable, as pilot flame makes the ignition occurs faster. However, a difference between these temperatures varies significantly between 21°C for polyurethane foam and 111°C for PVC film.

ASTM E 84, "Standard Test Method for Surface Burning Characteristics of Building Materials"

This ASTM fire test large-scale procedure describes determination of the relative burning behavior of the material in a form of a 24-ft long panel. The test is conducted with the specimen in the ceiling position with the surface to be evaluated exposed face down to the ignition surface. The specimen shall be either self-supporting by its own structural quality or supported from its front or the backside.

Two separate (and not necessarily related) readouts of the test are (a) flame spread along the surface of the specimen as a distance traveled by the boundary of a zone of flame over time and (b) smoke developed as a change in optical density (as a progress curve of light absorption percent) between the light source and the photoelectric cell mounted in the vent pipe. These data are used to calculate the respective FSI and SDI as described in the ASTM test procedure. The indexes are calculated as relative values to those of select grade oak (FSI arbitrarily set as 100) and inorganic reinforced cement board (FSI set as 0) surfaces under the specified conditions.

The procedure notices that some materials melt or drip to such a degree that they interfere with the continuity of the flame and result in an apparent low FSI.

The test method exposes a 24-ft (7.32-m) long by 20.25-in. (0.514-m) wide specimen to a controlled air flow and adjusts the observed flame spread to that with the select grade oak for which the flame spreads the entire length of the specimen in 5½ min. The specimen may consist of sections joined together.

The ASTM tests describe in detail the fire test chamber (the so-called 25-ft chamber), its insulation, dimensions, observation windows, lid assembly, gas burners, air intake, thermocouples, exhaust, and the photometer system (including a lamp and photocell) mounted in the vent pipe.

The FSI is calculated as follows. The end of the ignition fire—according to the ASTM procedure—shall be considered as being 4½ ft from the burners. Hence, flame spread distance shall be determined as the observed distance minus 4½ ft. Now, suppose the flame spreads 10 ft in 2.5 min, then remains for eight more minutes at the same 10-ft mark, and finally spreads again for the subsequent 2 min from 10-ft mark to 20-ft mark. The total area under that curve, given in the ASTM procedure, is equal to 90 ft min. The procedure says that in this case (an area is less than 97.5 ft min) the FSI is equal to $90 \times 0.515 = 46$ ft min.

If, suppose, this area under the curve is a little higher, say, 100 ft min (or any area higher than 97.5 ft min), the FSI in this case would be equal to $4900/(195–100) = 52$ ft min, where 4900 and 195 are constants, prescribed by the procedure.

The SDI is calculated in a similar manner, normalized by the area under the curve for red oak, multiplied by 100 and rounded to the nearest multiple of 5. For indexes of 200 and higher, the calculations shall be rounded to the nearest 50 points.

Precision of the test procedure is usually fair, with larger deviations for lower FSI figures. ASTM E 84–01 lists examples of FSI for six different materials, among them two samples of plywood (one was treated with flame retardant), one gypsum board, two plastic foams, and one composite panel. The determinations were made in eleven laboratories with four replicates of each material (Table 14.13).

"Repeatability" in this case is the difference (in FSI) between two averages, each one determined from four specimens of identical test material, using the same apparatus by the same analyst within a short time interval. "Reproducibility" in this case is the difference (in FSI) between two averages, each one determined from four specimens of identical test material, found by two operators working in different laboratories.

TABLE 14.13 Flame spread index (FSI). Within-laboratory and between-laboratory data

		Standard deviation	
Material	FSI, mean value	Repeatability	Reproducibility
Douglas Fir plywood	91	15	23
Rigid polyurethane foam	24	3	5
Fire retardant treated Douglas Fir plywood	17	3	6
Composite panel	17	2	4
Type X gypsum board	9	2	3
Rigid Polystyrene foam	7	1	4

ASTM E 1354, "Standard Test Method for Heat and Visible Smoke Release Rates for Materials and Products Using an Oxygen Consumption Calorimeter"

This standard test method is listed in ASTM D 7032 "Standard specification for establishing performance rating for WPC deck boards and guardrail systems (guards or handrails)" as a reference to "fire performance properties other than flame spread." The ASTM E 1354 fire test procedure describes determination of the ignitability, heat release rates, visible smoke development, and other related characteristics of materials and products. Specimens shall be exposed to heating fluxes in the range of 0–100 kW/m^2. The rate of heat release (i. e., the heat evolved from the specimen per unit of time) is determined by measuring the oxygen consumption, and smoke development is measured by decrease of optical density (or light transmission) of light by combustion product stream. Ignitability is determined as a measurement of time from initial exposure to time of sustained flaming (of at least 4 s), provided that ignition is initiated by electric spark. Specimens shall be tested in a horizontal position. Flaming of less than 4 s duration is identified in this ASTM procedure as flashing or transitory flaming.

In other part, this ASTM test is based on the observation that, generally, the net heat of combustion is directly related to the amount of oxygen required for combustion. The relationship is that approximately 13,100 kJ of heat are released per 1 kg (2.2 lb) of oxygen consumed. The procedure prescribes burning of specimens in ambient air conditions, while they are subjected to a specified external heat flux. Burning may be either with or without spark ignition. The primary measurements are oxygen concentrations and exhaust gas flow rate. The ASTM procedure describes in detail the test apparatus, specimen mounting, gas sampling, heat flux meter, and so on. Regarding smoke release measurements, they are conducted using a helium–neon laser, photodiodes as main beam and reference detectors, and appropriate electronics to derive the extinction coefficient and to set the zero reading. The smoke obscuration measuring system is attached to the exhaust duct.

The ASTM procedure specifies test size as 100 × 100 mm, up to 50 mm thick (4-in. × 4-in. up to 2-in. thick). Other sizes are also considered in the procedure in case of greater or smaller thickness of the tested material. The procedure describes calculations of heat release, mass-loss rate, effective heat of combustion, and smoke obscuration. For the latter, the extinction coefficient is calculated as

$$k = (1/L) \times \ln(I_0/I)$$

where

k = the smoke extinction coefficient (meter^{-1}),
L = optical beam path length (meter),
I = actual beam intensity,
I_0 = beam intensity with no smoke.

The ASTM test requires to report smoke obscuration as the average specific extinction area (m^2/kg) for each specimen. The average specific extinction area (σ, m^2/kg) is calculated as the volume exhaust flow rate (V, m^3/s), measured at the location of the laser photometer, multiplied by the smoke extinction coefficient (k, m^{-1}) and by the sampling time interval (Δt, s), divided by the specimen mass loss (Δm, kg), and averaged for repeated tests.

$$\sigma = Vk\Delta t/\Delta m$$

The ASTM procedure gives a range for average specific extinction areas for a number of different materials, which is between 30 and 2200 m^2/kg. Among those materials were fire retardant treated ABS, polyethylene, PVC, polyisocyanurate, polyurethane, and gypsum board.

E 162 "Standard Test Method for Surface Flammability of Materials Using a Radiant Heat Energy Source"

This ASTM test will be briefly described here. The test is not intended for use as a basis of rating for building code purposes, as, for example, ASTM E 84. The purpose of the test is to determine the relative surface flammability performance of various materials under specific test conditions under a radiant heat source.

A radiant 12 in. by 18 in. panel is preheated with a gas–air mixture to a radiant output equal to that obtained from a black body of the same dimensions operating at 670°C (1238°F). A test specimen 6 in. by 18 in. is suitably mounted facing the radian panel and being inclined as specified in the ASTM standard. The specimen is ignited at the top by a pilot flame, and the material burns downward. The operator records the flame progression at 3, 6, 9, 12, and 15 in. interval marks measured from the top of the sample. The operator also records the maximum temperature increase resulting from the burning sample measured by eight thermocouples located above the tested sample. The FSI is derived as

$$I_s = F_s \times Q$$

where
I_s = the flame spread index
F_s = the flame spread factor
Q = the heat evolution factor.

The ASTM procedure describes a calculation of the F_s, which is determined by the speed at which the flame front burns down the specimen. The higher the F_s value, the faster the specimen burns. The procedure also describes determination of the Q from the maximum temperature developed in the stack above the burning sample. The hotter the flame during the burning, the higher the heat evolution factor.

Note of the author: To illustrate values obtained according to the above ASTM procedure, the following table gives an example of two composite materials, one of which contains two-thirds of rice hulls and one-third of HDPE, another contains one-third of rice hulls, one-third of a mineral filler and one-third of HDPE (Table 14.14).

TABLE 14.14 Flame Spread Index (FSI), Flame Spread Factor (FSF), and Heat of evolution values (determined according to ASTM E 162) for two HDPE-based composite materials, one filled with rice hulls, another with rice hulls and a mineral filler. Detailed on the compositions are given on page 484. Data by the author.

Material	Flame spread index	Flame spread factor	Heat of evolution
Composite 1 (HDPE and rice hulls)	226±11	4.3±0.2	52.0±0.8
Composite 2 (HDPE, rice hulls, and a mineral filler)	164±15	3.6±0.4	46±3

E 662 "Standard Test Method for Specific Optical Density of Smoke Generated by Solid Materials"

This ASTM test will be briefly described here. The test is not intended for use as a basis of rating for building code purposes. The procedure covers measuring the smoke generated by solid materials and assemblies in thickness up to 1 in. The test is based on decrease of optical density of a light beam by smoke accumulated within a closed chamber due to nonflaming pyrolytic decomposition and flaming combustion. Both flaming and nonflaming exposures are conducted. Test results are expressed in terms of specific optical density.

The ASTM method uses an electrically heated radian energy source of a specified power, positioned over a vertically mounted specimen. This exposure provides the nonflaming exposure of the test. For the flaming condition, a specified flame burner is used.

The test specimens are exposed to the flaming and nonflaming conditions within a closed 18-ft^3 chamber, equipped with a photometric system.

Specimens of 3 in. by 3 in. size and up to 1 in. by thickness are predried for 24 h at 60°C (140°F) and then conditioned to constant weight at ambient temperature. The ASTM test describes calculations to obtain specific optical density of smoke.

Note of the author: To illustrate values obtained according to the above ASTM procedure, the following table gives an example of two composite materials, one of which contains two-thirds of rice hulls and one-third of HDPE, another contains one-third of rice hulls, one-third of a mineral filler and one-third of HDPE. During the flaming mode both samples ignited at about 10 s and burned for approximately 10 min (Composite 1) and 14 min (Composite 2). The data are average from three nonflaming and three flaming measurements (Table 14.15).

TABLE 14.15 Smoke Development Index (SDI) determined in flaming and nonflaming conditions according to ASTM E 662, for two HDPE-based composite materials, one filled with rice hulls, another with rice hulls and a mineral filler. Detailed on the compositions are given on page 484. Data by the author.

	Smoke developed index	
Material	Nonflaming	Flaming
Composite 1 (HDPE and rice hulls)	392±2	345±45
Composite 2 (HDPE, rice hulls, and a mineral filler)	298±100	351±72

FIRE PERFORMANCE OF COMPOSITE DECKS AND DECK BOARDS

UC Forest Products Laboratory (UCFPL) has undertaking fire performance testing of neat plastic and WPC deck boards. The testing was conducted in 2002. Table 14.16 and Figures 14.1 through 14.10 summarize and illustrate main results of the test.

(a)

(b)

Figure 14.1 Plastic lumber Eon (100% polystyrene), channeled deck board, before and after 2.5 min of a fire test (by permission from the University of California Forest Products Laboratory).

TABLE 14.16 Underdeck flame impingement test results (UCFPL)

Deck board	Material, shape	The time of board collapse or runaway combustion	Type of failure
Redwood	Wood	No degradation effects	No failure in 40 min
Eon	100% polystyrene, foamed, channeled, density 0.80 g/cm^3	Less than 4 min	Began dropping flaming debris; combustion accelerated
Maxituf	100% HDPE, solid, density 0.94 g/cm^3	Less than 4 min	Began dropping flaming debris; combustion accelerated
EverNew	100% PVC, hollow, density 1.44 g/cm^3	Less than 4 min	Collapsed
TimberTech	37% HDPE, channeled, 15% minerals, 48% wood fiber, density 1.22 g/cm^3	In less than 4 min began dropping flaming debris, collapsed in 9 min	Began dropping flaming debris; combustion accelerated; board collapsed
ChoiceDek	50% polyethylene, 50% wood fiber, slightly channeled, density 0.91 g/cm^3	In less than 4 min began dropping flaming debris, collapsed in 14 min	Began dropping flaming debris; combustion accelerated; board collapsed
Nexwood	42% HDPE, hollow, 46% rice hulls, 12% minerals, density 1.17 g/cm^3	In 10 min began dropping flaming debris, collapsed in 14 min	Began dropping flaming debris; board collapsed
Bedford #1	100% HDPE, solid, density 0.97 g/cm^3	In less than 2 min began dropping flaming debris, combustion accelerated in 17 min	Began dropping flaming debris; combustion accelerated
Ecoboard	100% polyethylene, foamed, density 0.85 g/cm^3	In less than 2 min began dropping flaming debris, combustion accelerated in 22 min	Began dropping flaming debris; combustion accelerated
Trex	50% LDPE/HDPE, 50% wood fiber, solid, density 0.92 g/cm^3	In 20 min began dropping flaming debris	Began dropping flaming debris
Rhino Deck	35% HDPE, 65% wood fiber, solid, density 1.08 g/cm^3	In 22 min began dropping flaming debris	Began dropping flaming debris
SmartDeck	35% polyethylene, 65% wood fiber, solid, density 1.10 g/cm^3	No degradation effect	No failure in 40 min
WeatherBest solid	28% HDPE, 60% wood flour, 12% talc + phenol-formaldehyde resin; solid, density 1.20 g/cm3	No degradation effect	No failure in 40 min
WeatherBest hollow	Same	No degradation effect	No failure in 40 min
Bedford #2	85% HDPE, 15% minerals, solid, density 1.06 g/cm^3	No degradation effect	No failure in 40 min

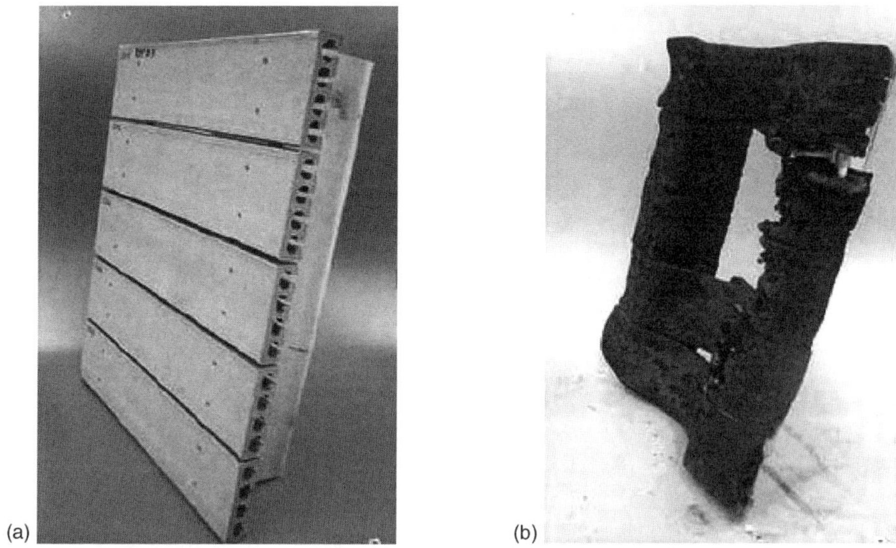

Figure 14.2 Wood-plastic composite TimberTech (37% HDPE, 15% minerals, 48% wood fiber, density 1.22 g/cm3), channeled deck board, before and after 9 min of a fire test (by permission from the University of California Forest Products Laboratory)

Figure 14.3 Wood-plastic composite ChoiceDek (50% polyethylene, 50% wood fiber, density 0.91 g/cm3), slightly channeled deck board, before and after 14 min of a fire test (by permission from the University of California Forest Products Laboratory).

Figure 14.4 Wood-plastic composite Nexwood (42% HDPE, 46% wood fiber, 12% minerals, density 1.17 g/cm3), hollow deck board, before and after 14 min of a fire test (by permission from the University of California Forest Products Laboratory).

Figure 14.5 Wood-plastic composite Trex (50% LDPE, 50% wood fiber, density 0.92 g/cm3), solid deck board, before and after 40 min of a fire test (by permission from the University of California Forest Products Laboratory).

Figure 14.6 Wood-plastic composite Rhino Deck (35% HDPE, 65% wood fiber, density 1.13 g/cm3), solid deck board, before and after 22 min of a fire test (by permission from the University of California Forest Products Laboratory).

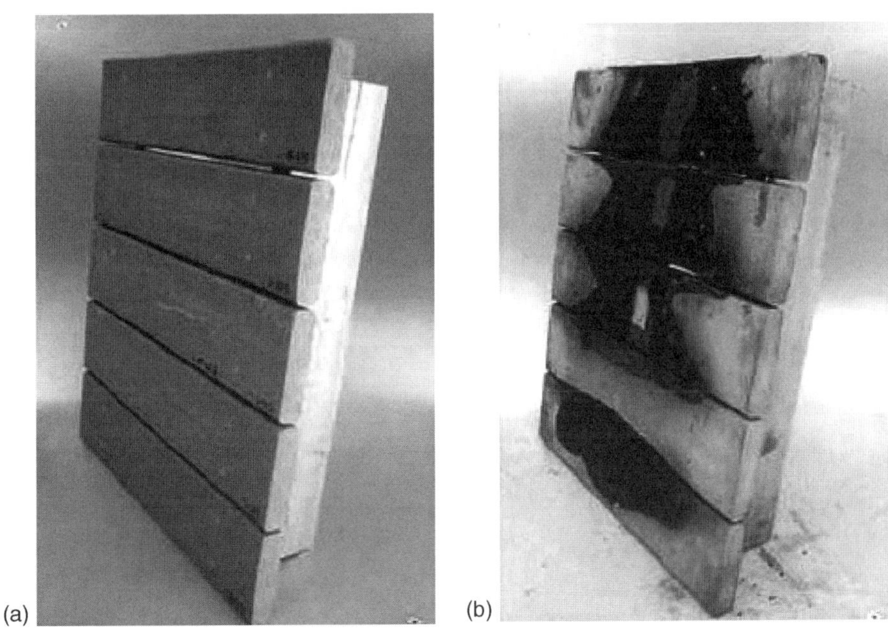

Figure 14.7 Wood-plastic composite SmartDeck (35% polyethylene, 65% wood fiber, density 1.10 g/cm3), solid deck board, before and after 40 min of a fire test (by permission from the University of California Forest Products Laboratory).

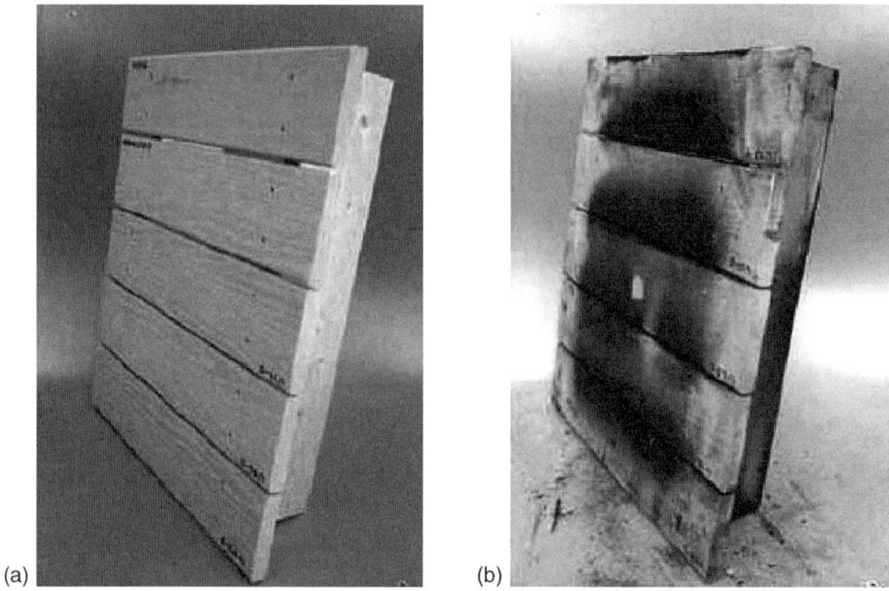

Figure 14.8 Wood-plastic composite WeatherBest (33% polyethylene, 60% wood fiber, 7% minerals, density 1.20 g/cm3), solid deck board, before and after 40 min of a fire test (by permission from the University of California Forest Products Laboratory).

Figure 14.9 Wood-plastic composite WeatherBest (33% polyethylene, 60% wood fiber, 7% minerals, density 1.20 g/cm3), hollow deck board, before and after 15 min of a fire test (by permission from the University of California Forest Products Laboratory).

Figure 14.10 Redwood (density 0.40 g/cm3), solid deck board, before and after 40 min of a fire test (by permission from the University of California Forest Products Laboratory).

REFERENCES

1. R.H. White and M.A. Dietenberger. Fire safety. In: *Wood Handbook*, Forest Products Society, Madison, WI, 1999, Chapter 17, p. 17–3.
2. C. Vasile and M. Pascu. *Practical Guide to Polyethylene.* Rapra Technology, Rapra Technology Ltd. UK, 2005, p. 93.
3. R.H. White and M.A. Dietenberger. Fire safety, In: *Wood Handbook*, Forest Products Laboratory, 1999, Chapter 17, p. 17–7.
4. J. Troitzsch. *Plastics Flammability Handbook. Principles, Regulations, Testing, and Approval.* 3rd edition, Hanser, *Munich-Cincinnati*, 2004, p. 18.
5. *Natural & Wood Fiber Composites,* Vol. IV, No. 9, Principia Partners, Cleveland, OH, September 2005, p. 2,
6. J. Troitzsch. *Plastics Flammability Handbook. Principles, Regulations, Testing, and Approval.* 3rd edition, Hanser, Munich-Cincinnati, 2004, p. 146.
7. J. Andrews. Flame retardants for the future today. In: *15th International Conference ADDITIVES 2006, Plastics Additives for Special Effects*, ECM, Plymouth, MI, Las Vegas, NV, January 30–February 1, 2006.
8. P. Moy. Recyclability of FR-PC/ABS composites using non-halogen flame retardants. In: *15th International Conference ADDITIVES 2006, Plastics Additives for Special Effects*, ECM, Plymouth, MI, Las Vegas, NV, January 30–February 1, 2006.

9. R.B. Dawson and S.D. Landry. The issues & regulatory status for flame retardants. In: *15th International Conference ADDITIVES 2006, Plastics Additives for Special Effects*, ECM, Plymouth, MI, Las Vegas, NV, January 30–February 1, 2006.
10. *Chem. Eng. News*, American Chemical Society, Washington, DC, June 26, 2006, p. 8.
11. R.L. Markezich. Clorine containing flame retardants. In: *15th International Conference ADDITIVES 2006, Plastics Additives for Special Effects*, ECM, Plymouth, MI, Las Vegas, NV, January 30–February 1, 2006.
12. J. Troitzsch. *Plastics Flammability Handbook. Principles, Regulations, Testing, and Approval*, 3rd edition, Hanser, Munich-Cincinnati, 2004, p. 147.
13. G. Beyer. Nanocomposites offer new way forward for flame retardants. *Plastic Addit. Compd.*, 2005, **7**(5), p. 32–35.
14. G. Guo, Y.H. Lee, Y.S. Kim, C.B. Park, and M. Sain, Flame retarding effects of nanoclay on wood-fiber composites, In: *The Global Outlook for Natural Fiber & Wood Composites 2004*, Intertech, Portland, ME, New Orleans, LA, December 8–10, 2004.
15. J. Innes, A. Innes, M. Wajer, D. Smith, and L. Granada. Nano materials as flame retardants in metal hydrate FR formulations. In: *15th International Conference ADDITIVES 2006, Plastics Additives for Special Effects*, ECM, Plymouth, MI, Las Vegas, NV, January 30–February 1, 2006.

15

THERMO- AND PHOTOOXIDATIVE DEGRADATION AND LIFETIME OF COMPOSITE BUILDING MATERIALS

INTRODUCTION. LIFETIME OF PLASTICS AND PLASTIC-BASED COMPOSITES: EXAMPLES

All plastics, as well as all organic materials, retain their functional properties during only a certain period of time. There are multiple events that shorten the lifetime of plastics—some of them are accidental, such as mechanical failure as a result of reaching an ultimate load, that is, a break load for a given profile, or a critical impact resistance, or a case of fire, and some of them are cumulative, spreading over a long time period, such as wear and tear, or thermo- and photooxidation.

In this chapter we will focus on thermo- and photooxidation and their combined effect on plastic-based fiber/mineral composite materials.

Thermooxidation of plastics is a relatively well-developed area of polymer chemistry. It is a dramatically underdeveloped area of composite building materials.

The resistance of commercial polymer grades to thermooxidation is commonly indicated by their UL temperature index (Table 15.1). It shows at which continuous temperature the polymer will serve for 100,000 h, that is, 11.4 years. "Will serve" in this context means until its impact strength, or strain at break, or other chosen mechanical property is reduced by 50%.

According to some other data [2], the service of unprotected polyethylene in the dark at 20°C (68°F) is 8–10 years, and that of stabilized polyethylene in moderate and tropical climates is 15–20 and 2–5 years, respectively.

Wood-Plastic Composites, by Anatole A. Klyosov

TABLE 15.1 UL temperature index for generic plastics [1]

Polymer	UL temperature index	
	(°C)	(°F)
LDPE	50	122
PVC (structural foam)	50	122
HDPE	55	131
PP (copolymer)	90	194
PVC (crosslinked)	95	203
PP (homopolymer)	100	212

Note of the author: Data in the table can describe plastics stabilized with different amounts of antioxidants and cannot be taken as the thermostability index for pure plastics.

That is why commercial plastics commonly contain added antioxidants, typically in amounts between 0.05 and 0.15% by weight, and 0.2–0.5% for stabilization of regrinds. This significantly extends the lifetime of plastics compared to unprotected ones, though a good portion of these antioxidants is consumed during the subsequent processing of plastics into shapes.

As a result, incoming plastics, as pellets, powder, or flakes, to a composite-making plant show variable life span, depending on the manufacturer of the plastic (see Table 15.2). This life span is often measured in the OIT values, or the oxidative induction time. This will be explained in detail below in this chapter, and essentially the OIT is a lifetime of the material in a chamber filled with pure oxygen and heated to a certain temperature, typically to 180–200°C. At this temperature the material is oxidized by about 1–10 million times faster than in ambient conditions.

TABLE 15.2 The lifetime coefficient as the OIT of HDPE obtained from different suppliers

HDPE	OIT (min)
HDPE-1	>100
HDPE-2	68 ± 1
Chevron CHVX891180	39 ± 2
Petromont PSPX7006	31 ± 2
HDPE, MFI 10	30 ± 2
Chevron CHVX896880	24 ± 1
HDPE-3	22 ± 2
HDPE-4	14 ± 4
HDPE-5	14 ± 2
HDPE-6	12 ± 2
An HDPE, recycled	4.4 ± 0.4
HDPE-7 (powder)	0.98 ± 0.08

The OIT values were measured at 190°C in pure oxygen. The origins of HDPE with the highest and lowest OIT values in the table are not disclosed.

In other words, if it takes 1 min for a material to complete its oxidation at 200°C, it might take between 2 and 20 years for complete oxidation of the same material in the natural environment. These are reasonable figures for many oxidation cases.

HDPE data in Tables 15.1 (at 55°C) and 15.2 (at 190°C) generally correspond to each other, despite (a) they were obtained at different temperature conditions; (b) HDPE in Table 15.1 is "generic"; and (c) the readout in Table 15.2 is not a loss of 50% of mechanical properties, but a certain degree of oxidative degradation. As it will be shown below, the temperature coefficient for thermooxidation of an HDPE-based composite (between 190°C and 130°C) is equal to 2.0–2.5/10°C. The last figure means that a temperature change by every 10°C results in an increase or decrease in the reaction rate by 2.0–2.5 times. This is not exact science in this particular case (as well as with temperature coefficients in general over a wide temperature range); however, here we look at a general picture.

Taking the last HDPE sample in Table 15.2 as apparently having the lowest amount (or none) of antioxidants, and using the temperature coefficient of 2.5/10°C as an approximation, we find that at 55°C the lifetime of the HDPE sample is approximately 90,000 h, that is, not far from 100,000 h in Table 15.1. This implies that we can use the temperature coefficient of 2.5 in order to make approximate temperature estimates for thermooxidation processes in rather wide temperature ranges, that is, over 100°C, or over 200°F.

Now, it is not commonly realized that by moving from a neat HDPE to a filled composite material, the lifetime of the resulting composite dramatically lowers, often by 50–100 times. That is, the OIT in Table 15.2 for composite materials can decrease to as low as 0.3–0.5 min, if no antioxidants are added in the composite manufacturing process. Actual drop in the OIT value largely depends on the compounding process (on shearing and heating effects first of all), on type of fillers, colorants, lubricants, and, of course, type and amount of added antioxidants, if any.

It can be calculated, using the above approach and the temperature coefficient of 2.5/10°C, that at the OIT of 0.3 min (at 190°C) the lifetime of a composite board on a deck at 70°C (158°F) will be 62 days, or only 2 months. This is when no antioxidants are added to the plastic. In fact, a temperature of 158°F is what I personally measured on a deck surface in Phoenix, AZ, in August, when air temperature was 108°F. Just for the record, 108°F is close to an average high temperature in July in that area. Typically, in the early afternoon, the deck surface temperature is 50°F higher than air temperature. This will be discussed later in this chapter.

The estimate of 2 months as a lifetime for a composite deck in Arizona climate is quite realistic and actually confirmed by direct observations—when hollow wood–plastic composite (WPC) deck boards did not contain added antioxidants, besides an amount that the incoming HDPE had. We have multiple observations of a lifetime of deck boards in Arizona as being between 4 months and a year, when no antioxidants were added to the composite formulation.

Industrial compounders cause severe shear and heating of WPC materials, particularly composites that contain abrasive ingredients. As a result, the OIT values can drop to much lower than those shown in Table 15.2. Examples will be given in subsequent sections of this chapter. Without added antioxidants, on top of what was already present in an incoming HDPE, the OIT for the composite can be as low as 0.3–1.0 min. This might result in an inadequate lifetime of composite decks, particularly in the South, that can be shorter than 3–5 years. Sometimes they last only for months.

THERMOOXIDATION, PHOTOOXIDATION, OXIDATIVE DEGRADATION, AND PRODUCT CRUMBLING AND FAILURE

Why some decks without an adequate amount of antioxidants might serve—in the South—for only a few months, or at best for only a few years? What happens after that (and in the course of it) and why?

Oxidative degradation of composite deck boards is the most common reason for board failure, except their mechanical failure. However, the oxidative degradation might cause a premature mechanical failure of composite materials. Oxidative degradation as a result of thermo- and photooxidation leads to board crumbling, when its surface can be easily scratched by a fingernail. Board surface becomes loose, powdery, and weak. After a while, it can be not only scratched, but also whole layers of the material can be easily removed by applying a little force. With hollow boards, thermooxidation leads to deterioration of profiles from both outside and inside, and sooner or later the board collapses. In the worst case, it can bring a bodily injury, particularly with an elevated deck.

Composite deck boards are much more sensitive to an oxidative degradation compared to pure plastics. Hence, they require much more antioxidants and other stabilizers.

Typically, composite deck boards are porous. The pores are formed by steam and by volatile organic compounds (VOC) during extrusion. Composite boards are partly foamed, and the pores are typically opened and connected to each other, forming chains of cavities. That is why composite materials absorb water, unlike many plastics. Air oxygen flows in, through these pores, and effectively oxidizes composite materials from inside, particularly at elevated temperatures, which often takes place on decks. Water, which is always present in composite materials, serves as a catalyst for the oxidation. Metals, which are often present in composites (as constituents of colorants, lubricants, biocides, fillers), also serve as efficient catalysts of oxidative degradation of composites. As a result, rates of oxidative degradation of composites are 50–100 times faster than those for their constituent plastics.

On top of thermooxidation, photooxidation speeds up an oxidative degradation of composite deck boards. The effect of photooxidation is dual. First, it adds to heating of the material, thereby speeding up the thermooxidation. Second, it causes an additional chemical damage by creating free radicals at the board surface, which propagate at some depth into the material. Hence, under hot sun, thermo- and photooxidation work in a synergism, reinforcing each other's action toward crumbling, deterioration, weakening, and finally board failure.

In a simplified manner, the chemical structure of polyethylene can be shown as follows:

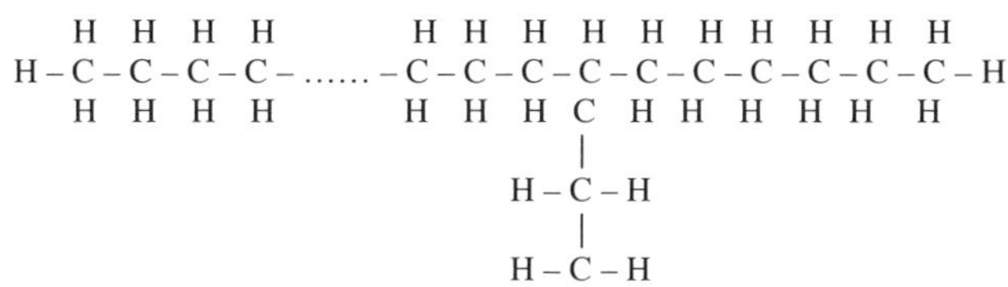

Hydrogen atoms in polyethylene are called

- Primary: at the very end of the chain;
- Secondary: along the chain;
- Tertiary: in branched polyethylene.

They show different sensitivity in abstracting by free peroxide radicals, in the ratio of 1:6:17.

Where the free peroxide radicals come from?

They are formed in the following chain of events. First, the initial step, a chain initiation, which happens as a result of thermal or photo-induced breakage of a chemical bond in the polymer. This often happens with the help of a catalyst (water, metal ion) in the vicinity of the reaction site, where the initiation step produces a hydrocarbon free radical, R^*:

$$RR \rightarrow R^* + R^* \text{ (primary macroalkyl radical)}$$

or/and

$$RH \rightarrow R^* + H^* \text{ (secondary macroalkyl radical)}$$

Scission of the backbone C—C bond requires seven times more energy compared to scission of the C—H bond.

Free radical R^* reacts with atmospheric oxygen, to form a peroxide free radical:

$$R_1^* + O_2 \rightarrow R_1OO^*$$

The peroxide free radical is very unstable and quickly acquires an electron by a radical mechanism via, for instance, the abstraction of hydrogen from an adjacent polymer molecule, thereby propagating further oxidation:

$$R_1OO^* + R_2H \rightarrow R_1OOH + R_2^*$$

Hence, it results in an unstable hydroperoxide R_1OOH and the new free radical, formed on another polymer molecule. In other words, the chain of events so far yields in fracture of the first polymer chain and a damage brought to the second polymer chain. The conversion of peroxides (ROO^*) to hydroperoxides (ROOH) is the rate-determining step for the chain reaction.

The new free radical R_2^* immediately reacts with oxygen:

$$R_2^* + O_2 \rightarrow R_2OO^*$$

and the chain reaction continues until it is quenched by, for example, an antioxidant.

Meanwhile, an unstable hydroperoxide R_1OOH collapses into two new free radicals:

$$R_1OOH \rightarrow R_1O^* + OH^*$$

By this moment the chain of events—since the initiation—results in fracture of the first polymer chain, forming of two free radicals ready to break two more polymer molecules, just-about fracture of a second polymer molecule, and a further propagation of free radicals-induced polymer breakage:

$$RO^* + RH \rightarrow ROH + R^*$$

$$OH^* + RH \rightarrow R^* + H_2O$$

This branched chain reaction of free radical polymer degradation leads to a rapid conversion of long polymer chains of the composite material to much shorter fragments of polymers and, therefore, makes the deck board loose, crumbling, weak, and finally collapsed, both visually and mechanically.

In principal, all the reactions leading to formation of free radicals are reversible and, indeed, many primary macroalkyl radicals recombine, either intramolecularly or intermolecularly, leading to branching and cross-linking. Often this results in decrease in the melt flow index in earlier steps of oxidation (the mechanical properties do not seem to be very sensitive to cross-linking in these steps), and then, with a progressive reduction in molecular weight of the polymer due to chain scission, and the melt index is increased. Concurrently, the mechanical properties of the plastic or the composite material decrease. Eventually the material reaches a step of catastrophical failure due to its brittleness, weakness, and practical disappearance of plastic.

Let us take a look on how thermal treatment causes fragmentation of plastic component in composite deck boards. Table 15.3 shows the average molecular weight of HDPE, changed from the raw material to compounded and extruded board to the same board but thermally treated at 190°F for several hours. No antioxidant was added besides the one that was (apparently) present in the incoming HDPE provided by a supplier.

As some readers might have a preference not to consider molecular weights of plastics but their chain lengths; in the number of methylene groups per chain, these average numbers are given in Table 15.4.

TABLE 15.3 Average molecular weight (number-average, weight-average, and viscosity-average) of incoming HDPE (Sherman HDPE) and deck boards—freshly extruded and extensively treated in the Q-Sun accelerated weathering box (cycle: 1:48 h light and 0:18 h light and water)

	Average molecular weight of HDPE		
Material	Number-average	Weight-average	Viscosity-average
Neat HDPE	42,000	350,000	2,408,000
Composite board, surface	21,000	169,400	728,000
Composite board, inside the hollow channel	29,400	161,000	616,000
Composite board after extensive weathering, upper surface	8960	89,600	364,000

One can see that processing of the composition significantly reduces average chain length (and its molecular weight, accordingly) of the incoming HDPE. The longest HDPE chains suffer the most, their average length drops by more than three times. Their subsequent weathering in the box (UV light, heat, and water spray) causes further fragmentation of the HDPE chains, with short molecules becoming almost five times shorter compared to those in the incoming plastic and half of those in the freshly made composite board; an average (weight-average) chain length becoming four times shorter than in the initial HDPE and almost half shorter compared to that in the fresh board. The most significant shift is observed with the longest HDPE chains—their average length is decreased by more than six times overall (between

TABLE 15.4 Average chain length (number-average, weight-average, and viscosity-average) of incoming HDPE (Sherman HDPE) and deck boards—freshly extruded and extensively treated in the Q-Sun accelerated weathering box (cycle: 1:42 h light and 0:18 h light and water)

	Average chain length of HDPE		
Material	Number-average	Weight-average	Viscosity-average
Neat HDPE	3000	25,000	172,000
Composite board, surface	1500	12,100	52,000
Composite board, inside the hollow channel	2100	11,500	44,000
Composite board after extensive weathering, upper surface	640	6,400	26,000

Chain length is shown in an average number of methylene groups per polymer molecule.

On weight-average, number-average, and viscosity-average molecular weights

With low molecular weight chemical compounds having a clearly defined formula, there is no problem with molecular weight determination and its description. For ethylene ($CH_2{=}CH_2$), for example, molecular weight is 28.

Not so with polymers. They consist of a mixture of long, medium, and short chains, with different molecular weights. Reasonably, one should determine the average molecular weight to characterize the polymer. But what is the average in each particular case?

As an example, let us consider a similar real estate situation. There is a district in which there are ten 50-thousand-dollar houses, and two 5-million-dollar houses. What is the average price of houses in this area?

The answer depends on how to count.

If we take a simple average, division of $10,500,000 by 12 gives $875,000 as an average price of each house in the area. Is it a useful figure? Apparently, not. It does not say anything on the fact that the houses are very different in value. It does not describe either cheap or expensive houses. That is why real estate businesses do not use this kind of "average" prices. In polymer science this average figures are used, however, but typically in comparison with other "averages." This first one, considered above, called "weight-average." For one particular HDPE material, the weight-average chain length (degree of polymerization) is 25,000, and the weight-average molecular weight is 350,000. That is, a "weight-average" HDPE molecule contains 25,000 elementary $-CH_2-$ units, which compose a whole "weight-average" HDPE chain.

What do the real estate folks do? They calculate the so-called "median," meaning the price of a house "in the middle," when half of the houses cost less and half of them cost more than the "median." In our case with 12 houses a median will be a 50-thousand-dollar house. Would it be a fair estimate? No, but it is probably better than the above "weight-average" value. At least it describes low-price-level houses, because numerically they prevail. That is why in polymer science this way of showing the average is called "number-average." For the above-mentioned HDPE the number-average chain length (degree of polymerization of the majority of shorter chains) is 3000, and number-average molecular weight is 42,000.

A ratio between "weight-average" and "number-average" molecular weights is called polydispersity. It shows how wide is an average distribution of chain length. The narrower the range, the lower is the polydispersity. For the said HDPE the polydispersity is 8.3.

However, in polymer science there is one more "average," called viscosity-average. It is related to the long chains. In a real estate business they would call it "upscale houses," and they would be among those 5-million-dolar houses. This is an "average" among the costly houses. In polymer science the "average" is called viscosity-average, because viscosity shows only long chains and ignores short chains. Thus for the above-described HDPE the viscosity-average chain length (degree of polymerization) is 172,000, and the average molecular weight is 2,408,000.

When one measures all three "averages" and the polydispersity in the course of oxidative degradation of a plastic, neat, or WPC, he/she has a better idea how the material degrades in terms of its molecular chains.

the initial plastic and the weathered composite material), and it decreased twice between the fresh and the weathered board.

Tables 15.5 and 15.6 show similar data, but for composite boards made with another HDPE (Equistar HDPE). The last board has a lower density (specific gravity)

TABLE 15.5 Average molecular weight (number-average, weight-average, and viscosity-average) of deck boards—freshly extruded and kiln treated

	Average molecular weight of HDPE		
Material	Number-average	Weight-average	Viscosity-average
Composite board	50,200	276,000	1,236,000
Composite board, $d = 1.12$ g/cm^3, kiln-treated	41,700	257,900	1,009,000
Composite board, $d = 1.09$ g/cm^3, kiln-treated	15,100	50,900	114,000

compared to the preceding one in the table, and after kiln treatment it showed distinct signs of thermal degradation (brittle, crumbled, mechanically weak).

One can see that a higher density board (the first row in Tables 15.5 and 15.6) showed only an insignificant drop in integrity of its plastic. However, a lower density board (the last row) revealed catastrophic damage in terms of both polymer chain length and physical properties of the board. Clearly, a lower density board provides more of its inner space for air oxygen to get in and to oxidize the material more rapidly. This phenomenon and its implications will be discussed in more detail below.

Similar effects have been also observed with HDPE-based composite roof tiles. Table 15.7 shows the impact of a progressive weathering of roof tiles on molecular weight of their HDPE.

Both tiles A and B described in the table contained a good amount of antioxidants. Tile A after its extensive weathering (the last tile in the table) did not show any mechanical weakening; however, its surface was covered with microcracks, seen under the microscope, and the surface turned almost white. As it is seen from the table, the longest (average) HDPE molecules at the surface, after weathering,

TABLE 15.6 Average chain length of deck boards—freshly extruded and kiln-treated, as described in Table 15.5

	Average chain length of HDPE		
Material	Number-average	Weight-average	Viscosity-average
Composite board	3600	19,700	88,000
Composite board, $d = 1.12$ g/cm^3, kiln-treated	3000	18,400	72,000
Composite board, $d = 1.09$ g/cm^3, kiln-treated	1000	3600	8100

TABLE 15.7 Average molecular weight (number-average, weight-average, and viscosity-average) of a very top layer of composite roof tiles in the progress of their weathering on roof top

	Average molecular weight of HDPE		
Material	Number-average	Weight-average	Viscosity-average
Tile A, freshly made	40,350	208,500	714,000
Tile B, freshly made	37,300	178,000	591,000
Tile A, weathered	16,900	140,300	456,000
Tile B, weathered	19,000	70,700	264,000
Tile A, progressively weathered	4400	11,400	25,000

became shorter by almost 30 times. In fact, HDPE at the tile surface was almost completely disintegrated.

FACTORS ACCELERATING THE OXIDATIVE DEGRADATION OF COMPOSITES

The oxidative degradation of plastics and plastic-based composites typically proceeds in two distinct stages: (a) the induction time, which may be regarded as the safe lifetime of the material and (b) a progressive degradation, which results in a rapid consumption of oxygen, fast decrease in molecular weight of plastic, increase in brittleness, and fast deterioration of mechanical properties of the material. In the process of the second stage the composite profile acquires a loose texture, becomes soft, structurally unsound, and eventually collapses under its own weight. Visually it looks like its plastic disappears and the material consists of only loose fillers (Fig. 15.1). The surface of the shape, such as a deck board, becomes so loose that it can be easily scratched by a fingernail or a corner of a credit card.

The induction time and, hence, durability, or the lifetime of composites depend mainly upon the following factors:

- Physical and chemical structure of the polymer;
- History of the plastic (virgin, recycled);
- The type and amount of cellulose fiber;
- The type and amount of mineral fillers;
- The presence of stress;
- The presence of metal catalysts (in colorants, lubricants, fillers, etc.);
- The presence of moisture;
- Antioxidants and their amounts;
- Amount of added regrind, if any;

Figure 15.1 Catastrophic failure of a composite deck board due to an oxidative degradation/crumbling (Chini Vallee, AZ)

- Density (specific gravity) of the composite;
- Temperature—heating during the processing; postmanufacturing heat treatment (annealing);
- Location of the deck and climatic conditions;
- Solar radiation (UV light).

We will consider these factors.

Density (Specific Gravity) of the Composite

Apparently, the effect of density in composite materials has been seriously underestimated in the literature, and, seemingly, by many manufacturers of composite building materials. Typically, density is considered as a factor determining a profile weight in terms of transportation expenses and convenience of handling during an installation of deck, and as a factor defining expenses for manufacturing and raw materials. However, density of a given composite material largely determines its lifetime also.

It was mentioned above that composite deck boards are porous by virtue of their manufacturing. The pores are typically open and interconnected, forming chains of cavities. Air oxygen flows in, through these pores, and effectively oxidizes composite materials from inside, particularly at elevated temperatures, which often take place on the surface of decks.

As a result, neat HDPE-made shapes, for example, are oxidized only from the surface, whereas HDPE-based composite profiles (as other composites) are oxidized largely from within, wherever air oxygen can penetrate through pores. Inner surface, available to oxygen, is sometimes hundreds of times larger, compared to the outer surface of plastics, available to oxygen. Pore volume of composite materials available to water, is 0.5–6% of the total volume of composites during 24-h immersion in water, and up to 20% and more of the total volume during 40-day immersion in water. This can be taken as an indication of the minimum inner volume of composites accessible to air, one-fifth of which is oxygen. For neat polyolefins, the inner volume, accessible to water or oxygen, is essentially zero.

Besides water absorption, another approach for estimations of an inner volume of composites is as follows. Let us consider two composite materials, Trex and GeoDeck. Trex composites consists of two major ingredients (50:50), that is, polyethylene (density 0.94 g/cm^3) and wood flour (density 1.30 g/cm^3). Hence, the overall specific gravity of Trex composites should be 1.12 g/cm^3 ($0.94 \times 0.5 + 1.30 \times 0.5$). In reality, it is 0.94 g/cm^3. If we assume that the difference between the two figures is due to porosity, the latter is equal to 19%. That is, about one-fifth of the whole volume of the composite material is open for oxygen to destroy plastic from within the material.

GeoDeck's overall density should be equal to 1.24 g/cm^3. Therefore, when GeoDeck board has specific gravity of 1.06 g/cm^3, its porosity is 17%, and when it is of 1.23 g/cm^3, its porosity is only 0.8%. Hence, the last material is more than 20 times less likely to suffer from oxidative degradation compared to the first one. In a simplified manner, it will serve 20 times longer. This was confirmed by direct experiments in the weathering box (Table 15.12 below).

Certainly, the density of a composite material is only one of the factors that determine the rate of the oxidative degradation of the composite material. However, all conditions equal, density plays a very significant role in determining the lifetime of a deck board.

Table 15.8 summarizes experimental results on the effect of density (specific gravity) of a WPC on its heat resistance (actually, on the velocity of its thermal oxidation) in an airflow oven. The control composite boards were not heat-treated. All other boards were kept in the oven for 87 h at 225°F. It was found that the lower the density, the faster the oxidation (Table 15.8).

TABLE 15.8 Thermooxidation of the "zero level" GeoDeck composite boards in an airflow oven at 107°C(225°F) after 87 h, measured by residual load at failure (flex and shear strength) after boards conditioning

Specific gravity (g/cm^3)	Residual load at failure compared to control (no heating) (%)
1.02	41
1.07	48
1.08	51
1.09	42
1.10	64
1.105	63
1.11	50
1.12	63
1.13	59
1.17	75

No antioxidant was added to composite materials ("zero level" of antioxidants).

Table 15.8 shows an obvious trend—the higher the density, the higher the durability of deck boards. The reason is rather obvious: the lower the board density, the higher its porosity, the faster air oxygen flows into the board and into its pores. The higher the temperature, the faster the oxidation of the composite material. The higher the oxidation, the weaker the material.

In this respect a WPC board is principally different from boards made from neat plastic. In neat plastic, oxidation of the material takes place only on the surface. In the bulk, oxidation goes in severe oxygen starving conditions. WPCs typically have porous matrix, hence, a rather significant water absorption. If water is absorbed so well, oxygen is absorbed much faster, because it has smaller, easily diffusible molecules.

The composite boards described in Table 15.8 did not have any added antioxidants. "Added" in this context means that only those antioxidants were introduced by plastic manufacturers into the plastic, that have been present in the initial material, if anything. No antioxidants were additionally added into the WPC formulation ("zero level of antioxidants").

Table 15.9 shows the data obtained with the same WPC boards, but some amount of an antioxidant was added into the formulation ("first level of antioxidant amount"). These data show the same trend—the higher the density, the higher the durability. Clearly, in the presence of a relatively small amount of an antioxidant ("first level"), the durability of the boards is significantly higher compared with that in Table 15.8. It is worth mentioning that in the presence of the antioxidant, boards with density below 1.125 g/cm^3 were not produced (compare with Table 15.8). Apparently, the antioxidant slowed down the decomposition of HDPE in the extruder, hence, effectively decreased the amount of VOC and, as the result, decreased porosity of the extruded deck board, and increased its density.

TABLE 15.9 Thermooxidation of the "first level" (with respect to antioxidants) GeoDeck composite boards in an airflow oven at 106°C (223°F) after 93 h, measured by residual load at failure (flex and shear strength) after boards conditioning

Specific gravity (g/cm^3)	Residual load at failure compared to control (no heating for control) (%)
1.125	74
1.14	84
1.14	86
1.15	83
1.16	84
1.16	90
1.17	93
1.18	94
1.19	95

Data in Table 15.10 were obtained at the tripled amount of the antioxidant, compared with that in Table 15.9 ("second level") and at a higher temperature and a much longer heating time. Again, these data show the same trend—the higher the density, the higher the durability. It is remarkable that at a noticeably higher temperature

TABLE 15.10 Thermooxidation of the "second level" (with respect to antioxidants) GeoDeck composite boards in an airflow oven at 111°C (232°F) after 264 h, measured by residual load at failure (flex and shear strength) after boards conditioning

Specific gravity (g/cm^3)	Residual load at failure compared to control (no heating) (%)
1.125	86
1.15	93
1.16	96
1.175	95
1.20	99

The "second level" contained triple amount of the antioxidant compared to the "first level" (Table 15.9).

and a much longer heating time, boards with a relatively high density (1.20 g/cm^3) showed practically no degradation, whereas at much more comfortable conditions, deck boards with low density (1.02–1.09 g/cm^3) but without added antioxidant lost 50% and more of their strength.

This phenomenon shows that the two powerful stabilizing factors, which prolong the lifetime of composite deck boards, are board density and added antioxidants. These two factors are in a way functionally interchangeable, but taken together they work in synergism. Antioxidants block propagation of free radicals, and density controls the amount of air oxygen flowing into the pores of the composite matrix. High density of a composite material effectively blocks access of oxygen and slows down oxidative degradation.

Data of Table 15.8 can be presented in the following convenient way (Table 15.11), after smoothing their fluctuations.

To show the effect of the antioxidant on lifetime of the composite board, Table 15.12 combines data for three levels of the antioxidant—"zero level," "first level,"

TABLE 15.11 Effect of board density on durability of GeoDeck composite boards with no added antioxidant in terms of half-life

Specific gravity (g/cm^3)	Time to reach 50% reduction of board strength at 107°C (225°F) (h)
1.02	68
1.07	82
1.08	90
1.09	98
1.10	106
1.105	110
1.11	116
1.12	131
1.13	146
1.17	210

Data are based on airflow experiments.

TABLE 15.12 Effect of board density on durability of GeoDeck composite boards with no antioxidant and two levels of the antioxidant in terms of half-life (h and months) at 107°C (225°F)

Specific gravity (g/cm^3)	No antioxidant	"First level"	"Second level" (tripled the first level)
1.02	68	—	—
1.07	82	—	—
1.08	90	—	—
1.10	106	—	—
1.12	131	195	2,100 (3 months)
1.15	—	390	4,200 (6 months)
1.17	210	810	7,040 (10 months)
1.19	—	1,200 (1.7 months)	—
1.20	—	—	28,600 (40 months)

Data are based on airflow experiments.

Determination of half-life time of a composite deck board at a certain temperature

In the course of airflow oven experiments (Tables 15.8–15.12) we have the initial and residual determined break strength of GeoDeck boards (in terms of break load for board sections exposed in the airflow oven at different temperatures for different time periods).

Considering thermooxidative degradation of composite deck boards as a first-order reaction, the following equation is applicable:

$$c/c_0 = e^{-kt}$$

where

c/c_0 = residual average break strength (after exposure of WPC samples for a certain time at a certain temperature),

k = first-order rate constant and

t = exposure time (336 h for the case below).

The most extended series of data that we have obtained contains 39 data points for exposure time of 336 h at 114°C (237°F) for GeoDeck deck boards containing 0.25% of an antioxidant (Irganox B225). An average residual break strength (c/c_0) for this series of data was equal to 0.67.

Because $\ln 0.67 = -k \times 336$ h, the first-order rate constant for oxidative degradation of GeoDeck boards (containing 0.25% of Irganox) at 114°C is equal to

$$k = 1.19 \times 10^{-3}\ h^{-1}$$

This oxidation rate constant is directly associated with the half-life time (that is, time at which strength is reduced to half of the initial one), by the following formula:

$$t_{1/2} = 0.693/\text{k}$$

and the half-life time of the GeoDeck boards at these conditions is equal to 582 h (24.3 days).

In other words, if a GeoDeck board is kept in the oven at 237°F for 24 days, its strength drops to 50% of the initial, presumably due to oxidation.

Similarly, it was found that at 107°C (225°F) the same GeoDeck board retained 93.6% of its strength after 112 h of exposure (for the second most extended series of data consisting of 29 data points), and its first-order rate constant at this temperature is equal to

$$k = 0.6 \times 10^{-3}\ \text{h}^{-1}$$

and the half-life time to 1155 h, that is, 48.1 days.

and tripled "second level." Temperature corrections are also made (temperature coefficient of 2.5/10°C, see above).

Table 15.12 shows a combined effect of density and antioxidants. The lifetime of the least and the most durable deck boards in the table differs by 420 times. In the real world this would correspond to 1 month and 35 years, respectively!

Hence, one can say that only the effect of density on lifetime of composite deck board can reach 10–20 times overall.

Temperature

Temperature accelerates oxidative degradation of composite materials following a common—for chemical reactions—temperature factor between 2 and 3. As it was described above, this means that by changing temperature by 10°C a reaction velocity changes 2–3 times.

Determination of the temperature coefficient for a WPC deck board due to oxidative degradation

It was shown in the preceding insert that at temperature lower than 7°C, or 12°F, the reaction rate decreases from 1.19×10^{-3} to $0.6 \times 10^{-3}\ \text{h}^{-1}$, that is, 1.98 times. This corresponds to a temperature coefficient (change in reaction rate per each 10°C) equal to 2.7.

This can be shown as follows:

Because a 7°C temperature difference resulted in 1.98-fold change in the reaction rate, therefore

$$x^7 = 1.98 \qquad (15.1)$$

where x is the change of the reaction rate caused by 1°C. Equation 15.1 is transformed to

$$7 \log x = \log 1.98 = 0.297$$

and $\log x = 0.0424$.

There are two basic ways to continue from here. One way is to take the antilogarithm of 0.0424, and find that $x = 1.103$. This means that the temperature change by 1°C causes the oxidation rate change by 10.3%. Hence, the temperature change by 10°C will cause the oxidation rate change by $1.103^{10} = 2.7$.

Another way is to consider that $\log x = 0.0424$ (see above); hence, $10 \log x = 0.424$. The antilogarithm of 0.424 equals to 2.7.

Note: The antilogarithm of 0.424 equals to 2.66. However, we round it up to 2.7, because the temperature coefficient of 2.66 would have an unrealistically high precision, pretending that it is determined with a 1% accuracy (and even better).

As one can see, this value of the temperature coefficient of 2.7 for the oxidative degradation of GeoDeck composite boards above 100°C (more accurately, between 107 and 114°C) is rather close to that of 2.5, used above as an educated guess.

In the preceding sections we have used a temperature coefficient of 2.5/10°C for oxidation reactions of HDPE and HDPE-based composite materials. It seemed to work well. As it is shown in the insert, the temperature coefficient, determined directly from experimental data, is equal to 2.7, which is rather close.

Determination of the activation energy and the temperature coefficient for the oxidative degradation of a WPC deck board

There is a third way to calculate the temperature coefficient from the above experimental data, using a concept of the activation energy of the reaction.

A formula for the temperature dependence of a rate constant is

$$k = A\mathrm{e}^{-E/RT} \tag{15.2}$$

or

$$\ln k = \ln A - E/RT$$

where A, E, and R are constants, values of which are irrelevant for temperature coefficient calculations, and T is the absolute temperature (273° + deg. C). For a pair of temperatures, as shown above,

$$\ln k_1/k_2 = E/R\,(1/T_2 - 1/T_1)$$

and for the above values of k (1.19×10^{-3} and 0.60×10^{-3} h) and T (387° and 380°K) we obtain

$$\ln 1.983 = E/R\,(2.63158 \times 10^{-3} - 2.58398 \times 10^{-3})$$

or

$$0.686 = E/R \times 4.76 \times 10^{-5}\ \mathrm{K}^{-1}$$

From this, $E/R = 14{,}414$ K, and E (activation energy) = 28.64 kcal/mol, because R (the universal gas constant) is equal to 1.987 cal/mol/K, or 8.3 J/mol/K.

For 10°C increase in temperature (for instance, from 104 to 114°C, or 377 and 387°K) this formula can be transformed to

$$\ln(\text{the temperature coefficient}) = E/R \times (2.6525 \times 10^{-3} - 2.5840 \times 10^{-3}),$$

or

$$\ln \text{(the temperature coefficient)} = E/R \times 6.85 \times 10^{-5}\ \text{K}^{-1}$$

From this,

$$\ln \text{(the temperature coefficient)} = 0.987,$$

and the temperature coefficient is equal to 2.7. It shows the increase in rate of oxidative degradation of the composite material with each 10°C increase in temperature.

Note: An antilogarithm (natural) of 0.987 equals to 2.68. However, as it was explained earlier (see insert), the temperature coefficient of 2.68 would pretend, precision wise, that it is determined with an accuracy of 1%. It is not so, of course.

At another insert (p. 508) it was shown that the half-life time (in terms of a drop of strength to 50% of the initial due to oxidation) of GeoDeck boards at 237°F equals to 582 h (24 days). At 225°F, only 12° lower, the half-life time of the boards doubles, to 48 days. This shows how sensitive WPC materials are to temperature, due to their oxidation.

An average high temperature difference between Phoenix, AZ, (106°F) and Boston, MA (81°F) is 25°F (13.9°C). With the temperature coefficient of 2.7/10°C, or 1.7/10°F, a difference in the oxidation rates between the two said geographical regions will be four times (see insert below). In reality, it is 4.5 times, as will be shown below. The values are pretty close to each other, taking into account the complexity of calculations of oxidation rates of decks in the real world, particularly when the calculations are based on amounts of warranty claims due to boards crumbling! Also, the average high temperature difference between Arizona and Massachusetts varies from year to year, along with the differences between the temperatures. However, taking the difference in the

Some useful calculations using the temperature coefficient (°C)

If the temperature coefficient is equal, say, 2.7/10°C, that is, by increasing the temperature by 10°C the reaction rate is increased by 2.7 times. A question: What would be the rate increase if temperature is increased by, say, 13.9°C?

This ratio can be described as follows:

$$x^{10} = 2.7$$

$$x^{13.9} = y$$

where

x is the increase in the reaction rate per 1°C

y is the increase in the reaction rate per 13.9°C.

In order to solve the ratio, let us take logarithm of both the parts:

$$10 \log x = \log 2.7$$

$$13.9 \log x = \log y$$

From the upper line, $\log 2.7 = 0.431$, and $\log x = 0.0431$

From the lower line, $13.9 \times 0.0431 = \log y$

Hence, $\log y = 0.60$,

and $y = 4.0$

An answer: If temperature is increased by 13.9°C, the rate of oxidative degradation increases by four times.

oxidative degradation between the two areas as 4.5 times at the average temperature difference of 13.9°C would result in an apparent temperature coefficient of 2.95. However, if the temperature difference is not 13.9°C, but, say, 14.9°C, the temperature coefficient of the natural oxidative degradation, resulting in boards crumbling, will be 2.78.

The last value can be obtained as follows:

$$x^{14.9} = 4.5$$

$$x^{10} = y$$

where
x is the increase in the reaction rate per each 1°C
y is the increase in the reaction rate per 10°C, that is, the temperature coefficient.

The value of 4.5 shows that with the temperature increase by 14.9°C, the reaction rate increased by 4.5 times (see above).

Solution of these two equations in the form

$$14.9 \log x = \log 4.5 = 0.6532$$

$$10 \log x = \log y = 0.438$$

gives $y = 2.74$ (the temperature coefficient), or, rather, 2.7, taking into account an accuracy of the data its calculation was based upon.

Some useful calculations using the temperature coefficient (°F)

The same approach as described in the preceding insert can be used to calculate the temperature coefficient in degrees Fahrenheit. If the temperature coefficient is equal to, say, 2.7/10°C, that is, the reaction rate increases 2.7 times per 10°C,

$$x^{10} = 2.7$$

and knowing that the temperature increase of 10°C is equal to the temperature increase of 18°F, we can write:

$$y^{18} = 2.7$$

$$y^{10} = z$$

where
x is the increase in the reaction rate per 1°C,
y is the increase in the reaction rate per 1°F and
z is the increase in the reaction rate per 10°F.

In order to solve the two last equations, let us take logarithm of both of them:

$$18 \log y = \log 2.7$$

$$10 \log y = \log z$$

From the upper equation, $\log y = 0.0240$.
From the lower equation, $\log z = 0.240$.
Hence, $z = 1.74$
Answer: If the temperature coefficient is equal to 2.7/10°C, it is equal to 1.7/10°F.

One more example of temperature dependence of the oxidative degradation of composite materials, now in pure oxygen, is given in Table 15.13. Here a specimen of GeoDeck, an HDPE-based composite material filled with rice hulls and Biodac® (a blend of calcium carbonate, kaolin/clay, and delignified cellulose fiber), was placed in pure oxygen atmosphere at temperatures between 130 and 190°C, and its oxidation was continuously monitored in a differential scanning calorimeter, following a modified ASTM D 3895.

One can see that for this particular material (GeoDeck), the temperature coefficient is close to 2.0 (energy of activation is 25.0 kcal/mol). At the doubled amount of the antioxidant, the OIT figures are 11.72 min at 190°C and 25.12 min at 180°C, and the temperature coefficient in this temperature range is 2.1. It is somewhat lower than the temperature coefficient for the same material oxidation in the airflow oven, but the latter test was conducted in a lower temperature range (the temperature coefficient as well as the activation energy depends on the temperature range), and the readout of the test was a break load, unlike heat evolution in the OIT test. Overall, the temperature coefficient is not a universal constant, and even for the same process it depends on many factors which are sometimes difficult, if possible, to control.

Because the temperature coefficient for the above process (Table 15.13) is practically the same in the presence of different amounts of the antioxidant, it is safe to say that the antioxidant does not change the chemistry of oxidative degradation of the composite material. It just slows down the process of degradation by 37 times in this particular case (from zero amount of added antioxidant to 0.5% by weight).

However, temperature dependencies in the presence of antioxidants can be misleading, when the antioxidants(s) is volatile at high temperatures. In this case the OIT will be progressively decreased with the increase in temperature not only because of a chemical depletion of the antioxidant(s) but also because of its physical depletion. This in turn would result in an unrealistically steep temperature dependencies of the OIT, as well as too high temperature coefficients and unrealistically high activation energies of the reaction. If at very high temperatures a volatile antioxidant(s) is completely volatilized, the temperature dependence would come to normal, reflecting only the chemical processes of oxidation.

TABLE 15.13 OIT for an HDPE-based composite material (GeoDeck, 0.25% of an antioxidant) in pure oxygen atmosphere

Temperature(°C)	OIT (min)	OIT increase per 10°C
190	0.32	—
180	0.50	1.56
170	1.03	2.06
160	2.10	2.04
150	3.99	1.90
140	8.17	2.05
130	17.14	2.10
Average		2.0 ± 0.2

Apparently, this can explain an unusually steep temperature dependence of the OIT of polyethylene degradation, described in "Application News T95" of Shimadzu DSC, entitled "Measurement of Oxidizing Induction Time of PE by DSC." The authors have obtained the following data for the oxidation of polyethylene in the presence of an antioxidant:

Temperature (°C)	OIT (min)
180	316
190	78.58
200	20.46
210	7.1

The data show that the temperature coefficient for the reaction is progressively decreased by each 10°C increase: 4.02 (180–190°C), 3.84 (190–200°C), 2.88 (200–210°C). Clearly, the temperature coefficient of about 4 is too high; it is practically unrealistic for chemical processes of this kind. The energy of activation for the numbers given in the table above is equal to 54 kcal/mole that is also unusually high for such a reaction. To add insult to the injury, the authors made a mistake in calculating the activation energy, by plotting decimal logarithm (log) of OIT instead of the natural logarithm (ln) of the OIT, following Arrhenius equation (Eq. 15.2), and did not consider the transition coefficient between the two logarithmic forms ($\ln x = 2.3 \log x$). Due to the mistake, the authors obtained an erroneous figure of the activation energy of 22.2 kcal/mole, which looked quite typical for oxidation reactions. In the presence of copper, which normally serves as a catalyst of the oxidation and depletes an antioxidant much faster, the authors have obtained the following data:

Temperature (°C)	OIT (min)
180	14.1
190	8.9
200	6.3

The temperature coefficient for the reaction is 1.41, both from 180 to 190°C and from 190 to 200°C, and the energy of activation is 14 kcal/mole. Therefore, as a result of the combination of methodological errors, experimental mistakes, misinterpretations, and miscalculations, the authors obtained data which are related—concurrently—to different phenomena.

It should be realized that damage to a composite material caused by the oxidative degradation is cumulative. During hot melt processing, thermoplastics degrade to some extent. If antioxidants are present in the material, they protect the material from oxidation, but the amount of the antioxidants is diminished accordingly. The subsequent annealing of the extruded shapes (if the annealing is a part of the processing) consumes some more antioxidants, depending on the annealing temperature and duration. Finally, when composite boards make onto a deck, the environment and elements (temperature, UV light, moisture as a catalyst of the oxidative degradation) continue to destroy antioxidants. The rest of the antioxidants determine the lifetime of the deck. If there were no added antioxidants in the composite material in the first place, the lifetime of the deck depends solely on resistance of the plastic to the elements, that is, on the amount of the antioxidants contained in the incoming plastics. It may be enough for, say, the warranty time (10–20 years), or it may be not. It all depends on this amount and on the location of the deck. Some specific examples will be given in the following section.

Generally, temperature extrapolations over wide temperature range can be very misleading. Theoretical temperature dependencies such as Arrhenius equations (Eq. 15.2) normally hold only in rather narrow temperature ranges. The same is applicable to linear plots in those narrow temperature ranges. Strictly speaking, not extrapolations outside this range, but only interpolations within this range, can be acceptable. Both activation energies (E in Eq. 15.2) and the temperature coefficients are generally changed with temperature. Unfortunately, it is very common when these "constants" are determined at high temperatures and then extrapolated to a low-temperature range. Such extrapolations are almost always wrong. Here is a typical, but striking example [3].

The OIT for a polyethylene material was determined over a wide range of temperatures, between 220 and 140°C, and it increased in this range from 3 to 5000 h (the temperature coefficient of 2.53). However, the slope for the Arrhenius plot was not linear and gave a transient activation energy in the temperature range from 212 kJ/mol (50.75 kcal/mol) at the highest temperatures to 108 kJ/mol (25.86 kcal/mol) at the lowest temperatures. The temperature coefficients were 2.92 and 2.11, respectively. The OIT figures obtained at the highest temperature of 210°C and being extrapolated to 90°C, using the respective calculated activation energy at high temperatures, predicted a lifetime for the profile of 384 years at 90°C (194°F). Linear extrapolation from 140°C to 90°C would, however, predict a lifetime of 24 years. However, if the decline in activation energy (hence, the temperature coefficient) is assumed to continue between 140 and 90°C, then the predicted OIT at 90°C would be only 5 years.

Generally, when high-temperature data are extrapolated to lower temperatures using the Arrhenius equation, overly optimistic (and grossly distorted) results are obtained. Another indicative example is shown by Ref. [4]. When polypropylene surface embrittlement was studied in a wide temperature range, the calculated activation energies were 16, 41, and 82 kJ/mol (3.8, 9.8, and 19.6 kcal/mol) at 30, 70, and 90°C, respectively. Again, at 90°C the temperature dependence of the reaction rate was steep, and extrapolation to lower temperatures would have given very low rates of embrittlement. In reality, they are much faster even at the lower temperatures.

The Physical and the Chemical Structure of the Polymer

The rate of abstracting hydrogen atoms by free peroxide radicals depends on the position of hydrogen atoms in a polymer chain (primary, secondary, tertiary), with the respective resistance to the abstraction as 17:6:1. For polyethylene, the main chain hydrogens are secondary, with an intermediate sensitivity to abstraction, but at chain branches they tend to be tertiary, the most sensitive to abstraction. Hence, highly branched LDPE offers less resistance to thermooxidation than the more linear HDPE. Nonbranched LLDPE offers intermediate resistance.

General resistance of several generic plastics to oxidation is shown in Table 15.1. LDPE is the least resistant and polypropylene homopolymer is the most resistant, whereas HDPE takes an intermediate place. Resistance of unsaturated C═C aliphatic polymers is quite low because of the ease of abstraction of hydrogen atom.

The OIT values for different polymers is hard to compare because plastics are manufactured, as a rule, with added antioxidants. Table 15.2 shows examples for some HDPE, with the lowest OIT number below 1 min, and the highest more than 100 min. Apparently, all the OIT values above less than 1 min reflected an increasing amount of antioxidants.

OIT for polypropylene that does not contain antioxidants is very low. Added antioxidants linearly increase the OIT of polypropylene. For example, it was found that a phenolic antioxidant increases the OIT of polypropylene following a simple formula:

$$\text{OIT (min)} = 71 \times [\%\text{AO}]$$

where the OIT was measured at 200°C. Thus, for 0.3% of the antioxidant (AO), the OIT was 21 min, for 0.1% AO the OIT was 7 min, and without the AO the OIT was close to zero [5]. The authors list data for commercial polypropylenes from three different suppliers, having the OIT values of 19, 18, and 12 min (at 200°C). Using the above formula, it might be conjectured that these polypropylenes contained ontioxidants equivalent to the above phenolic antioxidant in concentration of 0.27, 0.25, and 0.17%, respectively.

ABS (GE Cycolac GPM-4700) even in the presence of some added antioxidants had—at 187°C—the lowest OIT value of 4.7 min (with 0.2% Irganox 1010 and 0.2% of Irgafos 168). These amounts of the antioxidants are rather high, and in case of HDPE or polypropylene, each one of them increase its OIT values by an order of magnitude and more. However, the data were rather scattered, and another experiment with OIT without antioxidants has shown the OIT of 7.4 ± 4.9 min, and with 0.025% of either Cyanox 425 or Cyanox 2246 the OIT was of 2.6 min (in both cases). Eventually, a model has predicted that the ABS in the presence of 0.2% of Irgafos 168 has the OIT between 7 and 9 min, and different antioxidants increase the OIT up to 15–33 min. Besides, as it was noted, the ABS itself contained "a small amount of antioxidant as supplied from the manufacturer" [6].

Effect of antioxidants and the respective OIT values with styrenic block copolymers are not directly related to WPCs, at least known WPC on the current market; however, these figures can be compared with those for other polymers. Some data are listed in Table 15.14 as reference ones.

Table 15.14 shows that the styrenic block copolymers have the OIT values in the singular digits, if antioxidants are not added. Considering that the data were obtained at 161–165°C, the OIT would be an order of magnitude lower, in fractions of a second at 190°C, that is, close to the OIT for polyolefins.

TABLE 15.14 The OIT values for polybutadiene-based (SBS) and polyisoprene-based (SIS) styrenic block copolymers in the presence of some antioxidants

Polymer	Antioxidant	Antioxidant, concentration (%, w/w)	The OIT (min)
SBS	None	0	2.5 ± 1.2
	Mixed tocopherols	0.1	13 ± 5
		0.2	14 ± 7
		0.3	18 ± 10
		0.4	19 ± 7
SBS	Irganox 1010	0.1	12 ± 3
		0.2	11 ± 3
		0.4	19 ± 2
SIS	Irganox 1010	0.025	10.4 ± 0.6
		0.05	11.2 ± 0.8

The OIT were obtained at 161°C (for the first set of SBS data in the table) and 165°C(for the second set of SBS data and for SIS data). Data in the table were calculated from data of [7].

According to the data in literature, poly(phenylene sulfide), polycarbonate, and polysulfone are the most stable of the polymers studied, whereas PVC, both flexible and rigid, is the least oxidatively stable. HDPE was among the most oxidatively stable polymers under high-pressure oxygen at 175°C [8].

History of Plastic (Virgin, Recycled)

Virgin polyolefins contain, as a rule, some loaded antioxidants, often around 0.1% w/w. It is normally enough for neat polyolefins to extend their service up to, say, 10 years outdoors, with the length of the service varying with conditions of the exposure. However, it is often not enough for WPC deck boards or railing components, particularly if they are compounded in a rather aggressive twin screw compounder and contain abrasive minerals, if the WPC profile is hollow (with panels and ribs of 0.18–0.25 in. thin), and if the cellulose fiber does not possess good antioxidative properties. In those WPC profiles the OIT value can drop from 20–30 min (virgin incoming plastic) to 0.3—2.0 min, that is, by 10–100 times. Only added antioxidants can save such profiles from a rapid deterioration under direct sunlight, particularly in the South.

If the incoming plastic is recycled, in which the initial antioxidant is largely (or completely) depleted, WPC deck boards made from such materials are doomed, if not loaded with a good amount of antioxidants.

The Type and Amount of Cellulose Fiber

Wood flour or rice hulls contribute to the increase in stability of WPCs against thermooxidation (see, for example, data in Table 3.11 and 15.15).

Not only plastic but also cellulose contributes to free radical formation and adds their share to the system. Cellulose is degraded by temperature effects and UV light,

TABLE 15.15 Effect of rice hulls on the OIT (at 190°C) of an HDPE, processed on the Brabender laboratory extruder

Amount of rice hulls (%, w/w)	The OIT (min)
0 (neat HDPE)	4.4
Traces	7.4
10	15
30	22
60	37

which breaks the β-glycoside bond in cellulose molecules and causes an abstraction of hydrogen atom from various positions in glucose residues in the chain. These events typically lead to free radicals formation. Increased humidity or water presence in material increases rates of photochemical processes.

However, cellulose is a relatively stable material. In lignified cellulose, such as wood flour, the degradation of lignins dominates the photochemical processes. Most of the lignin is lost from the wood surface during UV exposure, and the depth of degradation is in the range of 0.5–2.5 mm (0.02″–0.1″) [9].

The thermal degradation rate of cellulose at 302°F (150°C) approaches the rate of UV-caused degradation produced by an unfiltered mercury lamp (wavelength 254 nm, energy 471 kJ/mol). In other words, these conditions are rather powerful because the solar cutoff is at 295 nm with energy 406 kJ/mol, and the mid-range UV (350 nm), which is close to a wavelength used in typical weathering boxes (340 nm), is characterized by energy of 341 kJ/mol.

The Type and Amount of Mineral Fillers

These effects will be considered below. In short, mineral fillers generally decrease the stability of WPC materials against thermooxidation. This overall effect is caused by two principle effects—attrition and the respective temperature rise during processing, and direct catalysis of oxidation. Sometimes small amounts of mineral fillers (traces or singular percentage amounts) cause some increase in the OIT, but above about 10% w/w minerals typically accelerate the oxidative degradation of WPCs, unless their negative effects are neutralized by loading of antioxidants.

The Presence of Stress

Stress in plastic composites and profiles can decrease the free energy of activation of oxidation processes and, hence, speed up the oxidation degradation of the materials in the stressed areas. Besides, stress decreases local densities (specific gravities) of composites and thereby increases porosity and provides space for oxygen from air to diffuse in and oxidize the material "from inside."

Chapter 6 gives many examples of density distribution across the profile shown in Figures 15.2 and 15.3. Typically, density distribution, hence, stress, covers a much

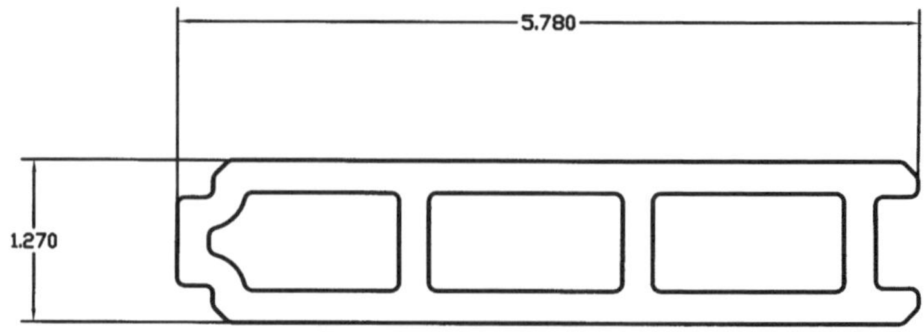

Figure 15.2 GeoDeck composite board profile, tongue-and-groove.

higher gradient across the profile of Figure 15.2, and vary between 1.03 and 1.145 (Fig. 6.4), or between 1.045 and 1.162 (Fig. 6.6), which is more than 0.1 g/cm^3. For the more symmetrical profile shown in Figure 15.3, variations of density are significantly less, such as between 1.185 and 1.24 (Fig. 6.8) and between 1.20 and 1.24 (Fig. 6.9).

More stressed tongue-and-groove HDPE-based composite deck boards are deteriorated due to the faster oxidative degradation compared to traditional boards. Seventy-six percent of all filed warranty claims regarding crumbling of GeoDeck due to severe oxidation were concerned with tongue-and-groove boards, and the rest 24% with traditional boards.

The board shown in Figure 15.4 was produced without added antioxidants. It was exposed for 75°C (167°F) for 1 month, and then transferred to a weathering box (0.35 W/m^2, 1:42/0.18 cycle). After a month in the oven, at 167°F, the board's surface looked good. After just 1 day in the weathering box, at about the same maximum temperature (actually, about 10° lower) but with water spray for 18 min every 2 hours, its surface became slightly but noticeably soft. On the second day the deterioration increased, and after 5 days the boards cracked along spider lines and fell apart.

The experiment was repeated with a similar board, but exposure in the oven at 75°C was for 3 months, and then the board was transferred to the weathering box as described above. After 72 h of exposure at 1:42/0.18 cycle, the board cracked along spider lines, and after 86 h the board broke itself in two parts. After 6 days the board was deeply crumbled.

Figure 15.5 provides an example of stress-induced splitting of composite boards as a step for a progressing crumbling. Soon after the picture was taken, the board had started to show degradation and crumbling.

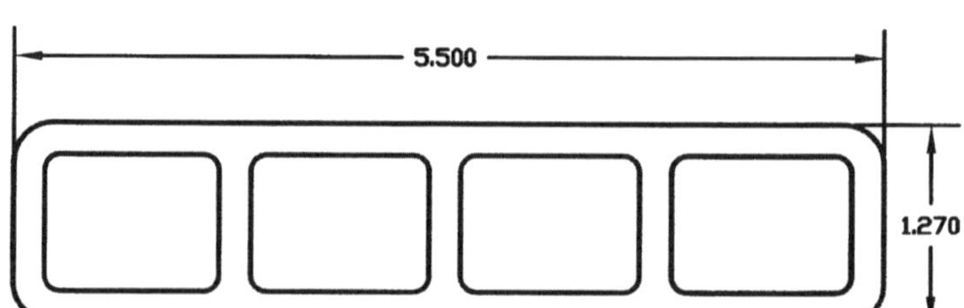

Figure 15.3 GeoDeck composite board profile, Traditional.

Figure 15.4 Tongue-and-Groove board manufactured without added antioxidants, after accelerated weathering for 1 month at 75°C and then 5 days in a weathering box.

Spider lines are essential elements in extruded composite *hollow* deck boards. Spiders in an extrusion die interrupt the melt flow, causing it to separate and then rejoin (see Figs. 15.6 and 15.7). This welding of the profiles creates the so-called spider lines along hollow composite deck board profiles.

Spider lines are usually visible to the naked eyes as linear surface depressions. They are stress concentrators. Figures 15.8–15.11 show the development of an oxidative crumbling of a composite deck board with clearly visible cracks along spider lines, and Figure 15.12 shows a catastrophic failure of WPC board due to oxidative decomposition.

Figure 15.5 Stress-induced through-thickness splitting as part of crumbling (oxidative degradation) of a composite board.

Figure 15.6 A hollow die, or a spider die with a fixed mandrel as an integrate part of the die to make the tongue-and-groove profile (see Fig. 15.2).

Figure 15.7 A hollow die or a spider die with a fixed mandrel as an integrate part of the die to make the Traditional profile (see Fig. 15.3).

Figure 15.8. An intermediate step of crumbling of Mahogany tongue-and-groove composite deck board.

Figure 15.9. An intermediate step of crumbling of Mahogany tongue-and-groove composite deck board.

Figure 15.10. An advanced step of crumbling of Mahogany tongue-and-groove composite deck board.

Figure 15.11. An advanced step of crumbling of Mahogany tongue-and-groove composite deck board.

Figure 15.12. Catastrophic failure of a tongue-and-groove composite deck board due to an oxidative degradation/crumbling.

The Presence of Metal Catalysts

Many metals or ions of metals are catalysts of oxidative degradation of plastics. The conversion of peroxides (R_1OO^*) to hydroperoxides (R_1OOH) via abstraction of hydrogen from an adjacent polymer molecule (R_2H) is typically the rate-limiting step for the propagating chain reaction of oxidation

$$R_1OO^* + R_2H \rightarrow R_1OOH + R_2^*$$

Metals apparently promote the above reaction:

$$R_1OO^* + R_2H + M^+ \rightarrow R_2^* + M^{++} + R_1O^* + OH^-$$

$$R_1OO^* + R_2H + M^{++} \rightarrow R_2^* + M^+ + R_1OO^* + H^+$$

thereby accelerating the oxidative degradation of the material. Metals are practically unavoidable in composite materials, hence, much lower stability of composites compared to the neat plastic.

The main sources of metal ions in composites are inorganic pigments, lubricants (such as metal stearates, except probably calcium stearate), fillers (silicates in rice hulls, typically as much as 17.5% by weight, and in wood flour, often about 0.5% by weight), contaminations during processing, such as metal particles and salts from compounders and extruders. Salts are formed as products of corrosion of the barrel resulting from acid-produced "steam explosion" of cellulosic materials as fillers at high temperature and pressure (see the insert). Actually, simple salts are the most damaging catalysts in oxidative degradation reactions of plastic and composites, compared to organometallic compounds.

"Steam explosion" is a process taking place when wet cellulosic material in a closed chamber is subjected to high temperature and pressure (at a certain combination of the two, when moisture in the cellulosic material does not boil), and then a sharp drop in pressure happens in the chamber. Moisture in the cellulosic filler momentarily "explodes" as a result of a violent boiling, and hemicellulosics, being the most chemically sensitive constituent of all natural cellulosic materials, break and release their multiple acetic residues in the form of acetic acid. Though a weak acid at ambient conditions, at high temperatures of hot melt processing, acetic acid becomes quite aggressive and corrodes the barrel, screw, and other parts of the extruder. This releases metal acetates into the cellulosic-filled composite material.

Inorganic pigments, such as iron oxide, often contain free iron. Its amount depends on where and how the iron oxide was mined. Some sources supply iron oxide with very little amount of free iron, some with a high amount of it. Generally, cheap iron oxide contains more free iron. Synthetic iron oxide typically contains less of the free iron.

Table 15.16 gives a few examples how some iron oxide industrial preparations affect an oxidation rate (as oxidative induction time) of HDPE.

One can see that with neat HDPE (no pigments) and at a rather common amount of a pigment, that is 3%, the pigments either slightly affect the OIT of the plastic or even increase it (pigment 2). However, at a further increase in some pigments (particularly 1, 4, and 5), the lifetime of the plastic drops 2–3 times. Because said pigments are commonly added to composite materials, their effect on the lifetime could not be ignored.

Now, let us consider effects of some of these inorganic pigments on the lifetime of some composite materials.

Table 15.17 shows that rice hulls significantly, more than 8-fold, increases the lifetime of the composite compared to neat HDPE. However, pigment 1, which at 3% did not show any noticeable effect on the OIT of neat HDPE (Table 15.16), in the presence of 60% of rice hulls drops the OIT of the composite by almost five times. It could have been not necessarily the result of a chemical interaction between them

TABLE 15.16 Effect of colorants on OIT (at 190°C) of an HDPE, processed on the Brabender laboratory extruder

	OIT for five iron oxide pigments (min)				
Amount of pigments (w/w)	1	2	3	4	5
0 (neat HDPE)			4.4 (no pigments)		
Traces	5.6	6.0	5.2	5.7	4.5
0.5%	6.0	7.2	—	—	—
3%	4.7	9.4	—	—	—
7%	2.4	9.7	—	—	—
12%	1.9	9.9	3.7	1.7	1.5

TABLE 15.17 Effect of colorants and fillers on OIT (at 190°C) of an HDPE, processed on the Brabender laboratory extruder

	OIT (min)				
Amount of pigments and fillers (w/w)	Filler 1	Filler 1 + Filler 2 (30%)	Filler 1 + Filler 2 (30%) + pigment 1 (2%)	Filler 1 + Filler 2 (30%) + pigment 2 (2%)	Filler 1 + Filler 2 (30%) + pigment 1 (2%) + antioxidant (0.5%)
0 (neat HDPE)			4.4		
Filler 1 (traces)	7.4	—	—	—	—
Filler 1 (10%)	15.0		—	—	—
Filler 1 (30%)	22.0	9.3	5.4	7.8	>90
Filler 1 (60%)	37.0	—	—	—	—
Filler 1 (60%) + colorant 1 (3%)	8.3	—	—	—	—

Filler 1 rice hulls; Filler 2 Biodac.

(the amount of rice hulls was much higher than that of the pigment), but an increase in shearing, hence, more heating, and the resulting drop in the OIT.

Furthermore, an addition of Biodac® almost completely neutralizes the positive effect of rice hulls on the OIT, and the latter came back, to 5.4 min. Pigment 2 is less damaging to the composite in terms of its OIT. Finally, an addition of 0.5% of an antioxidant dramatically—by 20-fold—increases the lifetime of the composite.

These effects play a very important role in the field, on real composite decks, containing these or other pigments. Of all warranty claims filed regarding GeoDeck (see next section), 61% were concerned Mahogany boards, 29% Cedar boards, and only 10% Driftwood boards. Clearly, these effects could not possibly be predicted with the present state of knowledge. However, after the described experiments were conducted, they provided guidelines for selecting fillers and/or pigments, and optimizing their amounts in composite materials.

Metals in metal stearates quite significantly effect the lifetime of plastics and composite materials. Table 15.18 illustrates these effects.

One can see that the most popular zinc stearate decreases the lifetime of the HDPE by more than three times, copper stearate by 12.5 times, and cobalt stearate by 25 times.

Titanium dioxide is a known catalyst of thermooxidation and photooxidation of polymers. At 0.5% amount of it in LDPE, durability of the latter drops to about 50% of the initial. There are some data that titanium dioxide destabilizes polypropylene against oxidation by 30 times [3]. These effects can be stopped by addition of antioxidants.

The Presence of Moisture

There is a common belief, supported by experimental data that water (rain in the field and water spray in a weathering box) washes out antioxidants and other additives,

TABLE 15.18 The effect of various metal stearates (0.5% by weight) on the OIT of polypropylene at 125°C [3]

Metal stearate	OIT (min)
None (neat HDPE)	125
Cadmium	50
Zinc	40
Titanium	28
Nickel	18
Vanadium	15
Iron	12
Copper	10
Manganese	8
Chromium	8
Cobalt	5

and thereby decreases the resistance of the plastic or the composite material to oxidative degradation. Because composite materials typically absorb much more water compared to plastics, this washing effect on the accelerated degradation should be much more pronounced in composites. Besides, water can hydrolyze some antioxidants and thereby cause their depletion.

However, water apparently plays a more direct role in oxidative degradation of composite materials. When a composite board, having no added antioxidants, and with the OIT close to zero, was placed in a weathering box (1:42–0:18 cycle, i.e., 1 h 42 min UV light and 18 min UV light + water spray), a noticeable crumbling of the board surface was observed after 5 days. When the same board was placed into the same box, but with water spray turned off (2:00 UV light cycle), there was no crumbling of the board after 30 days. The board did not have any oxidants to wash out. Apparently, two more factors could be considered: water acts as a catalyst of the oxidative degradation, and/or water spray induces a shock of the board forcing it to contract, and after it the board expands again. These consecutive expansions–contractions rock the board back and forth, which in turn induces stress or pressure at the interphase, because the filler(s) dimension is normally unaffected by the expansion–contraction. Generally, separation of surface of fillers and the matrix under stress causes permanent failure. On top of this, oxygen diffuses into cavities resulting from this preparation. All of this leads to a faster oxidative degradation.

The following experiments were conducted to clarify the role of water in the deterioration of composite deck boards. A board produced without added antioxidants was studied (GeoDeck, Mahogany hollow board). The board was exposed to 75°C (167°F) for 1 month and then transferred to a weathering box (0.35 W/m^2, 1:42–0:18 cycle). After a month in the oven, at 167°F, the board's surface looked good. After just 1 day in the weathering box, at about the same maximum temperature (actually, about 10° lower) but with water spray for 18 min every 2 h, its surface became slightly but noticeably soft. On the second day the deterioration increased, and after 5 days the board cracked along spider lines and fell apart.

The experiment was repeated with a similar board, but exposure in the oven at 75°C was for 3 months, and then the board was transferred to the weathering box as described above. After 72 h of exposure at 1:42–0:18 cycle, the board cracked along spider lines, and after 86 h the board broke itself in two parts. After 6 days the board was deeply crumbled.

It should be noted again that the damaging effect of water in this case could not be explained by washing off some antioxidants because there was not any. Either water spray produced temperature shock and the respective violent contraction with the subsequent heating and expansion, or water served as a catalyst of the oxidative degradation, or both.

To continue the experiment, the same board was placed into the weathering box, but water was shut off, and the board was irradiated with the UV light (0.35 W/m^2, 63° black panel) continuously for 14 days. There were no cracks, and no crumbling was observed. After 14 days the weathering regime was changed back to 1:42–0:18 cycle, and in less than 2 days the board showed clear signs of crumbling. On the third day the board cracked along the spider line, and on the fourth day it fell apart.

Antioxidants and Their Amounts

There are many types of antioxidants. The most popular in composite materials are "primary antioxidants" or "chain breaking antioxidants," and "secondary antioxidants" or "preventative antioxidants."

Generally, antioxidants either scavenge free radicals or prevent free radicals formation.

"Primary antioxidants" (AH) are represented by hindered phenolic compounds and secondary aromatic amines, or hindered amines. They provide easy hydrogen to peroxide radicals (ROO^*), and thereby prevent a hydrogen abstraction from a nearby polymer chain:

$$ROO^* + AH \rightarrow ROOH + A^* \quad (15.3)$$

where A^* is more stable than R^* and acts as a free radical scavenger:

$$A^* + ROO^* \rightarrow ROOA \quad (15.4)$$

This all takes place because the mobility of the primary antioxidant AH is greater than the mobility of the polymer (R_2H below).

In the absence of the antioxidant the above reaction would damage the polymer (R_2H) and produce a hyperactive free radical R_2^*, which would continue the chain reaction:

$$R_1OO^* + R_2H \rightarrow R_1OOH + R_2^* \quad (15.5)$$

Hence, this type of antioxidants indeed breaks the destructive chain reaction.

As a general rule, the lower molecular weight antioxidants are used to protect the plastic during processing (compounding and extrusion), whereas the higher molecular weight antioxidants are considered to be useful for long-term protecting during service, such as of composite deck boards. This is because the higher molecular weight antioxidants are less likely to be washed out from the profile with rains, freeze–thaw, and other similar events.

If "primary antioxidants" compete with free radicals for hydrogen abstraction from the polymer and then scavenge another free radical (Eqs. 15.3 and 15.4), "secondary antioxidants" directly decompose unstable hydroperoxides (Eq. 15.6), thereby preventing with free radicals formation (Eq. 15.7):

$$\mathrm{ROOH} + \mathrm{A} \rightarrow \mathrm{ROH} + \mathrm{AO} \tag{15.6}$$

$$\mathrm{ROOH} \rightarrow \mathrm{RO}^{*} + \mathrm{OH}^{*} \tag{15.7}$$

A division of antioxidants into primary (chain breaking) and secondary (preventative) is often rather arbitrary, and Irganox 1010 and Irganox 1076, for example, are considered both primary and secondary.

Antioxidants are not catalysts; they are consumed by protecting the polymer from oxidation. Antioxidants end up as a stable peroxide (Eq. 15.4), as an oxidized antioxidant (Eq. 15.6), or as other stable structures. It is practically impossible to exactly predict in a real situation for how long a given amount of antioxidants will hold in a particular shape, such as a composite deck board. If manufacturing conditions are relatively stable, conditions of natural exposure vary in a wide range of temperature, solar radiation, moisture, among others. What are they going to be is practically uncontrollable and unpredictable, when composite profiles leave the manufacturing plant. Hence, the principal goal regarding the choice of added antioxidants and their amounts is to protect the product in worst conditions in which the product might be installed—at least for the warranty time period, that is, typically for 10–20 years—and to minimize the cost for the added antioxidants with the same efficiency.

A common misunderstanding is that some antioxidants provide protection only during processing and are not effective in the course of service in the field, and, conversely, that some field-efficient antioxidants do not provide protection during processing. This is generally not true. If antioxidants are good in quenching free radicals (Eq. 15.4), in competing with polymers for hydrogen abstraction (Eqs. 15.3 and 15.5), and in preventing free radicals formation (Eqs. 15.6 and 15.7), they are good at it everywhere, both in processing at high temperatures and in the field, during a long-term service, at relatively low temperatures, but at a barrage of UV light, generating an avalanche of free radicals in the composite material.

It does not mean that all antioxidants are equal in their performance and in a potential residence time in the plastic or the composite material. Low-molecular weight antioxidants face a higher likelihood to be washed out of the material during a long-term service, particularly out of a composite material with a high porosity (low density or specific gravity, see above). Therefore, higher molecular weight antioxidants are generally preferred for a postmanufacturing service. Some antioxidants are more volatile at high temperatures; hence, they are less preferred at processing. It could be that some antioxidants have a low activation energy for their free radical quenching reactions, and although they can be efficient at low-temperature protection of, say, composite decks, they can be relatively slow at high processing temperatures. Some antioxidants (such as phenyl sulfides) have been reported to crystallize from their supersaturated solution in a matrix of the material after the melt is cooled down, and thereby greatly reduce their capacity to act as stabilizers.

In other words, the principal mechanisms of antioxidants action are the same at processing temperatures and in the field; however, some other factors might influence their actual performance, and these factors should be considered in making choice which antioxidant(s) to pick for a particular processing and for a particular type of a long-term service. Also, primary and secondary antioxidants often act in a synergism. This is the main reason why many plastic and composite manufacturers choose a combination of two or several antioxidants. One of the most popular combinations is Irganox B 225, which is a 1:1 blend of Irganox 1010 (primary, hindered phenolic antioxidant with low volatility, high molecular weight of 1178 Da) and Irgafos 168 (secondary antioxidant, phosphite, hydrolytically stable, low molecular weight of 647 Da). It is believed that Irgafos is protective mainly during the processing, and Irganox 1010 is protective during the long-term service; however, both of them are relatively effective in both the situations.

Popular blend compositions of Irganox and Irgafos types of antioxidants produced by Ciba Specialty Chemicals (Tarrytown, NY) are listed in Table 15.19.

TABLE 15.19 Irganox and Irgafos blend compositions manufactured by Ciba Specialty Chemicals

Product	Ratio	Components
Irganox B 215	1:2	Irganox 1010: Irgafos 168
Irganox B 220	1:3	Irganox 1010: Irgafos 168
Irganox B 225	1:1	Irganox 1010: Irgafos 168
Irganox B 311	1:1	Irganox 1330: Irgafos 168
Irganox B 313	1:2	Irganox 1330: Irgafos 168
Irganox B 501 W	1:1	Irganox 1425 WL: Irgafos 168
Irganox B 561	1:4	Irganox 1010: Irgafos 168
Irganox B 900	1:4	Irganox 1076: Irgafos 168
Irganox B 921	1:2	Irganox 1076: Irgafos 168
Irganox B 1411	1:1	Irganox 3114: Irgafos 168
Irganox B 1412	1:2	Irganox 3114: Irgafos 168
Irganox HP 2215	2:4:1	Irganox 1010: Irgafos 168: HP-136
Irganox HP 2225	3:3:1	Irganox 1010: Irgafos 168: HP-136
Irganox HP 2251	3:2:1	Irganox 1010: Irgafos 168: HP-136
Irganox HP 2411	3:3:1	Irganox 3114: Irgafos 168: HP-136
Irganox HP 2921	2:3:1	Irganox 1076: Irgafos 168: HP-136
Irganox XP 420	3:2:1	Irganox 1010: Irgafos P-EPQ: HP-136
Irganox XP 490	3:2:1	Irganox 1076: Irgafos P-EPQ: HP-136
Irganox XP 620	3:2:1	Irganox 1010: Irgafos 126: HP-136
Irganox XP 621	6:2:1	Irganox 1010: Irgafos 126: HP-136

TABLE 15.20 Ciba antioxidants sold under different names

Ciba antioxidants	Longchem (Taiwan)	Chitec (Taiwan)	Chemark(China)	Mayzo (Georgia, USA)
Irganox 1010	Longnox 10	Chinox 1010	Chemark AO-1010	BNX 1010
Irganox 1024	Longnox 24	Chinox 1024	Chemark AO-1024	BNX MD-1024
Irganox B 225	Longnox B1068	Chinox B225		BNX 1225
Irganox 1076	Longnox 76F	Chinox 1076	Chemark AO-1076	BNX 1076
Irganox 1098	Longnox 98	Chinox 1098		
Irgafos 168	Longnox AO-68	Chinox 168	Chemark AO-168	Benefos 1680

Irganox 1010 and Irgafos 168 are also known in the market by the names of Longnox 10 and Longnox AO-68, respectively, and their 1:1 blend (Irganox B 225) is Longnox B1068 (Longchem C & S International Corp., Taiwan). Another name for the same 1:1 blend is Chinox B225 (Chitec Chemical Co., Ltd, Taiwan). These and some other analogs are shown in Table 15.20.

One more reason to use a combination of antioxidants is that they often show a synergistic effect. Secondary antioxidants attempt to prevent formation of free radicals, and primary antioxidants are designed to prevent propagation. Generally speaking, every material, plastic and composite needs an optimized package of antioxidants, where optimization includes type of antioxidants, amount of each one in the package, and the total cost. It is rather unrealistic to have this task fully accomplished for each composite material; however, it is an R & D goal worth pursuing.

Besides synergism, some antioxidants (or, more generally, stabilizers) reveal an unexpected antagonism between each other. For example, carbon black is a weak antioxidant and an efficient UV stabilizer. When combined with some sulfur-based antioxidants, carbon black works in synergism. However, in combination with many hindered amines and some phenolic antioxidants, carbon black reduces the degree of protection of polyethylene [10]. For example, at 140°C the degradation induction time of polyethylene in combination with the antioxidants is

- 3% carbon black $\rightarrow$ 20 h;
- 0.1% DPPDA (*N,N'*-diphenyl-*p*-phenylendiamine) $\rightarrow$ 450 h;
- 3% carbon black + 0.1% DPPDA $\rightarrow$ 100 h.

This kind of antagonistic behavior of antioxidants practically cannot be predicted. Therefore, to avoid premature degradation of the product, potential packages of antioxidants should be thoroughly investigated, both separately and in combination.

It has been reported that for a number of different antioxidants and over a rather wide range of their concentrations, the following equation holds:

$$\mathrm{OIT} = k\sqrt{[\mathrm{AO}]}$$

where

[AO] = amount of antioxidant (w/w)

k = constant for the given material and test conditions.

The above equation typically holds above a certain (threshold) amount of the antioxidant. Below this amount the antioxidant is largely consumed during processing, and the residual amount, if any, does not contribute to the protection proportional to the square root of its initial amount.

For the HDPE-based composition containing rice hulls and Biodac® as principal fillers (GeoDeck), the above equation has the following form:

$$\mathrm{OIT\ (at\ 190°C)} = 28\sqrt{[\mathrm{AO}]}$$

which holds at the amounts of the antioxidant above 0.16%. For example, at 0.18% (w/w) AO, the average OIT at 190°C is 2.4 min; at 0.25% AO the OIT is 4.6 min; at 0.5% AO the OIT is 10.4 min; at 0.6% AO the OIT is 12.3 min. This is enough for GeoDeck composite deck boards to withstand the climatic conditions of Phoenix, AZ, for more than 20 years, as it was shown by laboratory experiments along with OIT measurements of composite deck materials in the field.

Composite deck boards are rather often considered for making a deck around a swimming pool. Obviously, chances are that boards in this case would come into contact with chlorinated water. To achieve a proper lifetime for such boards, they should be protected with a higher amount of antioxidants. Chlorine as water disinfectant reacts with water and forms free radicals as follows:

$$\mathrm{Cl_2 + H_2O = HCl + HOCl}$$

$$\mathrm{HOCl = HCl + O^*}$$

The higher the concentration of chloride in water, the higher the amount of the free radicals. This might reduce the lifetime of the pool composite deck, unless a sufficient amount of antioxidants is added to the composite material.

Generally, free radicals whose action destroys plastics and plastic-based composites can be quenched by antioxidants. As it was shown above, antioxidants either prevent free radical formation or convert free radicals to stable derivatives by providing them with a missing electron and, hence, forming a stable chemical bond. Though antioxidants cannot reverse a damage done to plastics (typically, shorter polymer chains, cross-linking between polymer chains, poorer mechanical properties, higher brittleness), they slow down or practically stop further development of the oxidative degradation and the resulting damages. Free radicals are short-lived chemical species, and they either instantly propagate or cease to be free radicals, generally following the mechanisms described above.

Solar Radiation (UV Light)

Solar radiation is normally accompanied by heating of the surface of materials. Very seldom "cold" solar radiation takes place, such as in high mountain areas. Therefore, effect of solar radiation is difficult to separate from thermal effects on plastics and plastic-based composites. Deck surface temperatures under direct sunlight and in a shade, are quite different from each other. In a shade, deck surface temperature is practically equal to air temperature, whereas under direct sunlight it is about 40°F higher in the north of the United States, and 50°F higher in the south. Hence, an effect of solar radiation practically always carries a thermal component.

The principal damaging element of solar radiation is UV light. When natural, solar UV light, which is highly energetic, hits a deck surface, it generates free radicals, the same way as thermal treatment does. The only difference is that thermal effects generate free radicals in the bulk material, whereas UV light generates free radicals only at the surface. UV light has, as it is known, a dual nature, that is, both a wave and a corpuscular at the same time. When those highly energetic waves or particles hit deck surface, they break polymer chains and form free radicals, as it was shown in chemical equations (Eqs. 15.3–15.7).

In fact, only about 6% of solar radiation consists of UV light with wavelength less than 400 nm, which carries enough energy to break covalent chemical bonds. The strongest covalent bonds can be ruptured by solar UV light with wavelength of 300 nm. The higher the sun above the horizon, the shorter the wavelength transmitted to a deck, and the higher the potential damage by both UV light and temperature. The higher the altitude, the stronger the UV light, the shorter its wavelength transmitted to the surface, and the higher its energy.

The broken polymer molecules and low molecular weight products (such as hydroperoxides) relay free radicals to their neighboring molecules. This propagates as free radical chain reactions (linear or branched) to the depth of composite board. In the case of "cold" UV light, free radicals are formed and propagate into the very thin "layer" of board surface, normally not more than 0.075 mm (3 mils), and progressively cease within and immediately below that layer. In the case of a normal solar radiation, particularly in the south, high temperature of deck board maintains the propagation of UV-generated free radicals at a distance into the bulk board, and additionally creates more free radicals by similar mechanisms of polymer chain scission (Eqs. 15.3–15.7). This makes the affected, damaged upper layer of the board to spread deeper and deeper, until the board practically collapses.

Figures 15.8–15.12 illustrate progressive steps of this process, leading to crumbling and eventually collapsing of a board. First, some microcracks appear at the board surface, and then the affected surface gradually changes its texture and after a while becomes soft and loose. One can easily scratch it with a fingernail or a corner of a credit card. This is the result of a triple action—UV light, temperature, and oxygen from air. A fourth factor, that is water, also plays an important part in the oxidative degradation.

The loose layer gradually becomes deeper and deeper, as if plastic disappears from the composite material, leaving only loose filler behind. In fact, plastic disappears indeed because its polymer chain becomes shorter and shorter, and after a while it is not a plastic anymore. If the board is hollow, soon the upper panel collapses (Figs. 15.11 and 15.12) exposing ribs and inner channels of the board. The bottom panel of the hollow board still holds, as it was subjected only to thermal effect, not to UV light.

Figure 15.13 shows the result of composite board deterioration from the top. This effect was reached by accelerated weathering of a composite board that did not contain added antioxidants. It took just a few days in a weathering box before the board's top layer became practically decomposed (notice a dark band of the loose, decomposed material at the board's top surface, compared with a control board).

This, in turn, leads to the board weakening. When tested for a break load (3-pt load test) at a span of 3 in. (small pieces of composite boards were weathered and tested), a control, hollow composite board showed a break point at an average load of 3433 ± 46 lb. A weathered, crumbled sample of the same size, shown in Figure 15.13, showed a partial break load at 1381 lb (40% of the initial), when the top panel of the hollow board yielded, resulting in large, catastrophic cracks. At continuing load, the bottom panel failed at 1968 lb (57% of the initial).

Another example can be given with a tongue-and-groove composite hollow board. A control board showed a break point at an average load of 2666 ± 32 lb. A weathered, crumbled sample of the same size showed a partial break load at 1408 lb (53%

Figure 15.13 A cross-cut of a hollow composite board (5/4″ x 5-1/2″): top—the board was exposed in a weathering box under UV light (0.35 W/m^2, 340 nm) and a periodic water spray (18 min every two h); bottom—control, not exposed board.

of the initial), when the top panel failed. At continuing load, the bottom panel failed at 1861 lb (70% of the initial).

This process of board oxidative degradation cannot be simulated in an air oven because temperature and oxygen evenly affect all sides of the board, and the oxidative deterioration does not proceed only "from the above." In fact, in a hollow board the thermal degradation in air oven starts and proceeds from inside, from the ribs, which, as a rule, have lower density (specific gravity), hence, higher porosity, that leads to a faster oxygen diffusion into the ribs and faster degradation of ribs. When broken, hollow profile after exposure in an air oven that led to a significant reduction in the board strength shows a clear visible pattern of specific gravity distribution. Those ribs that had the lowest density are the darkest ones, which reflects the highest degree of "burning," that is, oxidative degradation. The color of "burning" gradually disappears toward upper and lower panels of the board, that is, toward the surface, where density is the highest by virtue of melt flow pattern at the extrusion of the profile.

Pieces of hollow, composite deck boards were kept in an oven for 3 months at 75°C (167°F) and 100°C (212°F). In the first case, boards failed on average at 59% of the initial break load (control boards, kept at ambient temperature). A cross section of the board was all evenly dark colored, indicating a practically uniform oxidation of the material. A similar board, containing an antioxidant, after 3 months at 75°C showed break strength at the same level, as the control board. There was no change in break strength, hence, the oxidative degradation. A board, made without antioxidant and kept at 100°C for 3 months, decreased its break strength to 33% of the control value.

Because UV degradation along with heat creates the same—by nature—free radicals, as the heat alone (see above), the same antioxidants as described above can be used for board protection against oxidative degradation. In other words, it does not matter what causes free radicals, heat or UV light: When free radicals got into bulk of the composite board and propagate there, antioxidants stop them there, provided that antioxidants are used in sufficient amounts.

However, in order to shield the surface of the material from UV light, at least partially, UV absorbers (UVA) and light screens are also used. They do not convert free radicals into stable compounds by *chemical* means, or prevent free radicals formation by converting their predecessors (such as hydroperoxides) into stable compounds, but they absorb UV light, by converting it into low-frequency bond vibration, hence, heat, and thereby prevent free radicals formation by *physical* means. UVA, as follows from their name, absorb wavelengths primarily in the range 290–400 nm; hence, they do not color the material, whereas light screens absorb both UV and visible light (above 400 nm). Carbon black, for example, is a light screen, though it is a powerful absorber of UV light.

Table 15.21 shows an effect of a classical antioxidant, Irganox 1010, and a hindered amine light stabilizer (HALS) Tinuvin 770, as well as a combination of the two on UV embrittlement times (in hours) for polypropylene films containing various pigments in 1% amount. One can see that all three pigments protect the plastic from UV light. Tinuvin 770 shows much better protection from UV light compared with Irganox 1010. However, when combined, they show a certain, but various, antagonism with two pigments and a synergism with the third one.

TABLE 15.21 UV embrittlement times (in hours) for polypropylene films containing pigments, with and without stabilizing additives [11]

	Pigment (1%)			
Additive (0.1%)	None	Cadmium yellow	Green chromium oxide	Blue copper phthalocyanide
None	80	95	135	210
Irganox 1010	410	500	520	630
Tinuvin 770	2750	5620	5000	5750
Irganox + Tinuvin	2120	3500	3700	6750

One of the typical catalysts of thermo- (and photo-) oxidative processes is titanium dioxide. Being introduced as a pigment into WPCs, it always accelerates both fading and oxidative degradation.

HALS are the dominant means for protection of plastics and plastic-based composites from UV light. The mechanism of HALS action is not clear as yet; HALS are considered as primary antioxidants, free radical scavengers, and/or as oxygen trap with the subsequent conversion of it to water. HALS are consumed during service, showing an "induction period" depending on an amount of HALS added to the material. This consumption (induction period) is determined by both chemical depletion and washing the additive with rainfalls. That is why high molecular weight HALS hold longer and hence are more efficient during a long-term service.

Popular blend compositions of Tinuvin and Chimassorb types of HALS and UVA produced by Ciba Specialty Chemicals (Tarrytown, NY) are listed in Table 15.22.

For a comparison with UVA, HALS have much higher molecular weights Table 15.23.

TABLE 15.22 Tinuvin and Chimassorb blend compositions (HALS) and individual compounds (UVA) manufactured by Ciba Specialty Chemicals

Product	Ratio	Components
HALS		
Chimassorb 2030	1:1	Chimassorb 2020: Tinuvin 622
Chimassorb 2040	1:1	Chimassorb 2020: Tinuvin 770
Tinuvin 111	1:1	Chimassorb 119: Tinuvin 622
Tinuvin C 353	2:1	Chimassorb 119: Tinuvin 234
Tinuvin 492		Chimassorb 119: Oxides/Stearates
Tinuvin 494		Chimassorb 119: Oxides/Stearates
Tinuvin 783	1:1	Chimassorb 944: Tinuvin 622
Tinuvin 791	1:1	Chimassorb 944: Tinuvin 770
UVA		
Chimassorb 81, mol. weight 326		
Tinuvin 234, mol. weight 448		
Tinuvin 326, mol. weight 316		
Tinuvin 327, mol. weight 358		
Tinuvin 328, mol. weight 352		

TABLE 15.23 Molecular weight of some HALS

HALS	Molecular weight (Da)
Chimassorb 119	2286
Chimassorb 944	2000–3100
Chimassorb 2020	2600–3400
Tinuvin 123	737
Tinuvin 622	3100–4000
Tinuvin 765	509
Tinuvin 770	481

Tinuvin 234 and Tinuvin 326 are also known in the market by the names of Longsorb 90 and Longsorb 326, respectively. Another names for the same compounds are Chemark UV-234 and Chemark UV-326 (Chemark, China), Chisorb 234 and Chisorb 326 (Chitec Chemical Co., Ltd, Taiwan), Eversorb 234 and Eversorb 73 (Everlight, Taiwan). These and some other analogs are shown in Table 15.24.

In reality, the effect of solar radiation on plastics and composite materials can lead to many outcomes. Typically, but not always, UV light causes discoloration, or fading of plastic and plastic-based composite materials. That fading reflects oxidative degradation of the material at the very surface. It may lead to a damage of the bulk of the composite material, or it may not. This all depends on the amounts of antioxidants in the bulk of the material. Composite material with high amounts of

TABLE 15.24 Ciba UV absorbers and light stabilizers sold under different names

Ciba UVA and light stabilizers	Longchem (Taiwan)	Chitec (Taiwan)	Chemark (China)	Mayzo (GA, USA)	Cytec(NJ, USA)	Everlight (Taiwan)
Chimassorb 81	Longsorb BP-12	Chisorb BP-12	—	BLS 531	Cyasorb 531	Eversorb 12
Tinuvin 234	Longsorb 90	Chisorb 234	Chemark UV-234	—	—	Eversorb 234
Tinuvin 326	Longsorb 26	Chisorb 326	Chemark UV-326	—	—	Eversorb 73
Tinuvin 329	Longsorb 54	Chisorb 5411	Chemark UV-5411	BLS 5411	Cyasorb 5411	—
Tinuvin 328	Longsorb 28	Chisorb 328	Chemark UV-328	BLS 1328	Cyasorb 2337	Eversorb 74
Tinuvin 770	Longstab 70	Chisorb 770	Chemark UV-770	BLS 1770	—	Eversorb 90
Tinuvin 765	Longstab 92	Chisorb 292	Chemark UV-292	BLS 292	Cyasorb 3581	—
Tinuvin 622	—	Chisorb 622LD	—	—	—	—
Chimassorb 944	—	Chisorb 944LD	—	—	—	—

antioxidants can fade dramatically (particularly, if they do not contain colorants) but can retain their mechanical properties for an indefinite time period. Other composites can retain their color (particularly those with a high amount of pigment), but suffer from oxidative crumbling, if they do not have enough of antioxidants.

The most adequate laboratory tool to study the effect of UV light on plastic and composite materials is a weathering box, or an environmental chamber, weatherometer, among others. None of them matches precisely the spectrum of natural UV light; however, they provide all four major components of oxidative degradation of materials at conditions of natural exposure: UV light, heat, water, and oxygen.

A characteristic value of irradiance that is typically used in weathering boxes in practice is 0.35 W/m^2 at a given wavelength of 340 nm. The value 0.35 W/m^2 is a natural daylight irradiance measured at 340 nm at 80°F in Florida. Such energy being translated to Joules (W/s) and to a whole year is 11,038,600 J/m^2, that is, about 11 MJ/m^2 (at 340 nm). The energy of an integrated hemispherical annual solar radiation at 5° in Florida (Miami) is about 6500 MJ/m^2 for a total solar energy and 280 MJ/m^2 for a total UV energy (in the range of 295–385 nm). Such a difference between the UV irradiance at the selected wavelength (340 nm) in a weathering box and the integrated solar radiance in Florida does not mean that weathering in the box takes place much slower than an outdoor weathering in Florida. In fact, direct comparisons are difficult, albeit tempting.

Let us consider what is going on with composite deck boards in a weathering box.

Figure 15.14 shows that the top surface temperature of the composite board reaches 165°F (74°C) and continues at that level for at least 42 min. Between 20 min and 1 h 42 min, that is, for at least 82 min, the top surface temperature is above 150°F. On top of it the board is subjected to a high UV irradiation.

Besides, every 2 h the board receives a shock when it is sprayed with cold water. Figure 15.15 shows a temperature swing the board goes through, which is 84, 74, and 54°, respectively. The resulting expansion–contraction "rock" the board, unevenly from the top to the bottom, and contribute to stress and acceleration of the respective oxidative degradation of the board.

Table 15.25 gives the numerical data for this particular experiment.

A typical question that everyone talking about a weathering box has asked: What is the time period of a natural exposure that would correspond to a certain time in the weathering box, say, to 1 day or to 1 month or to 1000 h in the box?

Of course, for each specific material one can find an equivalency in a certain deterioration—a degree of fading, a loss in strength, an increase in water absorption—between a respective time period in the weathering box and a respective time period of the natural exposure. The trouble is that this "equivalency" each time will be different for any geographical location and for any color of the material under study.

Example: Two composite roof tiles, gray and green, were placed on a roof in Bedford, MA, until they faded by 1.0 unit (in Hunter Lab scale units). For the gray tile it took 4 months outdoors and 700 h (1 month) in the weathering box. Hence, it could have been concluded that 1 month in the weathering box is equivalent to 4 months outdoors (in Bedford, MA). This ratio would also suggest that 2000 h in the box is equivalent to about 1 year outdoors in Bedford.

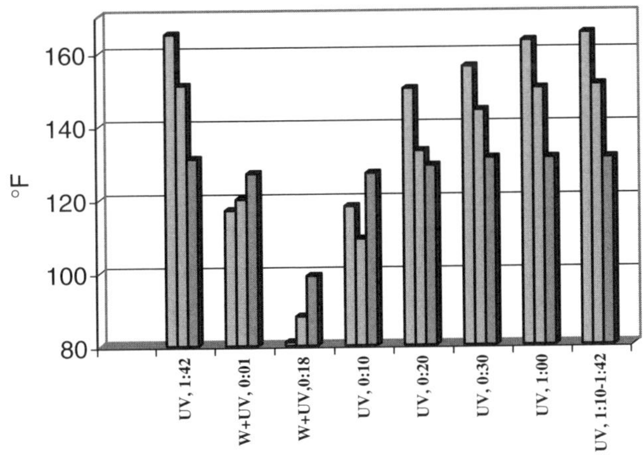

Figure 15.14 A GeoDeck Mahogany board temperature at three different points of the board in the course of one cycle of weathering (Q-Sun 3000, UV filter: Daylight, UV sensor: 340, 0.35 W/m^2, black panel 63°C, ASTM G155-97, Cycle 1: Light 1:42, Light + Spray 0:18). The first bar in each triple cluster—the top surface of the board; the second bar—the "inner" upper surface of the hollow channel; the third bar—the bottom surface of the board.

The first bar cluster—the end of the UV-only cycle, that is 1 hr 42 min of UV light; the second cluster—1 min into UV light 1 water spray; the third cluster—the end of water spray, that is 18 min into the cycle. The following five clusters: 10 min into the UV-only cycle; 20 min; 1 hr; the period between 1 h 10 min and 1 h 42 min (the same temperature).

For the green tile, however, for the same time period (4 months) on the roof but for only 300 h (12.5 days) in the weathering box, the tile faded by 2.0 units. Hence, the ratio is 1 month in the weathering box is equivalent to almost 10 months outdoors (in Bedford). In other words, the green tile was much more sensitive to the UV light

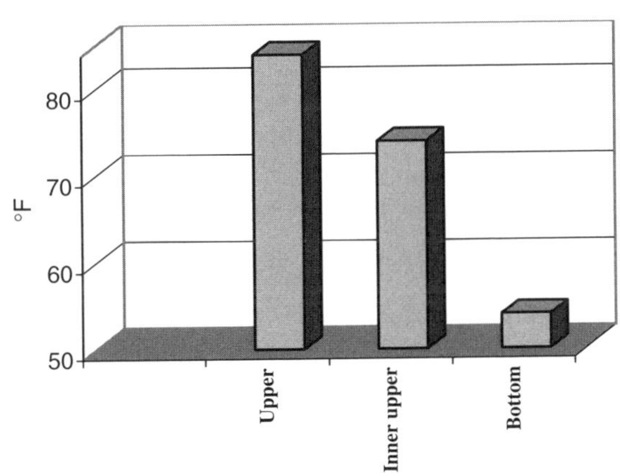

Figure 15.15 A temperature swing for a Mahogany GeoDeck board at three different points of the board in the course of one cycle of weathering (for the weathering condition, see Fig. 15.14 legend). "Upper"—the top surface of the board; "Inner upper"—the "inner" upper surface of the hollow channel; "Bottom"—the bottom surface of the board.

TABLE 15.25 Temperature of the upper panel, inner upper panel, and of the bottom panel of a composite hollow deck board in the weathering chamber

Side of the Board	UV, 1:42 (the very end of UV cycle)	Temperature,°F at the certain cycle time						
		Water spray + UV light		UV light				
		0:01 min	0:18 min	0:10	0:20	0:30	1:00	1:10–1:42
Upper	165	117	81	118	150	156	163	165
Inner-Upper	151	120	88	109	133	144	150	151
Bottom	131	127	99	127	129	131	131	131

Q-Panel, 0.35 W/m^2 at 340 nm, 2-h cycle: 1:42 UV light + 0:18 min UV light and water spray.

in the box compared to the natural sunlight. 1000 h in the box is now equivalent to more than a year (13 months) in Bedford.

Which set of data should be taken for calibration aiming at "standard" weathering? None, apparently. A calibration would be determined by color of the material; hence, it is related to this particular material only.

Another example: GeoDeck composite Mahogany board of a low density and with no added antioxidants started crumbling after 4 months in Phoenix, AZ. The same boards started crumbling after 4 days in a weathering box. Hence, one can say that 1 day in the weathering box is approximately equivalent—with respect to crumbling—to 1 month in Phoenix, and 1000 h in the box is equivalent to 3.5 years in Phoenix.

It would be not exactly thoughtful to say that the weathering conditions in Massachusetts are 3–4 times more severe than these in Phoenix, AZ. First, different readouts were considered, crumbling and fading; second, composites had different colors and different surface texture. This is important in weathering. If a readout is a drop in mechanical properties after weathering, the "calibration" can change again.

In terms of crumbling, as it would be explained below, taking a number of warranty claims as a readout, 1 year in Florida is approximately equal to 1.5 years in Pennsylvania, Maryland, Virginia, and New Jersey, and to 3.4 years in New England, Connecticut, New York State. One year in Phoenix, AZ, is approximately equal to 1.3 years in Florida and 4.5 years in Wisconsin, Minnesota, Michigan. More detailed comparisons can be obtained from Table 15.30.

Case study: HDPE-based roof slates (another main components were a flame retardant and an inorganic filler) made with an insufficient amount of both green chromium oxide pigment and a UV stabilizer after about 2 years on a roof in a southern state severely faded (L on the Hunter Lab units scale went from the initial 46 to 64 units). Inspection of their surface under a light microscope revealed a vast net of cracks (Fig. 15.16) compared with an unexposed tile (Fig. 15.17). A 10 times higher magnification shows cracks in more detail (Fig. 15.18).

At 43× magnification, one can visually track the cumulative mechanical UV damage by the appearance and subsequent growth of cracks in the tile surface. It turns out that fading and surface cracking phenomena always go together: a sample

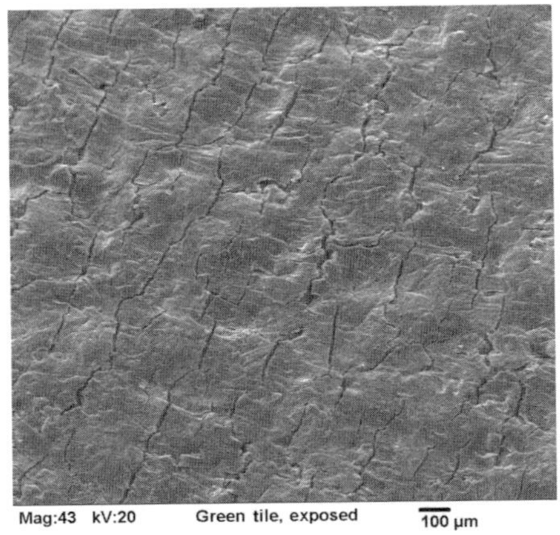

Figure 15.16 Green HDPE-based roof slate taken from a roof after 2 years of exposure. Magnification 43×.

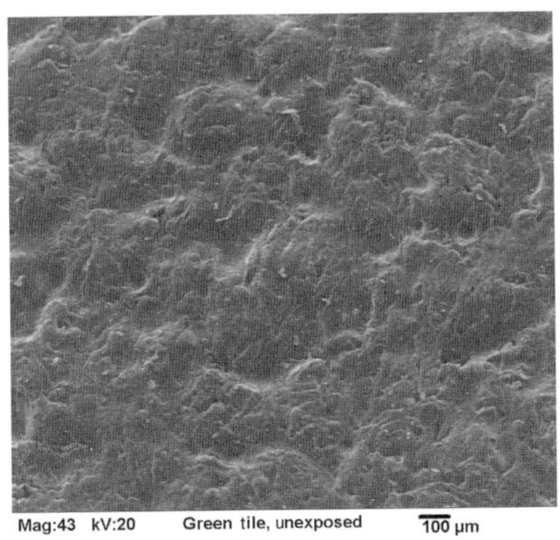

Figure 15.17 Green HDPE-based roof slate, unexposed. Magnification 43×..

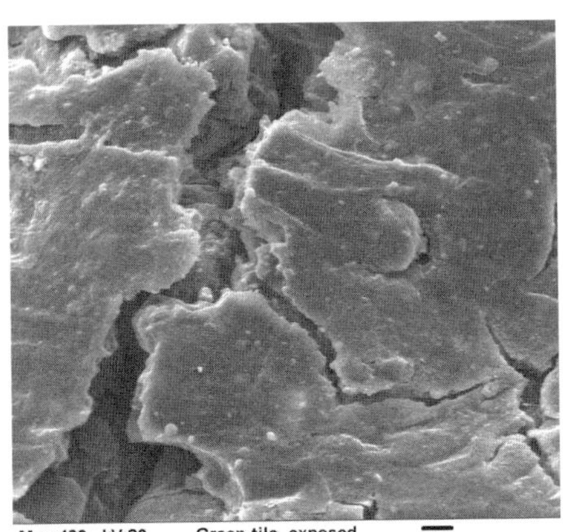

Figure 15.18 Scanning electron microscope image of a green-colored HDPE-based roof slate taken from a roof after 2 years of exposure. Magnification 430×.

that fades quickly also shows earlier and more rapid crack growth than a sample that fades more slowly. Commercial experience with HDPE shows that after many years exposure to UV radiation, surface crack growth eventually leads to a measurable reduction in tensile strength.

In a weathering box (0.35 W/m^2 at 340 nm, 2-h cycle @ 1:42 UV light and 0:18 water spray + UV light), cracks typically appeared between 2000 and 2500 h of exposure of gray-color tiles, and after as little as 600 h of exposure of green-color tiles. These cracks grew rapidly with additional exposure. After the pigment content and the UV stabilizer amount were increased, no fading and cracking occurred until after 4000 h in the weathering box.

Amount of Added Regrind, If Any

Adding regrind, often damaged to some extent by oxidative degradation during processing or in the field (when recalled composite material is used to make regrind), to "fresh" composite material is not adding a destructive element. Regrind does not carry free radicals because, as it was described above, free radicals are short-lived chemical species, and they either instantly propagate or cease to be free radicals. Added regrind can, however, decrease strength and stiffness (flexural modulus) of the resulting profile, if it is added in excessive amounts, but it would not accelerate oxidative degradation of the final composite material if a proper amount of antioxidants is added.

If, however, no antioxidants are added to a blend of plastic, fillers, additives, and regrind, an effective lifetime of the resulting composition might be lower compared to that of the same formulation without added regrind. A simplified formula looks like

$$\text{OIT final} = \text{OIT}_1 \times A + \text{OIT}_2 \times B$$

where
OIT final is the oxidative induction time of the final composition,
OIT_1 is the OIT of the composition without added regrind,
OIT_2 is the OIT of the regrind,
A is the fraction of the formulation without added regrind, and
B (=1.0 − A) is the fraction of the added regrind in the final formulation.

If, say, the OIT of plastic is 20 min, the OIT of the resulting composition without added regrind is 2 min (OIT_1), the OIT of a regrind is 0.5 min (OIT_2), and 20% of the regrind is added to the formulation ($B = 0.20$),

$$\text{OIT final} = 2 \times 0.8 + 0.5 \times 0.2 = 1.7 \text{ min}$$

Hence, the apparent lifetime of the composition with the added regrind is decreased by 15%. Of course, the real lifetime of the composition in the field would be determined by the temperature factor, and the above calculations give only a rough estimate of the effect of a regrind on oxidative degradation of composites.

If the antioxidant is added to the final blend in such an amount that it would increase the OIT of the initial formulation (without added regrind) to 20 min, the above formula would look like

$$\text{OIT final} = 20 \times 0.8 + 0.5 \times 0.2 = 16.1 \text{ min}$$

Though the apparent lifetime of the final formulation is still lower (by almost 20%) compared to that without added regrind, the final OIT is high enough to make the composite serve in the field for quite a long time, probably, well over the warranty time period.

To make the OIT of the final formulation equal to 20 min, 20% more antioxidant should be added to the blend, compared to that without added regrind.

Restabilization of an LDPE regrind with an added antioxidant was considered by Ref. [12]. The initial OIT of the LDPE was 24 min. After one cycle of processing the OIT was 18 min, after two, three, and four cycles of reprocessing the OIT was 14, 12, and 7 min, respectively. After that 0.2% of Recyclostab 421 was added into the regrind, and the OIT became 38 min. This means, with no surprise, that the OIT for the restabilized material was higher than for the material processed without restabilization.

In many cases a regrind requires higher amounts of an antioxidant added for restabilization. Here is an example of restabilization of an HDPE material that was stored for up to 18 years. Placed in a weathering box, the material began to deteriorate after 1000 h of exposure. With 0.2% of added Recyclossorb 550, tensile impact strength of the material increased to 260% of the initial. With 0.5% added Recyclossorb 550, tensile impact strength increased 6.4 times [12].

Hence, with added regrind there are two main issues that should be considered: (a) lifetime of the final product, which can easily be adjusted to the necessary value by just adding more antioxidant to the final formulation and (b) other properties of the final composite product, mechanical properties first of all; they cannot be adjusted by adding antioxidants, and they would be determined by superposition of those for the original formulation and for the regrind in added amounts.

Table 15.26 shows that with increase in the antioxidant, the OIT systematically increases, from the initial 5.3 to 10.9 min. Addition of 20% regrind invariably decreases the OIT (by 10–30%), but further increase in the antioxidant fully neutralizes the negative effect of regrind. 0.05–0.1% of the antioxidant on each step completely removes the negative action of 20% regrind.

ASTM RECOMMENDATIONS

ASTM Tests for Oxidative Induction Time

ASTM D 3895 "Standard Test Method for Oxidative Induction Time of Polyolefins by Differential Scanning Calorimetry" *Note of the author*: This procedure cannot be used for filled composite materials regarding "Sampling" section, particularly for those employing rather large filler particles. Instead of compression-molded test samples "into sheet format prior to analysis to yield consistent sample morphology and weight" and cut "specimen disks (6.4-mm diameter) from the sheet to have a

TABLE 15.26 Effect of regrind and an antioxidant on the OIT of WPC deck boards

Amount of regrind (%)	Amount of an antioxidant (%)	The OIT (min)
0	0.25	5.3
0	0.30	6.5
0	0.35	7.6
0	0.40	8.7
0	0.45	9.7
0	0.50	11.7
20	0.25	4.6
20	0.30	5.0
20	0.35	5.9
20	0.40	6.8
20	0.45	8.1
20	0.50	10.9

The OIT was determined at 190°C. Density (specific gravity) of all boards was 1.24 g/cm^3.

weight of approximately 5–10 mg, depending on sample density," a specimen of a filled composite material can be simply cut from or shave off from a sample (shape, profile, laboratory sample). A weight is not necessary to be exactly controlled, provided it is in a convenient range of about 5 mg to about 30 mg (usually 10–20 mg). The weight would affect the height of the peak of the heat flow of the oxidation isotherm, but not the OIT itself. Thus, a modified (in terms of sampling) ASTM D 3895 procedure is applicable to plastic/cellulose fiber/minerals composite materials.

OIT is a relative measure of a material's resistance to oxidative decomposition. It is determined by the thermoanalytical measurement of the time interval of exothermic oxidation of a material at a specified temperature (typically between 140 and 210°C) in an oxygen atmosphere. The procedure employs a differential scanning calorimeter (DSC). It is very practical to use an automatic sample, that is a carousel, which typically holds 50 specimens, and descriptions for 65 specimens can be programmed into the instrument before the runs. For low OIT numbers (less than 1–2 min), the procedure takes about 30 min per specimen; hence, 50 samples would run automatically for about 25 h, generating a wealth of data practically without an operator's involvement. For high OIT numbers, around 100 min, it would take almost 2 h per specimen. However, the instrument allows to cut the run after a specified time and automatically go to the next specimen.

The sample to be tested is heated at a constant rate (usually 20°C/min) under a nitrogen atmosphere. When the specified temperature has been reached, the specimen is maintained at this temperature for several minutes (usually 5 min) to reach a thermal equilibrium, and the nitrogen atmosphere is changed to oxygen maintained at the same flow rate (usually 50 cc/min). The specimen is then held at the specified constant temperature until the oxidation reaction is displayed on the thermal curve. The time interval from when the oxygen flow is first initiated to the oxidative reaction is referred to as the induction period. The end of the induction period is signaled by an abrupt increase in the specimen's evolved heat and may be observed by a DSC. The OIT is determined from the data recorded during the isothermal test, namely from the moment when the nitrogen in the chamber was replaced with the oxygen (it takes just a few seconds) until the time point at which the steepest slope on the heat flow curve is extrapolated to the flat baseline of the thermal curve. Contemporary DSC instruments perform the extrapolation and calculations of OIT automatically, on the DSC computer's screen (see Figs. 15.19 and 15.20, showing the OIT measurement for a WPC).

Precision of the OIT measurements is usually fair. ASTM D 3895-02 lists an example with four polyethylene samples [HDPE, LDPE (two samples) and LLDPE], tested in 11 different laboratories (a round robin test). They showed the OIT of 163 ± 8, 24 ± 3, 83 ± 9, and 120 ± 8 min, respectively, for within-laboratory standard deviations of the average, that is, within 5–13% of the average, and 163 ± 22, 24 ± 4, 83 ± 17, and 120 ± 15 min for between-laboratory standard deviations of the average, that is, within 13–20% of the average.

Note of the author: A recommended (modified ASTM D 3895) OIT procedure for filled composite materials. A convenient test temperature for cellulose–plastic

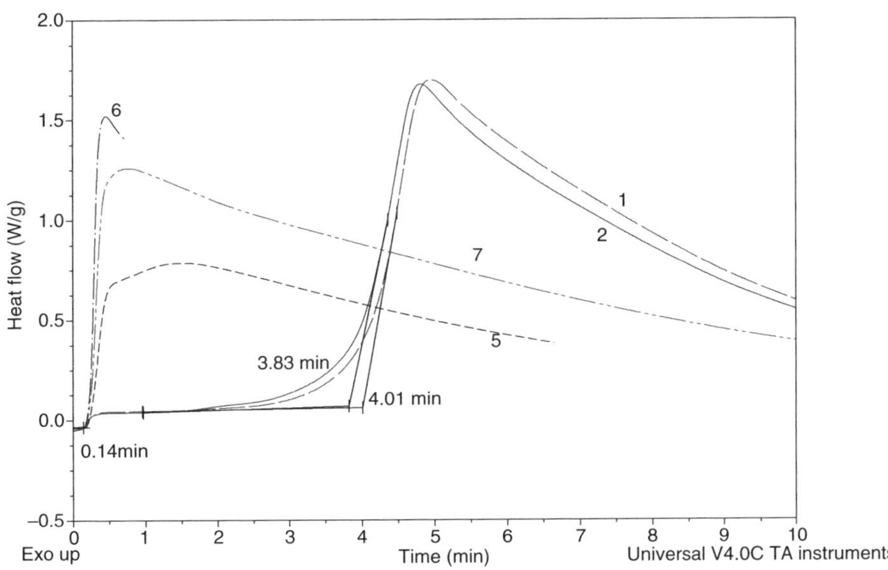

Figure 15.19 Examples of typical OIT measurements. The graph shows the OIT for wood–HDPE composite decking materials at 190°C. Samples 1 and 2 (containing an antioxidant) show the OIT equal to 4.01 and 3.83 min, respectively. Samples 5, 6, and 7 are the same material, except no antioxidant was added to the formulation. Besides, the last three samples were placed in different amounts into the DSC chamber. The respective heat produced as the result of their oxidation, is proportionally different. In all the last three cases, the OIT was equal to 0.14 min.

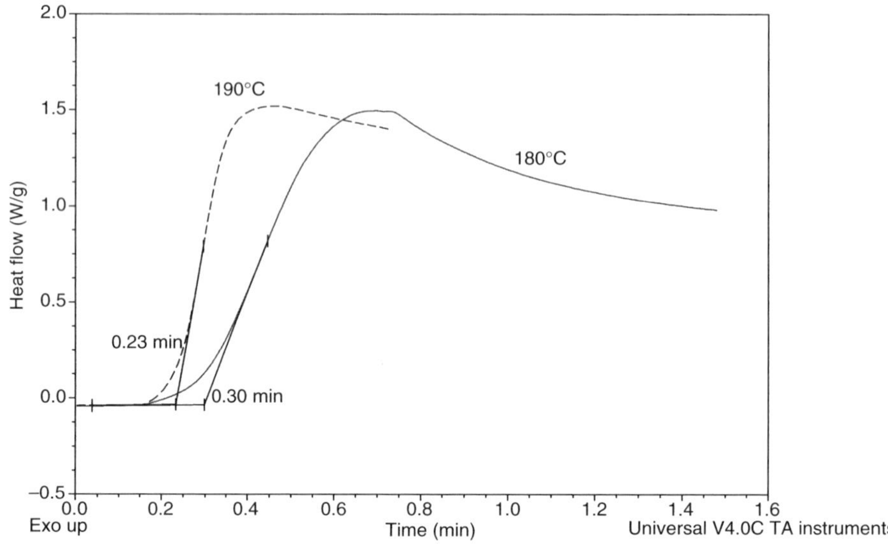

Figure 15.20 Examples of typical OIT measurements at two different temperatures. The graph shows the OIT for wood—HDPE composite decking materials at 180 and 190°C. The samples did not contain an antioxidant. They OIT values were equal to 0.30 and 0.23 min, respectively.

composites is 190°C. At this temperature most of the commercial composite building materials (such as deck boards) fall into the OIT range of 2–100 min, with a median of 20 min. Commercial HDPE samples typically fall into the OIT range of 1–100 min. A DSC instrument with an autosampler is recommended. The test sample is taken directly from the material to be analyzed (such as a deck board). A typical specimen is 10–20 mg by weight. A specimen is placed into an aluminum dish and placed onto a sampler (carousel), from which it is placed automatically into the cell (chamber). An identical empty pan is used as the reference. Indium is recommended to be used as the calibrant because its melting point (180°C) is close to the specified analysis temperature (190°C).

Nitrogen prepurge time prior to beginning the heating cycle is set for 5 min. Programmed heating of the specimen (under nitrogen flow of 50 mL/min) from ambient temperature to 190°C is at a rate of 20°C/min. When the set temperature (typically 190°C) is reached, programmed heating is discontinued and the sample is equilibrated for 5 min at the set temperature. After it the gas is changed to oxygen at a flow rate of 50 mL/min. This changeover point to oxygen flow is considered the zero time of the instrument.

Isothermal operation is continued until at least 5 min have elapsed after the steepest point of the exotherm was displayed. The test then is terminated, the gas selector is switched back to nitrogen, and the cell is cooled to 70°C to start a new test. The OIT is measured to within (0.01 min from zero time to the intercept point, obtained by extrapolation of the steepest linear slope of the exotherm onto the extended baseline.

Notes of the author: In many cases of determination of the OIT, the exotherm curve does not have a well-defined onset point. This particular question was studied by the ASTM D20 (plastics) committee. It was noted that the steepest linear slope (see Figs. 15.19 and 15.20) sometimes varies. An initial increase in the slope can be followed by a small decrease in the slope, and then followed by an abrupt increase leading to the steepest slope of the run. In that case it was unclear, if to follow the ASTM procedure instructions, which slope shall be used to determine the OIT—the first steepest point before the slope decrease, or the second steepest point, which is the steepest point of the run? Besides, the ASTM procedure says that a run can be stopped 2 min "after the steepest point of the exotherm has been displayed." However, because the steepest point can come later than expected, it was unclear whether to follow the ASTM procedure, whether to wait until the exotherm reaches a plateau, or whether to consider only the first steep portion of the plot?

In other words, the question was if the "steepest" slope, is equivalent to the term the "first" slope, or not, in case of multiple slopes? Or, when to stop the test?

The consensus among D20 members was that the testing should continue until the exotherm is completed, and the ballot called for a revision of Clause 10.2.5 to "Continue isothermal operation until the exotherm has peaked out for at least 2 min to allow a complete examination of the entire exotherm. At the tester's discretion, the test may be terminated if time requirements stated in the product's specification have been met."

However, this item has triggered another question, namely which linear slope of the exotherm should be extrapolated "to intercept the extended baseline" (Clause 10.3.1). The ASTM procedure defines it as the "steepest linear slope." However, in cases where multiple slopes are observed, the steepest slope could come at a later time when considerable amount of oxidation has already occurred. In other words, the question is when does oxidation take place? In the beginning, but with a small amount of the material, or later, but with a larger amount of the material, which represents the bulk of the material?

A comment of the D20 members was that it is up to the user what best represents the material for an application. If multiple slopes are due to the types of oxidation in a material, the definition of oxidation induction need to be made to fit the circumstances.

As a result, a note was suggested to be included into the ASTM as follows:

Note 14: If the multiple slopes are due to the type of oxidation, the definition of "oxidation induction" needed to be addressed to reflect the circumstance. It is up to the users which slope best represents the material property for an application. It must be noted, however, if the tangent line is not drawn from the first steep slope.

ASTM D 5885 "Standard Test Method for Oxidative Induction Time of Polyolefin Geosynthetics by High-Pressure Differential Scanning Calorimetry" Principally, this is the same test as the above and is performed using a DSC, except now with a high-pressure cell that can sustain a pressure of 5500 kPa. The test is designed for highly stabilized materials. It is applicable only to materials whose

OIT value under 3.4 Mpa (500 psi) of oxygen is greater than 30 min at 150°C. An elevated pressure of oxygen in this test is used to accelerate the reaction and to reduce analysis time.

Precision of the high-pressure OIT measurements is usually similar to that outlined in ASTM D 3895. ASTM D 5885 gives an example with two polyethylene samples tested in six different laboratories (a round robin test) at 150°C. They showed the high-pressure OIT of 231 ± 6 and 231 ± 18 min, respectively, for within-laboratory and between-laboratory standard deviations of the average, that is, 2.6 and 7.8%, respectively.

When the same polyethylene sample was tested at 200°C and 10 kPa (1.5 psi) oxygen pressure, that is, 50°C higher temperature but 340 times lower pressure, the OIT was 31.4 ± 1.6 and 31.4 ± 3.1 min, respectively (that is, 7.4 times lower), for within-laboratory and between-laboratory standard deviations of the average, that is, 5.1 and 9.9%, respectively.

Note of the author: It is doubtful that this procedure finds a widespread application for studying composite materials. First, composite materials are generally less stable (have lower OIT) than the respective plastics. Second, if at, say, 190°C the OIT is still high, such as 150 min (there is apparently no obvious reason to use such composite materials in building industry) then by increasing temperature to 210°C the OIT would decrease to about 40 min or even lower.

ASTM Tests for Determination of Phenolic Antioxidants in Plastics

There are a number of standard tests describing liquid chromatography procedures for determination of phenolic antioxidants such as Irganox 1010, Irganox 1076, Isonox 129, among others in polyolefins. These tests provide a means to directly evaluate a capacity of a given material to withstand oxidative degradation during processing and a subsequent long-term service. However, because these tests are developed for neat plastics, they are applicable in their current form only to incoming polyolefins.

Two subsequent events complicate the applicability of the said tests for composite materials. First, addition of fillers in high amounts, such as normally employed in wood fiber/mineral/plastic composites, significantly increases oxidative degradation of plastics during processing—due to increased shearing and, hence, overheating, and because of a number of accelerators/catalysts of oxidation that appeared in the system along with the fillers. Among them are metals (free metals, oxides, salts), moisture (almost unavoidably presenting in cellulosic fibers), air, and VOC microbubbles in the hot melt. Therefore, special studies need to be done with each formulation and for compounding/extrusion conditions to establish a correlation between the amounts of particular antioxidants in the incoming plastics and these in the extruded composite, and, if necessary, with performance-relevant properties of the extruded composite. It should not come as a surprise whether the amount of antioxidant(s) determined using the said ASTM procedures in the incoming plastic and in the extruded composite differ by orders of magnitude.

Second, the same approach has to be applied to establish a correlation between the amounts of particular antioxidants in the incoming plastics, in the extruded composite, and the performance properties of the composite product in the field, such as the lifetime of composite deck boards on a deck. This all was further complicated by lack of standard procedures for determination of phenolic (or any on that matter) antioxidants using liquid chromatography in cellulose fiber/minerals/plastic composite materials. All the said ASTM procedures are based on the extraction of ground plastic samples by refluxing with organic solvents such as isopropanol (2-propanol) and subsequent examination of the solvent extract by liquid chromatography using reverse phase (C-18 column) with UV detection at 200 nm. This works with neat polyolefins because the only extractable materials are plastic additives, among them antioxidants. With highly filled composite materials, extracts might often present a very complicated and rich mixture of compounds difficult to analyze using the said procedure. Sugars and lignin fragments, in particular, would strongly absorb UV light at 200 nm, shielding peaks from antioxidants. Hence, serious studies are in need in order to revise and modify the said procedures, or to develop new ASTM procedures applicable to cellulose/plastic composites.

In short, the following ASTM procedures are applicable to determine phenolic antioxidants in incoming plastics and to compare various incoming plastics in that regard.

ASTM D 1996 "Standard Test Method for Determination of Phenolic Antioxidants and Erucamide Slip Additives in Low-Density Polyethylene Using Liquid Chromatography" The procedure is described for the determination of Irganox 1010, Irganox 1076, and Isonox 129 in low-density polyethylene. The LDPE sample is ground to a 20-mesh particle size, and 5 g of it is extracted by refluxing with 50 mL of 2-propanol. The solvent extract is examined by liquid chromatography using reverse phase (C-18 column) in acetonitrile–water gradient with UV detection at 200 nm using Tinuvin P as an internal standard. The lowest level of detection for a phenolic antioxidant "under optimum conditions" is approximately 2 ppm, that is, $2 \times 10^{-4}\%$ (see, however, the next paragraph).

Precision of the measurements is usually good. ASTM D 1996–97 (Reapproved 2003) lists an example of six LDPE samples containing three antioxidants at high and low amount for each one, tested in 14 different laboratories (a round robin test). They showed the determined Irganox 1010 average concentration of 0.092 ± 0.004 and 0.0244 ± 0.0015% for within-laboratory tests ("high" and "low" Irganox concentration, respectively), and 0.092 ± 0.006 and 0.0244 ± 0.0025% for between-laboratory tests, respectively. Thus, standard deviation of the average was 4–6% for within-laboratory tests and 6–10% for between-laboratory tests. Tests for Irganox 1076 (at 0.10 and 0.025%) and Isonox (at 0.094 and 0.024%) gave similar results in terms of precision and bias.

ASTM D 5524 "Standard Test Method for Determination of Phenolic Antioxidants in High-Density Polyethylene Using Liquid Chromatography" The method technically is very similar to that described in ASTM D 1996, except HDPE samples are analyzed, and extraction is performed with cyclohexane. As the standard notes, besides Irganox 1010, Irganox 1076, and Isonox 129, the test method should be applicable for the determination of other antioxidants such as Cyanox 425, Cyanox 1790, Cyanox 2246, Ultranox 236, and Ultranox 246, but the applicability of this test method has not been investigated for these antioxidants. Precision of the procedure is slightly poorer compared to that of ASTM D 1996: 6–13% for within-laboratory tests and 9–19% for between-laboratory tests with concentrations of antioxidants in HDPE from approximately 0.02 to 0.08%.

ASTM D 5815 "Standard Test Method for Determination of Phenolic Antioxidants and Erucamide Slip Additives in Linear Low-Density Polyethylene Using Liquid Chromatography" This ASTM is very similar to ASTM D 1996, except that LLDPE samples are analyzed, and extraction is performed with either isobutanol or isopropanol prior to liquid-chromatographic separation. Isopropanol is recommended as the extraction solvent for lower crystallinity LLDPE (density 0.925 g/ cm^3 and below), and isobutanol is recommended as the extraction solvent for higher crystallinity LLDPE. As the standard notes, besides Irganox 1010, Irganox 1076, and Isonox 129, the test method should be applicable for the determination of other antioxidants such as Ultranox 626, Ethanox 330, Santanox R, and Topanol CA, but the applicability of this test method has not been investigated for these antioxidants. Precision of the procedure is somewhat poorer compared to that of ASTM D 1996: 3–7% for within-laboratory tests and 4–23% for between-laboratory tests with concentrations of antioxidants in HDPE from approximately 0.02 to 0.11%.

ASTM D 6042 "Standard Test Method for Determination of Phenolic Antioxidants and Erucamide Slip Additives in Polypropylene Homopolymer Formulations Using Liquid Chromatography" This ASTM is similar to in ASTM D 1996, except that polypropylene samples are analyzed, and extraction is performed with a cyclohexane–methylene chloride mixture using either reflux or ultrasonic bath prior to liquid-chromatographic separation. The test method is applicable for the determination of antioxidants such as Irganox 1010, Irganox 1076, Irganox 3114, and a phosphate antioxidant Irgafos 168. As the standard notes, this test should be applicable for the determination of other antioxidants, such as Ultranox 626, Ethanox 330, and Santanox R, but the applicability of this test method has not been investigated for these antioxidants. Precision of the procedure is slightly poorer compared to that of ASTM D 1996: 3–9% for within-laboratory tests and 9–19% for between-laboratory tests with concentrations of antioxidants in polypropylene of approximately 0.1%.

ASTM D 6953 "Standard Test Method for Determination of Antioxidants and Erucamide Slip Additives in Polyethylene Using Liquid Chromatography" This ASTM is technically similar to ASTM D 1996; however, it covers a wider set of procedures. Antioxidants are extracted with either isopropanol (densities less than

0.94 g/cm^3) or cyclohexane (densities higher than 0.94 g/cm^3) prior to liquid-chromatographic separation. Before refluxing, the polyethylene sample is ground to a 1-mm (~20 mesh) or 0.5-mm (~40 mesh) particle size. C-8 or C-18 column is recommended for reverse phase chromatography. The test method is applicable for the determination of antioxidants such as Irganox 1010, Irganox 1076, Isonox 129, and a phosphate antioxidant Irgafos 168. As the standard notes, this test should be applicable for the determination of other antioxidants, such as Ultranox 626, Ethanox 330, Santanox R, and Topanol CA, but the stability of these during extraction has not been investigated.

Precision of the procedure is similar to that of ASTM D 1996: 4–6% for within-laboratory tests and 6–12% for between-laboratory tests with concentrations of antioxidants in polypropylene approximately from 0.02 to 0.1%.

ASTM D 3012 "Standard Test Method for Thermal-Oxidative Stability of Polypropylene Using a Specimen Rotator Within an Oven"

Note of the author: This procedure is aimed at estimating the resistance of polypropylene, in molded form, to accelerated aging by heat in the presence of oxygen from air using an airflow oven. Technically, it can apparently be applied to polypropylene-based composite materials without any changes in the procedure. In case of HDPE-based composites, the test should be conducted at temperatures lower than 150°C, and apparently, not higher than 120°C. Also, in a modified procedure, an extruded specimen of a composite material can be employed.

According to the test, five specimens (each 50 mm long, 100 mm wide, and 1 mm thick) are conditioned at ambient temperature as specified in the test, and placed in a forced draft oven equipped with a biaxial rotator. Specimens should be attached to the rotator by metal slips lined with fluoropolymer film and should not directly contact with the metal clips or metal parts of the oven. The frequency of rotation about the horizontal and vertical axes of the rotator should be 1–3 min^{-1}. The time to failure is determined by regular visual examination of the specimens as the number of days after which the specimen shows localized crazing, crumbling, or discoloration, or a combination thereof. According to the standard procedure, the oven temperature shall be 150°C (302°F).

Precision of time-to-failure measurements in controlled conditions, when the said test procedure is closely followed, is rather fair. ASTM D 3012-00 lists an example with three polypropylene samples, apparently, of different origin, tested in seven different laboratories (a round robin test). They showed oxidative stability of the samples (time to failure) at 150°C as 14.0 ± 0.8, 35 ± 3, and 63 ± 5 days, respectively, for within-laboratory standard deviations of the average, that is, within 6–9% of the average, and 14 ± 3, 35 ± 7, and 63 ± 19 days, respectively, for between-laboratory standard deviations of the average, that is, within 20–30% of the average.

Note of the author: Obviously, with antioxidants as part of the composition, the time to failure will be delayed accordingly. This time, determined at 150°C, is not directly related to the lifetime of the material in the field and, according to the standard, shall not be used to predict performance. However, conducting the test at different temperatures, for example, in the range from 150°C to 100°C, with intervals of 10°C, and extrapolating data obtained to a desired temperature(s) through use of the Arrhenius relation or the respective temperature coefficients, reasonable estimates can be obtained. These estimates, though, can be still rather far away from reality, particularly if the extrapolation is used for estimations of real-life performance of a product at natural conditions, with seasonal, daily, and hourly fluctuation of temperature, rains, and other uncontrolled climatic conditions. Furthermore, even extrapolations to a desired temperature in controlled laboratory conditions using the Arrhenius plot can result in quite a deviation because these plots are, as a rule, nonlinear in a wide temperature range. The closer the studied temperature range would be to the desired temperature to be extrapolated to, the lesser would be the deviation. It is unwise to obtain data at, say, 150°C, 140°C, and 130°C, and extrapolate them to (say) ambient temperature.

ASTM D 5510 "Standard Practice for Heat Aging of Oxidatively Degradable Plastics" This practice defines the exposure conditions of plastic at various temperatures when exposed solely to hot air for an extended period of time. Any particular property of plastics can be chosen as a readout. Technically, any composite material, shape and profile, can be tested in accordance with this practice. Two methods of oven exposure are recommended in the practice, of which only procedure B is applicable to composite materials because it is intended for specimens having a nominal thickness greater than 0.01 in. Procedure A is recommended for film specimens.

The said practice recommends to use a minimum of four exposure temperatures in order to determine the relationship between temperature and a define property change. The lowest temperature should produce the desired level of property change or product failure in approximately 6 months. The second temperature should produce the same level of property change or product failure in about 1 month. The third and fourth temperatures should do it in approximately 1 week and 1 day, respectively. Using some assumptions, this practice provides several typical heat aging time schedules for selections of the exposure temperatures.

SURFACE TEMPERATURE OF COMPOSITE DECKING AND RAILING SYSTEMS

As in this chapter we often discuss events depending on deck temperature, it is reasonable to consider what is "deck temperature," how it depends on day time, on air temperature, what is the difference on top and bottom of a deck, and how the temperature of composite deck surface is related to a wooden lumber deck [such as pressure-treated lumber(PTL)] at the same day temperature, same solar radiation, and the same location.

Figure 15.21 shows the dynamics of temperature at the surface and the bottom of a PTL deck in the course of a summer day in Massachusetts. The midday air temperature was 83°F. Dark bars show top surface temperature, and white bars show the temperature at the bottom of the same deck.

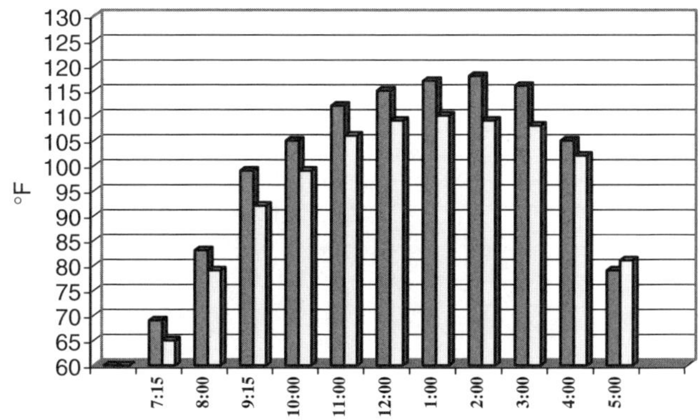

Figure 15.21 Temperature at the top surface (dark bars) and at the bottom (white bars) of a PTL deck in the course of a summer day in Bedford, M. A. The midday (12 P.M–2 P.M) air temperature was 83°F. Vertical axes: temperature, (degree Fahrenheit). Horizontal axes, daytime, A.M → P.M.

One can see that the maximum deck top surface temperature (118°F) was reached at 2 P.M., and that temperature was 35°F higher than the air temperature. Besides, in the midday the deck bottom temperature was 7–9°C lower than the deck top temperature.

Figure 15.22 shows surface temperatures of a composite GeoDeck (hollow profile), recorded on the same day as in Figure 15.21.

It can be seen that the Cedar GeoDeck maximum top surface temperature (116°F) of the deck was reached at 2–3 P.M. and that temperature was 33°F higher than the air temperature. Actually, Cedar GeoDeck top surface is about 2° cooler compared

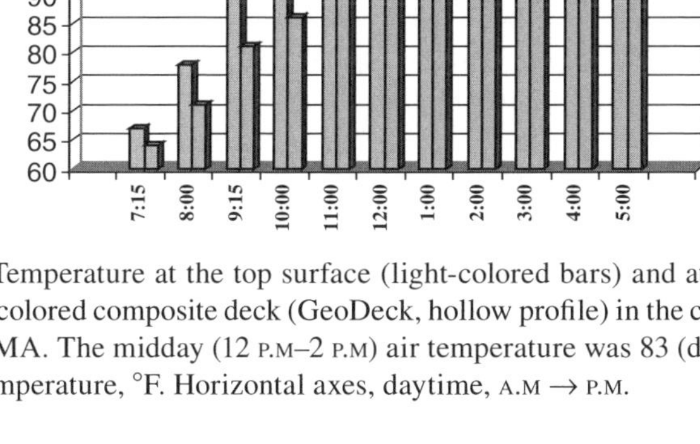

Figure 15.22 Temperature at the top surface (light-colored bars) and at the bottom (dark bars) of a Cedar-colored composite deck (GeoDeck, hollow profile) in the course of a summer day in Bedford, MA. The midday (12 P.M–2 P.M) air temperature was 83 (degree Fahrenheit). Vertical axes: temperature, °F. Horizontal axes, daytime, A.M → P.M.

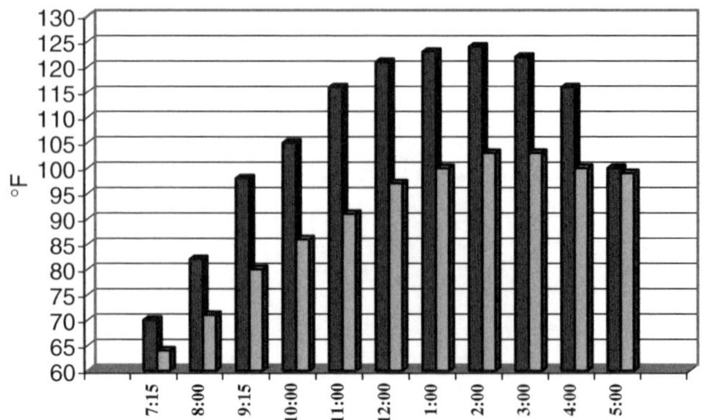

Figure 15.23 Temperature at the top surface (dark-colored bars) and at the bottom (lighter bars) of a Mahogany-colored composite deck (GeoDeck, hollow profile) in the course of a summer day in Bedford, MA. The midday (12 P.M–2 P.M) air temperature was 83°F. Vertical axes: temperature, (degree Fahrenheit). Horizontal axes, daytime, A.M → P.M.

to PTL deck. However, the bottom surface of the composite deck is much cooler than that of the wooden deck. In the midday the composite deck bottom temperature was 13–18° lower than the deck top temperature. The highest difference between top and bottom surfaces was at 11 A.M. (18°), when the top was already hot (109°) but the bottom was still cool (91°) because of a low thermal conductivity of the hollow composite profile.

The Mahogany GeoDeck maximum top surface temperature (124°F) was reached at 2 P.M. (Fig. 15.23), and that temperature was 41°F higher than the air temperature. Thus, Mahogany GeoDeck top surface is about 6 degrees warmer compared to PTL deck. The bottom surface of the composite deck is still much cooler than that of the wooden deck. In the midday the composite deck bottom temperature was 21–25° lower than the deck top temperature. The highest difference between top and bottom surfaces was again at 11 P.M. (25°), when the top was already hot (116°) but the bottom was still cool (91°) because of a low thermal conductivity of the hollow composite profile.

Notice that the maximum temperature at the bottom of the wooden deck (110°F) is noticeably higher compared to that of Cedar and Mahogany composite decks (103°F).

The Driftwood GeoDeck maximum top surface temperature (126°F) was reached at 2 P.M. (Fig. 15.24), and that temperature was 43°F higher than the air temperature. Thus, Driftwood GeoDeck top surface is about 8° warmer compared to PTL deck. The bottom surface of the composite deck is still much cooler than that of the wooden deck. In the midday the composite deck bottom temperature was 22–23° lower than the deck top temperature. The highest difference between top and bottom surfaces was at 12 p.m. (23°), when the top was already hot (121°) but the bottom was still cool (98°) because of a low thermal conductivity of the hollow composite profile.

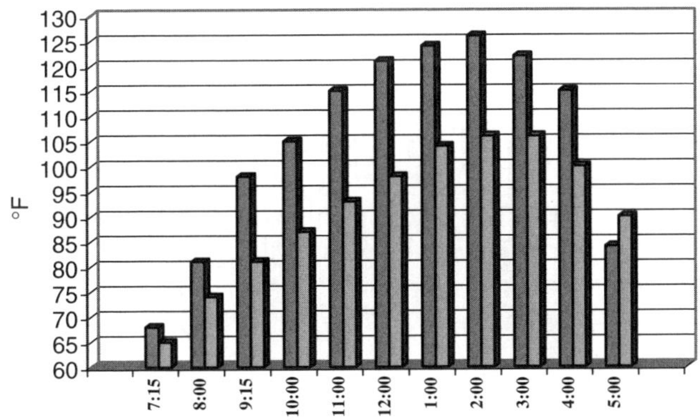

Figure 15.24 Temperature at the top surface (light-colored bars) and at the bottom (darker bars) of a Driftwood-colored composite deck (GeoDeck, hollow profile) in the course of a summer day in Bedford, MA. The midday (12 P.M–2 P.M) air temperature was 83°F. Vertical axes: temperature, (degree Fahrenheit). Horizontal axes, daytime, A.M → P.M.

A heating pattern is quite different with railing systems. Although a deck when exposed to sunlight all day is constantly under solar radiation though at different angles along the day, posts and handrails turn different sides to sunlight throughout the day. Hence, their heating profiles are quite different, such as shown in Figure 15.25.

One can see that at 7:15 A.M. only one, eastern, side of the post was much warmer (78° and 81° for two posts) compared to the other three sides, which were turned away from sun (65° and 66–67°, respectively). It was also much warmer than the top surface of the deck itself (68°), because the deck surface was at a low angle toward the sun that time.

By 9:15 A.M. the eastern side was still the warmest one (95° and 98° for the two posts), and the top deck surface reached the same temperature (98°). For the "colder" post, three other sides showed a gradient of temperatures: 85° on the southern side, 82° on the western, and 81° on the northern side.

Between 9:15 A.M. and 4 P.M. the temperature of the eastern side was steadily decreasing (from 98 to 95° for the two posts, respectively) and then stayed at about the same level (around 90–91°), because sunlight kept moving away from it to the southern and then western side. In turn, the southern and the western sides reached their temperature peaks by 1 P.M (98°) and 4 P.M. (111°) for one post, and 95 and 104° for another post, respectively. Meanwhile, the top surface of the same deck reached 126°, being constantly heated by the sun.

In other words, posts are always noticeably cooler throughout a hot day compared to the deck surface. The same is valid for balusters and handrails, as it is shown in Figure 15.26 for a handrail on the same deck as described above.

One can see that the data show the same pattern as that with the posts. By 10 A.M. the eastern side was much warmer compared to its opposite side. At about 12:30 P.M. their temperature became equal, and from 1 P.M through 4 P.M. the western side was warmer. By 5 P.M. temperature dropped for both of them. Overall, the handrail

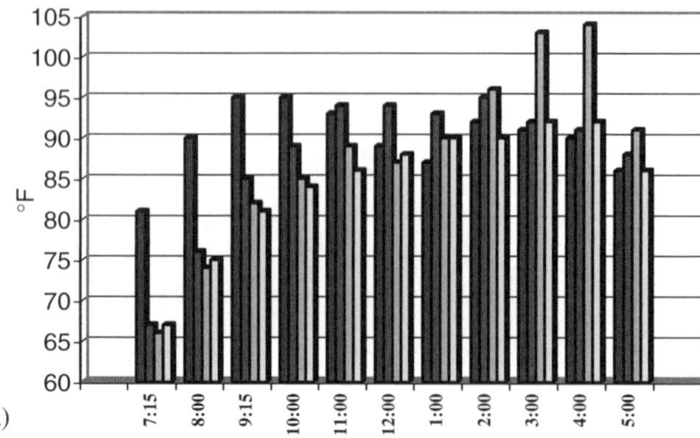

(a)

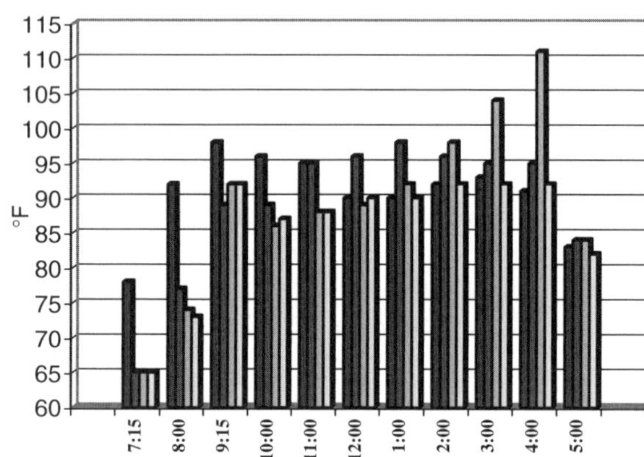

(b)

Figure 15.25 Temperature at four sides of a Driftwood post in a railing system (hollow profiles). In each cluster of bars the first one faced east, the second south, the third west, and the fourth north. The measurements were taken in the course of a summer day in Bedford, MA. The midday (12 P.M.–2 P.M.) air temperature was 83°F. Vertical axes: temperature, (degree Fahrenheit). Horizontal axes, daytime, A.M. → P.M. Data are shown for two Driftwood posts on the same deck but in different locations.

surface temperature was around 95° throughout the day (±5°), whereas the same deck surface reached 126° on the same day.

This explains why handrail systems are much more resistant to the environmental oxidative degradation compared to deck boards. As it is described in the next section, Kadant Composites had 1707 claims on crumbling of composite products, and among them there were only a few claims on components of railing system. Practically all of the claims were related to composite boards.

Both HDPE- and polypropylene-based composite deck boards are heated by direct sunlight to about the same surface temperature. As an example, the following

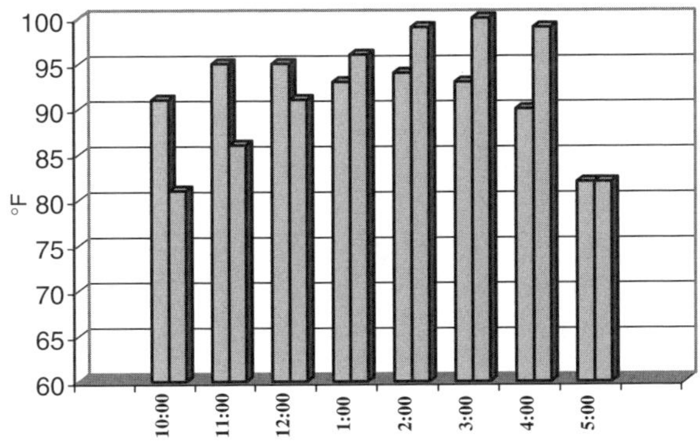

Figure 15.26 Temperature at two sides of a Driftwood handrail in a railing system (hollow profiles). In pairs of bars in the Figure the first one faced East, the second—opposite side. The measurements were taken in the course of a summer day in Bedford, MA. The midday (12 P.M.—2 P.M.) air temperature was 83°F. Vertical axes: temperature, (degree Fahrenheit). Horizontal axes, daytime, A.M. → P.M.

figures were obtained on a hot July day in Massachusetts, with 90° in shade, on the surface of CorrectDeck (polypropylene-based composite) and GeoDeck (HDPE-based composites) at noontime:

- CorrectDeck red, 119°
- CorrectDeck brown, 126°
- GeoDeck Mahogany, 122°.

At 2 P.M. the surface temperatures were as follows:

- CorrectDeck red, 127°
- CorrectDeck brown, 133°
- GeoDeck Mahogany, 128°.

One can see that the temperatures were close and mainly determined by the color rather than the nature of the plastic.

More detailed studies with the same boards were undertaken on a hot August day, between 7 A.M. and 12 P.M. (after that some clouds had introduced interfering factors), see Table 15.27.

One can see that the difference between air temperature and composite board surface temperature was equal to 2–4° early in the morning and reached 34–45° by noontime.

Again, there was no distinct effect of the nature of the plastic—polyethylene or polypropylene—on the surface temperature of composite boards. The effect of a colorant was much more noticeable.

TABLE 15.27 Air temperatures and composite board surface temperatures for CorrectDeck (red and brown) and GeoDeck (Mahogany)

Time of the day	Air temperature (°F)	Composite board surface temperature (°F)		
		CorrectDeck, red	CorrectDeck, brown	GeoDeck, Mahogany
7:00 am	66	68	70	70
8:00 am	73	84	90	92
9:00 am	80	98	106	107
10:00 am	87	108	118	118
11:00 am	89	123	133	129
12:00 pm	90	124	135	130

LIFE SPAN OF ZERO-ANTIOXIDANT GEODECK DECKS IN VARIOUS AREAS OF THE UNITED STATES

As it would be described below in detail, GeoDeck composite boards were manufactured with an insufficient amount of an antioxidant from April of 2002 through September of 2003, for more than a year. The amount of composite deck boards, made during those 17 months, was enough to make estimated 11,000 decks. By September of 2003 some of those decks had started to show signs of progressive oxidative degradation, that is, crumbling. Three years later, by September of 2006, the amount of crumbled decks on the record, with warranty claims filed, reached 1707 (Table 15.28).

Of course, not all of those 11,000 decks will crumble. Many of them are covered decks, many of them are kept in a shadow, and many of them have high enough density of the material, which makes them resistant to crumbling.

As those 1707 decks were installed practically in all states across the country, the analysis of the dynamics of the crumbling based on the warranty claims provides us with a unique opportunity to learn about the phenomenon of the oxidative degradation of the WPC material in different climatic conditions. We consider each warranty claim as a single experimental point obtained in the laboratory with a size of the whole country.

Multiple experimental observations have shown that GeoDeck composite boards when they do not contain added antioxidants began crumbling in the Phoenix, AZ, area after 4–24 months, depending on board density, location (full or partial exposure to direct sunlight), and apparently on conditions of the board manufacturing (extrusion speed, possible excessive heating during processing, duration of postmanufacturing heat treatment, etc.)

It would be safe to suggest that in the worst case scenario (low density of boards, such as 1.06–1.08 g/cm^3) and direct sunlight most part of the day, GeoDeck boards without added antioxidants begin crumbling after about 6 months in the Phoenix, AZ, area. The OIT of those boards at 190°C typically are in the range of 0.2–0.3 min. Thus, for those boards the difference between the OIT and their average "lifetime"

TABLE 15.28 Number of warranty claims regarding crumbling of GeoDeck boards in 2004–2006

State	Number of claims
MO	136
AZ	127
MD	100
PA	98
CA	80
CT, KS	71 each
TX	70
NC	69
IL	68
NJ	66
KY	63
FL	58
NY	57
IN	56
OH, VA	52 each
NM	50
MI	46
WI	42
TN	40
MN	32
SC	31
MA	28
CO	26
GA, LA	20 each
OR	15
WA	12
NE	11
AR	10
VT	6
NV, DE, WY, RI	4 each
NH	3
IA, OK	2 each
MS	1
ME, DC, WV, ND, SD	None
TOTAL	1707

is about 0.8–1.3 million times. In this context the lifetime can be defined as a time period after which crumbling becomes rather noticeable, and an owner with a good likelihood files a warranty claim.

The difference in the oxidative degradation of 0.8–1.3 million times between 190°C and, say, 40°C can be described with the temperature coefficient of 2.5–2.6, that is, close enough to 2.7, calculated above and based on very different premises.

Such a good fit is rather surprising taking into account an extrapolation over such a wide temperature range (190°–40°C) and a number of assumptions. At any rate, it shows that such an approach to oxidative degradation of composite materials based on the OIT and observation of crumbling makes sense.

This brings us to warranty claims as an instrument in determining the relative rate of boards failure due to crumbling in different regions of the United States (Table 15.28).

By themselves, these data do not tell us much. Clearly, the numbers of claims in some states greatly exceed those in other states, but maybe there are just a few GeoDeck decks in those "other" states in the first place. South Carolina has less crumbled decks than Connecticut, but maybe there are more decks in Connecticut.

In order to normalize these data with respect to the quantities of decks in the regions, the amount of sales in the regions in 2003 (when the majority of the defective decks was sold) was considered. The continental United States territory was subdivided into 10 regions, which essentially reflect distributorships of GeoDeck in the United States, and the ratios of warranty claims per sales are shown in Table 15.29.

TABLE 15.29 Number of warranty claims due to composite board crumbling (in 2004–2006) and number of claims per $100,000 of sales in 2003 (R)

States	Number of warranty claims, crumbling	R
AZ	127	25.1
TX, NM	120	23.0
AR, MO, IA	148	19.0
FL	58	16.9
CA, NV, CO, WA, OR, WY	141	15.8
KY, TN, NC, SC, GA, LA, MS	244	14.5
PA, MD, VA, NJ, DE	320	11.8
KS, OK, NE	84	9.1
NY, MA, VT, NH, CT, ME	169	6.1
MN, WI, IL, IN, MI, OH	296	5.0
Total	1707	

Total sales in 2003 was for $17,058,000 (see Tables 1.3 and 9.5).

Now, Table 15.29 clearly shows that the southern states produced many more warranty claims due to crumbling, compared with the Great Lakes states and New England states (besides state of New York). Can we say, based on data from Table 15.29, that crumbling in Arizona is five times faster than crumbling in the Great Lakes states?

Not quite so. This would be a statement describing not the rate of crumbling, but rather a number of warranty claims at a single time point, that is, the end of 2006. In reality, the dynamics of the crumbling follows a curve with an initial lag-time (Fig. 15.27).

Figure 15.27 shows that each curve differs from another by two parameters: the lag-time and the steady-state portion of the curve. Another feature is the decline of warranty claims, apparently due to depletion of the inventory. This is particularly noticeable for curves for Arizona and Florida. Seasonal (winter periods) slow down

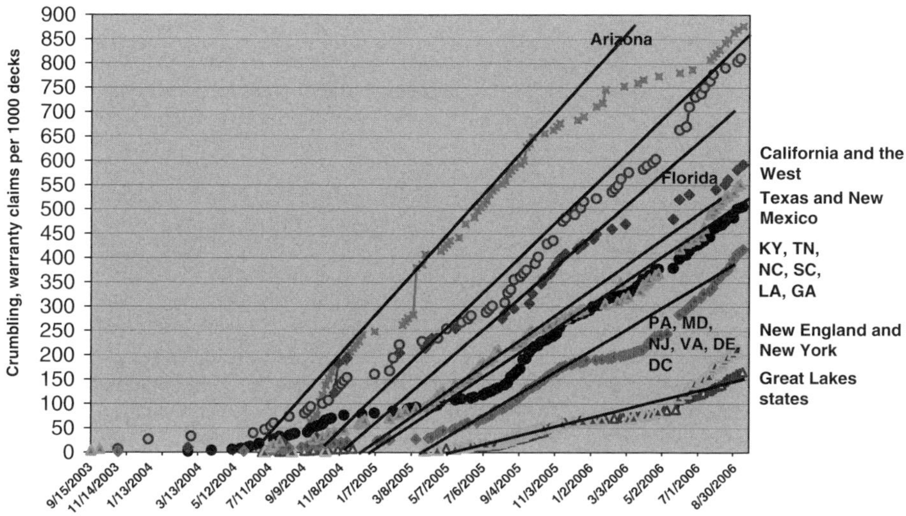

Figure 15.27 The dynamics of warranty claims accumulated over 2003–2006 in seven zones—Southern states: AZ; TX and NM; FL; KY, TN, NC, SC, LA, GA; Pacific and neighboring states: CA, WA, OR, CO, NV, WY; Atlantic states: PA, MD, NJ, VA, DE; Northern states: MN, WI, IL, IN, MI, OH; New England and NY. It was assumed that in terms of sales (Table 15.29) one deck is worth of \$3500 on average. Actual figures support this assumption.

of warranty claims are also noticeable. A lag-time reflects the antioxidant depletion from the composite, and after it burns out, the "defenseless" composite starts to get oxidized and finally crumbles. At some point the owner files the warranty claim; hence, a point appears in Figure 15.27.

Therefore, the data in the last column in Table 15.29 represent a mix of all three features: the lag-time, the steady-state period, and the decline (if it is observed).

This kind of kinetics, as shown in Figure 15.27, is described by the following equation:

$$P = At - B(1 - e^{-kt})$$

where (in this particular case)

P = the number of crumbling decks after a certain time period (t),

A = the crumbling rate constant for the composite in the given geographical area (see Fig. 15.27), when the antioxidant is completely depleted. The constant A is related to the linear portion of the respective curve,

B = constant, approximately equal to 400/1000 in our case (Fig. 15.27),

k = the rate constant describing depletion of the antioxidant in the first phase of the oxidative degradation.

When time (t) is long enough, e^{-kt} becomes much less than 1; in this particular case when (t) is more than a year, the above equation becomes

$$P = At$$

TABLE 15.30 Relative rates of crumbling of the composite boards in some geographical areas in the United States

States	Relative rate of composite board crumbling
AZ	4.5
TX, NM	3.8
FL	3.4
CA, NV, CO, WA, OR, WY	2.8
KY, TN, NC, SC, GA, LA, MS	2.7
PA, MD, VA, NJ, DE	2.4
NY, MA, VT, NH, CT, ME	1.0
MN, WI, IL, IN, MI, OH	1.0

The data are related to linear, steady-state portions of curves in Figure 15.27, when the antioxidant is completely depleted and the boards are "defenseless" with respect to oxidative degradation.

and crumbling is proportional to time span. This is related to the linear portion of the curve in Figure 15.27. Again, the linear portion describes the rate of boards crumbling when all antioxidants are depleted (Table 15.30).

In terms of crumbling, 1 year in Florida is equal to 3.4 years in New England, but only when all antioxidant is depleted from the board and the steady-state phase is reached. The shorter the time span from the deck installation is taken, the higher will be the above ratio. When we began only the crumbling observation by counting first batches of warranty claims, the ratio of crumbling claims in Florida and New England was 51. Half a year later the ratio decreased to 29. This was still a time-lag phase for the antioxidant depletion. This example shows that to take those ratios without considering which phase of crumbling development is analyzed can be misleading.

What the a rational behind the data in Table 15.30?

Clearly, the difference between the rates of crumbling should be explained by differences in climatic conditions in the respective areas. The principal parameters of these conditions apparently are air temperature, solar radiation, and precipitations. Each of these parameters contributes to the oxidative degradation of boards, though in an irregular and poorly predictable manner. These parameters are changes with high amplitudes between a hot day and a chilly night, overcasted and cloudless sky, summer and winter seasons, and rainy and dry periods.

At the first glance, the pattern of these factors is too complicated to try to identify any meaningful explanations of the observed dynamics of board crumbling. However, these complicated interferences and superpositions of various factors result nevertheless in a rather reproducible patterns of dynamics of a board oxidative degradation in any given area (Fig. 15.27).

It is reasonable to suggest that temperature is one of the most important parameters determining the dynamics of board crumbling. However, which temperature to consider? An average yearly temperature? Hardly, because the oxidative degradation exponentially depends on temperature and should be negligible during winter season compared to summer hot and sunny days. Hence, it makes sense to consider the "average high" temperature on, say, July, over a number of years. Table 15.31 shows the available data. These data are typically collected for at least a 30-year period and normalized as a statistical average.

TABLE 15.31 July average high temperature

Area	July average high temperature (°F)
Phoenix, AZ	106
Las Vegas, NV	105
Dallas, TX	98
Fresno, CA	98
Oklahoma City, OK	94
Albuquerque, NM	93
Salt Lake City, UT	93
Orlando, FL	92
Hattiesburg, MS	92
Pine Bluff, AR	92
Greensboro, AL	92
Baltimore, MD	92
New Orleans, LA	91
Topeka, KS	90
Nashville, TN	90
Jerome, ID	90
Lincoln, NE	90
Paducah, KY	89
Miami, FL	89
Durham, NC	89
Charleston, SC	89
Evansville, IN	89
St. Louis, MO	89
Raleigh, NC	88
Lewiston, ID	88
Denver, CO	88
Atlanta, GA	88
Washington, DC	88
Carbondale, IL	88
Dover, DE	88
Norfolk, VA	87
Casper, WY	87
Honolulu, HI	87
Des Moines, IA	86
Sioux Falls, SD	86
New Brunswick, NJ	86
Philadelphia, PA	86
New York, NY	85
Bismarck, ND	84
Atlantic City, NJ	84
Buckhannon, WV	84
Helena, MT	84
Madison, WI	83
Albany, NY	83

(*Continued*)

TABLE 15.31 (***Continued***)

Area	July average high temperature (°F)
Providence, RI	83
Chicago, IL	83
Spokane, WA	83
Hanover, NH	83
Detroit, MI	83
Minneapolis, MN	83
Los Angeles, CA	82
Cleveland, OH	82
Burlington, VT	82
Bedford, MA	81
Green Bay, WI	81
Portland, OR	80
Hartford, CT	80
Buffalo, NY	80
Caribou, ME	76
San Diego, CA	76
Seattle, WA	75
San Francisco, CA	71
Anchorage, AK	65

According to the data in Table 15.31, an average high temperature difference between Phoenix, AZ, and Bedford, MA, is 25°F (13.9°C). Is it enough to explain actual rates of composite boards crumbling in these two areas?

For the calculations see insert on p. 510. The temperature difference of 25°F would result in a difference in rate of oxidative degradation by four times, if the temperature coefficient is equal to 2.7 (per 10°C). Table 15.30 shows the difference of 4.5 times. Close enough. However, actual temperature that we have to consider is not air temperature, listed in Table 15.31, but the deck surface temperature. As it is described above, a typical composite deck surface temperature is 50°F higher than air temperature in Arizona, and 40°F higher in Massachusetts. Therefore, actual temperature difference between decks in Phoenix and Bedford on a hot sunny day, when the fastest oxidation occurs, would not be 25°F, but 35°F (19.4°C). At this temperature difference and the temperature factor of 2.7, the ratio of composite crumbling in Phoenix and Bedford will be 6.9 (see insert on p. 563).

Thus, for the temperature coefficient of 2.7, the difference in crumbling rate between Phoenix and Bedford can be between 4.0 and 6.9, for temperature differences of 13.9°C and 19.4°C, respectively. In reality this difference is equal to 4.5 (Table 15.30).

These calculations show that even temperature alone can explain differences in crumbling rate across the United States. We are not aiming here at exact figures; we just show the principal possibility to understand the phenomenon of crumbling in quantitative, or at least semiquantitative, terms.

In fact, the difference of 4.5 in crumbling rates between Phoenix and Bedford can be easily explained assuming the temperature coefficient of 2.7 and temperature difference of 15.1°C, that is, only 1.2° higher than that shown in Table 15.31. It is quite realistic.

Similarly, the difference between crumbling rates in Texas and New England (3.8 times, see Table 15.30) can be explained using the respective average high air temperatures in Dallas, TX (98°F = 36.7°C), and Bedford, MA (81°F = 27.2°C). For the air temperature difference of 9.5°C and the temperature coefficient of 2.7 the calculated difference in rates of crumbling is only 2.6. However, for composite board surface temperature 50°F higher in Dallas and 40°F higher in Bedford, the temperature difference is 15°C, and the ratio of crumbling rates should be 4.4 times. In reality it is 3.8 times. It can be fully explained if the temperature difference of board surfaces is not 15°C, but 13.5°C, that is, only 1.5°C lower. Again, it is quite realistic.

The above approach allows us to estimate the average lifetime of composite deck boards in practically any area of the United States, based on crumbling data available for any other area. The graph shown in Figure 15.27 can serve as a template for GeoDeck boards or other composite boards having the same oxidative degradation kinetics. The slope for any geographical area can be calculated from Table 15.31, taking into account the average high temperature difference between, say, Phoenix, AZ, and the city of interest, and introducing an additional temperature increment for the board surface (typically +40°F for the north and +50°F for the south, or an interpolated increment for intermediate areas), as it was shown above.

By doing this, the half-life of composite boards can be calculated and (tentatively) predicted. In Figure 15.27, the half-life of composite decks corresponds to the time period at which 500 decks are crumbled out of 1000 installed (the vertical axis).

At temperature difference of 19.4°C and the temperature coefficient of the reaction of 2.7, the following equations are applicable:

$$x^{10} = 2.7$$
$$x^{19.4} = y$$

where
x = increase in the reaction rate per 1°C,
y = increase in the reaction rate per 19.4°C

In order to solve the ratio, let us take logarithm of both the parts:

$$10 \log x = \log 2.7$$
$$19.4 \log x = \log y$$

From the upper equation, $\log 2.7 = 0.431$ and $\log x = 0.0431$
From the lower equation, $19.4 \times 0.0431 = 0.836 = \log y$
and $y = 6.9$
This means that when temperature is increased by 19.4°C and the temperature coefficient is 2.7, the reaction accelerates by 6.9 times.

TABLE 15.32 Average half-life time for GeoDeck WPC boards manufactured without added antioxidant

States	Half-life time for the deck (years)
AZ	1.8
TX, NM	2.2
FL	2.7
CA, NV, CO, WA, OR, WY	2.8
KY, TN, NC, SC, GA, LA, MS	3.0
PA, MD, VA, NJ, DE	3.4
NY, MA, VT, NH, CT, ME	6.2
MN, WI, IL, IN, MI, OH	6.2

Actual data for the first six rows, and extrapolated data for the last two rows

For GeoDeck boards without added antioxidants (manufactured between April 2002 and September 2003) the half-life times for different geographical areas are shown in Table 15.32.

Current GeoDeck boards have the OIT values more than 40 times higher compared with the boards described in Table 15.32. It is expected that service time for GeoDeck boards currently manufactured is increased even higher than 40 times.

THE OIT AND LIFETIME OF COMPOSITE DECK BOARDS

The oxidative degradation capacity of fillers, inorganic (metal oxides, particularly free metal-containing) colorants, and other additives in WPC is greatly underestimated by many researchers and manufacturers of composites. Composite materials generally require much more antioxidants compared to neat plastics, particularly when an energy-intense compounding is employed.

Our experience has shown that OIT is a good proxy test for an estimation of how long the board will serve on a deck in the real world in a given geographical area. One might think—at the first glance—that there should not be any meaningful correlation between OIT and actual deck lifetime, because an OIT test is conducted in pure oxygen at a very high temperature, whereas a board deterioration on a deck takes place under UV light, at much more moderate temperature (compared with that of an OIT test), in the air, and under some effect of moisture (rain, humidity). However, both OIT and natural oxidative degradation are controlled by the same principal single factor, that is, the amount of antioxidants in the composition. Other factors such as specific climatic conditions, location of the deck, among others will, of course, effect the deck lifetime, but again, an OIT value should be the first one to be taken into consideration. Specific correlations between the OIT and the deck lifetime will be different for each composite board brand and for each geographical area, and will be affected by an extent of exposition of decks to direct sunlight. Composite board density, if known, will also contribute to the rate of oxidative degradation, as it was described above.

All observations of actual decks in the southern United States have shown that the larger the OIT, the longer the lifetime of the decks before crumbling is noticeable (if it is ever noticeable). For GeoDeck decks in Arizona, OIT in the range of 0.2–0.3 min led to the boards crumbling in about 4–6 months. OIT in the range of 0.5–0.8 min resulted in the boards crumbling in about 12–20 months. Boards with the OIT around 1.0 min lasted for about 2 years. Boards with OIT above 2–3 have not been crumbling in a 3-year observation period. These data are related mainly to hollow composite boards, though solid boards with the same OIT values can behave similarly.

Tentatively, the lifetime of GeoDeck boards in Arizona in the worse case scenario (e.g., board of low density/specific gravity and exposed to direct sunlight most part of day) can be determined (in years of service) by multiplication of their OIT values (in minutes) by 2. For instance, boards with OIT of 5–6 min would serve in those conditions for a minimum of 10–12 years, and more likely 12–15 years and longer when in a shade. Similar GeoDeck boards would serve longer in other United States regions, as it was outlined above. In New England, in the Great Lakes region, and other northern regions, the average lifetime of composite boards will be up to 3–5 times longer with respect to oxidative degradation compared to that in Arizona. This means that in northern regions of the United States in the worse case scenario (boards of low density, direct exposure to sunlight most of the day), lifetime of composite boards will be (in years) the OIT (in min) multiplied by 10, and on an average the OIT multiplied by 20. Besides, the actual lifetime of boards will be affected by some other factors—wearing and tearing, for example, and other unrelated to the oxidation factors.

DURABILITY (IN TERMS OF OXIDATIVE DEGRADATION) OF WOOD–PLASTIC COMPOSITE DECK BOARDS AVAILABLE IN THE CURRENT MARKET

The OIT values of composite materials reflect two principal factors: (a) how much of the antioxidants is present in the material and (b) how resistant is the material itself to thermal oxidation. The OIT values of the commercial composite materials determine the sensitivity of the deck boards to oxidative degradation, hence, the lifetime of the boards, first of all in the south. If they are heat/UV stable in the south, no need to worry in the north.

Figure 15.28 provides a comparison of commercially available composite deck boards with respect to OIT values of their materials. The figure shows that the range of sensitivity of the boards to oxidative degradation is more than 50 times, from 2 min to more than 100 min. The median is at 22 min. Actually, it is a reasonably high number. However, a third of all commercial composite boards have OIT at 10 min and below, and several of them have the OIT below 5 min. This might be a time bomb for the respective companies, particularly if they sell their products to Arizona, Florida, Texas, California, and other regions where decks are quite popular.

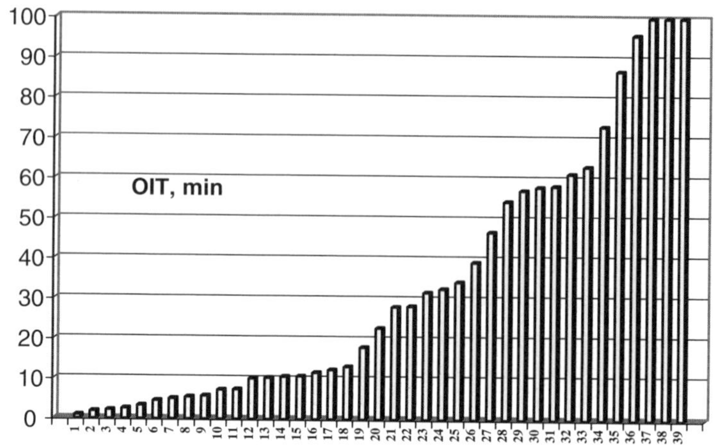

Figure 15.28 The OIT values for commercially available WPC deck boards. The manufacturers and board brand names are numbered in the increasing order of the OIT values.

It was described above what an OIT value below 1 min, and particularly in the range of 0.2–0.4 min, can bring to hollow WPC boards. 60%–90% of these boards, exemplified by GeoDeck deck boards, were crumbled in the south after a time span from 4 months to 3 years, and 15–20% of those boards crumbled in the north (see Fig. 15.27). In the weathering box, these boards crumbled—in worst cases—after 4 days. Generally, the ratio between crumbling in the weathering box and on a deck under direct sunlight most of the day was about 1:20 (in the south) and 1:100 (in the north).

Now, let us consider how long can a board serve whose OIT is "one step up," and is equal to about 3 min. Figure 15.28 shows five of such commercial WPC boards.

Two GeoDeck hollow boards with the OIT of 3 min were kept in the weathering box for 8640 h (360 days) and 9140 h (381 days) at "standard" conditions (0.35 W/m^2 at 340 nm, 102:18 min cycle, 63°C black panel temperature). Samples were taken from different parts of the boards, and results are shown in Table 15.33.

One can see that in both cases in the top layer of the boards the OIT decreased to 0.27–0.28 min, that is, the top layers were just about to crumble (see multiple examples above). The ribs, however, were still contained enough antioxidants to hold for a while, as well as the lower panel, contacting the hot surface in the weathering box. In reality the bottom layer of boards is typically well protected from crumbling.

Therefore, the top layer of hollow boards exemplified by GeoDeck, having initially the OIT of 3 min, can hold in the weathering box for over a year. This would translate to 20 years in the south, and much longer in the north, using the above coefficients. In reality their service time might be noticeably lower, considering other still overlooked factors. Therefore, the OIT of 3 min is barely acceptable. We would recommend the OIT at least 10 min for a good durability of WPC boards.

TABLE 15.33 The OIT values of GeoDeck hollow deck boards after long-term exposure in the weathering box

Exposure time in the weathering box	Samples taken from	The OIT (min)	The OIT change after the exposure (%)
360 days	Rib, bottom	1.95	35
	Rib, top	1.14	62
	Lower panel (exposed to lower temperature and moisture)	0.80	73
	Top panel (exposed to UV light and water spray)	0.27	91
381 days	Rib, bottom	2.18	27
	Rib, top	1.36	55
	Lower panel (exposed to lower temperature and moisture)	0.75	75
	Top panel (exposed to UV light and water spray)	0.28	91

The initial OIT was 3.0 min.

OXIDATIVE DEGRADATION AND CRUMBLING OF GEODECK DECK BOARDS: HISTORY OF THE CASE AND CORRECTION OF THE PROBLEM

The GeoDeck boards crumbling (oxidative degradation) issue resulting in warranty claims (225 claims on the record until May 2005, and 1707 claims until October 2006) was resolved in October, 2003. In December 2004, additional measures were taken to further increase the lifetime of GeoDeck boards beyond the reasonable doubt regarding 20-year warranty.

Suspect GeoDeck boards were manufactured between April 2002 through September 2003. Not all of these boards were defective because many of them had had high density (specific gravity) that, as it is shown above, significantly protected the boards against oxidative degradation. Some of these boards had sufficient amounts of an antioxidant, particularly at the end of 2002. However, because those, better boards could not have been identified among defective boards that were produced without added antioxidant and sold, all boards made between April 2002 and September 2003 had been recalled in 2004–2005 (see below).

Density, Porosity and Mechanical Properties of Geodeck before the Problem had Emerged

Porosity of composite materials is important because oxygen in the air diffuses through pores within the board that effectively increases the surface area of contact between oxygen and the plastic. That is why pure plastic (containing almost no pores) is much more resistant to oxidation compared to porous composite material, based on the same plastic compound. Boards of low-density have dozens of times more of the surface area available for oxidation compared with boards of high density. Hence, low density boards are oxidized dozens of times faster—in the same conditions—compared to high-density boards made of the same formulation.

For GeoDeck material, the highest density that can be achieved (by maximum compaction of the material, such as by compressive molding) is 1.24 g/cm^3. Theoretical calculations, taking into account specific gravity of plastic, the fillers and the additives also give the density value of 1.24 g/cm^3 for the shaped GeoDeck composite.

In 2001–2002 (first half of 2002) density of GeoDeck profiles was 1.13–1.16 g/cm^3. It was lower than the maximum because rice hulls were not dried (their typical moisture content was 9.5% w/w), and the resulting steam in the compounder produced microbubbles in pellets of the composite materials. Hence, these pellets were rather loose and carried air micropockets into the extruder, leading to slightly "puffed" composite material. Besides, compounding and extrusion typically lead to some amounts of VOC formed during the high-temperature processes in the hot melt. This also led to some "puffiness" and, hence, a decrease in density of the final composite compared to theoretical values (1.13–1.16 g/cm^3 compared to 1.24 g/cm^3). However, mechanical and other properties of GeoDeck profiles were quite satisfactory and well exceeded the building code (BOCA) requirements. Antioxidants in 2001–2002 were added to the formulation following the manufacturer's recommendations.

In April–May of 2002 a new twin-screw compounder was employed at the plant. It was much more efficient and more aggressive in terms of shearing compared with the former single-screw compounder. Besides, the manufacturing speed of GeoDeck profiles was significantly increased. This led to a decrease in density (specific gravity) of GeoDeck boards from 1.13–1.16 g/cm^3 to 1.10–1.06 and even to 1.05 g/cm^3. It was not a consistent trend. In some isolated cases the density was as low as 1.04 and 1.02 g/cm^3. In September of 2002 it was found that some boards (made in August of 2002) showed a fast deterioration in the weathering box (see below).

Overall, average densities of GeoDeck boards in June–December 2002 are listed in Table 15.34.

TABLE 15.34. Average (or isolated boards) densities of GeoDeck boards in June–December 2002

Month (2002)	Average density (g/cm^3)
June	1.12
July	1.17
August	1.06–1.03
September	1.16
December	1.12

Composite boards manufactured in September–December of 2002 were tested in the weathering box, as well as tests for flexural strength and modulus were conducted. The results were within norm. Accelerated weathering tests, conducted in February of 2003 by an independent testing company using boards made at the end of 2002, showed that flex strength and modulus of GeoDeck boards before and after accelerated weathering satisfied ICBO requirements, and that flex strength decreased by only 6% and flex modulus decreased by only 9%. The observations, reported to the ICC-ES in 2003 were as follows: *"There was no marked evidence of surface cracking, blistering, flaking, color change, or any other deleterious effects observed of the test specimens."*

Emerging of the Problem

The first sign of a problem was noticed in September of 2002. A board manufactured in August 2002 and exposed in the weathering box for only 860 h, showed cracks on the surface. Similar boards made in September 2002 showed cracks at the surface after 830 h in the box. These were isolated effects, and they were attributed to some technological deviations at the plant in Green Bay. However, based on those observations, in November of 2002 the amount of antioxidants was almost doubled compared to the original manufacturer's recommendations, and these boards with increased amount of the antioxidant were produced from November 2002 to January 2003. These boards were tested in the weathering box for 2000 h and were found quite satisfactory.

Starting January 2003 GeoDeck boards were made without adding antioxidant, as the HDPE supplier had assured that the plastic had an unusually high amount of antioxidants. These boards were made until March–April 2003, and after that the manufacturing continued with another HDPE source but the additions of the antioxidant was not resumed. This was a dramatic mistake. Manufacturing without the antioxidant was continued until the beginning of October 2003. The plant QC system could not detect the problem because there was no procedure in place that would have been able to detect a low amount (or a complete absence) of antioxidants in composite materials. The procedure was developed and implemented in the plant QC system later (see below).

In July of 2003 a new step was introduced at the plant, namely annealing of GeoDeck boards, in order to eliminate boards shrinkage. Initially, third-party kilns were employed to do the job, and in many cases boards were obviously overheated. They often became brittle and discolored. Being placed in the weathering box, they showed crumbling—in the worst case—even after 4 days. More detailed studies have shown that the lower the density, the less time it takes for board crumbling. Boards of density 1.06–1.07 g/cm^3 crumbled after 4 days. Boards of density 1.08–1.10 g/cm^3 crumbled after 7–12 days. Boards of density 1.15–1.16 g/cm^3 did not crumble after a month.

Density (Specific Gravity) of GeoDeck Boards in Pre-October 2003

Density of GeoDeck boards manufactured in January–September of 2003 was low (Table 15.35) compared to those manufactured in 2002 and 2004 (see below).

TABLE 15.35. Average densities of GeoDeck boards in 2003

Month	Average density (g/cm^3)
January	1.11
February	1.11
March	1.10
April	1.09
May	1.09
June	1.08
July	1.08
August	1.09
September	1.12
October	1.10
November	Plant was shut down
December	1.19

Data in Table 15.35 can be presented in a different format (Table 15.36). They show that in 2003 majority of all boards had density below 1.12 g/cm^3, and among those most of the boards had density in the range of 1.06–1.09 g/cm^3.

Tests of flex strength and flex modulus showed the boards (including those of density 1.06–1.08 g/cm^3) satisfactory and meeting the ICBO/ICC criteria.

Because of a higher board porosity (lower density), water absorption increased from 2% in 2002 to 3% (and in isolated cases to 4.5%) in 2003. It was within the building code requirements compared with that of other composite deck boards in the market.

As it was realized later, all these factors, that is, low board density, higher water absorption, and a faster crumbling in the weathering box, resulted in a low lifetime of boards in the southern areas of the United States. An R&D concept regarding crumbling and a fast deterioration of composite boards was developed, examined, and verified, and the problem was fixed in October–December 2003.

Correction of the Crumbling Problem

Antioxidant Level In October 2003 the former amount of the added antioxidant was resumed ("first level"). This continued for only 10 days, from October 4 to 14, and then the plant was shut down until December 5, 2003. When the plant restarted, the antioxidant level was more than doubled ("second level"). In December 2004, it was more than doubled again ("third level"). There were no particular reasons for the last increase (third level) of the antioxidant, because all laboratory data showed that the second level was enough to make GeoDeck boards lifetime to satisfy the warranty time (20 years). However, the third level of the antioxidant was employed for a still better protection of GeoDeck boards and taking into account the warranty situation.

TABLE 15.36 Distribution of GeoDeck board densities throughout 2003

Density (g/cm^3)	Number of tested boards having density showed in the left column	Percentage of boards tested throughout 2003 by the rank of density
1.01	1	1.06 g/cm^3 or less
1.02	1	4%
1.03	N/A	
1.04	2	
1.05	11	
1.06	31	
1.07	105	1.07–1.09 g/cm^3
1.08	266	43%
1.09	178	
1.10	170	1.10–1.12 g/cm^3
1.11	109	30%
1.12	98	
1.13	90	1.13–1.15 g/cm^3
1.14	80	17%
1.15	47	
1.16	14	1.16–1.18 g/cm^3
1.17	18	4%
1.18	26	
1.19–1.24	28	2%
Total	1275	100%

Density Along with increasing the antioxidant level, drying of rice hulls was introduced at the plant, and vented extruders were installed at several extrusion lines. As a result, density of GeoDeck profiles increased from 1.06–1.09 g/cm^3 (almost half of GeoDeck boards produced in 2003 had density in this range, see Table 15.36) to 1.18–1.22 g/cm^3, water absorption decreased from 3 to 1.0%, and lifetime of GeoDeck boards (observing crumbling) was increased from 4 days in the weathering box to more than 32 months, without any signs of crumbling.

Note: One day in the weathering box regarding crumbling is equivalent (on a conservative estimation) to 1 week of summer conditions in Phoenix, AZ. Hence, 32 months in the weathering box is equivalent to 960 weeks or 18 years of summer conditions in Arizona.

The OIT Procedure: Proxy of Lifetime at Accelerated Oxidation Three more procedures to study durability of boards were introduced in the company. One procedure was dealing with heating a sample of composite material in pure oxygen at a high temperature (typically 190°C) and measuring the time span before the sample is progressively oxidized. ASTM D3895 was used as a prototype: however, it could not be followed as recommended due to the nature of a filled composite material (ASTM D3895 is applicable to pure plastics). Details of the procedure are outlined below, in the Addendum section. Essentially, the procedure shows the lifetime of the composite material in pure oxygen and at a high temperature. The more the antioxidants in the formulation, the longer the lifetime of the composite material. This procedure can be considered as a proxy with respect to the lifetime of composite board in the real world.

For example, a typical GeoDeck composite profile made pre-October 2003 showed a lifetime (pure oxygen, 190°C) equal to 0.15–0.30 min. For the simplicity and following the above ASTM procedure, we will call this "lifetime" OIT. The "first level" material (see above) showed the respective OIT equal to 1.5–3 min, which is about 10 times higher. The "second level" GeoDeck (since December 2003, see above) showed OIT 4–5 min and then fluctuated but consistently increased in the range of 6 to 10 min between September and December 2004. In December 2004, the "third level" of antioxidants was employed, and the OIT reached 12–16 min.

As one can see, the OIT of the 2004–2005 GeoDeck materials has increased by 60–80 times compared to that of 2003. Considering that the lifetime of most of 2003 GeoDeck boards in Arizona is 4–24 months and higher, one can estimate that the lifetime of the "first level" boards would be 5–10 years in Arizona (conservative estimate; these boards were produced in 2002 and for 10 days, between October 4 and 14, 2003, see above), "second level" 20+ years in Arizona, and of the "third level" more than 20 years in Arizona; hence, much longer in other areas of the United States. In fact, if one adds the density factor of GeoDeck boards (currently 1.16–1.22 g/cm^3 compared to 1.06–1.09 g/cm^3 in 2003 for majority of boards, see above), the lifetime of current GeoDeck boards will be significantly longer than it is shown above.

Accelerated (Artificial) Weathering The second procedure was a well-known accelerated weathering using the weathering box. The readout was crumbling. As it was described above, for 2002 GeoDeck boards, crumbling was not observed for at least 2000 h (83 days) in the weathering box, as an independent study had shown. For pre-October 2003 boards, when density of the boards was low (1.06–1.09 g/cm^3 for majority of GeoDeck boards, see Table 15.36) crumbling was observed after only 4 days. Boards of this category crumbled after 4 months in Arizona. When density was higher (1.13–1.16 g/cm^3), crumbling was not observed even after a month in the box. Apparently, boards from this category crumbled only after 24 months in Arizona or did not crumble at all during the time of observation. Other boards, with intermediate densities crumble in the period between 4 and 24 months and more, with the life expectancy up to 5 years for the very dense boards, as calculations show.

As it was mentioned above, GeoDeck boards of the "first level," made at the end of 2002 and tested by an independent testing company in February 2003, were described in their ICC-ES report as *"There was no marked evidence of surface cracking, blistering, flaking, color change, or any other deleterious effects observed of the test specimens."*

The "second level" GeoDeck boards (OIT 3–5 min, made in December 2003) were placed in the weathering box in February 2004, and 32 months later, did not show any signs of crumbling. Again, the difference of those proxy lifetimes is high, in this case more than 100 times. One can suggest that compared with 2003 GeoDeck boards (4–24 months in Arizona before crumbling is noticeable), GeoDeck boards produced after December 2003 can have a lifetime more than 400 months in Arizona, that is, more than 30 years.

And in December of 2004 the amount of antioxidants was almost tripled.

Air-Flow Oven One more approach employed at the company was to study oxidative deterioration of GeoDeck boards in airflow oven. Preliminary experiments have shown that GeoDeck boards can stay in a vacuum oven indefinitely long even at high temperatures, above 230°F. At higher temperatures the board becomes deformed because the temperatures are too close to the melting point of HDPE.

Airflow oven provides oxygen for the oxidation and a high temperature to accelerate the process. However, there is no UV light, so the oven shows only the effect of temperature on the oxidation process. There is practically no crumbling even after a long exposure of boards in the airflow oven, and reduction of strength was chosen as a readout in these experiments. The testing was conducted in such a way that a combination of flex strength and shear strength was measured, as indicators of structural integrity of GeoDeck boards.

The experiment aimed at a comparison of GeoDeck boards made with no added antioxidants ("zero level"), "first level" boards (made for several days only in October 2003), "second level" boards (made between December 2003 and December 2004), and "third level" boards (made after December 2004). The results are listed in Tables 15.8, 15.9, and 15.10.

These tables show the obvious trend–the higher the density, the higher the durability. Clearly, in the presence of a relatively small amount of an antioxidant ("first level") the durability of the boards is significantly higher compared with that in Table 15.8. It is worth mentioning that in the presence of the antioxidant, boards with density below 1.125 g/cm^3 could not be produced (compare with Table 15.8). Apparently, the antioxidant slowed down the decomposition of HDPE in the extruder hence, effectively decreased the amount of VOC and, as the result, decreased the porosity of the extruded deck board and increased its density.

It is remarkable that at a noticeably higher temperature and a much longer heating time, boards with a relatively high density (1.20 g/cm^3) showed practically no degradation, whereas at much lower temperature deck board with low density (1.02–1.09 g/cm^3) but without added antioxidant lost 50% and more of their strength.

Experiments with "third level" GeoDeck boards (made in or after December 2004) could not detect any reduction in flex/shear strength at conditions given in Table 15.10. Residual load at failure was in the range of 99 ± 2%, and could not be reliably measured.

Even without considering the "third level" boards, a comparison of "zero level" and "second level" GeoDeck boards leads to striking conclusions: The lifetime of the least and the most durable deck boards (in terms of strength reduction resulting from oxidation of boards by heating them at the same temperature) differs by 420 times. In the real world this would correspond to 1 month and 35 years, respectively.

This phenomenon shows that two powerful stabilizing factors, which prolong the lifetime of composite deck boards, are board density and added antioxidants. These two factors are in a way functionally interchangeable, but taken together they work in synergism. Antioxidants block propagation of free radicals, and density controls the amount of air oxygen flowing into the pores of the composite matrix. High density effectively blocks the access of oxygen and slows down the oxidative degradation of composite materials.

Experiments in the airflow oven conducted at different temperatures and for different time periods have shown that the oxidative degradation of GeoDeck boards (leading to the reduction in board strength) had a steep dependence on temperature, with the temperature coefficient close to 3/10°C (see above).

ADDENDUM: TEST METHOD FOR OXIDATIVE INDUCTION TIME OF FILLED COMPOSITE MATERIALS BY DIFFERENTIAL SCANNING CALORIMETRY

This test procedure is instrumentally similar to ASTM D 3895 (see above). However, sampling is principally different because filled composite material cannot be made into sheet format prior to analysis.

DSC instrument was employed. OIT is a relative measure of a material's resistance to oxidative decomposition. It is determined by the thermoanalytical measurement of the time interval of exothermic oxidation of a material at a specified temperature (190°C typically in our case) in an oxygen atmosphere.

The sample to be tested is heated at a constant rate (20°C/min) in nitrogen. When the specified temperature has been reached, the nitrogen atmosphere is changed to oxygen maintained at the same flow rate. The specimen is then held at constant temperature until the oxidation reaction is displayed on the thermal curve. The time interval from when the oxygen flow is first initiated to the oxidative reaction is referred to as the induction period. The end of the induction period is signaled by an abrupt increase in the specimen's evolved heat and may be observed by a DSC. The OIT is determined from the data recorded during the isothermal test.

The test sample was taken directly from the material to be analyzed (such as a deck board). A typical specimen was of 10–20 mg by weight. A specimen was placed into an aluminum dish and placed onto a sampler (carousel), from which it was automatically placed into the cell. An identical empty pan was used as the reference. Indium was used as the calibrant, as its melting point (180°C) was close to the specified analysis temperature (190°C).

Nitrogen prepurge time prior to beginning the heating cycle was 5 min. Programmed heating of the specimen (under nitrogen flow of 50 mL/min) from ambient temperature to 190°C was at a rate of 20°C/min. When the set temperature (typically 190°C) has been reached, programmed heating was discontinued and the sample equilibrated for 5 min at the set temperature. After this the gas was changed to oxygen at a flow rate of 50 mL/min. This changeover point to oxygen flow was considered the zero time of the instrument.

Isothermal operation was continued until at least 5 min have elapsed after the steepest point of the exotherm has been displayed. The test was then terminated, the gas selector was switched back to nitrogen, and the cell was cooled to 70°C to start a new test. The OIT was measured to within ±0.01 min from zero time to the intercept point, obtained by extrapolation of the steepest linear slope of the exotherm onto the extended baseline. Typical examples of the OIT graphs are shown in Figures. 15.19 and 15.20.

Typical results are as follows:

- GeoDeck 2001 manufacturing: the OIT values are between 10 and 2.5 min;
- GeoDeck 2002 manufacturing: from 3.14 min (January) through 0.15–0.95 min (February–November) to 0.83–1.69 min (December);
- GeoDeck 2003 manufacturing: 0.3 ± 0.2 min (January–September), 0.45–0.82 min (October), 2.2–4.5 min (November–December);
- GeoDeck 2004 manufacturing: 4.0–6.5 min (January–August), 5.5–12.8 min (September), 7.5–13.1 min (October–November), 8.7–14.7 min (December);
- GeoDeck 2005 and 2006 manufacturing: 11.2–16.5 min.

Figure. 15.29 shows a profile of the OIT values for GeoDeck composite deck boards manufactured from 2001 through the spring of 2005.

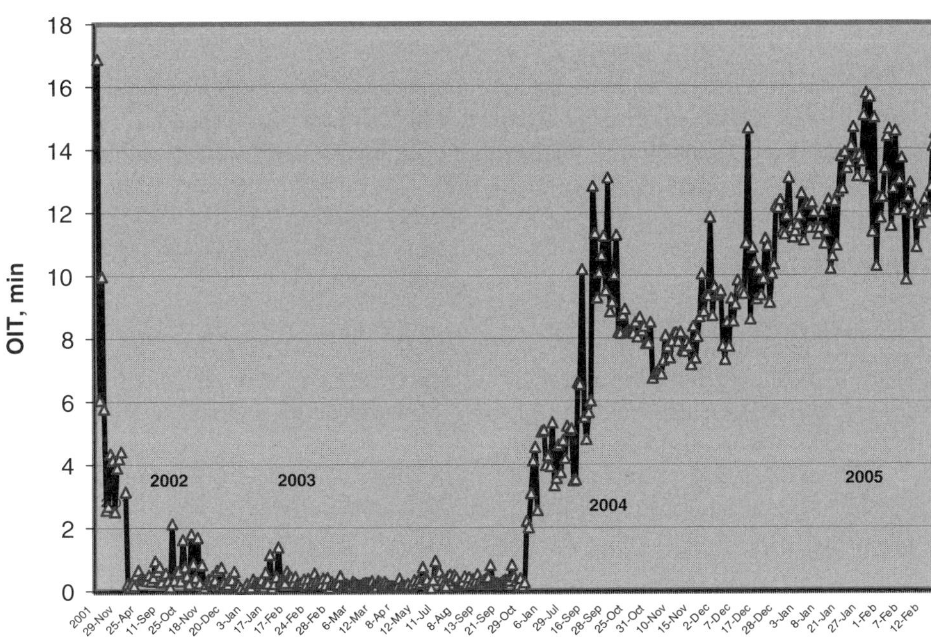

Figure 15.29. A profile of the OIT values for GeoDeck composite deck boards manufactured from 2001 through spring of 2005.

TABLE 15.37 Sensitivity of the composite boards to oxidative degradation as measured in terms of OIT, time before crumbling in a weathering box, loss of strength as a result of exposure at high temperature for a certain period in an airflow oven, and actual and expected lifetime of the composite boards on a deck in the Phoenix, AZ, area

OIT (190°C, pure oxygen) (min)	Weathering box (UV light + water spray, 1:42 min + 18 min cycle), days of exposure before crumbling	Airflow oven, days of exposure at the given temperature, residual break strength%	Lifetime in Arizona, actual and expected
0.3(pre-November 2003)	4 days	87 h (3.6 days) at 225°F, 40–50%	Actual: 4 months to 2 years
4-5 (November 2003—September 2004)	More than 32 months, no crumbling	112 h (4.7 days) at 226°F, 94–95%	Actual: More than 2 years with no sign of crumbling, expected—10–15 years and more
12–15(2005)	No crumbling	No change after 14 days at 226 264 h (11 days) at 232 F, 97–99%	Actual: No crumbling. Expected—more than 20 years

Expected lifetime was calculated as described above.

The described dynamics of the OIT increase in the composite deck boards has certainly reflected in their lifetime. This is shown in Table 15.37, which describes three "generations" of the composite boards with respect to their OIT and lifetime.

CASE STUDIES

GeoDeck Decks in Arizona

Table 15.28 shows that there were 127 warranty claims filed in Arizona regarding crumbled GeoDeck decks. Some pictures of these decks are shown in Figures 15.10–15.12 and 15.30–15.35.

Table 15.38 lists as examples the OIT values of actual decks installed in Arizona. The first several figures are related to boards manufactured after September 2003. The majority of the OIT values are related to boards manufactured in 2002–2003.

GeoDeck Decks in Massachusetts

There are 28 warranty claims on the record, filed regarding crumbling decks in Massachusetts (Table 15.28). The first one was filed in Chicopee, MA (Fig. 15.36).

Figure 15.30. Crumbling deck, Arizona.

Figure 15.31. Crumbled board, Arizona.

Figure 15.32. Crumbled handrail, Arizona.

Figure 15.33. Crumbling board, Arizona.

Figure 15.34. Crumbling deck, Arizona.

Figure 15.35. Crumbling composite stair and deck boards, Arizona.

TABLE 15.38 OIT figures (at 190°C) for actual composite decks in Arizona

Town (AZ)	OIT (min)
Jerome	13.4
Heber	11.9
Heber	8.72
Prescott	8.27
Prescott	5.28
Pinetop	3.41
Heber	2.62
Heber	2.60
Pine	0.95
Pinetop	0.86
Sedona	0.85 (in shade)
	0.57 (partial sun)
	0.19 (direct sun)
Sedona	0.82
Show Low	0.77
Ashfork	0.42
Show Low	0.35
Payson	0.34
Walker	0.34
Pinetop	0.32
Heber	0.30
Sedona	0.30
Pine	0.28
Prescott	0.27
Ashfork	0.27
Phoenix	0.26
Strawberry	0.26
Heber	0.25

(*Continued*)

TABLE 15.38 (*Continued*)

Show Low	0.22
Chino Valley	0.22
Prescott	0.22
Show Low	0.20
Sedona	0.19
Prescott	0.18
Walker	0.16
Prescott	0.16
Sedona	0.16
Pinetop	0.15
Sedona	0.15
Prescott	0.14
Strawberry	0.10

All the deck composite boards with OIT less than 1 min were manufactured in 2002–2003.

An inspection of the deck has shown that it began actively crumbling: Some boards produced cracks along spider lines, and board fragments between these cracks started to fall out. Top surface of the deck became soft and easily scratchy.

As it was the only warranty claim regarding crumbling in Massachusetts in 2005, it was subjected to a certain scrutiny. The deck was installed in July of 2003, almost 2 years before the damage was noticed and the claim filed. The OIT of the two spare boards stored in a garage showed figures of 0.29 and 0.59 min. Boards with such OIT values would have crumbled in, say, Arizona within the first 2 years; however, in Massachusetts they should stay much longer. The fact is that Chicopee is situated in Berkshire Mountains, on an elevation of about 250 ft. It is not too high, obviously, but high enough to get more UV light than on the plain; hence, higher temperature on the deck. Indeed, as the owner has informed, temperature on the deck is high to make snow melting rather fast, as well as water evaporation from the deck right after rain.

Figure 15.36. The crumbling deck in Chicopee, MA.

Figure 15.37. A crumbling deck in Massachusetts.

This case shows how even a relatively modest elevation can accelerate the oxidative degradation of a composite deck.

Figures 15.37 and 15.38 show some other examples of crambling decks in Massachusetts.

GEODECK VOLUNTARY RECALL

In 2003–2004 the manufacturing company has voluntarily recalled many trackloads of suspect composite boards from distributors of the product.

A major consequence was that in 2005 the manufacturing company had applied to the US Consumer Product Safety Commission (CPSC) informing them on the situation and announcing a safety recall of deck boards manufactured between April 2002 and October 2003. These products amount approximately 4.9 million ft^2 of

Figure 15.38. A crumbling composite deck, Massachusetts.

composite deck boards and approximately 340,000 lineal feet of railing materials, which represent approximately 11,000 decks based on average deck size of 456 ft^2. As a result, the product safety recall was put forward (see Safety Poster below).

Safety Recall

GeoDeck™ decking boards and railings

Manufactured between April 2002 and October 2003

Kadant Composites Inc. is voluntarily recalling certain GeoDeck™ Decking and Railing Materials manufactured between April 2002 and October 2003. These materials, if exposed to hot temperatures and strong sunlight, may degrade prematurely posing a risk of injury.

Consumers should check their GeoDeck™ material for (i) visible cracks on the surface of the decking materials and/or (ii) a surface that can be easily scratched with a fingernail or the corner of a credit card. Once it has been verified by Kadant Composites Inc. that a consumer possesses the recalled decking and/or railing materials, the company will replace such materials at no cost to the consumer and advise consumers to dispose of recalled materials in accordance with local law.

For additional information please contact GeoDeck™ Customer Service toll free at 1-800-545-1710
Monday through Friday, 9 A.M.–5 P.M. (US/Eastern Time)
www.kadantcompositesrecall.com

IN COOPERATION WITH THE U.S. CONSUMER PRODUCT SAFETY COMMISSION POST UNTIL November 18, 2005

PROBLEM GEODECK DECKS: INSTALLATION TIME AND WARRANTY CLAIMS

The number of warranty claims filed up to October 2006 was 1707 (Table 15.28). Practically all of them are related to GeoDeck boards manufactured in pre-October of 2003. This could be established by considering the defected boards directly, however, most of them lost the stamp with the manufacturing date. The stamps were commonly cut off during the deck installation, or masked by other boards on the deck, or faded, or could not be found on boards due to different reasons. It comes as no surprise that the deck owner has no knowledge when the boards were manufactured. However, it is much easier to learn when the deck was installed, and this information is requested on the warranty claim form.

Obviously, before a warranty claim is filed, several events should take place. First, the board should have been manufactured some time before purchasing. Then the deck should be installed. In many cases the time period between the manufacturing and installation is 1–2 years, sometimes more than that. Then, some time should pass until the deck begins crumbling. Depending on the geographical area, it might take between several months and a number of years. It might never happen, particularly if the deck is covered or located in the north. Finally, it takes some time between when the deck owner notices that something is wrong with the deck and when he/she files a warranty claim.

Figure 15.39 shows a number of total crumbling warranty claims related to the month in which the deck was installed. A total of 1099 claims are shown for which the installation time is on the record. Claim analysis shows that the average time between the defective deck installation and the respective claim is 13.3 ± 6.5 months, that is, approximately 13 months. The median of this time is also 13 months.

One can see that most of the problem decks were installed in 2003 and 2004 (88% total of warranty claims). Four percent of all problem decks were installed in 2005, and these were apparently made in 2002–2003. According to warranty claim forms, some deck owners had purchased GeoDeck boards 2–3 years before the deck was actually installed. There are only two isolated claims in 2006.

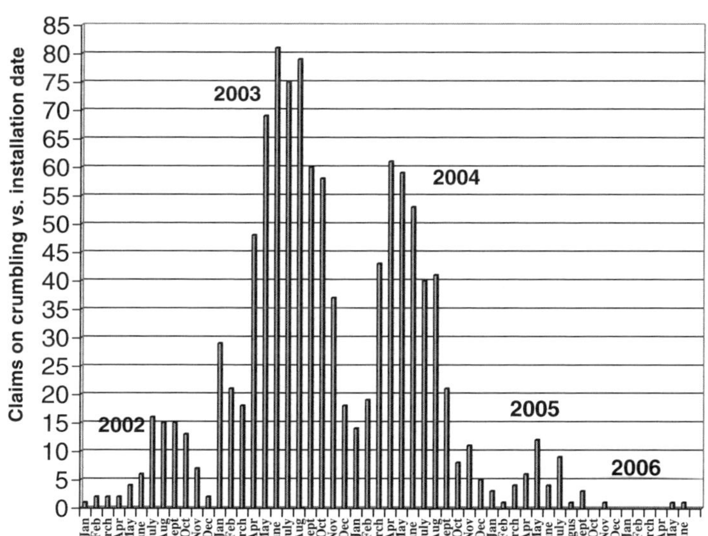

Figure 15.39. A total number of warranty claims filed regarding each month in which decks were installed. The graph shows data for 1099 warranty claims available in the record, for which deck installation date was reported.

OIT tests of numerous composite deck boards conducted by Kadant Composites have provided a massive database that serves as a part of their warranty claim system. In many cases the company has collected samples of "good" decks (just 10–20 mg samples are needed for the OIT tests) to determine an expected lifetime of the decks, and if the OIT was low, the company had offered to replace the deck free of charge, not waiting for the crumbling to come. This approach has met a very positive response from deck owners, who were impressed with such a proactive action by the company.

Since 2004, Kadant Composites and then LDI Composites are making GeoDeck composite boards with an excellent durability, which should withstand the southern US climate, such as in Arizona, Texas, Louisiana, Florida, California, among others for 20 and more years.

REFERENCES

1. D.C. Wright. *Failure of Polymer Products Due to Thermo-Oxidation,* Rapra Review Reports, Report 131, Vol. 11, No. 11, 2001, pp. 8–9.
2. C. Vasile and M. Pascu. *Practical Guide to Polyethylene.* RAPRA Technology, Rapra Technology Ltd., UK, 2005, p. 79.
3. D.C. Wright. *Failure of Polymer Products Due to Thermo-Oxidation.* Rapra Review Reports, Report 131, Vol. 11, No. 11, 2001, pp. 13–14.
4. S. Ding, M.T.K. Ling, A. Khare, and L. Woo, Activation energies of polymer degradation, In: J. Moalli (Ed.), *Plastic Failure: Analysis and Prevention*, Plastics Design Library, Norwich, NY, 2001, pp. 219–225.
5. L. Woo, S.Y. Ding, A Khare, and M.T.K. Ling. Failure progression and mechanisms of irradiated polypropylenes and other medical polymers. In: G. Wypych (Ed.), *Weathering of Plastics. Testing to Mirror Real Life Performance,* Plastic Design Library, 1999.
6. R.M. Suffield, S.H. Dillman, and J.E. Haworth. Performance of tocopherols as antioxidants in ABS. *ANTEC*, Society of Plastic Engineers, Brookfield, CT, Boston, MA, May 1–5, 2005.
7. R.M. Suffield, J.E. Kiesser, and S.H. Dillman. Antioxidant performance of mixed tocopherols in styrenic block copolymers. In: *Proceedings: ANTEC*, Society of Plastic Engineers, Brookfield, CT, Boston, MA, May 1–5, 2005.
8. K. Pielichowski and J. Njuguna. *Thermal Degradation of Polymeric Materials.* Rapra Technology, Rapra Technology Ltd., UK, 2005, p. 20.
9. G. Wypych. *Handbook of Material Weathering*, 3rd edition, William Andrew Publishing, Toronto, 2003, p. 316.
10. D.C. Wright. *Failure of Polymer Products Due to Thermo-Oxidation.* Rapra Review Reports, Report 131, Vol. 11, No. 11, 2001, p. 12.
11. N.S. Allen and A. Parkinson, *Polymer Degradation and Stability*, 1983, Vol. 5, No. 3, p. 189; cit. from D. Wright, *Failure of Plastics and Rubber Products. Causes, Effects and Case Studies Involving Degradation.* Rapra Technology Ltd., UK, 2001, pp. 84.
12. G. Wypych *Handbook of Material Weathering*, 3rd edition, William Andrew Publishing, Toronto, 2003, pp. 666–667.

16

PHOTOOXIDATION AND FADING OF COMPOSITE BUILDING MATERIALS

INTRODUCTION

Photooxidation of plastics and wood–plastic composites (WPCs) was described in principal detail in the preceding Chapter 15. It was emphasized that photooxidation acts in a synergism with thermooxidation of the materials, speeding up an oxidative degradation of WPC products, particularly being exposed to direct sunlight.

Along with thermooxidation, photooxidation causes an additional chemical damage by creating at the board surface free radicals, which propagate at some depth into the material. The respective chemical equations were also given in the preceding chapter, as well as a description of where the free radicals are coming from, what they do, and how to stop their action. UV absorbers and UV stabilizers were described in the preceding chapter as well, as they both slow down oxidative degradation of WPC materials.

However, photooxidation has its own specific manifestations, and, first of all, changing the material's color or fading (Fig. 16.1). Hence, we will focus in this chapter primarily on fading, what causes it, how to characterize it, and how to minimize it. We will also show how much commercial composite deck boards differ in terms of fading.

Despite many similarities between thermooxidation and photooxidation of plastics, on one hand, and WPCs, on the other, there are some important differences. Photodegradation of plastics typically occurs in oxygen starvation conditions, while photodegradation of WPC, which are always porous, occurs in the presence of plenty of oxygen in the

Wood-Plastic Composites, by Anatole A. Klyosov

Figure 16.1 Fading of a composite deck.

upper layers as well as in the bulk of the material. Hence, different kinetics and sometimes different chemistry of photodegradation of WPC compared with those of neat plastics.

How Fading is Measured

The most known among WPC researchers and engineers is the Hunter Lab color scale and a Hunter Lab Color scale meter. They were basically developed in the 1950s and 1960s, and the current formulas were finalized in the mid of 1960s. The most challenging question in that time and earlier was how to represent colors based solely on the numbers?

The Hunter Lab color space is organized in a rather simple way (Fig. 16.2). The *L* axis, as lightness, runs from top to bottom. The maximum for *L* is 100, the "absolute

Figure 16.2 The Hunter Lab color space in a cube form. It is based on the Opponent-Colors Theory. The theory assumes that the human eye perceive color as pairs of opposites, namely, light-dark, red-green, yellow-blue (with HunterLab permission). See color insert.

white," or, rather, a perfect reflecting diffuser. The minimum for L is 0, the "absolute black." The hundred units in the full scale have a good practical meaning. One L units, in a simplified manner, is what a naked eye can detect. That is, a difference between, say, $L = 45$ and $L = 46$ can be visually detected.

For most commercial WPC deck boards, the initial L value is between 40 and 55. For some, such as Trex Brasilia and Trex Madeira, it is around 32–35, and for Trex Winchester, Evergrain, and CorrectDeck, it is around 40. They are on a darker side of the color scale of composite deck boards, though for Trex Brasilia the color is mixed due to a dark surface "exotic" pattern. Among the most light-colored boards, with the initial L value close to or around 60, are Boardwalk, Fiberon Buff Cedar, Nexwood Red, Life Long, Procell, Millenium, TimberTech, and UltraDeck.

When commercial composite deck boards fade, some their lightness goes up to 85 units and higher (see below). Typically, for most composite deckboards the L value goes from the initial 40–55 to 60–70 after only 1000 h in the weathering box. This will be explained later in this chapter.

For solid TimberTech boards, the initial L value is close to 70, and after a relatively short exposure it is above 80, and then gets close to 90 (85.7 after 1000 h in the weathering box, and 87.2 after 1500 h). For hollow UltraDeck the L value changes from the initial 62.3 to 71.3 after 1000 h in the box to 72.0 after 1500 h. For the Austrian Fasalex, a change in L is from the initial 54.2 to 70.9 after 1000 h to 71.3 after 1500 h. More data are given in Table 16.5.

Now, back to the Hunter Lab color space diagram. There are two more axes there. The axis "between right and left" in Figure 16.2 is "a" axis. It goes from negative on the left (green) to positive on the right (red). It does not have specific numerical limits. Obviously, for the red-colored Trex Madeira, TimberTech, Xtendex, or Fiberon/Perfection the "a" value is rather high—in fact, it is in the 8–12 range. A light-brown Life Long also has the red component ("a") in the 10-s. In otherwise colored boards, the "a" figure shows just a "tone," a green or a red component in the "overall" board color. For some composite deck boards, the initial "tone" is even slightly greenish, such as for PVC-based Millenium and Procell, and their initial "a" values are in a slightly negative area (around -0.2 to -0.9). Typically, they do not change much during weathering.

The axis, perpendicular to "a" in Fig. 16.2 is the "b" axis. It goes from negative (blue) to positive (yellow). It also does not have specific numerical limits. Obviously, cedar-colored or brown deck boards have their "b" value high—in fact, in its 20-s, such as with UntraDeck, Life Long, Premier (brown), Xtendex (dark-colored), or GeoDeck Cedar. With bright-yellow colored Fasalex (Austria) the "b" value goes into the 40-s. In other cases boards show just a "tone," a blue (very seldom) or a yellow (very often) component in the "overall" board color.

Fading or other color change results in the change of L, a, and b (or some of them). Fading always results in the increase of lightness (L) and decrease of yellowness (b). These changes are expressed as ΔL, Δa, and Δb. The numbers are often used in quality control at manufacturing sites, along with the respective tolerances.

The overall change in color, or the total color difference, is often expressed as ΔE, which is a square root from a sum of each of the changes in the second power:

$$\Delta E = \sqrt{\Delta L^2 + \Delta a^2 + \Delta b^2}$$

Clearly, ΔE is the highest value compared with each of the three others, or at least equals to the highest of the individual color changes (when other two equal to zero).

Fading: Some Introductory Definitions

In this chapter, by the term "weathering" we mean fading because other aspects of weathering, first of all resulting from oxidative degradation of plastics and WPCs, were considered in the preceding chapter.

When subjected to UV light, composite materials fade. This happens because high-energy UV light (whether we consider it as waves in an electromagnetic field, or as a stream of massless particles called photons) breaks polymer chains and other organic compounds at the surface. They are often broken in such a manner that free radicals are formed and propagated, until the free radicals are terminated by themselves (e.g., due to recombination), or by added antioxidants.

If to take a look at fading surface under the light microscope (at a magnification of 20–100), one can see microcracks, which are developing along with fading developments. Examples of electron microscopy views on fading surfaces are given in the preceding chapter (Figs. 15.15–15.17). Reflection and scattering of light by edges of surface microcracks typically results in fading. Fading is also caused by chemical destruction of organic pigments, wood extractives, and other chromophores in the surface layer of the composite material, but typically fading of plastics and WPC products is caused (or accompanied) by plastic degradation. Examples of chromophores can be carbonyl groups often introduced during processing of plastics; lignin in WPC, photodegradation of which at 350–400 nm along with oxidation results in yellowing. However, color change of composite decks in reality is more complex, and after weathering of wood flour (or rice hulls) containing WPC, the "*b*" value is typically decreased, not increased (Table 16.1). Out of 21 commercial WPC deck boards listed in Table 16.1, only three showed an increase in "*b*" value after an exposure in the weathering box. Other 18 showed decrease in "*b*" (yellowness) in a range of 0.3–10 units after 1000 h irradiation and water spray in the conditions described in Table 16.1.

Fading of WPC materials can be decreased by introduction of antioxidants and stable colorants. Inorganic colorants, such as iron oxide, are more stable than organic colorants [1–3]. The irony is that inorganic pigments often accelerate WPC degradation due to formation of free radicals in the materials and on the material surface. It happens because inorganic pigments, such as iron oxide, typically contain free iron or other free metals, the powerful catalysts of plastic oxidation. Hence, inorganic pigments often protect the WPC surface from fading, but accelerate the WPC oxidative degradation in the bulk.

The protective effect of colorants against fading depends on the type of pigments and antioxidants, their sources, amounts, and, of course, climatic and other

TABLE 16.1 Shift in "*b*" value on the Hunter lab scale in the course of weathering of some WPC deck boards

		"*b*" value (yellowness)		
WPC board	Weathering (h)	Initial	After the weathering	Δb
GeoDeck Cedar	2400	21.9	23.5	+1.6
Life Long	1000	20.2	21.4	+1.2
	1500		22.1	+1.9
Xtendex	1000	20.3	21.1	+0.8
	1500		21.3	+1.0
Procell	1000	10.6	10.3	−0.3
	1500		10.3	−0.3
Millenium	1000	2.6	2.2	−0.4
	1500		2.1	−0.5
Boardwalk	1000	27.3	26.8	−0.5
Trex Saddle	1000	19.9	19.1	−0.8
GeoDeck Mahogany	35,000 (outdoors, Bedford, MA)	14.0 ± 0.3	12.9 ± 0.3	−1.1
GeoDeck, Mahogany	2400	15.0	13.8	−1.2
Cross Timbers	800	1.7	−0.6	−2.3
GeoDeck, Driftwood	2400	5.5	2.7	−2.8
Fiberon Buff Cedar	1000	17.4	14.3	−3.1
TimberTech (open profile)	500	13.2	9.3	−3.9
Premier, brown	1000	23.0	19.0	−4.0
	1500		19.0	−4.0
Rhino Deck, gray	700	7.9	3.6	−4.3
Rhino Deck, brown	700	14.3	9.0	−5.3
Premier, gray	1000	10.2	4.6	−5.6
	1500		4.3	−5.9
EverX	300	14.6	7.6	−7.0
UltraDeck	1000	22.5	14.7	−7.8
	1500		14.4	−8.1
Perfection	1000	18.0	10.0	−8.0
WeatherBest	1000	12.7	4.4	−8.3
	1500		4.3	−8.4
TimberTech	1000	13.1	3.1	−10.0
	1500		2.9	−10.2

Only the 12th row shows natural, outdoors weathering (35,000h); all other figures are related to accelerated weathering, in Q-Sun 3000 (0.35 W/m^2 at 340 nm, 102:18 cycle, 63°C black panel temperature). The data are arranged in the order of decreasing of the yellow component (b) compared to the initial value.

conditions. Moisture is also a powerful catalyst of oxidation and formation of free radicals in the material. Therefore, moisture accelerates fading dramatically.

Dynamics of fading can be expressed by ΔL, Δa, and Δb, as well as ΔE per given time. Some commercial composite deck boards fade very fast. On being placed outdoors and exposed to direct sunlight, some of them show clear fading after only a couple of weeks. Some of them almost do not fade after years of direct exposure to sunlight even in Arizona, Texas, or Florida. GeoDeck composite deck boards, for example, did not fade after being almost completely crumbled, as shown in the preceding chapter. This in turn illustrates that it was not antioxidants that would have prevented fading; the above said GeoDeck boards which did not have any antioxidants and showed progressive fading, particularly in the South, retained their colors even when almost completely crumbled.

This can be rephrased as follows: Antioxidants can decrease fading to some extent and in some situations, but they do not prevent it. On the contrary, lack of antioxidants does not necessarily accelerate fading. Antioxidants often disappear from the very surface of a composite deck board rather quickly; they would prevent oxidation of the bulk of the board but not the very surface. Without a good amount of colorants, the board would fade.

Brushed boards fade much slower than unbrushed boards. A thin layer or a film of plastic on a board surface fades much faster than a brushed board, from which this film was removed by brushing.

ACCELERATED AND NATURAL WEATHERING OF WOOD–PLASTIC COMPOSITE MATERIALS AND A CORRELATION (OR LACK OF IT) BETWEEN THEM: THE ACCELERATION FACTOR

Fading of composite materials is typically studied using accelerated weathering tests, or employing outdoors facilities, such as those provided by Atlas Weathering Services Group at their Direct Weathering test sites in Miami, FL and Phoenix, AZ, or by Q-Lab Test Service Division at their test sites in Homestead, FL and Buckeye, AZ, or at makeshift facilities that practically anyone can arrange. Conditions for these tests are outlined in the respective ASTM procedures (see below). Here I only mention that the most known procedure for accelerated weathering is a 2-h cycle of 102 min UV light (0.35 W/m^2 at 340 nm) and 18 min water spray along with a continuing UV light at 63 °C black panel temperature. Each one of these particular conditions and regimes can vary by choice, of course. In this chapter by tests in a weathering box I would refer to the above conditions, unless indicated otherwise.

Typical questions addressed by folks considering accelerated weathering data obtained in a weathering box are—“what really means 1000 hrs or whatever—in the weathering box with respect to the real world?” “To what time period outdoors is corresponds?” “What is an equivalency coefficient—or whatever—between the weathering box and outdoors conditions?”

A counterquestion typically is “where?” It depends. Where outdoors? In Arizona? In Minnesota?

O.K., they understand. Pick any area. Pick both, in Arizona *and* in Minnesota. What is a difference between weathering of materials in these areas and in the box?

Well, nobody can answer without some assumptions, conditions and restrictions, and a significant error margin. The most important factor—it depends—which product color we are considering. The thing is that the spectral power distributions of sunlight, on one hand, and artificial light sources, on the other, are different, hence, different responses of different materials.

In a way, however, the above questions make perfect sense. If there is no correlation between the weathering outdoors and in the accelerated weathering box, why bother? Why conduct those accelerated weathering tests that are pretty expensive, starting in the first place with that the costs for those weathering boxes are between $27,000 and $50,000, depending on amenities. Larger size weathering boxes cost around $100,000.

Of course, there are certain relationships between fading of the same material outdoors and in the weathering box in terms of time, required for the same degree of fading, or a color shift. The thing is that this time period is different for the same color shift for L (lightness), "a" (greenness and redness), "b" (blueness and yellowness), for ΔE ("total" color change), and also for each different color and for each geographic region. Hence, the initial question transforms into—how different are *those* relationships between each other?

Besides, fading of WPCs is different from that of neat plastics. Wood affects fading of composites. There is not enough data accumulated with WPC in order to answer those questions.

Let us consider what kind of relationships we can identify now. For this, let us consider some background first and then take a look at specific examples.

Generally, the spectrum of sunlight is predominantly visible and infrared. Only about 6% of UV light cases fading and other photodegradation effects in polymers and WPCs, and only wavelengths between 100 and 290 nm are the most destructive. In fact, UV radiation with these wavelengths is filtered out with the earth's ozone layer.

Overall, the solar radiation measured on the earth's surface consists of wavelength between 295 and 3000 nm. Wavelength between 295 and 400 nm are referred as UV light, and they make up about 7% of the total radiation. Visible part of the solar radiation (400–800 nm) make about 55% of the total, and the rest, about 38%, is related to infrared light of the solar radiation.

The most energetic and therefore the most damaging part of the spectrum is the UV region. It is often divided into three regions, UV-A (400–315 nm), UV-B (315–280 nm), and UV-C (below 280 nm). The UV-A light transmits through window glass. UV-B is filtered by most of window glasses. UV-C is filtered by the atmosphere and is irrelevant at outdoor weathering.

Hence, one of the most challenging tasks for constructors of accelerated weathering testing instruments is to match the spectrum of the artificial light with that of natural sunlight.

Let us give some comparative examples of natural and accelerated weathering.

Two experimental HDPE-based roof tiles, gray and green one, were placed each on a roof of a two-story high building in Bedford, MA, and Green Bay, WI (four tiles total), and kept there for 6 months. Meanwhile, the same tiles were exposed in the weathering box.

Gray tile: In Bedford the gray tile faded by 1.60 ± 0.15 units (ΔL), in Green Bay by 1.65 ± 0.20 units, which is practically the same. In the weathering box the same fading, by 1.60 L units, was observed after 955 h of exposure. Therefore, 1000 h in the box for the gray tile correspond to 1.7 L units of fading, and 1 day in the weathering box corresponds to 4.6 days outdoors. This is for both Bedford, MA, and Green Bay, WI.

Green tile: In Bedford the green tile faded by 2.60 ± 0.20 units (ΔL), in Green Bay by 2.50 ± 0.20 units, which is practically the same. In the weathering box the same fading, by 2.55 L units, was observed after 346 h of exposure. Therefore, 1000 h in the box for the green tile correspond to 7.4 L units of fading, and 1 day in the weathering box corresponds to 12.7 days outdoors. This is again for both Bedford, MA, and Green Bay, WI.

As one can see, products of two different colors, being exposed to both outdoors and in the weathering box, gave almost 300%; difference in their "correlations" of accelerated and natural weathering.

That is why such a correlation does not exist in a general case.

Another example. *Mahogany WPC deck board:* GeoDeck Mahogany deck boards increase the lightness (L) by 0.4 units after 1000 h in the weathering box, and by 0.8 units after 2230 h in the box. A GeoDeck Mahogany board, exposed on a building roof in Bedford, MA for 4 years, gave the following changes in the colors:

Unexposed board:

$L = 47.2 \pm 0.6$

$a = 9.1 \pm 0.3$

$b = 14.0 \pm 0.3$

Exposed board:

$L = 48.8 \pm 0.5$

$a = 8.7 \pm 0.3$

$b = 12.9 \pm 0.2$

After 4 years, lightness is increased by 1.6 units, "redness" decreased by 0.4 units, and "yellowness" decreased by 1.1 units. That is, 4 years outdoors (in Bedford, MA) approximately correspond (in terms of lightness) to 4460 h in the box, that is, 1 day in the box correspond to 7.9 days outdoors.

In a brief (and limited) summary of said experiments, we have that one day in the box might correspond—in terms of fading—to approximately 5–8–13 days outdoors in Massachusetts (and Wisconsin). As an order of magnitude of the value this may be acceptable, but not for quantitative predictions.

Let us introduce one more term, the acceleration factor. The acceleration factor—in this context—is the ratio of time for the same extent of fading outdoors to that in the accelerated weathering box. For example, an accelerated factor of 10 means that a one-month accelerated weathering test corresponds to 10 months

TABLE 16.2 Acceleration factors for some polymers, exposed in Atlas Ci5000 Xenon-arc weatherometer according to ASTM G26 (102 min UV:18 min UV + water spray; 340 nm, 0.35 W/m², black panel temperature of 63°C) and on a weathering rack in Milwaukee, WI, oriented to face south at an inclination of 45°

Material	Hours in the weathering box = 1 year outdoor exposure in Milwaukee	Acceleration factor
Polystyrene	1274 ± 552	6.9 ± 4.8
Vinyl (B-946)	684 ± 175	12.8 ± 2.6
Polyester (B-689)	1329 ± 1262	6.6 ± 3.2

Outdoors data for 0, 12, and 24 months were used in calculations [4].

outdoors. From the above example it is obvious that the accelerated factor would be quite different for *L*, *a*, and *b* color measurements. For lightness (*L*), the accelerated factor for the above examples of composite materials (roof tiles and a deck board) is between 5 and 13.

For a comparison, acceleration factors for several different classes of polymers are shown in Table 16.2

Note 1: Numerical data in the second column, and to some extent in the third column are listed incorrectly. It does not make much sense to show data with a pretended precision of better than 0.1% (four digits) while the error margin in fact reaches 40–95%. Data in the third column can be listed as 7 ± 5, 13 ± 3, and 7 ± 3.

Note 2: Color of the samples and color fade was measured using two criteria, that is, using a densitometer (the percent reduction in reflected optical density), and *L*, "*a*," and "*b*" units of the Hunter color space, along with color difference, ΔE. In general, better results were obtained when optical densities were measured.

One can see that acceleration factor between 5 and 13 (obtained for exposures in Massachusetts and Wisconsin) does not deviate too much from acceleration factors between 7 and 13 (Table 16.2), obtained by a different group of researchers in Wisconsin. The author [4] concluded that the acceleration factor of 11.0 ± 3.7 describes better the weathering in Milwaukee. For South Florida the same author cited the acceleration factor of 8.4, and those in South Florida and Arizona varied in the range of 4.5–15, depending upon the type of material. One year in Florida and Arizona is approximately equal to 2 years in Michigan [4].

The same author [4] noted that a year of outdoor weathering in Milwaukee corresponds to approximately 800 ± 400 h in the weathering box at standard conditions (see Table 16.2). An acceleration factor in South Florida was found to be 8.4 [5], or vary between 4.5 and 15 in South Florida and Arizona [6].

It might be good enough for semiquantitative considerations, but hardly for quantitative predictions of fading of plastics and composite materials.

Here are a few more examples of acceleration factors. A wood flour (50% w/w)-filled HDPE was subjected to the accelerated weathering under standard conditions

(ASTM G 155; 102:18 cycle, 340 nm) and to natural weathering (ASTM G7) at an undisclosed location [7]. ΔE factor was calculated and compared.

- For a Cedar-colored WPC, the acceleration factor was 6.1, and for a Redwood-colored it was 7.3.
- The same materials but with UV stabilizers and antioxidants introduced (as much as 5% total), the acceleration factors were equal to 5.3 and 7.7, respectively.
- When wood flour was first colored and then blended with HDPE, the respective acceleration factors were 8.6 and 7.7, respectively. In this case both natural and accelerated weathering were slower than those for nondyed wood flour, but the difference was not very significant for the accelerated weathering. Hence, a higher acceleration factor.
- With reprocessed HDPE the respective acceleration factors were 8.2 and 6.0.
- With other metal oxide pigments (but, apparently, the same color) the respective acceleration factors were 7.9 and 6.2.
- With concentrated color pigments the acceleration factors were equal to 5.9 and 13.5. In this case the natural weathering was significantly slower, but accelerated weathering was similar with that for the "standard" samples. Hence, a higher acceleration factor.

One can see that all acceleration factors in this study were in the range of 5.3–13.5, which is in the same range as those listed above. Not good enough for truly quantitative predictions, but enough to say that 1 day of accelerated weathering in a weathering box (at the indicated above conditions) is approximately equivalent to a week or two outdoors in the Midwest or the New England region.

As it was mentioned earlier, the acceleration factor depends on which color is measured. In some cases an inflection of the acceleration factor was observed, which is when until a certain time point of weathering the natural weathering results in higher fading than the accelerated weathering (the acceleration factor is less than 1) and after that the accelerated weathering becomes more damaging than the natural one (the accelerated factor is higher than 1). An example is given in Ref. [8]. A pine flour (58% w/w, 40 mesh)-filled HDPE (31% w/w) weathered in the Atlas UVA-2000 weatherometer according to ASTM D 6662, on one hand, and naturally weathered in Moscow, ID, showed a higher natural fading compared to the artificial ("accelerated") one in terms of ΔE until 400 h of exposure, but lower natural fading between 800 and 2000 h. The acceleration factor before 400 hrs was 0.55 and between 800 and 2000 h varied up and down between 1.5 and 2.0. When ΔL was measured, the acceleration factor was always more than 1 and gradually decreased between 400 and 2000 h from 9.4 to 1.5 [8].

Let us take another angle in this consideration and predict which values of acceleration factors can be anticipated, using direct comparisons of UV radiant exposures in the weathering box and the real world.

TABLE 16.3 The annual UV radiant exposures (295–385 nm) in Homestead, FL (latitude 25°27′ North, longitude 80°30′ West, elevation 7 ft) and Buckeye, AZ (latitude 33°23′ North, longitude 112°35′ West, elevation 1055 ft), 45° south exposure

	The annual radiant UV exposure at 45° (MJ/m^2)	
Year	Florida	Arizona
2000	314	333
2001	290	345
2002	325	348
2003	322	354
2004	339	345
2005	320	374
Average	318 ± 16	350 ± 14

Q-Lab data.

Table 16.3 shows examples of the annual UV radiant exposure (295–385 nm) in Florida and Arizona.

One can see that annual solar UV energy slightly varies, but in a rather narrow range (± 5% in Florida and ± 4% in Arizona). The respective annual data for the Florida site are available for a 10-year time period (data for 1996–1999 are 265, 254, 324, and 317 $MJ/^2$, respectively; Q-Lab data), and the average figure over the 10-year period is 307 ± 28 MJ/m^2. Let us calculate how much time it would take for a sample to get the same UV exposure in a weathering box under standard conditions, from which we need in this case only the value of 0.35 W/m^2 at 340 nm.

The first problem that we immediately encounter is that data for outdoor UV exposure (Table 16.3) are expressed in MJ/m^2 over 295–385 nm, which is integrated over the spectrum range and time (a year), but exposure in the weathering box is expressed in W/m^2 at only wavelength range of 340 nm, and per 1 s.

In order to directly compare these figures, we should consider that the energy contained in the 340 nm wavelength range is about 1% of that in the range of 295–385 nm. In this case 0.35 W/m^2 per 340 nm/s corresponds to 35 W/m^2 per the range of 295–385 nm/s, and to 126,000 J/m^2 per hour, or 3,024,000 per day.

TABLE 16.4 The theoretical acceleration factors for Florida and Arizona, for annual UV radiant exposures listed in Table 16.3

Year	Florida	Arizona
2000	3.52	3.31
2001	3.81	3.20
2002	3.40	3.17
2003	3.43	3.12
2004	3.26	3.20
2005	3.45	2.95
Average	3.5 ± 0.2	3.2 ± 0.1

Thus, the cumulative annual UV exposure of 318 MJ/m^2 over the said spectral range (over the years 2000–2005) will be reached in the weathering box for 318,000,000/3,024,000 days = 105 days. The acceleration factor is 365/105 = 3.5 (in Florida).

For other examples listed in Table 16.3, theoretical acceleration factors are given in Table 16.4

One can see that theoretical acceleration factors, related to the UV exposure in the weathering box (at 0.35 W/m^2) and to actual outdoor exposure in the test sites in Florida and Arizona, are rather consistent from year to year. In the real world they certainly would be deviating from the above figures due to difference in colors of samples, moisture content in samples, and so on. Nevertheless, considering that "real" acceleration factors in Wisconsin and New England are in the range of 5–13 or 11 ± 4 (see above), the theoretical acceleration factors of 3.5 in Florida and 3.2 in Arizona certainly make sense.

FADING OF COMMERCIAL WOOD-PLASTIC COMPOSITE MATERIALS

As the first commercial WPC deck boards were offered, it was recognized that they were fading with time. It was an undeniable feature of WPC. Nowadays practically every informational and marketing material of manufacturers, as well as their warranty policies, contain recognition that their products fade under exposure to sunlight. This recognition is often rather vague and elusive, and sometimes even used as a marketing tool, pretending that the fading of deck boards is a good and esthetically sound feature (some excerpts are given below).

Apparently, the most honest approach was taken by UltraDeck manufacturers, who placed in their marketing materials actual pictures showing the extent of fading of their product to a "natural color" (Fig. 16.3). In their materials the manufacturer suggests a customer to "allow the UltraDeck to weather at least 1 year."

The following are excerpts taken from published manufacturers materials. In these quotations direct marketing calls are omitted.

- *Trex:* Because Trex contains natural wood fiber, slight fading will occur with exposure to sunlight and moisture. Trex decking reaches its weathered color after 10–12 weeks. The "after weathering" color of the boards on a Trex deck may be different from the colors shown in the Trex Color Palette guide.

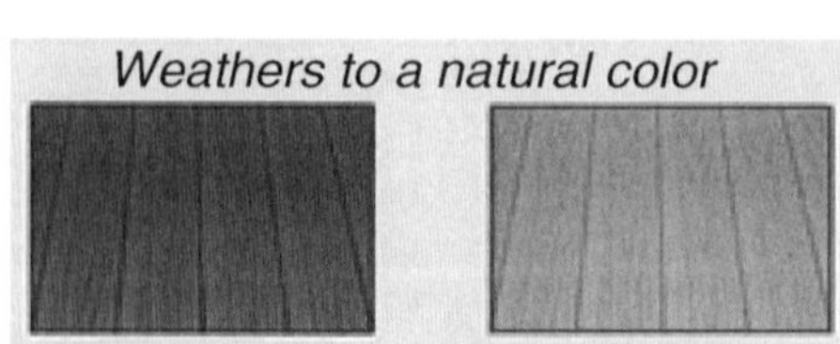

Figure 16.3 Fading of UltraDeck wood-plastic composite deck board, according to the manufacturer's data (© Midwest Manufacturing Extrusion). See color insert.

- *Rhino Deck:* Deck areas that have different exposures to the elements, for example, under an overhang or a table would have less exposure, therefore, will show less weathering to neighboring boards . . . Color variations and uneven fading may occur when exposed to the elements.
- *WeatherBest:* You will see some color variation from the original color due to the weathering process.
- *Monarch:* All Monarch™ products weather to some degree . . . decking and railing systems weather to a beautiful seasoned tone. The rate of color seasoning depends on exposure to sunlight, weathering, and other environmental factors . . . However, beginning with installation, Monarch . . . deck boards' seasoning process is usually complete within 6–12 months, similar to all-wood materials.
- *Oasis:* Yes, composite products will weather.
- *Premier:* Decking will experience less color fade and will "even out" to one consistent level of color within 8–12 weeks of installation.
- *CorrectDeck:* Standard CorrectDeck colors will fade a minimum of 10% during the first 12 weeks after installation.
- *Xtendex:* will fade to a lighter shade within weeks of exposure to ultraviolet rays.
- *Boardwalk:* When exposed to sunlight and water, Boardwalk weathers uniformly from medium tone color to a pleasant, light hue similar to that of weathered cedar.
- *Latitudes:* Latitudes Decking and Natural Railing will lighten slightly over time to a beautiful weathered tone. This color tone shift is dependent upon exposure to sunlight and other environmental factors. The weathering process begins upon installation and is generally complete within 60–90 days . . . At the end of the fading process the deck appears gray.
- *Cross Timbers:* Limited Warranty does not cover claims for variations or changes in the color of Decking.
- *EverGrain:* Each color will usually weather to a light shade after exposure to the environment, creating natural, beautiful look.
- *GeoDeck:* Most composite products are made with wood flour or chips and plastic. The lignins found in wood products are a major contributor to fading. GeoDeck uses Biodac® rather than wood chips as its fiber source. Biodac® has low amounts of lignins, and consequently its GeoDeck products are more resistant to fading.

The last statement differs in style from the above excerpts and is introduced here on the only reason: It fades indeed much less than all other WPC commercial products known to the author. Having said that, it is time to consider how much commercial deck boards fade.

Table 16.5 lists a number of WPC deck boards available in the market, along with ΔL values obtained after 1000 h of weathering at standard conditions. Only GeoDeck is named in the table. All other names are withheld after consulting with some of the manufacturers. Indeed, new modifications of WPC deck boards are introduced to the market, so the information listed in Table 16.5 may in some cases be outdated. Hence, the data should be considered as an indication to the range of fading of commercial composite deck boards.

TABLE 16.5 Comparison of commercial WPC deck boards in terms of ΔL values on the Hunter Lab color scale after 1000 h of weathering

Composite deck board	ΔL value after 1000 h of weathering	Initial L value
GeoDeck (Cedar)	0.4	52.2
GeoDeck (Mahogany)	0.4	47.2
GeoDeck (Driftwood)	0.4	52.8
4	1.2	64.0
5	3.5	61.2
6	4.2	57.3
7	4.6	48.8
8	4.9	50.5
9	5.0	56.9
10	5.1	52.0
11	5.3	45.8
12	6.0	61.4
13	6.9	34.6
14	7.7	53.0
15	7.8	47.8
16	9.0	62.3
17	9.1	56.9
18	9.3	40.8
19	9.4	56.4
20	10.2	55.0
21	10.4	42.6
22	12.1	50.2
23	12.5	32.4
24	12.8	38.3
25	13	53.8
26	16.7	61.3
27	16.7	54.2
28	18.0	56.5
29	18.3	67.4
30	18.4	39.7
31	19.5	36.8
32	20	51.1
33	20.4	53.2
34	23	53.6
35	24	45.1
36	35	52.6

Weathering was conducted in Q-Sun 3000 weathering chamber according to ASTM D 2565 (340 nm, 0.35 W/m^2, 102:18 cycle, 63°C black panel temperature). Data by the author.

Fading of GeoDeck boards after extensive time of accelerated weathering is shown in Table 16.6.

TABLE 16.6 Color (Hunter Lab color scale) of commercial GeoDeck composite deck boards before and after weathering for 2400 h in Q-Sun 3000 weathering chamber according to ASTM D 2565 (340 nm, 0.35 W/m², 102:18 cycle, 63°C black panel temperature)

	Color before and after 2400 h of accelerated weathering		
GeoDeck board	*L*	*a*	*b*
Cedar, initial	52.2	8.4	21.9
Cedar, after weathering	53.0	10.4	23.5
Mahogany, initial	47.2	9.1	15.0
Mahogany, after weathering	48.0	11.5	13.8
Driftwood, initial	52.8	−0.5	5.5
Driftwood, after weathering	53.8	0.1	2.7

Data by the author.

As it was indicated above, out of 21 commercial WPC deck boards, listed in Table 16.1, in 18 the yellow component decreases in its "*b*" value after exposure (UV light and water spray) in the weathering box. It seems this is rather common phenomenon, and the yellowing of lignin due to its photodegradation (see above) is certainly not a prevailing factor in fading of many wood-containing WPC. Plastic itself and wood extractives apparently contribute to fading the most.

Generally, color shift in WPC materials (as well as in neat polymers) is poorly understood. Very often there is no correlation between shifts in *L*, "*a*," and "*b*." Often only one of them changes in the process of weathering. Some samples show almost linear change of lightness during an extended time period, some show a significant change in the first 50–100 h, sometimes 200–400 h, and then attain an almost constant level.

For example, PVC-based Millenium deck board shows the main change of its lightness (*L*)—by 3.2 units—after only 120 h in the weathering box (standard conditions, see Table 16.5), and then lightness practically levels off, within decimals. Polypropylene-based Cross Timbers show a major jump in its lightness—by 8 units—after only 200 h in the box, and then the *L* levels off. HDPE-based Life Long in the first 300 h in the box increases its lightness by 4 units, and then practically levels off. Austrian Fasalex within only 400 h increases its lightness by 17 units (!), and then levels off.

Such a dynamics of the lightness change in the first period of the artificial weathering is not a common rule because, for example, HDPE-based TimberTech solid deck board increases its lightness in the first 200 h in the box by 14 units (!), but then continues to fade by 6 more units within the next more than 1000 h. HDPE-based

UltraDeck increases its lightness in the weathering box more slowly, within the first 600 h by 8.5 units, and then more slowly but steadily continues to fade.

Apparently, the main role in dynamics of fading plays pigments (or lack of them). For example, two apparently identical in composition HDPE-based Premier boards, but of different colors (brown and gray), fade quite differently. Gray board increases its *L* value within 300 h by 4.6 units and then levels off, while lightness of the brown board increases by 7.4 units within 500 h and then slowly continues to fade. If their compositions are indeed identical, this significant color change can be attributed mainly to the pigments in the boards and/or amounts of pigments, interaction of the pigments with other WPC ingredients, and similar pigment-related factors.

FADING OF COMPOSITE DECK BOARDS VERSUS THEIR CRUMBLING DUE TO OXIDATION

As it was noted above, antioxidants usually do not prevent fading. For instance, old (defective) GeoDeck boards which did not have any antioxidants and showed progressive deterioration, particularly in the South, retained their colors even when almost completely crumbled.

This can be rephrased as follows: Antioxidants can decrease fading to some extent and in some situations, but they do not prevent it. On the contrary, lack of antioxidants does not necessarily accelerate fading. Antioxidants often disappear from the very surface of a composite deck board rather quickly; they would prevent

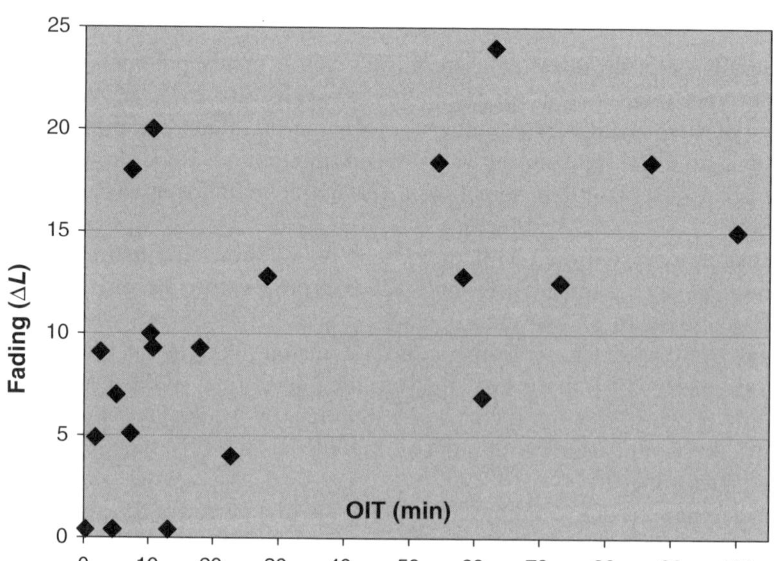

Figure 16.4 Fading of commercial wood-plastic composite deck boards (see Table 16.5) and the OIT (oxidative induction time) values for the same deck boards (see Chapter 15).

oxidation of the bulk of the board, but not the very surface. Without a good amount of colorants, the board would fade.

Having said this, it is not surprising to see that there is no clear correlation indeed between the amount of antioxidants in a WPC boards and the degree of its fading (Fig. 16.4).

It is tempting to see a sort of a correlation in Figure 16.4, such as

$$\text{Fading} = 7 + 0.11 \times \text{OIT}$$

However, this correlation does not have much of a sense. The OIT first of all depends on antioxidants present in the material. Some commercial additives or fillers, such as wood flour, can play a role of these antioxidants. Fading (rather, a lack of it) depends mainly on added pigments. There is no meaningful correlation between pigments and their antioxidant properties. Here is why:

- Often WPCs contain much of wood flour and, therefore, have a high OIT values, but have no colorants and, hence, have high fading. These WPC boards take the upper right area in Figure 16.4.
- WPC, containing a good amount of both antioxidants and pigments, move to the lower right area (high OIT and low fading). There are few of such WPC boards, as in Figure 16.4.
- Low OIT and low fading, as happened with an old GeoDeck, brings the respective WPC into the lower left quadrant as in Figure 16.4.
- When a WPC board does not contain antioxidants and pigments, it is prone to a fast oxidizing and high fading. This brings respective boards to the upper right area on the above graph. It is particularly not a good situation. There are several WPC boards, falling into this category, in the market.

At the same time, when a WPC contains a systematically increased amount of an ingredient, which increases the OIT of the material (that is, possesses an antioxidant properties) and concurrently increases its fading, one would observe a certain correlation between the two properties (Fig. 16.5).

One can see from Figure 16.5 that the added biopolymer significantly increased the OIT values of the HDPE-based composite and also increased its fading. These two properties are apparently unrelated, but as both of them were systematically and concurrently changed, they produced an illusion of interconnectivity between them. In fact, they just happened to be in the same place and in the same time.

FACTORS ACCELERATING OR SLOWING DOWN FADING OF COMPOSITES

Density (Specific Gravity) of the Composite

Density of the WPC does not accelerate fading, or at least such cases are not generally known. Accelerated weathering of GeoDeck with density (specific gravity) between

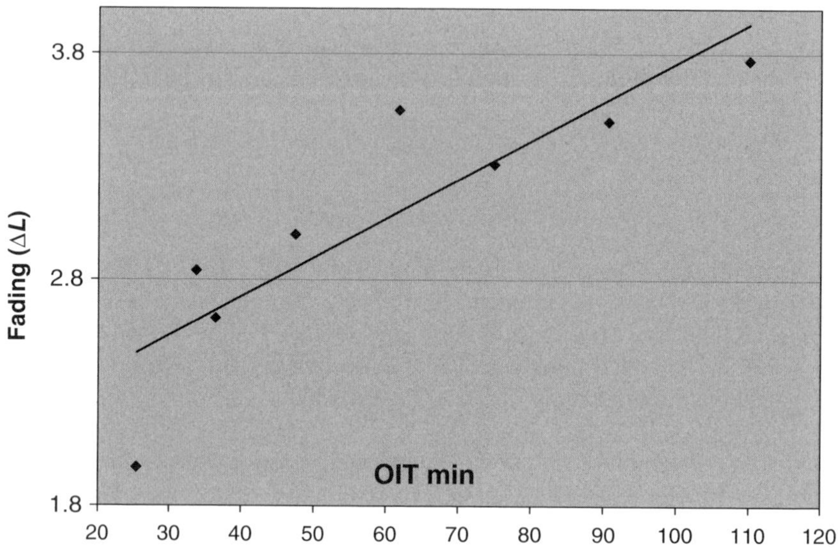

Figure 16.5 Fading of wood-plastic composite deck boards (in which a biopolymer was added at 2.5, 7, 10, 14, 17, 20, 22 and 30) and the OIT (oxidative induction time) values for the same deck boards.

1.24 and 1.05 g/cm^3 did not show any trend in fading. Natural weathering of DeoDeck on real decks across the country has not shown any trend either. Less dense boards, particularly those which did not contain antioxidants, oxidized faster, but this effect results in darkening of boards, not in their fading (Fig. 15.12 in the preceding chapter).

Temperature

It is rather difficult to separate effects of UV radiation from effects of temperature under typical experimental conditions. Systematic studies of the effects of "cold" UV radiation on WPC materials are not known to the author. However, observing the WPC decks in mountainous areas, where temperature is significantly lower compared to that at the bottom of mountains, I could not see any difference in the fading of composite deck boards and their plastic accessories (endcaps, etc.). It seems that fading does not depends on temperature, at least significantly. This makes sense because many photochemical reactions have their activation energy close to zero. In other words, temperature does not accelerate most of the photochemical reactions.

UV Absorbers and Their Amounts

UV absorbers generally decrease the oxidative degradation of plastics and plastic-based composites. However, they do not always significantly change the fading of WPC materials. For example, wood flour (50% w/w, 40 mesh)-filled HDPE (MFI 0.72) after an accelerating weathering in a box for 3000 h increased its *L* value by 115% after

TABLE 16.7 Effect of a light stabilizer Cyasorb UV-3853 (Cytec) on fading of a wood–HDPE composite material

	ΔE	
Exposure (h)	No UV stabilizers added	UV-3853, 0.1% in the composition
1500	1.5	0.4
4000	6.0	0.6
7500	9.5	0.6
10,000	10.5	0.6
11,000	11.0	1.5
12,500	11.2	2.0

Accelerated weathering in a Xenon-type weathering box [10].

0.5% (w/w) of a UV absorber was introduced into the formulation, its L decreased to +98% (from the +115%) above the control, and after doubling the amount of the UV absorber (to 1.0%), the L value increased again to +107% [9].

More often, however, UV absorbers (or UV stabilizers) decrease fading quite noticeably (see, for example, Tables 16.7 and 16.8).

Pigments and Their Amounts

Pigments typically slow down the fading of WPC materials. Pigments containing TiO_2 often increase fading because TiO_2 is an effective catalyst of the oxidative degradation of plastics, including oxidative processes on the surface.

For example, wood flour (50% w/w, 40 mesh)-filled HDPE (MFI 0.72) after an accelerating weathering in a box for 3000 h increased its L value by 115%; after 1% (w/w) of a pigment was introduced into the formulation, its final L decreased to +73% (from the +115%) above the control, and after doubling the amount of the pigment (to 2%), the final L value decreased further to +61%. A combination of a UV stabilizer (1%) and the pigment (2%) further decreased the fading (in terms of L) to +50% [9].

TABLE 16.8 Effect of a light stabilizer Cyasorb UV-3853 (Cytec) on fading of a wood–polypropylene composite material

	ΔE	
Exposure (h)	No UV stabilizers added	UV-3853, 0.2% in the composition
800	3	1.0
2400	12	1.5
3200	14	2.0
3800	16	2.1
5600	16	2.2
6800	—	2.3

Accelerated weathering in a Xenon type weathering box [10].

TABLE 16.9 Effects of pigments and antioxidants on fading of wood–polypropylene composites [11]

Wood flour, % (w/w)	Polypropylene, % (w/w)	IrganoxB-225, % (w/w)	Pigments, % (w/w)	ΔE after 1000 h of accelerated weathering
Effect of Irganox B-225				
46	39	0	0	21.2
48	48	0.1	0	35.9
45	45	0.3	0	32.8
Effect of both Irganox B-225 and/or pigments				
44	44	0	1.95	21.3
50	40	0.27	2	30.0
47	47	0.10	2.0	22.9
45	44	0.41	1.5	20.4
44	44	0.3	1.95	19.6
48	43	0.29	0.5	19.2
44	44	0.29	0.43	16.5
47	47	0.27	1.65	15.9
48	41	0.27	5.2	15.6
Effect of pigments at a constant amount of Irganox B-225				
46	46	0.1	2.0 + 2.0	14.0
46	46	0.1	2.0 + 2.0	13.6
46	46	0.1	0.5 + 1.5 + 2.0	13.4
46	46	0.1	1.0 + 1.0 + 2.0	13.3
45	45	0.1	2.0 + 4.0	13.0
46	46	0.1	1.0 + 1.0 + 2.0	12.4
46	46	0.1	2.0	12.2
46	46	0.1	2.0 + 2.0	10.9
46	46	0.1	2.0 + 2.0	8.7

Amounts of fiber and plastic were rounded to the nearest percent. Balance in the composition was made up by a wax. Pigments are different and they are not identified in the table.

Table 16.9 shows a good example when antioxidants did not decrease the fading of a number of WPCs, while introduction of pigments was much more effective.

The names of pigments are not indicated in Table 16.9, as it is not relevant to the subject under consideration. They are given in the reference in Table 16.9. The main result in the context of this consideration is that Irganox B-225 does not prevent fading, and, moreover, increases the fading at elevating concentrations of the antioxidant (the first three rows in the table). It is the pigments that decrease the fading (the second half of the table), and different pigments showed very different effects.

Antioxidants and Their Amounts

Antioxidants and their amounts effect oxidative degradation of plastics and plastic-based composite materials in the bulk, as it was shown in the preceding chapter. However, they do not seem to significantly effect fading of WPC materials, at least in many cases (see, for example, Table 16.9). GeoDeck deck boards, whether they do not contain antioxidants (the OIT of 0.2–0.4 min, see the preceding chapter) or are loaded with antioxidants (the OIT of 16–20) show the same fading, or rather a lack of it.

Cedar-colored composite containing wood flour (50% w/w) and HDPE without antioxidants and UV stabilizers (AO/UV) showed a ΔE shift of 8.4 and 6.8 units in a weathering box and outdoors, respectively, while the same composite loaded with AO/UV (5% w/w) showed a ΔE shift of 5.6 and 5.2, respectively. A decrease in fading was 50 and 30% in the box and at the natural weathering, respectively. However, with the same composition but colored differently (redwood), a ΔE shift of 10.0 and 6.8 units was observed in a weathering box and outdoors, respectively, while the same composite loaded with AO/UV (5% w/w) showed a ΔE shift of 9.2 and 5.9, respectively. A decrease in fading was 9 and 15% in the box and at the natural weathering, respectively [7]. Therefore, effect of antioxidants on fading significantly depends on the color of WPC materials.

History of Plastics (Virgin, Recycled)

This issue depends mainly on an amount of antioxidants in the WPC material and, hence, is related to the above section. As it was shown above, effect of recycled plastic (regrind) on fading depends mainly on the color of the WPC. For example, for a Cedar-colored WPC board, based on virgin HDPE, a ΔE shift of 8.4 and 6.8 units was observed in a weathering box and outdoors, respectively, while the same composite but based on recycled HDPE showed a ΔE shift of 11.7 and 7.1, respectively, that is, fading for the reprocessed HDPE was higher. However, for a Redwood-colored WPC board, the respective figures were 10.0 and 6.8 (artificial and natural weathering for a virgin-based composite) and 7.9 and 6.5 (for a reprocessed-based HDPE), respectively [7]. That is, fading was lower for the recycled-based composite material, both in a weathering box and outdoors.

It seems that both amount of antioxidants and recycled or virgin plastic do not effect the fading of composite materials. Some other factors prevail, such as the color of the composite.

Effect of Moisture in the Composite

This effect is apparently the most important one besides the UV light. As water is a catalyst of an oxidative degradation of composite materials (see the preceding chapter), it also catalyzes oxidative degradation and, hence, the fading, on the surface. With neat polyolefins effects of water on oxidative degradation is much less evident, as polyolefins practically do not absorb water. WPCs, because of their porosity, absorb much more (10–100 times more) water, and effect of moisture in oxidative degradation (and fading) of WPC is more noticeable.

TABLE 16.10 Effect of wood flour content on fading of wood flour (40 mesh pine) filled HDPE

Wood flour content, % (w/w)	Shift in ΔL (%)
10	2
20	16
30	33
40	38
50	51
60	102

Weathering was conducted under a 2-h cycle (108 min UV + 12 min UV/ water spray) [9].

Fading of WPC boards in a weathering box can be significantly decreased if to turn the water spray off. For example, for a wood flour (50% w/w, 40 mesh)–filled HDPE (the initial lightness of 48 units), exposure in a weathering box (UV + water) has led to the increase in ΔL (in percent) by 63, 80, and 87%, but when water was off, UV treatment only led to the increase in ΔL by only 9, 22, and 28% [9].

The Type and Amount of Cellulose Fiber

The higher the amount of the cellulose fiber in the composite, the higher is the fading, if all other conditions being equal (see, for example, Table 16.10).

Extruded Versus Injection-Molded Wood–Plastic Composite Materials

As the limited data show, there is not much of a difference between fading of the extruded and injection-molded profiles made of the same material. Density often differ between the two; however, as it was discussed above, density typically is not a factor influencing fading. An example is shown in Table 16.11.

TABLE 16.11 Shift of lightness (*L* in Hunter Lab color space) for an extruded (as is and planed) and injection-molded wood flour (50% w/w, 40 mesh)-filled HDPE (MFI 0.72)

	L value		
Exposure in the weathering box (h)	Extruded material	Injection molded material	Planed material
0 (Unexposed)	49	57	63
1000	79	85	85
2000	86	91	92
3000	90	93	93
Overall fading after 3000 h	41	36	30

Treatment was conducted in a xenon-arc weathering box at standard conditions according to ASTM D 2565 [12].

One can see that for the material under study ΔL was 30 after 1000 h for the extruded profile, and 28 for the injected-molded material. After 3000 h, L values for both the materials were close to each other. However, for the initial materials, the L values were higher for the injection-molded profile. For a planed profile, the initial L value was the highest one ($L = 63$), but after 3000 h, the L value was 93 [12]. This again confirms the observation (see above) that removing the upper polymer layer (by brushing or planning) increases the initial L value, and makes the overall fading (in terms of L shift) lower.

ASTM RECOMMENDATIONS

ASTM D 2565 "Standard Practice for Xenon-Arc Exposure of Plastics Intended for Outdoor Applications"

This ASTM procedure and practice forms a part of ICC AC 174 "acceptance criteria for deck board span rating and guardrail systems (guards and handrails)" and ASTM D 7032 "standard specification for establishing performance ratings for wood-plastic composite deck boards and guardrail systems (guards or handrails)." It describes test conditions, preparation of test specimens, and evaluation of test results for xenon-arc exposure of plastics intended for long-term use in outdoor applications.

There is no special requirement for specimens for weathering except that they should fit specimen holders and racks supplied with the exposure apparatus. Exposure of at least three replicate specimens of each test material and of control material is recommended. Thickness of test and control specimens should approximately be the same (within ± 10%).

Note 1 of the author: Thickness of the specimens can affect the temperature of the samples and, hence, their deterioration and other weathering characteristics.

The ASTM procedure lists test cycles commonly used for the xenon-arc (with daylight filters) exposure testing of plastics.

Note 2 of the author: Xenon-arc lamps usually provide a simulation of full spectrum sunlight for photostability testing.

Note 3 of the author: The first test cycle procedure (see below) is almost universally used for testing WPCs. Other cycle procedures (five of them are listed in the ASTM) are used mainly in automotive interior and exterior testing, and fabric testing. Only the first test cycle is described here as follows:

- 102 min light only is followed by 18 min of light with water spray (102/18 cycle). Uninsulated black panel temperature is 63 ± 2°C (during only the light-portion of the cycle), with an irradiance of 0.35 ± 0.02 W/m^2 at 340 nm.

The procedure notes that the irradiance of 0.35 W/m2 is related only to 340 nm, which typically corresponds to 41.5 ± 2.5 W/m2 between 300 and 400 nm. The ASTM prescribes, when possible, to maintain relative humidity at 50 ± 5% equilibrium during the light-only interval. The procedure also describes the required purity of water for water spray. The temperature of water used for specimen spray should be 16 ± 5°C (61 ± 9°F).

Note of the author: The procedure mistakenly recommends the temperature (see above) as 60.8 ± 9°F. Such a recommendation does not make much sense because when a deviation is recognized as being of ± 9°, one cannot pretend that the principal figure is measured with a precision of 0.1°. This notation is misleading and confusing. In fact, the recommendation simply suggests to have water spray temperature between 52 and 70°F. The precise figure of 60.8° is superficial and unnecessary.

The ASTM procedure does not specify duration of the weathering test.

Note of the author: AC 174 (Section 4.5) and ASTM D 7032 (Section 4.6) require flexural testing to be performed after exposure to a minimum of 2000 h accelerated weathering in accordance with this ASTM procedure.

ASTM D 1435 "Standard Practice for Outdoor Weathering of Plastics"

This standard considers the stability of plastics when they are exposed outdoors. The relative durability of plastics in outdoor use can be very different depending on the location of the exposure because of the differences in UV radiation, wetness, temperature, pollutants (acid rains, for example), and many other factors. Hence, exposure in several locations with different climates that represent a broad range of anticipated service conditions are recommended by this ASTM. Still, because of year-to-year climatological variations, results of exposure in 1 year cannot be used to predict the absolute rate at which a material degrades. Therefore, several years of repeated exposure to get an average test result for a given location is recommended.

According to the standard, unless otherwise specified, exposure racks should face the equator. The standard considers rack position at different latitudes, and the respective rack adjustments at certain periods of the year, and indicates that in most nondesert areas maximum annual UV exposure is provided by exposure at an angle of the latitude angle minus 10°. Besides, the standard considers at-latitude racks, 45° racks, 90° racks, horizontal racks (instead, 5° south exposure is recommended, to provide moisture runoff), and other angle racks. Materials and types of construction of test racks are also considered, as well as instruments for measuring solar radiation, ambient temperature and relative humidity, and their calibration procedures.

The procedure recommends to measure and express total solar irradiance and total UV irradiance in MJ/m^2. As an example, it lists typical average (based on several years of measurements) UV radiant exposure for 12 months to be 308 MJ/m^2 in subtropical climate at 5° exposure, and 333 MJ/m^2 in desert climate at latitude exposure, both between 295 and 385 nm.

ASTM D 4329 "Practice for Fluorescent UV Exposure of Plastics"

This practice covers specific procedures and test conditions for fluorescent UV exposure of plastics, including the preparation of test specimens and the evaluation of test results. The procedures are intended to induce property changes in the plastics intended for long-term use in outdoor applications associated with the effects of sunlight, moisture, and heat.

The practice recommends to use the spectral power distribution of the fluorescent UV lamp similar to that of UVA-340 fluorescent lamps.

Note of the author: Fluorescent lamps UVA-340 spectrally match *rather well* the sunlight between 295 and 340 nm. However, in the visible part of the spectrum, the irradiance of sunlight generally increases from 340 to 400 nm (from 0.6 to 1.2 $W/m^2/nm$, respectively), while UVA-340 lamps produce irradiance which decreases in the same range of wavelength from 0.6 to 0.1 $W/m^2/nm$.

The procedure recommends to employ Cycle A for most general application, and Cycle C for plastic building products:

Cycle A:
8 h UV at black panel temperature of 60 ± 3°C
4 h water spray at black panel temperature of 50 ± 3°C
Cycle C:
8 h UV at black panel temperature of 50 ± 3°C
4 h water spray at black panel temperature of 50 ± 3°C

ASTM D 4364 "Practice for Performing Outdoor Accelerated Weathering Tests of Plastics Using Concentrated Sunlight"

This practice describes the outdoor-accelerated-exposure testing of plastics and plastic-made products using Fresnel reflecting concentrator. The latter uses the sun as a source of UV and longer wavelength radiation and involves a system of plane mirrors focused on an air-cooled target board on which the test specimens are mounted.

The three basic exposure methods are as follows:

Cycle No. 1
Daytime (6 A.M. until 9 P.M.), fifteen 1-h consecutive cycles of water spray for 8 min, dry time 52 min.
Nighttime (9 P.M. until 6 A.M.), three 3-hr consecutive cycles of water spray for 8 min, dry time 172 min.
Cycle No. 2
No water spray.
Cycle No. 3
Daytime (5 A.M. until 7 P.M.), no water spray.
Nighttime (7 P.M. to 5 A.M.), forty cycles (four cycles per hour) of 3-min water spray and 12-min dry time.

ASTM D 4459 "Practice for Xenon-Arc Exposure of Plastics Intended for Indoor Applications"

As WPC materials are not commonly used in indoor applications, this standard is referenced here just for record. It covers specific procedures and test conditions applicable in window glass-filtered xenon-arc devices. The irradiance is recommended to be controlled at one of the following levels (at the temperature of an insulated black panel of 55 ± 2°C, that is, 131 ± 4°F):

- 0.30 ± 0.02 W/m^2 at 340 nm
- 0.80 ± 0.05 W/m^2 at 420 nm
- 36.5 ± 2.5 W/m^2 between 300 and 400 nm.

ASTM D 5071 "Practice for Exposure of Photodegradable Plastics in a Xenon-Arc Apparatus"

This practice is technically equivalent to ASTM D 2565 (see above) that covers xenon-arc exposures of plastics intended for long-term use in outdoor applications. This procedure covers testing of photodegradable plastics intended to deteriorate rapidly when exposed to solar radiation, heat, moisture, and other degrading elements of the weather.

The procedure recommends three basic test cycles commonly used for xenon-arc exposure of photodegradable plastics:

Cycle No. 1

Continuous light (no water spray), 0.35 ± 0.02 W/m^2 at 340 nm, 41.4 ± 2.5 W/m^2 from 300 to 400 nm, 365 ± 20 W/m^2 from 300 to 800 nm; black panel temperature of 63 ± 2°C.

Cycle No. 2

Light only for 102 min, and light and water spray for 18 min. Same irradiance and black panel temperature as in Cycle No. 1

Cycle No. 3

18 h of repeated Cycle No.2, and 6 h at dark with relative humidity of 95%, repeated continuously.

ASTM D 5208 "Practice for Fluorescent Ultraviolet (UV) Exposure of Photodegradable Plastics"

This practice is technically equivalent to ASTM D 4329 (see above) that covers UV exposure of plastics intended for long-term use in outdoor applications. This procedure covers testing of photodegradable plastics intended to deteriorate rapidly when exposed to solar radiation, heat, moisture, and other degrading elements of the weather.

Unless otherwise specified, the procedure recommends to control irradiance at 0.78 ± 0.02 $W/m^2/nm$ at 340 nm.

The procedure recommends to employ Cycle A, B, or C, the last one for materials to be used for testing toxicity after exposure (no water spray to wash away products of photochemical degradation).

Cycle A:

20 h UV at black panel temperature of 50 ± 3°C

4 h water spray at black panel temperature of 40 ± 3°C.

Cycle B:
4 h UV at black panel temperature of 50 ± 3°C
4 h water spray at black panel temperature of 40 ± 3°C.
Cycle C:
Continuous UV at black panel temperature of 50 ± 3°C.

ASTM D 5272 "Practice for Outdoor Exposure Testing of Photodegradable Plastics"

This practice is technically equivalent to ASTM D 1435 (see above) that covers outdoor exposure testing of plastics intended for long-term use in outdoor applications. This procedure covers testing of photodegradable plastics intended to deteriorate rapidly when exposed to solar radiation, heat, moisture, and other degrading elements of the weather. The procedure recommends the 5° exposure angle as the typical conditions for degradation experienced by photodegradable plastics when discarded as litter.

The practice includes Table 16.12 of average monthly solar-UV radiation (295–385 nm) on a 5° surface

TABLE 16.12 Average monthly solar-UV radiation on a 50 surface in Miami, FL and Phoenix, AZ

	Average solar-UV radiation (MJ/m^2, 295–385 nm)	
Month	Miami, FL(26°N latitude)	Phoenix, AZ(34°N latitude)
Jan	16.9	16.4
Feb	19.6	19.4
March	23.6	28.5
Apr	31.7	36.3
May	33.8	41.3
June	32.0	40.4
July	31.0	39.1
Aug	28.3	37.2
Sept	26.2	30.9
Oct	23.2	24.5
Nov	16.0	17.8
Dec	16.1	14.5
Annual	301.4	346.3

ASTM G 155 "Standard Practice for Operating Xenon-Arc Light Apparatus for Exposure of Nonmetallic Materials"

This is a more general standard compared to that described in ASTM D 2565 (in which ASTM G 155 is referred to, see above) and is related not to plastics, as ASTM D 2565 does, but to nonmetallic materials in general. It describes test conditions, preparation of test specimens, and evaluation of test results for xenon-arc exposure of nonmetallic materials intended for long-term use in outdoor applications.

Regarding apparatus, this procedure recommends to use xenon arc as laboratory light source, but equipped with one of the two different in kind filters: (a) daylight filters and (b) window glass filters. Daylight filters are used to filter xenon-arc lamp emission in a simulation of terrestrial sunlight. Window glass filters are used to filter xenon-arc lamp emission in a simulation of sunlight filtered through window glass. For example, between 301 and 320 nm, daylight filters passes 2.1–6.4% of total xenon-arc UV irradiance (5.6% of total UV sunlight irradiance), while window glass filters pass only 0–3.3% of the total xenon-arc UV irradiance (0.1–1.5% of the total sunlight UV irradiance). Between 321 and 340 nm, these figures equal to 11.1–15.0% and 1.9–14.3% for xenon-arc UV irradiance, and 18.5 and 9.4–14.8% for UV sunlight irradiance, respectively.

Compared to ASTM D 2565, which lists five test cycles commonly used for xenon-arc exposure testing of plastics, ASTM G 155 describes 12 tests. However, the mostly used Cycle 1 of ASTM D 2565 (0.35 W/m^2 at 340 nm, 102 min (UV light):18 min (UV light + water spray, 63°C black panel temperature) is identical to Cycle 1 of ASTM G 155.

ADDENDUM

Some definitions and technical terms used in descriptions of photodegradation of plastics and WPCs (from Atlas Weathering Testing Guidebook, with some modifications):

Absorption	A process by which light or other electromagnetic radiation is converted into heat or other radiation when incident on or passing through the material.
Accelerated outdoor weathering	Outdoor weathering using the sun as the source of irradiance, and where the rate of deterioration is accelerated over that of in-service exposure position, increasing one or more of the influencing parameters.
Ambient air temperature	The existing temperature of the air or of an object in thermal equilibrium with the surrounding atmosphere.
Black panel thermometer	A temperature measuring device consisting of a metal panel, having a black coating that absorbs all wavelengths uniformly, with a thermally sensitive element firmly attached to the center of the exposed surface. The black panel thermometer is used to control a laboratory weathering device and to provide an estimate of the maximum temperature of samples exposed to a radiant energy source.

Exposure, backed	A technique of weathering in which test specimens being exposed are mounted onto a solid backing material of sufficient strength to hold the specimen. When the specimen and the backing are in direct contact, the backing material must be of a type that will not contaminate the specimen. When two materials are intimately joined to form one composite, the materials below the top surface are not considered as a backing.
Exposure, open-backed	A technique of weathering in which the test specimens are exposed such that the portion of the specimen being evaluated is open to the effects of the weather on all sides.
Fading	A color change in a material that involves a weakening or lightening with time, usually as a result of exposure to light, weather, among others.
Fluorescence	The process by which electromagnetic radiation of one spectral region is absorbed and reradiated at other, usually longer, wavelengths.
Fluorescent UV lamp	A lamp in which the irradiance from a low-pressure mercury arc is transformed to a higher wavelength UV by a phosphor. The spectral power distribution of a fluorescent lamp is determined by the emission spectrum of the phosphor and the UV transmittance of the glass tube.
Fresnel reflector system	Flat mirrors arranged in an array such that they reflect onto a target, the illumination area of which simulates the size and shape of the flat mirror. Such an array simulates the ray-tracing of a parabolic trough of the same aperture angle.
Hue	The attribute of color perception by means of which an object is judged to be red, yellow, green, blue, purple, and so on.
Hunter *L*, *a*, *b* scales	A uniform color scale devised by Hunter in 1958 for use on a color difference meter, based on the opponent color theory of vision.

Irradiance, infrared	Radiation per unit area for which the wavelengths of the monochromatic components are greater than those for visible radiation, and less than about 1 mm. *Note*: There are some variations in quantitative definition of IR radiation, for example, between 780 nm and 1mm (1,000,000 nm); between 780 nm and 1400 nm; between 1.4 μm (14,000 nm) and 3.0 μm; between 3.0 μm (3000 nm) and 1 mm.
Irradiance, ultraviolet	Radiation per unit area for which the wavelengths of the monochromatic components are shorter than those for visible radiation. *Note:* There are some variations in quantitative definition of UV radiation, for example, between 100 nm and 400 nm; between 315 nm and 400 nm; between 280 nm and 315 nm; between 100 nm and 280 nm.
Irradiance, visible	Any radiation per unit area capable of causing a visual sensation. *Note*: There are some variations in quantitative definition of VIS radiation. The lower limit is generally taken between 380 and 400 nm, and the upper limit between 700 and 780 nm.
Irradiance	The rate at which radiant energy is incident on a surface per unit area (W/m^2).
Lightfastness	The resistance to color change when exposed to a light source.
Lightness	Perception by which white objects are distinguished from gray, and the light from dark-colored objects.
Monochromatic	Electromagnetic radiation of a single wavelength; object of a single color.
Photodegradation	Photochemically induced changes in the condition of the material.
Radiant energy	Energy traveling through space in the form of photons or electromagnetic waves of various lengths.
Radiation, direct	The solar radiation received from the sun without having been scattered by the atmosphere.

Radiation, diffuse	The solar radiation received from the sun after its direction has been changed by scattering in the atmosphere.
Radiation, total solar	The sum of the direct and the diffuse radiation on a surface. Total solar radiation is sometimes used to indicate quantities integrated over all wavelengths of the solar spectrum. The most common measurement of solar radiation is total radiation on a horizontal surface, often referred to as global radiation.
Solar radiant exposure	The incident energy per unit area on a surface, found by integration of irradiance over a specified time period (J/m^2).
Spectral power distribution	The variation of energy due to the source over the wavelength span of the emitted radiation.
Wavelength	The distance, measured along the line of propagation, between two points that are in phase on adjacent waves. Wavelength determines the color of light. Wavelengths of visible light range from about 400 to about 800 nm.
Weathering, direct	A technique of weathering in which the test specimens are exposed to all prevailing elements of the atmosphere.
Weathering, natural	Outdoor exposure of materials to unconcentrated sunlight, the purpose of which is to assess the effects of environmental factors on various functional and decorative parameters of interest.
Whiteness	Perception of high lightness, high diffusion, and absence of hue.
Yellowness	The attribute by which an object is judged to depart from a preferred white toward yellow.
Xenon arc	A light source produced by a high-intensity electrical current through a tube containing low-pressure xenon gas. The spectral energy distribution of this light source is generated by the electrical current arcing through the xenon gas plasma between the electrodes, and is modified by filters.

REFERENCES

1. J. Rediske. Improved look and appearance for wood-plastic composites after weathering. In: *Proceedings of Progress in Woodfibre-Plastic Composites.* Canadian Natural Composites Council, University of Toronto, Toronto, Canada, May 23–24, 2002.
2. K. O'Connor. WPC's: increasing need for color stability. In: *Proceedings of WPC Conference 2004 "Realizing the Full Potential,"* Baltimore, MD, October 11–12, 2004.
3. C. Vasile and M. Pascu. *Practical Guide to Polyethylene*, Rapra Technology, UK, 2005, p. 79.
4. B.M. Klemann. Correlations between xenon arc accelerated weathering tests and outdoor weathering. *Material Testing Product and Technology News,* Vol. 35, Issue 75, pp. 1–9.
5. L.S. Crump. *Proceedings of the 51st Annual Conference of the SPI Composites Institute*, Session 22, Society of the Plastic Industry, Washington, DC, 1996, pp. 1–34.
6. D.R. Bauer. *Polymer Degradation and Stability, Vol. 69*, Ingenta Connect Publications—Bath, Oxford, Providence, Cambridge, 2000, pp. 307–316.
7. D.S. Bajwa and D. Bruce. Improvements in weathering characteristics of wood-plastic composites. In: *8th International Conference on Woodfiber Composites*, Forest Products Society, Madison, WI, May 23–25, 2005.
8. J. Fabiyi, A. McDonald, M. Wolcott, and K. Englund. Chemical changes that occur during the weathering of wood plastic composites. In: *8th International Conference on Woodfiber Composites*, Forest Products Society, Madison, WI, May 23–25, 2005.
9. N.M. Stark. The effect of weathering variables on the lightness of HDPE/WF composites. In: *8th International Conference on Woodfiber Composites*, Forest Products Society, Madison, WI, Madison, May 23–25, 2005.
10. T. Steele and L. Davis. UV stabilization in wood fiber systems. In: *Proceedings of Progress in Woodfibre-Plastic Composites,* Candadian Natural Composites Council, University of Toronto, Toronto, Canada, May 23–24, 2002.
11. S. Goldstein. Stabilization of wood composites. In: *Proceeding of Additives 2006, 15th International Conference*, ECM, Plymouth, MI, Las Vegas, NV, January 30–February 1, 2006.
12. N.M. Stark. Factors influencing the weatherability of wood-polyethylene composites. In: *Proceedings of The Global Outlook for Natural Fiber & Wood Composites 2003*, Intertech, Portland, ME, New Orleans, LA, December 3–5, 2003.

17

RHEOLOGY AND A SELECTION OF INCOMING PLASTICS FOR COMPOSITE MATERIALS

INTRODUCTION: RHEOLOGY OF NEAT AND FILLED PLASTICS, COMPOSITE MATERIALS, AND REGRINDS

It is a very seldom situation when a composite manufacturer uses just one stream of its principal plastic. Typically, composite materials are manufactured using variable supplies of a polymer. Manufacturers are maneuvering making a choice between various suppliers, as they always consider cost of plastic and its available properties, listed in the material specification.

However, there is not much information available about plastic properties besides density and melt flow index (MFI). Having no consistent grade or/and source of plastics, a manufacturer faces from time to time a significant variation in the flowability of a polymer that is already purchased, sometimes in amounts of rail cars. These variations lead to improper process settings for production. This in turn results in a variety of defects on the surface of extruded profiles, such as sharkskin (fish skin), surface roughness, edge tearing, and so forth, and other property inconsistency in the final product.

Clearly, the problem originates from the poor melt flowability of the composition at the processing conditions and can be preventively identified by considering rheology of the filled composition, or maybe even of the plastic itself, at an earlier step of plastic selection.

Another similar situation happens when one of the principal ingredients of the composition is regrind, particularly when regrind is obtained from reclaimed composite profiles, returned from a dealer or a distributor due to some deviations

Wood-Plastic Composites, by Anatole A. Klyosov

in technological processes, earlier or current. In these cases a combined plastic (say, virgin and regrind) is characterized by a broader molecular weight distribution (MWD) of the plastic compared to the initial plastic, that is, a distribution often shifted to a shorter polymeric chains. The resulting formulation can have not only lower MFI but also different flowability in the extruder die.

Hence, the key questions are as follows: how to identify a plastic, which is appropriate for the given technological process before it gets to the industrial manufacturing of composite materials? Which supplier to choose? How to identify a "bad" regrind before it is mixed and processed along with the basic formulation?

As these questions are directly related to the melt flowability, rheology might be an answer. However, which rheological parameter to look at? At what conditions? There are too many of them.

This chapter will show that there are at least three rheological criteria to use:

- The power-law index, using either capillary or dynamic rheometers, or both;
- The storage modulus and the loss modulus, using a dynamic rheometer; and
- Melt fracture measurements, using a capillary rheometer.

In order to illustrate these approaches, we need to consider some basic concepts of melt rheology.

BASIC DEFINITIONS AND EQUATIONS

Shear Rate, Shear Stress, Shear Viscosity, Dynamic Viscosity, Apparent Viscosity, Limiting Viscosity

The fundamental feature of a plastic hot melt is that it moves differently in the direction of flow in a sense that it is characterized by a gradient in velocity per unit normal distance (normal, i.e., perpendicular to the direction of the flow).

The occurrence of velocity differences across the flow direction is called *shearing*, and the difference in velocity per unit normal distance (velocity gradient) is called *shear rate* or *rate of shear deformation* ($d\gamma/dt$, or $\dot{\gamma}$, see below).

This shearing in turn creates a hot melt deformation, or *stress*. The stress required to bring about a shearing type of a hot melt deformation (a required force divided by the area over which it works) is called *shear stress* (τ, see below). In case of capillary flow, apparent shear stress is the measured resistance to the flow through a capillary die.

The resistance to shear flow, equal to the ratio of shear stress to shear rate, is called *shear viscosity* or *dynamic viscosity* (η, see below).

This is expressed in one of Isaac Newton's many laws, which states that the shear stress τ is directly proportional to the shear rate, that is, to the rate at which the fluid is strained. The shear rate by definition is $d\gamma/dt$, where γ is the shear strain. Using the symbol $\dot{\gamma}$ for $d\gamma/dt$, we obtain

$$\tau = \eta\dot{\gamma} \tag{17.1}$$

where η is the constant of proportionality, called the *dynamic viscosity* or *shear viscosity*.

$$\eta = \tau/\dot{\gamma} \tag{17.2}$$

At a high shear viscosity, even low shear rate creates a high shear stress. On the contrary, if a high shear rate creates a low shear stress, this corresponds to a low shear viscosity.

Hence, when the shear is directly proportional to the shear rate, such a fluid is called "Newtonian." However, polymers are typically non-Newtonian fluids, and nonlinearity between shear rate and shear stress is typically observed. In these cases the viscosity in Eq. (17.2) is called *apparent viscosity*, as it does not reflect some nonlinearity between shear rate and shear stress.

In other words, *apparent* viscosity (as well as other *apparent* values in polymer rheology, such as apparent shear rate and apparent shear stress) is a value calculated assuming Newtonian behavior and considering all pressure drops within the capillary (when using a capillary rheometer). A nonlinearity between shear rate and shear stress is typically observed for polymer melts. The fluid may behave like Newtonian at a very low shear rates to give a *limiting viscosity* η_0.

The apparent viscosity is no longer a constant and becomes a function of the shear rate. The Eq. (17.2) still holds, however, in the following form:

$$\eta_c = d\tau/d\gamma \tag{17.3}$$

and η in Eq. (17.2) becomes the slope of the secant line from the origin to the shear stress at the given value of the shear rate.

In SI system viscosity is measured in Pascal-seconds (1 Pa.s = 10 poise). In English system of measurements, the load of 1 lb/in.^{2} (1 psi) is equal to 6895 Pa. One Pascal is 1 N/m^2 or 1.02×10^{-5} kg/cm^2 or 1.45×10^{-4} psi.

It is known that the viscosity of a filled system is increased proportionally to the volume fraction of the filler. This rule was described 100 years ago by Einstein [1] and since then is known as Einstein equation:

$$\eta_r = \eta_m (1 + 2.5\phi)$$

where
η_r is the relative viscosity,
η_m is the matrix viscosity, and
ϕ is the filler volume fraction.

There are some strict limitations of the above equation, and it is shown here mainly on historical reasons. Commercial composite materials typically have the viscosity higher than predicted by the above equation, and it is progressively higher with the increased filler loading. This is due to the fact that Einstein equation is strictly limited to dilute solution of rigid spheres, where buoyancy is not important.

Shear-Thinning Effect and the Power Law Equation

Flowability of plastics is largely determined by the dependence of viscosity on shear rate. Viscosity of water, for example, does not change with shear rate. When water moves through a capillary, fast or slow, its viscosity is same. In a forced oscillation rheometer, parallel plates immersed in water can move fast or slow, but the viscosity of water remains still the same. Therefore, a plot of viscosity against shear rate looks as a flat straight line, parallel to the horizontal axis (Fig. 17.1).

It is not so in polymer hot melts.

Below it will be shown that in polymer fluids the slope in (log viscosity vs. log shear rate) coordinates, such as in Figure 17.1, is always negative (see Fig. 17.2). The more negative the slope, the lower the coefficient, showing how much the polymer fluid is deviating from behavior of the Newtonian fluid. For a Newtonian fluid this coefficient, which is called "the power-law index," is equal to 1. For some "moderately" non-Newtonian fluids, the power-law index is equal to 0.9–0.8. For strongly non-Newtonian fluids, it is equal to 0.2–0.3.

As will be shown in a few paragraphs, the slope in (log viscosity vs. log shear rate) coordinates is equal to $n - 1$, where n is the power-law index of shear viscosity. As one can see from Figure 17.1, for water, this slope $(n - 1) = 0$, and $n = 1$. That is why water is called Newtonian fluid.

These polymer fluids are typically called non-Newtonian or nonlinear fluids, as they show a decrease of viscosity with increasing fluid velocity (shear rate). This is also known as shear thinning. This behavior results from the fact that the polymer molecules are long and have many contact points interacting with each other, or entanglements. These molecular interactions determine the viscosity of polymers. When one moves them slowly, viscosity is still relatively high. For example, it is

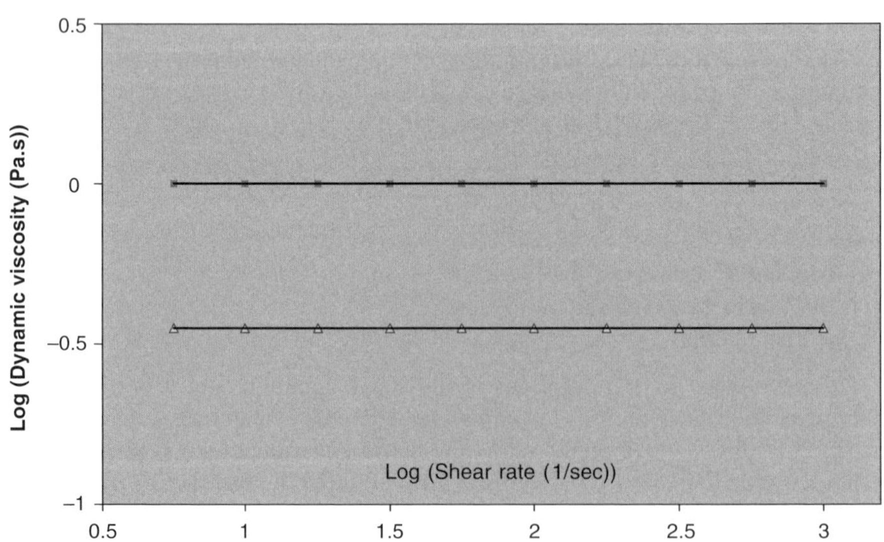

Figure 17.1 Dynamic viscosity of water at 20°C (top line) and 80°C (bottom line).

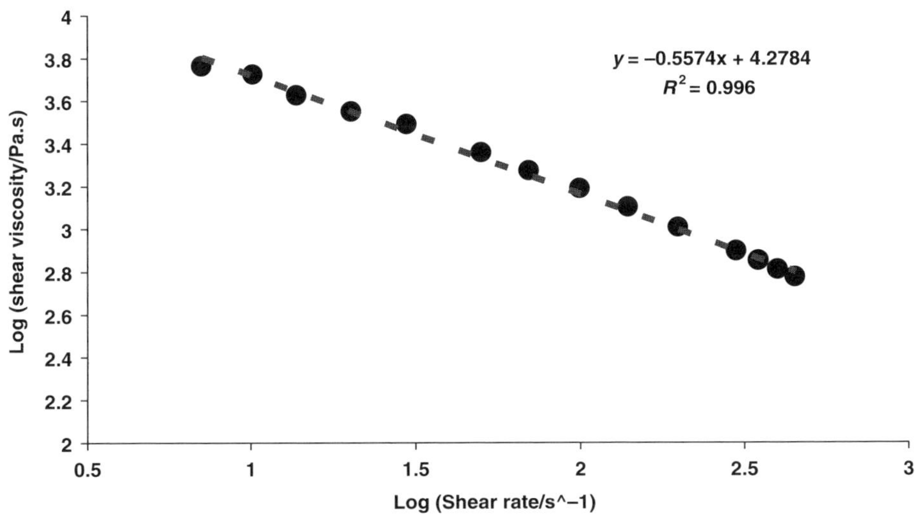

Figure 17.2 Dependence of the apparent shear viscosity on the shear rate for HDPE (melt index 0.5 g/10 min).

observed with a thick gel of starch when still, or stirred slowly. When the polymer system moves faster, entanglements of the long molecules are progressively disturbed, the number of contacts along molecules decreases, and with it, viscosity decreases. This is a reversible process, and when the shear rate decreases, the viscosity again increases.

Plot of log viscosity versus log shear rate, or a double logarithmic plot, has a stretched concaved down shape (or stretched s-shape), where a relatively long linear middle part that ends with nonlinear (shorter) parts on both sides of the straight line, at too low and too high shear rates (Fig. 17.2). The slope of the plot is negative (the higher the shear rate, the lower the viscosity) and equal to $n - 1$. In Figure 17.2 the slope is (approximately) equal to -0.56, and $n = 0.44$. The power-law index n for polymers is typically in the range of below 0.1 to about 0.8 (see below).

The linear part of the double logarithmic plot is described by the following equation (the power law equation):

$$\eta = m\dot{\gamma}^{n-1}, \tag{17.4}$$

or

$$\log \eta = \log m - (1 - n) \log \dot{\gamma} \tag{17.5}$$

where η is viscosity, $\dot{\gamma}$ is shear rate, m is the consistency index (the viscosity at unit shear rate, i.e., at $\dot{\gamma}=1\ s^{-1}$) and n is the power-law index.

As

$$\tau = \eta\dot{\gamma},$$

in this linear range where the power law equation holds, the shear stress is equal to

$$\tau = m\dot{\gamma}^n, \tag{17.6}$$

For Newtonian fluids, such as water ($n = 1$),

$$\eta = m\dot{\gamma}^0, \text{ or } \eta = m,$$

that is, the viscosity is constant at a given temperature and does not depend on a flow velocity.

Hence, the plot log viscosity versus log shear rate for a non-Newtonian fluid, which is derived from the power law equation, gives two important parameters: the consistency index m and the power-law index n. The consistency index numerically shows the viscosity at a shear rate of 1.0 s^{-1} and is determined from the interception point with the vertical axis at log $\dot{\gamma} = 0$; in Figure 17.2 the log (consistency index) is equal to 4.278, and the consistency index is 18,985 Pa.s^n. The power-law index, which is calculated from a slope, as in Figure 17.2, is a measure of the degree of shear thinning.

As a rule, viscosity versus shear rate data in a wide range of shear rates do not accurately fit the simple power law Eq. (17.4), that is, do not accurately follow linear relationship for more than two to three logarithmic units in the shear rate. The neat plastics exhibit a Newtonian plateau in the low shear rate (or frequency) region, showing high apparent power-law index and a more pronounced shear thinning effect at higher shear rates or frequencies. Such pattern can be fitted to the Cross model [2]:

$$\eta = \eta_0/[1 + (\lambda\dot{\gamma})^{1-n}] \tag{17.7}$$

that is very close to the Carreau equation [3]:

$$\eta = \eta_0/[1 + (\lambda\dot{\gamma})^2]^{(1-n)/2} \tag{17.8}$$

Here λ is a time constant and is often interpreted as the relaxation time of the polymer, particularly in the filled composites. It significantly increases with wood flour content in filled high density polyethylene (HDPE) [4]. The authors suggested that it is a result from the increased relaxation time of the reduced amount of polymer in the filled composite.

The relaxation time in rheology, and particularly in rotational rheometry, is a measure of the rate at which the viscoelastic fluid changes in response to the change in flow due to the oscillatory movements of the fluid. Typically, an apparent relaxation time is defined as the time for the disturbances to decrease by a factor of $1/e$, that is, 0.368.

These movements induce anisotropy across the fluid by rearrangement of its microstructure. The anisotropy varies during the oscillatory cycle, and its degree depends on the period of oscillation and microstructure of the fluid, which in turn depends on the nature of the polymer, its molecular characteristics, the filler, and its amount.

When λ is small (less than 1–2 s), and the shear rate is low, the denomination in Eqs. (17.7) and (17.8) is attending unity, and viscosity is attending the limiting shear viscosity η_0. For example, when λ is 1.6 s (as of neat polyethylene in some experiments [4], $n = 0.5$, and the shear rate is 0.1 s^{-1}, the Eq. (17.7) is

$$\eta = \eta_0/(1 + 0.4) = \eta_0/1.4$$

Equation (17.8) gives $\eta = \eta_0/1.006$, that is, practically equal to the limiting shear viscosity.

With a highly filled HDPE, when λ is, say, 16 s, the shear rate is 10^4 s^{-1} and $n = 0.5$, Eq. (17.7) becomes

$$\eta = \eta_0/(1 + 400)$$

Equation (17.8) gives practically the same value of $\eta = \eta_0/400$.

With a lower power-law index, say, $n = 0.2$, the apparent viscosity would still be close to the limiting shear viscosity at λ of 1.6 s and the shear rate of 0.1 s^{-1}. For these conditions, Cross equation (17.7) gives $\eta = \eta_0/1.23$, and Carreau equation (17.8) gives $\eta = \eta_0/1.01$. However, at λ of 16 s and the shear rate of 10^4 s^{-1} (see above), the apparent viscosity would be much lower, that is, the shear thinning effect would be much more pronounced. Both Cross and Carreau equations give $\eta = \eta_0/14{,}500$, that is, 36 times lower apparent viscosity compared to that at $n = 0.5$.

This means that at high shear rates (even at moderate λ values), both Cross and Carreau equations are transformed to

$$\eta = \eta_0 \lambda \dot{\gamma}^{\,n-1},$$

or

$$\eta = m\dot{\gamma}^{n-1},$$

that is,

$$\log \eta = \log m + (1-n) \log \dot{\gamma}$$

where η is viscosity, $\dot{\gamma}$is shear rate, m is the consistency index (a product of the limiting shear viscosity η_0 and the time constant λ, apparently the relaxation time of the polymer), and n is the power-law index.

Volumetric Flow Rate and a Pressure Gradient Along the Capillary

It should be noted that the shear rate is related to the volumetric flow rate. In a capillary rheometer, for example, the shear rate at the capillary wall is related to the volumetric flow rate as shown in the equations:

$$\dot{\gamma} = 32Q/\pi D^3 \tag{17.9}$$

or

$$\dot{\gamma} = 4Q/\pi r^3 \tag{17.10}$$

where
$\dot{\gamma}$ is the apparent shear rate (s^{-1}),
D or r is the diameter or the radius of the die, respectively (mm), and
Q is a volumetric flow rate, that is, quantity of fluid extruded per time (mm^3/s).

Let us give some examples, using two different systems, in millimeters and inches.

Example one: At the volumetric flow rate of 1 cm^3 (1000 mm^3)/s and a die diameter of 1 mm (0.1 cm), the shear rate is 1.02×10^4 s^{-1}. Note that the shear rate is measured in reciprocal seconds (1/s, or s^{-1}).

Another example: At the linear flow rate of 8.3 in./min and the die diameter of 1 mm (0.0394 in.), the volumetric flow rate is $R^2 \times 8.3$ in.^{3}/min, that is, 0.01 in.^{3}/min, or 1.683×10^{-4} in.^{3}/s. Using formula (17.8), we obtain that the shear rate is 28 s^{-1}.

With the die of 2 mm in diameter, at the same linear flow rate of 8.3 in./min, the shear rate will be 14 s^{-1}.

Using V_L as a linear flow rate, formulas (17.9) and (17.10) become

$$\gamma = 4V_L/r \tag{17.11}$$

Obviously, the volumetric flow rate in a tube is simply the volume swept by the piston in unit time,

$$Q = \pi R^2 V,$$

where R and V are the radius and velocity of the piston, respectively.

Combining this equation with (17.10), we obtain

$$\dot{\gamma} = 4R^2 V/r^3 \tag{17.12}$$

where R is the radius of the piston and r is the radius of the die.

Another way to describe the volumetric flow rate is by using the Poiseuille (or Hagen–Poiseuille) equation:

$$Q = \Delta P \pi r^4/8\eta L \tag{17.13}$$

Where ΔP is the pressure drop (pressure gradient) along the length of the tube (capillary), between the inlet and exit; r and L are the radius and length of the capillary, and η is the viscosity of the fluid.

As

$$\eta = \tau/\dot{\gamma}, \quad (17.14)$$

$$\dot{\gamma} = 4Q/\pi r^3, \quad (17.10)$$

and (17.13) can be presented as

$$\eta = (\Delta Pr/2L)\,(\pi r^3/4Q), \quad (17.15)$$

then, combining (17.10) and (17.14) with (17.13), we obtain

$$\tau = \Delta Pr/2L. \quad (17.16)$$

In other words, the apparent shear stress is determined by the pressure drop along the length of the capillary (i.e., pressure at the entrance of the measuring capillary) and the radius and length of the capillary.

The relationships (17.10), (17.14), and (17.16) are very important for the generation of flow curves for hot melts using capillary rheometry.

These and other equations shown in this chapter assume that the velocity at the wall of the capillary is zero, that is, there is no slip at the wall. Other typical assumptions are as follows: the flow is laminar and isothermal, it is time-independent, steady along the capillary, the velocity of any fluid element is a function of the radius only, and the pressure gradient along the tube is constant. These assumptions will be discussed later.

Walls Slip Phenomenon

It is well known that HDPE and PVC, as well as many polymer suspensions and filled composites, exhibit slip at the capillary wall. Considering slip at the wall, the volumetric flow rate Q through the die is given by

$$Q = \dot{\gamma}_r \pi D^3/32 + \pi R^2 V_s \quad (17.17)$$

where

$\dot{\gamma}_r$ is the true shear rate (which differs from the apparent shear rate as shown in Eq. (17.18) and equals to the apparent shear rate at a very large value of R) and

V_s is the slip velocity.

The first term in Eq. (17.17) is the flow rate due to shearing, and the second one is the flow rate due to slip. The apparent shear rate in case of wall slip is given by the Mooney equation [5]:

$$\dot{\gamma} = \dot{\gamma}_r + 4V_s/R \quad (17.18)$$

Hence, a plot of $\dot{\gamma}$ against $1/R$ gives $\dot{\gamma}_r$ as the intercept and $4V_s$ as the slope. This works only if the shear stress is not a function of R; that is, why L/D ratio of the die should be kept constant [6].

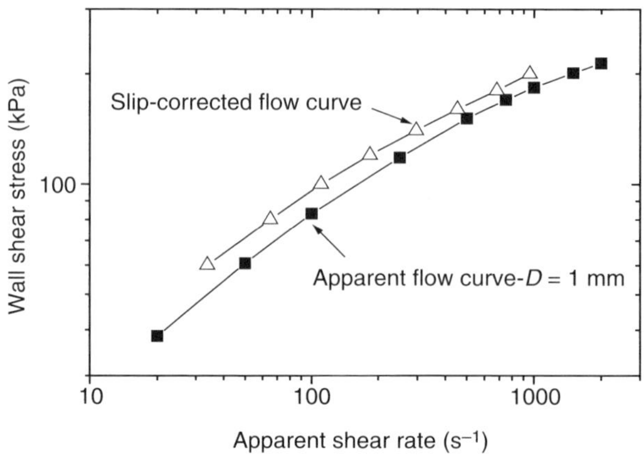

Figure 17.3 Slip corrected flow curve of 50%-wood-flour-filled polypropylene.

Figure 17.3 shows slip corrected flow curve for 50% wood flour-filled polypropylene. One can see the importance of the wall slippage for filled polymers, which is typically ignored in viscosity calculations in the literature.

The Rabinowitsch Correction

The Rabinowitsch correction accounts for the fact that the true shear rate is often larger (because of shear thinning) than the *apparent* shear rate for non-Newtonian materials. Hence, for any non-Newtonian fluid the expression for the wall shear rate is given not by Eq. (17.11), but as

$$\gamma_w = 4Q/\pi r^3(3/4 + 1/4\ [\mathrm{d}\ \ln Q/\mathrm{d}\ \ln\ \tau_w]) \tag{17.19}$$

where τ_w is the wall shear stress. If the non-Newtonian fluid obeys the power law, in which the shear stress depends upon the shear rate to the nth power (see Eq. 17.6), Eq. (17.19) takes the form

$$\gamma_w = 4Q/\pi r^3(3n + 1)/4n \tag{17.20}$$

where n is the power-law index. Clearly, for Newtonian fluids with n equals unity, Eq. (17.20) becomes Eq. (17.10). The quantity $(3n + 1)/4n$ is often used for transformation between the apparent and true (dynamic) viscosities, and it is known as the *Rabinowitsch correction*. The ratio $4Q/\pi r^3$ is called the apparent wall shear rate.

Note that the shear rate is always maximum at the wall, and the fluid velocity is always zero at the wall. On the contrary, the shear rate is zero and the velocity is maximum at the center of the capillary (Fig. 17.4).

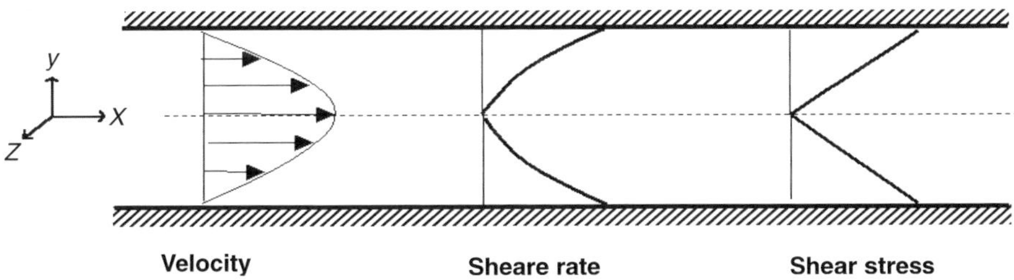

Figure 17.4 Fluid velocity, shear rate, and shear stress profiles in capillary flow [7].

Just to give an idea how different are the apparent and true shear rates, we will notice that the higher the shear thinning effect, the larger the difference. For example, at the power-law index $n = 0.8$, the difference in shear rates and, hence, apparent and true viscosities is 6.25%. For $n = 0.2$, the difference is 100%.

The following figure (Fig. 17.5) shows apparent and corrected shear rates.

The upper limit on the shear rate in capillary viscometers is about 10^5–10^6 s^{-1}, and the lower limit is about 1-10 s^{-1}.

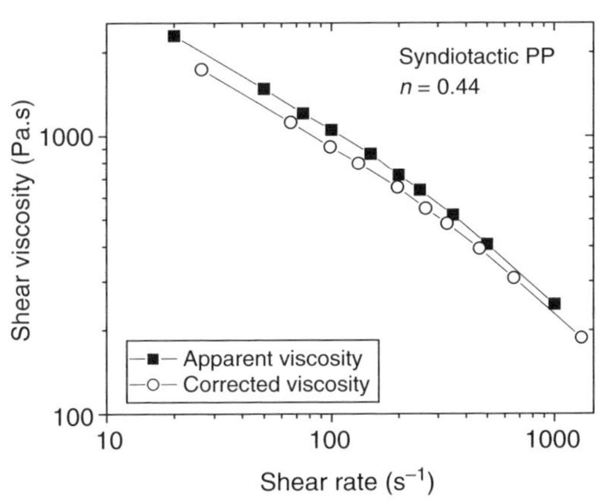

Figure 17.5 Data for syndiotactic polypropylene at 180°C, $n = 0.44$, correction factor $(3n+1)/4n=1.32$.

ASTM RECOMMENDATIONS IN THE AREA OF CAPILLARY RHEOMETRY

There are five principal ASTM standards related to melt flow rates of thermoplastics. All of them deal with capillary rheometers one way or another. The capillary rheometer is the most common device for measuring viscosity of polymer melts. In deriving the viscosity relations, the following important assumptions should be taken into consideration:

1. The flow is fully developed, steady, isothermal, and laminar.
2. The flow is unidirectional, that is, fluid velocity has only one component in the direction of the capillary axis (cylindrical coordinate system).
3. No slip at the wall.
4. The fluid is incompressible and viscosity is independent of pressure.

A simpler version of a capillary rheometer is known as melt indexer, or an extrusion plastometer.

ASTM D 1238-04, "Standard Test Method for Melt Flow Rates of Thermoplastics by Extrusion Plastometer"

This standard deals with the melt flow indexer, which is a simplified version of the capillary rheometer. The MFI is a popular parameter in the plastic processing industry and serves for specification of materials. Melt index is inversely proportional to the viscosity. The test procedure describes the measurement of the weight in grams of a material extruded in 10 min through a die of length 8 mm and diameter 2.095 mm, using a standard load of 2.16 kg or 5 kg on a piston 9.55 mm in diameter. The result is a mass flow rate at a single stress, that is, a single point on a flow curve.

For polyethylene, a standard test designation is 190/2.16, which means that the test is conducted at 190°C using 2.16-kg load (including piston). Besides, the following conditions have been found satisfactory for polyethylene: 125/0.325, 125/2.16, 250/1.2, 190/0.325, 190/10, 190/21.60, and 310/12.5. However, 190/2.16 is the most often employed conditions.

For polypropylene, standard test conditions are 230/2.16.

Most of plastics used in composite building materials have MFI that fall generally between 0.15 and 50 g/10 min, the range that is considered by ASTM D 1238 ("Procedure A"). Plastics that have MFI below 1 g/10 min (MFI 1.0) are commonly called "fractional melt" plastics.

As the shear rate and shear stress at standard tests conditions in the melt flow indexer are very low (τ_w = 19,401 Pa at 2.16-kg load for instance) compared to the range usually observed in plastic processing, the MFI is of very limited value for plastics processing behavior. Furthermore, the flow rate obtained with the extrusion plastometer is not a fundamental polymer property. MFI is an empirically defined parameter. Many plastics can have the same MFI, but otherwise they can be very different in their chemical structure, molecular weight and MWD, the power-law index, and melt flowability in general. The same MFI just means that the melts show the same approximate volume flow rate through the die at given conditions. In other words, they must all have a common point on the flow curve. Other parts of the curve, over the whole shear rate range, can be totally different.

Nevertheless, the melt flow indexer is a very useful instrument besides measuring MFI. It can be used for studying plastic thermal degradation, by consecutive running of the same material or by running hot melt slow and recording MFI over time. By doing so, various antioxidants and their various amounts can be tested and optimized. It can be used for studying flowability of plastics at different temperatures, hence, temperature dependence of flowability. It can be used for studying flowability of plastics at different loads. Overall, it can be used for developing an elaborated QC for incoming plastics by creating a database of patterns for different plastics (or for "same" plastics but from different vendors), having in particular the same MFI but different flowability at various test conditions.

In order to overcome the shortcomings of the melt index method, high load MI (HLMI) has been introduced to indicate the grams of polymer pushed out of the barrel under the action of a heavier weight, usually 10 kg (or 21.6 kg). The ratio of the HLMI/MI indicate how shear thinning the resin is. The higher the ratio, the more shear thinning the polymer [8].

Precision of MFI measurements is usually quite good. ASTM D 1238-04 lists several examples. For instance, three polyethylene samples tested in nine different laboratories (a round robin test) showed MFI of 0.270 ± 0.008, 0.400 ± 0.012, and 2.040 ± 0.026 g/10 min for within-laboratory standard deviations of the average, that is, within 1–3% of the average, and 0.270 ± 0.022, 0.400 ± 0.038, and 2.040 ± 0.079 g/10 min for between-laboratory standard deviations of the average, that is, within 4–10% of the average, respectively.

ASTM D 3835-02, "Standard Test Method for Determination of Properties of Polymeric Materials by Means of a Capillary Rheometer"

This standard covers measurement of the rheological properties of polymers with both stable and unstable melt viscosity parameters at various temperatures and shear rates. The test procedure lists typical test temperature conditions: for polyethylene 190°C, for polypropylene 230°C, for poly(vinyl chloride) 170–205°C, however, this indicates that the most useful data are generally obtained at temperatures consistent with processing experience. The test method also prescribes using the Rabinowitsch shear rate correction (see above) and indicates that the basic rheology equations (17.10), (17.15) and (17.16) yield true shear rate and true viscosity for Newtonian fluids only; for non-Newtonian fluids only the *apparent* shear rate and viscosity are obtained.

The ASTM test method gives several examples of precision of rheological measurements, conducted by 13 laboratories (a round-robin test). The precision is usually very good. For instance, for polypropylene copolymer at the highest set rate of 3162 s^{-1} and stress of 150 kPa, apparent viscosity was 47.3 ± 0.6 Pa.s for within-laboratory standard deviation of the average and 47.3 ± 1.8 Pa.s for between-laboratory standard deviation of the average, that is, 1.3 and 3.9%, respectively. For the lowest set rate of 3.2 s^{-1} and stress of 5 kPa, apparent viscosity was 1550 ± 120 Pa.s for within-laboratory standard deviation of the average and 1550 ± 220 Pa.s for between-laboratory standard deviation of the average, that is, 8 and 14%, respectively.

In a similar round-robin test performed for low-density polyethylene at the highest set rate of 3162 s^{-1} and stress of 320 kPa, apparent viscosity was 101.2 ± 0.7 Pa.s for within-laboratory standard deviation of the average, and 101.2 ± 5.2 Pa.s for between-laboratory standard deviation of the average, that is, 0.65 and 5.1%, respectively. For the lowest set rate of 3.2 s^{-1} and stress of 15 kPa, apparent viscosity was 4652 ± 126 Pa.s for within-laboratory standard deviation of the average and 4652 ± 803 Pa.s for between-laboratory standard deviation of the average, that is, 2.7 and 17.3%, respectively.

ASTM D 5422-03, "Standard Test Method for Measurement of Properties of Thermoplastic Materials by Screw-Extrusion Capillary Rheometer"

The standard aims at description of the material behavior that is more likely to occur in factory processing compare to that obtained with a piston-type capillary rheometer (ASTM D 3835).

The test method addresses two most important corrections to measure rheological values, Rabinowitsch (see above) and Bagley. The Bagley correction factor is applied to the apparent wall shear stress, in order to obtain the actual shear stress at the wall of the capillary die. The Bagley correction compensates for energy losses at the entrance and exit of the die and also removes the influence of any static pressure in the system that does not vary with die length. It is applied as though it were an additional length of capillary; hence, it is often termed "end effect." The details of the Bagley correction are outside of the scope of this chapter and are described in ASTM D 5422, as well as details of the standard test method procedure. The procedure allows for the determination of apparent shear rate, apparent shear stress, apparent viscosity, corrected shear stress, corrected shear rate, corrected viscosity, shear sensitivity (the power-law index), and entrance/exit effects. All these flow parameters are sensitive to molecular weight and molecular-weight distribution of polymers; therefore, the test method can distinguish differences between lots of the "same" plastic. Furthermore, these flow parameters are sensitive to the type and amount of filler in composite materials, and additives and other ingredients of composites.

The ASTM test method gives an example of precision of rheological measurements, based on a single laboratory repeatability study using one LDPE homopolymer resin. The laboratory used a computerized single-screw rheometer having three heated zones. The precision/repeatability was fairly good. For instance, at the highest fixed corrected shear rate of 12,500 s^{-1}, the corrected average shear stress was 251,900 ± 2,676 Pa and corrected viscosity 20.2 ± 0.2 Pa.s, with repeatability of 3 and 2.9%, respectively. For the lowest fixed corrected shear rate of 1000 s^{-1}, the corrected average shear stress was 120,300 ± 2,617 Pa and corrected viscosity 120.3 ± 2.6 Pa.s, with repeatability of 6.1 and 6.1%, respectively.

ASTM RECOMMENDATIONS IN THE AREA OF ROTATIONAL RHEOMETRY

Although the flow in a capillary rheometer is regarded as pressure-driven flow, namely, the shear is generated by the pressure difference along the capillary length, in rotational rheometers the shear is generated between a moving and a fixed surface [9].

ASTM D 4440-01, "Standard Test Method for Plastics: Dynamic Mechanical Properties Melt Rheology"

This test method employs nonresonant forced vibration techniques for determining the *complex viscosity* (see below) and viscoelastic characteristics of thermoplastic resins as a function of frequency, strain amplitude, temperature, and time. A wide range of frequencies can be used, typically from 0.01 to 100 Hz.

In these cases the relative velocity of the shearing plates is not constant but varies in a sinusoidal manner so that the shear strain and the rate of shear strain are both cyclic, and the shear stress is also sinusoidal. For non-Newtonian fluids, the stress is out of phase with the rate of strain. In this situation a measured *complex viscosity* (η^*) contains both the shear viscosity, or dynamic viscosity (η'), related to the ordinary steady-state viscosity that measures the rate of energy dissipation, and an elastic component (the imaginary viscosity η'' that measures an elasticity or stored energy):

$$\eta^* = \eta' + i\eta'' \qquad (17.21)$$

These two viscosities are calculated using the following equations:

$$\eta' = G''/\omega$$

$$\eta'' = G'/\omega$$

where ω is the fixed angular frequency of the oscillations in radians per second ($\omega = 2\pi f$, where f is the frequency in hertz). G' is the storage (elastic) modulus, and G'' is the loss (viscous) modulus. The ratio of the loss modulus to the storage modulus, G''/G', is known as tan δ and is an important characteristics of viscoelastic materials. In perfectly elastic materials G'' is zero, and in perfectly viscous materials (Newtonian fluids) G' is zero. Typically, both G' and G'' are functions of frequency and temperature.

The zero-shear viscosity, or the limiting viscosity, can be calculated from the loss modulus using the following equation:

$$\eta_0 \rightarrow G''(\omega)/\omega \qquad (17.22)$$

at the value ω approaching zero.

An instrument employed in the test method is either a cone-and-plate rheometer or a parallel-plate rheometer. In the first case, a flat, circular plate and a linearly concentric cone are rotated relative to each other. The cone is normally truncated so that there is no physical contact between the two. The fluid is in the space between the plate and cone. Either of the two members can be rotated or oscillated, and one measures the torque needed to keep the other member stationary. For both Newtonian and non-Newtonian fluids, the basic equations for cone-and-plate rheometers are as follows:

$$\tau = 3M/2\pi r^3 \tag{17.23}$$

$$\gamma = \omega/\alpha \tag{17.24}$$

$$N_1 = 2N/\pi r^2 \tag{17.25}$$

where
M is the torque,
N is the normal force needed to hold the apex of the truncated cone at the center of the disk (N is equal zero for a Newtonian fluid),
N_1 is first normal stress difference, which is a measure of the polymer elasticity,
ω is angular rotation in radians per second,
r is the plate radius (typically about 25 mm),
α is the cone angle (usually a few degrees).

Parallel-plate rheometers are often more useful for studying rheology of filled polymers or composite materials, particularly when the size of the fillers is comparable to the distance between the truncated cone and the surface of the plate. Again, the torque M and the normal force N tending to separate the two plates are measured. In steady-shear flow, the shear rate and the shear stress at the edge of the disks located at $r = R$ are given by

$$\gamma(R) = \omega R/h \tag{17.26}$$

and

$$\tau(R) = 3M/2\pi R^3\ [1 + (d \ln M)/3\, d \ln \gamma(R)] \tag{17.27}$$

where h is the vertical gap between the two plates [10].

ASTM D 4065-01, "Standard Practice for Plastics: Dynamic Mechanical Properties: Determination and Report of Procedures"

The standard describes laboratory practice and gives a summary of techniques and calculations used to determine dynamic mechanical properties, among them elastic and loss moduli of hot melts as indicative of the viscoelastic characteristics of a plastic. The standard practice indicates that elastic modulus decreases rapidly with increasing temperature (at constant or near constant frequency) or increases with increasing frequency (at constant temperature). A maximum is observed for loss modulus. These parameters are influenced by fillers and additives, their amounts, as well as by certain processing treatment.

TABLE 17.1 Zero-shear viscosity and weight average molecular weight (M_w) for linear polyethylenes at 150°C [11]

Molecular weight, M_w (Da)	Zero-shear viscosity (Pa.s)
35,000	100
100,000	6000
200,000	80,000
250,000	100,000

COMMON OBSERVATIONS

Neat Plastics

Molecular Weight of Polyethylenes and Viscosity of Their Hot Melts What are typical viscosity values for HDPE? Table 17.1 shows some data. Viscosity is often measured in SI system, in Pascal-seconds (1 Pa.s = 10 poise). In English system of measurements, the load of 1 lb/in.^{2} (1 psi) is equal to 6895 Pa. One Pascal is 1 N/m^2, or 1.02×10^{-5} kg/cm^2, or 1.45×10^{-4} psi.

Similar data, but at a higher temperature, are given in Table 17.2

Data in Table 17.2 for commercial grade HDPE can be described using a formula

$$\eta_0 = C\, M_w^{3.4} \tag{17.28}$$

where $C = 8.1 \times 10^{-13}$ g/cm s [12, 13].

The power of 3.4 still represents one of the major unresolved problems in polymer physics. It is known, however, that for most polymers the formula (17.28) is valid for their molecular weights higher than 10,000–40,000.

Effect of Temperature on Viscosity Increase of temperature *decreases* the shear viscosity in such a way that the viscosity follows the "reverse" Arrhenius equation (17.29), to a good approximation. The Arrhenius equation in rheology is "reverse" because the common Arrhenius equation is typically applied to velocities

TABLE 17.2 Zero-shear viscosity and weight-average molecular weight for HDPE at 190°C [12]

Molecular weight, M_w (Da)	Zero-shear viscosity (Pa.s)	Log η_0
40,000	40	1.60
55,000	100	2.00
100,000	1,000	3.00
200,000	10,000	4.00
400,000	70,000	4.85

of chemical reactions, which *increase* with temperature. Therefore, "rheological" Arrhenius equation has an unusual (for chemists) positive power form:

$$\eta_0 = K e^{E/RT} \tag{17.29}$$

Of course, the activation energy in Eq. (17.29) has a principally different meaning from that in equations for temperature dependencies of rates of chemical reactions, such as in chemical kinetics.

In the above equation, K is a constant characteristic of the polymer and its molecular weight and its distribution, branching, and so forth, E is the "activation energy" of the flow, R is the universal gas constant (1.987 cal/mol grad or 8.314 J/mole grad), and T is the absolute temperature. As it was noticed above, this equation works to a good approximation, primarily because the temperature range in which it is typically applied is rather narrow. Below it, the plastic is solid, and above it, the plastic decomposes. Hence, this range commonly does not exceed 50–60°K (or Celsius, on that matter).

Compared to the velocities of chemical reactions, the viscosity of polyolefins changes with temperature much less, hence, lower energy of activation values for hot melts. The so-called temperature coefficient for chemical reactions, that is, a change of the velocity by each 10°C, is typically between 2 and 3 that corresponds to energy of activation between 28 and 44 kcal/mol (116–183 kJ/mol), if measured between 170 and 180°C. This can be compared to some typical (generic) values of the energy of activation for the shear viscosity of HDPE (6.3–7.0 kcal/mol, or 26–29 kJ/mol), LDPE (11.7 kcal/mol, or 49 kJ/mol), polypropylene (9–10 kcal/mol or 38–42 kJ/mol) ([13], p. 46). In other words, the temperature coefficient for the viscosity of said plastics is equal to 1.18 for HDPE, 1.34 for LDPE, and 1.26 for PP.

The equation that expresses the temperature coefficient through the activation energy (E) and the temperature range is as follows:

$$E = R \ln k/(1/T_1 - 1/T_2)$$

where

k is the temperature coefficient,

T_1 and T_2 (in °K) differ by 10°C (or 10°K, i.e., the same temperature range).

As a difference of reciprocal T_1 and T_2 depends on the temperature (e.g., it is 4.98×10^{-5} between 170 and 180°C [443 and 453°K] and 3.33×10^{-5} between 270 and 280°C [543 and 553°K]), the temperature coefficient and the activation energy in turn depend on temperature.

The above generic values of temperature coefficients for molten plastics are even slightly higher than those observed in reality. For example, Table 17.3 lists data for eight HDPE preparations of various origins (obtained from different suppliers). One can see that even over a 30° temperature span, the viscosity does not changes much. In fact, between 170 and 200°C, the average difference in viscosity is 1.33 ± 0.04, and between 200 and 230°C, the average difference is 1.31 ± 0.05. In both the cases

TABLE 17.3 The consistency index of HDPE of various origins at three different temperatures

HDPE	Consistency index (Pa.s)		
	170°C	200°C	230°C
Chevron CHVX891180	28,900	22,900	16,600
Chevron CHVX896880	25,700	19,500	15,800
Equistar EQUX621048	17,000	12,300	9570
Petromont PSPX7006	19,100	13,800	10,500
Equistar EQUX631675	17,800	13,200	9770
Equistar GPLX74618	24,600	18,700	14,100
Equistar EQUX632225	18,200	14,100	10,700
Dupont/Sherman Tyvek Repro, Lot 45104	26,000	20,000	15,800

Galaxy V capillary rheometer (Kayness Inc., Model 8025). Consistency index is the viscosity at unit shear rate, that is, at $\dot{\gamma} = 1\ s^{-1}$.

the HDPE viscosity temperature coefficient is equal to 1.10/10°C. This corresponds to the activation energy of approximately 4 kcal/mol (17 J/mol).

The Power-Law Index of Some Neat Plastics As it was explained in the section "Shear-thinning Effect and the Power Law Equation," the higher the deviation of a molten plastic from Newtonian behavior, the lower its power-law index and the steeper the dependence of its viscosity on shear rate. The lower the power-law index, the more sensitive the shear viscosity to flow speed. As it will be shown in this chapter later, the lower the quality of a regrind, the lower (often) its power-law index. As can be seen from Tables 17.4 and 17.5, PVC, PS, ABS, and PMMA have particularly low power-law index compared to PET or PE, hence, that lowers the window of processability for the former plastics in terms of speed. Both the

TABLE 17.4 The power-law index, shear viscosity, and other useful properties of a number of neat generic polymers [14]

Polymer	The power-law index (n)	Density (g/cc)	Glass transition (°C)	Melting point (°C)
PA-6.6	0.75	1.14	55	265
PA-6	0.70	1.13	50	225
PET	0.60	1.35	70	275
LLDPE	0.60	0.92	−120/−90	125
HDPE	0.50	0.95	−120/−90	130
LDPE	0.35	0.92	−120/−90	120
PP	0.35	0.91	−10	175
PVC	0.30	1.40	80	Decomp.
PS	0.30	1.06	101	Decomp.
ABS	0.25	1.02	115	Decomp.
PMMA	0.25	1.18	105	Decomp.

TABLE 17.5 The power-law index of certain neat polymers [15]

Polymer	The power-law index (n)
HDPE	0.15–> 0.5
LDPE	0.25–0.50
PP	0.19–0.44
PVC	0.11–0.38
ABS	0.20–0.37
Nylon	0.04–> 0.5
PS	0.15–0.42
PMMA	0.16–0.37

tables show similar information with respect to power-law index but from different sources. Hence, they show how different may be the data that are supposed to be the same.

The Power-Law Index and Molecular Weight Distribution For HDPE, n hardly depends on MWD M_w/M_n. Between M_w/M_n 4 and 80, the n values are scattered between 0.68 and 0.25. The n values for the lowest and the highest M_w/M_n ratios are equal to 0.36 and 0.38, respectively [16]. For PP the pattern is similar, and between M_w/M_n 3.5 and 25, n values are located in the narrow range of 0.19–0.44, though in a less scattered manner [16]. There was no correlation found between shear stress and HDPE molecular weight distribution ([16], p. 80).

Composite Materials

Rheology of Filled Plastics and Wood–Plastic Composites There are very few data published on rheology of filled plastics or multicomponent plastic-based composite materials. Most of them, in addition to those cited above, are related to long glass or graphite fiber-filled plastic [17–20].

The above references mainly describe rheological behavior of graphite fiber (length 0.5–16 cm) in poly-ether-ketone-ketone (PEKK) at 370°C. Authors conclude that the transient and steady-state rheological properties of these materials are different from the unfilled melt.

With such a few works published on rheology of composite materials, particularly on rheology of wood–plastic composites (WPCs), to use words "common knowledge" here is somewhat risky. However, there are some observations commonly known to researchers and manufacturers working in the field. For example, it is commonly known that fillers increase the viscosity of hot melts. It is generally known that composites have lower power-law index compared to their neat plastics. It is known that neat plastics swell more than the respective filled plastics. On the contrary, there are issues in rheology of composites and their plastics that remain to be solved, leaving alone those that are important for the industry but have not been addressed as yet in the literature. Those issues will be described below in the section "Almost Uncharted Areas of Composite and Plastic Rheology."

TABLE 17.6 Effect of filler and additives in HDPE (MFI 0.48 g/10 min) on the consistency index (η_0) [zero viscosity] and power-law index (n) determined in the parallel plate (dynamic shear) rheometer

HDPE neat and filled	Consistency index (Pa.s^n)	Power-law index	Consistency index (Pa.s)^n	Power-law index
	160°C		170°C	
Neat	12,300	0.54	12,700	0.53
Filled with 70% rice hulls	163,700	0.41	135,500	0.42
Filled with 60% rice hulls	90,700	0.41	66,700	0.44
Same plus 1% coupling agent	59,900	0.46	57,300	0.47
Same plus 1% lubricant	72,600	0.42	54,100	0.45
Same plus 1% coupling agent plus 1% lubrient	68,500	0.46	61,450	0.46

Rice hulls was ground, 20–80 mesh [21].

Filler Increases the Dynamic Viscosity Table 17.6 and 17.7 show examples of dynamic viscosity of rice hulls- and wood flour-filled HDPE, respectively.

One can see that the consistency index increases from 12,300 Pa.s^n for the neat HDPE (MFI 0.48) to 163,700 Pa.s^n for the 70% filled material (20–80 mesh ground rice hulls), and from 44,000 Pa.s^n for the neat HDPE (MFI 0.3) to 150,000,000 Pa.s^n for the 70% filled plastic (120-mesh wood flour). Overall, the viscosity increases by 11–13 times on transition from neat HDPE to rice hulls-filled HDPE and by 3400 times on that to wood flour-filled HDPE). The sharp increase of viscosity in the last case is a result, among other factors, of a sharp drop in the power-law index, from 0.33 to 0.12.

TABLE 17.7 Effect of filler and additives in HDPE (MFI 0.3 g/10 min) on the consistency index (m) and power-law index (n) determined at 180°C in parallel plate rheometer

HDPE neat and filled	Consistency index (Pa.s^n)	Power-law index	λ (Cross model parameter, see Eq. 17.7) (s)
Neat	44,000	0.33	1.60
Filled with 50% wood flour	1,100,000	0.28	3.82
Filled with 60% wood flour	12,100,000	0.20	7.5
Filled with 70% wood flour	150,000,000	0.12	—
Same plus 3% coupling agent plus 3% lubricant	30,000,000	0.14	—

Wood flour from maple, 120 mesh, vacuum dried. The coupling agent was maleated polyethylene containing 1% (w/w) grafted maleic anhydride (Epolene G-2608, Eastman Chemicals). The lubricant was alkene bis fatty amid (Glycolube WP-2002, Lonza Inc.) [4].

TABLE 17.8 Effect of fillers in HDPE (MFI 0.50 g/10 min) on the apparent viscosity, consistency index (apparent viscosity at the shear rate of 1 s^{-1}) and power-law index, determined in the capillary rheometer (at 170°C) and the parallel plate rheometer (at 177°C)

	Apparent viscosity (Pa.s) at the shear rate (s^{-1})				Power-law
HDPE neat and filled	1	10	100	500	index
Neat[a]	18,020	5,217	1,445	532	0.44
Composite Deck board[a]	45,930	10,312	2,253	718	0.34
Composite Deck board[b]	32,860	8,936	2,156	n/d	0.41
Composite material (regrind)[a]	40,150	9,340	2,134	671	0.36

The fillers were rice hulls (20 mesh) and Biodac (50% cellulose fiber, 50% minerals).

[a] Capillary rheometer.

[b] Parallel plate rheometer.

Similarly, Table 17.8 shows data for neat HDPE and a composite material, containing rice hulls and Biodac® (29% each) as principal fillers. Because the power-law index is lower for the composite material (0.34 compared to 0.44 for the neat HDPE), the effect of filler on viscosity is increased with decrease of the shear rate. At the shear rate of 500 s^{-1}, fillers increase the viscosity by only 35%; at the shear rate of 1 s^{-1}, fillers increase the viscosity by 150%, when melts of a composite deck board and the neat HDPE are compared.

Generally, the increase in viscosity in filled materials is larger at low shear rates; that is, at low shear rates the composites become more shear thinning, and their power-law index decreases. For example, the power-law index decreases from 0.54 for the neat HDPE to 0.41 for the 70% filled HDPE–rice hulls (20–80 mesh ground), see Table 17.6 (160°C data). This is typically an indication that at lower shear rates the interaction of the filler particles with each other and the particle–matrix interaction throughout the matrix is stronger and results in longer relaxation times [4, 21–23].

Viscosity and the Power-Law Index of Wood–Plastic Composite Materials Let us consider in more detail how fillers tend to make the system more shear thinning, that is, to decrease the power-law index. At lower shear rates or frequencies, neat plastics often exhibit a Newtonian plateau, that is, a "higher" apparent power-law index, and in the presence of fillers, the plateau often turns upward or even disappears. In other words, the addition of filler often makes the power-law plot more steep, that is, shifts it to a more uniform straight line dependence of viscosity verses shear rate (or frequency).

Besides, the increased shear thinning in wood-filled composites could probably be due to the higher local shear rate in the polymer occupying the space between wood particles. In this case, the same macroscopic shear induces higher local shear rate and lower viscosity in the thin polymer layer between wood flour particles [24].

TABLE 17.9 Shear viscosity of some neat HDPE and PP, a commercial composite material (GeoDeck) and its regrind in a range of shear rates

	HDPE $n = 0.47$ (170°C)	Polypropylene[a] $n = 0.26$	GeoDeck composition (cellulose fiber and minerals) $n = 0.38$ (170°C)	GeoDeck regrind, partially oxidized material $n = 0.19$ (170°C)
Shear rate (s^{-1})	Shear viscosity (Pa.s)			
1	25,000*	10,400	36,000*	130,000*
10	7,500	2,750**	9,000	20,000*
100	2,200	800	2,150	3,300
1000	480	140	480*	540*
10000	100*	20	110*	70*

Capillary rheometry. Extrapolated or interpolated viscosity values are marked * and **, respectively.
[a] Reference [25].

Tables 17.9 and 17.10 show the magnitude of shear thinning behavior for some neat polymers (HDPE and PP) and composites.

One can see that when the shear rate increases by 10,000 times and, the shear viscosity in a capillary rheometer decreases quite dramatically: for HDPE in 250 times, for polypropylene in 520 times, for GeoDeck composition in 327 times, and for GeoDeck regrind almost in 2000 times. In a graphical presentation, data of Table 17.9 are given in Figure 17.6.

Similar to Table 17.9 (capillary rheology), Table 17.10 (dynamic rheology) shows that a change of the shear rate by 100 times led to a decrease in the shear viscosity in 25 times for HDPE and approximately in 60, 70, and 160 times for three different regrinds.

Steady Shear Viscosity and Dynamic Viscosity Data *Neat* HDPE rheology data fairly well correspond to each other when obtained by both capillary and rotational rheometers. This actually means that HDPE melt obeys the Cox–Merz rule [26]. The

TABLE 17.10 Complex melt viscosity of some commercial composite materials (GeoDeck) and their regrind in a range of shear rates

Shear rate (rad/s)	GeoDeck composite, $N = 0.53$	GeoDeck regrind, $n = 0.41$	GeoDeck regrind, $n = 0.39$	GeoDeck regrind, n = 0.27
	Complex melt viscosity (Pa.s)			
0.1	14,000	97,000	164,000	236,000
1	4,500	24,500	40,000	34,000
10	1,700	6,600	11,000	8,200
100	560	1,600	2,500	1,500

Dynamic parallel-plate rheometry, 177°C.

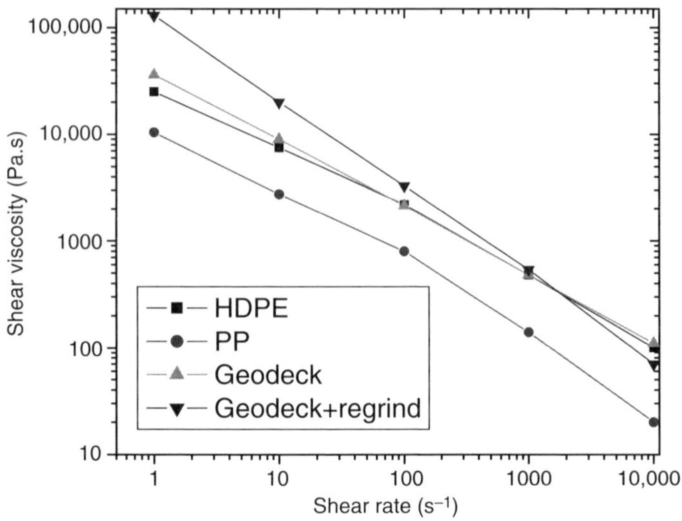

Figure 17.6 Shear rate–shear (complex) viscosity data for the four materials. Numerical data are given in Table 17.9.

rule states that the shear rate dependence of the steady-state viscosity is similar to the frequency dependence of the dynamic viscosity [9]; that is,

$$\eta(\dot{\gamma}) = |\eta^{*}(\omega)| \text{with} \dot{\gamma} = \omega$$

However, HDPE-based wood-filled *composites* often show very different rheometry data using these two approaches. They typically show much higher dynamic viscosity compared to the steady shear viscosity at the same shear rate values [27]. It seems that studying the melt rheological properties of composite materials, particularly highly filled polymer systems under dynamic steady shearing conditions (such as using parallel-plate rheometers above shear rates of 0.1 s^{-1}), could produce erroneous results due to slip and subsequent sample breakage at high speed of rotation [27]. For neat polymers, the viscosity data are usually taken up to shear rate of 5 s^{-1}. At higher speeds, the polymer melt starts exhibiting secondary flow and edge instabilities [28].

Table 17.11 shows data for neat and a rice-hulls-filled HDPE in the steady shear flow (a cone die) and the dynamic shear. One can see that the difference in the consistency index for neat HDPE between these two experimental approaches was only 14%. However, that for the filled composite was 32%. In the presence of the coupling agent, the difference was much higher: 116,400 Pa.s^n (cone die) and 50,600 Pa.s^n (ARES rheometer), that is, 130% difference.

The power-law index measurements were in a reasonable agreement between the two modes (Table 17.11) and show the same trends: for the neat HDPE 0.60 and 0.54 (cone die and ARES, respectively), for the 60% filled composite 0.39 and 0.41, and for the 60% filled composite in the presence of 1% coupling agent 0.40 and 0.47, respectively.

In general, capillary (steady shear) and parallel plate (dynamic shear) viscosity data of cellulose-filled composite materials (HDPE- or PP-based) at high filler

TABLE 17.11 The consistency index (m) and power-law index (n) determined for the neat and the filled HDPE using the cone die (steady shear flow) and the parallel-plate (dynamic shear) rheometer (ARES) at 160°C

HDPE neat and filled	Consistency index (Pa.s^n)	Power-law index	Consistency index (Pa.s^n)	Power-law index
	Cone die (steady shear)		ARES (dynamic shear)	
Neat	14,000	0.60	12,300	0.54
Filled with 10% rice hulls	16,200	0.59	14,000	0.53
Filled with 50% rice hulls	55,700	0.48	37,800	0.47
Filled with 70% rice hulls	—	—	163,700	0.41
Filled with 60% rice hulls	119,900	0.39	90,700	0.41
Same plus 1% coupling agent	116,400	0.40	50,600	0.47
Same plus 1% lubricant	—	—	72,600	0.42
Same plus 1% coupling agent plus 1% lubricant	89,300	0.49	76,600	0.46

25-mm parallel plate, 2-mm gap, 5% strain. Rice hulls was ground, 20–80 mesh [21].
ARES = Advanced Rheometric Expansion System.

loadings significantly differ from each other [4, 21, 29, 30]. An example of such a discrepancy is shown in Figure 17.7. This is typically explained by a different alignment of cellulose particles in steady shear (in a capillary rheometer and in extrusion) and dynamic shear and also by a possible slip at the wall in the steady-shear

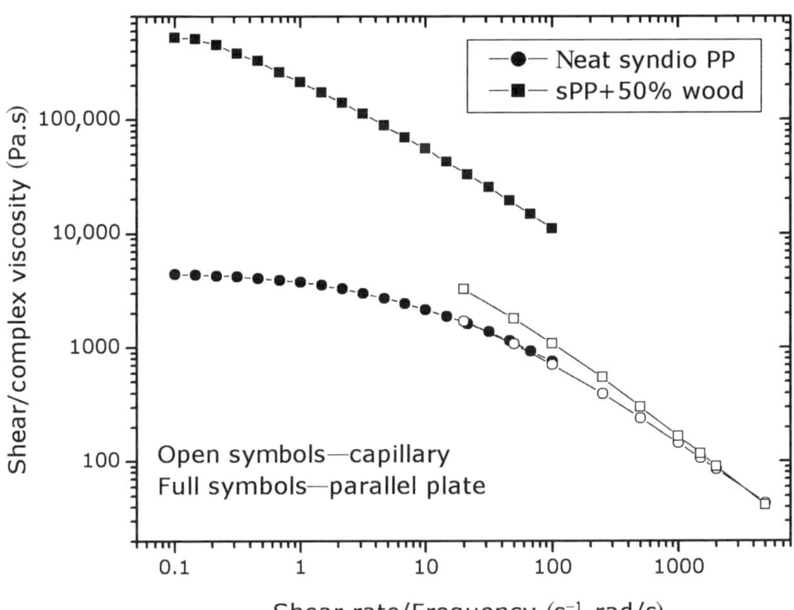

Figure 17.7 A comparison of data for a neat polypropylene and the polypropylene filled with 50% wood flour (w/w) in terms of capillary viscometry and dynamic rheometry (courtesy of V. Hristov).

mode. Another possible reason could be the enhanced filler–matrix and filler–filler interactions resulting in much higher viscosity values compared to those from capillary. Capillary flow is of different nature in which filler orientation effects due to combined shear and extensional forces lead to less resistance to flow [30].

Figure 17.7 shows that shear rate and frequency are almost identical for neat polymers; that is, the Cox–Merz rule (see above) is valid for these systems. The power-law index n, calculated from the slope of the two viscosity curves in Figure 17.7, would be the same (excluding data at very low frequencies, at so-called Newtonian plateau). But for filled plastics this rule is not applicable, and one will get different values of the power-law index from the two viscosity curves. In other words, one will obtain much higher viscosity data from a parallel plate rheometer.

Typically, WPC based on polypropylene and polyethylene show deviation from the Cox–Merz rule. This is due to the different nature of flow. Capillary flow is a pressure-driven flow, including entrance and exit effects, wall slip, friction in the barrel, and orientation effects. Parallel-plate flow is pure drag shear flow, in which particle–particle and matrix–particle interactions result in higher viscosities for filled polymers. In other words, a straightforward question "is a 100-fold increase in shear rate and 100-fold increase in frequency result in the same effects?" the answer would be "yes" for neat polymers, and "no" for wood–filled composites.

Table 17.12 shows data for a series of regrinds all derived from the same composition, that is, HDPE (MFI 0.5) filled with rice hulls and Biodac®. The limiting shear viscosity (for a parallel-plate rheometer) and the consistency index (for a capillary rheometer) were estimated using eq. (17.22) and (17.5), respectively.

Certainly, one can see that there is practically no correlation between the zero-shear viscosity determined on a parallel-plate rheometer and the consistency index, determined using a capillary rheometer. Regrinds that resulted in a bad quality boards (roughness, sharkskin), namely regrinds B, G, and I, do not show any certain pattern neither in their loss modulus, nor in their consistency index.

TABLE 17.12 The loss modulus (parallel plate rheometer) and the consistency index (capillary rheometer) for regrinds of various degradation (oxidation) degree

	Parallel-plate rheometer	Capillary rheometer
Regrind	G'' (Pa) at $\omega = 0.1$ rad/s as an estimated η_0 (Pa.s)	Consistency index (Pa.s^n) at $\gamma = 1\ s^{-1}$
A	1.09×10^3	4.0×10^4
B	4.16×10^3	3.9×10^4
C	4.58×10^3	3.5×10^4
D	4.79×10^3	13.2×10^4
E	7.79×10^3	4.0×10^4
F	11.1×10^3	3.9×10^4
G	13.3×10^3	9.2×10^4
H	14.6×10^3	5.4×10^4
I	22.2×10^3	5.2×10^4

All regrinds were derived from the same HDPE-based composition, filled with rice hulls (29%) and Biodac (29%).

Dynamic rheometry was not (and, apparently, cannot be) employed for studying melt fracture of neat plastics and composite materials. This has been done so far only using capillary rheometry. However, dynamic oscillatory measurements can produce the most reliable rheological data on filled polymers [2, 4]. It should be noted that measurements at dynamic oscillatory conditions bellow frequency of 0.1 rad/s likely produce erroneous results due to the increased time for reaching steady state at low frequencies [4].

Capillary Rheometer and an Extruder: Are They in Agreement? One can ask a reasonable question: If rheology data are obtained in a capillary rheometer, are they applicable to an extruder? To answer this question, at least for a specific set of conditions, a direct comparison was made [31]. It was found that the capillary rheometer and extruder are in good agreement for neat plastics (polystyrene and polypropylene), but extruder systematically measures lower viscosities in glass-fiber-filled plastics.

However, data points for both neat and filled plastics accurately fit the same curves in double-log plots (log η vs. log$\cdot\dot{\gamma}$). Furthermore, the rheometer data of the melt collected from the die exit agree with the extruder data and also show lower viscosities in glass-fiber-filled plastics. This implies that the rheological properties of the filled polymer are affected by the intense shearing in the extruder, which leads to fiber breakage and structural changes in the filled polymer. Interestingly, according to the data [31], these changes led to higher power-law index for hot melts in an extruder. Hence, these composite materials were closer to melt fracture compared to neat polymers, as it will be shown in the last section. Their hot melts are moving at much higher values of the shear rate.

Extrudate Swell The extrudate swell of wood-filled plastics typically increases with the flow rate (shear rate), decreases with filler loading, and practically does not depend on the melt temperature. A nonfibrous filler, such as calcium carbonate, also suppresses swell of polyethylene, and in the case of medium-density polyethylene, a swell suppression was maximal at 30% of calcium carbonate (0.4 μm particles) [32]. A similar effect was shown using HDPE filled with rice hulls (Table 17.13).

The swell *increase* with the flow rate for neat HDPE was attributed to the higher shear that resulted in larger elastic forces to be released at the die exit.

On the contrary, the swell *decrease* with the increase in filler content was attributed to the lesser amount of elastic polymer chains in the system to recover and swell at the exit. As the filler content was increased, the swell was approaching the Newtonian value of 13% [21].

TABLE 17.13 Increase of swell with the increase of flow rate at extrusion of neat HDPE and that filled with rice hulls

Amount of rice hulls in HDPE	Increase of swell with increase of flow rate
0 (neat HDPE)	Increased by 76–89%
10%	Reduced to 57–70%
50%	Reduced to 17–27%
60%	Reduced to 8–19%

The extrusion was from the cone die, the die exit diameter of 0.300 in. Increase of flow rate was from 110 to 170 g/min (neat HDPE and 10%-filled HDPE), from 90 to 190 g/min (50%-filled HDPE), and from 10 to 180 g/min (60%-filled HDPE) [21]).

TABLE 17.14 Effect of particle size of maple filler in 40% wood fiber–60% HDPE on rheological properties of the composite system

Maple fiber size (mesh)	The power-law index	Log viscosity, Pa.s at shear rate 1 s^{-1}(log $\gamma = 0$)
40	0.53	4.32
60	0.53	4.28
80	0.53	4.29
100	0.54	4.28

Wood flour as a filler decreases elasticity of polyethylene [33]. In other words, the shear stresses in the system of polyethylene highly filled with wood flour dominated over normal stresses. This was shown using a parallel-plate rotational rheometer under steady shearing conditions. The decreased elasticity was expressed as a decrease of the stress ratio with increasing wood flour loading at the same shear rate.

ALMOST UNCHARTED AREAS OF COMPOSITE AND PLASTIC RHEOLOGY

Effect of Particle Size of Filler on Rheology of Wood–Plastic Composites Very few studies were published regarding the effect of particle size of wood fillers on rheology (e.g., a conference poster [34]). Tables 17.14–17.16 show the data.

As one can see, there is no dependence of rheological behavior of the hot melts on species of wood fiber in this particular case. As a result of transition from 60% HDPE to 40% HDPE, viscosity of the hot melt increased by 2.6–3.2-fold, and the composite became significantly more sensitive to a change of the shear rate (the power-law index has noticeably decreased).

Again, with the increase of the wood fiber content, viscosity increases and the power-law index decreases. The viscosity of the hot melt becomes more sensitive to fluctuations of the shear rate; hence, the higher might be fluctuations of pressure of the hot melt, the easier the hot melt can overshoot the critical shear rate, leading to rough profiles, sharkskin, and surface rupture (see below).

TABLE 17.15 Effect of origin (species) of wood filler (40 mesh in all cases) in 40% wood fiber–60% HDPE and 60% wood fiber–40% HDPE on rheological properties of the composite system

Composite	The power-law index	Log viscosity, Pa.s at shear rate 1 s^{-1}(log $\gamma = 0$)
40% maple–60% HDPE	0.48	4.39
40% pine–60% HDPE	0.49	4.39
60% maple–40% HDPE	0.35	4.90
60% pine–40% HDPE	0.36	4.80

TABLE 17.16 Effect of percentage of maple fiber (40 mesh in all cases) in HDPE on rheological properties of the composite system

Composite	The power-law index	Log viscosity, Pa.s at shear rate 1 s^{-1} (log $\gamma = 0$)
30% fiber–70% HDPE	0.51	4.15
40% fiber–60% HDPE	0.49	4.39
50% fiber–50% HDPE	0.40	4.65
60% fiber–40% HDPE	0.35	4.90
70% fiber–30% HDPE	0.33	4.95

Effect of Coupling Agents, Lubricants, and Polymer Processing Additives Coupling agents are often used in WPCs, as wood fillers are hydrophilic whereas polyolefins are hydrophobic, and they need to be compatibilized for better performance of the material. Generally, functionalized polymers such as maleic anhydride grafted polyolefins are employed as coupling agents in filled plastics. In order to avoid segregation of the coupling agents as a separate phase within the matrix, it is preferential for the agent to be compatible with the matrix [35].

High molecular weight polymer containing relatively low concentration of grafted maleic anhydride is desirable for a coupling agent. The high molecular weight ensures co-crystallization with the matrix, and low level of grafted maleic anhydride units will prevent interactions between maleic anhydride groups from the same polymer chain. This in turn can lead to the formation of aggregates well dispersed within the matrix. The amount of the coupling agent is also of crucial importance [36–38]. It is generally accepted that the amount of the coupling agent should be kept relatively low (not more than 3–5 wt %).

The available data for the influence of coupling agents on the rheological properties of wood-filled composites are rather controversial. It is not surprising, considering the variety of effects of coupling agents on mechanical properties of WPCs, described in Chapter 5. It was suggested there that different experimental and industrial conditions might principally change a mode of interaction of fillers, plastic, and coupling agents.

Therefore, it would have been expected that the effects of coupling agents on rheology of WPCs were different. That is what was observed. One set of data is shown in Table 17.6. Adding of a coupling agent decreased the consistency index and increased the power-law index in both studies. The authors [21] attributed the effect to an improved dispersion of filler in the hot melt in the presence of the coupling agent. However, they noticed that in their particular case the coupling agent had a higher melt index than the matrix material. In another two sets of experiment, adding of coupling agents increased the shear viscosity [4, 22]. These results were attributed to the effective coupling between matrix and filler. Further increasing of the amount of the coupling agent however led to reduction of the viscosity.

Lubricants significantly increase the output, widen the processing window, and lower the melt temperature. It was logical to suggest that lubricants would reduce viscosity and probably not affect the power-law index. That is what was observed (Table 17.6). Adding lubricants to the WPC system with the coupling agent increased viscosity and did not change the power-law index (Table 17.6). However, as the coupling agent was a maleated polyolefin and the lubricant contained metal, they certainly interacted with each other, as described in Chapter 5. Apparently, this interaction resulted in increase of the viscosity.

A few words on polymer processing additives (PPAs). They are widely used in the extrusion of linear polyolefins to prevent sharkskin, flow instability, and melt fracture [39]. In typical industrial usage, fluoropolymers are added to the polymer in small quantities (less than 0.1% of mass fraction of PPA in polymer). To be effective, the PPA must do two things. First, they must coat the die wall, particularly the die exit. Second, they must induce slippage between themselves and the polymer melt. As a result of slippage, the shear stress is reduced, and consequently the power requirements are lowered while keeping high production efficiency and high product quality. Most popular commercial grades PPAs are Dynamar, Viton, Teflon, and Kynar. Examples of applications of fluoropolymers in wood-plastic extrusion are given in Ref. [40].

Thermoplastic silicone elastomers (TPSE) are also promising materials for using as lubricants in WPCs. A particular example is a high molecular weight polydimethylsiloxane–urea copolymer that combines the properties of thermoplastics and silicones in one material. This material has low viscosity and could readily migrate to the die wall initiating slippage. Moreover, because of its high melting temperature, TPSE will not lubricate the solids conveying extruder zone, a problem typical for liquid silicone lubricants. Furthermore, TPSE is inert to the coupling agent and improves the impact strength, surface smoothness, and abrasion resistance. An illustration of WPC surface morphology obtained by using the TPSE is given in Figures 17.8 and 17.9. It was shown that a coupling agent and TPSE provided smooth extrudate surface at any studied output rate [42].

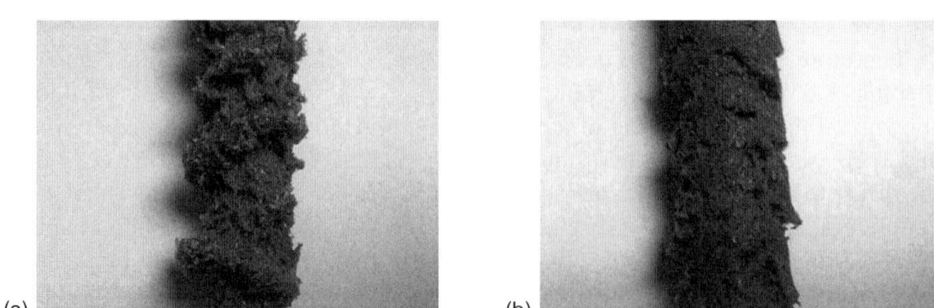

Figure 17.8 Extrudate surface morphology of mPP/wood-filled composites, obtained at 25 rpm (courtesy of V. Hristov): (a) 50% wood flour. (b) 50% wood flour + coupling agent.

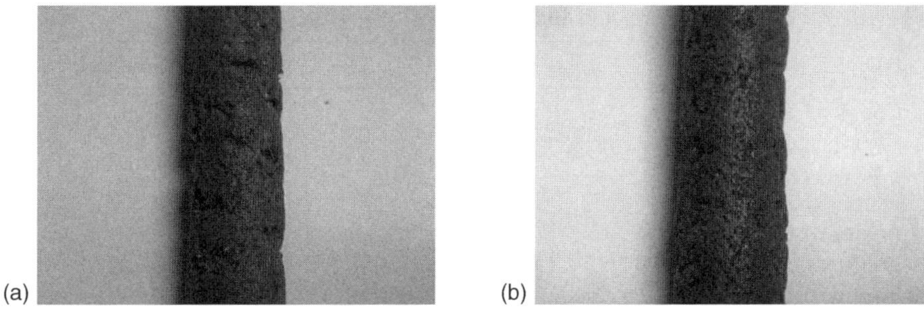

Figure 17.9 Extrudate surface morphology of mPP50 wood flour + coupling agent + TPSE (courtesy of V. Christov): (a) 25 rpm. (b) 50 rpm.

Varying Plastic Sources—Which to Choose for Composite Materials? Movement of different fluid layers of the plastic-based composites sometimes has a different pattern with and without added regrind. It is often described as a difference in shear rate at the same viscosity. This might lead to a more narrow window for a "proper" flowability of the hot melt before a melt fracture is observed and result to a roughness, sharkskin effect, or other kinds of defective extruded profiles at conditions that would normally give good quality products.

Hence, a question: What kind of test should be conducted as a part of QC for incoming plastics, in order to prevent such a situation? What minimum set of parameters should be measured in order to reliably decline any particular plastic, either pure plastic, or regrind, or a final composition?

Clearly, this minimum set of parameters should be defined by rheology of plastics. The question can be rephrased—how to find a simpler connection between viscosity-related numbers, on one hand, and the extruder runability, on the other?

Table 17.17 shows the power-law index data for eight HDPE materials. The first four give good quality composite boards (filled with rice hulls and minerals); the last four polymers result in sharkskin and tear profiles (hollow deck boards). In all the eight cases, a formulation of the composite was the same, including 39% (w/w) of each of the HDPE. Densities of all the HDPE were practically the same.

Table 17.18 shows data for four more HDPE materials; all of them resulted in good quality boards.

Clearly, the power-law index alone of a pure HDPE cannot serve as a factor pointing at an expected quality of an extruded composite profile.

Let us take a look at temperature dependence of the power-law index (Table 17.19).

These are the same HDPE samples described above. It seems that temperature dependencies of the power-law index of neat HDPE also cannot serve as a factor pointing at an expected quality of an extruded composite profile. On a temperature range of 60°C, the power-law index differs by a range of 5–10%.

TABLE 17.17 The power-law index of HDPE of various origin

HDPE	The power-law index (n)
Chevron CHVX891180[a]	0.392 ± 0.003
Chevron CHVX896880[a]	0.400 ± 0.009
Equistar EQUX621048[a]	0.472 ± 0.005
Petromont PSPX7006[a]	0.455 ± 0.008
Equistar EQUX631675[b]	0.442 ± 0.003
Equistar GPLX74618[b]	0.423 ± 0.009
Equistar EQUX632225[b]	0.442 ± 0.007
Dupont/Sherman Tyvek Repro, Lot 45104[b]	0.468 ± 0.003

Galaxy V capillary rheometer (Kayness Inc., Model 8025), the capillary diameter 0.10 cm, 200°C.
[a] The plastics gave a good quality composite deck boards (hollow profiles, plastic filled with rice hulls and minerals).
[b] The plastics gave a poor quality composite deck boards, sharkskin inside the hollow profile, tear surface (plastic filled with rice hulls and minerals).

On the contrary, the zero-shear viscosity, or consistency index, that is, viscosity of an almost non moving polymer melt, depends on temperature quite significantly, namely by 74±7% on average for all eight points in Table 17.20 between 170 and 230°C.

However, it seems that temperature dependencies of the consistency index of neat HDPE also cannot serve as a factor pointing at an expected quality of an extruded composite profile.

Now let us consider the so-called polymer processability, which is represented by the MFI ratio measured at 10 and 2.16 kg loads, namely HLMI/MI (high load melt index over melt index) (Table 17.21)

Overall, it seems that neither the power-law index, nor the consistency index, nor temperature dependencies of the both, nor polymer processability ratios (in terms of MFI) is a good or appropriate indicator of the expected quality of extruded composite profiles.

Higher molecular weight of polyethylene often improves the physical properties of products (though, see below). However, increasing of the molecular weight, hence, the viscosity of the polymer, usually increases the extrusion pressure and torque on

TABLE 17.18 The power-law index of commercial grades HDPE

HDPE	The power-law index
CCBX58758 + CCBX72914 (Trademark Plastics)	0.44
CCBX057340 (Dow Chemical)	0.45
CCBX73082 (Trademark Plastics)	0.44
CCBX71026	0.45

Galaxy V capillary rheometer (Kayness Inc., Model 8025), the capillary diameter 0.10 cm, 170°C. All extruded composite deck boards were of a good quality.

TABLE 17.19 The power-law index of commercial grades HDPE at three different temperatures

	The power-law index		
HDPE	170°C	200°C	230°C
Chevron CHVX891180[a]	0.383 ± 0.004	0.392 ± 0.003	0.422 ± 0.003
Chevron CHVX896880[a]	0.390 ± 0.004	0.400 ± 0.009	0.409 ± 0.016
Equistar EQUX621048[a]	0.447 ± 0.007	0.472 ± 0.005	0.483 ± 0.003
Petromont PSPX7006[a]	0.428 ± 0.003	0.455 ± 0.008	0.473 ± 0.005
Equistar EQUX631675[b]	0.421 ± 0.005	0.442 ± 0.003	0.464 ± 0.003
Equistar GPLX74618[b]	0.395 ± 0.003	0.423 ± 0.009	0.437 ± 0.003
Equistar EQUX632225[b]	0.422 ± 0.003	0.442 ± 0.007	0.447 ± 0.003
Dupont/Sherman Tyvek Repro, Lot 45104[b]	0.454 ± 0.005	0.468 ± 0.003	0.481 ± 0.003

Galaxy V capillary rheometer (Kayness Inc., Model 8025).

[a] The plastics gave a good quality composite deck boards (hollow profiles, plastic filled with rice hulls and minerals).

[b] The plastics gave a poor quality composite deck boards, sharkskin inside the hollow profile, tear surface (plastic filled with rice hulls and minerals).

the screw, and as a result, such polymers are more difficult to process. To overcome this, resin producers often broaden the MWD (molecular weight distribution) concurrently by increasing its molecular weight. This indeed makes the polymer easier to extrude or mold than a polymer with a narrow distribution. However, sometimes this leads to unpredictable rheological behavior of polyethylene composites.

TABLE 17.20 The consistency index (as log of viscosity) of commercial grades HDPE at three different temperatures

	Log viscosity (Pa.s)		
HDPE	170°C	200°C	230°C
Chevron CHVX891180[a]	4.46 ± 0.01	4.36 ± 0.01	4.22 ± 0.01
Chevron CHVX896880[a]	4.41 ± 0.01	4.29 ± 0.02	4.20 ± 0.04
Equistar EQUX621048[a]	4.23 ± 0.01	4.09 ± 0.02	3.98 ± 0.01
Petromont PSPX7006[a]	4.28 ± 0.01	4.14 ± 0.01	4.02 ± 0.02
Equistar EQUX631675[b]	4.25 ± 0.01	4.12 ± 0.01	3.99 ± 0.01
Equistar GPLX74618[b]	4.39 ± 0.01	4.27 ± 0.02	4.15 ± 0.01
Equistar EQUX632225[b]	4.26 ± 0.01	4.15 ± 0.03	4.03 ± 0.01
Dupont/Sherman Tyvek Repro, Lot 45104[b]	4.43 ± 0.01	4.30 ± 0.01	4.20 ± 0.01

Galaxy V capillary rheometer (Kayness Inc., Model 8025).

[a] The plastics gave a good quality composite deck boards (hollow profiles, plastic filled with rice hulls and minerals).

[b] The plastics gave a poor quality composite deck boards, sharkskin inside the hollow profile, tear surface (plastic filled with rice hulls and minerals).

TABLE 17.21 Polymer processability ratio for a series of commercial grades HDPE

	Melt Flow Index at 190°C		
HDPE	MFI at 2.16 kg	MFI at 10 kg	$MFI_{10}/MFI_{2.16}$
Chevron[a] CHVX891180	0.28	—	—
Chevron CHVX896880[a]	0.36	—	—
Equistar EQUX621048[a]	0.89	12.60	14.16
Petromont PSPX7006[a]	0.47	9.20	19.57
Equistar EQUX631675[b]	0.58	11.83	20.40
Equistar GPLX74618[b]	0.56	10.99	19.63
Equistar EQUX632225[b]	0.25	6.51	26.04
Dupont/Sherman Tyvek Repro, Lot 45104[b]	0.33	—	—

[a] The plastics gave a good quality composite deck boards (hollow profiles, plastic filled with rice hulls and minerals).

[b] The plastics gave a poor quality composite deck boards, sharkskin inside the hollow profile, tear surface (plastic filled with rice hulls and minerals); speed change and temperature change did not improve the situation.

Typically, polymers with broad MWD exhibit non-Newtonian flow at low shear rates compared to that of narrow MWD polymers. In addition, shear thinning is generally much more pronounced in polymers with broader MWD. As a result, output rate of polyethylenes with different molecular weights and MWD is influenced by the power-law index in a rather complex way. This was illustrated in a detailed study of four HDPE samples (Table 17.22).

It should be mentioned that the relationships between average molecular weights, MWD, and the power-law index of the respective polymer melts are not clear and completely unexplored in case of wood-filled composites. For example, increasing viscosity does not always improve physical properties of products. It was found that the increase of MFI of polypropylene from 3 to 30 g/10 min did not alter the efficiency of wood fiber dispersion and did not result in an improvement of any measured property of WPC. On the contrary, a change of MFI for HDPE from 0.15 to 7.0 led to better wetting of wood fiber and superior mechanical properties of the WPC.

TABLE 17.22 The power-law index for neat HDPE samples with various molecular characteristics [42]

			The power-law index		
HDPE	M_w (g/mol)	M_w/M_n	180°C	210°C	240°C
A	200,000	28	0.35	0.37	0.40
B	320,000	33	0.28	0.31	0.34
C	250,000	15	0.28	0.29	0.32
D	450,000	58	0.30	0.26	0.25

It is possible that such inconsistencies reflect different interfacial interactions between plastics and cellulosic filler. For example, loading of 50% (w/w) of wood flour to two different polyethylenes can increase the system viscosity 100-fold or only 25-fold. In the first case polyethylene was metallocene PE (MFI 4 g/10 min, MWD = 2.5, power-law index $n = 0.79$); in the second case it was HDPE (MFI = 0.3 g/10 min, MWD = 5, power-law index $n = 0.45$). It is likely that the less viscous metallocene polyethylene melt, which contains fewer branches, is capable of deeper penetration into the porous structure of the wood. This would increase the mechanical adhesion contacts between the matrix and filler, expressed in an increased resistance to shearing [43, 44].

HDPE-D in this table, having the highest weight-average molecular weight and the widest MWD ($M_w = 450{,}000$, $M_n = 7760$), shows an unusual temperature dependence of its power-law index, that is, its n value decreases with temperature, unlike three other HDPE samples in the Table 17.22. As a result, viscosities of these HDPE hot melts are higher or lower compared to each other, depending on temperature and shear rate. Generally, HDPE-D, having the lowest (overall) power-law index, showed the lowest output rate among the four studied HDPE. The authors [42] concluded that the output rate increases with the power-law index, all else being constant, and the dependence on n becomes more significant as the screw rpm (shear rate) is increased.

Some plastic engineers use the term "Coefficient of shear sensitivity," relating it to a ratio of viscosity of a hot melt at 100 s^{-1} to that at 1000 s^{-1}. However, this coefficient is based on two experimental points only and often is much less reliable compared to the power-law index. In fact, this coefficient is related to the power-law index as

$$n = 1 - \log \eta_1/\eta_2$$

where n is the power-law index, η_1 and η_2, respectively, are viscosities at 100 and 1000 s^{-1}. Clearly, determination of n from a series of experimental points would give more reliable index.

Rheology of Regrinds of Wood–Plastic Composites As it will be shown below, the presence of a partially degraded (oxidized) regrind often decreases the power-law index, for example, from 0.50 to 0.20–0.30. This leads to a higher shear thinning at a higher velocity of the hot melt, to a higher difference in viscosity across the extruder, and to a melt fracture at a lower shear rate compared to a composition without "bad" regrind. In other words, at the same extrusion conditions (temperature, rpm), "normal" composition is extruded into a good quality profile, whereas the same composition but containing, say, 20% of a "bad" regrind (having noticeably lower average molecular weight) shows pronounced sharkskin (Figs. 17.10–17.14).

Now, let us consider rheology parameters for a composite material and the same material but subjected to a partial oxidative destruction by heating it in a kiln for extended period of time (no antioxidants were added to the

Figure 17.10 A composite deck board to be discarded.

formulation). The data are shown in Figures 17.15 and 17.16 and reviewed in Table 17.23.

Figure 17.15 shows data for a "good" regrind. It has the power-law index of 0.39 and the consistency index of 35,230 Pa.s^n. Its mixture (at 20% w/w) with the freshly made formulation results in a good runability and a good quality profile (a composite deck board).

Figure 17.11 A composite deck board to be discarded.

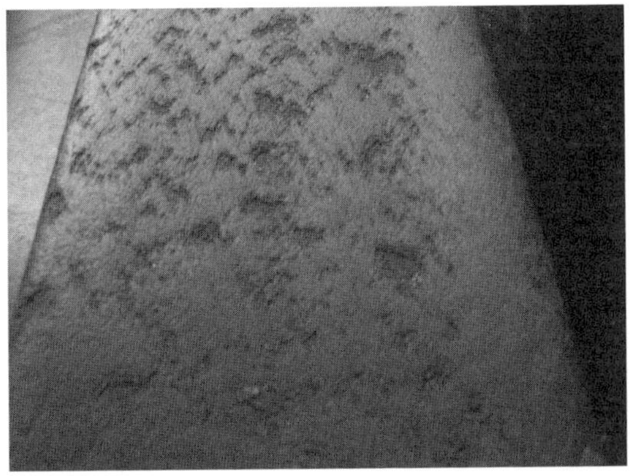

Figure 17.12 A composite deck board to be discarded.

Figure 17.13 A composite deck board to be discarded.

Figure 17.14 A composite deck board to be discarded.

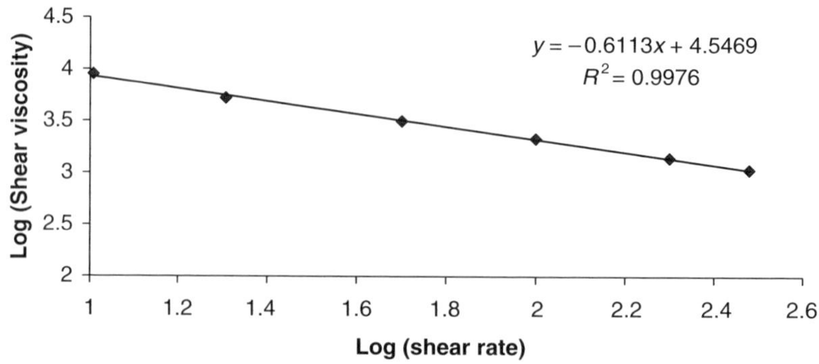

Figure 17.15 Double logarithmic plot for a "good" regrind (HDPE-based composite material). The power-law index 0.39.

On the contrary, Figure. 17.16 shows data for a "bad" regrind. It has the power-law index of 0.19 and the consistency index of 132,000 Pa.s^n. Its mixture (at 20% w/w) with the freshly made formulation results in a tear board, with a rough skin.

Hence, an indication to a potentially troublesome regrind can be a low power-law index, below 0.30 for this particular formulation and using this particular extruder with this particular screw at this specific processing conditions.

It can be seen from Table 17.23 that degraded regrind significantly deviates by its power-law index from the parent GeoDeck composition. Most of regrinds from recalled boards in Table 17.23 that resulted in rough, distorted boards had low OIT (oxygen induction time) values, between 0.17 and 0.40 min (regrinds 6, 8, and 10). Regrinds that resulted in good boards have generally higher OIT values, between 0.37 and 1.28 min (regrinds 1, 2, 4, 5, and 9).

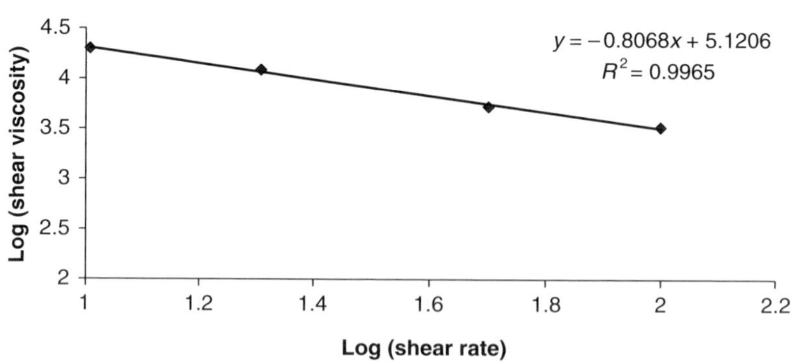

Figure 17.16 Double logarithmic plot for a "bad" regrind (the same HDPE-based composite material shown in Figure 17.15, but partially degraded after exposure at high temperature in a kiln). The power-law index 0.19.

TABLE 17.23 The power-law index and the consistency index (as log viscosity) for GeoDeck composition (compounded pellets) and the same GeoDeck composite material subjected to a partial oxidative destruction (regrind from recalled boards containing no added antioxidants)

Composition	The power-law index	The consistency index as log viscosity
GeoDeck, pellets	0.55	4.15
Regrind 1 (recalled boards, OIT = 0.37 min)[a]	0.53	3.67
Regrind 2 (recalled boards, OIT = 0.97 min)[a]	0.48	5.20
Regrind 3 (pellets made with 5% regrind from recalled boards)[a]	0.43	5.68
Regrind 4 (recalled boards, OIT = 1.29 min)[a]	0.42	5.56
Regrind 5 (recalled boards, OIT = 0.71 min)[a]	0.41	5.38
Regrind 6 (recalled boards, OIT = 0.17 min)[b]	0.41	5.14
Regrind 7 (recalled boards)[b]	0.39	5.61
Regrind 8 (OIT = 0.40 min)[b]	0.325	5.78
Regrind 9 (OIT = 0.54 min)[a]	0.32	5.42
Regrind 10 (recalled boards, OIT = 0.35 min)[b]	0.28	5.61
Regrind 11 (recalled boards, brittle, weak)[c]	0.28 (No continuous flow in a capillary rheometer, n could not be determined, estimate gave n is close to zero)	4.35
Regrind 12 (recalled boards, good quality, OIT = 1.47 min)[a]	0.24	5.38
Board 1 (board in a good shape)[d]	0.41	5.51
Board 6 (rough inside)[e]	0.41	5.50
Board 8 (significant roughness inside)[f]	0.40	5.44
Board 4 (good shape, but pitted)[g]	0.38	5.55
Board 10 (distorted board)[h]	0.38	5.85

Rheometrics System FOUR*, equipped with 25-mm diameter parallel plates, a forced oscillatory strain of 10%, and isothermal (177°C) frequency sweeps from 0.01 to 100.0 rad/s (ASTM D4440). Regrinds 1–5 in the amount of 20% (w/w) in the GeoDeck formulation brought about good quality profiles (hollow deck boards), regrinds 6–9 in amounts of 5–20% gave boards rough inside. Boards 1, 4, 6, and 8 were made from the regular GeoDeck formulation with an addition of 20% of the respective regrind. The OIT is given in a few cases.

[a]Good board resulted with this regrind.

[b]Rough, unaccepted board resulted from this regrind.

[c]Board could not be made with this regrind.

[d]Board made with 20% (w/w) regrind 1 (see the table).

[e]Board made with 20% (w/w) regrind 6 (see the table).

[f]Board made with 20% (w/w) regrind 8 (see the table).

[g]Board made with 20% (w/w) regrind 4 (see the table).

[h]Board made with 20% (w/w) regrind 10 (see the table).

The OIT values in this case correspond to amounts of an antioxidant present in the material. When the OIT values are below 1 min, the plastic often is partially oxidized, degraded, its molecular weight shifted to lower values. This often leads to lower values of the power-law index, and for the first group of the (lower) OIT values, the average power-law index is 0.34; for the second (higher) OIT values, the average power-law index is 0.43. Hot melt in the first case is more sensitive to speed variations, that is, it is more likely to result in melt fracture (see the next section).

The power-law index plays an important role in melt flow. It is obvious that the high flow rate results in the increase of the die pressure. However, for a Newtonian fluid (such as water, $n = 1$), a 10 × increase in pressure is accompanied by a 10 × increase in flow rate. For a non-Newtonian fluid with $n = 1/2$, a 10 × increase in pressure is accompanied by a 100 × increase in flow rate. For $n = 1/3$, a 10 × increase in pressure results is accompanied by a 1000 × increase in flow rate. For $n = 1/4$, a 10 × increase in pressure is accompanied by a 10,000 increase in flow rate. For $n = 1/5$, as in the case of a "bad" regrind above, a 10× increase in pressure results in a 100,000 increase in flow rate! This is essentially applicable for any extrusion die. Hence, the power-law index of a hot melt essentially determines its extrusion behavior.

It follows from the above that compared to GeoDeck hot melt composition without an added regrind ($n = 1/2$), regrind hot melt ($n = 1/5$) flows 1000 times faster at a 10 × increase in die pressure. No wonder that at the same extrusion conditions adding a "bad" regrind (with low value of the power-law index) to otherwise "good" hot melt leads to melt fracture (too high shear rate) and, as a result, to tearing of the extruded profiles, sharkskin and roughness.

Table 17.24 shows data for a series of regrinds all derived from the same composition, that is, HDPE (MFI 0.5) filled with rice hulls and Biodac®, 29% of each filler. The storage and loss moduli were plotted against the frequency in double logarithmic coordinates, and the respective slopes are shown in Table 17.24. In all cases the storage modulus and the loss modulus increased with frequency. The same data were analyzed earlier, in Table 17.12, in terms of their zero-shear viscosity.

It looks like regrinds that result in bad quality boards (roughness, sharkskin), namely regrinds B, G, and I, show a slower (less steep) increase in the storage modulus with frequency, compared with "good" regrinds. Generally, it makes sense because the storage modulus describes the elastic component of the composite material, and degradation of the matrix should reduce its elasticity. At the same time the loss modulus, describing viscous properties of hot melt in the composite material, also increases slower with frequency for "bad" regrinds (see also Table 17.25). Certainly, this phenomenon warrants further studies.

Melt Fracture of Plastics and Their Composites and Regrinds: Surface Tearing

From the preceding section it seems unlikely that potentially "good" or "bad" neat *HDPE* could be reliably identified in terms of runability and accompanying defects (sharkskin, roughness) of HDPE-based composite materials based only on their

TABLE 17.24 Storage and loss modulus (parallel plate rheometer, 177°C) of regrinds of various degradation (oxidation) degree as a function of frequency

	A slope of Log G′ and Log G″ versus log ω	
Regrind	The storage modulus G′	The loss modulus G″
A	0.53	0.53
C	0.52	0.47
D	0.51	0.45
H	0.46	0.40
E	0.45	0.38
F	0.45	0.40
B	0.44	0.39
G	0.44	0.36
I	0.37	0.23
Powder regrind, significantly degraded	Nonlinear plot. Slope 0.22 between 0.1 and 1 rad/s, 0.40 between 1 and 10 rad/s	Nonlinear plot. Slope 0.18 between 0.1 and 1 rad/s, 0.32 between 1 and 10 rad/s

All plots had a positive slope; a value of each one is listed in the table. All regrinds were derived from the same HDPE-based composition, filled with rice hulls (29%) and Biodac (29%).

power-law index and/or consistency index. It is more likely that *regrinds* can be assigned as "good" or "bad" by determining their power-law index and consistency index, and/or the storage modulus and the loss modulus as a function of the angular frequency. When all four of these parameters are lower compared with those of other

TABLE 17.25 Storage and loss modulus (parallel-plate rheometer, 177°C) of composite deck board material as a function of frequency

	A slope of Log G′ and Log G″ versus Log ω		
Composite deck board material	The storage modulus, G′	The loss modulus, G″	Comments from the manufacturing plant QC
1	0.49	0.40	Runs well
8	0.44	0.40	Some roughness, pitted board
6	0.43	0.36	Rough inside of the hollow board; boards were rejected
10	0.42	0.35	Distorted board; apparently, bad plastic
4	0.41	0.37	Pitted board

All plots had a positive slope; a value of each one is listed in the table. All composite boards were derived from the same HDPE-based composition, but origins of HDPE were different (see Table 17.23). HDPE was filled with rice hulls (29%) and Biodac (29%). The resulting boards are shown in Figs. 17.10–17.14 and 17.19–17.23.

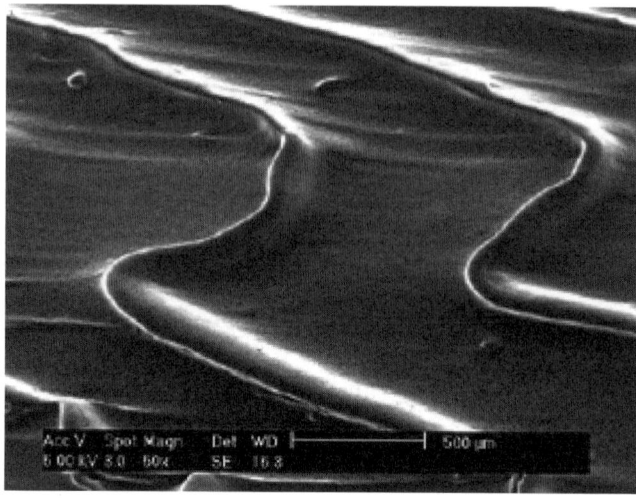

Figure 17.17 Scanning electron microscopy of sharkskin surface on HDPE (Source: Dawn Arda, Polymer Fluids Group, Department of Chemical Engineering, University of Cambridge; with permission).

("fresh") regrind materials, it is likely an indication that a critical shear stress can be reached at lower shear rate that would result in melt fracture (see, e.g., Fig. 17.17) and defective extrusion profiles.

Let us consider melt fracture features and manifestations.

Melt fracture is a flow instability phenomenon occurring when hot melts are extruded through a die at a shear rate, which exceeds a critical shear stress. It typically shows three regular melt fracture patterns, which at further increase of the shear rate turn into irregular shapes. These three melt fracture, patterns are sharkskin, helix, and spiral in sequence, with an increase of the shear rates [45]. The manifestations of melt fracture have been assigned to a variety of causes including irregular flow at the die entrance, stick-slip movement along the die walls, and failure by excessive strain. The initiation of the phenomena seems to occur at the die entrance with some polymers, and within the die with others [46]. Apparently, certain defects occur when the tensile stress in the extensional flow region at the die entrance exceeds the strength of the material [47].

It should be noted that melt fracture is an elastic phenomenon, and it has been reported only for non-Newtonian fluids. At the melt fracture, the extrudate appears distorted in a periodic or random manner (Fig. 17.18). It reflects interference between fluid rheology and fluid adhesion that depends on the surface properties of the die [48].

An important element in melt fracture is also wall slip phenomenon [5, 49]. It is related to the so-called sharkskin, or sharkskin melt fracture, which is also called surface melt fracture. It is a low amplitude surface distortion of extruded polymer. Sharkskin is generally observed in case of linear polymers with narrow MWD, below the oscillating stick-slip transition. Sometimes (but not always), there is a change

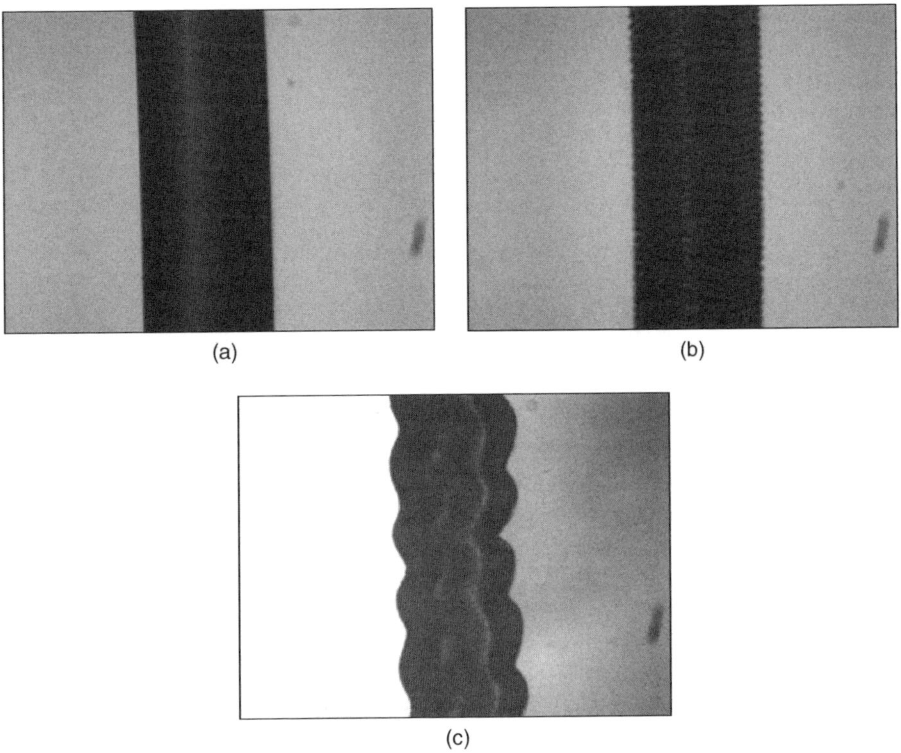

Figure 17.18 Typical evolution of melt facture of narrow MWD metallocene HDPE with increasing shear rate: (a) 250 s^{-1}, (b) 750 s^{-1}, and (c) 10,000 s^{-1}. Capillary diameter 1 mm, length 32 mm; temperature 180(C (Courtesy of V. Hristov).

in the slope of the flow curve at the onset of sharkskin [49]. Some researchers believe that the rough surface is caused by the tearing that occurs just past the die. This flow discontinuity at the exit occurs due to high stress levels, and the plastic very near the surface undergoes great stretching as its velocity rapidly increases from near zero inside the die to a high value outside the die [47, 50]

A specific pattern of sharkskin depends on polymer chemistry, molecular weight and MWD, and polymer–wall surface interaction [50, 51]. With a further increase of the shear rate, a spurt flow can be observed that results from pressure oscillations in the extruder [52]. At higher shear rates, gross melt fracture eventually occurs, with the extrudate coming out with an irregular pattern [45].

It is recognized that HDPE with a high molecular weight typically does not show sharkskin and enters directly into the spurt stage when shear rate increases above a critical value. On the contrary, the sharkskin for narrow MWD LLDPE is a common phenomenon when the shear stress exceeds a critical value of 0.14 MPa, and at a further increase of the shear rate, the hot melt enters the spurt stage of melt fracture. It is noticeable that LDPE does not exhibit slip-stick even at high shear rates.

Sharkskin, when the surface of the extrudate becomes visibly opaque, occurs at a wall shear stress level that is typically of the order of 0.1 MPa. At higher wall shear stress, typically of the order of 0.3 MPa, the flow becomes unsteady and the extrudate alternates between sharkskin and smooth segments (stick-slip, spurt flow, or cyclic melt fracture) [52, 53].

The onset of melt fracture can be determined using one of any combinations of the following criteria: (a) surface defects as regular surface patterns; (b) swagging extrudate coming out of the die; (c) periodical pressure oscillations with a magnitude greater than the doubled machine noise; and (d) changing of slope in the flow curve (wall shear stress vs. apparent shear rate). For the virgin HDPE (a) or (d) criteria are typically sufficient. For the filled HDPE criterion (c) is typically applied.

From the preceding rheological tests of problem runs, it was tentatively concluded that for the plant extruders the following rheological parameters might explain good or problem runs (Table 17.26).

TABLE 17.26 A tentative summary of rheological behavior of HDPE and HDPE-based regrind materials (applicable to a certain extruders, employed at LDI Composite plant, Green Bay, WI)

	Critical share rate (1/s)	Power-law index (n)	Consistency Index (Pa.s^n)
"Good" Neat HDPE	500–600	0.40–0.55	20,000–30,000
"Good" regrind	200–400	0.36–0.39	30,000–40,000
"Questionable" regrind	200–400	0.33–0.37	40,000–50,000
"Bad" regrind	Below 200, down to 20–100	0.19–0.30	Higher than 50,000

"Good" and "bad" profiles are related to hollow GeoDeck boards (5/4 × 6); rheological parameters in the table are given in tables of this chapter.

Table 17.27 describes neat HDPE and some regrinds (scrap and returns from the field) that were combined to make composite deck boards. Four combinations of said ingredients were used (as listed in Table 17.27), and all four resulted in problem extrusion runs. Boards were slightly distorted, pitted, and rough inside. Besides, a few more assorted ingredients are listed in Table 17.27, which might have been resulted in problem runs.

One can see that in Run 1 a mixture of two HDPE was used, both of which had rather low power-law index values (0.40 and 0.42), and the second one had a significantly lower value of the critical shear rate (300 s^{-1}). What is worse, the regrind, used in the amount of 15% in the final mix, had an exceptionally low critical shear rate (50 s^{-1}) along with a rather low value of the power-law index (0.32). This means that the processing window for this combination was quite narrow, and a slight increase in a flow rate would result in melt fracture, board roughness, sharkskin, and other profile defects (Figs. 17.10–17.14, and 17.19–17.23).

Run 2 was similar to Run 1. Again, the principal HDPE (77% of total neat HDPE in the formulation) was rheologically good enough (except maybe a lower power-law index of 0.39 compared to a conventional range of 0.40–0.55); the second neat

TABLE 17.27 The consistency index (apparent viscosity at the shear rate of 1 s^{-1}) and power-law index for some HDPE materials and HDPE-based composite regrinds (scrap and returns from the field), determined in the capillary rheometer (at 170°C)

Neat HDPE and composite regrinds	Consistency index (Pa.s^n)	Power-law index	Critical shear rate (s^{-1})
Run 1			
Neat HDPE 1	22,200	0.40	600
Neat HDPE 2	21,900	0.42	300
Regrind 1	49,800	0.32	50
Run 2			
Neat HDPE 3	24,500	0.39	600
Neat HDPE 4	22,600	0.42	400
Regrind 2	41,800	0.35	50
Run 3			
Neat HDPE 5	20,100	0.42	600
Neat HDPE 6	17,500	0.45	600
Regrind 3	45,900	0.35	200
Run 4			
Neat HDPE 7	24,400	0.40	600
Neat HDPE 8	25,500	0.42	400
Regrind 4	38,400	0.37	300

The fillers were rice hulls (94% smaller than 20 mesh, and 10% smaller than 80 mesh) and Biodac (50% cellulose fiber, 50% minerals)

HDPE had a lower critical shear rate of 400 s^{-1} compared to that in the primary plastic (600 s^{-1}), and the added regrind (15%) had a very low critical shear rate, making melt fracture likely to happen in the composite processing.

Run 3 was unstable; however, rheological parameters of all the main ingredients were within norm. The regrind was added in the amount of only 6%, than 4%, then 2%, and the run was still equally unstable. Clearly, the regrind was not a culprit in the problem run. However, both neat HDPE employed in the run had lowest viscosities among all other HDPE materials normally used in the formulation. Instead of a typical MFI of 0.5, these two HDPE materials had MFI equal to 0.66 (77% of total HDPE in the formulation) and 0.60. This was probably a reason of the process instability in this particular run.

Run 4 showed signs of melt fracture at otherwise normal speed. All lines had to slow down. The rheological properties of all plastic and plastic-based (regrind) ingredients were almost within norm, except for the power-law index of the principal HDPE (65% of total), which was on the low side (0.40). Normally, it is as high as above 0.50. The regrind was "good," and it was used in the amount of only 2% of the formulation. Actually, the regrind was a ground product from the preceding run (not shown here).

Table 17.28 continues Table 17.27 and also lists HDPE and regrinds from some other problem runs.

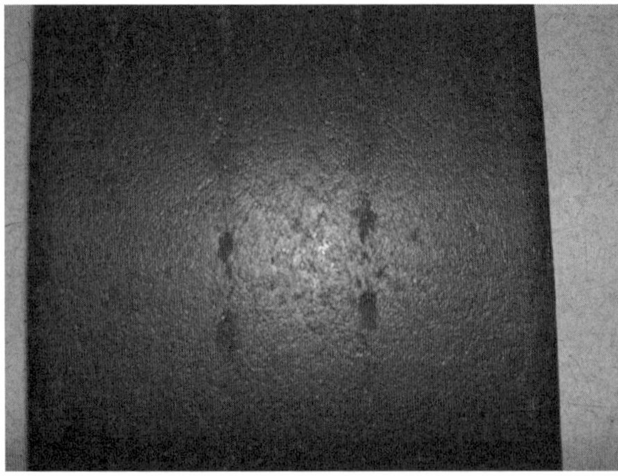

Figure 17.19 A composite deck board of a poor surface quality.

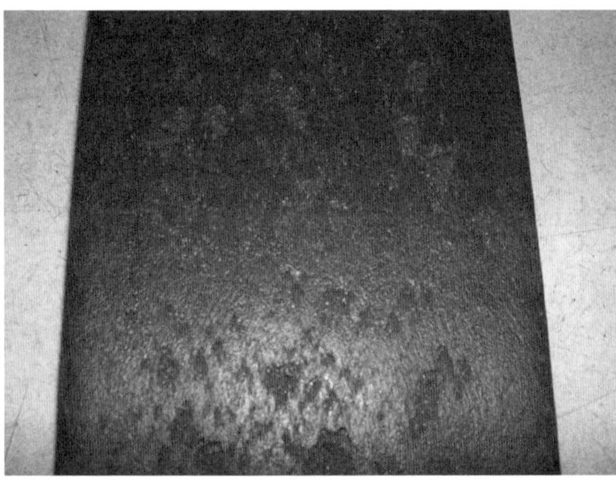

Figure 17.20 A composite deck board of a poor surface quality.

Figure 17.21 A composite deck board of a poor surface quality.

Figure 17.22 A composite deck board of a poor surface quality.

Neat HDPE 9, used at 23% of the total HDPE in the formulation, had exceptionally low critical shear rate (100 s^{-1}) and the power-law index (0.34). It apparently should make the processing window narrower compared to a normal run. Regrinds 4 (Table 17.27), 5, and 6 were within norm and could hardly create serious problems. Probably, the HDPE materials were culprits in the respective runs (see HDPE 9 in Table 17.28 with a very low critical shear rate of 100 s^{-1}). Regrind 7 also showed very low critical shear rate (100 s^{-1}) and power-law index (0.33) and could create problems with run stability and output.

Table 17.29 shows that critical share rate is the best in identifying "problem HDPE" materials. For plastics resulted in a "good run" and good extruded profiles, critical shear rate was close to 500 s^{-1} and above it, up to 750 s^{-1} (the first four HDPE

Figure 17.23 A composite deck board of a poor surface quality (a "washing board").

TABLE 17.28 The consistency index (apparent viscosity at the shear rate of 1 s^{-1}) and power-law index for some HDPE materials and HDPE-based composite regrinds (scrap and returns from the field), determined in the capillary rheometer (at 170°C)

Neat HDPE and composite regrinds	Consistency index ($Pa.s^n$)	Power-law index	Critical shear rate (1/s)
Neat HDPE 9	33,300	0.34	100
Regrind 5	37,600	0.37	200
Regrind 6	42,500	0.36	200
Regrind 7	44,300	0.33	100

The fillers were rice hulls (94% smaller than 20 mesh, and 10% smaller than 80 mesh) and Biodac (50% cellulose fiber, 50% minerals)

samples in the table). For "problem plastics" critical shear rate was around or below 300 s^{-1} down to about 100 s^{-1} (the last four HDPE samples in the table). It is quite understandable, as low critical shear rate is associated with a relatively low extrusion speed at which melt fracture is observed.

The processing window for the first four HDPE materials was wider in terms of extrusion speed, stress, and temperature. The last four HDPE materials were closer to melt fracture and the resulting sharkskin, roughness, tearing, and rupture of the extruded shape. The corresponding composite materials formulated with the last four resins had a lower throughput.

Let us consider shear rate for a wider selection of plastics, HDPE-composite materials, and regrinds (Table 17.30).

TABLE 17.29 The critical shear stress and the critical shear rate for HDPE of various origin at 170°C

HDPE	Log critical shear stress (Pa)	Log critical shear rate (1/s)	Critical shear rate (1/s)
Chevron CHVX891180[a]	5.48	2.74	550
Chevron CHVX896880[a]	5.42	2.67	460
Equistar EQUX621048[a]	5.48	2.88	750
Petromont PSPX7006[a]	5.42	2.71	510
Equistar EQUX631675[b]	5.20	2.19	150
Equistar GPLX74618[b]	5.32	2.32	210
Equistar EQUX632225[b]	5.34	2.49	310
Dupont/Sherman Tyvek Repro, Lot 45104[b]	5.37	2.05	112

Galaxy V capillary rheometer (Kayness Inc., Model 8025), 1 mm diameter capillary.

[a]The plastics gave a good quality composite deck boards (hollow profiles, plastic filled with rice hulls and minerals).

[b]The plastics gave a poor quality composite deck boards, sharkskin inside the hollow profile, tear surface (plastic filled with rice hulls and minerals); speed change and temperature change did not improve the situation.

TABLE 17.30 The critical shear stress and the critical shear rate for HDPE of various origin at 170°C

HDPE, HDPE-based composite material or regrind	Log critical shear stress (Pa)	The power-law index	critical shear rate (1/s)
Neat HDPE, CCBX58758 + CCBX72914 (Trademark Plastics)[a]	5.45 (280,000)	0.44	600
Neat HDPE, CCBX057340 (Dow Chemical)[a]	5.45 (280,000)	0.45	500
Neat HDPE, CCBX73082 (Trademark)[a]	5.43 (270,000)	0.44	500
Neat HDPE, CCBX71026[a]	5.46 (290,000)	0.45	550
GeoDeck, pellets[b]	5.54 (330,000)	0.19 (Stickslip, very high viscosity)	100
Regrind 1 (recalled boards)[a]	5.43 (270,000)	0.36	200
Regrind 2 (recalled boards)[a]	5.51 (330,000)	0.39	300
Regrind 3 (pellets made with 5% regrind from recalled boards)[a]	5.52 (330,000)	0.33	200–300
Regrind 4 (recalled boards)[a]	5.48 (300,000)	0.38	200–300
Regrind 5 (recalled boards)[a]	5.52 (330,000)	0.37	300–400
Regrind 6 (recalled boards)[b]	5.43 (270,000)	0.37	200
Regrind 7 (recalled boards)[b]	5.45 (280,000)	0.25	100–200 (Flow discontinuous until load above 250 lb, typically 177–184 lb)
Regrind 8 (recalled boards)[b]	5.54 (350,000)	0.34	300
Regrind 11 (recalled boards, brittle, weak)[b]	5.24 or lower (175,000 or lower)	Could not be determined, close to zero	20 (no continuous flow)
Board 1 (board in a good shape), made with 20% (w/w) of regrind 1	5.49 (310,000)	0.34	300

Galaxy V capillary rheometer (Kayness Inc., model 8025), 1-mm diameter capillary.

[a] Good board resulted with this plastic or regrind.

[b] Rough, unaccepted board resulted from this composite material or regrind.

One can see that the first four HDPE samples that ran well in the extrusion of the GeoDeck composition have a higher critical shear stress and higher critical shear rate. Their critical shear rate was in the range of 500–600 s^{-1}. The extrusion "window" for them was wider in terms of extrusion speed, stress, and temperature.

"Good" regrind, the addition of which in the amount of 20% to the GeoDeck formulation has resulted in good run and good boards, had the critical shear rate lower than that of neat HDPE, and it was in the range of 200–400 s^{-1}. "Bad" regrind, resulting in distorted boards, showing sharkskin, roughness, tearing, and rupture of the extruded shape, had the critical shear rate of 200 s^{-1} and below, to 20 s^{-1}. Regrind 8, which had the critical shear rate of 300 s^{-1} but resulted in rough boards, was mixed with a low critical shear rate (150 s^{-1}) neat HDPE.

These examples show that both neat HDPE and a regrind should be controlled by measuring their critical shear rate.

Table 17.31 shows a temperature dependence of the critical shear rates. One can see that increase of temperature widens the window of a good runability of HDPE regarding their critical shear rate.

Melt fracture for highly filled composite materials has a more complex character compared to that of neat plastics. And this is, of course, due to the effect of the filler. For example, using a cone die (the entrance diameter 2.5 in., the exit diameter 0.300 in., the length of the die 12 in.) it was shown for the neat HDPE that there was no visual signs of extrudate distortion for any of the flow rate tested (5.5–177 g/min), unlike that for filled plastics. Similarly, the 10%-filled (ground rice hulls) HDPE did not exhibit any observable extrudate distortion. However, the 60%-filled composite

TABLE 17.31 The critical shear stress and the critical shear rate (as logarithms) for HDPE of various origin at 170, 200, and 230°C

HDPE	Log critical shear stress (Pa)	Log Critical shear rate (1/s)		
		170°C	200°C	230°C
Chevron CHVX891180[a]	5.48	2.74	2.90	3.07
Chevron CHVX896880[a]	5.42	2.67	2.84	3.05
Equistar EQUX621048[a]	5.48	2.88	3.01	3.19
Petromont PSPX7006[a]	5.42	2.71	2.86	3.02
Equistar EQUX631675[b]	5.20	2.19	2.43	2.53
Equistar GPLX74618[b]	5.32	2.32	2.49	2.66
Equistar EQUX632225[b]	5.34	2.49	2.73	2.96
Dupont/Sherman Tyvek Repro, Lot 45104[b]	5.37	2.05	2.25	2.43

Galaxy V capillary rheometer (Kayness Inc., Model 8025).

[a] The plastics gave a good quality composite deck boards (hollow profiles, plastic filled with rice hulls and minerals).

[b] The plastics gave a poor quality composite deck boards, sharkskin inside the hollow profile, tear surface (plastic filled with rice hulls and minerals); speed change and temperature change did not improve the situation.

material showed a tearing distortion (sharkskin) that appeared to form at the die exit. The magnitude of the distortion was quite large and regular [21]. The same phenomenon was observed with HDPE filled with 50% wood flour [29].

It might come as a surprise that at higher flow rates the magnitude of the sharkskin effect can become less severe and even disappear at even more higher flow rates. When the flow rate was reduced, the tearing would appear again [21]. The tearing would persist and grow as the flow rate was reduced to zero. Hence, the concept of a critical wall shear stress increasing with the shear rate does not work here. It seems that in the described case a critical wall shear stress was *increased* with the *reduction* of the shear rate, at least the calculated shear rate, using conventional approaches. In this case a tearing defect could be caused by the rearrangement of the velocity profile at the exit of the die in such a way that a stretching flow is formed at the die surface leading to cohesive failure of the melt. Higher extrusion rates and higher melt temperatures might increase slip at the wall or very near the die exit and reduce the stretching at the exit, leading to the disappearance of the defect [21].

Indeed, increase of the die temperature from 140 to 180°C led in this case to a progressive disappearance of the sharkskin. This temperature effect is expected according to the prevailing theories on sharkskin, as the higher melt temperature reduces the viscosity of the melt near the die surface and thus reduce the wall shear stress and also allows stresses to relax more quickly. However, the suggested way to reduce sharkskin by increasing throughputs needs to be more carefully examined and verified in which cases it is justified and in which not, and why so. An alternative suggestion is that the tearing defect may be caused by tensile forces at the exit of the die [47].

It is recognized that coupling agents, that provide better interaction between hydrophilic cellulose fiber and hydrophobic polymer matrix, can make the material more resistant to high shear stress in the die and to excessive tensile forces at the die exit, hence, more resistant to melt fracture [41]. Besides, the usage of a thermoplastic silicone elastomer induces slip at the die wall leading to disappearance of the extrudate surface tearing (see Fig. 17.24 and comments above). It seems that slip at the wall is crucial for obtaining extrudates with acceptable surface quality.

Surface tearing is a multiparameter feature of both neat plastics and wood-filled composite materials. For example, moderate concentration of wood flour in the polyethylene melt results in characteristic tearing, which resembles sharkskin of the neat HDPE. Higher amounts of filler lead to more exaggerated sharkskin effect, most pronounced at 50 wt% maple wood flour loading [43]. The extrudate surface tearing diminished both with increasing *L/D* ratio of the die and increasing the shear rate (Fig. 17.24).

This behavior is similar for the 60%-filled HDPE as well; however, the diagonal line in Figure 17.24 is shifted to the left. This means that increasing the wood flour loading decreases the critical shear rate of obtaining smooth extrudates, and this shear rate decreases with increasing *L/D* of the die.

Effects of coupling agents on flowability of composite materials are, of course, much more complicated than just to provide better interaction between the filler and

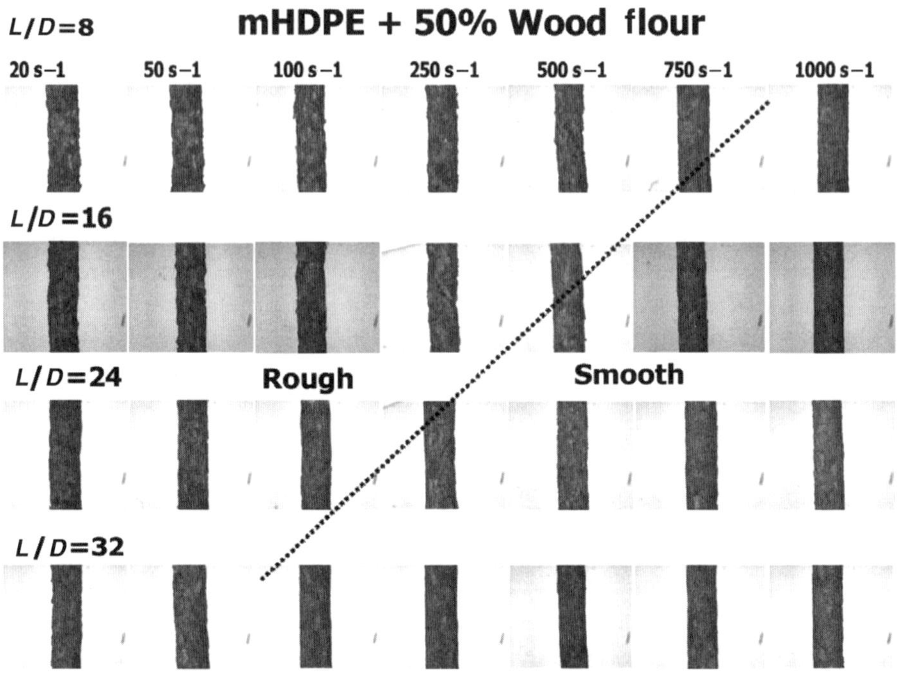

Figure 17.24 Extrudate surface morphology of 50-wt% wood-filled metallocene HDPE (MWD = 2.5) composites in dependence on L/D of the die and shear rate; D =1 mm (courtesy of V. Hristov, to be published in *Rheologica Acta*).

polymer matrix, hence, making the material more resistant and able to withstand the high shear stress in the die. Coupling agents themselves can change the flowability quite noticeably. It was observed with GeoDeck hollow boards that addition of a relatively small amount of Integrate (a maleic anhydride functionalized HDPE, see Chapter 5) results in a severe sharkskin surface in the hollow channel, whereas without the coupling agent both outer and inner surfaces of the board were just fine. All attempts to vary the concentration of Integrate, switching to a similar Polybond, changing lubricants, including Neustrene 060, and an ester-type lubricant recommended by the manufacturers, were unsuccessful, and a sharkskin of a more or less intensity was quite visible. At the same time small solid samples extruded from the same material with the same additives, including those made by Integrate and Polybond manufacturers, were smooth and of a good quality. Clearly, something was fundamentally different in the flowability of small solid samples and industrial size hollow boards.

When dynamic rheology was applied, the following data were obtained (Fig. 17.25).

One can see that the coupling agent though led to a drop in viscosity, at the same time resulted in a significant drop in the power-law index, from an apparent 0.65 to

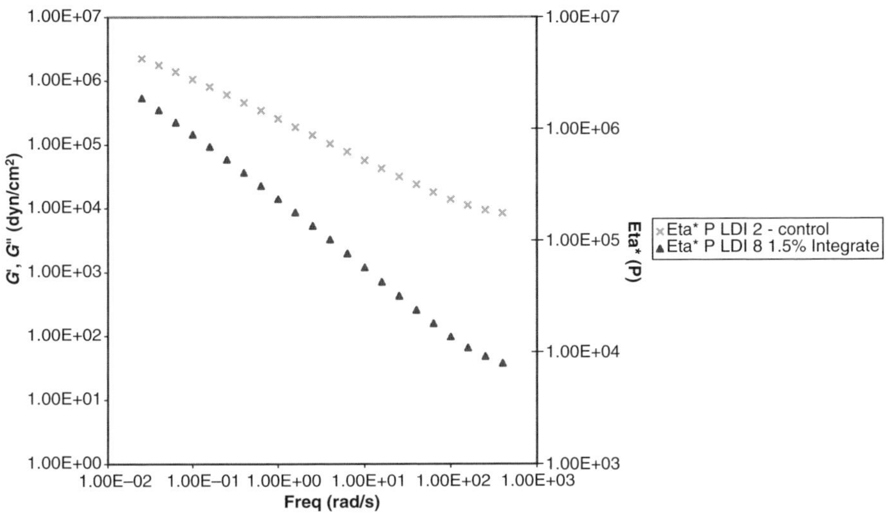

Figure 17.25 Dynamic rheology of GeoDeck formulation (upper curve) and the same formulation in the presence of 1.5% Integrate (coupling agent, maleated HDPE) and the respective lubricant (lower curve).

an apparent 0.40. This is an indicator of the flow instability of the system. The introduction of the coupling agent led to a much steeper slope on the graph, that is, to a much higher shear thinning effect of the melt with increase in frequency. This means that even a small change in pressure in the barrel, that is, in speed of the flow, would immediately (and steeply) result in a change in viscosity of the melt, in its shear rate, and, hence, shear stress in the melt.

Once the surge in the shear rate of the melt into the critical shear rate area happens, it will result in reaching and exceeding of the critical shear stress, hence, melt fractures, and one would likely see the sharkskin or much more severe surface defects. That was exactly what has happened when Integrate was introduced into the GeoDeck formulation.

Therefore, a graph such as in Figure 17.25 can be considered as a readout of an expected trouble, when the slope with a coupling agent is steeper compared to that for the control (no coupling agent). It means that either the lubricant is wrong or the lubricant is not in a right amount. One should try to change either one or both until both the slopes are close to each other.

As melt fracture and surface tearing are complicated phenomena, many factors, besides those discussed here, are also important, including die design, geometry of the extruder, formulation, and the presence of coupling agents in particular, the interfacial interactions between filler and matrix, and even *modus of operandi* of workers on the line. Overall, though, it might be suggested that three parameters, a low critical shear rate (lower than 200 s^{-1} for GeoDeck boards), a low power-law index (below 0.30 for GeoDeck boards), and a high consistency index (higher than 50,000 Pa.s^n for GeoDeck boards), are indicative of a "bad" regrind. Naturally, for each extruder and for every set of processing conditions, these critical parameters would be different. However, manifestations of "bad" profiles due to melt fracture and surface tearing are generally the same, making the product unacceptable.

REFERENCES

1. A. Einstein. *Ann. Phys.* 1906, **19**, 289–306.
2. A.V. Shenoy. *Rheology of Filled Polymer Systems*, Kluwer Academic Publishers, 1999, p. 339.
3. R.B. Bird, R.C. Armstrong, and O. Hassager. *Dynamics of Polymeric Liquids*, Vol. 1. 2nd edition, John Wiley & Sons, Inc., New York, 1987, p. 171.
4. V. Hristov, E. Takacs, and J. Vlachopoulos. Viscoelastic behavior of highly filled HDPE/wood flour composites. *ANTEC*, Boston, May 1–5, 2005.
5. M. Mooney. Explicit formulas for slip and fluidity. *J. Rheol.* 1931, **2**, 210–222.
6. S. Hatzikiriakos, and J.M. Dealy. Wall slip of molten high density polyethylenes. II. Capillary rheometer studies. *J. Rheol.* 1992, **36**, 703–741.
7. R.J. Castillo, D. Strutt, and J. Vlachopoulos. *Proceedings of Experiments and Simulations with Barrier Screws*, ANTEC, Society of Plastic Engineers, Brookfield, CT, San Francisco, CA, 2002, pp. 318–322.
8. J. Vlachopoulos, SPE Extrusion Division, Society of Plastic Engineers, Brookfield, CT, *Newsletter* 1999, **23**, 3.
9. C. Macosko. *Rheology: Principles, Measurements, and Applications*, Wiley-VCH Publication, New York, 1994, p. 181.
10. R.K. Gupta. *Polymer and Composite Rheology*, 2nd edition, Marcel Dekker, New York, 2000.
11. S.J. Park and R.G. Larson. Modeling the linear viscoelastic properties of metallocene-catalyzed high density polyethylenes with long-chain branching. *J. Rheol.* 2005, **49**, 523–536.
12. A.I. Isayev (Ed.), *Injection and Compression Molding Fundamentals* Marcel Dekker, New York, 1987, p. 9, 87.
13. R.K. Gupta. *Polymer and Composite Rheology*, 2nd edition, Marcel Dekker, New York, 2000, p. 38.
14. C. Rauwendaal, *Polymer Extrusion*, 3rd edition, Hanser Publishers, Munich, 1994.
15. C.A. Hieber. Melt-viscosity characterization and its application to injection molding. In: A.I. Isayev (Ed.), *Injection and Compression Molding Fundamentals,* Marcel Dekker, New York, 1987, p. 6.
16. A.I. Isayev (Ed.), *Injection and Compression Molding Fundamentals*, Marcel Dekker, New York, 1987, p. 78.
17. A.V. Shenoy, *Rheology of Filler Polymer Systems*, Kluwer Academic Publishers, Norwel, MA, 1999, p. 492.

18. T.S. Creasy, S.G. Advani, and R.K. Okine, Transient rheological behavior of a long discontinuous fiber-melt system. *J. Rheology*, 1996, **40**, 497–519.
19. T.S. Creasy and S.G. Advani, A model long-discontinuous-fiber filled thermoplastic melt in extensional flow. *J. Non-Newtonian Fluid Mech.* 1997, **73**, 261–278.
20. T.S. Creasy and S.G. Advani, and R.K. Okine, Nonlinear response of long-discontinuous-fiber/melt system in elongational flows. *Rheol. Acta* 1996, **35**, 347–355.
21. Z. Charlton, D. Suwanda, and J. Vlachopoulos. Extrusion of heavily filled HDPE profiles. In: *17th Meeting of the Polymer Processing Society*, PPS, Melville, NY, Montreal, May 21–24, 2001.
22. M. Botros, *Intertech Conference, The Global Outlook for Natural and Wood Fiber Composites*, New Orleans, LA, December 3–5, 2003.
23. M. Gahleitner. Melt rheology of polyolefins, *Prog. Polym. Sci.* 2001, **26**, 895–944.
24. T.D. Papthanasiou and D.C. Guell (Eds.), *Flow-Induced Alignment in Composite Materials*, Woodhead Publishing, Cambridge, UK, 1997, p. 24.
25. P. Prentice. *Rheology and its role in plastic processing.* Rapra Review Reports, Report 84, Vol. 7, No. 12, 1995, p. 9.
26. W.P. Cox and E.H Merz, *J. Polym. Sci.* 1958, **28**, 619.
27. V. Hristov, E. Takács, and J. Vlachopoulos. Surface tearing and wall slip phenomena in extrusion of highly filled HDPE/wood flour composites, *Polym. Eng. Sci.*, 2006, **46**, 1204–1214.
28. J. Dealy and K.F. Wissburn. *Melt Rheology and its Role in Plastics Processing*, Van Nostrand Reinhold, New York, 1990.
29. J. Vlachopoulos, V. Hristov, and Z. Charlton. Rheological aspects of wood polymer composites extrusion. In: *Americas Regional Meeting Proceedings*, PPS, Melville, NY, Quebec City, August 15–19, 2005.
30. V. Hristov, E. Takács, and J. Vlachopoulos, Surface tearing and wall slip phenomena in extrusion of highly filled HDPE/wood flour composites, *Polym. Eng. Sci.*, 2006, **46**, 1204–1214.
31. A.I. Isayev and R.K. Upadhyay, Flow of polymeric melts in juncture regions of injection molding. In: A.I. Isayev (Ed.), *Injection and Compression Molding Fundamentals*, Marcel Dekker, New York, 1987, pp. 167–169.
32. F. Mijangos-Santiago and J.M. Dealy. Effect of filler content and additives on the extrudate swell of polyethylene pipe resin. *Polym. Eng. Sci.* 1991, **31**(16), 1176–1181.
33. V. Hristov, and J. Vlachoupoulos. Viscoelasticity and extrudate surface tearing of natural fiber composites. *Rheol. Acta*, in press.
34. M.W. Chastagner, M.P. Wolcott and K.R. Englund. *Characterizing the Rheological Properties of Wood-Plastic Composite Formulations.* Wood Materials and Engineering Laboratory, Washington State University, 2005.
35. V. Hristov, M. Krumova, and G. Michler. Influence of the excess coupling agent on the microdeformation processes and mechanical properties of polypropylene/wood flour composites. *Macromol. Mater. Eng.*, 2006, **291**, 677–683.
36. A.K. Bledzki, O. Faruk, M. Huque. *Polym.-Plast. Technol. Eng.* 2002, **41**, 435.
37. R. Karnani., M. Krishnan, and R. Marayan. *Polym. Eng. Sci.* 1997, **37**, 476.
38. H. Nitz, P. Reichert, H. Römling, and R. Mülhaupt. *Macromol. Mater. Eng.* 2000, **276/277**, 51.

39. S. Kharchenko, K.B. Migler, and S. Hatzikiriakos. Conventional polymer processing additives. In: S.G. Hatzikiriakos and K.B. Migler (Eds.), *Polymer Processing Instabilities—Control and Understanding*, Dekker, New York, 2005.
40. L.M. Sherman. *Plastics Technology*, July 2005, p. 58.
41. V. Hristov, E. Takács, and J. Vlachopoulos. Extrudate surface tearing in extrusion of wood plastic composites, In: *ANTEC 2006*, Charlotte, NC, May 7–11, 2006.
42. R.K. Krishnaswamy, D.C. Rohlfing, A.M. Sukhadia, and K.R. Slusarz. Extrusion of broad-molecular-weight-distribution polyethylenes, *Polymer. Eng. Sci.* 2004, **44**(12).
43. V. Hristov and J. Vlachopoulos. Viscoelasticity and extrudate surface tearing of natural fiber composites. *Rheolog. Acta*, in press.
44. V. Hristov. *Polym. Eng. Sci.* in press.
45. Z. Tao and J.-C. Huang, Study on the melt fracture of metallocene poly(ethylene-octene) in capillary flow. *J. Appl. Polym. Sci.* 2005, 98, 903–911.
46. J.P. Tordella. In: F.R. Eirich (Ed.), *Rheology*, Vol. 5, Academic Press, New York, 1969, Chapter 2.
47. F.N. Cogswell. Stretching flow instabilities at the exits of extrusion dies. *J. Non-Newtonian Fluid Mech.* 1977, **2**, 37–47.
48. A.V. Ramamurthy. Wall slip in viscous fluids and influence of materials of construction. *J. Rheol.*, 1986, **30**, 337–357.
49. K.B. Migler. Sharkskin instability in extrusion. In: S.G. Hatzikiriakos and K.B. Migler (Eds.), *Polymer Processing Instabilities—Control and Understanding*, Dekker, New York, 2005.
50. K.B. Migler, Y. Son, F. Qiao, and K. Flynn. Extensional deformation, cohesive failure, and boundary conditions during sharkskin melt fraction. *J. Rheol.*, 2002, **46**, 383–400.
51. A.V. Ramamurthy. Extrudate irregularities and the polymer-metal interface connection, In: *Proceedings of 10th International Congress on Rheology*, Australian Society of Rheology, Sidney, 1988, pp. 85–87.
52. G.V. Vinogradov and L.I. Ivanova. Wall slippage and. elastic turbulence of polymer in the rubbery state. *Rheol. Acta*, 1968, **7**(3), 243–255.
53. M.M. Denn. Extrusion instabilities and wall slip. *Ann. Rev. Fluid Mech.,* 2001, **33**, 265–287.

SUBJECT INDEX